New Research on Detection and Removal of Emerging Pollutants—Volume I

New Research on Detection and Removal of Emerging Pollutants—Volume I

Editors

Avelino Núñez-Delgado
Zhien Zhang
Elza Bontempi
Mario Coccia
Marco Race
Yaoyu Zhou

MDPI

Basel • Beijing • Wuhan • Barcelona • Belgrade • Novi Sad • Cluj • Manchester

Editors

Avelino Núñez-Delgado
University Santiago de Compostela
Lugo
Spain

Zhien Zhang
University of Cincinnati
Cincinnati
USA

Elza Bontempi
University of Brescia
Brescia
Italy

Mario Coccia
National Research Council of Italy (CNR)
Turin
Italy

Marco Race
University of Cassino and Southern Lazio
Cassino
Italy

Yaoyu Zhou
Hunan Agricultural University
Changsha
China

Editorial Office
MDPI
St. Alban-Anlage 66
4052 Basel, Switzerland

This is a reprint of articles from the Topic published online in the open access journals *Processes* (ISSN 2227-9717), *Applied Sciences* (ISSN 2076-3417), *Materials* (ISSN 1996-1944), *Sustainability* (ISSN 2071-1050), and *Toxics* (ISSN 2305-6304) (available at: https://www.mdpi.com/topics/Emerging_Pollutants).

For citation purposes, cite each article independently as indicated on the article page online and as indicated below:

Lastname, A.A.; Lastname, B.B. Article Title. *Journal Name* **Year**, *Volume Number*, Page Range.

Volume I
ISBN 978-3-7258-0825-0 (Hbk)
ISBN 978-3-7258-0826-7 (PDF)
doi.org/10.3390/books978-3-7258-0826-7

Set
ISBN 978-3-7258-0793-2 (Hbk)
ISBN 978-3-7258-0794-9 (PDF)

Contents

About the Editors

Avelino Núñez-Delgado

Avelino Núñez-Delgado, Ph.D., was born in O Barco de Valdeorras (Ourense province, Galicia, Spain). He obtained a Ph.D. in the Department of Soil Science and Agricultural Chemistry, at USC, in 1993. Between 1993 and 1996, he was a Post-doc Researcher in France (University of Montpellier) and Spain (USC), at the end of which he became a Professor in the Department of Soil Science and Agricultural Chemistry, Engineering Polytechnic School, Campus Lugo, University of Santiago de Compostela (USC), Spain, a position which he has covered to this day. He has nine patents, several research awards, and more than 400 publications to date (December 2023), around 200 of which are in D1 and Q1 JCR journals. He was the Principal Investigator and/or a collaborator in more than 40 research projects. He was listed among the top 2% of world researchers by the Stanford ranking system and among the world top researchers by Researchgate, Expertscape, Web of Sciences, Scopus, and other world research classifications. Currently, he is collaborating with a variety of research teams from various countries around the world. He is a Book Editor for Springer Nature, Elsevier, and other top scientific publishers. He is a Book Series Editor for Springer Nature, an Editor for various top research journals (covering roles such as Chief Editor, Associate Editor, Special Issues Editor, Managing Guest Editor, and Guest Editor), and a Reviewer for national and international research projects.

Zhien Zhang

Zhien Zhang is currently a Visiting Scientist at the University of Cincinnati. Prior to this position, he was a Research Assistant Professor at West Virginia University and a Senior Researcher at the Ohio State University. His research interests are in the following areas: carbon capture, utilization, and storage (cCuS); gas separation; absorption; membrane; gas hydrate; process modeling and simulation; and optimization. To date, he has published more than 120 peer-reviewed journal articles, 20 journal editorials, two books, and six book chapters (h-index of 53) and has been invited to and delivered more than 30 talks and seminars. He is an Editor in journals such as *Applied Energy, Environmental Chemistry Letters, Gas Science and Engineering,* and *Chemical Papers* and serves as a committee member in several international conferences. He was recognized as a Highly Cited Researcher by Clarivate in 2021 and 2022.

Elza Bontempi

Elza Bontempi has a permanent position at the University of Brescia, where she is currently a Full Professor teaching courses on the Fundamentals of Chemistry for Technology. She is responsible for the research line concerning eco-materials at the Chemistry for Technologies Laboratory. She has been responsible for several national and international research projects developing new technologies and sustainable materials from waste and by-products. In recent years, her scientific activity has focused on the recovery of critical raw materials from exhausted batteries, such as lithium and cobalt, in the context of the circular economy. She is the author of more than 300 peer-reviewed papers and several patents in the field of material recovery. She is included in the list of Unstoppable Women (the 1000 women who are changing Italy) and Top Italian Scientists in the field of natural and environmental sciences and in the list of 100 experts.

Mario Coccia

Mario Coccia is a social scientist acting as the Research Director in the National Research Council of Italy and as a visiting scholar at the Arizona State University (USA). He has been a researcher at the Max Planck Institute of Economics and a visiting professor at the Polytechnic of Torino and at the University of Piemonte Orientale (Italy). He has carried out scientific research at the Georgia Institute of Technology, Yale University, UNU-Maastricht Economic and Social Research Institute on Innovation and Technology (United Nations University-MERIT), RAND Corporation (Washington D.C.), University of Maryland (College Park), Bureau d'Économie Théorique et Appliquée (Strasbourg, France), Munk School of Global Affairs (University of Toronto, Canada), and the Institute for Science and Technology Studies (University of Bielefeld, Germany). He investigates via statistical analyses, models, experiments, and observational studies with an interdisciplinary scientific perspective in order to explain the evolutionary properties of science and technology in society, emerging research fields and the related scientific development, new technological trajectories, processes of coevolution between technologies, and the measurement of scientific and technological advances over time and space. He is a member of the Editorial Board of many international journals, and his research publications include more than 350 international papers on several disciplines.

Marco Race

Marco Race (Associate Professor) was born in Napoli, graduated with an M.Sc. in Environmental Engineering at the Università degli Studi di Napoli Federico II in May 2012, and obtained his Ph.D. in Environmental Systems Analysis at UNINA in 2016. Since 2022, he has been an Associate Professor at the University of Cassino. His main research fields concern the treatment of waste or wastewater treatments, the remediation of soil and groundwater, novel contaminant (bio)monitoring and risk assessment approaches, and trace metals and organics in biogeochemical cycles. He is the author of more than 100 papers published in international journals, conferences proceedings, and books chapters. He obtained an international award on soil reclamation.

Yaoyu Zhou

Dr. Yaoyu Zhou is a Full Professor in the College of Environment and Ecology, at the Hunan Agricultural University, Changsha, Hunan province, China. Prof. Zhou's academic background covers waste management and the decontamination of aqueous effluents. Prof. Zhou also has experience in fundamental soil science and the remediation of various contaminants in soils and sediments. Prof. Zhou is listed in Stanford's list of the top 2% of scientists in the world (2020). Together with some of his graduate students and colleagues, Prof. Zhou has published over 220 academic papers, 23 of which were ranked as ESI top papers (17 nominated as "Highly Cited Papers" and 8 nominated as "Hot Papers"). After being supported by the Hong Kong Scholar Program, Prof. Zhou worked in the Hong Kong Polytechnic University from 2018 to 2020. He also holds some international positions: he is a Member of the Editorial Board of Environmental Technology, Carbon Research, and Biochar and the Guest Editor of the *Journal of Environmental Management* (JCR Q1, New Research on Soil Degradation and Restoration) and *Science of the Total Environment* (JCR Q1, Antibiotics and Heavy Metal; and JCR Q1, BEEM conference 2019).

materials

MDPI

Editorial

Editorial on the Topic "New Research on Detection and Removal of Emerging Pollutants"

Avelino Núñez-Delgado [1,*], Zhien Zhang [2], Elza Bontempi [3], Mario Coccia [4], Marco Race [5] and Yaoyu Zhou [6]

1 Department of Soil Science and Agricultural Chemistry, Engineering Polytechnic School, University Santiago de Compostela, 27002 Lugo, Spain
2 Department of Chemical and Biomedical Engineering, West Virginia University, Morgantown, WV 26506, USA
3 INSTM (National Interuniversity Consortium of Materials Science and Technology), Department of Mechanical and Industrial Engineering, University of Brescia, Via Branze, 38, 25123 Brescia, Italy
4 Research Institute on Sustainable Economic Growth, National Research Council of Italy (CNR), Turin Research Area of the CNR, 10135 Turin, Italy
5 Department of Civil and Mechanical Engineering, University of Cassino and Southern Lazio, Via Di Biasio 43, 03043 Cassino, Italy
6 College of Resources and Environment, Hunan Agricultural University, Changsha 410128, China
* Correspondence: avelino.nunez@usc.es

With the Topic "New Research on Detection and Removal of Emerging Pollutants" (https://www.mdpi.com/topics/Emerging_Pollutants) closed to new submissions, the Editors would like to share some comments on it.

The journals involved in the Topic were Materials (with 23 papers finally published), Processes (with 21 papers published), Sustainability (with 13 papers published), Applied Sciences (with 7 papers published), and Toxics (with 2 papers finally published).

To date, with the Topic just closed for submissions, the most cited papers have received 22 citations [1], 21 citations [2], 15 citations [3], 10 citations [4], 9 citations [5,6], 8 citations [7], 7 citations [8,9], and 6 citations [10–12], while the other papers included in the Special Issue received between 5 and 0 citations at the time of writing of this editorial piece.

Overall, the Editors think that the Topical Issue has provided very interesting and high-quality contributions to the broad field of research on emerging pollutants. The removal of emerging pollutants is a challenging topic that is receiving increasing attention at the level of investigation and risk concern perceived by the society. In fact, improving the means for both quantification and removal of toxic substances is clearly relevant in the current situation of environmental stress affecting the different environmental compartments [13–17].

In addition, the Editors consider useful the experience of combining the five journals involved in the Topic which promotes a wider diffusion of this Special Issue, covering a broader spectrum of researchers and potential readers.

This field of research needs continuous and higher efforts, so it is expected that additional issues and Topics focused on it will be developed in the coming future.

Conflicts of Interest: The authors declare no conflict of interest.

Citation: Núñez-Delgado, A.; Zhang, Z.; Bontempi, E.; Coccia, M.; Race, M.; Zhou, Y. Editorial on the Topic "New Research on Detection and Removal of Emerging Pollutants". *Materials* **2023**, *16*, 725. https://doi.org/10.3390/ma16020725

Received: 9 January 2023
Accepted: 10 January 2023
Published: 11 January 2023

References

1. Pop, C.-E.; Draga, S.; Măciucă, R.; Niță, R.; Crăciun, N.; Wolff, R. Bisphenol a Effects in Aqueous Environment on *Lemna minor*. *Processes* **2021**, *9*, 1512. [CrossRef]
2. Sui, X.; Wang, X.; Li, Y.; Ji, H. Remediation of Petroleum-Contaminated Soils with Microbial and Microbial Combined Methods: Advances, Mechanisms, and Challenges. *Sustainability* **2021**, *13*, 9267. [CrossRef]
3. Alghamdi, W.M.; El Mannoubi, I. Investigation of Seeds and Peels of *Citrullus colocynthis* as Efficient Natural Adsorbent for Methylene Blue Dye. *Processes* **2021**, *9*, 1279. [CrossRef]
4. Xue, S.; Cao, S.; Huang, Z.; Yang, D.; Zhang, G. Improving Gas-Sensing Performance Based on MOS Nanomaterials: A Review. *Materials* **2021**, *14*, 4263. [CrossRef] [PubMed]
5. Cui, Y.; Zhang, H.; Zhu, J.; Peng, L.; Duan, Z.; Liu, T.; Zuo, J.; Xing, L.; Liao, Z.; Wang, S.; et al. Unstimulated Parotid Saliva Is a Better Method for Blood Glucose Prediction. *Appl. Sci.* **2021**, *11*, 11367. [CrossRef]
6. Foday Jr, E.H.; Bo, B.; Xu, X. Removal of Toxic Heavy Metals from Contaminated Aqueous Solutions Using Seaweeds: A Review. *Sustainability* **2021**, *13*, 12311. [CrossRef]
7. Gao, Y.; Huang, J.; Li, M.; Dai, Z.; Jiang, R.; Zhang, J. Chemical Modification of Combusted Coal Gangue for U(VI) Adsorption: Towards a Waste Control by Waste Strategy. *Sustainability* **2021**, *13*, 8421. [CrossRef]
8. Cardei, P.; Nenciu, F.; Ungureanu, N.; Pruteanu, M.A.; Vlăduț, V.; Cujbescu, D.; Găgeanu, I.; Cristea, O.D. Using Statistical Modeling for Assessing Lettuce Crops Contaminated with Zn, Correlating Plants Growth Characteristics with the Soil Contamination Levels. *Appl. Sci.* **2021**, *11*, 8261. [CrossRef]
9. Mao, Y.; Xiong, R.; Gao, X.; Jiang, L.; Peng, Y.; Xue, Y. Analysis of the Status and Improvement of Microalgal Phosphorus Removal from Municipal Wastewater. *Processes* **2021**, *9*, 1486. [CrossRef]
10. Huynh, A.T.; Chen, Y.-C.; Tran, B.N.T. A Small-Scale Study on Removal of Heavy Metals from Contaminated Water Using Water Hyacinth. *Processes* **2021**, *9*, 1802. [CrossRef]
11. Mansour, A.M.; Almutairi, A.; Alyami, S.; Obeidat, M.A.; Almkahles, D.; Sathik, J. A Unique Unified Wind Speed Approach to Decision-Making for Dispersed Locations. *Sustainability* **2021**, *13*, 9340. [CrossRef]
12. Cela-Dablanca, R.; Nebot, C.; Rodríguez López, L.; Fernández-Calviño, D.; Arias-Estévez, M.; Núñez-Delgado, A.; Fernández-Sanjurjo, M.J.; Álvarez-Rodríguez, E. Efficacy of Different Waste and By-Products from Forest and Food Industries in the Removal/Retention of the Antibiotic Cefuroxime. *Processes* **2021**, *9*, 1151. [CrossRef]
13. Delgado, A.N. Our Environment: Everything Is Natural on Earth, but … Editorial Piece on Current and Future Soil and Environmental Research. *Processes* **2023**, *11*, 6. [CrossRef]
14. Wang, S.; Zhang, W.; Jia, F.; Fu, H.; Liu, T.; Zhang, X.; Liu, B.; Núñez-Delgado, A.; Han, N. Novel Ag_3PO_4/boron-carbon-nitrogen photocatalyst for highly efficient degradation of organic pollutants under visible-light irradiation. *J. Environ. Manag.* **2021**, *292*, 112763. [CrossRef] [PubMed]
15. Haider, F.U.; Ejaz, M.; Cheema, S.A.; Khan, M.I.; Zhao, B.; Liqun, C.; Salim, M.A.; Naveed, M.; Khan, N.; Núñez-Delgado, A.; et al. Phytotoxicity of petroleum hydrocarbons: Sources, impacts and remediation strategies. *Environ. Res.* **2021**, *197*, 111031. [CrossRef] [PubMed]
16. Ahmed, W.; Mehmood, S.; Núñez-Delgado, A.; Ali, S.; Qaswar, M.; Khan, Z.H.; Ying, H.; Chen, D.Y. Utilization of Citrullus lanatus L. seeds to synthesize a novel $MnFe_2O_4$-biochar adsorbent for the removal of U(VI) from wastewater: Insights and comparison between modified and raw biochar. *Sci. Total Environ.* **2021**, *771*, 144955. [CrossRef] [PubMed]
17. Barreiro, A.; Cela-Dablanca, R.; Nebot, C.; Rodríguez-López, L.; Santás-Miguel, V.; Arias-Estévez, M.; Fernández-Sanjurjo, M.; Núñez-Delgado, A.; Álvarez-Rodríguez, E. Occurrence of Nine Antibiotics in Different Kinds of Sewage Sludge, Soils, Corn and Grapes After Sludge Spreading. *Span. J. Soil Sci.* **2022**, *12*, 10741–10753. [CrossRef]

processes

MDPI

Article

Efficacy of Different Waste and By-Products from Forest and Food Industries in the Removal/Retention of the Antibiotic Cefuroxime

Raquel Cela-Dablanca [1], Carolina Nebot [2], Lucia Rodríguez López [3], David Fernández-Calviño [3], Manuel Arias-Estévez [3], Avelino Núñez-Delgado [1,*], María J. Fernández-Sanjurjo [1] and Esperanza Álvarez-Rodríguez [1]

1 Department Soil Science and Agricultural Chemistry, Engineering Polytechnic School, University Santiago de Compostela, 27002 Lugo, Spain; raquel.dablanca@usc.es (R.C.-D.); mf.sanjurjo@usc.es (M.J.F.-S.); esperanza.alvarez@usc.es (E.Á.-R.)

2 Department of Analytical Chemistry, Nutrition and Bromatology, Faculty of Veterinary Medicine, University of Santiago de Compostela, 27002 Lugo, Spain; Carolina.nebot@usc.es

3 Soil Science and Agricultural Chemistry, Faculty Sciences, University Vigo, 32004 Ourense, Spain; lucia.rodriguez.lopez@uvigo.es (L.R.L.); davidfc@uvigo.es (D.F.-C.); mastevez@uvigo.es (M.A.-E.)

* Correspondence: avelino.nunez@usc.es

Abstract: Environmental pollution due to antibiotics is a serious problem. In this work, the adsorption and desorption of the antibiotic cefuroxime (CFX) were studied in four by-products/residues from the forestry and food industries. For this, batch-type experiments were carried out, adding increasing concentrations of CFX (from 0 to 50 μmol L^{-1}) to 0.5 g of adsorbent. The materials with a pH higher than 9 (mussel shell and wood ash) were those that presented the highest adsorption percentages, from 71.2% (23.1 μmol kg^{-1}) to 98.6% (928.0 μmol kg^{-1}). For the rest of the adsorbents, the adsorption was also around 100% when the lowest concentrations of CFX were added, but the percentage dropped sharply when the highest dose of the antibiotic was incorporated. Adsorption data fitted well to the Langmuir and Freundlich models, with R^2 greater than 0.9. Regarding desorption, the materials that presented the lowest values when the highest concentration of CFX was added were wood ash (0%) and mussel shell (2.1%), while pine bark and eucalyptus leaves presented the highest desorption (26.6% and 28.6%, respectively). Therefore, wood ash and mussel shell could be considered adsorbents with a high potential to be used in problems of environmental contamination by CFX.

Keywords: antibiotics; eucalyptus leaves; mussel shell; pine bark; pine needles; retention/release; wood ash

Citation: Cela-Dablanca, R.; Nebot, C.; Rodríguez López, L.; Fernández-Calviño, D.; Arias-Estévez, M.; Núñez-Delgado, A.; Fernández-Sanjurjo, M.J.; Álvarez-Rodríguez, E. Efficacy of Different Waste and By-Products from Forest and Food Industries in the Removal/Retention of the Antibiotic Cefuroxime. *Processes* **2021**, *9*, 1151. https://doi.org/10.3390/pr9071151

Academic Editor: Zhien Zhang

Received: 16 June 2021
Accepted: 29 June 2021
Published: 1 July 2021

1. Introduction

From 2000 to 2015, the worldwide consumption of antibiotics in humans and in veterinary medicine increased by almost 65% [1,2], due to the increase in the world population, and to the higher demand for protein, which intensified animal production, requiring a higher use of antibiotics [3,4].

Among these drugs, cephalosporins, belonging to the group of beta-lactams, are widely used in the treatment of bacterial infections, as they have good tolerance and few side effects [5,6]. In human medicine, they are used mainly in specific infections of the human genital tract, as well as in serious infections such as meningitis [7,8]. In veterinary medicine, cephalosporins are widely used in infections of the respiratory tract and mammary glands [8–10].

Among the second-generation cephalosporins, cefuroxime (CFX) is the most prescribed, representing more than 50% of the total administration of cephalosporins in most European countries [11]. These antimicrobials are poorly absorbed by the intestine and a significant proportion (up to 90%) are excreted through feces and urine as the parent compound [12–14], being incorporated into wastewater in the case of humans, and passing to slurry pits or directly into the environment in the case of farm animals. These wastewaters reach treatment plants, but many of these facilities have not been designed to eliminate antibiotics [14,15], removing between 20 and 90% of the pollutants, either through its accumulation in sewage sludge [16], or by degradation processes, which affect antibiotics such as penicillin [17]. However, other antimicrobials such as cephalosporins, fluoroquinolones, and tetracyclines are more resistant to natural degradation [17,18].

As an example, a study conducted in Greece on the presence of antibiotics in wastewater detected high concentrations of amoxicillin, clarithromycin, CFX, and ciprofloxacin [19]. The incorporation of antibiotics into the soil, through wastewater or fertilization with sewage sludge or livestock manure, can cause the appearance of bacterial resistance, contamination of underground or surface water bodies, and the passage of these pollutants to the food chain, both through drinking water and animal or vegetable products, as different crops can absorb the antibiotics present in the soils [20], constituting a threat to human health [21].

Recently, Cela-Dablanca et al. [22] studied the retention of CFX in various soils, finding adsorption values between 40.8% and 99.6% (between 54.39 and 125 μmol kg^{-1}) in those devoted to agricultural production, and between 74.6% and 93.5% (between 109.9 and 116.71 μmol kg^{-1}) in forest soils. However, taking into account that the presence of antibiotics both in soils and other environmental compartments is considered a matter of real concern, growing research is focusing on the design of systems and procedures to remove/retain these molecules, with special emphasis on wastewater, which is often used as irrigation waters [2,23], and also in soils [24–26].

The methods generally used for the removal of antibiotics include advanced oxidation, biological technology, and membrane separation, but they can be excessively expensive and even produce toxic by-products [14,24,27]. Adsorption technologies have the advantages of being relatively simple, low-cost, long-lasting, and renewable, and they do not generate toxic by-products [25,26]. The most common adsorbents include activated carbon, mineral materials, and biological materials [13]. Within these, the most used for the removal of antibiotics is activated carbon, but its high cost and difficulty of regeneration are considered relevant disadvantages [28]. In view of that, it would be clearly interesting to determine the adsorption capacity and potential effectivity of alternative low-cost adsorbent materials. In this line, bio-adsorbents derived from a wide variety of sources, many of them residual materials that need to be recycled, can be a viable option for the retention/removal of antibiotics present as contaminants in environmental compartments [29].

In view of the above background, this work focused on studying the adsorption/desorption characteristics of the antibiotic CFX when it interacts with different residues/by-products generated by the forestry industry (eucalyptus leaf, pine bark, pine needles, and wood ash), and of a waste from the food industry (mussel shell), in order to evaluate their potential suitability to be used in processes of removal/retention of this antibiotic, which could be very relevant for environmental protection and preservation of public health.

2. Materials and Methods

2.1. Sorbent Materials

The following materials were used: (a) four residues and by-products derived from the forestry industry, specifically eucalyptus leaves from plantations in the province of Lugo (Spain), pine bark (a commercial product of Geolia, Madrid), pine needles (from plantations located in the province of Lugo, Spain), and wood ash from a combustion boiler in Lugo (Spain); and (b) a waste/by-product from the food industry: crushed mussel shell (<1 mm in diameter), provided by Abonomar S.L. (Pontevedra, Spain). It should be noted that Galicia is one of the geographic areas with a higher production related to the forest industry, as well as in relation to mussel processing, which generates high amounts of by-products and wastes needing recycling.

The characterization of these sorbents was carried out following the methods detailed in the Supplementary Material.

2.2. Chemical Reagents

The CFX used (purity $\geq$ 95%) was supplied by Sigma-Aldrich (Barcelona, Spain). Phosphoric acid (85% extra pure) and acetonitrile (purity $\geq$ 99.9%) used for HPLC were supplied by Fisher Scientific (Madrid, Spain), and Ca_2Cl (95% purity) by Panreac (Barcelona, Spain). To carry out HPLC determinations, all solutions were prepared with milliQ water (Millipore, Madrid, Spain).

2.3. Adsorption and Desorption Experiments

Batch-type experiments were carried out to study the adsorption/desorption of CFX on/from the different bio-adsorbents. For this, 0.5 g of bio-adsorbent was weighed, and 10 mL of a solution with different concentrations of the antibiotic (2.5, 5, 5, 10, 20, 30, 40, and 50 μmol L^{-1}) was added, also containing 0.005 M of $CaCl_2$ as a background electrolyte. The suspensions were shaken for 48 h in the dark by means of a rotary shaker (this time being enough to reach equilibrium, according to previous kinetic tests, data not shown). These suspensions were then centrifuged at 4000 rpm for 15 min. The resulting supernatants were filtered through 0.45 μm nylon syringe filters, and the antibiotic concentration in the filtered liquids was determined by HPLC-UV with LPG 3400 SD equipment (Thermo-Fisher, Waltham, MA, USA).

For this, a Luna C18 column (150 mm long, 4.6 mm internal diameter, 5 μm particle size) was used, provided by Phenomenex (Madrid, Spain), as well as a pre-column (4 mm long, 2 mm in diameter, 5 μm particle size) packed with the same material as the column. The injection volume was 50 μL and the flow rate was 1.5 mL min^{-1}. The mobile phase consisted of acetonitrile (phase A) and 0.01 M of phosphoric acid (phase B). A linear gradient was used varying from 5% to 32% of phase A and from 95% to 68% of phase B. The initial conditions were re-established in 2 min and maintained for 2.5 min. The total analysis time was 15 min, with a retention time of 8.69 min, and the wavelength used for detection was 212 nm. The amounts of antibiotic adsorbed were calculated by the difference between the concentrations initially present in the added solutions and those remaining in the solutions at equilibrium. All determinations were made in triplicate. Figure S1 (Supplementary Material) shows some selected chromatograms corresponding to the quantification of CFX.

Once the adsorption experiments were carried out, desorption was studied, allowing the evaluation of the reversibility of the process. For this, a volume of 10 mL of 0.005 M of $CaCl_2$ (without antibiotic) was added to the material resulting from the adsorption process, and then the subsequent procedure carried out for adsorption was repeated. All determinations were made in triplicate.

2.4. Data Treatment

The experimental data obtained in the adsorption/desorption tests were adjusted to the Freundlich (Equation (1)), Langmuir (Equation (2)), and Linear (Equation (3)) models.

$$q_e = K_F C_{eq}{}^n \quad (1)$$

$$q_e = \frac{q_m K_L C_{eq}}{1 + K_L C_{eq}} \quad (2)$$

$$Kd = q_e / C_{eq} \quad (3)$$

where q_e is the amount of antibiotic retained by the bio-sorbent (calculated as the difference between the concentration added and that remaining in the equilibrium solution); C_{eq} is the concentration of antibiotic present in the solution at equilibrium; K_F is the Freundlich constant related to the adsorption capacity; n is a parameter of the Freundlich model associated with the degree of heterogeneity of the adsorption; K_L is the Langmuir adsorption constant; q_m is the maximum adsorption capacity according to the Langmuir model; and K_d is the partition coefficient in the linear model.

The SPSS Statistics 21 software was used to carry out the adjustment of the data derived from the adsorption experiments to the Langmuir, Freundlich, and Linear models, as well as statistical correlation studies among parameters of the bio-adsorbent materials and adsorption.

3. Results and Discussion

3.1. Characteristics of the Sorbent Materials

The bio-adsorbents used in this study showed some marked differences in their physicochemical properties (Table 1). The pH in water ranged from highly acidic values, such as that of pine needles (pH = 3.68), to clearly alkaline values, such as that of wood ash (pH = 11.31). The total C content also presented a wide range, going from 13.26% in wood ash to more than 50% in eucalyptus leaves and pine needles. Regarding the total N content, it ranged between 0.08% in pine bark and 1.35% in eucalyptus leaves.

Table 1. Characteristics of the bio-adsorbent materials. C: total carbon; N: total nitrogen; Ca_e, Mg_e, Na_c, K_e, Al_e: elements in the exchange complex; Sat. Al: Al-saturation in the exchange complex; eCEC: effective cation exchange capacity; X_T: total content of the element (X); Al_o, Fe_o: noncrystalline Al and Fe. Average values (n = 3), with coefficients of variation always <5%.

Parameter	Unit	Eucalyptus Leaves	Pine Bark	Pine Needles	Wood Ash	Mussel Shell
C	%	53.05	48.70	50.31	13.23	11.43
N	%	1.35	0.08	1.08	0.22	0.21
C/N		39.18	608.75	46.76	60.14	55.65
pH_{water}		4.88	3.99	3.68	11.31	9.39
pH_{KCl}		4.81	3.42	3.51	13.48	9.04
Ca_e	$cmol_c\ kg^{-1}$	7.95	5.38	2.13	95.0	24.75
Mg_e	$cmol_c\ kg^{-1}$	8.53	2.70	7.15	3.26	0.72
Na_e	$cmol_c\ kg^{-1}$	1.36	0.46	1.42	12.17	4.37
K_e	$cmol_c\ kg^{-1}$	12.93	4.60	11.09	250.65	0.38
Al_e	$cmol_c\ kg^{-1}$	0.13	1.78	2.15	0.07	0.03
eCEC	$cmol_c\ kg^{-1}$	30.90	14.92	23.94	361.15	30.25
Sat Al	%	0.42	11.93	8.98	0.02	0.10
Available-P	$mg\ kg^{-1}$	262.84	70.45	217.95	462.83	54.17
Na_T	$mg\ kg^{-1}$	242.31	68.92	271.54	2950	5174.00
MgT	mg/kg	840.96	473.55	653.40	26,171	980.66
Al_T	$mg\ kg^{-1}$	80.58	561.50	246.95	14,966	433.24
K_T	$mg\ kg^{-1}$	4464.10	737.84	4123.44	99,515	202.07
Ca_T	$mg\ kg^{-1}$	2262.96	2318.81	538.96	136,044	280,168
Cr_T	$mg\ kg^{-1}$	0.13	1.88	0.74	36.28	4.51
Mn_T	$mg\ kg^{-1}$	614.92	30.19	356.28	10,554	33.75
Fe_T	$mg\ kg^{-1}$	43.13	169.78	47.27	12,081	3535
Co_T	$mg\ kg^{-1}$	0.03	0.20	0.38	17.25	1.02
Ni_T	$mg\ kg^{-1}$	2.17	1.86	0.93	69.25	8.16
Cu_T	$mg\ kg^{-1}$	2.80	<LD	3.81	146.33	6.72
Zn_T	$mg\ kg^{-1}$	7.66	6.98	5.78	853.00	7.66
As_T	$mg\ kg^{-1}$	0.02	<LD	0.02	8.36	1.12
Cd_T	$mg\ kg^{-1}$	0.0	0.13	0.05	19.93	0.07
Al_o	$mg\ kg^{-1}$	45.0	315.0	169.0	8323	178.33
Fe_o	$mg\ kg^{-1}$	77.0	74.0	15.0	4233	171.0

The effective cation exchange capacity (eCEC) values also varied greatly depending on the type of bio-adsorbent, from 23.94 $cmol_c\ kg^{-1}$ for pine needles to 361.15 $cmol_c\ kg^{-1}$ for wood ash. In pine bark and mussel shell, the predominant exchangeable cation was Ca^{2+}, while it was K^+ in wood ash, eucalyptus leaves, and pine needles. In addition, noteworthy are the high values of available P present in wood ash, followed by eucalyptus leaves and pine needles (always higher than 200 mg kg^{-1}). The total contents of heavy metals were clearly higher in wood ash compared to the other bio-adsorbents, also presenting the highest concentrations of noncrystalline Fe and Al (Fe_o, Al_o), followed by mussel shell (Table 1).

3.2. Adsorption and Desorption of CFX

3.2.1. Adsorption

Figure 1 shows the adsorption curves of CFX for the different adsorbents. Table S1 (Supplementary Material) shows the values of the adsorbed amounts and adsorption percentages as a function of the concentration of antibiotic added. The maximum adsorption values were between 551.62 μmol kg^{-1} of pine bark and 927.95 μmol kg^{-1} of wood ash.

A statistical analysis showed that the maximum adsorption values for CFX were significantly and positively correlated with the eCEC of the adsorbents ($r = 0.927$, $p < 0.05$), with its Ca contents ($r = 0.918$, $p < 0.05$), its K contents ($r = 0.927$, $p < 0.05$), and its Na contents ($r = 0.903$, $p < 0.05$).

Comparing CFX adsorption data for the five adsorbents, wood ash showed the highest adsorption percentages, retaining practically 100% of the amount added, even for the highest concentration (50 μmol L^{-1}). For pine bark and pine needles, percentage adsorption values decreased as the CFX concentration added increased, going from 100% (for the lowest concentration added) to around 55% (for the highest concentration added). Eucalyptus leaves also presented a higher CFX adsorption percentage (94.26%) when low concentrations of the antibiotic were added, compared with a value of 67.29% reached when higher CFX concentrations were used, which is probably due to the progressive saturation of the adsorption sites. In the case of mussel shell, adsorption percentages varied to a lesser extent depending on the concentration of antibiotic added, ranging between 71.19% and 85.02%.

Comparing the results obtained for the different adsorbents, it is clear that the materials with the highest pH (mussel shell and wood ash) are those with the highest adsorption values, especially when the highest concentrations of the antibiotic are added. Similarly, Fakhri and Adami [30] pointed out that the parameter that most influences the adsorption of cephalosporins is pH, as it simultaneously affects the chemical speciation of the antibiotics and the adsorbent surfaces, making the adsorption of cephalosporins low at pH < 6, while a basic or alkaline medium favors the adsorption processes for these compounds. These same authors also studied the adsorption of CFX in different materials (magnesium oxide nanoparticles and carbon nanotubes), finding that sorbents with pH > 8 presented a higher adsorption, as has occurred in the present study with wood ash. The acid dissociation constant (pK_a) is an important parameter that determines the mobility/retention of organic compounds, and therefore their distribution in environmental compartments [8,31]. Cephalosporins have at least two dissociation constants ($pK_{a1} = 3.15$ and $pK_{a2} = 10.97$) [32,33], and their molecules can have a positive or negative charge, as well as behave as a zwitterion, depending on the pH of the medium. At pH above 10.97, they are in anionic form, while they are cations when the pH is below 3.15, and they will be zwitterions between these two pH values. In view of that, CFX would be a zwitterion in all the adsorbents used in this work (except wood ash), so the positively charged groups (NH_4^+) of the antibiotic will interact electrostatically with the surfaces of the negatively charged adsorbents, which are more abundant in adsorbents with a higher pH, while the anionic groups (COO^-) of CFX will bind to these surfaces through a cationic bridge [34]. In the case of wood ash, the pH value is higher than that of the pK_{a2} value of the antibiotic (10.97), so CFX would be negatively charged, as would certain components of the ash, specifically noncrystalline Fe and Al minerals (Table 1), favoring adsorption taking place through a cationic bridge, an interaction that will be also favored by the high concentration of changeable Ca^{2+} present in the ash.

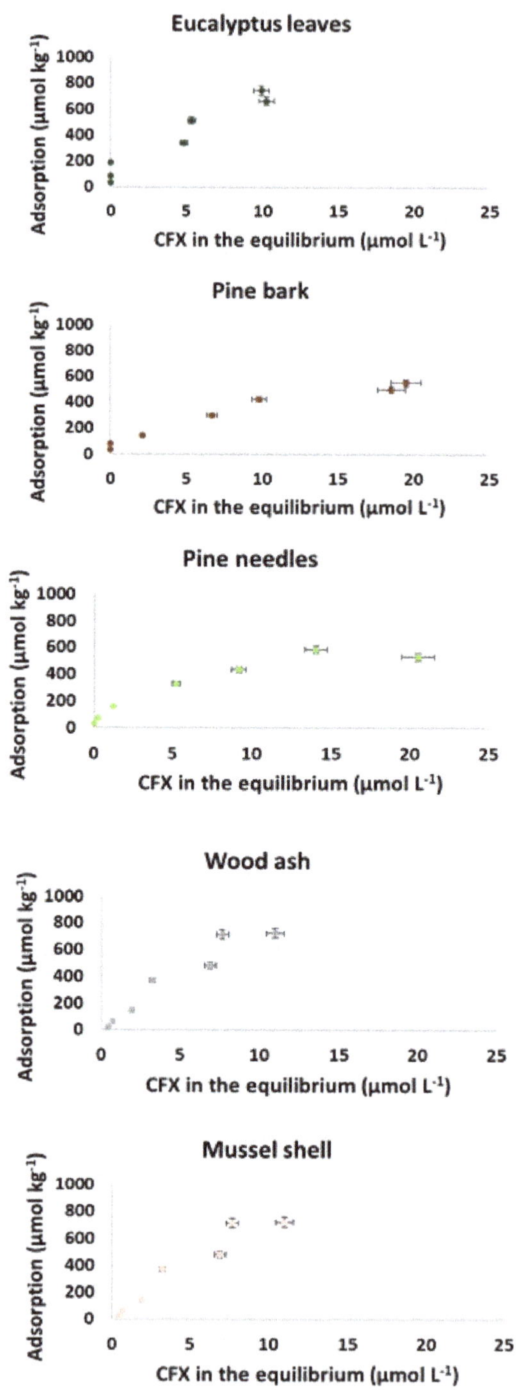

Figure 1. Adsorption curves for CFX and the five sorbent materials used.

When all five adsorbents were considered together, significant correlations were obtained between the amount of CFX adsorbed when the highest concentration of this antibiotic was added and the exchangeable cations (with values of r = 0.918, 0.903, and 0.927, for Ca, Na, and K, respectively; $p < 0.05$). This suggests their participation in bonds taking place by means of cationic bridges.

For three of the materials used in this study (wood ash, pine bark, and mussel shell), Conde-Cid et al. [35,36] studied in previous works their adsorption capacity for three tetracycline antibiotics (oxytetracycline, chlortetracycline, and tetracycline) and for three sulfonamides (sulfadiazine, sulfamethazine, and sulfachloropyridazine). In the case of tetracyclines, a high adsorption capacity was also obtained for wood ash, with pine bark also showing a high capacity to retain tetracyclines (unlike what was observed in the present study for CFX and pine bark), while the mussel shell was not suitable for the removal of these substances. Regarding the results of sulfonamide adsorption on those materials, Conde-Cid et al. [36] also obtained results showing differences in relation to the current work, with pine bark retaining practically 100% of the added sulfonamide, while wood ash and mussel shell were not effective in retaining these antibiotics.

Several authors have studied the adsorption of different cephalosporins on other materials [37–41]. For example, activated carbon obtained from different plant remains showed an adsorption capacity higher than 80% for cephalexin present in an aqueous solution [38,39,41]. In a study by Samarghandi et al. [42], the natural zeolites used retained 28% of the added cephalexin at pH 7, but the adsorption increased to 89% when these zeolites were coated with manganese oxide nanoparticles. Biochar made from pine wood showed a very high adsorption for different antibiotics and, therefore, can be considered promising in terms of treating contamination by these substances [14,43,44]. This kind of biochar reaches and even exceeds the adsorption capacity of powdered activated carbon biochar, a product that is commercially available but is very expensive [14,44]. It should be noted that, in the present study, residual materials such as mussel shell and especially wood ash showed high potentials to adsorb CFX, without performing additional modification treatments.

3.2.2. Modeling of Adsorption Data

The Langmuir isotherm assumes that the adsorption is homogeneous and in a single layer, with no interaction among the molecules of the sorbate. It is recognized that each sorbate molecule occupies a site, and no further adsorption can take place on it [45]. Unlike Langmuir's model, the Freundlich's model assumes that adsorption can occur in multiple layers, and that the adsorption sites are heterogeneous, with those with the highest energy being the first to be occupied, which means that there may be different functional groups involved in adsorption on the surface of the sorbent, with the intervention of different binding energies [45].

In the present study, the experimental adsorption data obtained were adjusted to the Freundlich (Equation (1)), Langmuir (Equation (2)), and Linear (related to the Henry's isotherm equation) (Equation (3)) models. Table 2 shows the parameters of the adsorption equations obtained from the adjustments to the three models. Figure 2 shows graphically the fitting to the adsorption models used.

Table 2. Fitting of the adsorption data to the Freundlich, Langmuir, and Linear models. K_F (L^n μmol^{1-n} kg^{-1}); K_L (L μmol^{-1}); n (dimensionless); q_m (μmol kg^{-1}); K_d (L kg^{-1}). -: standard error values too high for fitting.

	Freundlich					Langmuir					Linear Model		
Adsorbent	K_F	**Error**	n	**Error**	R^2	K_L	**Error**	q_m	**Error**	R^2	K_d	**Error**	R^2
Eucalyptus leaves	66.94	18.00	0.72	0.13	0.955	0.067	0.056	862.05	458.17	0.951	36.41	2.45	0.929
Pine bark	110.86	27.94	0.54	0.09	0.965	0.096	0.038	817.67	140.82	0.971	30.41	2.70	0.867
Pine needles	161.98	27.35	0.43	0.07	0.967	0.202	0.067	700.09	75.87	0.974	34.25	4.31	0.748
Wood ash	-	-	-	-	-	-	-	-	-	-	597.15	181.51	0.145
Mussel shell	93.25	34.91	0.77	0.18	0.909	0.077	0.067	1223.47	666.70	0.923	56.83	5.09	0.884

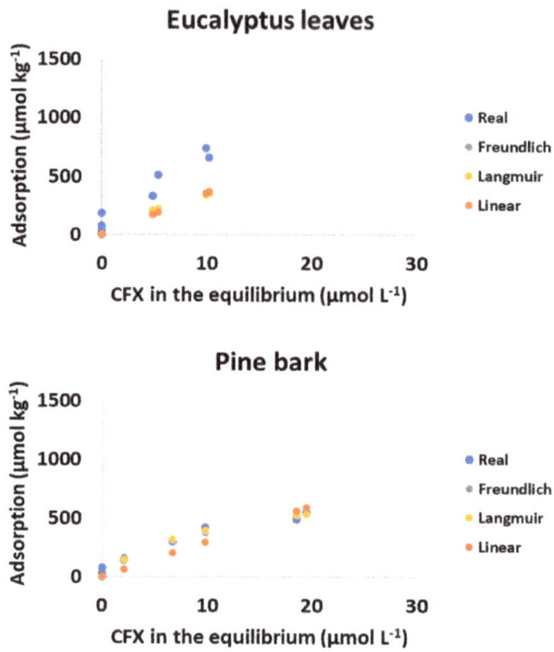

Figure 2. Adsorption curves for CFX also including data of fitting to the adsorption models used.

For the linear model, the R^2 values were lower than 0.90 in all materials (with the exception of eucalyptus leaves, $R^2 = 0.929$), with R^2 being very low in the case of wood ash. On the contrary, in the Freundlich and Langmuir models, R^2 values were higher than 0.9, with the exception of wood ash (bearing in mind that for this material, it was not possible to adjust both models, due to the existence of too high an error). In fact, CFX adsorption onto wood ash did not fit any model, as it adsorbed 100% of the concentrations of CFX added, which prevents the adjustment of the experimental data to the adsorption models used.

The Freundlich's K_F parameter, related to the adsorption capacity, varied between 66.9 and 162.0 L^n μmol^{1-n} kg^{-1}. The dimensionless Freundlich's n parameter, related to the heterogeneity of the active sites of the sorbent [46], showed values between 0.43 and 0.77, with the highest corresponding to mussel shell. All the adsorbents used in this study presented values of $n < 1$, which indicates a heterogeneous adsorption surface, where the highest energy sites are occupied first. This means that the adsorption energy decreases exponentially as the available surface is smaller [46,47]. It is also relevant that, the closer n is to zero, the more heterogeneous the adsorption surface will be [46,48,49].

3.2.3. Desorption

Table 3 shows data corresponding to CFX desorption from the different adsorbents, expressed in µmol kg^{-1} (and as a percentage, between parentheses). The adsorbents presenting the highest desorption values were eucalyptus leaves (ranging between 6.4 and 212.2 µmol kg^{-1}) and pine bark (ranging between 8.4 and 157.0 µmol kg^{-1}). Regarding the rest of the adsorbents (pine needles, mussel shell, and wood ash), the desorbed quantities did not exceed 20.7 µmol kg^{-1} in any case.

With desorption data expressed as percentages, it is confirmed that eucalyptus leaves and pine bark were the materials showing the highest desorption, with values ranging between 19.5 and 28.6% for eucalyptus leaves, and between 25.9 and 59.1% for pine bark. In the rest of the materials, desorption was less than 20.6% for pine needles, 12.3% for mussel shell, and 0% for wood ash. The lower desorption values of mussel shell and wood ash correspond to the presence of higher contents of noncrystalline Al and/or Fe compounds (Al_o, Fe_o) in these bio-adsorbents (Table 1).

Table 3. CFX desorption, expressed in μmol kg^{-1} (and in a percentage, between brackets), for the five different sorbents, as a function of the concentration of antibiotic added (C_0, in μmol L^{-1}). -: no data. Average values (n = 3), with coefficients of variation always <5%.

			C_0 (μmol L^{-1})				
Sorbent	**2.5**	**5**	**10**	**20**	**30**	**40**	**50**
Eucalyptus leaves	6.4(19.8)	16.3(20.2)	41.1(22.2)	65.9(19.5)	141.1(27.5)	164.5(24.9)	212.2(28.6)
Pine bark	8.4(25.9)	36.2(44.8)	84.0(59.1)	133.3(44.5)	157.0(37.1)	141.8(28.6)	146.5(26.6)
Pine needles	0(0)	15.4(20.5)	20.7(13.0)	19.5(5.9)	-	-	-
Wood ash	0(0)	-	-	0(0)	-	0(0)	0(0)
Mussel shell	0.3(12.2)	0.6(8.4)	1.6(10.7)	1.5(3.9)	1.4(2.9)	1.5(2.1)	1.5(2.1)

In three of these materials (oak ash, pine bark, and mussel shell), previous works studied the desorption of other groups of antibiotics, specifically three tetracyclines and three sulfonamides. For the former, Conde-Cid et al. [35] also obtained a low desorption from wood ash, but (unlike what was observed in the present study with CFX) these authors found that pine bark adsorbed tetracyclines in a way that was practically irreversible, while desorption from mussel shell was high (up to 44% of what was adsorbed). In the case of sulfonamides, Conde-Cid et al. [36,50] (2021, 2020) reported that only pine bark retained irreversibly high concentrations of these antibiotics, while wood ash and mussel shell had a low adsorption capacity and desorbed a high percentage of what was previously retained.

Considering together data on CFX adsorption and desorption for the five different adsorbents used, wood ash presented the best results, with the highest adsorption and the lowest desorption values. In previous studies, this material was also found to be very effective for the irreversible adsorption of tetracyclines [35]. Mussel shell also has potential utility for retaining CFX present in polluted media. However, pine bark could not be recommended for the adsorption of CFX, as it had low adsorption and high desorption, especially for the highest concentrations added, despite the fact that in previous studies, it showed great effectiveness to strongly retain tetracyclines and sulfonamides [35].

4. Conclusions

Among the five sorbents evaluated, the most effective for the adsorption of the antibiotic cefuroxime (CFX) were wood ash and mussel shell. Both materials were those with the highest pH values and were also characterized by their richness in noncrystalline compounds. Both sorbents would be of high interest for being used in CFX retention/removal processes, which would contribute to their recycling. The rest of the materials studied (pine bark, pine needles, and eucalyptus leaves) could not be recommended for retention/removal of this antibiotic, as they showed low adsorption and high desorption when the highest concentrations of CFX were added.

CFX adsorption generally showed a good fit to the Langmuir isotherm, and especially to the Freundlich model. Furthermore, in these materials, the Freundlich's n values were always lower than 1, which would indicate the relevance of heterogeneous adsorption sites, with those with the highest energy being the first to be occupied. Future additional studies could focus on delving into the retention mechanisms of CFX and other antibiotics (especially cephalosporins) in the sorbents that showed better results in the current work. Likewise, the influence of the simultaneous presence of several antibiotics, or of antibiotics and other contaminants, both organic and inorganic, could be evaluated, as well as the impact of modifying the values of different variables that could affect the final effectivity in the retention/release of the pollutants. Globally, the results of this study can be considered relevant at an environmental level, in relation to the potential promotion of waste and by-product recycling, protection against pollution, and its potential repercussions on aspects that affect public health.

Supplementary Materials: The following are available online at https://www.mdpi.com/article/10.3390/pr9071151/s1, Table S1: Adsorption amounts (μmol kg^{-1}) and percentages for the five adsorbents used, after adding the different initial concentrations (C_0) of the antibiotic CFX; Figure S1. Selected chromatograms corresponding to the detection of CFX after adding various concentrations of the antibiotic to the different adsorbents used in the study.

Author Contributions: Conceptualization, E.Á.-R., M.J.F.-S., A.N.-D. and M.A.-E.; methodology, E.Á.-R., M.J.F.-S., A.N.-D., M.A.-E., R.C.-D. and C.N.; software, E.Á.-R. and R.C.-D.; validation, E.Á.-R., M.J.F.-S., A.N.-D., M.A.-E., D.F.-C. and C.N.; formal analysis, R.C.-D., C.N. and L.R.L.; investigation, E.Á.-R., M.J.F.-S. and R.C.-D.; resources, E.Á.-R., M.J.F.-S. and M.A.-E.; data curation, E.Á.-R., M.J.F.-S., A.N.-D. and C.N.; writing—original draft preparation, E.Á.-R., M.J.F.-S. and R.C.-D.; writing—review and editing, A.N.-D.; visualization, E.Á.-R., M.J.F.-S., A.N.-D., M.A.-E., D.F.-C., R.C.-D., C.N. and L.R.L.; supervision, E.Á.-R., M.J.F.-S. and C.N.; project administration, E.Á.-R., M.J.F.-S. and M.A.-E.; funding acquisition, E.Á.-R., M.J.F.-S. and M.A.-E. All authors have read and agreed to the published version of the manuscript.

Funding: This research was funded by SPANISH MINISTRY OF SCIENCE, INNOVATION AND UNIVERSITIES, grant numbers RTI2018-099574-B-C21 and RTI2018-099574-B-C22.

Conflicts of Interest: The authors declare no conflict of interest. The funders had no role in the design of the study; in the collection, analyses, or interpretation of data; in the writing of the manuscript, or in the decision to publish the results.

References

1. Klein, E.Y.; Vanboeckel, T.P.; Martinez, E.M.; Pant, S.; Gandra, S.; Levin, S.A.; Goossens, H.; Laxminarayan, R. Global increase and geographic convergence in antibiotic consumption between 2000 and 2015. *Proc. Natl. Acad. Sci. USA* **2018**, *115*, 3463–3470. [CrossRef]
2. Wei, M.; Lv, D.; Cao, L.; Zhou, K.; Jiang, K. Adsorption behaviours and transfer simulation of levofloxacin in silty clay. *Environ. Sci. Pollut. Res.* **2021**. [CrossRef]
3. Zhao, R.X.; Feng, J.; Liu, J.; Fu, W.J.; Li, X.Y.; Li, B. Deciphering of microbial community and antibiotic resistance genes in activated sludge reactors under high selective pressure of different antibiotics. *Water Res.* **2019**, *151*, 388–402. [CrossRef]
4. Kovalakova, P.; Cizmas, L.; McDonald, T.J.; Marsalek, B.; Feng, M.; Sharma, V.K. Occurrence and toxicity of antibiotics in the aquatic environment: A review. *Chemosphere* **2020**, *251*, 126351–126361. [CrossRef]
5. Ren, X.; Liu, D.; Ding, N.; Huang, K.; Xiong, Y.; Du, G.; Zeng, F. Safety evaluation of cephalosporins based on utilization and adverse drug events: Analysis of two databases in china. *Expert Opin. Drug Saf.* **2012**, *11*, 689–697. [CrossRef]
6. Hu, X.; Sun, T.; Jia, L.; Wei, J.; Sun, Z. Preparation of metal-organic framework based carbon materials and its application to adsorptive removal of cefepime from aqueous solution. *J. Hazard. Mater.* **2020**, *390*, 122190–122198. [CrossRef]
7. Dancer, S.J. The problem with cephalosporins. *J. Antimicrob. Chemother.* **2001**, *48*, 463–478. [CrossRef] [PubMed]
8. Ribeiro, A.R.; Sures, B.; Schmidt, T.C. Cephalosporin antibiotics in the aquatic environment: A critical review of occurrence, fate, ecotoxicity and removal technologies. *Environ. Pollut.* **2018**, *241*, 1153–1166. [CrossRef] [PubMed]
9. Ray, P.; Knowlton, K.F.; Shang, C.; Xia, K. Development and validation of a UPLC-MS/MS method to monitor Cephapirin excretion on dairy cows following intramammary infusion. *PLoS ONE* **2014**, *9*, 112343–112355. [CrossRef] [PubMed]
10. EMA (European Medicines Agency). European Surveillance of Veterinary Antimicrobial Consumption. Sales of Veterinary Antimicrobial Agents in 29 European Countries in 2014. EMA/61769/2016. Available online: https://www.ema.europa.eu/en/documents/report/sixth-esvac-report-sales-veterinary-antimicrobial-agents-29-european-countries-2014_en.pdf (accessed on 10 June 2021).
11. Versporten, A.; Coenen, S.; Adriaenssens, N.; Muller, A.; Minalu, G.; Faes, C.; Vankerckhoven, V.; Aerts, M.; Hens, N.; Molenberghs, G.; et al. European Surveillance of Antimicrobial Consumpetion (ESAC): Outpatient cephalosporin use in Europe (1997–2009). *J. Antimicrob. Chemother.* **2011**, *66*, 25–35.
12. Sarmah, A.K.; Meyer, M.T.; Boxall, A.B.A. A global perspective on the use, sales exposure pathways, occurrence, fate and effects of veterinary antibiotics (VAs) in the environment. *Chemosphere* **2006**, *65*, 725–759. [CrossRef]
13. Duan, H.; Hu, X.; Sun, Z. Magnetic zeolite imidazole framework material-8 as an effective and recyclabe adsorbent for removal of ceftazidime from aqueous solution. *J. Hazard. Mater.* **2020**, *384*, 121406–121414. [CrossRef]
14. Russell, J.N.; Yost, C.K. Alternative, environmentally conscious approaches for removing antibiotics from wastewater treatment systems. *Chemosphere* **2021**, *263*, 128177–128187. [CrossRef] [PubMed]
15. Gupta, A.; Garg, A. Degradation of ciprofloxacin using Fenton's oxidation: Effect of operating parameters, identification of oxidized by-products and toxicity assessment. *Chemosphere* **2018**, *193*, 1181–1188. [CrossRef] [PubMed]
16. Perini, J.A.L.; Tonetti, A.L.; Vidal, C.; Montagner, C.C.; Nogueira, R.F.P. Simultaneous degradation of ciprofloxacin, amoxicillin, sulfathiazole and sulfamethazine, and disinfection of hospital effluent after biological treatment via photo-Fenton process under ultraviolet germicidal irradiation. *Appl. Catal. B* **2018**, *224*, 761–771. [CrossRef]

17. Becker, D.; Giustina, S.V.D.; Rodríguez-Mozaz, S.; Schoevaart, R.; Barceló, D.; de Cazes, M.; Belleville, M.; Sanchez-Marcano, J.; de Gunzburg, J.; Couillerot, O.; et al. Removal of antibiotics in wastewater by enzymatic treatment with fungal laccase—Degradation of compounds does not always eliminate toxicity. *Bioresour. Technol.* **2016**, *219*, 500–509. [CrossRef]
18. Guo, R.; Chen, J. Application of alga-activated sludge combined system (AASCS) as a novel treatment to remove cephalosporins. *Chem. Eng. J.* **2015**, *260*, 550–556. [CrossRef]
19. Iatrou, E.I.; Stasinakis, A.S.; Thomaidis, N.S. Consumption-based approach for predicting environmental risk in Greece due to the presence of antimicrobials in domestic wastewater. *Environ. Sci. Pollut. Res.* **2014**, *21*, 12941–12950. [CrossRef]
20. Richmond, E.K.; Rosi, E.J.; Walters, D.M.; Fick, J.; Hamilton, S.K.; Brodin, T.; Sundelin, A.; Grace, M.R. A diverse suite of pharmaceuticals contaminates stream and riparian food webs. *Nat. Commun.* **2018**, *9*, 4491. [CrossRef] [PubMed]
21. Marchant, J. When antibiotics turn toxic. *Nature* **2018**, *555*, 431–433. [CrossRef] [PubMed]
22. Cela-Dablanca, R.; Nebot, C.; López, L.R.; Ferández-Calviño, D.; Arias-Estévez, M.; Núñez-Delgado, A.; Álvarez-Rodríguez, E.; Fernández-Sanjurjo, M.J. Retention of the Antibiotic Cefuroxime onto Agricultural and Forest Soils. *Appl. Sci.* **2021**, *11*, 4663. [CrossRef]
23. Pan, M.; Chu, L.M. Occurrence of antibiotics and antibiotic resistance genes in soils from waste water irrigation areas in the Pearl River Delta region, southern China. *Sci. Total Environ.* **2017**, *12*, 145–152.
24. Ding, H.; Wu, Y.; Zou, B.; Lou, Q.; Zhang, W.; Zhong, J.; Lu, L.; Dai, G. Simultaneous removal and degradation characteristics of sulfonamide, tetracycline, and quinolone antibiotics by laccase-mediated oxidation coupled with soil adsorption. *J. Hazard. Mater.* **2016**, *307*, 350–358. [CrossRef]
25. Azhar, M.R.; Abid, H.R.; Periasamy, V.; Sun, H.; Tade, M.O.; Wang, S. Adsorptive removal of antibiotic sulfonamide by UiO-66 and ZIF-67 for wastewater treatment. *J. Colloid Interface Sci.* **2017**, *500*, 88–95. [CrossRef]
26. Malakootian, M.; Yaseri, M.; Faraji, M. Removal of antibiotics from aqueous solutions by nanoparticles: A systematic review and meta-analysis. *Environ. Sci. Pollut. Res.* **2019**, *26*, 8444–8458. [CrossRef] [PubMed]
27. Ata, R.; Sacco, O.; Vaiano, V.; Rizzo, L.; Tore, G.Y.; Sannino, D. Visible light active N- doped TiO_2 immobilized on polystyrene as efficient system for wastewater treatment. *J. Photochem. Photobiol. A* **2017**, *348*, 255–262. [CrossRef]
28. Crisafully, R.; Milhome, M.A.L.; Cavalcante, R.M.; Silveira, E.R.; De Keukeleire, D.; Nascimiento, R.F. Removal of some polycyclic aromatic hydrocarbons from petrochemical wastewater using low-cost adsorbents of natural origin. *Bioresour. Technol.* **2008**, *99*, 4515–4519. [CrossRef] [PubMed]
29. Núñez-Delgado, A.; Álvarez-Rodríguez, E.; Fernández-Sanjurjo, M.J.; Nóvoa-Muñoz, J.C.; Arias-Estévez, M.; Fernández-Calviño, D. Perspectives on the use of by-products to treat soil and water pollution. *Microporous Mesoporous Mat.* **2015**, *210*, 199–201. [CrossRef]
30. Fakhri, A.; Adami, S. Adsorption and thermodynamic study of Cephalosporins antibiotics from aqueous solution onto MgO nanoparticles. *J. Taiwan Inst. Chem. Eng.* **2014**, *45*, 1001–1006. [CrossRef]
31. Kümmerer, K. The presence of pharmaceuticals in the environment due to human use—present knowledge and future challenges. *J. Environ. Manag.* **2009**, *90*, 2354–2366. [CrossRef]
32. Ribeiro, A.R.; Schmidt, T.C. Determination of acid dissociation constants (pKa) of cephalosporin antibiotics: Computational and experimental approaches. *Chemosphere* **2017**, *169*, 524–533. [CrossRef]
33. Evagelou, V.; Tsantili-Kakoulidou, A.; Koupparis, M. Determination of the dissociation constants of the cephalosporins cefepime and cefpirome using UV spectrometry and pH potentiometry. *J. Pharmac. Biomed. Anal.* **2003**, *31*, 1119–1128. [CrossRef]
34. Legnoverde, M.S.; Simonetti, S.; Basaldella, E.I. Influence of pH on cephalexin adsorption onto SBA-15 mesoporous silica: Theoretical and experimental study. *Appl. Surf. Sci.* **2014**, *300*, 37–42. [CrossRef]
35. Conde-Cid, M.; Ferreira-Coelho, G.; Arias-Estévez, M.; Álvarez-Esmorís, C.; Nóvoa-Muñoz, J.C.; Núñez-Delgado, A.; Fernández-Sanjurjo, M.J.; Álvarez-Rodríguez, E. Competitive adsorption/desorption of tetracycline, oxytetracycline and chlortetracycline on pine bark, oak ash and mussel shell. *J. Environ. Manag.* **2019**, *250*, 109509–109519. [CrossRef]
36. Conde-Cid, M.; Cela-Dablanca, R.; Ferreira-Coelho, G.; Fernández-Calviño, D.; Núñez-Delgado, A.; Fernández-Sanjurjo, M.J.; Arias-Estévez, M.; Álvarez-Rodríguez, E. Sulfadiazine, sulfamethazine and sulfachloropyridazine removal using three different porous materials: Pine bark, "oak ash" and mussel shell. *Environ. Res.* **2021**, *195*, 110814–110820. [CrossRef]
37. Liu, H.; Liu, W.; Zhang, J.; Zhang, C.; Ren, L.; Li, Y. Removal of cephalexin from aqueous solutions by original and Cu (II)/Fe (III) impregnated activated carbons developed from stalks kinetics and equilibrium studies. *J. Hazard. Mater.* **2011**, *185*, 1528–1535. [CrossRef]
38. Ahmed, M.J.; Theydan, S.K. Adsorption of cephalexin onto active carbons from *Albizia lebbeck* seed pods by microwave induced KOH and K_2CO_3 activations. *Chem. Eng. Sci.* **2012**, *211–212*, 200–207. [CrossRef]
39. Pouretedal, H.R.; Sadegh, N. Effective removal of Amoxicillin, Cephalexin, Tetracycline and Penicillin G from aqueous solutions using activated carbon nanoparticles prepared from vine wood. *J. Water Process. Eng.* **2014**, *1*, 64–73. [CrossRef]
40. Mitchell, S.M.; Subbiah, M.; Ullman, J.L.; Frear, C.; Call, D.R. Evaluation of 27 different biochars for potential sequestration of antibiotic residues in food animal production environments. *J. Environ. Chem. Eng.* **2015**, 3, 162–169. [CrossRef]
41. Nazari, G.; Abolghasemi, H.; Esmaieli, M.; Pouya, E.S. Aqueous phase adsorption of cephalexin by walnut shell-based activated carbon: A fixed-bed column study. *Appl. Surf. Sci.* **2016**, *375*, 144–153. [CrossRef]
42. Samarghandi, M.R.; Al-Musawi, T.J.; Mohsensi-Bandpi, A.; Zarrabi, M. Adsorption of cephalexin from aqueous solution using natural zeolite and zeolite coated with manganese oxide nanoparticles. *J. Mol. Liq.* **2015**, *211*, 431–441. [CrossRef]

43. Taheran, M.; Naghdi, M.; Brar, S.K.; Knystautas, E.J.; Verma, M.; Ramirez, A.A.; Surampalli, R.Y.; Valero, J.R. Adsorption study of environmentally relevant concentrations of chlortetracycline on pinewood biochar. *Sci. Total Environ.* **2016**, *571*, 772–777. [CrossRef]
44. Shimabuku, K.K.; Kearns, J.P.; Martinez, J.E.; Mahoney, R.B.; Moreno-Vasquez, L.; Summer, R.S. Biochar sorbents for sulfamethoxazole removal from surface stormwater, and wastewater effluent. *Water Res.* **2016**, *96*, 236–245. [CrossRef]
45. Jafari, M.; Aghamiri, S.F.; Khaghanic, G. Batch Adsorption of Cephalosporins Antibiotics from Aqueous Solution by Means of Multi-Walled Carbon Nanotubes. *World Appl. Sci. J.* **2011**, *14*, 1642–1650.
46. Foo, K.Y.; Hameed, B.H. Insights into the modeling of adsorption isotherm systems. *Chem. Eng. J.* **2010**, *156*, 2–10. [CrossRef]
47. Behnajady, M.A.; Bimeghdar, S. Synthesis of mesoporous NiO nanoparticles and their application in the adsorption of Cr (VI). *Chem. Eng. J.* **2014**, *239*, 105–113. [CrossRef]
48. Kong, W.; Li, C.; Dolhi, J.M.; Li, S.; He, J.; Qiao, M. Characteristics of oxytetracycline sorption and potential bioavailability in soils with various physical–chemical properties. *Chemosphere* **2012**, *87*, 542–548. [CrossRef]
49. De Arsénio, S.; Abreu, A.S.; Moura, I.; Vera-Machado, A. Polymeric materials for metal sorption from hydric resources. *Water Purif.* **2017**, 289–322. [CrossRef]
50. Conde-Cid, M.; Fernández-Calviño, D.; Núñez-Delgado, A.; Fernández-Sanjurjo, M.J.; Arias-Estévez, M.; Álvarez-Rodríguez, E. Influence of mussel shell, oak ash and pine bark on the adsorption and desorption of sulfonamides in agricultural soils. *J. Environ. Manag.* **2020**, *261*, 110221–110231. [CrossRef]

processes

MDPI

Article

Thermodynamic Analysis of Brownian Motion-Induced Particle Agglomeration Using the Taylor-Series Expansion Method of Moments

Qing He [1,*] and Mingliang Xie [2,*]

[1] Guangdong Provincial Key Laboratory of Distributed Energy Systems, School of Chemical Engineering and Energy Technology, Dongguan University of Technology, Dongguan 523808, China
[2] State Key Laboratory of Coal Combustion, School of Energy and Power Engineering, Huazhong University of Science and Technology, Wuhan 430074, China
* Correspondence: heqing@dgut.edu.cn (Q.H.); mlxie@mail.hust.edu.cn (M.X.)

Abstract: On the basis of binary perfectly inelastic collision theory, the time evolutions of kinetic energy and surface area for a particle agglomerate system, due to Brownian motion, are investigated by using the Taylor series expansion technology. The asymptotic behaviors over a long time period show a significantly negative power function of time. The thermodynamic constraints of this system are then obtained according to the principle of maximum entropy, which establishes a relationship of inequality between the first three particle moments and some physical parameters (i.e., surface tension and temperature). In the thermodynamic equilibrium state, this function provides a new approach for estimating the effect of molecular structure on surface tension of liquid polymers.

Keywords: thermodynamic equilibrium; entropy criterion; Brownian agglomeration; moment method; asymptotic solutions; population balance equation

Citation: He, Q.; Xie, M. Thermodynamic Analysis of Brownian Motion-Induced Particle Agglomeration Using the Taylor-Series Expansion Method of Moments. *Processes* **2021**, *9*, 1218. https://doi.org/10.3390/pr9071218

Academic Editor: Avelino Núñez-Delgado

Received: 19 June 2021
Accepted: 13 July 2021
Published: 15 July 2021

1. Introduction

Particle agglomeration is a common phenomenon in both nature and industrial applications, such as particle synthesis and soot formation processes. It plays a significant role in these aerosol processes by profoundly affecting the size distribution of a particle system [1], which strongly determines the physical properties of aerosol particles, such as light scattering, toxicity, deposition rate and diffusion. Nowadays with the escalation of fine particle pollution, the agglomeration processes are also widely used in the field of contamination control to improve removal efficiency, especially for the particles whose diameters are less than 2.5 μm [2,3]. The main principle is that through physical or chemical action, particles can coagulate with each other to form particles with larger particle size and then be removed efficiently. An appropriate approach for investigating the time evolution of particle size distribution (PSD) due to agglomeration is typically called the population balance equation (PBE) or the classic Smoluchowski equation (SE), which can be expressed as the following form [4]:

$$\frac{\partial n(v,t)}{\partial t} = \frac{1}{2}\int_0^v \beta(v_1, v - v_1)n(v_1,t)n(v - v_1,t)dv_1 - \int_0^\infty \beta(v_1, v)n(v,t)n(v_1,t)dv_1 \quad (1)$$

where $n(v, t)$ is the number density function of the particles with volume from v to $v + dv$ at time t; $\beta(v, v_1)$ is the collision frequency function between particles with volume v and v_1.

Due to the strong non-linear integro-differential structure, the PBE is difficult to solve analytically. By trading off between accuracy and computational cost, three main numerical methods are proposed and developed, including the method of moments (MOM) [5,6], sectional method (SM) [7] and Monte Carlo method (MCM) [8]. It can't be ignored that the analytical solutions show great merit in computational cost and direct physical insights into agglomeration mechanisms. Thus, some researchers focus on the asymptotic or analytical solutions of the moments of PSD by converting the original PBE to a system of ordinary differential equations (ODEs). Due to the complexity of its kernel function and the universality of Brownian motion, the study on the solution of the PBE for Brownian agglomeration is considered to be important but one of the most difficulties. Mainly using the log-normal method of moments (LG-MOM) [6] or the Taylor series expansion method of moments (TEMOM) [9], the asymptotic behavior of moments, due to Brownian coagulation (for spherical particles) and agglomeration (for agglomerates) over the entire particle size regimes [10–13], the analytical solution for Brownian coagulation in the free-molecule and the continuum regime [14], and so on, are obtained. These articles reveal that the geometric standard deviation will reach a constant for a long period of time, namely, the self-preserving size distribution theory [15].

Particle coalescence upon collisions subject to conservation of mass and momentum is called ballistic aggregation, but it is well known that the kinetic energy of this system decreases with time [16]. However, the loss of particle kinetic energy after collisions is rarely taken into account in the framework of PBE. Nowadays, with an assumption of a perfectly inelastic collision process, the rate of change for kinetic energy is correlated with that of particle number density, and the relationship of inequality between particle moments and some physical parameters (i.e., surface tension and temperature) for Brownian coagulation have been firstly proposed by Xie and Yu based on the principle of maximum entropy [17,18]. In this paper, we will extend their efforts to Brownian agglomeration, and the asymptotic behaviors of kinetic energy, surface area and entropy over a long period of time are obtained.

2. Theory and Model

2.1. Brownian Agglomeration

Particle agglomeration due to thermal motion is called Brownian agglomeration. Unlike spherical particles, agglomerates are not rigid structures and can be described as fractal morphology statistically. They are clusters of primary particles, which are ideally considered to be spherical with point contacts and uniform size. Considering the case of monodisperse primary particles, which form power law agglomerates, the Brownian agglomeration kernels β are represented as [1]:

$$\beta_{FM} = B_1(v_i^{-1} + v_j^{-1})^{1/2}(v_i^{1/D_f} + v_j^{1/D_f})^2 \tag{2}$$

$$\beta_{CR} = B_2(v_i^{1/D_f} + v_j^{1/D_f})(v_i^{-1/D_f} + v_j^{-1/D_f}) \tag{3}$$

Here, the subscripts FM and CR stand for agglomeration in the free molecular and continuum regimes, respectively; the constants $B_1 = \left(\frac{6k_BT}{\rho_p}\right)^{\frac{1}{2}}(3/4\pi)^{2/D_f-1/2}a_{p0}^{2-6/D_f}$ and $B_2 = 2k_BT/3\mu$, with k_B the Boltzmann's constant; T is the temperature; μ is the gas viscosity; ρ_p is the particle density; a_{p0} is the radius of a primary particle; v is the particle volume; D_f is called the fractal dimension, which can be related to the arrangement of the primary particles within an agglomerate. It should be noted that $D_f < 2$ is not applicable for Equation (2) in physics [1], thus the following discussions are limited to a range of $2 \leq D_f \leq 3$.

2.2. Taylor Series Expansion Method of Moments

With the definition of k-th order moment M_k,

$$M_k = \int_0^\infty v^k n(v)dv \tag{4}$$

Equation (1) can be converted into a system of original differential equations by multiplying both sides with v^k and then integrating over all particle sizes:

$$\frac{dM_k}{dt} = \frac{1}{2}\int_0^\infty \int_0^\infty \left[(v+v_1)^k - v^k - {v_1}^k\right]\beta(v,v_1)n(v,t)n(v_1,t)dvdv_1 \tag{5}$$

The main objective of all MOMs is to achieve the closure of Equation (5). In the classic TEMOM, this is accomplished in two procedures [9,19]: (1) the collision kernel is directly approximated by a two-variable third-order Taylor series expansion, for example, the power function $\left(v_i^{-1} + v_j^{-1}\right)^{1/2}$ in Equation (2) can be expanded with respect to mean volume $u = M_1/M_0$; (2) and all the higher and fractional moments are approximated by the polynomial equation with respect to the first three moments:

$$M_k = \frac{M_1^k}{M_0^{k-1}}\left[1 + \frac{k(k-1)(M_C-1)}{2}\right] \tag{6}$$

Here the dimensionless moment $M_C = M_0M_2/{M_1}^2$ is the function of geometric standard deviation σ which can be noted as $\ln(M_C)/9 = \ln^2\sigma$ [6]. The moment equations based on TEMOM in the free molecule regime are obtained [11]:

$$\begin{aligned} \left.\frac{dM_0}{dt}\right|_{FM} &= -\frac{\sqrt{2}B_1{M_0}^2}{64D_f^4}\left(\frac{M_1}{M_0}\right)^{\frac{4-D_f}{2D_f}}\left(a_1{M_C}^2 + a_2M_C + a_3\right) \\ \left.\frac{dM_1}{dt}\right|_{FM} &= 0 \\ \left.\frac{dM_2}{dt}\right|_{FM} &= -\frac{\sqrt{2}B_1{M_1}^2}{32D_f^4}\left(\frac{M_1}{M_0}\right)^{\frac{4-D_f}{2D_f}}\left(b_1{M_C}^2 + b_2M_C + b_3\right) \end{aligned} \tag{7}$$

where the coefficients $a_1, a_2, a_3, b_1, b_2, b_3$ are:

$$\begin{aligned} a_1 &= D_f^4 - 24D_f^3 + 70D_f^2 - 48D_f + 16 \\ a_2 &= 54D_f^4 - 144D_f^3 + 52D_f^2 + 96D_f - 32 \\ a_3 &= 73D_f^4 + 168D_f^3 - 122D_f^2 - 48D_f + 16 \\ b_1 &= 3D_f^4 + 16D_f^3 + 10D_f^2 - 16D_f - 16 \\ b_2 &= 2D_f^4 - 96D_f^3 - 212D_f^2 + 32D_f + 32 \\ b_3 &= -133D_f^4 + 80D_f^3 + 202D_f^2 - 16D_f - 16 \end{aligned} \tag{8}$$

Now the most important moments for describing the particle dynamics, namely, the particle number density M_0, total particle volume M_1 and a polydispersity variable M_2, can be obtained. Here, M_1 remains constant due to the rigorous mass conservation requirement. The corresponding moment equations in the continuum regime are:

$$\begin{aligned} \left.\frac{dM_0}{dt}\right|_{CR} &= -\frac{B_2{M_0}^2}{4D_f^4}\left(p_1{M_C}^2 + p_2M_C + p_3\right) \\ \left.\frac{dM_1}{dt}\right|_{CR} &= 0 \\ \left.\frac{dM_2}{dt}\right|_{CR} &= \frac{B_2{M_1}^2}{2D_f^4}\left(p_1{M_C}^2 + p_2M_C + p_3\right) \end{aligned} \tag{9}$$

where the coefficients p_1, p_2, p_3 are:

$$p_1 = 1 - D_f^2;\ p_2 = -2 + 6D_f^2;\ p_3 = 1 - 5D_f^2 + 8D_f^4 \tag{10}$$

2.3. Principle of Maximum Entropy

As a characteristic function composed of internal energy U, total particle volume M_1, and particle number M_0, the rate of change for entropy S of a disperse system can be expressed as:

$$\frac{dS}{dt} = \frac{\partial S}{\partial M_0}\frac{dM_0}{dt} + \frac{\partial S}{\partial U}\frac{dU}{dt} + \frac{\partial S}{\partial M_1}\frac{dM_1}{dt} \tag{11}$$

According to the thermodynamic analysis [18], the rate of change for S can be arranged and then correlated with that of M_0, the particle kinetic energy k_e, and the particle specific surface area s:

$$\frac{dS}{dt} = -k_B \ln(M_0 \lambda_{th}^3)\frac{dM_0}{dt} + \frac{1}{T}\left(\frac{dk_e}{dt} + \gamma\frac{ds}{dt}\right) \tag{12}$$

in which λ_{th} is the thermal wavelength and γ is the surface tension. Thus, the focal point is to determine dk_e/dt and ds/dt. With the assumption of simplified physical model according to the binary perfectly inelastic collision theory, the loss of particle kinetic energy after collision for two colliding particles and the whole system are [18]:

$$\Delta k_e = -\frac{k_b T}{2}\left(1 - \frac{2\sqrt{v_1 v_2}}{v_1 + v_2}\right) \leq 0 \tag{13}$$

$$\frac{dk_e}{dt} = -\frac{k_B T}{4}\int_0^\infty \int_0^\infty \left(1 - \frac{2\sqrt{v_1 v_2}}{v_1 + v_2}\right)\beta(v_1, v_2) n(v_1, t) n(v_2, t) dv_2 dv_1 \tag{14}$$

Assuming that v_1 is the larger particle, the relative loss of k_e increases with a larger ratio of v_1 to v_2, which is illustrated in Figure 1. This shows that a wider range of PSD, namely, a larger M_C, would lead to a more rapid reduction in k_e. Substituting Equation (2) into the above equation and then using the Taylor series expansion technology, we can get the rate of change for kinetic energy in the free molecular regime:

$$\left.\frac{dk_e}{dt}\right|_{FM} = k_B T \frac{x_1 M_C^2 + x_2 M_C + x_3}{a_1 {M_C}^2 + a_2 M_C + a_3}\frac{dM_0}{dt} \tag{15}$$

in which x_1, x_2, x_3 are:

$$\begin{aligned} x_1 &= \left(D_f^4 - 12D_f^3 + 8D_f^2\right)/2 \\ x_2 &= 15D_f^4 + 12D_f^3 - 8D_f^2 \\ x_3 &= \left(-31D_f^4 - 12D_f^3 + 8D_f^2\right)/2 \end{aligned} \tag{16}$$

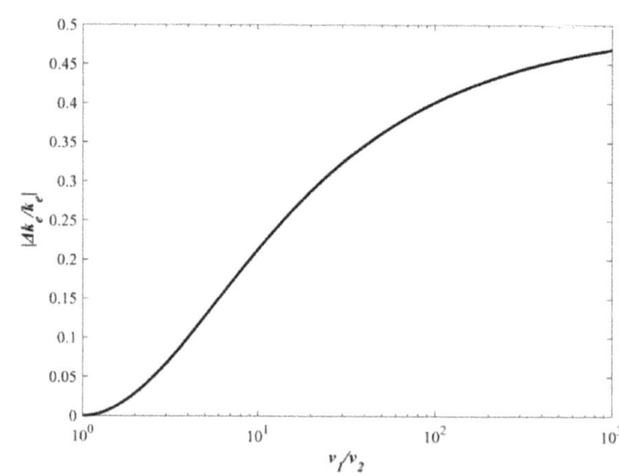

Figure 1. The relationship between $|\Delta k_e/k_e|$ and v_1/v_2.

As the structure of agglomerate is complex, modeling of its surface area is even more difficult, given the scarcity of experimental data. It is also very difficult to numerically determine which part of primary particles are the boundary particles and which part of the surface of these boundary particles forms the agglomerate surface. For an ideal case that k primary particles agglomerate with point contacts, its surface area equals $4k\pi a_{p0}^2$, where a_{p0} is the radius of primary particle [20]. Obviously, it is more suitable for chain-like structures with $D_f \rightarrow 1$ but not compact aggregates with $D_f \rightarrow 3$. In a statistical sense, the collision radius of agglomerates composed of k monomer is [1]:

$$r = Aa_{p0}\left(\frac{v}{v_0}\right)^{1/D_f} = Aa_{p0}k^{1/D_f} \tag{17}$$

where A is the dimensionless proportionality constant and can be assumed to be in unity to simplify calculations. In this paper, we will use this collision radius to calculate the surface area approximately:

$$s = 4\pi r^2 = 4\pi k^{2/D_f} a_{p0}^2 = B_3 v^{2/D_f} \tag{18}$$

in which the constant $B_3 = 4\pi(3/4\pi)^{2/D_f} a_{p0}^{2-6/D_f}$. Thus, for chain-like structures with $D_f = 2$, $s = 4k\pi a_{p0}^2$ equals to that of an agglomerate without necking and for compact aggregates with $D_f = 3$, $s = (36\pi)^{1/3} v^{2/3}$ equals to that of a spherical particle. Apparently, the agglomerates composed of the same number of primary particles with smaller D_f would have larger specific surface area and collision radius. Now the total surface area of this system can be expressed as:

$$s = \int_0^{\infty} sn(v,t)dv = B_3 M_{2/D_f} \tag{19}$$

and the rate of change for s can be written as:

$$\frac{ds}{dt} = B_3 \frac{dM_{2/D_f}}{dt} \tag{20}$$

$$\frac{ds}{sdt} = \frac{dM_{2/D_f}}{M_{2/D_f}dt} \tag{21}$$

where the fractional moment $M_{2/Df}$ is approximated by using Equation (6):

$$\frac{ds}{sdt} = \frac{dM_{2/D_f}}{M_{2/D_f}dt} \tag{22}$$

and its derivative with time t can be achieved:

$$\frac{dM_{2/D_f}}{dt} = \frac{(1-2/D_f)}{D_f}\frac{M_1^{2/D_f}}{M_0^{2/D_f}}\left[(1/D_f - 1)(2M_C - 2 - D_f)\frac{dM_0}{dt} - \frac{M_0^2 dM_2}{M_1^2 dt}\right] \tag{23}$$

Combining the first and third equations in Equation (7) gives:

$$\left.\frac{dM_2}{dt}\right|_{FM} = \frac{2M_1^2}{M_0^2}\frac{b_1 M_C{}^2 + b_2 M_C + b_3}{a_1 M_C{}^2 + a_2 M_C + a_3}\frac{dM_0}{dt} \tag{24}$$

Then Equation (23) can be rearranged as:

$$\left.\frac{dM_{2/D_f}}{dt}\right|_{FM} = \frac{(1-2/D_f)}{D_f}\frac{M_1^{2/D_f}}{M_0^{2/D_f}}\left[\begin{array}{c}(1/D_f - 1)(2M_C - 2 - D_f) - \\ \frac{2(b_1 M_C{}^2 + b_2 M_C + b_3)}{a_1 M_C{}^2 + a_2 M_C + a_3}\end{array}\right]\frac{dM_0}{dt} \tag{25}$$

Substituting the above equation into Equation (21), the rate of change for s in the free molecular regime has the following form:

$$\left.\frac{ds}{dt}\right|_{FM} = \frac{(1-2/D_f)\left[(1/D_f-1)(2M_C-2-D_f) - \frac{2(b_1M_C{}^2+b_2M_C+b_3)}{a_1M_C{}^2+a_2M_C+a_3}\right]}{D_f+(2/D_f-1)(M_C-1)}\frac{sdM_0}{M_0dt} \tag{26}$$

Analogously, the rate of change for particle kinetic energy and surface area in the continuum regime can be calculated as:

$$\left.\frac{dk_e}{dt}\right|_{CR} = k_BT\frac{q_1M_C{}^2+q_2M_C+q_3}{p_1M_C{}^2+p_2M_C+p_3}\frac{dM_0}{dt} \tag{27}$$

$$\left.\frac{ds}{dt}\right|_{CR} = \frac{(1-2/D_f)\left[(1/D_f-1)(2M_C-2-D_f)+2\right]}{D_f+(2/D_f-1)(M_C-1)}\frac{sdM_0}{M_0dt} \tag{28}$$

where q_1, q_2, q_3 are noted as:

$$\begin{aligned} q_1 &= (-9D_f^4+12D_f^2)/16 \\ q_2 &= (34D_f^4-24D_f^2)/16 \\ q_3 &= (-25D_f^4+12D_f^2)/16 \end{aligned} \tag{29}$$

Finally, the rate of change for S can be found:

$$\frac{dS}{dt} = \frac{1}{T}\left(-k_BT\ln(M_0\lambda_{th}^3)+k_BTC_2+\frac{\gamma s}{M_0}C_1\right)\frac{dM_0}{dt} \tag{30}$$

in which C_1, C_2 are functions of the dimensionless moment M_C and fractal dimension D_f:

$$C_1|_{FM} = \frac{(1-2/D_f)\left[(1/D_f-1)(2M_C-2-D_f) - \frac{2(b_1M_C{}^2+b_2M_C+b_3)}{a_1M_C{}^2+a_2M_C+a_3}\right]}{D_f+(2/D_f-1)(M_C-1)} \tag{31}$$

$$C_1|_{CR} = \frac{(1-2/D_f)\left[(1/D_f-1)(2M_C-2-D_f)+2\right]}{D_f+(2/D_f-1)(M_C-1)} \tag{32}$$

$$C_2|_{FM} = \frac{x_1M_C^2+x_2M_C+x_3}{a_1M_C{}^2+a_2M_C+a_3} \tag{33}$$

$$C_2|_{CR} = \frac{q_1M_C{}^2+q_2M_C+q_3}{p_1M_C{}^2+p_2M_C+p_3} \tag{34}$$

From the viewpoint of the second law of thermodynamics, the entropy of an isolated system will never decrease: $dS/dt \geq 0$. Moreover, the total particle number M_0 will decrease with time due to agglomeration: $dM_0/dt < 0$. Thus, the thermodynamics constraints for Brownian agglomeration at a certain temperature and pressure can be obtained:

$$\frac{\gamma s}{k_BT} \leq \frac{M_0(\ln(M_0\lambda_{th}^3)-C_2)}{C_1} \tag{35}$$

The equality would hold in the thermodynamic equilibrium state, and the critical time to reach this state can be determined. Moreover, the growth of the mean particle size M_1/M_0 will tend to a limit depending on the operating temperature and specific surface energy for thermal agglomeration technology.

3. Results

According to the self-preserving size distribution theory, the dimensionless particle moment M_C will tend to a constant at long time periods, and the asymptotic solutions of particle moments based on the TEMOM model can be found [11]. Here, the results are listed in the Appendix A. In the free molecular regime, the asymptotic solution of kinetic energy can be solved by directly integrating Equation (15) with respect to t:

$$k_e|_{FM} = C_3 + k_b T C_2 M_0 \rightarrow k_b T C_2 [g_1 t]^{-\frac{2D_f}{3D_f-4}} \quad (36)$$

where $C_3 = k_e(t_1) - k_e T C_2(t_1) M_0(t_1)$ is the integral constant, t_1 is the critical time in which the particle size distribution approaches self-preserving and the definition of g_1 is shown as Equation (A7). In our previous work [21], a criterion to calculate this critical time has been given based on the asymptotic solution of M_0 in the continuum regime, which can also be available in the free molecular regime. Now the effect of primary particle size a_{p0} on k_e can be obtained, which is as the same as that on M_0:

$$k_e|_{FM} \propto a_{p0}^{\frac{12-4D_f}{3D_f-4}} \quad (37)$$

Thus, its relative dissipative rate becomes:

$$\left.\frac{dk_e}{k_e dt}\right|_{FM} = \frac{dM_0}{M_0 dt} = -\frac{2D_f}{3D_f-4} t^{-1} \quad (38)$$

The asymptotic solution of surface area and the effect of primary particle size can be expressed as the following forms after substitution of Equations (A6) and (22) into Equation (19):

$$s|_{FM} \rightarrow B_3 M_1^{2/D_f} \left[1 + \frac{(2/D_f - 1)(M_C - 1)}{D_f}\right] [g_1 t]^{-\frac{2D_f-4}{3D_f-4}} \quad (39)$$

$$s|_{FM} \propto a_{p0}^{-\frac{6-2D_f}{3D_f-4}} \quad (40)$$

and its relative dissipative rate becomes:

$$\left.\frac{ds}{s dt}\right|_{FM} = \frac{D_f - 2}{D_f}\frac{dM_0}{M_0 dt} = -\frac{2D_f - 4}{3D_f - 4} t^{-1} \quad (41)$$

For simplification and without loss of generality, the calculation can be non-dimensionalized through the following relations: $M_0{}^* = M_0/M_{00}$, $M_1{}^* = M_1/M_{10}$, $M_2{}^* = M_{C0}M_2/M_{20}$, $t^* = tB_1 M_{00}^{(3Df-4)/2Df} M_{10}^{(4-Df)/2Df}$, $k_e{}^* = k_e/(M_{00}k_b T)$, $s^* = sM_{00}^{2/Df-1} M_{10}^{-2Df}/B_3$. Then the Equation (7) coupling with Equations (15) and (19) can be solved numerically by means of fourth-order Runge–Kutta method with the initial dimensionless conditions set as $M_{00} = 1$, $M_{10} = 1$, $M_{20} = 4/3$, $k_{e0} = 1/2$ (the star symbol '*' is omitted thereafter). The numerical and asymptotic solutions of kinetic energy and surface area are shown in Figure 2. According to the principle of equipartition of energy, the agglomerates share the molecular thermal motion of the fluid and have the same initial kinetic energy, thus the dissipative rate of k_e strongly depends on the collision rate. The evolutions of kinetic energy with time show the larger descent at lower fractal dimension because of the larger collision radius, which result in a more rapid agglomerated rate [22]. Oppositely, the contacting surface between primary particles in an agglomerate with smaller fractal dimension is less than that in an agglomerate with larger fractal dimension, thus the decay of surface area shows the reverse trend in the dual role of the higher specific surface area and more rapid agglomerated rate.

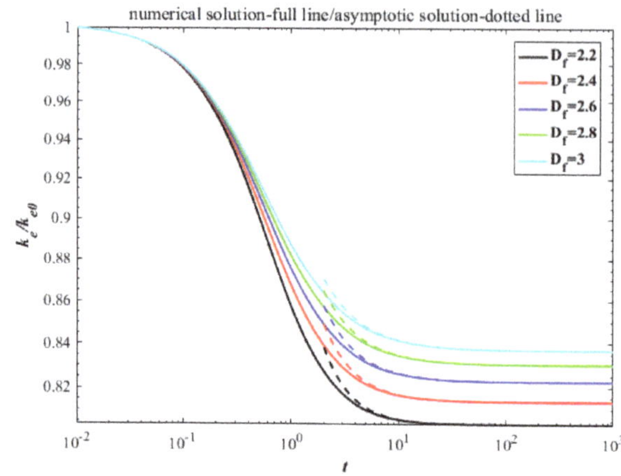

(a)

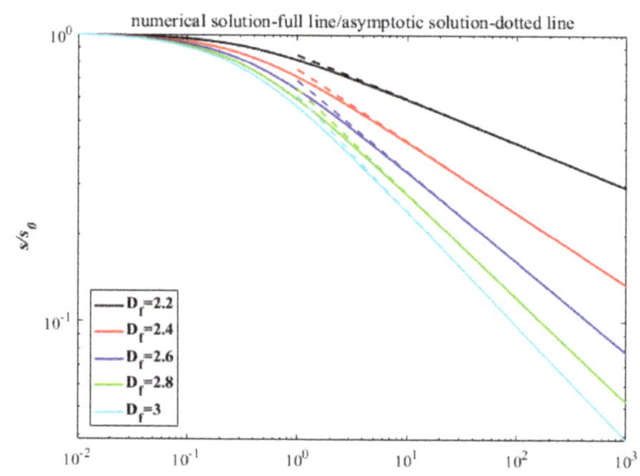

(b)

Figure 2. The decay with time between numerical solutions and asymptotic solutions in the free molecular regime: (**a**) kinetic energy; (**b**) surface area.

Analogously, the asymptotic solutions of kinetic energy and surface area, as well as their relative dissipative rates, in the continuum regime can be expressed as:

$$k_e|_{CR} \rightarrow k_b T C_2 M_0 = k_b T C_2 g_2^{-1} t^{-1} \tag{42}$$

$$\left.\frac{dk_e}{k_e dt}\right|_{CR} = \frac{dM_0}{M_0 dt} = -t^{-1} \tag{43}$$

$$s|_{CR} \rightarrow B_3 M_1^{2/D_f}\left[1 + \frac{(2/D_f - 1)(M_C - 1)}{D_f}\right][g_2 t]^{2/D_f - 1} \tag{44}$$

$$\left.\frac{ds}{sdt}\right|_{CR} = \frac{D_f - 2}{D_f}\frac{dM_0}{M_0 dt} = -\frac{D_f - 2}{D_f}t^{-1} \tag{45}$$

where g_2 is a function of D_f showed as Equation (A12). The results are showed in Figure 3. These allow us to simplify the rate of change for S as the asymptotic form in both the free molecular and continuum regime:

$$\frac{dS}{dt} = \frac{1}{T}\left(-k_B T M_0 \ln(M_0 \lambda_{th}^3) + k_e + \frac{D_f - 2}{D_f}\gamma s\right)\frac{dM_0}{M_0 dt} \tag{46}$$

And the corresponding thermodynamics constraints are:

$$\frac{\gamma s}{k_B T} \leq \frac{M_0(\ln(M_0 \lambda_{th}^3) - C_2)}{1 - 2/D_f} \tag{47}$$

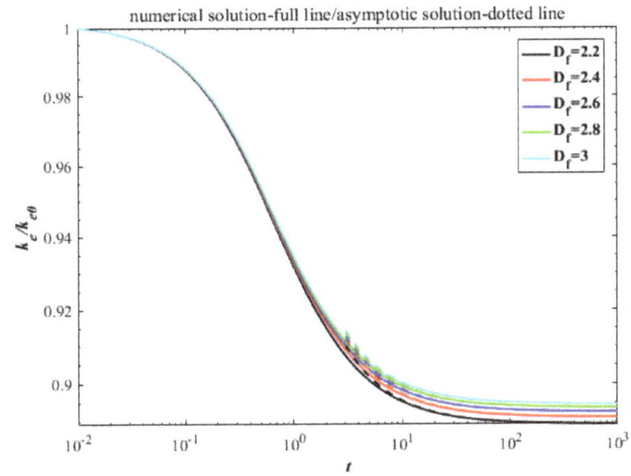

(a)

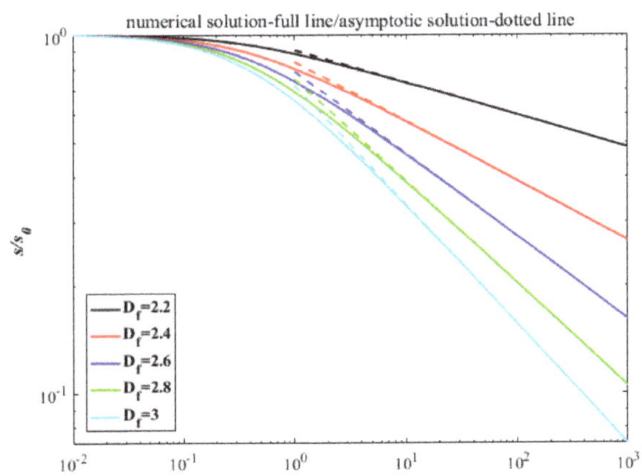

(b)

Figure 3. The decay with time between numerical solutions and asymptotic solutions in the continuum regime: (**a**) kinetic energy; (**b**) surface area.

4. Discussion

The above equation establishes an inequality relationship between moments and some physical parameters, such as temperature and specific surface energy, and the equality holds if and only if the system reaches the thermodynamic equilibrium. Obviously, increasing temperature leads to decreasing particle number density and greater mean volume, which can be useful for dust collection efficiency. It also shows the effect of molecular structure on surface tension. By substituting the expression of surface area into this equation, we can get the following formula in the thermal equilibrium state:

$$\gamma B_3 \frac{M_1^{2/D_f}}{M_0^{2/D_f}} \left[1 + \frac{(2/D_f - 1)(M_C - 1)}{D_f} \right] = \frac{(\ln(M_0 \lambda_{th}^3) - C_2)}{1 - 2/D_f} k_b T \tag{48}$$

For liquid pure substance, $M_0 \lambda_{th}{}^3$ usually takes the value as V_m in the free molecular regime and $M_1/M_0 = V_m / \mathrm{N_A}$, where V_m is the molar volume and $\mathrm{N_A}$ is the Avogadro constant. A modification coefficient γ_∞, which is equal to the value of surface tension at infinite molar volume, should be introduced because the surface tension decreases almost linearly with the increase of temperature. Then a correction function of molar volume can be constructed:

$$\gamma = \gamma_\infty - \frac{k_1}{V_m{}^{2/3}} \tag{49}$$

where k_1 is a function of the temperature and fractal dimension:

$$k_1 = \frac{(\ln V_m - C_2)}{V_m{}^{2/D_f - 2/3}} \frac{\mathrm{N}_A^{2/D_f} D_f}{B_3 \left(1 - 2/D_f\right) \left[D_f + (2/D_f - 1)(M_C - 1)\right]} k_b T \tag{50}$$

The surface tension increases as molar volume increases and tends toward the constant γ_∞ at infinite molar volume, and it decreases monotonously with increasing temperature and tends to zero at the critical temperature. Unfortunately, it should be noted that some important factors, i.e., the effect of end groups, cannot be considered. It also can be written as the molecular weight-surface tension relationship with $V_m = M/\rho$,

$$\gamma = \gamma_\infty - \frac{k_2}{M^{2/3}} \tag{51}$$

$$k_2 = \frac{(\ln(M/\rho) - C_2)}{M^{2/D_f - 2/3}} \frac{(\rho N_A)^{2/D_f} D_f}{B_3 \left(1 - 2/D_f\right) \left[D_f + (2/D_f - 1)(M_C - 1)\right]} k_b T \tag{52}$$

where M is the molecular weight and ρ is the density. Some research shows that the correlation with molecular weight is better than that with molar volume for alkanes and perfluoro alkanes, but the discrepancy can be ignored for the siloxanes [23]. The effect of fractal dimension on the slope k_2 is illustrated in Figure 4 for n-alkanes at 0 °C, where a_{p0} = 0.2 nm is equal to the radius of methane and n is the number of carbons. Compared to bulks, the surface tension of molecules with long-chain structure generally increases due to the large contact areas and intermolecular forces, which leads to a small slope k_2. In the range of 2.7 to 2.8, the result is mostly close to the value k_2 = 360 of least-squares fitting based on experimental data [24].

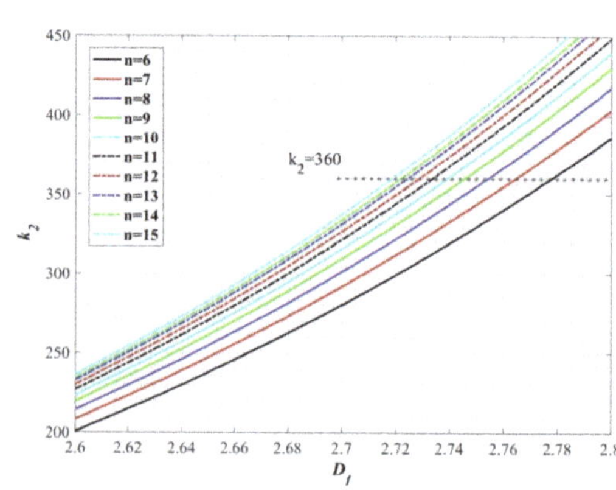

Figure 4. The effect of fractal dimension on the slope k_2 for n-alkanes.

5. Conclusions

On the basis of the theory of maximum entropy and binary perfectly inelastic collision, the thermodynamic constraints of Brownian coagulation for spherical particles are extended to agglomerates, and a relationship of inequality between particle moments and some physical parameters is established using the TEMOM. Meanwhile, the evolutions of kinetic energy and surface area with time are presented, as well as their asymptotic behaviors. While some of our present simplifying assumptions will have to be relaxed, even our present results are of potential interest for a number of applications, for example, the estimation of surface tension of liquid polymers and the enhancement of dust collection efficiency.

Author Contributions: Conceptualization, Q.H. and M.X.; methodology, Q.H. and M.X.; validation, M.X.; investigation, Q.H. and M.X.; resources, Q.H. and M.X.; data curation, Q.H.; writing—original draft preparation, Q.H.; writing—review and editing, Q.H.; supervision, M.X.; funding acquisition, Q.H. All authors have read and agreed to the published version of the manuscript.

Funding: This work was funded by the National Natural Science Foundation of China Grant Number 11902075.

Institutional Review Board Statement: Not applicable.

Informed Consent Statement: Not applicable.

Conflicts of Interest: The authors declare no conflict of interest.

Appendix A. The Asymptotic Solutions of TEMOM Model for Brownian Agglomeration

The self-preserving size distribution theory implies that the dimensionless moment M_C approaches a constant as time advances, thus we have:

$$\frac{dM_C}{dt} = \frac{M_2}{M_1^2}\frac{dM_0}{dt} + \frac{M_0}{M_1^2}\frac{dM_2}{dt} = 0 \tag{A1}$$

Substituting the first and the third equations of Equation (7) into Equation (A1) leads to a third-order algebraic equation of M_C in the free molecular regime:

$$c_1{M_C}^3 + c_2{M_C}^2 + c_3 M_C + c_4 = 0 \tag{A2}$$

In which c_1, c_2, c_3 and c_4 are functions of D_f:

$$\begin{aligned} c_1 = a_1 &= D_f^4 - 24D_f^3 + 70D_f^2 - 48D_f + 16 \\ c_2 = a_2 + 2b_1 &= 60D_f^4 - 112D_f^3 + 72D_f^2 + 64D_f - 64 \\ c_3 = a_3 + 2b_2 &= 77D_f^4 - 24D_f^3 - 546D_f^2 + 16D_f + 80 \\ c_4 = 2b_3 &= -266D_f^4 + 160D_f^3 + 404D_f^2 - 32D_f - 32 \end{aligned} \tag{A3}$$

and for a given value of D_f, the solution of M_C, which is also an invariant constant, can be solved as [11]:

$$\begin{aligned} M_{C1}|_{FM} &= \tfrac{1}{6c_1}\left((d_1 + d_2)^{1/3} - \tfrac{4d_3}{(d_1+d_2)^{1/3}} - 2c_2\right) \\ M_{C2}|_{FM} &= \tfrac{1}{12c_1}\left(-(d_1 + d_2)^{1/3} + \tfrac{4d_3}{(d_1+d_2)^{1/3}} - 4c_2\right) \\ M_{C3}|_{FM} &= \tfrac{-c_3+\sqrt{c_3^2-4c_2c_4}}{2c_2} \end{aligned} \tag{A4}$$

where d_1, d_2 and d_3 are:

$$d_1 = 12c_1\sqrt{3(27c_1^2c_4^2 - 18c_1c_2c_3c_4 + 4c_1c_3^3 + 4c_2^3c_4 - c_2^2c_3^2)}$$
$$d_2 = -108c_1^2c_4 + 36c_1c_2c_3 - 8c_2^3 \quad \text{(A5)}$$
$$d_3 = 3c_1c_3 - c_2^2$$

Then the asymptotic solution of M_0 can be obtained:

$$M_0|_{FM} \to (g_1t)^{-\frac{2D_f}{3D_f-4}} \quad \text{(A6)}$$

where g_1 is a function of D_f:

$$g_1 = \frac{\sqrt{2}B_1(3D_f-4)}{128D_f^5}M_1^{\frac{4-D_f}{2D_f}}\left(a_1M_C^2 + a_2M_C + a_3\right) \quad \text{(A7)}$$

And its relative agglomerate growth rate is:

$$\left.\frac{dM_0}{M_0dt}\right|_{FM} \to -\frac{2D_f}{3D_f-4}\frac{1}{t} \quad \text{(A8)}$$

Analogously, the asymptotic solution of M_C in the continuum regime is:

$$M_C|_{CR} = 2 \quad \text{(A9)}$$

And the asymptotic solution of M_0 and its relative growth rate are:

$$M_0|_{CR} \to (g_2t)^{-1} \quad \text{(A10)}$$

$$\left.\frac{dM_0}{M_0dt}\right|_{CR} \to -\frac{1}{t} \quad \text{(A11)}$$

where g_2 is a function of D_f:

$$g_2 = \frac{B_2(1+3D_f^2+8D_f^4)}{4D_f^4} \quad \text{(A12)}$$

References

1. Friedlander, S.K. *Smoke, Dust and Haze: Fundamentals of Aerosol Dynamics*, 2nd ed.; Oxford University Press: London, UK, 2000.
2. Hu, B.; Yi, Y.; Liang, C.; Yuan, Z.L.; Szczepan, R.; Yang, L.J. Experimental study on particles agglomeration by chemical and turbulent agglomeration before electrostatic precipitators. *Powder Technol.* **2018**, *335*, 186–194.
3. Yan, J.P.; Chen, L.Q.; Lin, Q. Removal of fine particles in WFGD system using the simultaneous acoustic agglomeration and supersaturated vapor condensation. *Powder Technol.* **2017**, *315*, 106–113. [CrossRef]
4. Müller, H. Zur allgemeinen theorie ser raschen koagulation. *Kolloidchem. Beih.* **1928**, *27*, 223–250. [CrossRef]
5. Dürr, R.; Bück, A. Approximate moment methods for population balance equations in particulate and bioengineering processes. *Processes* **2020**, *8*, 414. [CrossRef]
6. Pratsinis, S.E. Simultaneous nucleation, condensation and coagulation in aerosol reactors. *J. Colloid Interface Sci.* **1988**, *124*, 416–427. [CrossRef]
7. Gelbard, F.; Tambour, Y.; Seinfeld, J.H. Sectional representations for simulating aerosol dynamics. *J. Colloid Interface Sci.* **1980**, *76*, 541–556. [CrossRef]
8. Tandon, P.; Rosner, D.E. Monte Carlo simulation of particle aggregation and simultaneous restructuring. *J. Colloid Interface Sci.* **1999**, *213*, 273–286. [CrossRef] [PubMed]
9. Yu, M.Z.; Lin, J.Z.; Chan, T. A new moment method for solving the coagulation equation for particles in Brownian motion. *Aerosol Sci. Technol.* **2008**, *42*, 705–713. [CrossRef]
10. Xie, M.L. Asymptotic behavior of TEMOM model for particle population balance equation over the entire particle size regimes. *J. Aerosol Sci.* **2014**, *67*, 157–165. [CrossRef]
11. Xie, M.L. Asymptotic solution of moment approximation of the particle population balance equation for Brownian agglomeration. *Aerosol Sci. Technol.* **2015**, *49*, 109–114. [CrossRef]

12. Chen, Z.L.; Lin, J.Z.; Yu, M.Z. Asymptotic behavior of the Taylor-expansion method of moments for solving a coagulation equation for Brownian particles. *Particuology* **2014**, *14*, 124–129. [CrossRef]
13. Xie, M.L.; Wang, L.P. Asymptotic solution of population balance equation based on TEMOM model. *Chem. Eng. Sci.* **2013**, *94*, 79–83.
14. Xie, M.L.; He, Q. Analytical solution of TEMOM model for particle population balance equation due to Brownian coagulation. *J. Aerosol Sci.* **2013**, *66*, 24–30.
15. Friedlander, S.; Wang, C. The self-preserving particle size distribution for coagulation by Brownian motion. *J. Colloid Interface Sci.* **1966**, *22*, 126–132. [CrossRef]
16. Trizac, E.; Krapivsky, P.L. Correlations in Ballistic Processes. *Phys. Rev. Lett.* **2003**, *91*, 218–302. [CrossRef]
17. Xie, M.L. On the growth rate of particle surface area for Brownian coagulation. *J. Aerosol Sci.* **2017**, *113*, 36–39.
18. Xie, M.L.; Yu, M.Z. Thermodynamic analysis of Brownian coagulation based on moment method. *Int. J. Heat Mass Transf.* **2018**, *122*, 922–928. [CrossRef]
19. Yu, M.Z.; Lin, J.Z. Taylor series expansion scheme applied for solving population balance equation. *Rev. Chem. Eng.* **2017**, *34*, 561–594. [CrossRef]
20. Cao, L.N.Y.; Chen, S.C.; Fissan, H.; Asbach, C.; Pui, D.Y.H. Development of a geometric surface area monitor (GSAM) for aerosol nanoparticles. *J. Aerosol Sci.* **2017**, *114*, 118–129. [CrossRef]
21. He, Q.; Shchekin, A.K.; Xie, M.L. New analytical TEMOM solutions for a class of collision kernels in the theory of Brownian coagulation. *Phys. A* **2015**, *428*, 435–442. [CrossRef]
22. Wu, M.; Friedlander, S. Enhanced power law agglomerate growth in the free molecule regime. *J. Aerosol Sci.* **1993**, *24*, 273–282. [CrossRef]
23. Legrand, D.G.; Gaines, G.L. The molecular weight dependence of polymer surface tension. *J. Colloid Interface Sci.* **1969**, *31*, 162–167. [CrossRef]
24. Legrand, D.G.; Gaines, G.L. Surface Tension of Homologous Series of Liquids. *J. Colloid Interface Sci.* **1973**, *42*, 181–184. [CrossRef]

sustainability

MDPI

Article

A Study on Indoor Particulate Matter Variation in Time Based on Count and Sizes and in Relation to Meteorological Conditions

Marius Bodor

Faculty of Engineering, Department of Materials and Environmental Engineering, "Dunarea de Jos" University of Galati, 47 Domneasca Street, 800008 Galati, Romania; marius.bodor@ugal.ro; Tel.: +40-746-070-322

Abstract: An important aspect of air pollution analysis consists of the varied presence of particulate matter in analyzed air samples. In this respect, the present work aims to present a case study regarding the evolution in time of quantified particulate matter of different sizes. This study is based on data acquisitioned in an indoor location, already used in a former particulate matter-related article; thus, it can be considered as a continuation of that study, with the general aim to demonstrate the necessity to expand the existing network for pollution monitoring. Besides particle matter quantification, a correlation of the obtained results is also presented against meteorological data acquisitioned by the National Air Quality Monitoring Network. The transformation of quantified PM data in mass per volume and a comparison with other results are also addressed.

Keywords: particulate matter; counting; variation

Citation: Bodor, M. A Study on Indoor Particulate Matter Variation in Time Based on Count and Sizes and in Relation to Meteorological Conditions. *Sustainability* **2021**, *13*, 8263. https://doi.org/10.3390/su13158263

Academic Editor: Avelino Núñez-Delgado

Received: 7 June 2021
Accepted: 19 July 2021
Published: 23 July 2021

1. Introduction

Particulate matter (PM) is mentioned among the important pollutants that are necessary to be permanently monitored in highly populous cities in the European Union [1]. According to the mean diameter of the measured particulates, PM is classified in various categories, PM_{10} being one of the most researched and regulated [2]. Although already suspected to be dangerous, when present in great quantities in breathable air [3–6], PM of all kinds were even more so cataloged, in a new study, after using a physicochemical kinetic model [7] that proved the higher extent of the hazardous potential of these pollutants for human health. These later studies further emphasize the necessity to pinpoint the sources of particulate matter generators and to determine the variability in time and space of these pollutants in preparation towards measures to hinder their presence, at least in highly populated regions. Towards this purpose, an important factor to be considered is the relation of PM concentration to meteorological conditions, high amounts present in breathable air usually being in direct correlation with temperature, wind speed and precipitation [8,9].

The legislation currently in force [2] aims at the outdoor monitoring of PM, although a vast majority of the population carries out most of its activities indoors [10]. This is perhaps why, lately, a variety of studies have concentrated on indoor PM measurements [10–15]. Each mentioned study had in common the measurement of PM_{10}, and the general conclusion is that the indoor PM concentration is obviously strictly related to the outdoor PM concentration; a possible measure in the case of registering high PM values could be represented by air purification filters, especially on premises where a large and young part of the population is situated, namely in schools [11]. As a provisionary measure, the amount of indoor particulate matter may be reduced, with moderate expense [16] and high efficiency [17], through utilization of air purifiers [16], although, as specified in [18], the effectiveness of this process is highly dependent on the resuspension factors that can vary from the surface and purpose of the indoor space, the number of people using that space, heating, cooling or ventilation means, etc. All these factors can be utilized in concordance with data related to meteorological conditions, and the possible utilization of purifiers might thus become more efficient.

The acquisition of different data related to air pollution and meteorological data is constantly underway in large cities in most countries, and in addition to these existing systems, some pilot projects [19–22], among which are independent ones such as [23,24], are already using relatively cheap sensors that can be integrated in a network able to provide real-time data on pollutant concentration and/or meteorological data. The valuable results of all this research lacks only the will of decision makers in order to create an efficient system that could autonomously activate, when necessary, any air purifiers in sensible indoor locations.

The present work was intended as a case study towards creating a clearer view on particulate matter circulation in a city that seems to lack an efficient air pollution monitoring system. This could represent an aid to solving the problems that air pollution can create for people's health in agglomerated areas, and the present study can be of such aid through results regarding the variation of PM concentration in an indoor space during a specific time of the year. These data, correlated with meteorological conditions and compared with official results from the National Air Quality Monitoring Network, for the same area, exposed the limitations of the existing capabilities and the necessity to enhance their number and efficiency. Additionally, as a novelty of the present work, the architecture of the measuring campaign emphasized the difference in PM concentration relative to different positions of the same indoor location.

2. Materials and Methods

The location used in the present study is the same as that used in a previous study [25], and the timing was almost similar: namely, January, February and March months were used for the present study measurement campaign. These months are of greater interest, compared to the rest of the year, since the transition from the colder to warmer season offers the possibility to determine particle matter presence unhindered by the vegetation in its path. Additionally, similarities were encountered from the meteorological point of view between the two sampling campaigns, in both cases, little to no precipitation being registered.

An airborne particle counter apparatus, model Fluke 985 (Figure 1), was used to quantify the particulate matter by different types of dimensions. Some specifications are as follows: (i) 6 channels for measuring 0.3 μm, 0.5 μm, 1.0 μm, 2.0 μm, 5.0 μm and 10 μm sized airborne particles; (ii) flow rate: 2.83 L/min; (iii) light source: 775 nm to 795 nm, 90 mW class 3B laser. As a novelty for the indoor PM concentration measurement, the measuring apparatus used in this study was placed in three different spots of the same location, namely: S1—in the furthest point relative to the living room tilt-opened window; S2—near the living room tilt-opened window and S3—near the balcony's tilt-opened window. At all times, both the balcony and the living room windows were tilt-opened, and the height of the sampler's position was kept constant at 1200 mm. For each sampling spot, the same sampling program was used consisting of 168 sessions of airflow intake lasting 15 min each followed by a 45 min break and resulting in a 7-day period for each sampling spot. The architecture used for the measuring campaign was intended to emphasize the differences of PM concentration in three spots of the same location, starting from the furthest to the closest distance relative to the window that was seen as the major source of particulate matter.

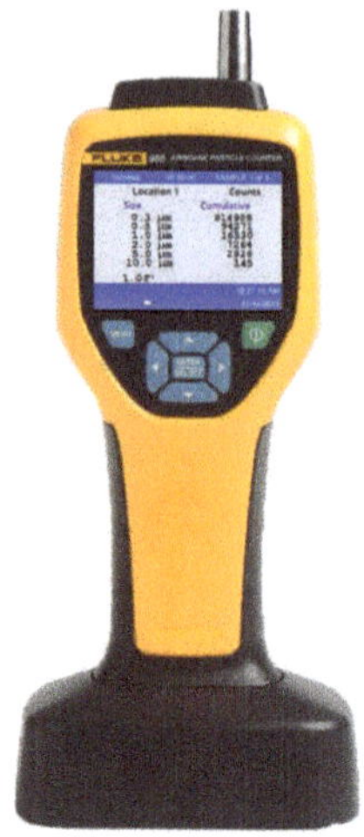

Figure 1. Fluke 985—the apparatus used to count the particle matter of different sizes [26].

The data related to meteorological conditions during sampling were collected from the official website of the Romanian National Air Quality Monitoring Network [27], the values being correlated with the exact timing of particle matter sampling, making the analysis of all data easier to compare.

Conversion of data related to counts of particle matter in mass per volume (as utilized in legislation regarding PM measurements) was done using a method that takes into account a mean of different densities that usually characterize these types of particles [28]. This method, although it presents a great deal of error margin, is nonetheless useful in the present study to demonstrate the indoor variation of particulate matter against time, the particulate matter dimension used for this purpose being 10 μm. This size allows for a better highlighting of PM evolution during an entire week, since the greater particulate mass also implies a more rapid deposition. Moreover, using the values for PM_{10}, another novelty of the present study is the comparison of PM evolution data from the national monitoring network against data of the present study, during the same period of time and from the same area.

3. Results and Discussion

The measuring of the data obtained using the apparatus and procedure described in the previous section was organized so that a specific time was the same for each measuring spot. Thus, the starting time was 2 p.m., and a graphic was drawn for each particulate matter dimension, resulting in six graphical representations with three curve series (for each measuring spot: S1, S2 and S3) each, presented in Figure 2.

As expected, the amount of smaller particulate matter was the highest registered during measurements. Although no reference was found in the literature, another confirmation was obtained related to the expectation that in the furthest measuring spot relative to the windows, the number of airborne particulates was mostly lower than the other measuring spots. These higher PM values near the windows are not necessarily expected to be present in other indoor sites such as schools with numerous individuals, as referred to in other studies [11,12]. Additionally, some unusually high values were registered, most of them for the S3 measuring spot, the majority being confirmed for all particulate matter dimensions. Another characteristic of smaller particulates (being airborne for longer) is easily distinguishable from graphics depicting 0.3 and 0.5 μm PM count evolution, high values being registered throughout the measuring period for all three measuring spots. Different, clearer patterns begin to unveil for the data related to 1 μm PM count evolution, higher numbers of particulates being registered starting from 8 a.m. until, usually, 12 a.m., when these values fall towards zero. The same pattern is seen even clearer for heavier particulates, the data for 10 μm dimensions being the most representative in this respect, these reaching nil values for all measuring spots during the still time, a trend also similar and confirmed in other studies [11,17,29].

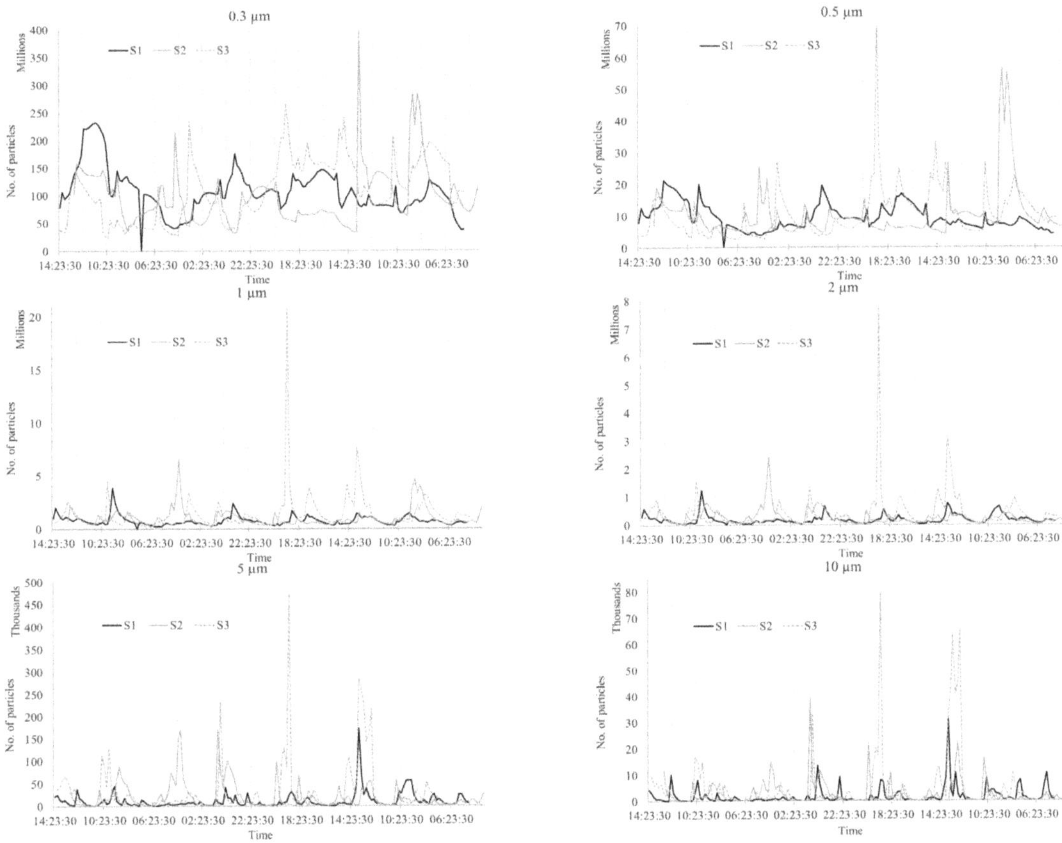

Figure 2. Results regarding the number of particulate matter measured in three different spots (S1—in the furthest point relative to the living room tilt-opened window; S2—near the living room tilt-opened window and S3—near the balcony's tilt-opened window) and for six different dimensions (0.3; 0.5; 1; 2; 5 and 10 μm).

Regarding the meteorological data, these are presented in Figure 3, which contains three graphics corresponding to the measuring spot that was utilized continuously, each one for a week. The measuring campaign started from the measuring spot farthest away from the windows; thus, the first graphic (from 30 January 2021 to 6 February 2021) corresponds to S1, the second graphic (from 8 February 2021 to 15 February 2021) to S2 and the last graphic (from 9 March 2021 to 16 March 2021) to S3. During the entire measuring period, the temperature varied from −7 °C to 17 °C, a normal variation for this time of the year. Higher temperatures were registered during the day period, as expected, and one of the direct consequences was the increase in registered wind speeds due to promoting the convection of air [30]. These two meteorological factors each present a similar curve to those of the PM number variation in Figure 2, highlighting the direct contribution of high temperature and wind speeds to a higher amount of particulate matter in the air of the present study's measuring spots (a pattern similar to some studies [31,32] and contrary to others [33,34]), confirming, as mentioned in [8], the spatial heterogeneity of concentration and meteorological factors. Opposite to this correlation, a higher value for precipitation was expected, as reported in [31], to decrease the number of airborne particulates. However, the lack of long timespans with precipitation deprived the present study of such confirmation. Two small precipitation periods were registered during measurements, and as mentioned, a clear result in PM number variation was not noticed, mostly since the measurements took place indoors, and a longer precipitation period would be needed to visibly affect the registered amount of airborne particulate matter.

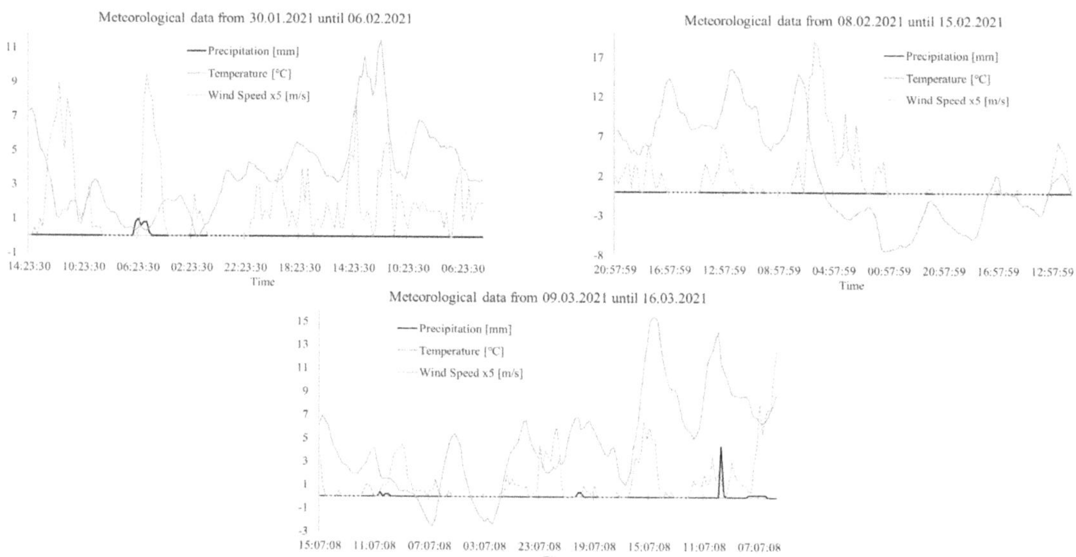

Figure 3. Meteorological data from the Romanian National Air Quality Monitoring Network during the particulate matter sampling period.

Overall, the correlation of meteorological conditions with PM concentration is an important aspect to be considered concerning air quality prediction, although, as mentioned in [8], other factors should be taken into account, such as the sources and components of PM, the local climate and the terrain [8,35,36], which cause the influencing degree of different processes to vary by region. All these factors should be thoroughly monitored for each city or region that might implement any type of decision making based on scientific results.

In order to place the data of the present work in context with the national legislation limits, one batch of data was transformed using a relatively trustworthy method [28] into data expressed in mass per volume. This data translation was realized in order to ease a comparison with results from the National Air Quality Monitoring Network (data selected from the closest measuring station, named GL-2), and the data batch was for PM_{10} (particulate matter with diameter lower than 10 μm), since this is one of the air quality indices regulated by the national air quality law [37] and was also that presenting a clearer variation in number during the measuring period for the present study.

Figure 4 presents the data for PM_{10} expressed in μg/m^3 against the maximum limit that, according to the national law, should not be exceeded 35 times throughout an entire year [37]. Considering that the transformation procedure from the PM_{10} number into the PM_{10} expressed in mass per volume is close to reality, the obtained data present a large number of maximum limit (50 μg/m^3) overruns, at least 17 times, during the week taken into account. Although the transformation of data is not exact, since it is impossible to precisely appreciate the density of counted airborne particulates, the results presented in Figure 4 are believed to be not far from reality, especially since, in a previous work, higher than normal values of airborne particulates were registered in the same location [25], as was the case in another paper dealing with PM_{10} measurements in the same city of Galați [38]. Aside from the city on which the present work focuses, other studies also reported higher than the maximum admissible values in indoor places, such as 280.6 μg/m^3 in Warsaw, Poland [11]; 578 μg/m^3 in Beijing, China [39] and 322 μg/m^3 in Castellón, Spain [12].

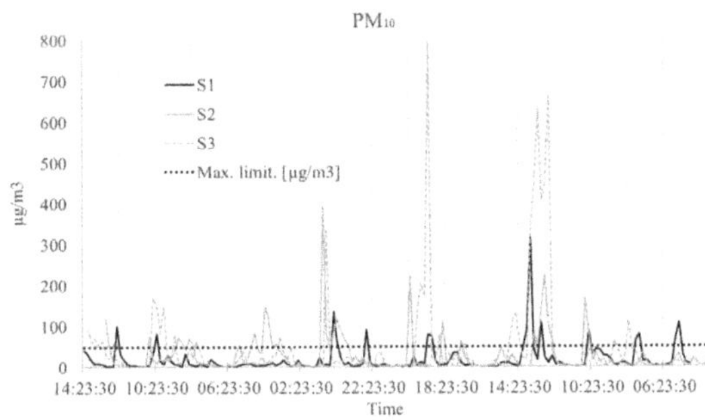

Figure 4. Hourly mean values of PM_{10}, expressed in μg/m^3, registered in the three measurement spots (S1, S2 and S3), compared with the maximum admissible value according to the national legislation.

Using the same results of PM_{10} transformed into μg/m^3, and in order to level out any extreme values, daily averages were calculated for each measuring spot and presented against the maximum regulated limit. All these and the averages of results gathered by the GL-2 measurement station of the National Air Quality Monitoring Network are presented in Figure 5. The results from the GL-2 station are represented as curves, each of them being calculated using the available data and corelated to the same time period as the measuring period of this study; thus, the abbreviation in the figure represents the Romanian National Air Quality Monitoring Network (RN) during the weeks when measurements took place in the three mentioned spots (S1, S2 and S3) for the present study.

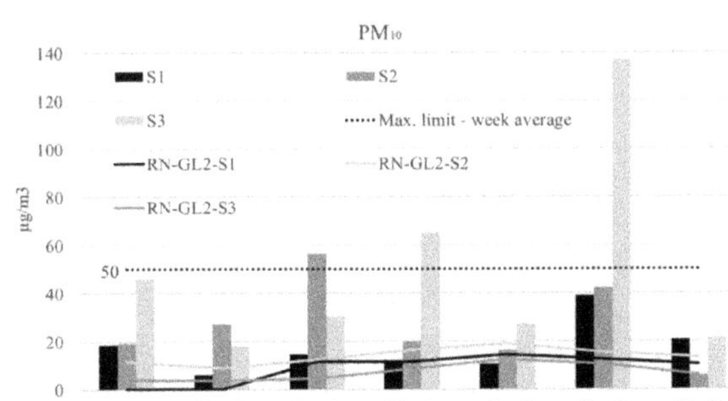

Figure 5. Daily mean values of PM_{10}, expressed in μg/m^3, registered in the three measurement spots (S1, S2 and S3), compared with the maximum admissible value according to the national legislation and the weekly mean values registered by the closest measuring station of the national measuring system (RN-GL2-S1, RN-GL2-S2 and RN-GL2-S3) during the same time as this study measurement period.

The trends in Figure 5 are somewhat similar when comparing the results from the two different sources of data, yet higher values are still registered for data from the present study compared to data from the national measuring network. Similar situations are reported in [11,12], where some indoor measurements present higher PM values compared to outdoor PM concentrations, although these studies are focused on schools as indoor measuring sites, where the number of individuals might represent the main reason for these results. Additionally, in [40], some cases are presented where higher indoor values were registered in households with smoking individuals. On the contrary, in [10,41], the values for outdoor PM concentration measurements were always higher than the indoor ones; for that study, only households were used as indoor measuring sites. For the present work, this difference might mainly be the result of the positioning of the measuring sites. As already suggested in a previous work [25], some air quality monitoring locations are not representative for all situations that might occur in a vast and industrialized city such as Galați. Thus, a more realistic view on the variation in time of particulate matter could be obtained in the future by diversifying and increasing the number of measuring stations used by the National Air Quality Monitoring Network. A possibility in this respect might be represented by small, cheap and resilient measuring units that already exist on the open market throughout the world [23,42,43] and their potential integration into the existing hardware and software infrastructure.

4. Conclusions

The present study demonstrates the direct dependance of particulate matter amount on meteorological conditions, a clear correlation being emphasized for temperature and wind speed, of which high values also correspond to high number of airborne particulates for the indoor location used in this work. These data were also presented against the maximum allowable by legislation and the ones registered by the National Measurement Network. Aside from similar trends between the two PM related datasets, high values and numerous overruns of the maximum admissible were also registered for the indoor measurements from this study, while the values from the National Measurement Network were always well below this maximum. The validation of the methodology used in the present work is realized by a comparison with other studies in the literature, confirming the correlation between indoor PM concentration and meteorological conditions.

All this being said, it is clear that a more thorough monitoring of these pollutants in large cities might be useful for decision makers and for the general population, which might be affected. In this respect, broader research could be undertaken, focusing on an entire year in the same indoor location and also using a more precise and complete instrumentation. This method of data acquisition might offer some clarifications regarding the degree of inertia of meteorological conditions towards indoor PM quantity variation. Additionally, the intermittent utilization of an air purifier in the same indoor location could offer some valuable data about the utility of such an apparatus in terms of obtaining a cleaner breathable air.

Funding: This research received no external funding.

Institutional Review Board Statement: Not applicable.

Informed Consent Statement: Not applicable.

Data Availability Statement: Publicly available data, related to the National Air Quality Monitoring Network, used in this study for comparison reasons was obtained from the following web-site: https://www.calitateaer.ro/public/monitoring-page/reports-reports-page/?__locale=ro.

Conflicts of Interest: The author declares no conflict of interest.

References

1. Directive 2008/50/EC of the European Parliament and of the Council of 21 May 2008 on Ambient Air Quality and Cleaner Air for Europe. 2008. Available online: https://eur-lex.europa.eu/legal-content/en/ALL/?uri=CELEX%3A32008L0050 (accessed on 5 July 2021).
2. Normativ 02/07/2007, Portal Legislativ, Monitorul Oficial. 2007. Available online: http://legislatie.just.ro/Public/DetaliiDocumentAfis/83993 (accessed on 5 July 2021).
3. Lelieveld, J.; Evans, J.S.; Fnais, M.; Giannadaki, D.; Pozzer, A. The contribution of outdoor air pollution sources to premature mortality on a global scale. *Nat. Cell Biol.* **2015**, *525*, 367–371. [CrossRef]
4. Dockery, D.W.; Pope, C.A. Acute Respiratory Effects of Particulate Air Pollution. *Annu. Rev. Public Health* **1994**, *15*, 107–132. [CrossRef]

5. Arangio, A.; Tong, H.; Socorro, J.; Pöschl, U.; Shiraiwa, M. Quantification of environmentally persistent free radicals and reactive oxygen species in atmospheric aerosol particles. *Atmos. Chem. Phys. Discuss.* **2016**, *16*, 13105–13119. [CrossRef]
6. Shiraiwa, M.; Selzle, K.; Pöschl, U. Hazardous components and health effects of atmospheric aerosol particles: Reactive oxygen species, soot, polycyclic aromatic compounds and allergenic proteins. *Free Radic. Res.* **2012**, *46*, 927–939. [CrossRef]
7. Alpert, P.A.; Dou, J.; Arroyo, P.C.; Schneider, F.; Xto, J.; Luo, B.; Peter, T.; Huthwelker, T.; Borca, C.N.; Henzler, K.D.; et al. Photolytic radical persistence due to anoxia in viscous aerosol particles. *Nat. Commun.* **2021**, *12*, 1769. [CrossRef]
8. Yang, Q.; Yuan, Q.; Li, T.; Shen, H.; Zhang, L. The Relationships between PM2.5 and Meteorological Factors in China: Seasonal and Regional Variations. *Int. J. Environ. Res. Public Health* **2017**, *14*, 1510. [CrossRef]
9. Paraschiv, S. Effects of wind speed, relative humidity, temperature and air pressure on PM10 concentration for an urban background area. *IOP Conf. Ser. Mater. Sci. Eng.* **2019**, *595*, 012059. [CrossRef]
10. Jodeh, S.; Hasan, A.R.; Amarah, J.; Judeh, F.; Salghi, R.; Lgaz, H.; Jodeh, W. Indoor and outdoor air quality analysis for the city of Nablus in Palestine: Seasonal trends of PM10, PM5.0, PM2.5, and PM1.0 of residential homes. *Air Qual. Atmos. Health* **2017**, *11*, 229–237. [CrossRef]
11. Rogulski, M. Indoor PM10 concentration measurements using low-cost monitors in selected locations in Warsaw. *Energy Procedia* **2018**, *147*, 137–144. [CrossRef]
12. Pallarés, S.; Gómez, E.; Martínez, A.; Jordán, M.M. The relationship between indoor and outdoor levels of PM10 and its chemical composition at schools in a coastal region in Spain. *Heliyon* **2019**, *5*, e02270. [CrossRef]
13. Nishihama, Y.; Jung, C.-R.; Nakayama, S.F.; Tamura, K.; Isobe, T.; Michikawa, T.; Iwai-Shimada, M.; Kobayashi, Y.; Sekiyama, M.; Taniguchi, Y.; et al. Indoor air quality of 5,000 households and its determinants. Part A: Particulate matter (PM2.5 and PM10–2.5) concentrations in the Japan Environment and Children's Study. *Environ. Res.* **2021**, *198*, 111196. [CrossRef] [PubMed]
14. Yoda, Y.; Tamura, K.; Shima, M. Airborne endotoxin concentrations in indoor and outdoor particulate matter and their predictors in an urban city. *Indoor Air* **2017**, *27*, 955–964. [CrossRef]
15. Argunhan, Z.; Avci, A.S. Statistical Evaluation of Indoor Air Quality Parameters in Classrooms of a University. *Adv. Meteorol.* **2018**, *2018*, 4391579. [CrossRef]
16. Liu, Y.; Zhou, B.; Wang, J.; Zhao, B. Health benefits and cost of using air purifiers to reduce exposure to ambient fine particulate pollution in China. *J. Hazard. Mater.* **2021**, *414*, 125540. [CrossRef]
17. Fermo, P.; Artíñano, B.; De Gennaro, G.; Pantaleo, A.M.; Parente, A.; Battaglia, F.; Colicino, E.; Di Tanna, G.; Junior, A.G.D.S.; Pereira, I.G.; et al. Improving indoor air quality through an air purifier able to reduce aerosol particulate matter (PM) and volatile organic compounds (VOCs): Experimental results. *Environ. Res.* **2021**, *197*, 111131. [CrossRef]
18. Kim, M.Y.; Jung, Y.G.; Park, J.C.; Yang, Y.K. The impact of airflow and air purification on the resuspension and removal of deposited particulate matter. *J. Build. Eng.* **2021**, *41*, 102367. [CrossRef]
19. Penza, M.; Suriano, D.; Villani, M.G.; Spinelle, L.; Gerboles, M. Towards air quality indices in smart cities by calibrated low-cost sensors applied to networks. *IEEE* **2014**, 2012–2017. [CrossRef]
20. Castell, N.; Kobernus, M.; Liu, H.-Y.; Schneider, P.; Lahoz, W.; Berre, A.J.; Noll, J. Mobile technologies and services for environmental monitoring: The Citi-Sense-MOB approach. *Urban. Clim.* **2015**, *14*, 370–382. [CrossRef]
21. Gulia, S.; Prasad, P.; Goyal, S.; Kumar, R. Sensor-based Wireless Air Quality Monitoring Network (SWAQMN)—A smart tool for urban air quality management. *Atmos. Pollut. Res.* **2020**, *11*, 1588–1597. [CrossRef]
22. Lu, Y.; Giuliano, G.; Habre, R. Estimating hourly PM2.5 concentrations at the neighborhood scale using a low-cost air sensor network: A Los Angeles case study. *Environ. Res.* **2021**, *195*, 110653. [CrossRef]
23. Global PM10 Map. Available online: https://www.uradmonitor.com/ (accessed on 30 May 2021).
24. World's Air Pollution: Real-Time Air Quality Index. Available online: https://waqi.info/ (accessed on 5 July 2021).
25. Bodor, M.; de Jos, D. Indoor Concentration and Composition of Fine Particulate Matter (PM2.5) Near an Intense Circulated Road in an Industrial City. *Ann. Dunarea Jos Univ. Galati. Fascicle IX Met. Mater. Sci.* **2020**, *43*, 5–10. [CrossRef]
26. Fluke 985 Particle Counter. Available online: https://test4less.co.uk/fluke-985-particle-counter.html (accessed on 2 April 2021).
27. Available online: https://www.calitateaer.ro/public/monitoring-page/reports-reports-page/?__locale=en (accessed on 7 April 2021).
28. Dust Collection Research–Dylos. Available online: https://billpentz.com/woodworking/cyclone/dylos.php (accessed on 14 May 2021).
29. Lu, X.; Sha, Y.H.; Li, Z.; Huang, Y.; Chen, W.; Chen, D.; Shen, J.; Chen, Y.; Fung, J.C. Development and application of a hybrid long-short term memory–three dimensional variational technique for the improvement of PM2.5 forecasting. *Sci. Total Environ.* **2021**, *770*, 144221. [CrossRef] [PubMed]
30. Luo, J.; Du, P.; Samat, A.; Xia, J.; Che, M.; Xue, Z. Spatiotemporal Pattern of PM2.5 Concentrations in Mainland China and Analysis of Its Influencing Factors using Geographically Weighted Regression. *Sci. Rep.* **2017**, *7*, 40607. [CrossRef] [PubMed]
31. Chen, T.; He, J.; Lu, X.; She, J.; Guan, Z. Spatial and Temporal Variations of PM2.5 and Its Relation to Meteorological Factors in the Urban Area of Nanjing, China. *Int. J. Environ. Res. Public Health* **2016**, *13*, 921. [CrossRef] [PubMed]
32. Li, J.J.; Wang, G.H.; Wang, X.; Cao, J.J.; Sun, T.; Cheng, C.L.; Meng, J.J.; Hu, T.F.; Liu, S.X. Abundance, composition and source of atmospheric PM2.5 at a remote site in the Tibetan Plateau, China. *Tellus B Chem. Phys. Meteorol.* **2013**, *65*. [CrossRef]
33. Chen, Z.; Cai, J.; Gao, B.; Xu, B.; Dai, S.; He, B.; Xie, X. Detecting the causality influence of individual meteorological factors on local PM2.5 concentration in the Jing-Jin-Ji region. *Sci. Rep.* **2017**, *7*, 40735. [CrossRef]

34. Zhang, H.; Wang, Y.; Hu, J.; Ying, Q.; Hu, X.-M. Relationships between meteorological parameters and criteria air pollutants in three megacities in China. *Environ. Res.* **2015**, *140*, 242–254. [CrossRef]
35. Wang, Y.; Ying, Q.; Hu, J.; Zhang, H. Spatial and temporal variations of six criteria air pollutants in 31 provincial capital cities in China during 2013–2014. *Environ. Int.* **2014**, *73*, 413–422. [CrossRef]
36. Daly, C.; Gibson, W.; Taylor, G.; Johnson, G.; Pasteris, P. A knowledge-based approach to the statistical mapping of climate. *Clim. Res.* **2002**, *22*, 99–113. [CrossRef]
37. LEGE 104 15/06/2011-Portal Legislativ. Available online: http://legislatie.just.ro/Public/DetaliiDocument/129642 (accessed on 28 May 2021).
38. Bodor, M.; Baltă, Ș; Pintilie, Ș.C.; Lazăr, A.L.; Buruiană, D. Studies regarding the distribution and composition of particulate matters in the air of an industrialized city. *SGEM* **2016**, *2*, 587–591. [CrossRef]
39. Pan, S.; Du, S.; Wang, X.; Zhang, X.; Xia, L.; Liu, J.; Pei, F.; Wei, Y. Analysis and interpretation of the particulate matter (PM10 and PM2.5) concentrations at the subway stations in Beijing, China. *Sustain. Cities Soc.* **2019**, *45*, 366–377. [CrossRef]
40. Stranger, M.; Potgieter-Vermaak, S.; Van Grieken, R. Comparative overview of indoor air quality in Antwerp, Belgium. *Environ. Int.* **2007**, *33*, 789–797. [CrossRef] [PubMed]
41. Jeong, C.-H.; Salehi, S.; Wu, J.; North, M.L.; Kim, J.S.; Chow, C.-W.; Evans, G.J. Indoor measurements of air pollutants in residential houses in urban and suburban areas: Indoor versus ambient concentrations. *Sci. Total Environ.* **2019**, *693*, 133446. [CrossRef] [PubMed]
42. Portable Particulate Monitor-Measure PM2.5/PM10. Available online: https://www.aeroqual.com/product/portable-particulate-monitor (accessed on 30 May 2021).
43. Air Quality Meter | PCE Instruments. Available online: https://www.pce-instruments.com/english/measuring-instruments/test-meters/air-quality-meter-kat_150925.htm (accessed on 30 May 2021).

processes

Article

Investigation of Seeds and Peels of *Citrullus colocynthis* as Efficient Natural Adsorbent for Methylene Blue Dye

Wafa Mohammed Alghamdi [1] and Ines El Mannoubi [1,2,*]

1 Chemistry Department, Faculty of Science, Albaha University, Al Bahah 65731, Saudi Arabia; fofe.88@hotmail.com
2 Department of Chemistry, Faculty of Sciences of Gabes, University of Gabes, Gabes 6072, Tunisia
* Correspondence: ines.mannoubi@gmail.com or ielmannoubi@bu.edu.sa

Abstract: Natural adsorbents as low-cost materials have been proved efficient for water remediation and have significant capacity for the removal of certain chemicals from wastewater. The present investigation aimed to use *Citrullus colocynthis* seeds (CCSs) and peels (CCPs) as an efficient natural adsorbent for methylene blue (MB) dye in an aqueous solution. The examined biosorbents were characterized using surface area analyzer (BET), scanning electron microscope (SEM), thermogravimetric analyzer (TGA) and Fourier transform infra-red (FT-IR) spectroscopy. Batch adsorption experiments were conducted to optimize the main factors influencing the biosorption process. The equilibrium data for the adsorption of MB by CCSs were best described by the Langmuir isotherm followed by the Freundlich adsorption isotherms, while the equilibrium data for MB adsorption by CCPs were well fitted by the Langmuir isotherm followed by the Temkin isotherm. Under optimum conditions, the maximum biosorption capacity and removal efficiency were 18.832 mg g^{-1} and 98.00% for MB-CCSs and 4.480 mg g^{-1} and 91.43% for MB-CCPs. Kinetic studies revealed that MB adsorption onto CCSs obeys pseudo-first order kinetic model (K_1 = 0.0274 min^{-1}), while MB adsorption onto CCPs follows the pseudo-second order kinetic model (K_2 = 0.0177 g mg^{-1} min^{-1}). Thermodynamic studies revealed that the MB biosorption by CCSs was endothermic and a spontaneous process in nature associated with a rise in randomness, but the MB adsorption by CCPs was exothermic and a spontaneous process only at room temperature with a decline in disorder. Based on the obtained results, CCSs and CCPSs can be utilized as efficient, natural biosorbents, and CCSs is promising since it showed the highest removal percentage and adsorption capacity of MB dye.

Keywords: biosorption; *Citrullus colocynthis*; methylene blue; isotherms; kinetic modeling; thermodynamic study

Citation: Alghamdi, W.M.; El Mannoubi, I. Investigation of Seeds and Peels of *Citrullus colocynthis* as Efficient Natural Adsorbent for Methylene Blue Dye. *Processes* **2021**, *9*, 1279. https://doi.org/10.3390/pr9081279

Academic Editor: Avelino Núñez-Delgado

Received: 14 June 2021
Accepted: 21 July 2021
Published: 25 July 2021

1. Introduction

Dyes are considered one of the most concerning industrial contaminants that are discharged into the environment in large amounts. They significantly spoil pleasant water characteristics and damage the aquatic environment [1]. Therefore, there are increasing research efforts, as it is vitally important to develop suitable and efficient techniques for the treatment of dye pollutants and their effective removal from wastewater to guarantee the safe discharge of treated liquid wastes into watercourses and water bodies [2]. Many conventional chemical, physical, and biological methods have been developed for the decontamination of water, but unfortunately, most of these methods are not applicable for large scale industries because of the limitations posed by the majority of them, such as high cost, low efficiency at low dye concentrations, and sludge production [3,4]. Adsorption is considered an efficient and prevalent process used for wastewater remediation, as it is capable of eliminating hazardous pollutants and color [5]. It has been reported as an attractive method, due to its multiple benefits, such as the operation facileness, the design simplicity and the high-quality treated water production [6,7]. Activated carbon is the most commonly used effective adsorbent for several kinds of dyes, owing to its high surface area, but it is restricted by its high cost and regeneration problems [8]. Various low-cost adsorbents, including agricultural and industrial by-products, have been developed as an alternative to commercial adsorbents and examined for their ability to remove dyestuffs [6]. The advantage of using these materials is largely by virtue of their omnipresent availability and low cost, and their regeneration is unnecessary [9]. Numerous studies reported the successful adsorption of dyes, using different parts of plants as adsorbent materials, such as *Calotropis Procera* [10], *Artocarpus heterophyllus* leaf [11], banana peels [12], almond gum [13], Oleander plant tissues [14], apple peels [15], *Vigna Trilobata* pod [16], *Cucumis sativus* peel [17], *Antigonon leptopus* leaf [18], *Lathyrus sativus* husk [19], etc.

Citrullus colocynthis is a species of the genus *Citrullus* of the *Cucurbitaceae* family known as handal in Arabic. It is native to tropical Africa and Asia and largely distributed in the arid regions of the Mediterranean basin [20]. The fruit is rich in biologically active constituents and used in the treatment of different ailments [21]. Recent studies have focused on the efficiency of *Citrullus colocynthis* waste ash and seed ash as phenol biosorbents [22,23].

In this work, MB was adopted as a model dye, which is widely used by a huge number of industries, such as the textile, pharmaceutical and food industries; exposure to it can cause different adverse health effects [24]. Several studies have highlighted the use of ecofriendly low-cost adsorbents based on natural materials for the removal of this contaminant from water systems.

To the best of our knowledge, CCSs and CCPs untreated forms have not been used for such a purpose. Hence, this work is dedicated to investigating the effectiveness of CCSs and CCPs as adsorbent materials for removing MB dye from aqueous solutions. The impact of the contact time, MB initial concentration, adsorbent particle size, adsorbent dosage, pH and temperature factors on the biosorption process were examined. Additionally, the biosorption kinetics, isotherms and thermodynamic properties were analyzed.

2. Materials and Methods

2.1. Materials and Solutions

MB dye (molecular formula: $C_{16}H_{18}ClN_3S$) was purchased from Sharlau (Spain) and was used without further purification. A total of 1000 mg/L MB dye stock solution was prepared in the laboratory. The required standard solutions were obtained by diluting the stock solution, using distilled water, and were used to prepare the calibration curve.

Citrullus colocynthis plant fruits were collected from the Al-Aqiq region in Al-Baha province located in the southwestern part of Saudi Arabia in November 2019. The collected fruits were rinsed thoroughly with tap water and air dried in the shade for two weeks. Then, CCSs and CCPs were separated, dried in the oven at 50 °C for about 24 h, crushed in a mixer grinder, screened through a set of sieves to obtain three fractions viz. <600 µm, 600–1180 µm and >1180 µm, and were stored in plastic containers for further use as adsorbent materials.

2.2. Characterization of Biosorbents

The point of zero charge (pH_{PZC}) of the biosorbents was measured as described by Miyah et al. [25]. The surface area, pore volume and pore radius were measured by the Surface Area and Pore Size Analyzer (model Nova touch-Quantachrome instruments, Anton Paar QuantaTech, New York, NY, USA). The TGA measurements were performed on thermogravimetric Analyzer (Model Pyris 1, Perkin Elmer, Waltham, MA, USA). The morphological features of the biosorbents and MB-loaded biosorbents were determined by SEM (model: Nova-Nano SEM-600, FEI, Hillsboro, OR, USA) at 15.0 kV and 500× magnification. Infrared spectra of the biosorbents and MB-loaded biosorbents were recorded over 4000–400 cm^{-1}, using Thermo Scientific Nicolet iS50 FT-IR spectrophotometer in the attenuated total reflectance (ATR) mode.

2.3. Batch Biosorption Experiments

The batch biosorption experiments were performed in a shaking incubator (JSSI-100C, JSR, Tokyo, Japan) at 140 rpm. The experiments were performed at different parameters, including contact time (0–240 min), initial MB dye concentration (2–7 mg/L), particle size of biosorbents (<600 μm, 600–1180 μm and >1180 μm), adsorbent dosage (0.5 g/L–10 g/L), pH (2–12), and temperature (20–60 °C). The solution pH was adjusted by adding 0.1M NaOH or 0.1M HCl before adding the adsorbent. Finally, the adsorbent was separated from the solution, using a HPLC syringe filter. The residual dye concentration was determined by monitoring the optical density for the dye solution at 665 nm, using a UV-Vis spectrophotometer (PD 303S, APEL, Japan).

The removal percentage R (%) was determined using Equation (1):

$$R\% = \frac{C_i - C_t}{C_i} \times 100 \tag{1}$$

where C_i and C_t are the MB dye concentration (mg/L) before and after the adsorption, respectively.

The equilibrium biosorption loading q_e (mg g^{-1}) of MB adsorbed by CCSs and CCPs was assessed using Equation (2):

$$q_e = \left(\frac{C_i - C_e}{m}\right) \times V \tag{2}$$

where C_e is MB dye concentrations (mg/L) at equilibrium, V is the volume of aqueous solution (L) and m is the mass of the biosorbent (g).

2.4. Adsorption Isotherms

Different isotherm equations are available to analyze the equilibrium data. Langmuir, Freundlich and Temkin equations, which are the most common ones, were adopted in the present investigation. Regression methods are generally used to assess the coefficients of the isotherm equations.

The used Langmuir equation was used in linearized form as shown in Equation (3) [14]:

$$\frac{1}{q_e} = \frac{1}{q_m K_L C_e} + \frac{1}{q_m} \tag{3}$$

In addition, the linear form of Freundlich equation is described by Equation (4) [26]:

$$\ln q_e = \ln K_f + \frac{1}{n} \ln C_e \tag{4}$$

The used Temkin isotherm linear expression is given in Equation (5) [19]:

$$q_e = \frac{RT}{b_t} \ln K_t + \frac{RT}{b_t} \ln C_e \tag{5}$$

where qe is adsorbent-phase concentration of adsorbate (mg g^{-1}) and C_e is the aqueous-phase concentration of adsorbate (mg/L) at equilibrium. q_m is the maximum adsorption capacity (mg g^{-1}) and K_L is the Langmuir constant (L/mg) related to the free adsorption energy. K_f (mg g^{-1}) (L $g^{-1})^{1/n}$ is the Freundlich constant related to the adsorption capacity and 1/n is an indicator of the adsorption strength. R is the ideal gas constant (8.314 J mol^{-1} K^{-1}), T is the absolute temperature (K), and b_t (J mol^{-1}) and K_t (L mg^{-1}) are the Temkin constants linked with the equilibrium binding constant and adsorption heat.

The Langmuir model is helpful in calculating the separation factor abbreviated as R_L, which is a dimensionless factor, by using Equation (6):

$$R_l = \frac{1}{1 + K_L C_i} \quad (6)$$

According to R_L values, the biosorption process can be either favorable ($0 < R_L < 1$), linear ($R_L = 1$), irreversible ($R_L = 0$) or unfavorable ($R_L > 1$) [27].

2.5. Biosorption Kinetics

The pseudo-first order kinetic model, developed by Lagergren in 1898, suggests the existence of physisorption, which is represented by Equation (7) [19]:

$$\ln(q_e - q_t) = \ln q_e - k_1 t \quad (7)$$

where q_e is the adsorption capacity (mg g^{-1}) at equilibrium time and q_t is the adsorption capacity at any time t (min), k_1 (min^{-1}) is the rate constant for the pseudo-first-order adsorption process.

The pseudo-second order kinetic model presumes that chemisorption is the rate-limiting step [28]. It is described by Equation (8):

$$\frac{t}{q_t} = \frac{1}{k_2 q_e^2} + \frac{1}{q_e} t \quad (8)$$

where q_e is the adsorption capacity (mg g^{-1}) at equilibrium and q_t is the adsorption capacity at any time t (min), k_2 (g mg^{-1} min^{-1}) is the rate constant for the pseudo-second-order adsorption process.

2.6. Thermodynamic Studies

Thermodynamic parameters' determination is required to study the spontaneity and feasibility of the adsorption process. These thermodynamic properties were assessed at a given temperature, using Equations (9) and (10) [27]:

$$\Delta G^\circ = -RT\ln K_d \quad (9)$$

$$\ln K_d = \frac{-\Delta H^\circ}{RT} + \frac{\Delta S^\circ}{R} \quad (10)$$

where $k_d = q_e/C_e$ is the apparent equilibrium adsorption constant at temperature T (K), ΔG° is Gibb's free energy, ΔH° is the enthalpy and ΔS° is the entropy.

3. Results and Discussions

3.1. Characterization of Biosorbents

3.1.1. Surface Area

The surface area, pore volume and pore radius values of CCSs and CCPs were 24.6794 and 11.9878 m^2/g, 0.023 and 0.013 cc/g, and 14.90 and 15.60 Å, respectively. It is worth noting that the surface area was approximately two-fold higher in the case of CCSs than in CCPs, suggesting that CCSs had better adsorbent capacity than CCPs. These observed values were higher compared with other reported biosorbents, such as orange and banana peels [29], jackfruit leaf [11], pinus durangensis sawdust [30], tea waste [31] and Lathyrus sativus husk [19], but were smaller than other adsorbents, such as activated carbon [32]. As the pore volumes were found to be less than 20 Å, CCSs and CCPs are considered predominately microporous materials, according to IUPAC classification [33].

3.1.2. TGA Analysis

The thermal stability of CCSs and CCPs was determined by TGA. The plot of the percent weight loss versus temperature over time is portrayed in Figure 1. Thermogravimetric

(TG) experiments were performed at a temperature range starting from 25 °C to 800 °C with 10 °C/min as heating rate. The TG pyrolysis plots exhibit the typical reverse s-shaped curves distinctive of the thermal degradation of lignocellulosic biomass under an inert atmosphere [34]. Moreover, the TG curves indicate that the thermal degradation of CCSs and CCPs occurred in two stages. During the first stage, the % weight loss was 2.80% and 3.12%, which occurred at 88.28 °C and 91.91 °C for CCSs and CCPs, respectively, which may be ascribed to the release of moisture. In the second stage, for both biosorbents, a continuous mass loss was observed with the rising of temperature, up to 610 °C and 670 °C for CCSs and CCPs, respectively. The highest mass % loss during the second stage was estimated to be 76.74% and 73.05%, which occurred at 351.28 °C and 303.18 °C for CCSs and CCPs, respectively. These findings may be associated with the removal of light molecules, including cellulose, hemicellulose and lignin, representing the essential constituents of the biomass among which hemicellulose is the most thermally unstable component [25]. It was revealed that the degradation temperatures of these components were in the range of 200–350 °C for hemicellulose, 250–400 °C for cellulose and 150–1000 °C for lignin [35]. At the end of this stage, it was observed that CCSs and CCPs were completely pyrolyzed, followed by a gradual drop in weight loss. The mass of biosorbents remained constant at 15.67% for CCSs and 17.14% for CCPs with the increasing of temperature, which is linked to the ash content of the biosorbents.

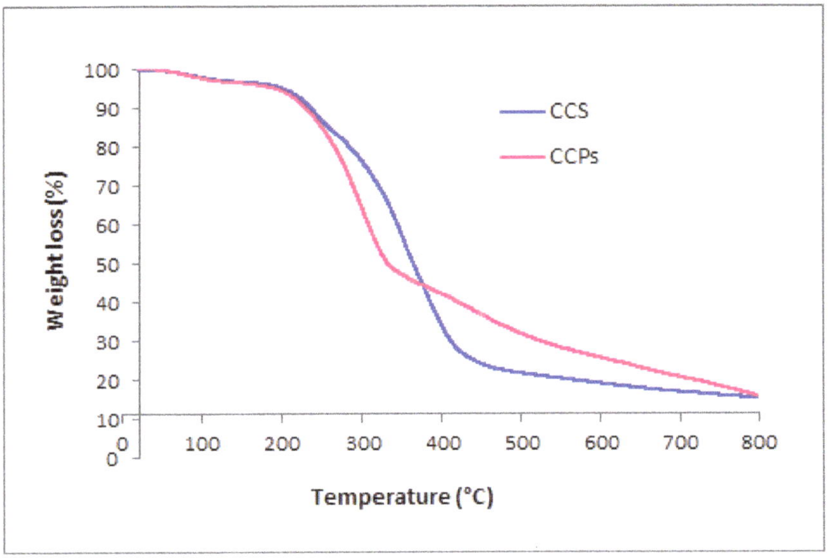

Figure 1. TG pyrolysis curves of CCSs and CCPs.

3.1.3. FT-IR Spectra

FT-IR spectroscopy is a useful technique for obtaining information on the nature of interactions between the adsorbent and adsorbate and for identifying the functional groups involved in these interactions [26]. The FT-IR spectra of CCSs and CCPs pre- and post adsorption are illustrated in Figure 2. The main FT-IR peaks were identified, assigned to their appropriate functional groups and the percent difference in band intensities (% ΔI) pre- and post-absorption was calculated. The findings are presented in Table 1. The FT-IR spectra for CCSs and CCPs showed the main peaks fundamentally linked to lignocellulosic biomass components [34], which were attributable to -OH groups (broadband), the -CH aliphatic group (sharp), the carbonyl group of ester and carboxylate, C=C of aromatic ring and C-O of carboxylate groups. It was observed that the positions of the bands before and after adsorption remained unchanged, while some major decrease in the band intensity of some specific peaks' characteristic of -OH groups, C=C, C=O and C-O bonds appeared in the IR spectra of MB-loaded CCSs and CCPs. These observations suggest that there is involvement of the above-named functional groups in the MB biosorption process. Similar findings were obtained by Shakoor and Nasar [26] who concluded significant engagement of the corresponding functional groups, due to the peak intensity reduction instead of the peak shifts.

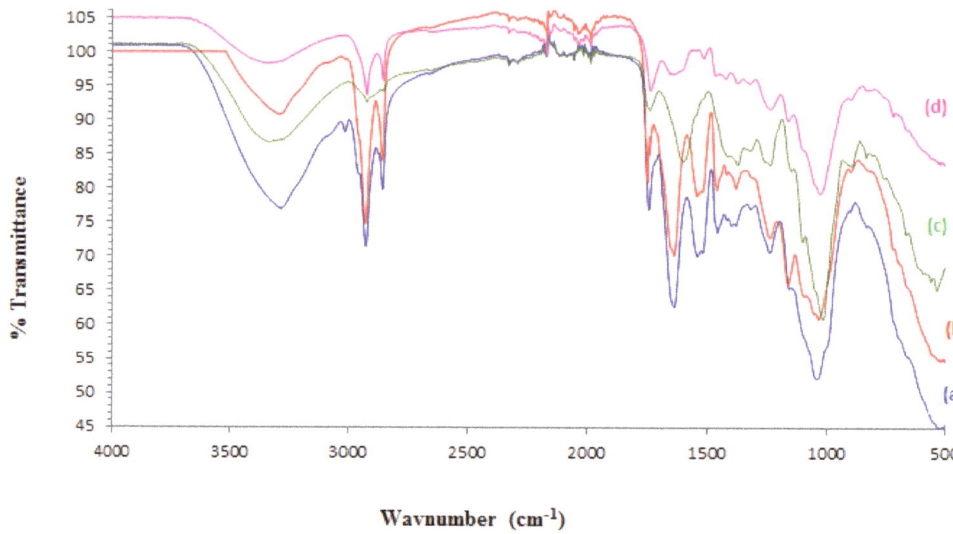

Figure 2. FT-IR spectra of (**a**) CCSs pre- adsorption, (**b**) CCSs post adsorption, (**c**) CCPs pre- adsorption and (**d**) CCPs post adsorption.

Table 1. FT-IR band position (cm^{-1}) and function groups assignments of CCSs and CCPs before and after adsorption.

CCSs	CCSs-MB	% ΔI	CCPs	CCPs-MB	% ΔI	Suggested Assignments	Reference
3292	3292	−13.66	3328	3328	−11.57	stretching vibration of -OH (alcohols and phenols)	[36]
2923	2924	−3.42	2917	2917	−1.05	asymmetric stretching of -CH group	[37]
2854	2854	−3.8	2847	2847	−2.4	symmetric stretching vibrations of -CH group	
1739	1739	−7.68	1735	1731	−6.25	C=O stretching vibration of carboxylate (-COO$^-$) / (-COOR)	[38]
1635	1636	−7.93				stretching vibrations of aromatic -C=O and -C=C bonds in carboxylic acid anions.	[39] [40]
1538	1540	−8.9	1596	1596	−13.58	aromatic ring vibrations	[41]
1455	1456	−6.23				phenolic -OH and -C=O stretching of carboxylates	[42]
			1370	1370	−12.04	stretching vibration of -COO	
1237	1236	−2.19	1235	1232	−8.24	C=O stretching vibration	[43]
1036	1032	−8.76	1027	1027	−17.06	C-O stretching of carboxylate groups	[44]

3.1.4. SEM

The pre- and post-adsorption SEM micrographs of CCSs and CCPs with MB, recorded at 500 magnifications, are illustrated in Figure 3. The SEM of CCSs exhibited the presence of irregular polygonal parenchymal cells with different sizes and shapes [45], while the SEM of CCPs exhibited a highly uneven, rough and porous surface, displaying clearly visible pockets with channels between them. The porous aspect and the surface heterogeneity are considered beneficial for any adsorbent [31]. It was noticed that these surfaces were completely modified after adsorption, indicating the filling of the pores and irregularities with the MB dye molecules.

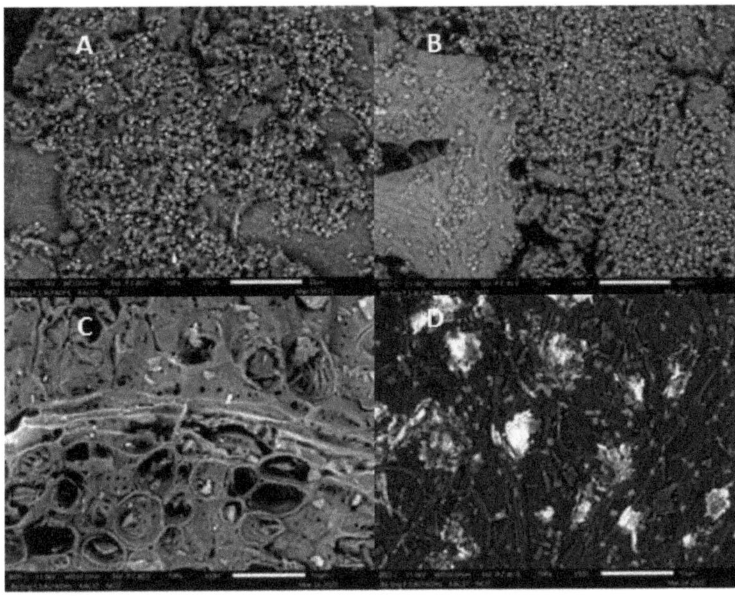

Figure 3. SEM micrograph of (**A**) CCSs pre- adsorption (**B**) CCSs post adsorption (**C**) CCPs pre-adsorption (**D**) CCPs post adsorption.

3.2. *Effect of Contact Time*

The effect of contact time on the MB removal by CCSs and CCPs as biosorbents over the time range of 0–240 min was studied; the changes in the MB removal efficiency are given in Figure 4. The removal rates of MB rose rapidly with both adsorbents during the first 20 min for CCSs and the first 45 min for CCPs, which was followed by a very slow adsorption rate. The equilibrium was reached after 60 min (R% = 95.57) for CCSs and 180 min for CCPs (R% = 85.55). This trend could be explained by the fact that a huge number of abundant unoccupied binding sites on the surface of both adsorbents were accessible at the adsorption process onset. In the middle stage, these sites were progressively covered, resulting in the adsorbent surface saturation. In the final stage, when all sites were fully occupied, the removal efficiency values remained constant, as no further adsorption occurs after the mentioned equilibrium times [26]. Hence, 60 min and 180 min were selected as the MB-CCS and MB-CCP optimum times, respectively, for subsequent batch experiments.

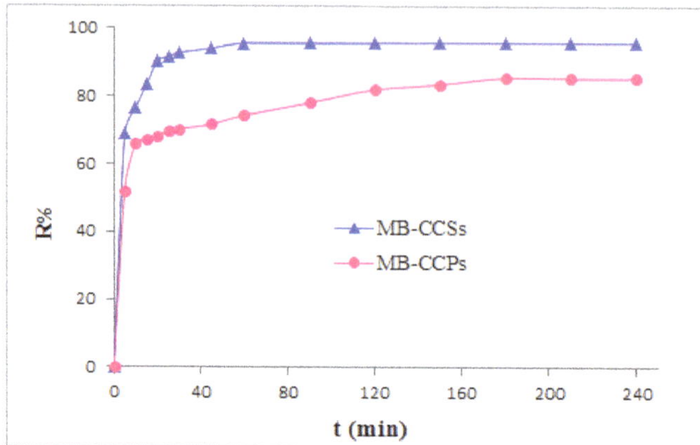

Figure 4. Effect of contact time on MB dye removal efficiency using CCSs and CCPs.

3.3. Effect of Initial MB Dye Concentration

The effect of the initial MB dye concentration on the removal of MB dye by CCSs and CCPs adsorbents was investigated at various concentrations, ranging from 2 mg/L–7 mg/L. Figure 5 represents the dye removal efficiency (R%) versus concentration. The obtained results revealed that the removal efficiency of dye by CCSs increased from 78.36% to 98.09% as the MB dye concentration varied from 2 mg/L to 6 mg/L but insignificant changes were observed in the removal efficiency with further increases in the dye concentration. The rapid uptake of MB at low concentrations could be linked to the availability of a huge count of free active sites on CCSs particles, compared to the initial count of MB molecules, enhancing the collisions between MB and CCSs. A higher concentration provides a vital driving force to surmount the mass transfer resistance of dye molecules between the aqueous solution and the solid surface [14].

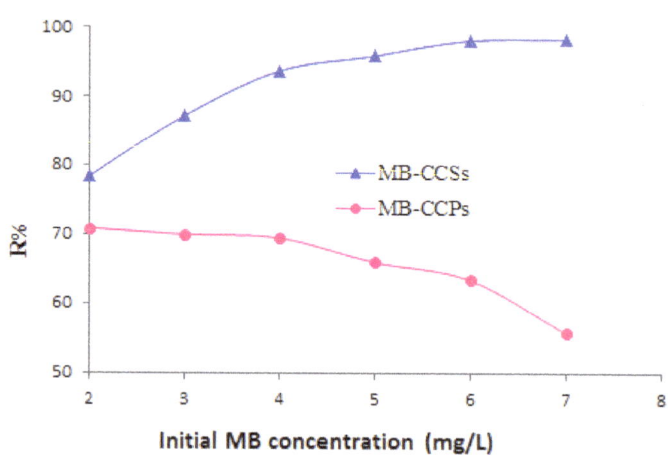

Figure 5. Effect of initial MB dye concentration on the removal efficiency of MB dye using CCSs and CCPs.

On the other hand, the dye removal efficiency by CCPs insignificantly decreased from 70.88% to 69.49% as the MB dye varied from 2 mg/L to 4 mg/L, followed by a quick decline to reach 55.82% at 7 mg/L. This could be explained by the availability of a limited number of active sites on the surface of CCPs, which would have become saturated at a definite concentration. At higher MB concentrations, the majority of the molecules are still unadsorbed, producing a decrease in the removal efficiency [11]. To make the biosorption process effective, MB dye concentrations of 6 mg/L and 4 mg/L were chosen as optimum values with CCSs and CCPs, respectively.

3.4. Impact of Adsorbent Particle Size

The impact of three different particle sizes of CCSs and CCPs on MB dye removal was investigated, and the results obtained are manifested in Table 2. The increase in particle size of CCSs decreased the MB dye removal efficiency (97.80% for <600 μm, 85.64% for 600 μm–1180 μm and 75.62% for >1180 μm). This may be explained by the fact that the adsorbent surface area with the smallest particles is the largest and contained the highest number of active sites and pores, which are more accessible [46]. Similar behavior was observed in using CCPs. Indeed, as the particle size decreased from >1180 μm to <600 μm, the MB removal efficiency was increased from 54.17% to 68.20%. However, the particles of the size 600–1180 μm were selected for further batch experiments, due to problems in the handling of CCPs smaller particles during the adsorption process.

Table 2. Effect of adsorbent particle size on MB removal efficiency.

Particle Size (μm)	R% (CCSs)	R% (CCPs)
<600	97.80	68.20
600 μm–1180 μm	85.64	63.50
>1180	75.62	54.17

3.5. Effect of Adsorbent Dosage

The effect of CCSs and CCPs dosage on MB dye removal was examined under various doses, ranging from 0.5 to 10 g/L. The obtained results are portrayed in Figure 6. It is obvious that a rise in the dosage of both biosorbents resulted in an increase in the MB removal efficiency. This may be associated with the attainability of unfilled active binding sites over a greater adsorbent surface area [18,47]. The removal efficiency increased promptly from 88.34% to 97.38% as the CCSs dosage increased from 0.5 to 3 g/L, followed by insignificant changes in R% with further enhancement in CCSs dosage. The removal of MB dye using CCPs adsorbent increased with a slow pattern from 15.35% to 81.14% by varying the CCPs dose from 0.5 to 9 g/L. Thereafter, R% remained unchanged, despite the further increase in the CCPs dosage. Based on these observations, 3 g/L and 9 g/L were concluded as the optimum CCSs and CCPs dosages, respectively.

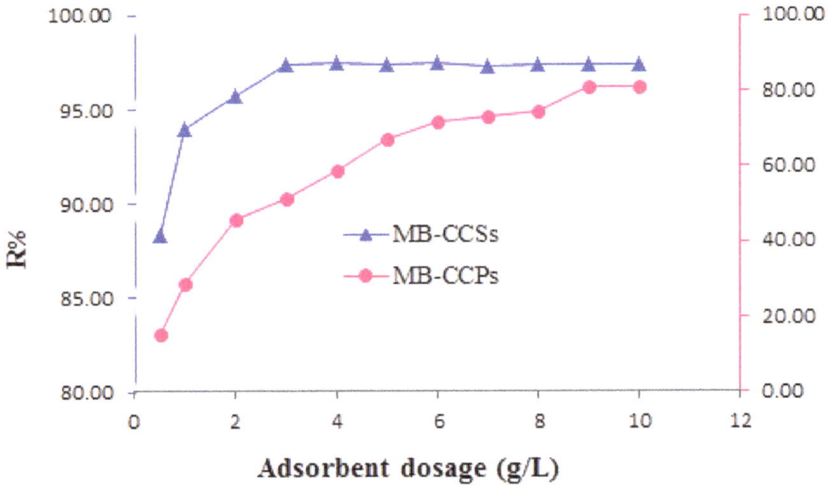

Figure 6. Effect of adsorbent dosage on MB removal efficiency using CCSs and CCPs.

3.6. Effect of PH

Batch experiments were executed by changing the pH of solutions from 2 to 10, while other parameters were maintained constant. Figure 7 showed the dye removal efficiency variation versus the initial solution pH. The percent removal of MB on CCSs and CCPs increased significantly by increasing pH from 2 to 8 and thereafter, no considerable change in MB efficiency removal was observed for further increases in the solution pH. The lowest values of MB adsorptive elimination were observed at pH < 8 because of the repulsion between the positively charged surface of both biosorbents and the cationic dye [47]. Moreover, the lower sorption of MB dye in acidic media may be associated with the competition from excess H^+ ions with the cationic dye for the binding sites [4,14]. MB dye removal efficiency under slightly alkaline conditions was found to be favorable. Indeed, at pH 8, the optimum values were 98.02% and 86.63% on CCSs and CCPs, respectively. This could be explained by the reduction in H^+ ions on the free sites and the predominance of negative charges on the surface of the biosorbents, enhancing the electrostatic attraction between the negatively charged adsorbent particles and the cationic dye [47,48]. Therefore, all further experiments were carried out at initial pH = 8 for MB-CCSs and MB-CCPs adsorption experiments.

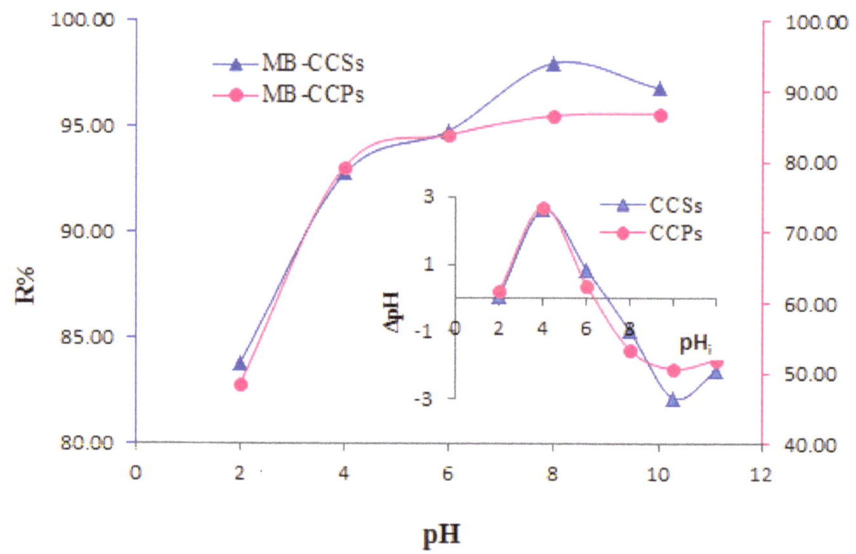

Figure 7. Effect of pH on MB removal efficiency using CCSs and CCPs (Inset–pH_{PZC}).

The impact of pH on the biosorption process could be interpreted based on pH_{PZC}. The values of pH_{PZC} were determined as 7 and 6.4 for CCSs and CCPs, respectively (Figure 7–inset). The surface of biosorbents is positively charged at pH < pH_{PZC}, owing to the protonation in the presence of extra hydrogen ions, negatively charged at pH > pH_{PZC} as a result of the deprotonation by the excess of OH^- ions neutral at pH = PH_{PZC} and [26]. Hence, the MB cationic dye should be adsorbed on a negatively charged biosorbent surface; in other terms, the pH of the medium should exceed pH_{PZC}, which approves our selection of pH = 8 as the optimum value.

3.7. Impact of Temperature

The impact of temperature on the MB dye uptake percentage by CCSs and CCPs was investigated between 293 and 333 K. As shown in Figure 8, the MB percent uptake by CCSs was increased from 93.58 to 98.00% upon increasing the temperature, which is indicative of the nature of the endothermic process. This could be linked to the rise in the kinetic energy of MB dye ions, resulting in increasing the chemical interaction with the adsorbent surface as well as an enhancement in the diffusion rate of the adsorbate molecules on the adsorbent particles [13,14].

An adverse trend was detected for the uptake of MB by CCPs, which was decreased from 91.43% to 82.52%, suggesting the exothermic nature of the adsorption process. This could be attributable to the proneness of MB molecules to run away from the biosorbent surface to the aqueous solution upon increasing the temperature, or could be assignable to the bonds' weakness between the biosorbent binding sites and the adsorbate molecules [11,41].

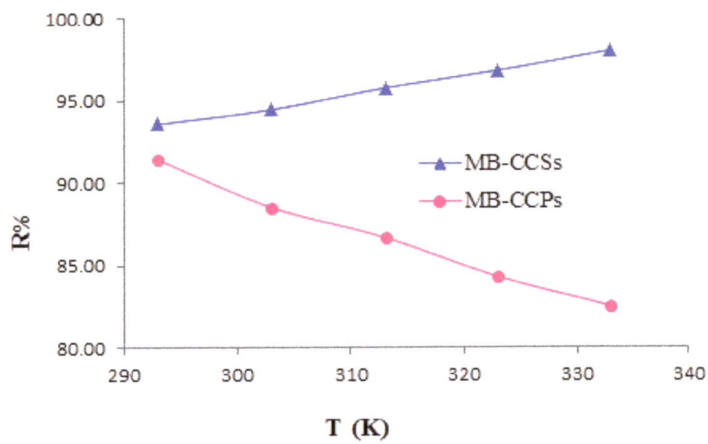

Figure 8. Impact of temperature on MB removal efficiency using CCSs and CCPs.

3.8. Isotherm Studies

Adsorption isotherms generally describe the nature of interactions between adsorbates and adsorbents by establishing the relationship linking the quantity of the dye adsorbed and the dye concentration remaining in the solution at equilibrium (q_e and C_e respectively) at a constant temperature and pH to evaluate the adsorbents' capacities [33].

In this investigation, three isotherm models were adopted to analyze the equilibrium data of MB dye loaded onto CCSs and CCPs (Langmuir, Freundlich and Temkin). The isotherm parameters of each model were calculated from the intercept and the slope of the appropriate linear curves portrayed in Figure 9a–c; the values are listed in Table 3. The correlation coefficients are also given in Table 3 to confirm the applicability of the adsorption isotherm models.

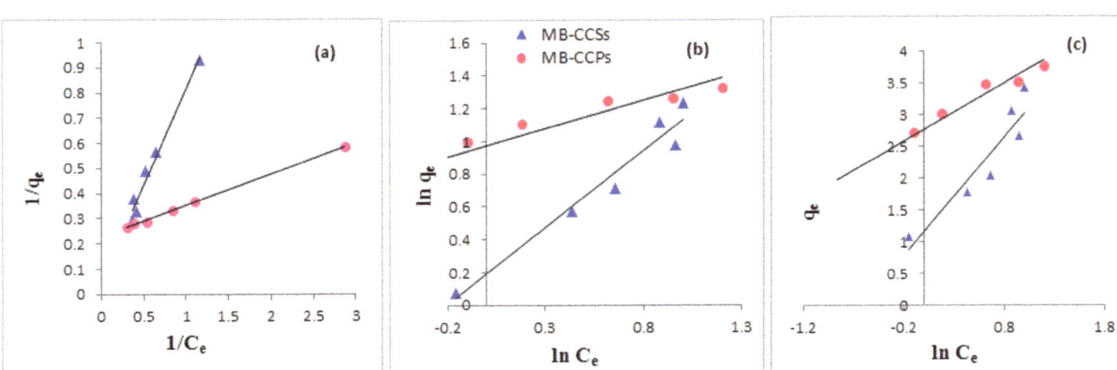

Figure 9. Adsorption Isotherm models: (**a**) Langmuir, (**b**) Freundlich, (**c**) Temkin.

By analyzing these results, it was noted that the Langmuir model presented the best fitness for MB-CCSs system pursued by the Freundlich model (R^2 = 0.9753 and R^2 = 0.9503 respectively). Unfortunately, the Temkin model is not well suited to the equilibrium data because of the moderate value (R^2 = 0.8642). The fitness of both Langmuir and Freundlich isotherms implied that CCSs and CCPs may display monolayer coverage and an irregular surface [38]. The Langmuir model presented the best fitness for the MB-CCPs system followed by the Temkin model (R^2 = 0.9982 and 0.9776, respectively). The Freundlich model could not fit well to the adsorption of MB onto CCPs based on the R^2 value (0.9456). The high value of the Langmuir model correlation coefficient postulates that the adsorption process was monolayer, reversible at the homogenous adsorbent surface with no lateral interaction between the adsorbate molecules [33]. Furthermore, the Temkin model correlation coefficient was greater than 0.95, suggesting that electrostatic interactions describing a chemical process are involved in the adsorption mechanism between MB and CCPs [31].

Table 3. Different parameters of isotherm models for MB adsorption onto CCSs and CCPs.

Sample	MB-CCSs	MB-CCPs
Langmuir		
q_m (mg g^{-1})	18.832	4.367
K_L (L/mg)	0.0698	1.8438
R^2	0.9753	0.9982
Freundlich		
K_f (mg g^{-1}) (L $g^{-1})^{1/n}$	1.2054	2.6485
n	1.0625	2.8910
R^2	0.9503	0.9456
Temkin		
K_t (L/mg)		20.884
b_t (kJ mol^{-1})		2.682
R^2	0.8642	0.9776

The Freundlich constants K_f and n that represent the adsorption capacity and the strength of adsorption were estimated. The values of n were larger than 1 for both CCSs and CCPs, which indicates that the adsorption processes were favorable and followed the normal Langmuir isotherm [19,26]. The values of K_t and b_t from the Temkin isotherm model were 20.884 L/mg and 2.682 kJ mol^{-1} for MB-CCPs, suggesting a strong interaction between the MB dye and the biosorbent surface, and that the sorption process is of a physio–chemical nature [38,49]. The obtained values of R_L listed in Table 4 varied between 0 and 1, suggesting the favorable MB biosorption on CCSs and CCPs.

Table 4. Dimensionless separation factor R_L.

MB (C_i, mg/L)	2	3	4	5	6	7
MB-CCSs	0.878	0.827	0.782	0.741	0.705	0.672
MB-CCPs	0.213	0.153	0.119	0.098	0.083	0.072

The maximum monolayer biosorption capacities (q_m) of Langmuir isotherms were equal to 18.832 and 4.480 mg g^{-1} for MB-CCSs and MB-CCPs, respectively. It is clear that CCSs was 4.2 times more efficient as a MB biosorbent than CCPs. On comparing the maximum adsorbed quantity of MB obtained in this study with other biosorbents reported in the literature (Table 5), it was clear that the biosorption capacity of CCSs was greater than those of six materials out of fifteen biosorbents and larger than that of activated carbon, while the adsorption capacity of CCPs was the lowest one.

Table 5. Comparison of biosorption capacities of different biosorbents for MB dye elimination.

Biosorbent	Q_m (mg g^{-1})	Reference
Brazil nut shells	7.81	[50]
Water hyacinth root powder	8.04	[51]
Spent rice	8.13	[52]
Sugar extracted spent rice biomass	8.13	[52]
Bio-char from pyrolysis of wheat straw	12.03	[53]
Pine sawdust	16.75	[54]
Cucumis sativus peels	21.459	[55]
Sunflower seed husk (*Helianthus annuus*)	23.20	[56]
Orange waste	30.3	[57]
Marula seed husk	33	[58]
Palm tree waste	39.47	[59]
Ginkgo biloba leaves	48.07	[44]
Vigna Trilobata pod	71.42	[16]
Coconut leaves	112.35	[60]
Tea waste	113.1461	[31]
Activated carbon	8.77	[61]
CCSs	18.832	present study
CCPs	4.480	present study

3.9. Adsorption Kinetics

The investigation of adsorption kinetics is of remarkable importance for the determination of the equilibrium time and the rate of adsorption [62]. In this work, biosorption kinetics of MB dye was assessed, using pseudo-first and pseudo-second order models. The parameters of each model were calculated from the intercept and the slope of the appropriate linear curves shown in Figure 10a,b. These values are summarized in Table 6 in addition to the correlation coefficients in order to check the fitness of the kinetic models with the experimental data.

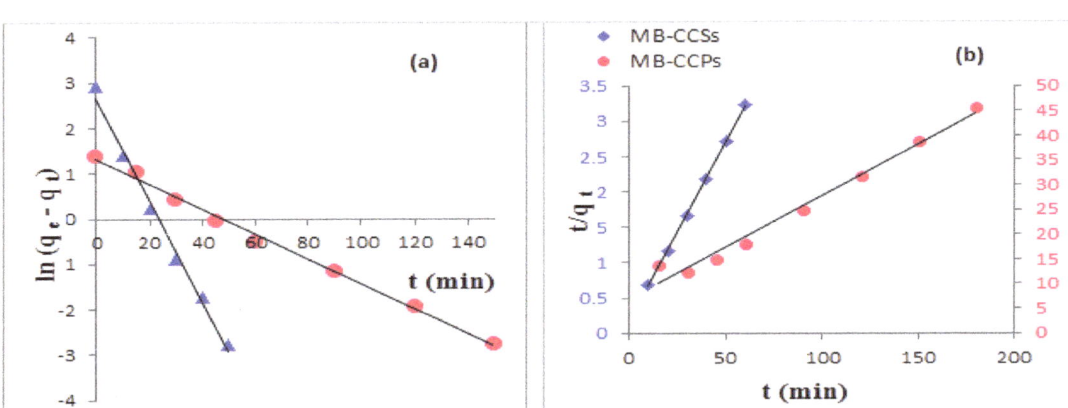

Figure 10. Adsorption kinetic models: (**a**) pseudo-first order, (**b**) pseudo-second order.

Table 6. Kinetic parameters for the biosorption of MB dye onto CCSs and CCPs.

Sample	MB-CCSs	MB-CCPs
MB (C_i, mg/L)	6	4
$q_{e,\,exp}$ (mg g^{-1})	18.832	4.367
Pseudo-First Order		
K_1 (min^{-1})	0.1113	0.0274
$q_{e1,\,cal}$ (mg g^{-1})	14.036	3.681
R^2	0.991	0.9947
Pseudo-Second Order		
K_2 (g mg^{-1} min^{-1})	0.0177	0.0061
$q_{e2,\,cal}$ (mg g^{-1})	19.531	4.801
R^2	0.9994	0.9803

It was observed that the experimental data of MB adsorbed by CCSs fitted perfectly to the pseudo-second order (R^2 = 0.9994) rather than pseudo-first order (R^2 = 0.991). Moreover, there was only a small difference between $q_{e,exp}$ (18.832 mg g^{-1}) and $q_{e,cal}$ (19.531 mg g^{-1}), confirming the suitability of the pseudo-second order mechanism process, while the pseudo-first order was the model that best fitted the experimental data of MB sorption by CCPs, which was due to the high value of R^2 (0.9947), compared to pseudo-second-order (R^2 = 0.9803). Kinetic data results revealed that the MB adsorption by CCSs and CCPs followed chemisorption and physisorption, respectively [63].

3.10. Thermodynamic Studies

Thermodynamic studies are useful for the estimation of the feasibility and the adsorption process nature. Hence, the thermodynamic parameters were calculated by examining the impact of temperature on the biosorption processes under optimum conditions. The values of ΔH° and ΔS° were deduced from the slope and the intercept of the linear plot of Van't Hoff, i.e., ln K_d versus 1/T, and were further used to calculate ΔG° at a given temperature. The mentioned plots are presented in Figure 11 and the estimated thermodynamic parameters are outlined in Table 7.

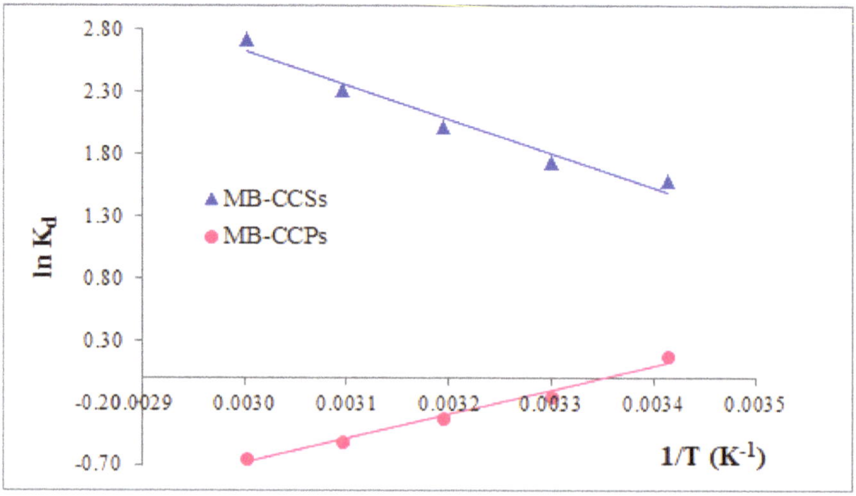

Figure 11. The plot of ln K_d as a function of 1/T for the removal of MB dye by CCSs and CCPs.

Table 7. Thermodynamic parameters for MB dye adsorption by CCSs and CCPs.

Adsorbent	ΔH°(kJ mol^{-1})	ΔS°(J mol^{-1} K^{-1})	Temperature (K)	ΔG°(KJ mol^{-1})
CCSs	23.09	91.17	293	−3.62
			303	−4.53
			313	−5.44
			323	−6.36
			333	−7.27
CCPs	−16.25	−54.47	293	−0.29
			303	0.25
			313	0.80
			323	1.34
			333	1.89

The negative values of ΔG° in the temperature range 293–333 K are indicative of the spontaneity and the feasibility of the biosorption of MB onto CCSs. These values decreased when temperature increased, indicating that adsorption is favored at high temperature values [17]. The positive value of ΔH° (23.09 kJ mol^{-1}) indicated that MB biosorption by CCSs is an endothermic and chemical process [48]. The positive value of ΔS° is an indication of the increase in disorder at the solid–liquid interface during the adsorption process, which could be associated with structural modifications of MB molecules and the adsorbent surface during the process [28]. The biosorption process is likely to take place spontaneously at any temperature since ΔH° > 0 and ΔS° > 0 [13].

MB dye adsorption by CCPs was favorable at a normal temperature and was non-spontaneous at higher temperatures, as the only negative value of ΔG° was obtained at 293 K, and it was observed that ΔG° increased positively between 303 and 333 K. The negative value of ΔH° shores up the exothermic comportment of the adsorption process. The negative value ΔS° suggests the decline of disorder at the solid/solution interface during the biosorption process, resulting in decreasing the adsorption capacity because of dye molecules escaping from the biosorbents [19].

3.11. Adsorption Mechanism

The results of the different parameters, physical characterization, kinetic and isotherm modeling were adopted to understand the adsorption mechanism. Based on the obtained results, MB biosorption onto CCSs and CCPs was found to be a complicated process involving different types of interactions. The removal of MB by CCSs and CCPs was rapid at the first stage and then reached an equilibrium. The MB uptake by CCSs could be considered a physicochemical process since the MB biosorption by CCSs follows pseudo-second order, which suggests that chemisorption interactions play a crucial role in this operation. It is best suited by Langmuir and Freundlich models, indicating a physisorption process, while the pseudo-first order best fits the biosorption of MB by CCPs, suggesting that physical interactions play a fundamental role in the MB-CCPs adsorption process, while electrostatic interactions are also involved between MB and CCPs since the Temkin model was well applicable ($R^2 > 0.95$).

The effect of pH on MB loading is another indicator on the importance of electrostatic interaction between the dye and biosorbents. Indeed, it was found that the electrostatic interaction between the cationic MB dye dissociated into MB^+ and Cl^- groups and adsorbents increased at pH > pH_{PZC} since the biosorbents' surface became more negatively charged. At the surface, the carboxylic groups were deprotonated and the carboxylate anions (-COO^-) were electrostatically attracted to the positively charged quaternary ammonium group of MB cationic dye [40]. Furthermore, dipole–dipole hydrogen bonds might be formed between hydroxyl groups acting as proton donors on the surface of the biosorbents and nitrogen atom acting as proton acceptors in the MB dye [64]. The n–π interactions could be formed between oxygen atoms in the carbonyl groups (electron donors) of the biosorbent surface and aromatic rings (electron acceptors) in the MB dye [19]. This mechanism is supported by the FT-IR results, which revealed that the intensity of the IR bands of the above-named functional groups decreased after MB adsorption. Accordingly, organic functional groups of CCSs and CCPs played a significant role in the MB biosorption. Hence, the adsorption of MB onto CCSs and CCPs could occur via three possible mechanisms, viz. hydrogen bonding, n–π interactions and electrostatic attraction, which prove that physisorption and/or chemisorption could be involved in the biosorption process. All these interactions are shown in Figure 12.

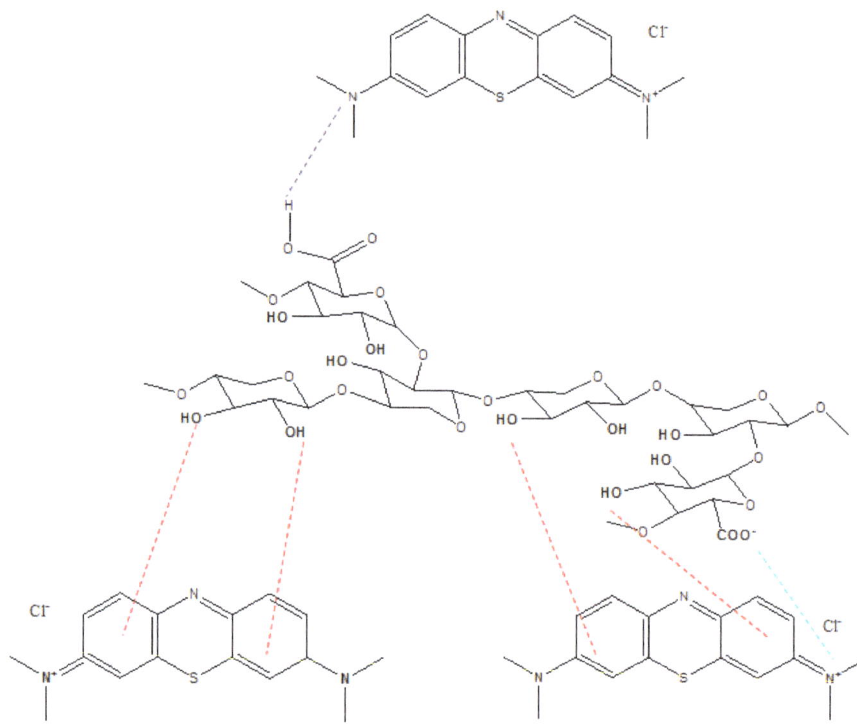

Figure 12. Proposed mechanism for MB adsorption by CCSs and CCPs.

4. Conclusions

MB dye adsorption onto CCSs and CCPs was studied concerning equilibrium, kinetics and isotherms. Batch experiments were used to determine the effect of different parameters on the MB removal efficiency. The results indicated that all parameters affected significantly the adsorption process. Maximum removal percentage was achieved, using 6 mg/L of MB dye, 3 g/L of adsorbent at 60 min and 333 K for CCSs and using 4 mg/L of MB dye, 9 g/L of adsorbent at 180 min and 293 K for CCPs under alkaline conditions. Under optimum conditions, 98.00% and 91.43% of MB removal was reached, using CCSs and CCPs, respectively. Kinetic studies revealed that biodsorption of MB by CCSs follows pseudo-second order, while MB biosorption onto CCPs is well described by pseudo-first order, suggesting chemisorption and physisorption processes, respectively. The equilibrium data fitted well with Langmuir and Freundlich models, indicating monolayer coverage at the heterogeneous surface for the MB-CCSs system, while Langmuir and Temkin models for MB-CCPs suggest that the adsorption process was monolayer and reversible at the homogenous adsorbent surface. MB maximum adsorption by CCSs was found to be 18.832 mg g^{-1}, which is considered relatively high, compared with many biosorbents, while MB maximum adsorption by CCPs was 4.480 mg g^{-1}. The thermodynamic results revealed that the biosorption process using CCSs was endothermic, feasible, spontaneous at any temperature and favorable at higher temperature values with increasing the randomness at the solid–liquid interface, while the adsorption process using CCPs was exothermic and spontaneous only at a normal temperature of 293 K with decreasing the randomness at the solid–liquid interface of the adsorbent

Author Contributions: Conceptualization, methodology, W.M.A. and I.E.M.; formal analysis, software, resources, writing—original draft preparation, W.M.A.; validation, investigation, data curation, visualization, supervision, writing—review and editing, I.E.M. All authors have read and agreed to the published version of the manuscript.

Funding: Not applicable.

Institutional Review Board Statement: Not applicable.

Informed Consent Statement: Not applicable.

Data Availability Statement: Data are contained within the article.

Conflicts of Interest: The authors declare no conflict of interest.

References

1. Tan, K.B.; Vakili, M.; Horri, B.A.; Poh, P.E.; Abdullah, A.Z.; Salamatinia, B. Adsorption of dyes by nanomaterials: Recent developments and adsorption mechanisms. *Sep. Purif. Technol.* **2015**, *150*, 229–242. [CrossRef]
2. Wahlström, N.; Steinhagen, S.; Toth, G.; Pavia, H.; Edlund, U. Ulvan dialdehyde-gelatin hydrogels for removal of heavy metals and methylene blue from aqueous solution. *Carbohydr. Polym.* **2020**, *249*, 116841. [CrossRef] [PubMed]
3. Bhatia, D.; Sharma, N.R.; Singh, J.; Kanwar, R.S. Biological methods for textile dye removal from wastewater: A review. *Crit. Rev. Environ. Sci. Technol.* **2017**, *47*, 1836–1876. [CrossRef]
4. Kuang, Y.; Zhang, X.; Zhou, S. Adsorption of Methylene Blue in Water onto Activated Carbon by Surfactant Modification. *Water* **2020**, *12*, 587. [CrossRef]
5. Chikri, R.; Elhadiri, N.; Benchanaa, M.; El Maguana, Y. Efficiency of Sawdust as Low-Cost Adsorbent for Dyes Removal. *J. Chem.* **2020**, *2020*, 8813420. [CrossRef]
6. De Gisi, S.; Lofrano, G.; Grassi, M.; Notarnicola, M. Characteristics and adsorption capacities of low-cost sorbents for wastewater treatment: A review. *Sustain. Mater. Technol.* **2016**, *9*, 10–40. [CrossRef]
7. Katheresan, V.; Kansedo, J.; Lau, S.Y. Efficiency of various recent wastewater dye removal methods: A review. *J. Environ. Chem. Eng.* **2018**, *6*, 4676–4697. [CrossRef]
8. Holkar, C.; Jadhav, A.; Pinjari, D.V.; Mahamuni, N.M.; Pandit, A.B. A critical review on textile wastewater treatments: Possible approaches. *J. Environ. Manag.* **2016**, *182*, 351–366. [CrossRef]
9. Saini, R.D. Textile Organic Dyes: Polluting effects and Elimination Methods from Textile Waste Water. *Int. J. Chem. Eng. Res.* **2017**, *9*, 121–136.
10. Oyelude, E.O.; Owusu, U.R. Adsorption of Methylene blue from aqueous solution using acid modified *Calotropis procera* leaf powder. *J. Appl. Sci. Environ. Sanit.* **2011**, *6*, 477–484.
11. Das, P.; Chakraborty, S.; Chowdhury, S. Batch and continuous (fixed-bed column) biosorption of crystal violet by *Artocarpus heterophyllus* (jackfruit) leaf powder. *Colloids Surf. B Biointerfaces* **2012**, *92*, 262–270. [CrossRef]
12. Moubarak, F.; Atmani, R.; Maghri, I.; Elkouali, M.; Talbi, M.; Latifa, M. Elimination of Methylene Blue dye with natural adsorbent "banana peels powder". *Glob. J. Sci. Front. Res.* **2014**, *14*, 39–44.
13. Bouaziz, F.; Koubaa, M.; Kallel, F.; Chaari, F.; Driss, D.; Ghorbel, R.E.; Chaabouni, S.E. Efficiency of almond gum as a low-cost adsorbent for methylene blue dye removal from aqueous solutions. *Ind. Crops Prod.* **2015**, *74*, 903–911. [CrossRef]
14. Abu-El-Halawa, R.; Zabin, S.A.; Abu-Sittah, H.H. Investigation of Methylene Blue Dye Adsorption from Polluted Water Using Oleander Plant (Al Defla) Tissues as Sorbent. *Am. J. Environ. Sci.* **2016**, *12*, 213–224. [CrossRef]
15. Enniya, I.; Jourani, A. Study of Methylene Blue Removal by a biosorbent prepared with Apple peels. *J. Mater. Environ. Sci.* **2017**, *8*, 4573–4581. [CrossRef]
16. Saibaba, K.V.N.; Kandisa, R.V. Adsorption Isotherm Studies on Methylene Blue Dye Removal Using Naturally Available Biosorbent. *Rasayan J. Chem.* **2019**, *12*, 2176–2182.
17. Shakoor, S.; Nasar, A. Utilization of *Cucumis Sativus* Peel as an Eco-Friendly Biosorbent for the Confiscation of Crystal Violet Dye from Artificially Contaminated Wastewater. *Anal. Chem. Lett.* **2019**, *9*, 1–19. [CrossRef]
18. Devi, V.S.; Sudhakar, B.; Prasad, K.; Sunadh, P.J.; Krishna, M. Adsorption of Congo red from aqueous solution onto *Antigonon leptopus* leaf powder: Equilibrium and kinetic modeling. *Mater. Today Proc.* **2020**, *26*, 3197–3206. [CrossRef]
19. Ghosh, I.; Kar, S.; Chatterjee, T.; Bar, N.; Das, S.K. Removal of methylene blue from aqueous solution using *Lathyrus sativus* husk: Adsorption study, MPR and ANN modelling. *Process Saf. Environ. Prot.* **2021**, *149*, 345–361. [CrossRef]

20. Degola, F.; Marzouk, B.; Gori, A.; Brunetti, C.; Dramis, L.; Gelati, S.; Buschini, A.; Restivo, F.M. *Aspergillus flavus* as a Model System to Test the Biological Activity of Botanicals: An Example on *Citrullus colocynthis* L. Schrad. Organic Extracts. *Toxins* **2019**, *11*, 286. [CrossRef]
21. Ahmed, M.; Ji, M.; Qin, P.; Gu, Z.; Liu, Y.; Sikandar, A.; Iqbal, M.F.; Javeed, A. Phytochemical screening, total phenolic and flavonoids contents and antioxidant activities of *Citrullus colocynthis* L. and *Cannabis sativa* L. *Appl. Ecol. Environ. Res.* **2019**, *17*, 6961–6979. [CrossRef]
22. Qasemi, M.; Afsharnia, M.; Zarei, A.; Najafpoor, A.A.; Salari, S.; Shams, M. Phenol removal from aqueous solution using *Citrullus colocynthis* waste ash. *Data Brief* **2018**, *18*, 620–628. [CrossRef]
23. Salari, S.; Afsharnia, M.; Moteallemi, A.; Ghasemi, M. Evaluation of removal efficiency of phenol from synthetic aqueous solutions by *Citrullus colocynthis* seed ash. *Environ. Health Eng. Manag.* **2018**, *5*, 49–55. [CrossRef]
24. Hasdemir, Z.M.; Simsek, S. Removal of Cationic Dye in Aquatic Medium by Using a New Composite Material. *Cumhur. Sci. J.* **2018**, *39*, 181–191. [CrossRef]
25. Miyah, Y.; Lahrichi, A.; Idrissi, M.; Boujraf, S.; Taouda, H.; Zerrouq, F. Assessment of adsorption kinetics for removal potential of Crystal Violet dye from aqueous solutions using Moroccan pyrophyllite. *J. Assoc. Arab Univ. Basic Appl. Sci.* **2017**, *23*, 20–28. [CrossRef]
26. Shakoor, S.; Nasar, A. Adsorptive decontamination of synthetic wastewater containing crystal violet dye by employing *Terminalia arjuna* sawdust waste. *Groundw. Sustain. Dev.* **2018**, *7*, 30–38. [CrossRef]
27. Ray, S.S.; Gusain, R.; Kumar, N. (Eds.) Adsorption equilibrium isotherms, kinetics and thermodynamics. In *Micro and Nano Technologies, Carbon Nanomaterial-Based Adsorbents for Water Purification*; Elsevier: Amsterdam, The Netherlands, 2020; pp. 101–118. [CrossRef]
28. Hamoudi, S.A.; Hamdi, B.; Brendlé, J. Removal of Ions Pb^{2+} and Cd^{2+} from Aqueous Solution by Containment Geomaterials. In *Exergetic, Energetic and Environmental Dimensions*; Dincer, I., Colpan, C.O., Kizilkan, O., Eds.; Academic Press: Cambridge, MA, USA, 2018; pp. 1029–1043. [CrossRef]
29. Kamsonlian, S.; Suresh, S.; Balomajumder, S.; Chand, S. Characterization of banana peels and orange peels: Biosorption mechanism. *Int. J. Sci. Technol. Manag.* **2011**, *2*, 1–7.
30. Salazar-Rabago, J.J.; Ramos, R.L.; Utrilla, J.R.; Perez, R.O.; Cordova, F.J.C. Biosorption mechanism of Methylene Blue from aqueous solution onto White Pine (*Pinus durangensis*) sawdust: Effect of operating conditions. *Sustain. Environ. Res.* **2017**, *27*, 32–40. [CrossRef]
31. Liu, L.; Fan, S.; Li, Y. Removal Behavior of Methylene Blue from Aqueous Solution by Tea Waste: Kinetics, Isotherms and Mechanism. *Int. J. Environ. Res. Public Health* **2018**, *15*, 1321. [CrossRef]
32. Saleem, J.; Bin Shahid, U.; Hijab, M.; Mackey, H.; McKay, G. Production and applications of activated carbons as adsorbents from olive stones. *Biomass Convers. Biorefin.* **2019**, *9*, 775–802. [CrossRef]
33. Sahoo, T.R.; Prelot, B. Chapter 7—Adsorption processes for the removal of contaminants from wastewater: The perspective role of nanomaterials and nanotechnology. In *Micro and Nano Technologies, Nanomaterials for the Detection and Removal of Wastewater Pollutant*; Bonelli, B., Freyria, F.S., Rossetti, I., Sethi, R., Eds.; Elsevier: Amsterdam, The Netherlands, 2020; pp. 161–222. [CrossRef]
34. Nyakuma, B.; Oladokun, O.; Dodo, Y.; Wong, S.; Uthman, H.; Halim, M. Fuel Characterization and Thermogravimetric Analysis of Melon (*Citrullus colocynthis* L.) Seed Husk. *Chem. Chem. Technol.* **2016**, *10*, 493–497. [CrossRef]
35. Díez, D.; Urueña, A.; Piñero, R.; Barrio, A.; Tamminen, T. Determination of Hemicellulose, Cellulose, and Lignin Content in different Types of Biomasses by Thermogravimetric Analysis and Pseudocomponent Kinetic Model (TGA-PKM Method). *Processes* **2020**, *8*, 1048. [CrossRef]
36. Silva, F.; Nascimento, L.; Brito, M.; da Silva, K.; Paschoal, W., Jr.; Fujiyama, R. Biosorption of Methylene Blue Dye Using Natural Biosorbents Made from Weeds. *Materials* **2019**, *12*, 2486. [CrossRef]
37. Prasad, A.L.; Santhi, T. Adsorption of hazardous cationic dyes from aqueous solution onto *Acacia nilotica* leaves as an eco friendly adsorbent. *Sustain. Environ. Res.* **2012**, *22*, 113–122.
38. Bouras, H.D.; Yeddou, A.R.; Bouras, N.; Hellel, D.; Holtz, M.D.; Sabaou, N.; Chergui, A.; Nadjemi, B. Biosorption of Congo red dye by *Aspergillus carbonarius* M333 and *Penicillium glabrum* Pg1: Kinetics, equilibrium and thermodynamic studies. *J. Taiwan Inst. Chem. Eng.* **2017**, *80*, 915–923. [CrossRef]
39. Koyuncu, H.; Kul, A.R. Removal of methylene blue dye from aqueous solution by nonliving lichen (*Pseudevernia furfuracea* (L.) Zopf.), as a novel biosorbent. *Appl. Water Sci.* **2020**, *10*, 72. [CrossRef]
40. Dinh, V.-P.; Huynh, T.-D.-T.; Le, H.M.; Nguyen, V.-D.; Dao, V.-A.; Hung, N.Q.; Tuyen, L.A.; Lee, S.; Yi, J.; Nguyen, T.D.; et al. Insight into the adsorption mechanisms of methylene blue and chromium(iii) from aqueous solution onto pomelo fruit peel. *RSC Adv.* **2019**, *9*, 25847–25860. [CrossRef]
41. Khattri, S.D.; Singh, M.K. Removal of malachite green from dye wastewater using neem sawdust by adsorption. *J. Hazard. Mater.* **2009**, *167*, 1089–1094. [CrossRef] [PubMed]
42. Barka, N.; Abdennouri, M.; El Makhfouk, M.; Qourzal, S. Biosorption characteristics of cadmium and lead onto eco-friendly dried cactus (*Opuntia ficus indica*) cladodes. *J. Environ. Chem. Eng.* **2013**, *1*, 144–149. [CrossRef]
43. Sharma, M.; Kaushik, A.; Kaushik, C.P. Waste biomass of *Nostoc linckia* as adsorbent of crystal violet dye: Optimization based on statistical model. *Int. Biodeterior. Biodegrad.* **2011**, *65*, 513–521.

44. Singh, R.; Singh, T.S.; Odiyo, J.O.; Smith, J.A.; Edokpayi, J.N. Evaluation of Methylene Blue Sorption onto Low-Cost Biosorbents: Equilibrium, Kinetics, and Thermodynamics. *J. Chem.* **2020**, *2020*, 8318049. [CrossRef]
45. Ambi, A.A.; Nuru, F.G.; Ibrahim, M.M.; Mora, A.T.; Abubakar, M.S. Pharmacognostic Studies and Elemental Analysis of the Surface Structures of the Seeds of *Colocynthis citrullus* (Thunb.). *Int. J. Glob. Sustain.* **2017**, *4*, 1–10.
46. Ikenyiri, P.N.; Ukpaka, C.P. Overview on the Effect of Particle Size on the Performance of Wood Based Adsorbent. *J. Chem. Eng. Process Technol.* **2016**, *7*, 1000315.
47. Amin, M.T.; Alazba, A.A.; Shafiq, M. Comparative study for adsorption of methylene blue dye on biochar derived from orange peel and banana biomass in aqueous solutions. *Environ. Monit. Assess.* **2019**, *191*, 735. [CrossRef]
48. Sivarajasekar, N.; Baskar, R.; Ragu, T.; Sarika, K.; Preethi, N.; Radhika, T. Biosorption studies on waste cotton seed for cationic dyes sequestration: Equilibrium and thermodynamics. *Appl. Water Sci.* **2017**, *7*, 1987–1995. [CrossRef]
49. Pathania, D.; Sharma, S.; Singh, P. Removal of methylene blue by adsorption onto activated carbon developed from *Ficus carica* bast. *Arab. J. Chem.* **2017**, *10*, S1445–S1451. [CrossRef]
50. de Oliveira Brito, S.M.; Andrade, H.M.C.; Soares, L.F.; de Azevedo, R.P. Brazil nut shells as a new biosorbent to remove methylene blue and indigo carmine from aqueous solutions. *J. Hazard. Mater.* **2010**, *174*, 84–92. [CrossRef] [PubMed]
51. Soni, M.; Sharma, A.K.; Srivastava, J.K. Adsorptive Removal of Methylene Blue Dye from an Aqueous Solution Using Water Hyacinth Root Powder as a Low Adsorbent. *Int. J. Chem. Sci.* **2012**, *3*, 338–345.
52. Rehman, M.S.U.; Kim, I.; Han, J.-I. Adsorption of methylene blue dye from aqueous solution by sugar extracted spent rice biomass. *Carbohydr. Polym.* **2012**, *90*, 1314–1322. [CrossRef]
53. Liu, Y.; Zhao, X.; Li, J.; Ma, D.; Han, R. Characterization of bio-char from pyrolysis of wheat straw and its evaluation on methylene blue adsorption. *Desalin. Water Treat.* **2012**, *46*, 115–123. [CrossRef]
54. Cheng, G.; Sunb, L.; Jiaob, L.; Peng, L.; Leia, Z.; Wanga, Y.; Lina, J. Adsorption of methylene blue by residue biochar from copyrolysis of dewatered sewage sludge and pine sawdust. *Desalin. Water Treat.* **2013**, *51*, 37–41. [CrossRef]
55. Shakoor, S.; Nasar, A. Adsorptive treatment of hazardous methylene blue dye from artificially contaminated water using *Cucumis sativus* peel waste as a low-cost adsorbent. *Groundw. Sustain. Dev.* **2017**, *5*, 152–159. [CrossRef]
56. Ong, S.T.; Keng, P.S.; Lee, S.L.; Leong, M.H.; Hung, Y.T. Equilibrium studies for the removal of basic dye by sunflower seed husk (*Helianthus annuus*). *Phys. Sci. Int. J.* **2010**, *5*, 1270–1276.
57. Irem, S.; Khan, Q.M.; Islam, E.; Hashmat, A.J.; Haq, M.A.U.; Afzal, M.; Mustafa, T. Enhanced removal of reactive navy blue dye using powdered orange waste. *Ecol. Eng.* **2013**, *58*, 399–405. [CrossRef]
58. Edokpayi, J.N.; Ndlovu, S.S.; Odiyo, J.O. Characterization of pulverized Marula seed husk and its potential for the sequestration of methylene blue from aqueous solution. *BMC Chem.* **2019**, *13*, 10. [CrossRef] [PubMed]
59. Belala, Z.; Jeguirim, M.; Belhachemi, M.; Addoun, F.; Trouvé, G. Biosorption of basic dye from aqueous solutions by Date Stones and Palm-Trees Waste: Kinetic, equilibrium and thermodynamic studies. *Desalination* **2011**, *271*, 80–87. [CrossRef]
60. Jawad, A.H.; Rashid, R.A.; Mahmuod, R.M.; Ishak, M.A.M.; Kasim, N.N.; Ismail, K. Adsorption of methylene blue onto coconut (*Cocos nucifera*) leaf: Optimization, isotherm and kinetic studies. *Desalin. Water Treat.* **2016**, *57*, 8839–8853. [CrossRef]
61. Ijagbemi, C.O.; Chun, J.I.; Han, D.H.; Cho, H.Y.; O, S.J.; Kim, D.S. Methylene Blue adsorption from aqueous solution by activated carbon: Effect of acidic and alkaline solution treatments. *J. Environ. Sci. Health* **2010**, *45*, 958–967. [CrossRef] [PubMed]
62. Lou, S. Applied Adsorption Kinetics Model for Removal of Hazards from Aqueous Solution: More Informational Parameters for Industrial Design. *J. Environ. Sci. Public Health* **2017**, *1*, 228–239. [CrossRef]
63. Kandisa, R.V.; Kv, N.S.; Gopinadh, R.; Veerabhadram, K. Kinetic Studies on Adsorption of Methylene Blue Using Natural Low Cost Adsorbent. *J. Ind. Pollut. Control* **2018**, *34*, 2054–2058.
64. Basharat, S.; Rehman, R.; Mahmud, T.; Basharat, S.; Mitu, L. Tartaric Acid-Modified *Holarrhena antidysenterica* and *Citrullus colocynthis* Biowaste for Efficient Eradication of Crystal Violet Dye from Water. *J. Chem.* **2020**, *2020*, 8862167. [CrossRef]

sustainability

MDPI

Article

Chemical Modification of Combusted Coal Gangue for U(VI) Adsorption: Towards a Waste Control by Waste Strategy

Yuan Gao [1,†], Jiandong Huang [2,†], Meng Li [2], Zhongran Dai [3], Rongli Jiang [1] and Jixiong Zhang [2,*]

[1] School of Chemical Engineering and Technology, China University of Mining & Technology, Xuzhou 221116, China; y.gao@cumt.edu.cn (Y.G.); ronglijcumt@163.com (R.J.)
[2] School of Mines, China University of Mining & Technology, Xuzhou 221116, China; huang@cumt.edu.cn (J.H.); limeng77521@126.com (M.L.)
[3] Key Discipline Laboratory for National Defense for Biotechnology in Uranium Mining and Hydrometallurgy, University of South China, Hengyang 421001, China; dzr1122@126.com
* Correspondence: cumtzjxiong@163.com
† Yuan Gao and Jiandong Huang have equal contribution to this work.

Abstract: Uranium mining waste causes serious radiation-related health and environmental problems. This has encouraged efforts toward U(VI) removal with low cost and high efficiency. Typical uranium adsorbents, such as polymers, geopolymers, zeolites, and MOFs, and their associated high costs limit their practical applications. In this regard, this work found that the natural combusted coal gangue (CCG) could be a potential precursor of cheap sorbents to eliminate U(VI). The removal efficiency was modulated by chemical activation under acid and alkaline conditions, obtaining HCG (CCG activated with HCl) and KCG (CCG activated with KOH), respectively. The detailed structural analysis uncovered that those natural mineral substances, including quartz and kaolinite, were the main components in CCG and HCG. One of the key findings was that kalsilite formed in KCG under a mild synthetic condition can conspicuous enhance the affinity towards U(VI). The best equilibrium adsorption capacity with KCG was observed to be 140 mg/g under pH 6 within 120 min, following a pseudo-second-order kinetic model. To understand the improved adsorption performance, an adsorption mechanism was proposed by evaluating the pH of uranyl solutions, adsorbent dosage, as well as contact time. Combining with the structural analysis, this revealed that the uranyl adsorption process was mainly governed by chemisorption. This study gave rise to a utilization approach for CCG to obtain cost-effective adsorbents and paved a novel way towards eliminating uranium by a waste control by waste strategy.

Keywords: combusted coal gangue; chemical modification; uranium; adsorption

Citation: Gao, Y.; Huang, J.; Li, M.; Dai, Z.; Jiang, R.; Zhang, J. Chemical Modification of Combusted Coal Gangue for U(VI) Adsorption: Towards a Waste Control by Waste Strategy. *Sustainability* **2021**, *13*, 8421. https://doi.org/10.3390/su13158421

Academic Editor: Avelino Núñez-Delgado

Received: 5 July 2021
Accepted: 24 July 2021
Published: 28 July 2021

1. Introduction

Even with the growth of renewables and natural gas, coal resources are still the primary fuel to provide energy for daily life and industry. However, coal mining activities lead to various wastes, including CO_2 emissions [1,2], heavy metal-induced acid mine drainage [3], and coal gangue heap [4]. Among them, the accumulation of coal gangue (CG) and combusted coal gangue (CCG) without appropriate utilization has resulted in land occupation, landscape destruction, as well as heavy metal pollutions [5]. Currently, coal gangue is mainly used in the field of backfilling, building materials, agriculture, energy generation, soil improvement, and other high-added applications [6–8].

Nuclear power can partially substitute fossil energy and provide low-cost electrical power [9]. However, uranium used in the nuclear reactor is radiative, which causes both environmental and health issues [10]. Considering the rapid growth in uranium demand, efficient methods are necessary to eliminate or reduce uranium from contaminated media. Numerous methods, such as ion exchange, chemical precipitation, chromatographic extraction, and electrochemical techniques, have been developed to deal with uranium contaminations [11]. Among them, adsorptions show the most promising method due to their easy-handling and efficiency [12,13]. The designed synthesized materials, such as polymers, geopolymers, zeolites, and MOFs, have been applied in the removal of U(VI) [14]. These materials feature high surface area, uniform pore structure, excellent stability, and thus high sorption capacities [15]. The functional species, such as the hydroxyl, phosphoryl oxygen and metal oxide, were found to have strong interactions with U(VI). However, the associated high cost of precursors limits their practical applications.

In general, the CG and CCG contain natural mineral adsorbents, such as quartz, kaolinite, illite, chlorite, feldspar, and calcite. Due to the huge production and low-cost, the adsorption on constituents of CG and CCG was tested in environmental related remediation in the early stage. The main challenge of using them as a commercial adsorbents is due to a low surface area, pores, low ion exchange, and as a result low adsorption capacity [16]. Recently, chemical activations by mixing CG and CCG with strong alkali (e.g., Na_2SiO_3, NaOH and KOH) or and acidic solutions (e.g., HCl) has been investigated, which can open the pores increase the surface area of particles through disaggregation, elimination of mineral impurities, and dissolution of the external layers. An alkali solution dissolves aluminosilicate-reactive materials and forms AlO_4 and SiO_4 tetrahedral units that polysialate disiloxo (Al-O-Si-Si-Si chain), Al-O-Si chain, and siloxo Al-O-Si-Si-Si chain. Some studies also believe that instead of forming polymers, $SiO_4{}^{4-}$ and $AlO_4{}^{5-}$ can be dissolved by breaking the skeleton of aluminosilicate, depending on the alkali type and concentrations [17,18]. Using modified CG as promising adsorbents has been performed on adsorptions of methylene blue and heavy metal ions, including Cu^{2+}, Cd^{2+} Zn^{2+}, and Pb^{2+} [19,20].

Despite the studies mentioned above, a systematic analysis of the chemical activation of CG and CCG, with a wide characterization of the obtained solids and a complete discussion of their structural differences, is lacking in the literature. The increasing interest in adsorption applications of natural, quite common, and cheap materials justifies such a systematic study of this process [19–25]. Regarding that radioactivity U(VI) and coal gangue are coexistent wastes in certain cases, the target of this work is not only to eliminate them by a low-cost strategy, but also to deeply understand the active adsorption component, under chemical treatments, of CCG [26,27]. Herein, we report the results of uranyl adsorption under various experimental conditions using a CCG-based adsorbents, HCG and KCG. The roles played by surface properties and constitutions were explained based on characterizations and mechanistic studies.

2. Results and Discussion

2.1. Characterizations

The details of the synthetic approach are described in the Supplementary Materials and illustrated in Scheme 1. To confirm the successful chemical-modification of CCG, detailed characterizations of all the materials were performed by FTIR, PXRD, TGA-DSC, XPS, and SEM measurements. As shown in Figure 1 (left), the PXRD patterns of the solids treated with acid at room temperature did not show significant variations with respect to quartz and kaolinite, which are the major mineralogical composition of CCG and HCG. The acid treatment produced the dissolution of Ca^{2+}, Mg^{2+} cations, provoking a relatively increased crystallinity in the HCG constitutions due to the increased purity and concentrations of quartz and kaolinite. In contrast, treatment with KOH at room temperature led to weak alternations in the composition and structure of the solids. The first strong difference observed in PXRD is that the characteristic patterns of kaolinite phase decreased and the patterns with new peaks were observed. The comparison of the new diffractograms with the PDF data file shows that the feldspathoid kalsilite ($KAlSiO_4$) was formed under this reaction condition, which has a framework of linked (Si, Al)O_4 tetrahedra, and corresponding PXRD peaks are located at 19°, 21°, and 28° [28–30].

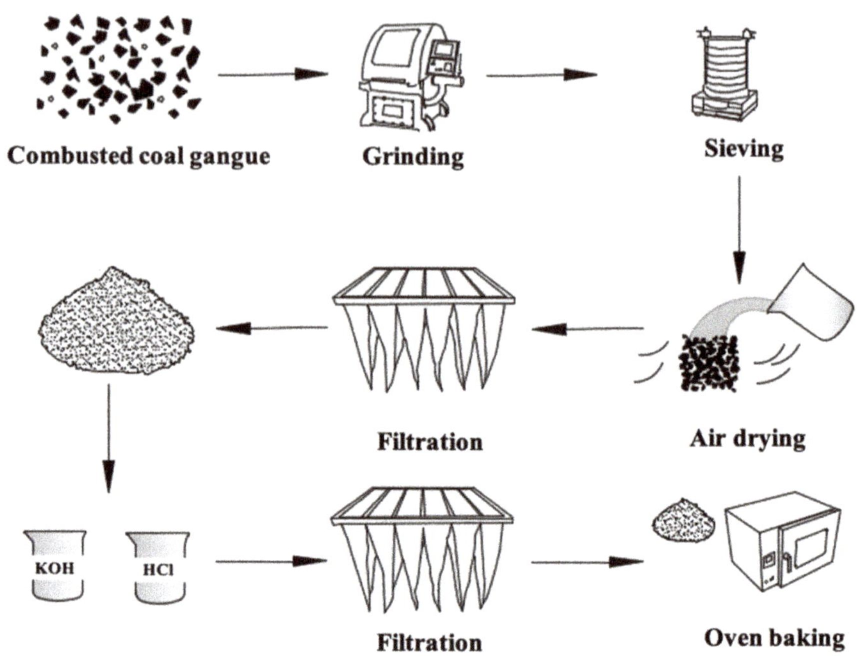

Scheme 1. The modification approach of CCG.

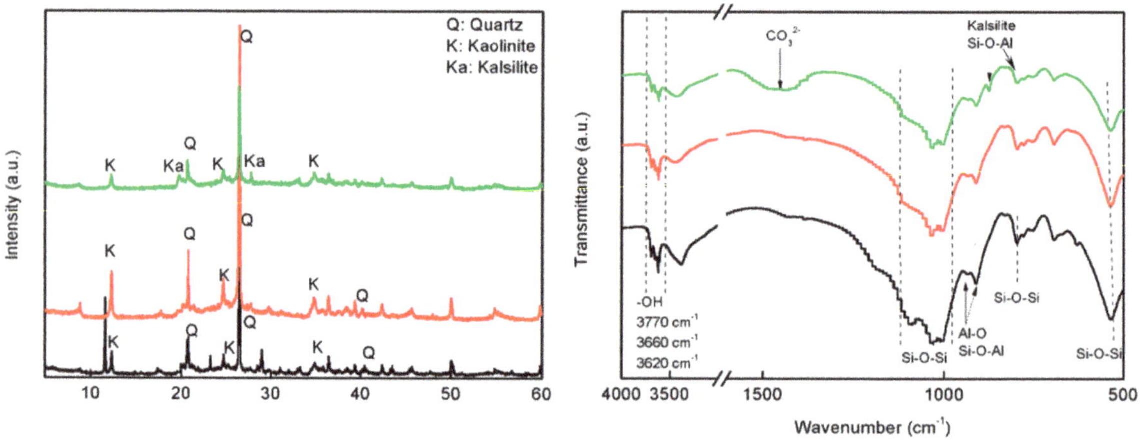

Figure 1. The PXRD (**left**) and FTIR (**right**) of CCG (black), HCG (red) and KCG (green).

FTIR spectrum showed the typical bands of quartz and kaolinite in CCG and HCG (Figure 1 (right)). Thus, the bands assigned to O-H stretching vibration in kaolinite were found at 3700, 3660, and 3620 cm^{-1} [31]. Meanwhile, the Al-O and Si-O-Al vibration from kaolinite can be found at 915 and 900 cm^{-1}. In this region of low wavenumbers, there are characteristic peaks of quartz: 1098, 1035, 1010, and 781cm^{-1} (Si-O-Si vibration). FTIR spectra of CCG treated with KOH showed the characteristic patterns of kalsilite, where the appearance of 987 cm^{-1} and 689 cm^{-1} peak assigned to Si-O-Al asymmetric stretching vibration in the linked (Si, Al)O_4 tetrahedra [32].

To further characterize the surface property of the prepared samples, the binding energies of C1s, O1s, Al2p, and Si2p as well as trace elements were determined by XPS

(Supplementary Figure S1), a magnified view of which is shown in Figure 2. There were no significant differences in the binding energy of Al2p, Si2p in CCG and HCG, while these peaks shifted to lower binding energy in KCG. Distinct variation in peak area also revealed changes in chemical structure [33]. As compared with CCG (1/1.40) and HCG (1/1.46), the increased ratio of Al/Si (1/1.26) in KCG was caused by the minor kalsilite formation. This suggests the alkaline treatment promoted an aluminum enrichment surface, while acid modification is on the contrary. These results are consistent with reported studies [34]. Moreover, the aluminum on the surface is always regarded as the active site compared to Si, which can influence adsorption as well as related physical properties [33].

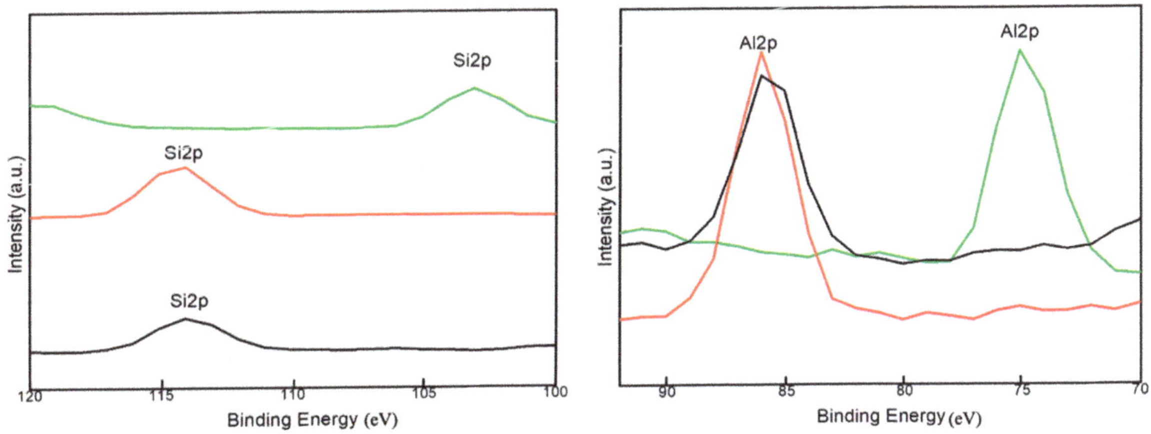

Figure 2. The zoom in XPS analysis of Si2p and Al2p in CCG (black), HCG (red) and KCG (green).

The thermogravimetric analysis showed that higher decomposition temperature was observed in all samples between 450 and 500 °C, whilst the KCG exhibited a dramatical weight loss below 200 °C as compared with CCG and HCG (Supplementary Figure S2). This can be related to the evaporation of lattice water, and explained by a larger pores and relatively high surface area in the KCG sample [31,34]. The accurate surface area and pore distributions of CCG, HCG, and KCG were investigated by nitrogen adsorption-desorption measurements, and the results are shown in Supplementary Figure S3. The CCG already showed certain porosity and activity, with a surface area of 21.34 m^2/g and a pore volume of 0.03 cm^3/g. This is larger than the reported mechanical activation coal gangue [34]. After chemical modifications, the BET surface area and a pore volume of samples increased conspicuously as compared to CCG (Table 1). Combined with the above characterizations, the increased surface in the HCG can be explained by the dissolution of metal ions. The improved surface property of KCG could be attributed to kalsilite formation.

Table 1. The calculation of BET surface area, pore volume and pore size by the Barrett-Joyner-Halenda analysis.

Sample	BET Surface Area (m^2/g)	Pore Volume (cm^3/g)	Pore Size (nm)	Standard Deviation of Fit (cm^3/g)
CCG	21.34	0.03	7.43	0.112
HCG	44.7	0.063	5.36	0.340
KCG	62.33	0.088	5.98	0.450

The surface features also related to the morphology of particles, therefore we conducted SEM measurements (Figure 3). Briefly, the structure of as-synthesized HCG and KCG became relatively loose after chemical modification. The surface of fresh CCG was relatively dense, and there were no large pores or fractures. The activated HCG and KCG presented decreased particle size and some irregular layered structures, which contribute to increased surface areas.

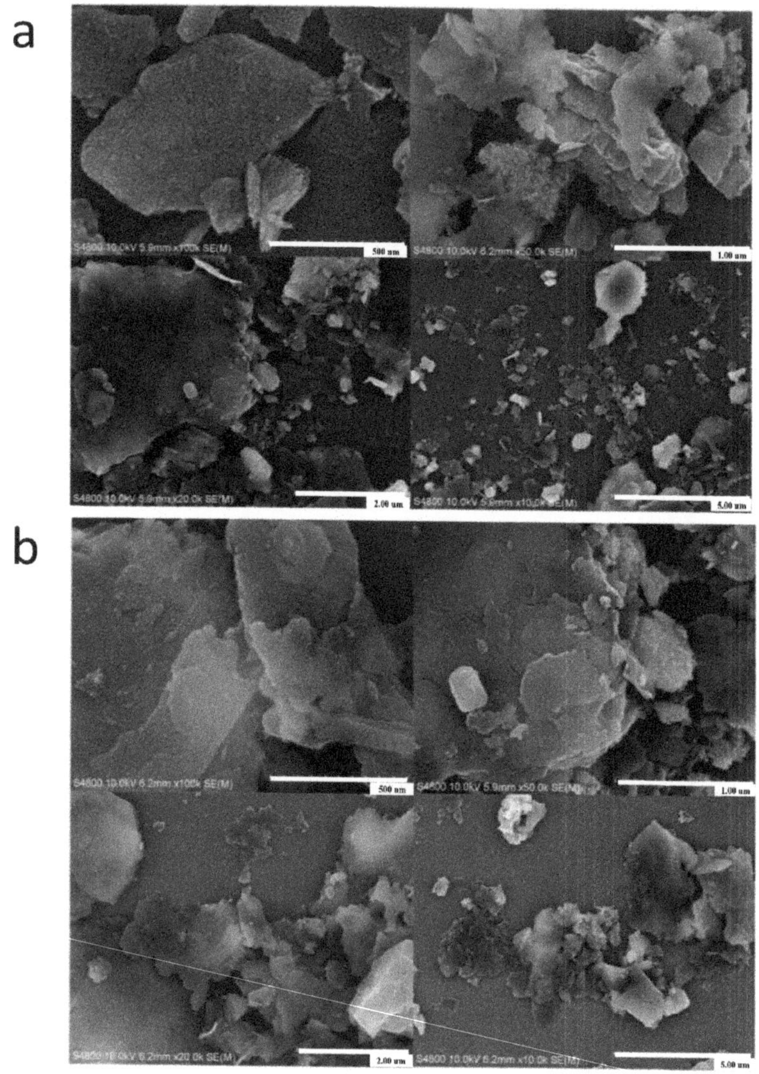

Figure 3. *Cont.*

Figure 3. The SEM images of CCG (**a**), HCG (**b**) and KCG (**c**) at 500 nm, 1.00 um, 2.00 um, and 5.00 um.

2.2. Adsorption Performance

The above characterizations, including crystallinity, functional groups, surface area, and pore volume, are key factors affecting adsorption process. Based on these, the batch experiments were carried out to investigate the UO_2^{2+} adsorption on CCG, HCG, and KCG samples.

2.2.1. Effect of pH

The acidity not only determines the existing forms of the targets, but influences the surface charge of the adsorbents, due to the repulsion forces and electrostatic attractions occurring between adsorbents and the U(VI) ions [35,36]. The U(VI) is in the style of cationic with a positive charge in solution. Thus, the surface charge of the modified coal gangue is strongly influenced by the pH of the solution. It can be noticed that uranyl carbonate starts to form from pH 6.5 as the predominant species, and the negative charge of uranyl carbonate will compromise the adsorption on the coal gangue surface. Therefore, we selected the batch experiments conducted from pH 2 to 6.

As depicted in Figure 4, CCG showed the removal efficiency from 2.1% to 15.2% under the pH from 2 to 6. The corresponding adsorption capacity increased from 4.2 to 39.6 mg/g. Generally, a more acidic solution can induce the protonation of reactive sites. For the kaolinite and quartz, the number of negatively charged sites, including Al-O^- and Si-O^-, increases under an increase in pH conditions [37,38]. As a result, the electrostatic attractions between the positive U(VI) and negative adsorbents became stronger and the removal percentage increased. These pH effects can explain the increasing adsorption trends of all samples (Figure 4 and Supplementary Table S1). Although HCG and KCG have improved porosity and surface area, the removal percentage of uranium by HCG slightly increased to 14.6% (43.8 mg/g) at pH 6. Meanwhile, KCG reflected a much higher adsorption performance than that of HCG, reaching the adsorption capacity of 140 mg/g. The improved adsorption properties can be explained by the fact that Al of lower binding energy has a higher affinity towards U(VI), and therefore KCG with a higher Al/Si (1/1.26) ratio leads to a better adsorption performance [39].

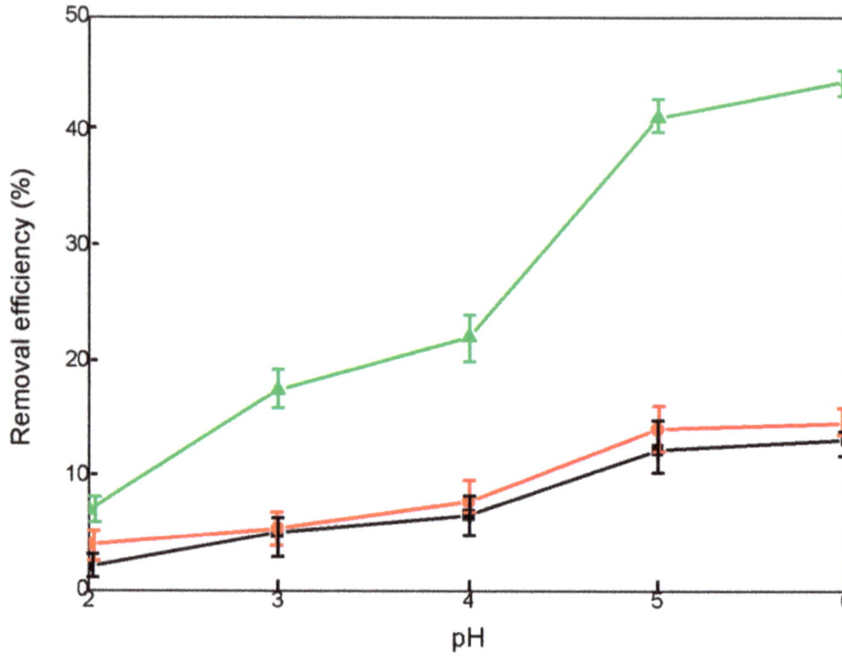

Figure 4. Removal efficiency (%) with 5 mg CCG (black), HCG (red) and KCG (green) adsorbents in 30 mL U(VI) (50 mg·L^{-1}) solution under variant pH conditions (from 2 to 6).

2.2.2. Effect of Adsorbent Dosage

Generally, the adsorption capacity of metal ions is determined by the amount of available adsorption sites on the adsorbents depends on the size and free energy of the hydrated ions, the activity of the metal ions, and other intrinsic reasons. As adsorbent dosage increases keeping all the other parameters at constant value, the removal efficiency of HCG increased continuously. The KCG sample, however, showed first increases, reached maximum and then decreased to zero, due to the precipitation of uranium by basic sites (Figure 5).

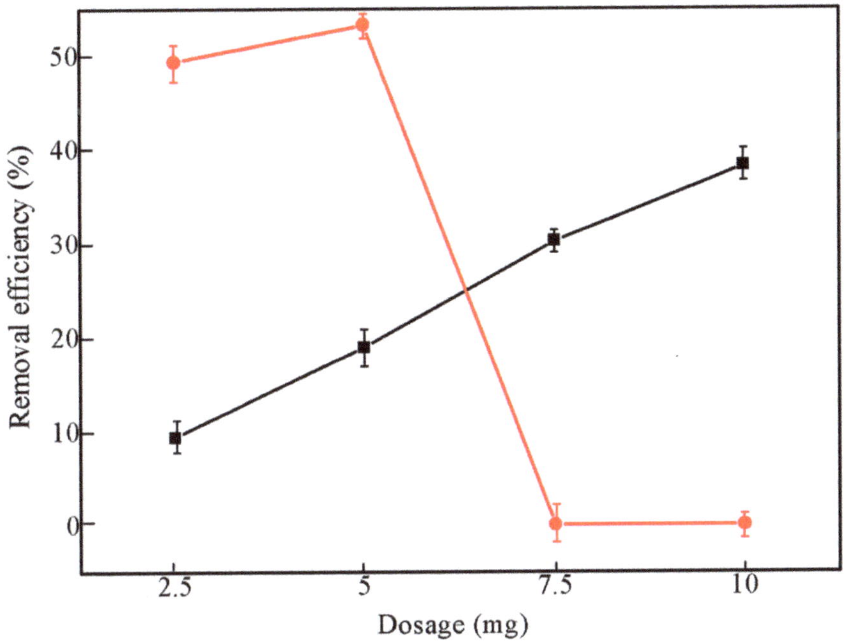

Figure 5. The effect of adsorbent dosage of HCG (black), and KCG (red) that influence the U(VI) adsorption.

2.2.3. Influence of Contact Time and Adsorption Kinetics

Supplementary Figure S4 shows the UO_2^{2+} evolution with the contact time. The removal efficiency of U(VI) adsorption onto HCG reached to 15% within 120 min, and then increased slightly with contact time and reached 15.6% at 120 min. The high removal efficiency of KCG reached equilibrium 41% within 120 min. To evaluate the adsorption rates of uranium ions onto HCG and KCG, the kinetics experiments were fitted by pseudo first-order and pseudo second-order kinetic models. The specific kinetic parameters as well as the calculated adsorption rates are shown in Table 2.

Table 2. Parameters of the pseudo-first-order and pseudo-second-order kinetic models for adsorption of U(VI) on HCG and KCG under pH 6.

Adsorbents	Pseudo-First-Order			Pseudo-Second-Order		
	R^2	q_e (mg/g)	K_1 (g/mg/min)	R^2	q_e (mg/g)	K_2 (g/mg/min)
HCG	0.985	22.8	−0.03	0.999	47.6	0.00412
KCG	0.955	126.8	−0.038	0.999	144.9	0.000667

As depicted in Figure 6, the pseudo second-order model provided a relatively high correlation coefficient (R^2 = 0.999). Moreover, the calculated q_e values of 47.6 mg/g and 145 mg/g from second-order was closed to the experimental data (46.8 mg/g and 140 mg/g, respectively), which suggested that the UO_2^{2+} ion adsorption onto HCG and KCG was mainly governed by chemisorption, which included diffusion, adsorption, and interior pore diffusion. The two-step immobilization process was assumed in chemical adsorption: (i) UO_2^{2+} moved to the surface of the HCG and KCG; (ii) UO_2^{2+} ions diverted from their surface to the active sites on the interior pore.

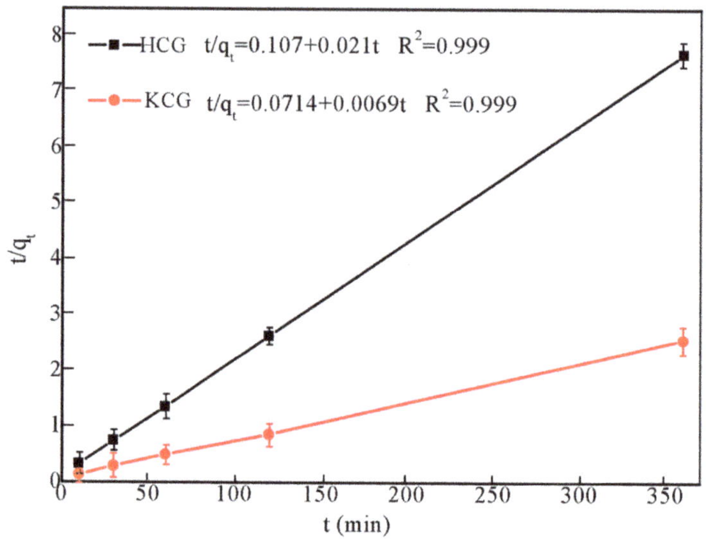

Figure 6. Adsorption kinetics of U(VI) on HCG (black) and KCG (red) fitted to pseudo-second-order kinetic model under pH 6.

2.2.4. Adsorption Mechanism

The structural characterizations and batch adsorption experiments showed that the structure optimization improved the adsorption properties of KCG controlled by chemical adsorption. In order to clarify the nature of the interaction of uranyl with adsorbents, the structure changes after adsorption were measured and analyzed by FTIR, PXRD, SEM, and XPS.

The comparison FTIR spectra before and after adsorption are shown in Figure 7 (left). The Al-O, Si-O, and Al-O-Si from kaolinite and kalsilite played an important role in the creation chemical bonding between adsorbents and U(VI), due to the disappear and/or shift of these bands together with the observation of SiO-UO_2^{2+}, AlO-UO_2^{2+}, and UO_2^{2+}-Si-O-Al-UO_2^{2+} [40]. The kalsilite formed in KCG had strong interactions with U(VI), and thus prominently improved the adsorption capacity, which can be confirmed by PXRD.

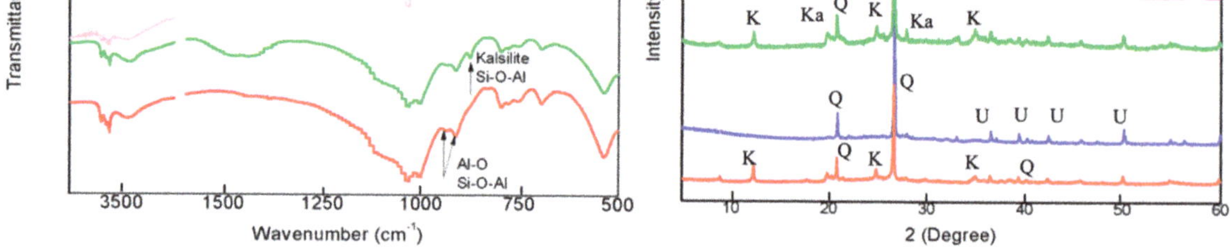

Figure 7. The FTIR (**left**) and PXRD (**right**) before and after UO_2^{2+} adsorption of HCG (red) and after adsorption (blue), KCG (green) and after adsorption (pink).

In Figure 7 (right), the crystalline phases in HCG are retained, while obvious changes can be observed in KCG after uranium adsorption, indicating the creation of new chemical bonding. XPS showed obvious U4f peaks at 382 eV in Figure 8 (left). Especially the atomic percentage of Si2s, Si2p, Al2s, and Al2p in KCG had a sharp decrease due to the higher loading of U(VI). Indeed, additional confirmation of chemical adsorption came from SEM, which displayed that the surface of adsorbents before uranium adsorption is smooth, but became rough and irregularly shaped after uranium adsorption (Figure 9). All these results show that the improved adsorption property is controlled by strong chemical interactions in KCG.

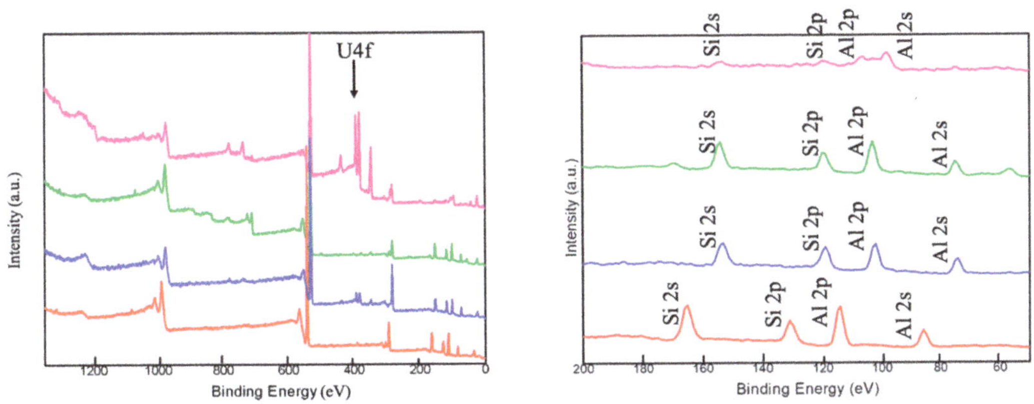

Figure 8. The XPS analysis before and after UO_2^{2+} adsorption of HCG (red) and after adsorption (blue), KCG (green) and after adsorption (pink). The right is the zoom in the analysis of Si and Al elements.

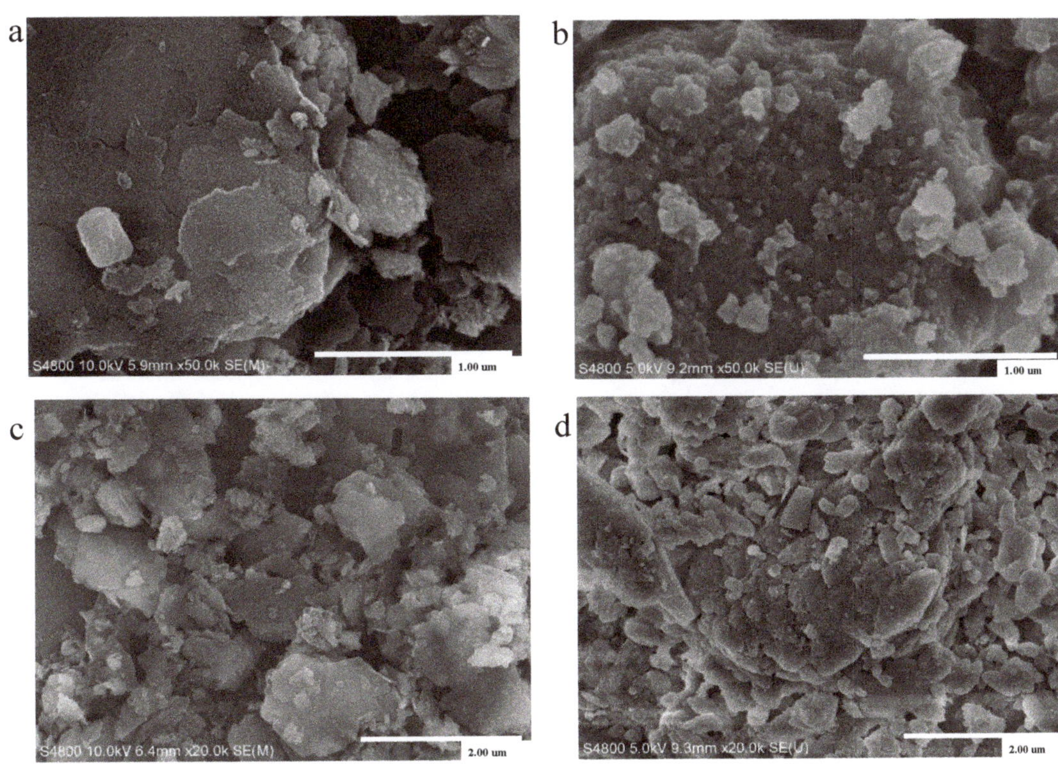

Figure 9. SEM images of HCG before (**a**) and after adsorption (**b–d**) are the SEM of KCG before and after adsorption of U(VI).

3. Conclusions

We successfully synthesized cost-effective adsorbents, namely HCG and KCG, from CCG by a simple chemical modification. Besides the optimized porosity and surface area, the kalsilite formation in KCG conspicuously enhanced the affinity towards U(VI). As we showed here, KCG exhibited an excellent adsorption capacity (140 mg/g) under pH 6 within 120 min. The pseudo second-order model of the kinetic study indicated a chemical interaction dominated process. Making use of the solid waste, the elimination of radioactivity U(VI) was realized by a waste control by waste strategy. As most of the research is still in the early stage, obtaining an in-depth understanding of the controlled synthesis of kalsilite from minerals and the reusability of natural adsorbents remains ongoing.

Supplementary Materials: The following are available online at https://www.mdpi.com/article/10.3390/su13158421/s1, Figure S1: The XPS of CCG (black), HCG (red) and KCG (green), showing the characteristic peaks, Figure S2: The TGA (Line) and DSC (dots) of CCG (black), HCG (red) and KCG (green), Figure S3: The N2 adsorption isotherms at 77K (Left) and the pore size distributions using BHJ analysis (Right). CCG (black), HCG (red) and KCG (green), Figure S4: Effects of adsorption contact time on the removal efficiency of U(VI) onto HCG (black) and KCG (red), Figure S5: The isotherm linear plots of CCG (a), HCG(b), and KCG(c), Table S1: Adsorption capacity under variant pH condition (from 2 to 6) (mg/g).

Author Contributions: Data curation, Z.D.; Funding acquisition, M.L.; R.J. and J.Z.; Investigation, Y.G. and J.H.; Methodology, Y.G.; Resources, J.Z.; Supervision, R.J. All authors have read and agreed to the published version of the manuscript.

Funding: This research was funded by the National Science Fund for Distinguished Young Scholars (No. 51725403), the National Natural Science Foundation of China (No. 52004271), and Hunan Provincial Natural Science Foundation for Excellent Young Scholars (No. 2020JJ3028).

Institutional Review Board Statement: Not applicable.

Informed Consent Statement: Not applicable.

Data Availability Statement: All data generated or analysed during this study are included in this published article.

Acknowledgments: This work is supported by the National Science Fund for Distinguished Young Scholars (No. 51725403), the National Natural Science Foundation of China (No. 52004271), and Hunan Provincial Natural Science Foundation for Excellent Young Scholars (No. 2020JJ3028). The "Qihang Plan" fund of the China University of Mining and Technology is acknowledged.

Conflicts of Interest: The authors declare no conflict of interest.

References

1. Burnham, A.; Han, J.; Clark, C.E.; Wang, M.; Dunn, J.B.; Palou-Rivera, I. Life-Cycle Greenhouse Gas Emissions of Shale Gas, Natural Gas, Coal, and Petroleum. *Environ. Sci. Technol.* **2011**, *46*, 619–627. [CrossRef]
2. Huang, J.; Shiva Kumar, G.; Ren, J.; Sun, Y.; Li, Y.; Wang, C. Towards the potential usage of eggshell powder as bio-modifier for asphalt binder and mixture: Workability and mechanical properties. *Int. J. Pavement Eng.* **2021**, 1–13. [CrossRef]
3. Akcil, A.; Koldas, S. Acid Mine Drainage (AMD): Causes, treatment and case studies. *J. Clean. Prod.* **2006**, *14*, 1139–1145. [CrossRef]
4. Jabłońska, B.; Kityk, A.; Busch, M.; Huber, P. The structural and surface properties of natural and modified coal gangue. *J. Environ. Manag.* **2017**, *190*, 80–90. [CrossRef]
5. Huang, Y.; Zhang, J.; Yin, W.; Sun, Q. Analysis of overlying strata movement and behaviors in caving and solid backfill-ing mixed coal mining. *Energies* **2017**, *10*, 1057. [CrossRef]
6. Zhang, J.X.; Ju, Y.; Zhang, Q.; Ju, F.; Xiao, X.; Zhang, W.Q.; Li, M. Low ecological environment damage technology and method in coal mines. *J. Min. Strata Control Eng.* **2019**, *1*, 013515.
7. Li, J.; Wang, J. Comprehensive utilization and environmental risks of coal gangue: A review. *J. Clean. Prod.* **2019**, *239*, 117946. [CrossRef]
8. Huang, J.; Duan, T.; Zhang, Y.; Liu, J.; Zhang, J.; Lei, Y. Predicting the permeability of pervious concrete based on the beetle antennae search algorithm and random forest model. *Adv. Civ. Eng.* **2020**, *2020*, 8863181.
9. Huang, J.; Sun, Y.; Zhang, J. Reduction of computational error by optimizing svr kernel coefficients to simulate concrete compressive strength through the use of a human learning optimization algorithm. *Eng. Comput.* **2021**, 1–18. [CrossRef]

10. Karakosta, C.; Pappas, C.; Marinakis, V.; Psarras, J. Renewable energy and nuclear power towards sustainable develop-ment: Characteristics and prospects. *Renew. Sustain. Energy Rev.* **2013**, *22*, 187–197. [CrossRef]
11. García-Balboa, C.; Baselga-Cervera, B.; García-Sanchez, A.; Igual, J.M.; Lopez-Rodas, V.; Costas, E. Rapid adaptation of microalgae to bodies of water with extreme pollution from uranium mining: An explanation of how mesophilic organisms can rapidly colonise extremely toxic environments. *Aquat. Toxicol.* **2013**, *144–145*, 116–123. [CrossRef] [PubMed]
12. Duan, S.; Xu, X.; Liu, X.; Wang, Y.; Hayat, T.; Alsaedi, A.; Li, J. Highly enhanced adsorption performance of U (VI) by non-thermal plasma modified magnetic Fe_3O_4 nano-particles. *J. Colloid Interface Sci.* **2018**, *513*, 92–103. [CrossRef] [PubMed]
13. Huang, J.; Zhang, Y.; Sun, Y.; Ren, J.; Zhao, Z.; Zhang, J. Evaluation of pore size distribution and permeability reduction behavior in pervious concrete. *Constr. Build. Mater.* **2021**, *290*, 123228. [CrossRef]
14. Veliscek-Carolan, J. Separation of actinides from spent nuclear fuel: A review. *J. Hazard. Mater.* **2016**, *318*, 266–281. [CrossRef]
15. Abney, C.W.; Mayes, R.T.; Saito, T.; Dai, S. Materials for the Recovery of Uranium from Seawater. *Chem. Rev.* **2017**, *117*, 13935–14013. [CrossRef] [PubMed]
16. Vellingiri, K.; Kim, K.-H.; Pournara, A.; Deep, A. Towards high-efficiency sorptive capture of radionuclides in solution and gas. *Prog. Mater. Sci.* **2018**, *94*, 1–67. [CrossRef]
17. Wang, X.; Chen, L.; Wang, L.; Fan, Q.; Pan, D.; Li, J.; Wang, X. Synthesis of novel nanomaterials and their application in efficient removal of radionuclides. *Sci. China Chem.* **2019**, *62*, 933–967. [CrossRef]
18. Carboni, M.; Abney, C.; Liu, S.; Lin, W. Highly porous and stable metal–organic frameworks for uranium extraction. *Chem. Sci.* **2013**, *4*, 2396–2402. [CrossRef]
19. Nekhunguni, P.M.; Tavengwa, N.T.; Tutu, H. Sorption of uranium (VI) onto hydrous ferric oxide-modified zeolite: As-sessment of the effect of pH, contact time, temperature, selected cations and anions on sorbent interactions. *J. Environ. Manag.* **2017**, *204*, 571–582. [CrossRef] [PubMed]
20. Munir, M.A.M.; Liu, G.; Yousaf, B.; Mian, M.; Ali, M.U.; Ahmed, R.; Cheema, A.I.; Naushad, M. Contrasting effects of biochar and hydrothermally treated coal gangue on leachability, bioavailability, speciation and accumulation of heavy metals by rapeseed in copper mine tailings. *Ecotoxicol. Environ. Saf.* **2020**, *191*, 110244. [CrossRef]
21. Mohammadi, R.; Azadmehr, A.; Maghsoudi, A. Fabrication of the alginate-combusted coal gangue composite for simul-taneous and effective adsorption of Zn(II) and Mn(II). *J. Environ. Chem. Eng.* **2019**, *7*, 103494. [CrossRef]
22. Li, D.; Song, X.; Gong, C.; Pan, Z. Research on cementitious behavior and mechanism of pozzolanic cement with coal gangue. *Cem. Concr. Res.* **2006**, *36*, 1752–1759. [CrossRef]
23. Moghadam, M.J.; Ajalloeian, R.; Hajiannia, A. Preparation and application of alkali-activated materials based on waste glass and coal gangue: A review. *Constr. Build. Mater.* **2019**, *221*, 84–98. [CrossRef]
24. Selvam, T.; Inayat, A.; Schwieger, W. Reactivity and applications of layered silicates and layered double hydroxides. *Dalton Trans.* **2014**, *43*, 10365–10387. [CrossRef]
25. Jabłońska, B.; Siedlecka, E. Removing heavy metals from wastewaters with use of shales accompanying the coal beds. *J. Environ. Manag.* **2015**, *155*, 58–66. [CrossRef]
26. Zhou, L.; Zhou, H.; Hu, Y.; Yan, S.; Yang, J. Adsorption removal of cationic dyes from aqueous solutions using ceramic adsorbents prepared from industrial waste coal gangue. *J. Environ. Manag.* **2019**, *234*, 245–252. [CrossRef] [PubMed]
27. Li, H.; Zheng, F.; Wang, J.; Zhou, J.; Huang, X.; Chen, L.; Liu, J.L. Facile preparation of zeolite-activated carbon composite from coal gangue with enhanced adsorption perfor-mance. *Chem. Eng. J.* **2020**, *390*, 124513. [CrossRef]
28. Qiu, R.; Cheng, F.; Huang, H. Removal of Cd2+ from aqueous solution using hydrothermally modified circulating fluid-ized bed fly ash resulting from coal gangue power plant. *J. Clean. Prod.* **2018**, *172*, 1918–1927. [CrossRef]
29. Yan, S.; Pan, Y.; Wang, L.; Liu, J.; Zhang, Z.; Huo, W.; Yang, J.; Huang, Y. Synthesis of low-cost porous ceramic microspheres from waste gangue for dye adsorption. *J. Adv. Ceram.* **2017**, *7*, 30–40. [CrossRef]
30. Zhao, H.; Huang, X.; Liu, F.; Hu, X.; Zhao, X.; Wang, L.; Gao, P.; Li, J.; Ji, P. Potential of a novel modified gangue amendment to reduce cadmium uptake in lettuce (Lactuca sativa L.). *J. Hazard. Mater.* **2021**, *410*, 124543. [CrossRef] [PubMed]
31. Meng, F.; Yu, J.; Tahmasebi, A.; Han, Y. Pyrolysis and combustion behavior of coal gangue in O_2/CO_2 and O_2/N_2 mixtures using thermogravimetric analysis and a drop tube furnace. *Energy Fuels* **2013**, *27*, 2923–2932. [CrossRef]
32. Zhou, J.; Zheng, F.; Li, H.; Wang, J.; Bu, N.; Hu, P.; Gao, J.-M.; Zhen, Q.; Bashir, S.; Liu, J.L. Optimization of post-treatment variables to produce hierarchical porous zeolites from coal gangue to enhance adsorption performance. *Chem. Eng. J.* **2020**, *381*, 122698. [CrossRef]
33. Becerro, A.I.; Escudero, A.; Mantovani, M. The hydrothermal conversion of kaolinite to kalsilite: Influence of time, temperature, and pH. *Am. Miner.* **2009**, *94*, 1672–1678. [CrossRef]
34. Purbasari, A.; Samadhi, T.W.; Bindar, Y. Thermal and Ash Characterization of Indonesian Bamboo and Its Potential for Solid Fuel and Waste Valorization. *Int. J. Renew. Energy Dev.* **2016**, *5*, 95–100. [CrossRef]
35. Novembre, D.; Gimeno, D.; D'Alessandro, N.; Tonucci, L. Hydrothermal synthesis and characterization of kalsilite by using a kaolinitic rock from Sardinia, Italy, and its application in the production of biodiesel. *Mineral. Mag.* **2018**, *82*, 961–973. [CrossRef]
36. Novembre, D.; Gimeno, D. The Solid-state Conversion of Kaolin to Kalsio4 Minerals: The Effects of Time and Temperature. *Clays Clay Miner.* **2017**, *65*, 355–366. [CrossRef]
37. Su, S.-Q.; Ma, H.-W.; Yang, J.; Zhang, P.; Luo, Z. Synthesis of kalsilite from microcline powder by an alkali-hydrothermal process. *Int. J. Miner. Met. Mater.* **2014**, *21*, 826–831. [CrossRef]

38. Zhang, N.; Ejtemaei, M.; Nguyen, A.V.; Zhou, C. XPS analysis of the surface chemistry of sulfuric acid-treated kaolinite and diaspore minerals with flotation reagents. *Miner. Eng.* **2019**, *136*, 1–7. [CrossRef]
39. Guo, Y.; Yan, K.; Cui, L.; Cheng, F. Improved extraction of alumina from coal gangue by surface mechanically grinding modification. *Powder Technol.* **2016**, *302*, 33–41. [CrossRef]
40. Shao, D.; Hou, G.; Li, J.; Wen, T.; Ren, X.; Wang, X. PANI/GO as a super adsorbent for the selective adsorption of uranium(VI). *Chem. Eng. J.* **2014**, *255*, 604–612. [CrossRef]

sustainability

MDPI

Article

Monitoring Study on Dust Dispersion Properties during Earthwork Construction

Qiming Luo [1,2], Lepeng Huang [1,2], Yuhong Liu [1,2], Xuanyi Xue [1,2], Fengbin Zhou [1,2] and Jianmin Hua [1,2,*]

1 School of Civil Engineering, Chongqing University, Chongqing 400045, China; luoqiming@cqu.edu.cn (Q.L.); huang_lepeng@cqu.edu.cn (L.H.); 201816021030@cqu.edu.cn (Y.L.); xuexuanyi@cqu.edu.cn (X.X.); 20141613158@cqu.edu.cn (F.Z.)

2 Key Laboratory of New Technology for Construction of Cities in Mountain Area, Ministry of Education, Chongqing University, Chongqing 400045, China

* Correspondence: huajianmin@cqu.edu.cn; Tel.: +86-133-3020-3691

Abstract: Dust generated in earthwork construction activities can seriously affect the air quality at a construction site and have adverse effects on the health of construction workers. To accurately and quantitively analyze the distribution characteristics of construction dust and the effect of dust prevention measures during earthwork construction under normal construction and construction with dust control measures, multiple collection points and one meteorological parameter collection point were placed at the construction site. From half an hour before the construction to half an hour after the construction, the particle concentration was recorded once every minute. The monitoring results indicated that there was a significant positive correlation between dust concentration during earthwork construction and the number of soil shipments. The dust concentration was highest at the earth excavation site, followed by the area of the waste truck's transportation path. Earth excavation primarily resulted in the generation of many coarse particles, the concentration of which was the highest near the excavation site. The average concentration increments of $PM_{2.5}$ and TSP (total suspended particulate) caused by earthwork construction were 55.06 and 375.17 $\mu g/m^3$ at the construction site, respectively. The concentration increment of $PM_{2.5}$ and TSP decreased by 72.01% and 40.16%, respectively, when a spray system and artificial sprinkling were adopted. Through the methodology and results of this study, construction companies can systemically plan their construction work by considering the key equipment to be used and can effectively manage the pollutants found within construction sites.

Keywords: earthwork; construction dust; real-time monitoring; dust sensor; concentration increment

Citation: Luo, Q.; Huang, L.; Liu, Y.; Xue, X.; Zhou, F.; Hua, J. Monitoring Study on Dust Dispersion Properties during Earthwork Construction. *Sustainability* **2021**, *13*, 8451. https://doi.org/10.3390/su13158451

Academic Editors: Avelino Núñez-Delgado, Zhien Zhang, Elza Bontempi, Mario Coccia, Marco Race and Yaoyu Zhou

Received: 15 July 2021
Accepted: 26 July 2021
Published: 28 July 2021

1. Introduction

Air pollution has become a vital issue worldwide, especially in areas in which infrastructure construction is carried out on a large scale. The construction industry is one of the main causes of air pollution [1]. Construction projects discharge a large amount of particulate matter into the environment, which causes harm not only to construction workers but also to the surrounding residents. The adverse effects of air pollution are extensive, especially in the area of public health. Studies have shown that an increase in the PM concentration in the air will cause harm to human bodies [2], and inhalable particulate matter will penetrate the lungs and cause lung diseases [3]. Therefore, air pollution has attracted increasing attention in various industrial fields.

Outside the construction industry, many scholars have conducted correlated studies, which have primarily focused on the dispersion rules of dust during tunnel construction and mining construction [4,5], diseases caused by dust [6,7], dust prevention measures [8,9] and health risk assessments of involved employees [10–12]. However, dust pollution in the construction industry should be emphasized equally. Over the years, urban construction activities have continued, especially due to the development and construction of new real estate. The development of large-scale construction has led to an increasingly serious problem of dust pollution [13–15]. As shown in the comparative estimates, for every increment of 3×10^4–$4 \times 10^4 m^2$ in the amount of construction in the urban center, the average contribution of dust generated from construction projects to the urban TSP (total suspended particulate) increases by 1 $\mu g/m^3$ [16]. Because of the large number of employees, as well as high labor intensity, outdoor production and poor working conditions in the construction industry, dust pollution greatly threatens the health of related employees and nearby residents [17,18]. With the increase in construction activities, the pollution of construction dust is predicted to become more serious in the future. Therefore, it is necessary to study the dispersion of building dust pollution.

As shown in previous studies [10,19,20], construction dust pollution is predominantly derived from the following four sources: (1) excavation dust; (2) field handling, stacking and processing of building materials; (3) dust from clearing and stacking construction waste and (4) primary dust caused by scattering during the transportation of building materials by vehicles and the secondary dust caused when the dust on the road surface is raised by the vehicles and then dispersed with the wind. The above four sources of construction dust pollution mostly occur during the foundation excavation and backfilling stages, which are primarily correlated with earthwork construction. Therefore, during the major construction stage of a project, dust pollution from earthwork is the most significant factor [21].

However, to date, there has been insufficient research on dust dispersion during earthwork construction. Most studies have been implemented on the premise that the level of dust generated at a construction site is kept consistent under a certain construction period and a certain construction area. This assumption mainly focuses on the analysis of the overall pollution contribution made by construction dust to the entire city, but the variations in the construction dust at the site are omitted [10,21–24]. Moreover, project construction is divided into several different construction stages, and the construction activities and site conditions of each stage differ significantly. The dust emission amount and intensity during different construction stages also vary. Some studies have emphasized the health hazards of construction dust to related employees as well as control measures [1,10,11,13,20,25,26]. In other studies, the meteorological factors and corresponding construction conditions at the location of the case have been ignored [27], the data collection points were limited, the field data coverage was incomplete or the data could not be collected in real-time [11,28].

Therefore, in this study, we chose a commercial-complex residential project located in a typical city in the Jianghan Plain of China to study the space–time distribution of dust diffusion under two construction conditions: normal construction and dust removal measures. Targeting earthwork construction, the research set up a plurality of dust concentration collection points inside and outside the workplace for real-time monitoring to record meteorological and construction data. First, Pearson's correlation coefficient was used to analyze the degree of correlation between the dust generated by earthwork construction activities and the influencing factors. Then, based on an analysis of the changes in the dust concentration at each collection point during earthwork construction, the distribution of the dust concentration at the construction site and changes in the dust particle composition during the middle and late stages of the construction were explored in different states. On this basis, the background concentration value of the dust was used for comparison, and the increment of the dust concentration at the construction site as well as the retention of dust at the construction site after completion were discovered to have different states.

2. Methods

2.1. Monitoring Location

The actual on-site monitoring was carried out on a housing construction project in Wuhan, Hubei Province, China (Figure 1). The total construction area of the project is approximately 280,000 m^2, consisting of 18 residential buildings, affiliated commercial buildings and basements. This site is a typical urban commercial-residential complex project. The Jianghan Plain, in which the project is located, is subject to a subtropical monsoon climate. The earth's surface constituents are predominantly composed of modern river alluvium and lake silt, classified as fine sand, silt and clay.

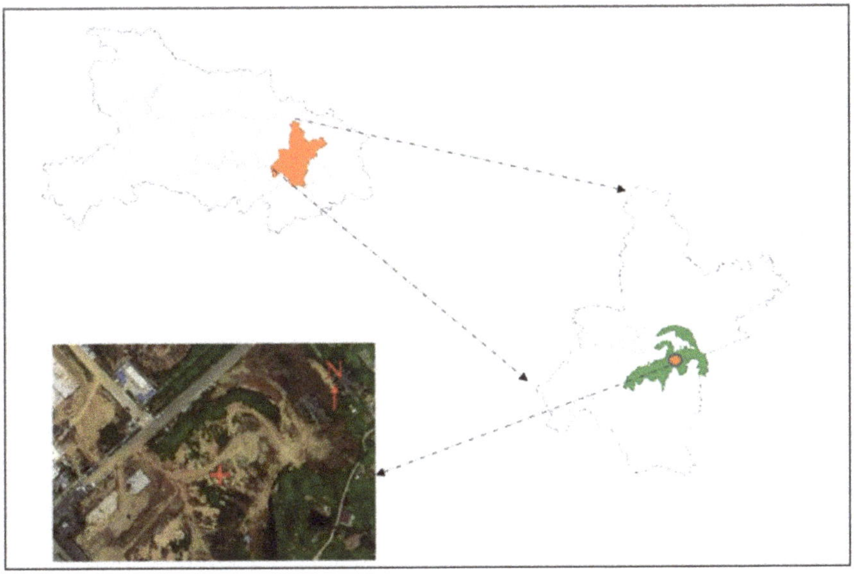

Figure 1. The geographic location of the project.

2.2. Monitoring Equipment

2.2.1. Meteorological Observation

To study the influence of meteorological factors on the distribution characteristics of construction dust, it is necessary to measure these factors on the construction site accurately. In this study, a fixed meteorological dust monitor (JXBS-3000-ZSYC) was selected. The monitor can accurately record meteorological factors such as wind speed, air humidity, atmospheric pressure and temperature in real-time. In addition, it is equipped with a built-in data logger to store the real-time monitoring data of the meteorological factors and promote an analysis of the collected data. Because of the large volume of the fixed meteorological dust monitor, such monitors are supposed to occupy large spaces. As a result, the instalment of a fixed meteorological dust monitor near a construction site may cause some inconvenience to the construction workers. As shown in Figure 2, the overall area of the construction site accounts for 18,750 square meters, within which meteorological factors such as wind speed, air humidity, atmospheric pressure and temperature in real-time were the same, so the fixed meteorological monitor was placed at the exit of the construction site (A1), as shown in Figure 2. During the research, the meteorological data were recorded once every 3 min.

2.2.2. Particulate Matter Measurement

A dispersion-type dust monitoring station, PH-YC01, and the inhalable dust continuous tester, PC-3A and PC-3B were used to measure the concentration of particulate matter. The PH-YC01 dispersion-type dust monitoring station is manufactured by Wuhan Xinpuhui Technology Co., Ltd. Given the fact that such a dispersion-type dust monitoring station needs to be connected to a 220 V power supply, it had to be fixed at the monitoring point. Each dispersion-type dust monitoring station was connected to an environmental data collector, and the data were read by the collector. The inhalable dust continuous tester is a portable instrument used to continuously measure the mass and concentration of particulate matter, the manufacturer of which is Qingdao Loobo Jianye Environmental

Protection Technology Co. Ltd. (Qingdao, China) Among the inhalable dust continuous testers, PC-3A was used to monitor the concentration of $PM_{2.5}$, while PC-3B was used to monitor the concentration of TSP.

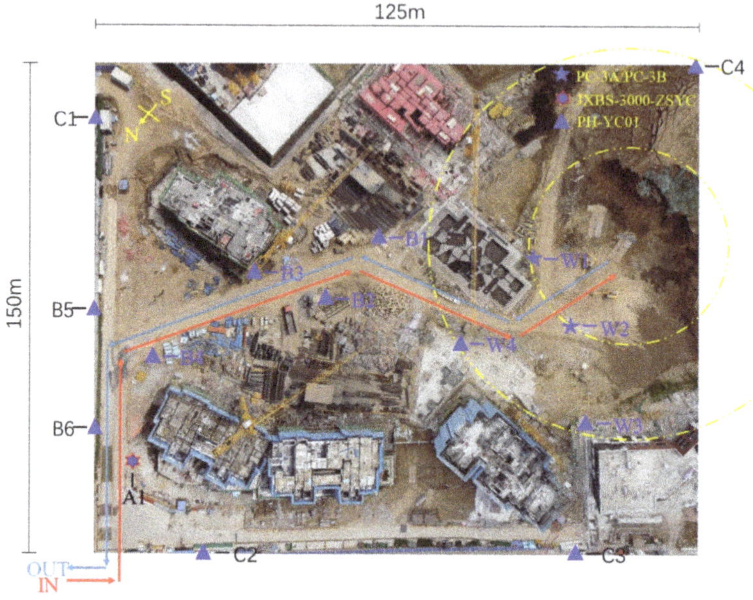

Figure 2. Schematic diagram of sampling points.

Both tools can continuously measure the mass and concentration of particulate matter. Inside the measuring cell, the particles in the sampled air were detected using the laser scattering method [29,30]. The interval ranged from 1 to 99 min; thus, the interval for the above dust monitoring stations was fixed at 1 min and the installation height of the particulate matter measurement sensor was 2.5 m.

2.3. *Monitoring Plan*

2.3.1. Particulate Matter Measurement

The study was conducted on 25–30 October 2020. The earthwork margin at the construction site was approximately 100,000 m^3. The construction site is located in the central urban area; thus, the municipal regulations stipulated that earthwork could only be transported at night. Therefore, the study was primarily conducted between 19:00 on the first day and 04:00 on the second day. The specific start and end times are presented in Table 1. The project management personnel were notified approximately 30 min before the start of construction, and they turned on the equipment for monitoring and recorded the time as the monitoring start time. The construction start time was when the excavator began working, and the end of the construction was when the excavator was shut down. The monitoring ended approximately 30 min after the end of construction, namely the shutdown time of the first monitoring device. It should be noted that during this period, the construction of the main structure was completely stopped, and only earth excavation and transportation were carried out. Therefore, the influences of the construction of the main structure were excluded, and the monitoring data in this study could accurately reflect the influence of the earthwork construction on the dust distribution characteristics.

Table 1. Research time interval table.

Date	Monitoring Interval	Construction Section	Monitoring Time (min)	Construction Time (min)
Start time 1 End time 1	25 October 20:30 26 October 03:38	25 October 21:05 26 October 03:04	428	369
Start time 2 End time 2	26 October 19:30 27 October 02:45	26 October 20:05 27 October 02:14	435	365
Start time 3 End time 3	27 October 20:00 28 October 03:41	27 October 20:40 28 October 03:10	461	390
Start time 4 End time 4	28 October 20:30 29 October 03:13	28 October 20:50 29 October 02:43	403	353
Start time 5 End time 5	29 October 19:00 30 October 03:19	29 October 19:33 29 October 23:37	499	244

2.3.2. Monitoring Indicators

At present, detection indicators aimed at the concentration of dust in construction projects mainly include dust fall, $PM_{2.5}$, PM_{10} and TSP [31,32]. The sampling frequency of dust fall is too low to accurately reflect the change in dust concentration in a short time [33]; therefore, this indicator was not selected for the study. In previous research, the correlation between $PM_{2.5}$ and PM_{10} was very high [34,35], and the changing trend and the change range were almost the same; thus, $PM_{2.5}$ and TSP were used in this study. The air humidity, temperature, atmospheric pressure and wind speed at the construction site were obtained using the meteorological monitor. In addition, the total numbers of excavators and earth shipments on that day were recorded.

2.3.3. Monitoring Points and Sampling Frequency

In this study, four types of monitoring points were considered. Here, the specific meanings of the four types of monitoring points and the significance of their selection are introduced in detail. W1, W2, W3 and W4 were Type 1 monitoring points, which were all located near the excavation site. Among them, W1 and W2 were approximately 25 m and W3 and W4 were approximately 50 m from the earthwork excavation site. There were no obstacles between the four points and the earthwork excavation site. B1, B2, B3, B4, B5 and B6 were Type 2 monitoring points, which were located on the transportation path of the truck. Of these, B1, B2, B3 and B4 were located near the unhardened pavement, and B5 and B6 were set on the hardened pavement. A car wash pool was located near B6, where the waste transport vehicle must be washed for mud removal before leaving the construction area. C1, C2, C3 and C4 were Type 3 monitoring points located at the four corner points of the construction site. O1 and O2 were Type 4 monitoring points. Point O1 was located at the entrance of the project department and along the path of the truck. Point O1 was located on the roadway outside the project; therefore, it is not shown in Figure 2. The O2 point was located near the residential area with a straight-line distance of 200 m from the project, and it was used as the background concentration value sampling point in the research.

2.4. *Meteorological Observation and Working Conditions*

Monitoring was carried out on five nights. Table 2 presents the working conditions of each day, and Table 3 presents the meteorological data of each day. In the following text, October 25 to October 26 is called the first day, October 26 to October 27 is called the second day, and so on; October 29 to October 30 is called the fifth day.

Table 2. Actual working condition.

Time	Day 1	Day 2	Day 3	Day 4	Day 5
Number of excavators	3	3	3	3	3
Total number of soil shipments	161	140	183	144	101
Prevention and control measures	×	×	×	√	√
Weather	Sunny	Sunny	Sunny	Sunny	Rain

Table 3. Weather condition.

<table>
<tr><td>Air Humidity</td><td>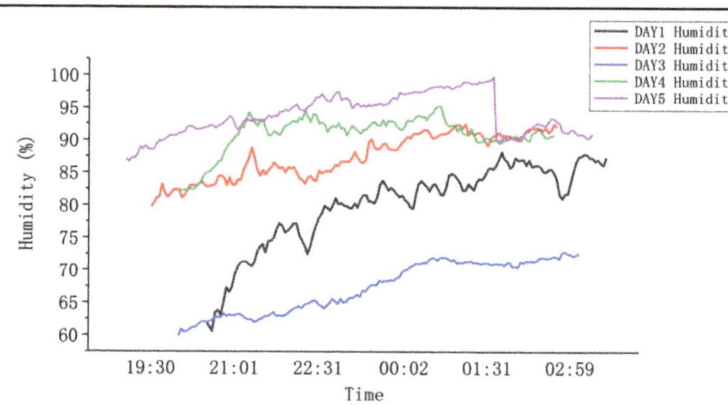
</td></tr>
<tr><td>Temperature</td><td>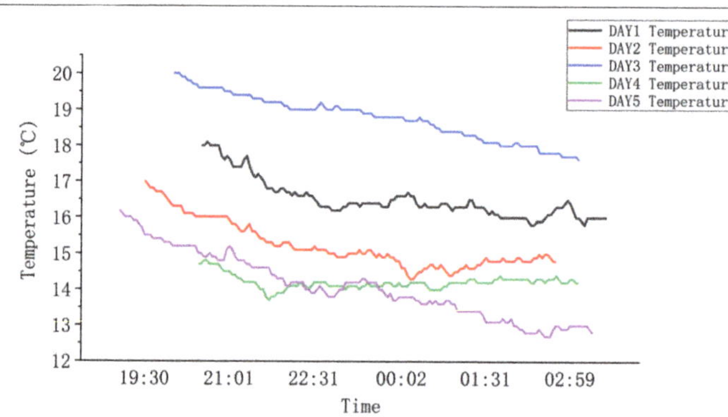
</td></tr>
<tr><td>Wind Velocity</td><td>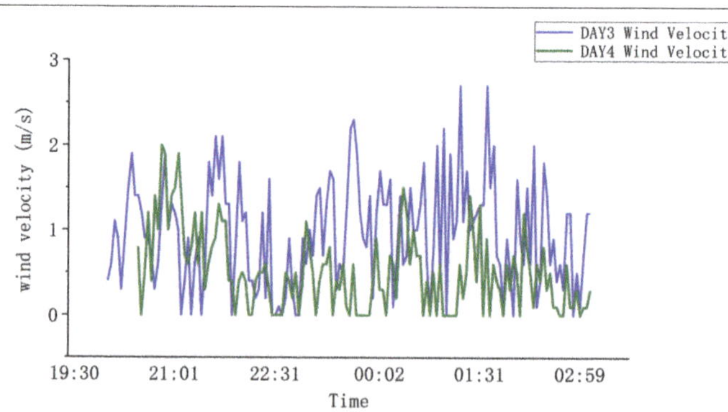
</td></tr>
</table>

This study aims to present the real temporal and spatial changes in the dust concentration at the construction site caused by the earthwork construction, and the construction party did not take relevant dust prevention measures during monitoring three days before the monitoring. On the fourth and fifth days, sprinkling systems and ground sprinkling were adopted for dust control. Figure 3 is a schematic diagram of dust prevention measures at the construction site. The light blue colored blocks indicate the locations where the spray device was installed, while the blue dots show the area where the road was sprayed. In general, the spray de-vice was installed above the enclosure of the construction site to prevent the construction dust from spilling out. Due to the complexity of the construction site, it was impossible to fix the spray device in a certain place, so the dust removal treatment was carried out at the excavation site and the earthwork transport road by artificial water spraying.

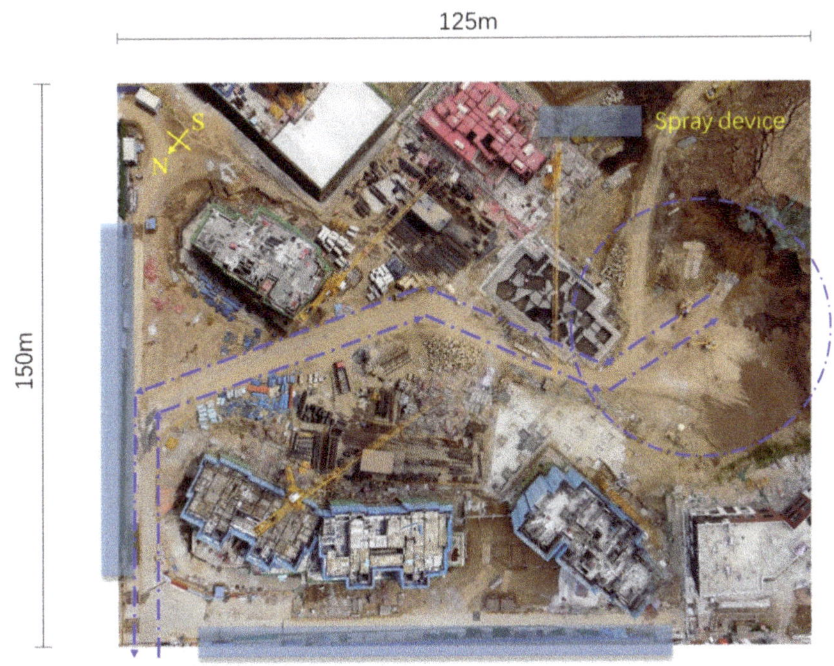

Figure 3. Schematic diagram of dust prevention measures.

According to the records, it rained on the construction site from 23:32 on the fifth day to 01:46 on the next day. According to the local regulations, the construction on the site was stopped when it rained. Therefore, the construction monitoring time on the fifth day was short, but relevant data after the rain were also collected. In the five days of earthwork construction, three excavators were working continuously at the construction site, but the total number of excavated waste trucks daily was different from the first day to the fifth day, at 161, 140, 183, 144 and 101, respectively.

As can be seen from the variation of the air humidity chart over time, the air humidity on the fifth day was almost always higher and rose as time went on. On the fifth day, the precipitous drop of the humidity occurred because the rain stopped. The construction site is located in the Jianghan Plain near the Yangtze River, and the overall humidity was at a relatively high level during the study period. Regarding the graph of temperature change over time, the temperature reduced over the five days as time progressed, and the temperature of the third day was the highest. It should be noted that the graph of wind speed change with time shows only the wind speed on the third and fourth days, because the wind speed on the first, second and fifth days was low, and the wind speed measured by the instrument in the above days was 0.

3. Results and Discussion

3.1. Analysis of Main Influencing Factors of Dust Emission from Earthwork

The Pearson correlation coefficient in the SPSS statistical analysis software was used to analyze the degree of correlation between the dust generated from earthwork construction activities and the related influencing factors, and then whether the relevant factors were the main influential factors of dust emissions during earthwork construction activities was determined [34]. Using a meteorological monitor, weather data such as the wind speed, temperature, humidity and atmospheric pressure at each construction site were obtained. The number of excavators on-site and the total number of soil shipments every day were recorded manually.

To compare the influence of various factors on dust emissions before and after the earthwork construction, data without dust prevention measures in the first three days were selected for analysis. The dust concentration values of the data collection points in the first three days were averaged as a whole, and the monitoring data were divided into three parts: before construction, during construction and after construction. The data were averaged over time at all times before and after construction. The data generated during construction were averaged every 0.5 h. Finally, 44 sets of data were obtained for the correlation analysis. It should be noted that if a shorter period were selected for averaging, the difference in the changes in various indicators would be too small, and the calculation results would be almost the same if more computing resources were used. If a longer period were selected, the changes in the indicators could not be accurately mastered. Since the wind speed was 0 in the first two days of the monitoring period, only 15 groups remained after the data processing of the third day. There were too few data samples, so wind speed and dust concentration were not analyzed in the correlation analysis. Relevant literature has proved that the two are closely related [11,34].

As shown in Table 4, the significance level between the average concentration of $PM_{2.5}$, TSP and each meteorological factor was greater than 0.01, failing to pass the significance test, and the correlation coefficient was less than 0.3, indicating that the dust emissions during earthwork construction were not significantly correlated with any single meteorological factor in Table 3. This result is also in line with those of previous studies [28,36]. The average concentrations of $PM_{2.5}$ and TSP were significantly positively correlated with the number of soil shipments, with a correlation coefficient of 0.814 and a significance level of less than 0.01, passing the significance test [2,28,36]. Obviously, the more soil shipments per unit time, the higher the average dust concentration at the construction site.

Table 4. Correlation analysis of the dust concentration in earthwork construction and potential factors.

		Temperature	Humidity	Atmospheric Pressure	Number of Soil Shipments
$PM_{2.5}$	Pearson correlation	−0.193	0.238	0.185	0.814 **
	Sig. (two-tailed)	0.210	0.120	0.230	0.000
	Number of cases	44	44	44	44
TSP	Pearson correlation	−0.202	0.211	0.075	0.843 **
	Sig. (two-tailed)	0.188	0.169	0.630	0.000
	Number of cases	44	44	44	44

** The correlation is significant at the level of 0.01 (two-tailed).

3.2. Increase in Dust Concentration Caused by Excavation and Transportation

Based on the high-density data collection, this study accurately quantified the dust concentration at the construction site. Based on striking a balance between the representativeness and accuracy of the data, the difference was used as the discriminant index to select the measured data to confirm the true situation of the data source.

As shown in Figure 4, as construction activities continued, the values of each monitoring point increased to different degrees. For the background concentration value, the value of the sampling point was relatively stable, while the values of other points fluctuated greatly. The numerical ranges of the different points varied considerably. For example, the concentrations of $PM_{2.5}$ and TSP at W1 near the earthwork excavation site were higher than the concentration at C2, which was located at the corner of the construction site.

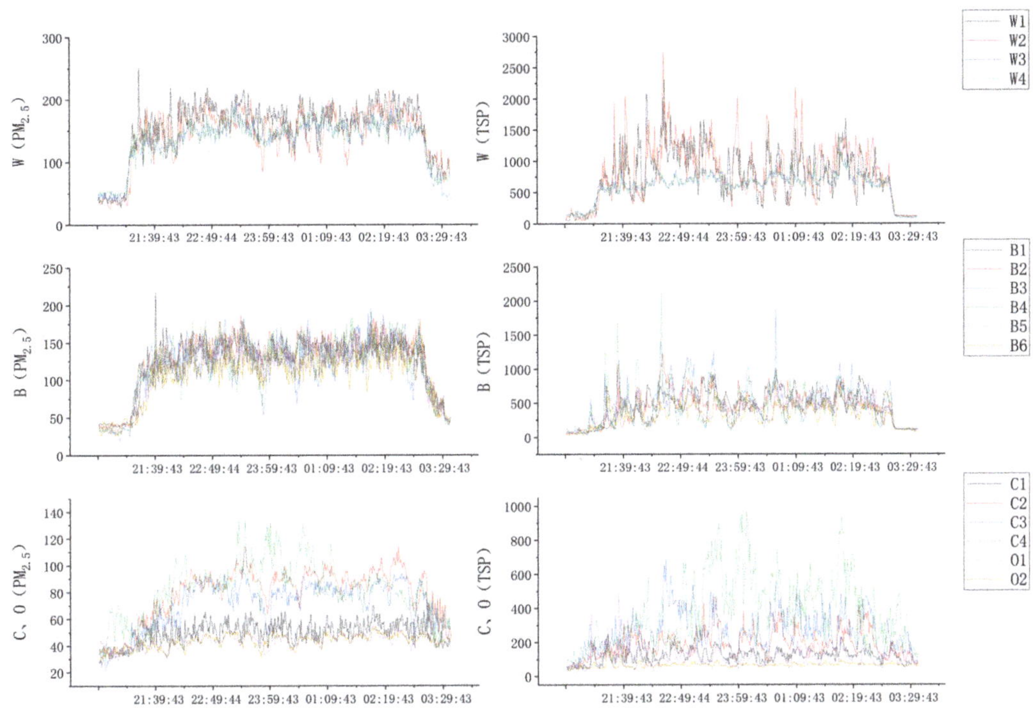

Figure 4. Concentration change diagram on 25 October at characteristic points in the monitoring interval.

Based on the above diagram, only the numerical order of each point could be determined. To further reveal the underlying causes, the concentration data of $PM_{2.5}$ and TSP during the start and end periods of the construction activities were processed to obtain the average concentration and medians at each point during the construction activities. In Table 4, each collection point corresponds to the three average concentrations. From left to right, the data represent the average concentration on the first, second and third days during construction, and the median index can be deduced similarly.

According to the data in Table 5, the following information can be obtained. During construction, the average concentration of dust at all the sampling points was greater than the background concentration, and all the positions of the construction site were affected to varying degrees. The order of the degree of influence (from large to small) was Type 1, Type 2 and Type 3. Furthermore, as the earthwork construction activities continued, the concentration values varied greatly across the construction site, and the data values of one or two collection points could not be used to characterize the dust concentration value of the entire construction site. If the scope of the construction site is substantial, it is necessary to analyze data from multiple collection points to characterize the concentration value of the construction site. For similar studies in the future, it is recommended to select more than four characteristic points and one background concentration value point for the research.

Table 5. Average and median dust concentration during construction for three days at each collection point.

Sites	Average Concentration during Construction ($\mu g\ m^{-3}$)		Median Concentration during Construction ($\mu g\ m^{-3}$)	
	$PM_{2.5}$	TSP	$PM_{2.5}$	TSP
W1	173	881	172	871
W2	168	924	170	862
W3	152	701	150	653
W4	149	691	148	653
B1	135	567	136	564
B2	138	543	135	514
B3	126	530	126	498
B4	130	529	128	515
B5	121	450	121	439
B6	114	408	114	397
C1	64	117	65	113
C2	89	220	91	208
C3	86	260	87	241
C4	92	434	92	404
O1	60	140	60	134
O2	62	92	62	91

Earthwork operations are mainly carried out using large machinery. The increase in the dust concentration is attributed to the waste transporting and loading behaviors of the excavators, the exhaust generated by the excavator and waste trucks during use, the dust emissions formed when the vehicles pass dust-covered surfaces and the secondary dust caused by the wind. These four types of behaviors occur continuously near the first type of points, so its average and median values are both at the highest level. The numerical comparison of the sampling points can also reflect the significant variations in the dust pollution in different areas at the construction site. By comparing the data from Table 5, the following analysis can be carried out. According to the comparison results of W1, W2, W3 and W4, for W1 and W2, the concentration values of $PM_{2.5}$ and TSP were relatively close and those for W3 and W4 were relatively close, but because the two points W3 and W4 were relatively far away from the excavation site, the $PM_{2.5}$ and TSP concentrations were lower than those of W1 and W2, indicating that the closer the point was to the earth excavation site, the higher the dust concentration value.

With regard to the second type of point, the increase in dust concentration was mainly attributed to the emissions of dust formed when the vehicles passed the dust-covered surface. The closer this was to the excavation site, the higher the concentrations of $PM_{2.5}$ and TSP of the sampling point. The distance between the excavation site and point B4 in the diagram was more than 100 m, indicating that the impact scope of dust dispersion caused by earthwork excavation was greater than 100 m. According to the comparison results of the dust concentration values of the sampling points on hardened and non-hardened pavements, the hardened pavement had a significant effect on dust suppression. The formation of road dust is caused by vehicles rolling over the dust on the road and dispersing it into the air. Regardless of whether the road is hardened or not, there will be a large amount of dust on the surface during earthwork, but the mass concentration of the dust on the hardened road will be lower than that on the unhardened pavement.

The third type of point was located at the four corners of the construction site. Because point C4 was close to the excavation site, its dust concentration value was relatively high. However, because C4 was behind a small hillside, the dust was stopped by the intermediate obstacle. The concentration of C4 was much smaller than that of W3 and W4. The other three points were also affected by on-site construction and truck transportation, and the dust concentration value was greater than the background concentration value. Point O1 was located on the path of the waste transport vehicle outside the construction site. However, because a car washing pool is set at the entrance of the construction site, the dust on the surface of the vehicle needs to be washed off before the vehicle leaves the construction site, so the dust concentration is relatively low. In addition, the results also indicate that the car wash pool can prevent the waste truck from taking pollutants out of the construction site.

3.3. Time for Dust Concentration to Reach Steady State

Construction sites have widely adopted measures using enclosure spraying or fog cannons to reduce dust. There is no scientific basis for the start-up time of spraying or fog gun machines, and they are often operated continuously. This mode will cause a significant waste of water resources. According to the observation of the former data, the increase in dust concentration during the construction process will reach a stable state after a period of time. In this state, the concentration value of the collection point will be close to the average concentration value of the collection point throughout the construction. Therefore, it is necessary to accurately predict the time at which the dust concentration reaches a stable state in order to organize dust reduction measures at the construction site and save water resources, which is helpful for the intelligent control of the construction process. Therefore, it is necessary to study these characteristic times. Direct observations of the change in the curve value per minute are insufficient to determine these factors, and detailed changes in dust concentrations could not be reflected by averaging the concentration value over a long period of time. In this study, the collected data were averaged every 10 min and sorted based on the time sequence. It is assumed that when Equations (1) and (2) are satisfied simultaneously, the dust concentration at the sampling point is deemed to have reached a steady state.

$$\mathrm{Avg}\{T_i\} > \mathrm{Min}\{\mathrm{Med},\ \mathrm{Avg}\} \times 80\% \tag{1}$$

$$\mathrm{Avg}\{T_{i+1}\} > \mathrm{Min}\{\mathrm{Med},\ \mathrm{Avg}\} \times 80\% \tag{2}$$

where T_i denotes the average value of the dust concentration for 10 min in a certain period, T_{i+1} represents the average value of the dust concentration for 10 min in the next period, Med represents the median of the dust concentration throughout the construction, and Avg represents the average value of the dust concentration throughout the construction. It should be noted that in most cases, the value of the dust concentration fluctuates greatly, and the median can better reflect the state of the dust concentration than the average.

Table 6 shows the time interval required for the first and second collection points to reach a stable state. As shown in the table, the time for the concentration of the two pollutants $PM_{2.5}$ and TSP to reach the steady-state level is inconsistent, and the time required for the concentration of $PM_{2.5}$ to reach the steady-state level is earlier than that of the TSP. This is because the TSP emissions are greater than that of $PM_{2.5}$ during the construction operation, so it takes a considerable amount of time to reach a higher concentration value. According to the time required for the above pollutants to reach a stable state, the construction organization could turn on the enclosure spray or fog gun machine for dust suppression for 5 to 15 min after the earth excavation and start of transportation, as recommended. The key dust reduction area of the fog gun machine is near the earth excavation site and on the transportation path of the truck.

As shown by the comparison of the values on the first, second and third days, it takes longer for the pollutant concentration to reach the pollution state during the first two days than on the third day. As measured by the meteorological observation instruments, the wind speed during the construction operation during the first two days was 0, and the windy state remained during the construction operation on the third day. It can be inferred that the wind accelerated the dispersion of pollutants, causing the concentration of pollutants to reach a pollution state level earlier.

As shown by the comparison of the first and second types of collection points, the time for the dust concentration of the first type of collection points to reach the pollution state level was shorter than that of the second type of collection points. The reason for this phenomenon was consistent with the reason why the average dust concentration of the first type of collection points was higher, which will not be repeated here. It takes a long time for the TSP concentration at the second type of points to reach the pollution level. Because of their large weight, TSP particles cannot float in the air for a long time and drift over substantial distances. This dispersion requires the movement of large construction equipment at the construction site, which can increase the regional TSP concentration in other areas.

Table 6. Time for the concentration values of the first and second types of collection points to reach a steady state.

Site	25–26 October		26–27 October		27–28 October	
	Time Required to Reach Construction Pollution State					
	PM2.5 (min)	TSP (min)	PM2.5 (min)	TSP (min)	PM2.5 (min)	TSP (min)
W1	15–25	15–25	5–15	25–35	0–10	10–20
W2	15–25	15–25	5–15	15–25	10–20	10–20
W3	5–15	5–15	25–35	25–35	10–20	10–20
W4	5–15	5–15	15–25	25–35	10–20	10–20
B1	25–35	35–45	15–25	35–45	0–10	10–20
B2	25–35	45–55	15–25	35–45	10–20	10–20
B3	15–25	25–35	25–35	35–45	10–20	20–30
B4	25–35	25–35	15–25	45–55	10–20	10–20
B5	25–35	15–25	5–15	45–55	10–20	20–30
B6	25–35	15–25	5–15	45–55	20–30	20–30

3.4. Dust Removal Rate of Construction Site under the Action of Dust Prevention Measures

To more intuitively understand the dust removal effect caused by dust control measures, the dust concentration data of the first three days without dust control measures were compared with the dust concentration data of the next two days with measures taken. The dust removal rate of each monitoring point can be calculated according to formulas (3) and (4):

$$R_{xPM2.5} = \frac{Avg(x_{1PM2.5} + x_{2PM2.5} + x_{3PM2.5}) - Avg(x_{4PM2.5} + x_{5PM2.5})}{Avg(x_{1PM2.5} + x_{2PM2.5} + x_{3PM2.5})} \quad (3)$$

$$R_{xTSP} = \frac{Avg(x_{1TSP} + x_{2TSP} + x_{3TSP}) - Avg(x_{4TSP} + x_{5TSP})}{Avg(x_{1TSP} + x_{2TSP} + x_{3TSP})} \quad (4)$$

where $R_{xPM2.5}$ and R_{xTSP} are, respectively, the $PM_{2.5}$ and TSP dust removal rate of monitoring point x. $x_{1PM2.5}$ and x_{1TSP} are the average dust concentrations of $PM_{2.5}$ and TSP at monitoring point x during the construction of the first day, respectively.

Table 7 gives the average dust concentration in the construction stage two days after each collection point and gives the dust removal rate at each monitoring point according to Formulas 3 and 4. O1 and O2 were outside the construction site, hence this paper did not calculate the values for those points.

Comparing the dust removal rates of TSP and $PM_{2.5}$ in Table 7. Overall, the dust removal efficiency of TSP was slightly higher than that of $PM_{2.5}$, indicating that manual water sprinkling and dust removal by spraying had a greater inhibitory effect on large particles. Points B1~B4 in the second type of monitoring points and the first type of monitoring points were mainly influenced by the dust removal by manual sprinkling, indicating that the manual sprinkling measures had a positive effect on dust reduction. Of the above points, the dust removal efficiency of Type 1 monitoring points was slightly higher than that of B1~B4, but the overall value was close to that of B1~B4. The $PM_{2.5}$ dust removal efficiency was 23.7–42.3%, and the TSP dust removal efficiency was 35.7–47.2%. B5 and B6 were close to the spray device and were located on the hardened road treated by sprinkler measures. Therefore, the dust removal efficiency of B5 and B6 was at the highest level in the collection point. The dust removal efficiency of $PM_{2.5}$ and TSP could both reach about 50%. Although C2 and C3 were close to the spraying device, they were not greatly affected due to being at a certain distance from the earth excavation site and the earth transportation road. Therefore, the dust removal efficiency was lower than the monitoring points located on the earth excavation site and the earth transportation road. C1 and C4 were located at the far east and south of the construction site. Various dust removal measures were located far from these points, but they still had a certain dust removal efficiency, indicating that dust control measures not only had an impact on the monitoring points near the measures taken but also could inhibit dust travel to a certain extent. In summary, artificial sprinkling or spraying for dust removal showed a good dust removal effect in the earthwork excavation site and earthwork transportation road, the dust removal efficiency for TSP was higher than $PM_{2.5}$, and the dust removal efficiency was greater than 30%. However, the dust removal efficiency was relatively low in positions where the dust itself was not greatly affected. The use of dust prevention measures could effectively prevent the jump and migration of dust.

Table 7. Average dust concentration in the second two days of construction at each collection point and overall dust removal efficiency.

Sites	Average Concentration during Construction ($\mu g\ m^{-3}$)		Dust Removal Efficiency (%)	
	$PM_{2.5}$	TSP	$PM_{2.5}$	TSP
W1	111	479	34.8	44.3
W2	103	458	37.9	47.2
W3	87	406	42.3	40.5
W4	89	405	39.5	39.7
B1	87	355	35.5	35.7
B2	89	340	35.1	35.8
B3	88	322	29.6	37.6
B4	98	331	23.7	35.8
B5	61	211	49.6	52
B6	58	202	49.4	49.5
C1	60	113	7	2.3
C2	60	173	33.3	21.3
C3	65	169	24.7	34.3
C4	84	327	8.2	23.5
O1	72	150	-	-
O2	50	81	-	-

3.5. Changes in the Composition of Dust

To more intuitively understand the changes in the composition of the particulate matter caused by the construction, the data from the entire monitoring period were divided into three parts: before construction (A), during construction (B), and after construction (C). Equations (5) and (6) were used to process the data:

$$\mathbf{K_{1A}} = \frac{\mathbf{C_{1AavePM2.5}}}{\mathbf{C_{1AaveTSP}}} \tag{5}$$

$$\mathbf{K_{A}} = \frac{\mathbf{K_{1A}} + \mathbf{K_{2A}} + \mathbf{K_{3A}}}{3} \tag{6}$$

where $C_{1AavgPM2.5}$ is the average value of all the $PM_{2.5}$ concentration values before construction and $C_{1AaveTSP}$ is the average value of all the TSP concentration values before construction. K_{1A} indicates the percentage of $PM_{2.5}$ in the TSP before the first day of construction. K_A indicates the proportion of $PM_{2.5}$ in the TSP in all monitoring data before construction. In the same way, according to the above rules, the proportions of $PM_{2.5}$ in the TSP, K_B and K_C during and after construction can be obtained. Notably, because the construction operation will affect the dust concentration index for a period of time after the construction is completed, during the calculation of K_C, the last 10 min average value of the monitoring data was used. It should be mentioned here that dust control measures were adopted on the fourth day. To compare the impact of dust fall on the change of dust composition, the data of the fourth day were calculated separately in this paper. In the case of sudden rain on the fifth day, the data on the fifth day would not be calculated. The data were processed in the above manner, and the K value broken line graph was used to show the change rule of each sampling point.

The following key information can be obtained by analyzing Figure 5:

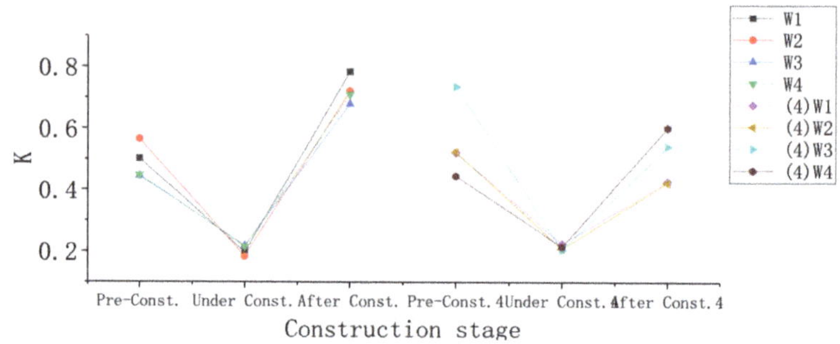

(a)

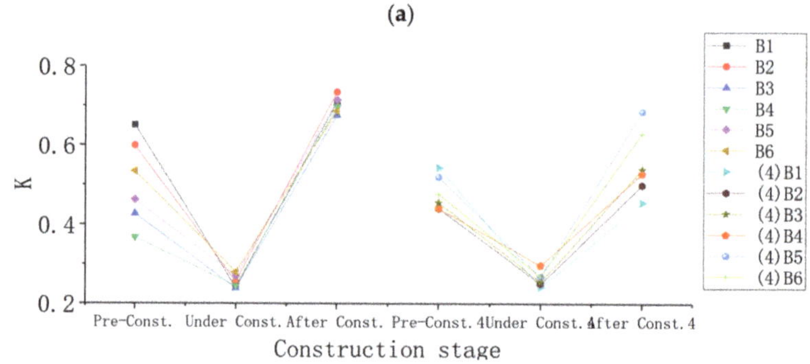

(b)

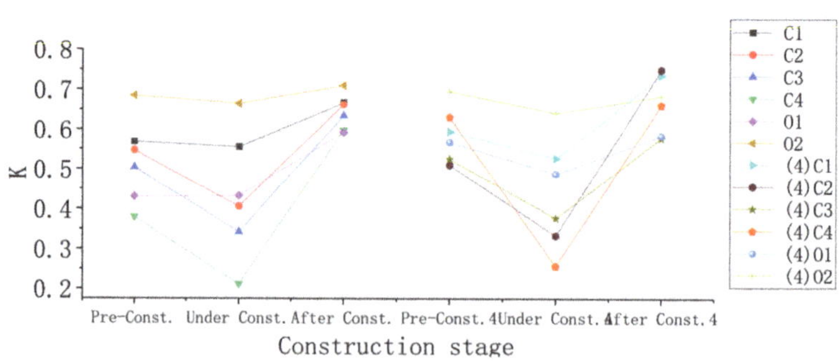

(c)

Figure 5. K value diagram of each sampling point: (**a**) K value change diagram for the first type of sampling point; (**b**) K value change diagram for the second type of sampling point; (**c**) K value change diagram for the third and fourth types of sampling points.

(1) Except for monitoring point O1, the K value of all sampling points tended to first decrease and then increase. This result indicates that in the non-construction stage, $PM_{2.5}$ accounted for a higher proportion of the TSP concentration, and the main pollution source was fine particles ($PM_{2.5}$), with fewer coarse particles ($PM_{2.5-100}$). However, coarse particles ($PM_{2.5-100}$) increased sharply and were significant pollution sources due to the impacts of the construction operations. Therefore, in dust fall

research, it is necessary to study the coarse and fine particles separately and determine the dust reduction measures for particles of different sizes. In earthwork construction, the dust reduction method of coarse particles is mainly taken into account;

(2) Before construction, the K value of each sampling point in the construction site differed significantly, varying between 0.35 and 0.6. Although other areas in the site were in a shutdown stage before sampling, the construction operations during the day still had a significant impact on those at night. The K value of each sampling point was less than the K value of the O2 point, indicating that construction operations during the day led to an increase in the concentration of coarse particles;

(3) During the construction, the K value of the first type of sampling point varied between 0.18 and 0.22, and the K value of the second type of sampling point varied between 0.24 and 0.28. This suggests that the closer the location to the excavation site, the higher the proportion of coarse particles in the pollutants. The weight of coarse particles was larger than that of fine particles, and they could not be suspended in the air for a long time [36]; the closer the point was to the place where the disturbance was greater, the higher the proportion of coarse particles. The K values of the third and fourth types of sampling points varied between 0.21 and 0.66. Because point C4 was located near the earth excavation site, the K value was small. The values at all other points decreased but remained different. As for points C1 and O1, C1 was located at the easternmost side of the construction site, avoiding the path of the waste truck, and was the farthest point from the excavation point relative to other sampling points and is the least affected. The K value of point O1 during the construction phase further illustrates the necessity of setting up a car wash pool;

(4) After construction, the K values of all the sampling points increased. The K value range of the first type point was 0.68–0.78. The K value range of the second type point was 0.68–0.73. This indicates that the coarse particles settled at places that were significantly affected by construction operations. However, their K value was higher than those of the third and fourth sampling points (0.59–0.71), because the fine particles were still suspended in the air due to the impact of construction;

(5) By comparing the variation rule of dust composition after the use of dust control measures, the following noteworthy points were found. First, during the construction stage, the K value of the site sampling points did not change significantly, indicating that the sprinkler system and the artificial road sprinkler made no significant difference regarding the influence on the coarse and fine particles during the construction stage. However, after the completion of construction, the K values of the sampling points of Type 1 and 2 were smaller than those without dust removal measures. This phenomenon indicated that after the construction, the use of dust removal measures would make the proportion of TSP in the dust lower; that is, after the construction, the coarse particles in the air would be relatively reduced. After the use of dust removal measures, coarse particles were more likely to be captured by water fog, which could achieve particle sedimentation or inhibit the migration of particles.

3.6. Increment of Dust Concentration on Construction Site

Based on the collected data, Equations (7)–(10) were used to calculate the dust concentration increments at the sampling point, the relative increments of the dust concentration at the sampling point, the average concentration increments of dust at the construction sites and the relative increments of the average dust concentrations at the construction sites. In this study, the above formula was used to analyze the emissions of dust from construction quantitatively. The increments in the dust concentration at the sampling point indicate the absolute value of that value caused by earthwork construction, and the relative increase in dust concentration at the sampling point indicates the percentage of the concentration increment caused by construction. In addition, the increment in the average concentration of dust at the construction site and the relative increment in the average concentration of the dust at the construction site were used to indicate an increase in the absolute value and percentage of the dust concentration generated by the earthwork construction for the entire construction site. It should be noted that because the construction operation will affect the concentration value for a period of time after the construction is stopped, the dust concentration at each sampling point is the average value of the dust concentration from the beginning of the construction until 30 min after stopping the construction.

$$\begin{gathered}\text{Dust concentration increment at sampling point} = \\ \text{Avg}\{\text{dust concentration value at each sampling point}\} - \text{background concentration value}\end{gathered} \tag{7}$$

$$\begin{gathered}\text{Relative increment of dust concentration at sampling point} = \\ \frac{\text{dust concentration increment at sampling point}}{\text{background concentration value}} \times 100\%\end{gathered} \tag{8}$$

$$\begin{gathered}\text{Average concentration increment of dust at constrution site} = \\ \frac{\sum\sum \text{increment of dust concentration at sampling points}}{\text{number of sampling points} \times \text{number of monitoring days}}\end{gathered} \tag{9}$$

$$\begin{gathered}\text{Relative increment of average dust concentration at construction site} = \\ \frac{\sum\sum \text{relative increment of dust concentration at sampling points}}{\text{number of sampling points} \times \text{number of monitoring days}}\end{gathered} \tag{10}$$

Figure 6 more intuitively shows the increase in the concentration of particulate matter caused by the construction operations at each sampling point. According to Equations (7) and (8), the average concentration increment of $PM_{2.5}$ is 55.06 $\mu g/m^3$ at the construction site, and the average TSP concentration increment is 375.17 $\mu g/m^3$, respectively, without taking any dust control measures. According to Formula (9) and (10), the average relative increment of the $PM_{2.5}$ concentration is 0.99 at the construction site, and the average relative increment in TSP concentration is 4.25. This means that in the earthwork construction phase, the concentration of $PM_{2.5}$ almost doubled compared with the shutdown period, and the TSP concentration increased by 4.25 times. $PM_{2.5}$/TSP = 0.147, indicating that approximately 14.7% of the particles in the dust emissions during earthwork construction had a particle size of less than 2.5 μm. These particles will remain suspended in the air for a long time when there is no precipitation [35]. After being inhaled, these particles can cause harm to humans [37]. In general, the content of particles with large diameters in earthwork construction dust is relatively large, and the content of particles with small diameters is relatively small. This conclusion has also been mentioned in similar studies [38]. The average concentration increment of $PM_{2.5}$ was 32 $\mu g/m^3$ and that of TSP was 224.67 $\mu g/m^3$ when the dust-proof measures of a spray system and artificial water sprinkling were adopted. Compared with situations without dust control measures, the pollutant value was greatly reduced. At this time, the average relative increment of the $PM_{2.5}$ concentration in the construction site was 0.64, and the average relative increment of TSP concentration was 2.88. According to the above values, the $PM_{2.5}$ concentration increment and TSP concentration increment were reduced by 72.01% and 40.16%, respectively.

According to the classification of environmental air function zones in the Chinese *Ambient Air Quality Standards* (GB3095-2012) [39], the first-class areas include nature reserves, scenic spots and other areas in need of special protection; the second-class areas are business, traffic, and resident mixed districts, cultural districts, industrial districts and rural areas. The construction site monitored in this work belongs to the second-class area of the ambient air function zone, so it is subjected to the second-level concentration limit. The daily average concentration limit of TSP is 300 $\mu g/m^3$, and the daily average concentration limit of $PM_{2.5}$ is 75 $\mu g/m^3$. According to the monitoring results, during the earthwork construction, if no means were used to suppress the generation and diffusion of dust, the sum of the background concentration value and the increment of the average concentration of dust in the construction site would be far higher than the concentration limit. The concentration level of dust particles could be greatly reduced by using a spray system and artificial sprinkling. Therefore, in this stage, the construction organization should take proper protective measures, and the staff at the construction site should take safety precautions to avoid bodily harm caused by dust pollution.

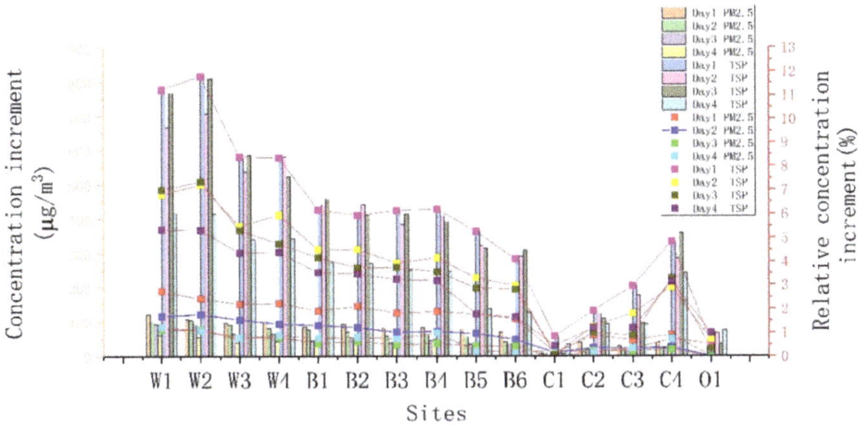

Figure 6. Increment and relative increment of dust concentration at each sampling point.

3.7. Remaining Rate of Dust in Construction Site

Due to local government regulations, most earthworks are completed at night. From the end of construction on the first day to the beginning of other projects on the second day, a sufficient time interval was reserved during this period. In this study, the data collected 30 min after the construction stopped at each collection point were monitored to calculate the retention rate of the dust. This parameter could be used to measure the residual amount of dust in the workplace after the construction stops. Project managers can thus more clearly determine the degree of dust retention at the site after the completion of the construction, which can promote construction organization and arrangement. The retention rate of the dust at the construction site refers to the ratio of the increment of the dust concentration generated by the construction activities to the dust concentration value caused by the construction activity remaining in the construction site after the construction activities have stopped for 30 min. The area under the concentration curve was used (Figure 7), and the retention rate was calculated as follows:

$$\text{Retention fraction} = \frac{B}{A + B} \times 100\% \tag{11}$$

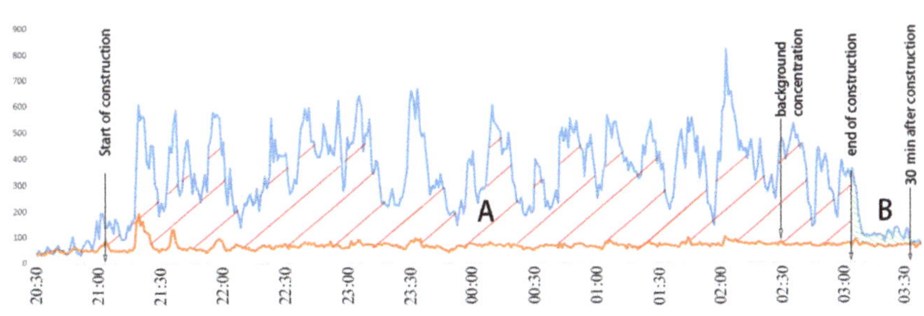

Figure 7. Schematic diagram of the concentration curve.

According to the data shown in Table 8, 0.82~3.76% of $PM_{2.5}$ and 0.81~1.9% of the TSP generated by the construction operations at the construction site would remain 0.5 h after the completion of the construction. The coarse particles produced during construction settle faster than the fine particles, and thus the fine particles will have a higher retention fraction. According to the comparison results of the data from the second and third days, the dust retention rate is higher under static wind conditions. Under the influence of wind, although the construction site is more likely to reach a polluted state, it is also easier to reduce the retention fraction of the dust at the construction site. According to the calculation results, the retention rates of $PM_{2.5}$ and TSP on the fourth and fifth days were relatively large because the denominator in Equation (11) was greatly reduced after the adoption of relevant dust prevention measures. By comparing the data of the fourth and fifth days, it could be seen that due to the occurrence of rain, the value of the retention fraction decreased, and the residual amount of particulate pollutants in the construction site decreased to a certain extent.

Table 8. Calculation of retention fraction.

Date	Retention Fraction	
	$PM_{2.5}$	TSP
25–26 October	0.0339	0.0158
26–27 October	0.0215	0.0107
27–28 October	0.0082	0.0081
28–29 October	0.0376	0.019
29–30 October	0.031	0.0154

4. Conclusions and Recommendations

4.1. Conclusions

This paper presents the results of a study on dust dispersion during the earthwork construction of a typical urban commercial-residential complex project in Jianghan Plain, China. The concentration values of $PM_{2.5}$ and TSP in each area of the construction site were taken as the research objects. The study provided estimates of the critical parameters related to dust dispersion during earthwork construction. These parameters could be used to reveal the distribution and changes in the dust at construction sites during earthwork construction. At the same time, the positive effect of dust control measures on pollutant concentration reduction was also presented. This study can be used as a reference for related earthwork monitoring in other areas of the Jianghan Plain. The following conclusions were drawn from this study:

(1) The aforementioned dust emissions were related to the average concentrations of $PM_{2.5}$ and TSP. The correlation coefficient with the number of unearthed trips was 0.814, and the significance level was less than 0.01, which showed a strong positive correlation;

(2) During the earthwork construction stage, all parts of the construction site were affected, and there were obvious differences. The data of one or two collection points could not be used to characterize the dust concentration value of the entire construction site. The earth excavation area was the most affected, and the impact scope of the dust dispersion caused by it was greater than 100 m, followed by the impact on the transportation path of the truck. The closer the point was to the earth excavation site, the higher the dust concentration;

(3) The dust removal efficiency was more than 30% when artificial sprinkling or spraying was used in the earth excavation site and the earth transport road, and the dust removal efficiency of TSP was higher than $PM_{2.5}$. Where the two measures were used at the same time, the dust removal efficiency of $PM_{2.5}$ and TSP could reach about 50%;

(4) The time required for the concentration of the two pollutants$PM_{2.5}$ and TSP—to reach the steady-state level was inconsistent. The time for the $PM_{2.5}$ concentration to reach the steady-state level was earlier than that of the TSP. The closer to the earthwork excavation, the shorter the time to reach the steady level. In the non-construction stage, $PM_{2.5}$ accounted for a higher proportion of the TSP particle concentration.

Under the influence of earthwork construction, coarse particles occupy a dominant position as a pollution source. The closer to the excavation site, the higher the proportion of coarse particles in the pollutants. The use of dust control measures did not change the proportion of pollutants in the construction process, but after the completion of construction, the coarse particles in the air of the construction site were relatively reduced;

(5) The average concentration increment of $PM_{2.5}$ at the construction site caused by earthwork construction was 55.06 $\mu g/m^3$, and the average concentration increment of TSP was 375.17 $\mu g/m^3$. The sum of the background concentration value and the average incremental dust concentration at the construction site was much higher than the concentration limit. The average concentration increment of $PM_{2.5}$ was 32 $\mu g/m^3$ and that of TSP was 224.67 $\mu g/m^3$ when the dust-proof measures of the spray system and artificial water sprinkling were adopted. Due to the use of dust control measures, the $PM_{2.5}$ concentration increment and TSP concentration increment were reduced by 72.01% and 40.16%, respectively. With or without dust control measures, the ratio of the concentration of particles with different diameters was close to the following: TSP:$PM_{2.5}$ = 1:0.147;

(6) Half an hour after the completion of construction, 0.82–3.76% of $PM_{2.5}$ and 0.81–1.9% of TSP remained at the construction site because of earthwork construction operations. The coarse particles generated during construction work settle faster than fine particles. Under static wind conditions, the dust retention rate is higher. Under the influence of wind, although the construction site was more likely to reach a steady state, the retention rate of dust at the construction site could be reduced more easily. Rainwater would reduce the number of particulate pollutants in the construction site.

4.2. Recommendations

The research results presented in this paper were based on the earthwork construction of a typical urban commercial-residential complex project located in the Jianghan Plain. The results generated from the selection of different construction sites from those used in this study might be slightly different. Therefore, it is necessary to conduct a long-term study on earthwork construction in more engineering cases to better understand dust dispersion in workplaces during earthwork construction.

Based on the conclusions drawn in this paper, our research may enable construction practitioners to create much more targeted countermeasures for different construction activities based on their dust emission and to resolve the concerns raised by dust exposure on a macro scale. The specific dust prevention measures implemented could include the deployment of a dust sensor network. The collated information could be stored in a repository in order to create much more targeted control measures for construction activities. Furthermore, this approach may be able to expeditiously solve the unpredictable civil complaints of residents who live near construction sites through the effective management of and reduction in construction pollutants.

Author Contributions: Conceptualization, J.H., L.H. and Q.L.; project administration, J.H.; formal analysis, Q.L.; validation, Y.L., F.Z. and X.X.; investigation, Q.L., Y.L., F.Z. and X.X. writing—original draft preparation, Q.L. All authors have read and agreed to the published version of the manuscript.

Funding: This research was funded by the National Key Research and Development Plan of the 13th Five-Year Plan (Grant No.2016YFC0702105).

Institutional Review Board Statement: Not applicable.

Informed Consent Statement: Not applicable.

Data Availability Statement: The data presented in this study are available on request from the corresponding author. The data are not publicly available due to data are applied to follow-up studies.

Acknowledgments: The instruments used in this study were provided by the Key Laboratory of New Technology for Construction of Cities in the mountain area (Chongqing University).

Conflicts of Interest: The authors declare no conflict of interest. We wish to confirm that we do not have any commercial or associative interest that represents a conflict of interest in connection with the work submitted. We confirm that we have given due consideration to the protection of intellectual property associated with this research and that there are no impediments to publication with respect to intellectual property.

References

1. Li, C.Z.; Zhao, Y.Y.; Xu, X.X. Investigation of dust exposure and control practices in the construction industry: Implications for cleaner production. *J. Clean Prod.* **2019**, *227*, 810–824. [CrossRef]
2. Gautam, S.; Patra, A.K.; Prusty, B.K. Opencast mines: A subject to major concern for human health. *Int. J. Geol. Min.* **2012**, *2*, 25–31.
3. Finkelman, R.B.; Orem, W.; Castranova, V.; Tatu, C.A.; Belkin, H.E.; Zheng, B.S.; Lerch, H.E.; Maharaj, S.V.; Bates, A.L. Health impacts of coal and coal use: Possible solutions. *Int. J. Coal Geol.* **2002**, *50*, 425–443. [CrossRef]
4. Kanaoka, C.; Furuuchi, M.; Inaba, J.; Ohmata, K.; Myojo, T. Flow and dust concentration near working face of a tunnel under construction. *J. Aerosol Sci.* **2000**, *31*, 31–32. [CrossRef]
5. Sun, Z.; Su, Z. Study on dust diffusion regularity in tunnel slag process. *Ind. Saf. Environ. Prot.* **2017**, *43*, 100–103.
6. Bakke, B.; Stewart, P.; Eduard, W. Determinants of Dust Exposure in Tunnel Construction Work. *Appl. Occup. Environ. Hyg.* **2002**, *17*, 783–796. [CrossRef]
7. Galea, K.S.; Mair, C.; Alexander, C.; de Vocht, F.; van Tongeren, M. Occupational Exposure to Respirable Dust, Respirable Crystalline Silica and Diesel Engine Exhaust Emissions in the London Tunnelling Environment. *Ann. Occup. Hyg.* **2016**, *60*, 263–269. [CrossRef]
8. Liu, J.; He, N.; Wen, W.Y.; Liu, J. Water Jet Vacuum Dust Suppression Device Used to Tunnel Development. *Adv. Mater. Res.* **2012**, *594–597*, 1188–1192. [CrossRef]
9. Xu, Y.; Yang, D.H.; Zhang, Y.W. Tunnel Construction Dust Monitoring and Dust Control Technology of High Altitude and High and Cold Area. *Appl. Mech. Mater.* **2014**, *578–589*, 703–708. [CrossRef]
10. Li, X.; Su, S.; Huang, T. Health damage assessment model for construction dust. *J. Tsinghua Univ. (Sci. Technol.)* **2015**, *055*, 50–55.
11. Huang, T. The Monitoring and Analysis of the Health Damage Caused by Fugitive Dust in the Building Construction Phase. Master's Thesis, Tsinghua University, Beijing, China, 2013.
12. Chen, X.F.; Guo, C.; Song, J.X.; Wang, X.; Cheng, J.H. Occupational health risk assessment based on actual dust exposure in a tunnel construction adopting roadheader in Chongqing, China. *Build Environ.* **2019**, *165*, 106415. [CrossRef]
13. Wu, Z.Z.; Zhang, X.L.; Wu, M. Mitigating construction dust pollution: State of the art and the way forward. *J. Clean Prod.* **2016**, *112*, 1658–1666. [CrossRef]
14. Gangolells, M.; Casals, M.; Gasso, S.; Forcada, N.; Roca, X.; Fuertes, A. A methodology for predicting the severity of environmental impacts related to the construction process of residential buildings. *Build Environ.* **2009**, *44*, 558–571. [CrossRef]
15. Zhu, W.N.; Feng, W.; Li, X.D.; Zhang, Z.H. Analysis of the embodied carbon dioxide in the building sector: A case of China. *J. Clean Prod.* **2020**, *269*, 122438. [CrossRef]
16. Programme, T.U.N.D. *Urban Air Pollution Control in China*; Science and Technology of China Press: Beijing, China, 2001.
17. Liu, J.G.; Diamond, J. Science and government—Revolutionizing China's environmental protection. *Science* **2008**, *319*, 37–38. [CrossRef] [PubMed]
18. Chan, C.K.; Yao, X. Air pollution in mega cities in China. *Atmos. Environ.* **2008**, *42*, 1–42. [CrossRef]
19. Wen, L. Numerical Simulation of the Spatial Migration Rule of Fugitive Dusts at Urban Building Construction Sites. Master's Thesis, Lanzhou University, Lanzhou, China, 2011.
20. Xie, Z. *Research on Formation, Diffusion and Controlling of Construction Dust*; Chongqing University: Chongqing, China, 2018.
21. Huang, T.; Li, X.; Su, S.; Liu, S. Monitoring and Analysis of Dust Pollution during Earthwork Construction. *J. Saf. Environ.* **2014**, *14*, 317–320.
22. Zhang, X.L.; Shen, L.Y.; Zhang, L. Life cycle assessment of the air emissions during building construction process: A case study in Hong Kong. *Renew. Sustain. Energy Rev.* **2013**, *17*, 160–169. [CrossRef]
23. Zhao, X.; Cheng, S.; Tian, G.; Li, G.; Wang, H. Construction Fugitive Dust Pollution and Control in Beijing. *J. Beijing Univ. Technol.* **2007**, *10*, 1086–1090.
24. Zhang, Z.; Wu, F. Evaluation of health damage caused by construction dust pollution. *J. Tsinghua Univ. (Sci. Technol.)* **2008**, *6*, 922–925.
25. Nlj, E.T.; Hilhorst, S.; Spee, T.; Spierings, J.; Steffens, F.; Lumens, M.; Heederik, D. Dust control measures in the construction industry. *Ann. Occup. Hyg.* **2003**, *47*, 211–218.
26. Fan, S.C.; Wong, Y.W.; Shen, L.Y.; Lu, W.S.; Wang, T.; Yu, A.; Shen, Q.P. The effectiveness of DustBubbles on dust control in the process of concrete drilling. *Saf. Sci.* **2012**, *50*, 1284–1289. [CrossRef]
27. Shou-bin, F.; Gang, T.; Gang, L.; Yu-hu, H.; Jian-ping, Q.; Shui-yuan, C. Road fugitive dust emission characteristics in Beijing during Olympics Game 2008 in Beijing, China. *Atmos. Environ.* **2009**, *43*, 6003–6010. [CrossRef]
28. Gautam, S.; Prusty, B.K.; Patra, A.K. Dispersion of respirable particles from the workplace in opencast iron ore mines. *Environ. Technol. Innov.* **2015**, *3*, 11–27. [CrossRef]

29. Li, J.Y.; Li, H.R.; Ma, Y.H.; Wang, Y.; Abokifa, A.A.; Lu, C.Y.; Biswas, P. Spatiotemporal distribution of indoor particulate matter concentration with a low-cost sensor network. *Build. Environ.* **2018**, *127*, 138–147. [CrossRef]
30. Cheriyan, D.; Choi, J.H. Estimation of particulate matter exposure to construction workers using low-cost dust sensors. *Sustain. Cities Soc.* **2020**, *59*, 102197. [CrossRef]
31. Zhao, P.; Feng, Y.; Jin, J.; Han, B.; Zhu, T.; Zhang, X. Characteristics and control indicators of fugitive dust from building construction sites. *Acta. Sci. Circumstantiae* **2009**, *29*, 1618–1623.
32. Lee, C.H.; Tang, L.W.; Chang, C.T. Modeling of fugitive dust emission for construction sand and gravel processing plant. *Environ. Sci. Technol.* **2001**, *35*, 2073–2077. [CrossRef]
33. China, Ministry of Ecology and Environment of People's Republic of China. *Ambient Air–Determination of Dustfall–Gravimetric Method*; Ministry of Ecology and Environment of People's Republic of China: Beijing, China, 1994.
34. Hong, J.; Kang, H.; Jung, S.; Sung, S.; Hong, T.; Park, H.S.; Lee, D.E. An empirical analysis of environmental pollutants on building construction sites for determining the real-time monitoring indices. *Build. Environ.* **2020**, *170*, 106636. [CrossRef]
35. Cheriyan, D.; Hyun, K.Y.; Jaegoo, H.; Choi, J.H. Assessing the distributional characteristics of PM10, PM2.5, and PM1 exposure profile produced and propagated from a construction activity. *J. Clean Prod.* **2020**, *276*, 124335. [CrossRef]
36. Ding, G. *Study on the Emission Characteristics and Modeling of Building Construction Dust*; South China University of Technology: Guangzhou, China, 2020.
37. Cohen, A.J.; Anderson, H.R.; Ostro, B.; Pandey, K.D.; Krzyzanowski, M.; Kunzli, N.; Gutschmidt, K.; Pope, A.; Romieu, I.; Samet, J.M.; et al. The global burden of disease due to outdoor air pollution. *J. Toxicol. Env. Health A* **2005**, *68*, 1301–1307. [CrossRef] [PubMed]
38. Yan, H.; Ding, G.L.; Li, H.Y.; Wang, Y.S.; Zhang, L.; Shen, Q.P.; Feng, K.L. Field Evaluation of the Dust Impacts from Construction Sites on Surrounding Areas: A City Case Study in China. *Sustainability* **2019**, *11*, 1906. [CrossRef]
39. China, Ministry of Ecology and Environment of People's Republic of China. *Air Quality Standard of China*; China, Ministry of Ecology and Environment of People's Republic of China: Beijing, China, 2012.

materials

Article

Nano-Porous-Silicon Powder as an Environmental Friend

Marwa Nabil [1,*], Kamal Reyad Mahmoud [2], Raghda Nomier [3], El-Maghraby El-Maghraby [3] and Hussien Motaweh [3]

1 Department of Electronic Materials Researches, Advanced Technology and New Materials Research Institute, City for Scientific, Research and Technology Applications, New Borg El-Arab City 21934, Egypt
2 Department of Physics, Faculty of Science, Kafrelsheikh University, Kafr El Sheikh 33516, Egypt; kamalreyad@gmail.com
3 Department of Physics, Faculty of Science, Damanhour University, Damanhur 22511, Egypt; raghda.mhmd91@yahoo.com (R.N.); maghrabym@yahoo.com (E.-M.E.-M.); prof_motaweh@yahoo.com (H.M.)
* Correspondence: marwamoh2000@yahoo.com

Abstract: Nano-porous silicon (NPS) powder synthesis is performed by means of a combination of the ultra-sonication technique and the alkali chemical etching process, starting with a commercial silicon powder. Various characterization techniques {X-ray powder diffraction, transmission electron microscopy, Fourier Transform Infrared spectrum, and positron annihilation lifetime spectroscopy} are used for the description of the product's properties. The NPS product is a new environmentally friendly material used as an adsorbent agent for the acidic azo-dye, Congo red dye. The structural and free volume changes in NPS powder are probed using positron annihilation lifetime (PALS) and positron annihilation Doppler broadening (PADB) techniques. In addition, the mean free volume (VF), as well as fractional free volume (Fv), are also studied via the PALS results. Additionally, the PADB provides a clear relationship between the core and valence electrons changes, and, in addition, the number of defect types present in the synthesized samples. The most effective parameter that affects the dye removal process is the contact time value; the best time for dye removal is 5 min. Additionally, the best value of the CR adsorption capacity by NPS powder is 2665.3 mg/g at 100 mg/L as the initial CR concentration, with an adsorption time of 30 min, without no impact from temperature and pH. So, 5 min is the enough time for the elimination of 82.12% of the 30 mg/L initial concentration of CR. This study expresses the new discovery of a cheap and safe material, in addition to being environmentally friendly, without resorting to any chemical additives or heat treatments.

Keywords: microporous materials; positron annihilation spectroscopy; X-ray diffraction

Citation: Nabil, M.; Mahmoud, K.R.; Nomier, R.; El-Maghraby, E.-M.; Motaweh, H. Nano-Porous-Silicon Powder as an Environmental Friend. *Materials* **2021**, *14*, 4252. https://doi.org/10.3390/ma14154252

Academic Editor: Avelino Núñez-Delgado

Received: 25 June 2021
Accepted: 19 July 2021
Published: 30 July 2021

1. Introduction

The major threat, for the time being, which must be dealt with on a global level is toxic and carcinogenic environmental pollutants. In particular, the new technologies developed for the easier decolorization of different compound types have attracted widespread interest [1]. Many industries produce residual dyes (i.e., dye intermediates, textile, paper, and pharmaceutical industries, etc.). Wastewater treatment systems have to deal with a wide range of organic pollutants. Pollution with dyes is undesirable, as many of the dyes released are toxic and carcinogenic [2]. In order to remove the wastewater color, several physical and chemical experiments have been performed. Therefore, it was found that the process of de-pigmentation using physical adsorption technology is the most effective and economically appropriate [3]. So, the adsorption technique is one of the best techniques for water reuse, as a result of its economic cost, simple design, ease of operation and non-toxicity [4]. Accordingly, many porous adsorbent materials, such as activated carbon [5], peat, chitin, and silica, are used for testing the possibility of dye removal [6]. However, intraparticle diffusion associated with porous adsorbents may lessen the rate and capacity of adsorption [7]. Therefore, the adsorption process is a surface process; its adsorption value and its specific surface area are directly proportional to each other [8]. The ratio between the rising surface area and nano-adsorbent mass of materials can promote the sorbent material's adsorption capacity.

Generally, NPS material is a network containing a homogenous mixture of air and silicon. From the optical point of view, NPS is specified as an effective medium, and it is considered environmentally suitable for use as an adsorbent material. Its optical properties rely on the silicon prorated volumes, and the pore filling medium [9]. The ultra-sonication technique is one of the most famous materials processing techniques that is widely used for powder technology, as a result of its simplicity and effectiveness, as shown in previous research [10].

The PALS is an important tool and non-destructive technique that has been used for the characterization and investigation of the microstructural properties of different materials. Positron experiments confirmed the sensitivity of PALS to the studied defects in metals/alloys, and free volume/pores in molecular solids [11]. Additionally, in the case of porous materials, the formation of positrons are implanted from a radioactive source in the molecular solids, and each pore of them annihilates with e^- of the material's atoms and for the formation of a positronium (Ps), s shown in previous studies. Thus, the pluck annihilation rate (also known as the lifetime) is correlated with the pore size in the simple free-volume model size according to the simple free-volume model [12–16].

Doppler broadening spectroscopy (PADB) supplies valuable information regarding the inner electronic shells' contribution and provides valuable data about chemical annihilation. The S-parameter is defined as the ratio of counts in the central part of the Doppler broadened spectrum to the area below the annihilation line completely. It depends on the average density of volume defects, which is open. On the other hand, the ratio between the area below the annihilation line fixed-wing region and the area under the whole annihilation line is defined as W [17]. This is related to the positron annihilation with deeply bound core electrons, which provides information about the chemical environment of the defect. Thus, the PALS technique obtains the e^- density data at the positron annihilation site, and the PADB methods provide information on the momentum distribution of electrons. All of them are widely used in modern materials science. Several studies, in particular on solids and porous systems, have included NPS and nonporous SiO_2 via the PALS technique [18–25].

As shown in our previous research [26–30], the solid's nano-scale microstructure is studied. We report herein the application of the PALS for tracking the free volume size changes for synthesized NPS powder via a combination of the alkali chemical etching process and the ultra-sonication technique, starting with commercial silicon powder. Additionally, this work is targeted at studying acidic Congo red dye removal from aqueous solutions using synthesized NPS powder via the adsorption process.

2. Materials and Methods

2.1. NPS Powder Production and Characterization

The combination of two techniques (ultra-sonication and alkali etching process) for NPS powder production was performed, starting with commercially available Si-powder (Silicium, Pulver—99%, Burlington, VT, USA), as shown in previous studies. A suitable amount of Si powder was dispersed in n-propanol and KOH was dispersed in distal H_2O. The product powders were filtrated, washed, and then dried overnight.

The construction and crystallization of the synthesized NPS were analyzed using XRD (X-ray 7000 Shimadzu diffractometer, Kyoto, Japan). (Fourier transform infrared spectrophotometry (FTIR-Shimadzu FTIR-8400 s) was used to determine the NPS powder forming chemical bonds. In addition, high-resolution transmission electron microscopy (HR-TEM, Tecnai G20, FEI, Eindhoven, The Netherlands) was utilized in the description process of the NPS powder product's morphology.

2.2. Positron Annihilation Lifetime Measurements

This work used the spectrometer due to its fast spectrometry [31], with a resolution of ~350 ps via the ^{60}Co source at room temperature for measurements of the lifetime of the positron. To study the activity of 15 μCi of ^{22}Na, the sample was deposited and dehydrated upon Kapton foil (7.6 μm thick), and then glued using epoxy glue. During the measurements, this assembly was sandwiched between two similar samples as a positron source. The measurement of each sample was repeated at least 2–3 times, and the total number of elementary annihilation events was approximately 1–2 million. The LT computer program from Kansy was used to resolve the collected spectra. [32]

As a result of measured spectra analysis, there are 3 lifetime items (τ1, τ2, and τ3). The 1st lifetime item τ_1 is produced due to the P-positronium (p-Ps) atom (fixed at 0.125 ns). The 2nd lifetime item τ_2 is produced during the positron annihilation via free electrons inside the material. Finally, the 3rd lifetime item (τ_3), which is the longest lifetime component, depends on the ortho-positronium (o-Ps) annihilation via the "pick-off" mechanism in the amorphous regions free volume sites. All items were determined via the fit's variance (1.005 to 1.18). So, τ_3 provides valuable data regarding the free volume cavities' mean size when probed by o-Ps.

For the free-volume model [12], the o-Ps lifetime focused inside a spherical solid potential well (radius = R_o) and the free volume of radius R, and no electrons were found below it, as shown in the following equation [33,34]:

$$\tau_o - P_s = 0.5 \times [1 - (R/R_O) + (1/2\pi)\, \mathrm{Sin}\, (2\pi R/R_o)]^{-1} \quad (1)$$

where $\delta R = R_o - R = 1.656$ Å is the fitted empirical electron layer thickness. With this value of δR, the free volume radius (R) was calculated from Equation (1), and the average size of the free volume holes (V_f) was calculated as $V_f = (4/3)\ \pi R_3$ (in Å^3).

Furthermore, the free volume hole fraction, f_v, can be estimated using the empirical equation [35]:

$$f_y = CV_f I_3 \quad (2)$$

where V_f is in angstrom cube, I_3 in percent, and C is an arbitrarily chosen scaling factor for a spherical cavity.

2.3. Doppler Broadening Measurements

Using a Ge-detector (Ortec, p-type high-purity, GEM series, Oak Ridge, TN, USA), Doppler broadening was measured. Its energy resolution was FWHM = 1.6 keV for the γ-line (1.33 MeV) of ^{60}Co. A relative efficiency of 25% was applied for defining S and W as the line-shape parameters of the Doppler broadening. Ortec 570 was used for magnifying the detector output signals, which were then obtained via an Ortec 919 multichannel analyzer (MCA). A 5 μCi ^{22}Na sample was prepared via a droplet of NaCl solution that was dried on 2 congruent Kapton foils, and then glued using epoxy glue. Both samples (disks) were coordinated with the ^{22}Na source in a 4π configuration. The energy was calibrated (~68 eV/channel) via the ^{133}Ba source. The Doppler broadening spectra were taken until more than two million counts had accumulated in the peak. These measurements were performed in air at room temperature. The obtained Doppler broadening spectra were analyzed using the SP ver. 1.0 program. The calculation of the S- and W-parameters, depending on the centroid channel with maximum counts of the 511 keV peak, was accurately determined. The input data were fixed for all spectra of the studied samples.

2.4. Dye Decolorization Using the Batch Procedure

The wastewater was synthesized by dissolving the acidic Congo red dye in distilled water to gain the required waste solution concentrations; this was applied to monitor the efficiency of the NPS product's adsorption. Then, 30 mL of wastewater solution (10 and 50 ppm) was mixed with 0.15, 0.3, 0.6 and 1 g of the NPS product for 15 min using the orbital shaker. The solid phase was separated from the solvent phase using a centrifugation technique (600× *g* rpm for 15 min). The remaining acidic Congo red concentration was analyzed using a UV-Visible spectrophotometer at a wavelength of 486 nm.

The residual mass comprised adsorbed metal ions; the collected filtrate was exposed to metal ion assessment via the UV-Visible Spectrophotometer Double Auto Cell (Labomend. INC, Los Angeles, CA, USA). We then calculated the percentage of metal uptake, using the sorption efficiency, and then the amount of metal ions that were adsorbed [36]. The tests were executed to determine the impact of contact time (20–60 min) and the temperature of the waste solution (25 °C).

$$\text{Sorption efficiency} = (C_i - C_f)/C_i \times 100 \quad (3)$$

$$\text{Amount Adsorbed } (Q_e) = (C_i - C_f)/W \times V \quad (4)$$

where C_i is the initial metal ion concentration in the solution (mg/L), C_f is the final metal ion concentration in the solution (mg/L), W is the adsorbent weight (g), V is the solution volume (L), and Q_e is the amount of metal ions that adsorbed per gram of adsorbent.

3. Results and Discussion

3.1. NPS Powder Characterization

The chemically prepared NPS powder material was examined via various physico-chemical techniques in order to investigate its structure and properties.

3.1.1. X-ray Diffraction Analyses

Figure 1 describes the diffraction peaks of the NPS powder product which is perfectly reported in the cubic phase NPS (JCPDS Card No. 01-079-0613 and 00-027-1402). The strongest peak appears at 2θ = 28.23°, which corresponds to (111), while other peaks appeared at 2θ = 47.193°, 56.023°, 76.261°, 87.9382° and 94.8370°, which correspond to (220), (311), (312), (422) and (511), respectively. It is also noted that limited silica formation occurs at 2θ = 23.128°. There are no impurity peaks in the pattern, meaning that it is corroborative of the high purity of the prepared NPS powder.

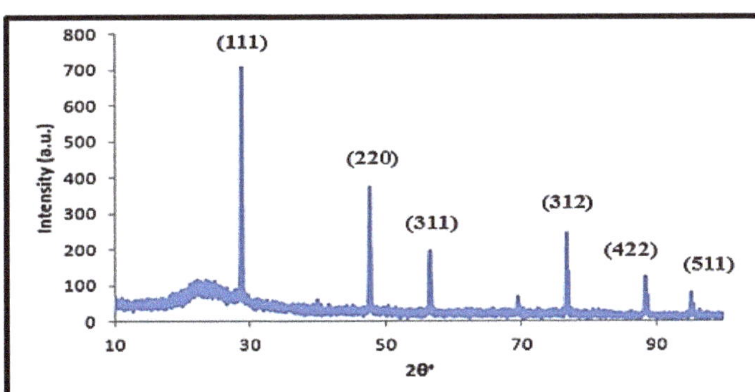

Figure 1. X-ray diffraction pattern of the prepared NPS powder.

3.1.2. Fourier Transform Infrared Spectroscopy (FTIR)

Figure 2 presents the NPS product's FTIR spectrum in the range 400–4000 cm^{-1}. The peaks within the wavelength range of 1000–1300 cm^{-1} are assigned to Si–O asymmetric stretching in Si–O–Si, and the peak at 449 cm^{-1} corresponds to Si–O bending. In addition, the formation of an NPS product peak is recorded at 1072 cm^{-1}. The broad peak at 3449 cm^{-1} corresponds to the presence of interstitial water and the hydroxyl group. The peak at 1662 cm^{-1} corresponds to the free water molecules' deformation vibration [37]. Therefore, the FTIR spectrum agrees that the product is pure NPS with no pollutants due to oxidizing and wetting agents, which were utilized in the preparation step.

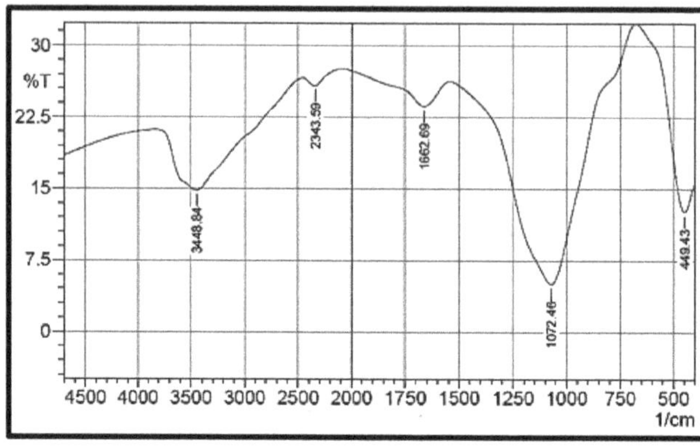

Figure 2. FTIR spectrum of the prepared NPS powder.

3.1.3. Transmission Electron Microscopic Analyses (TEM)

TEM images of the prepared NPS powder are shown in Figure 3. In the preparation conditions—7 g of commercial Si powder, 3 wt.% KOH at sonication times of 3 and 4 h—the morphological construction of the NPS powder product as illustrated in Figure 3A has a spherical NPS morphology covered with a nano-silica layer. In the case of Figure 3B, the TEM image presents the cubic shape with good crystallinity. Figure 3 shows that the NPS product is in the nano range. These results provide a prediction of the NPS powder product, which has a huge surface area, which is useful for enhancing its dye pollutant removal affinity.

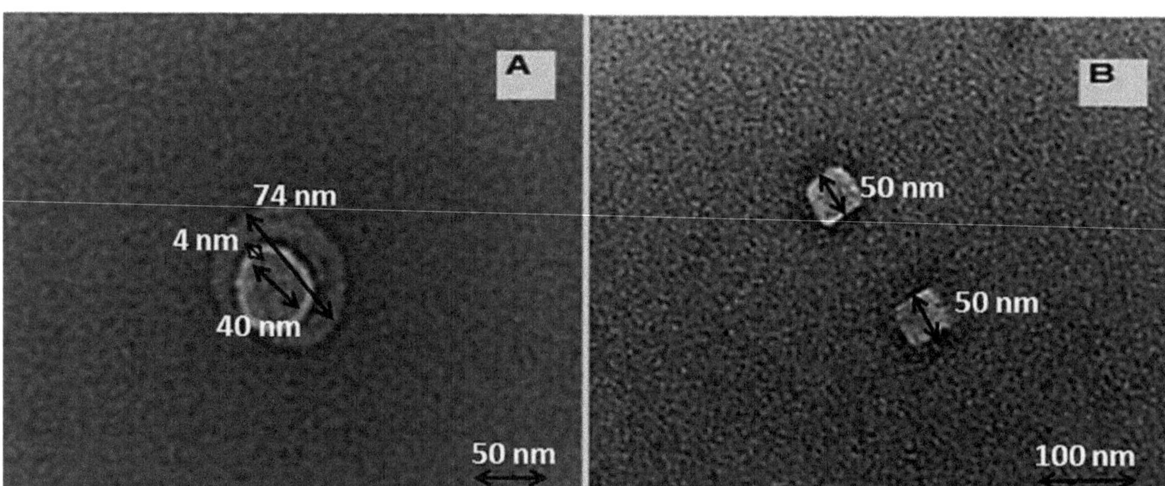

Figure 3. TEM images for NPS and nano porous silica powders in the preparation conditions—7 g of commercial Si powder, 3 wt.% KOH, and at sonication times; (**A**) 3 h, and (**B**) 4 h.

3.2. Positron Annihilation Lifetime (PAL) Parameters

The spectra of positron lifetime are classified in terms of three items of positron lifetime, τ_1, τ_2, τ_3, while the intensities are I_1, I_2, I_3 for the NPS powder product, respectively. Due to the poor resolution time of the apparatus (≈350 ps), the short-lived item data for the p-Ps are unreliable. For the minimization of the scatter of the other parameters, τ_1 was fixed at 125 ps. Accordingly, the suitable quality of the spectra did not change and the derived parameters were close to those obtained when the analysis was made without any restrictions.

For all measured samples, the intermediate-lifetime component (τ_2 = 0.351–0.497 ns) and its relative intensities (I_2) ranged from 29.00% to 44.2%, as shown in Table 1. Additionally, it may arise from the interaction of positrons with e^-s placed in higher negative charge density. The τ_2 and I_2 values are found to be in the same order as those commonly seen in the literature [20,24]. The longest-lived one, τ_3, may be attributed to the o-Ps annihilation localized in nano-regions, within the silica matrix, which is very sensitive to the microstructural changes. In molecular systems, the o-Ps localized in a cavity annihilate via a pickoff annihilation technique with an antiparallel electron spin from the cavity wall surroundings. The τ_3 determination provides valuable information on the mean size of free volume cavities probed by o-Ps.

Table 1. The results of lifetime components with the corresponding intensities at 7 and 5 g of commercial silicon powder in different weight of KOH (wt.%) at sonication times of 2, 3, and 4 h).

Sample Composition	Sonication Time (h)	τ_2 (ns)	τ_3 (ns)	I_1 (%)	I_2 (%)	I_3 (%)
7 g Si Powder + 6 wt.% KOH	2	0.359 ± 0.004	1.780 ± 0.055	57.22 ± 0.50	40.56 ± 0.50	2.21 ± 0.10
	3	0.378 ± 0.002	1.648 ± 0.019	68.20 ± 0.17	29.00 ± 0.17	2.79 ± 0.043
	4	0.449 ± 0.004	1.334 ± 0.012	41.33 ± 0.19	39.61 ± 0.20	18.3 ± 0.59
	4 + No. H *	0.008 ± 0.456	1.510 ± 0.013	40.01 ± 0.94	42.79 ± 0.86	17.2 ± 0.42
7 g Si Powder + 4.5 wt.% KOH	2	0.351 ± 0.004	1.78 ± 0.055	57.2 ± 0.50	40.6 ± 0.50	2.21 ± 0.10
	3	0.369 ± 0.006	1.73 ± 0.043	63.3 ± 0.93	34.5 ± 0.93	2.18 ± 0.10
	4	0.372 ± 0.002	2.19 ± 0.026	55.8 ± 0.53	41.7 ± 0.53	2.45 ± 0.03
	4 + No. F *	0.470 ± 0.009	1.54 ± 0.014	36.6 ± 0.83	44.2 ± 0.70	19.3 ± 0.43
7 g Si Powder + 3 wt.% KOH	2	0.376 ± 0.009	1.58 ± 0.080	61.0 ± 1.40	36.0 ± 1.5	2.79 ± 0.20
	3	0.358 ± 0.004	1.97 ± 0.038	58.9 ± 1.00	38.8 ± 1.0	2.32 ± 0.08
	4	0.373 ± 0.007	1.86 ± 0.050	64.4 ± 1.00	33.3 ± 1.0	2.33 ± 0.11
	4 + No. F *	0.497 ± 0.011	1.42 ± 0.015	43.9 ± 0.88	37.1 ± 0.8	18.9 ± 0.50
5 g Si Powder + 3 wt.% KOH	2	0.351 ± 0.003	2.15 ± 0.05	57.7 ± 0.95	40.06 ± 0.95	2.28 ± 0.07
	3	0.353 ± 0.004	1.63 ± 0.05	62.7 ± 0.48	35.66 ± 0.48	1.63 ± 0.07
	4	0.382 ± 0.015	2.26 ± 0.026	64.4 ± 0.42	32.59 ± 0.42	2.99 ± 0.04
	4 + No. F *	0.481 ± 0.014	1.31 ± 0.014	41.0 ± 0.89	37.49 ± 0.91	21.47 ± 0.6

* No. H: without heat treatment, No. F: without filtration.

Table 1 contains the calculated values of the τ_3 and its relative intensity I_3 that classifies the annihilation parameters of the o-Ps as a function of sonication time (2, 3, and 4 h) at the preparation conditions (commercial silicon weight (5 and 7 g) and several $KOH_{conc.}$ (3, 4.5 and 6 wt.%)). Additionally, samples at a sonication time of 4 h and special conditions without heat treatment (No. H) or without filtration (No. F) are presented. The range of the longest-lived item, τ_3, is 1.31–2.19 ns, and its corresponding intensity (I_3) in within the range 1.63–21.5%, for all the measured samples. The values of t_3 and I_3 are the smallest [20,21,38,39].

It is clear from Table 1 that there is a directly proportional relationship between the sonication time and the values of I_3%). This is a logical relationship, which is a result of the enhancement of the porosity percentage in Si powder for NPS formation. One can notice that a surprising enhancement of the I_3% values was shown in samples with 4 h sonication time with special conditions of no heat treatment (No. H) or no filtration (No. F). In these processes, a longer period time of the oxidized agent was achieved as a result of the NPS surface oxidation process. Consequently, the porous silica layer was formed on the NPS core, which has a larger surface area, increasing the porosity percentage and decreasing the o-Ps lifetime (τ_3 ns), and consequently decreasing the size of the free volume, as shown in Table 1. This result was proven and in good agreement with the TEM measurements (see Section 3.1.3).

The calculated values of the o-Ps lifetime (τ_3 ns) in Table 1 were used to calculate the radius R of the free volume V_f (Å^3) according to the free-volume model [12]. Figure 4 shows the variations of mean free volume V_f (Å^3) as a function of sonication time (2, 3, and 4 h) for the measured samples at weight of commercial silicon powder (5 and 7 g) in different KOH concentrations (6, 4.5, and 3 wt.%). In addition to samples at a sonication time of 4 h in special conditions and with a slow drying process (without heat treatment (4 h + No. H)) and without the separation process (without filtration (4 h + No. F)). It is clear from Figure 4 that the effect of sonication time on the mean free volume V_f (Å^3) has the same trend as the o-Ps lifetime, τ_3 (ns) (as shown in Table 1), and also the same explanation can be suggested. It can be concluded that a severe reduction in mean free volume was observed in samples at a sonication time of 4 h in the special conditions and with a slow drying process, without filtration as a result of the formation of a silica layer on the NPS material.

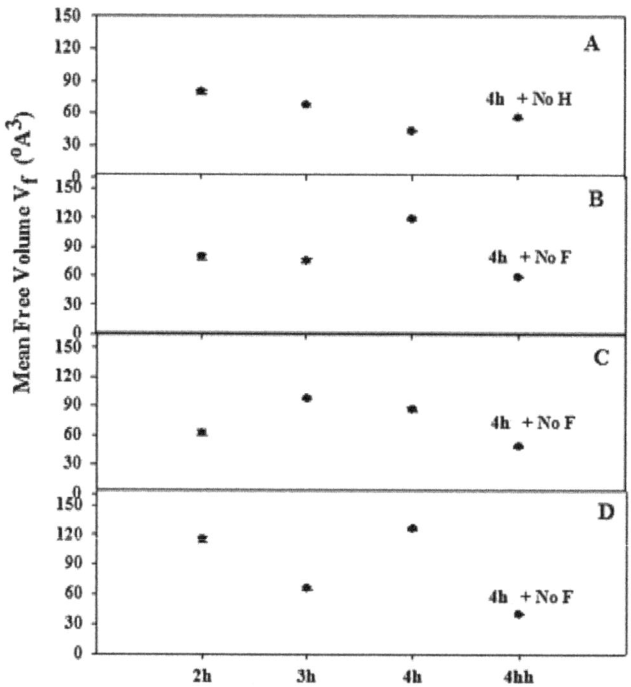

Figure 4. The variation of mean free volume, Vf (Å3), as a function of sonication time (2, 3 and 4 h) in special conditions; 7 g commercial silicon powder in different KOH concentrations (wt.%): (**A**) at 6 wt.% KOH, (**B**) 4.5 wt.% KOH, (**C**) 3 wt.% KOH and (**D**) 5 g commercial silicon powder and 3 wt.% KOH.

The variation of the fractional free volume (F_v) of the NPS samples as a function of sonication time (2, 3 and 4 h) for all the measured samples are also shown in Figure 5. The results show a small increase in the values of F_v with a sonication time of 2, 3 and 4 h for all different KOH concentrations, then a steep increase at a sonication time of 4 h in the special conditions and with a slow drying process (without heat treatment (4 h + No. H)) and without the separation process (without filtration (4 h + No. F)). This enhancement may be attributed to the formation of a silica layer on the NPS material.

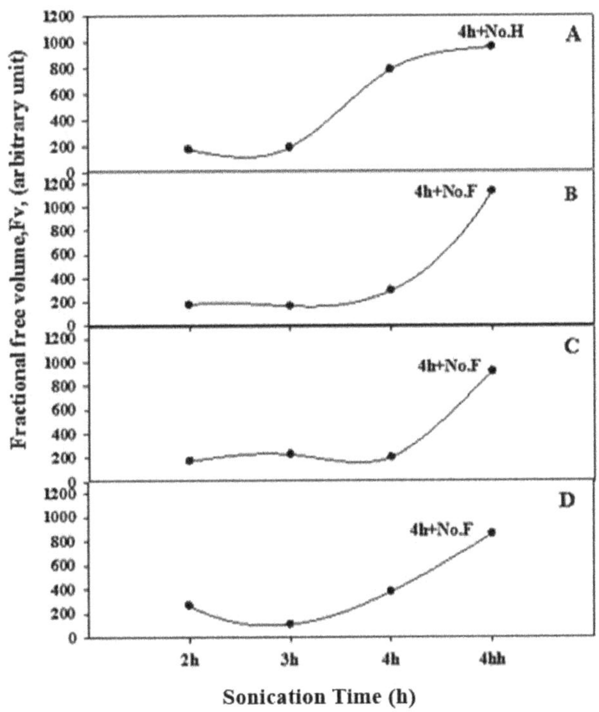

Figure 5. The variation of fractional free volume, Fv, as a function of sonication time (2, 3 and 4 h) in special conditions; 7 g commercial silicon powder in different KOH concentrations (wt.%): (**A**) at 6 wt.% KOH, (**B**) 4.5 wt.% KOH and (**C**) 3 wt.% KOH and (**D**) 5 g commercial silicon powder and 3 wt.% KOH.

3.3. Doppler Broadening Spectroscopy Measurements

Although the PALS results are strongly indicative of a long-lifetime component task Ps, the results show this conclusion via another distinct technique, such as Doppler broadening of annihilation radiation (DBAR). The sharpness of an annihilation peak of 511 keV can be measured by the so-called S-parameter, which is an indicator of the fraction of positrons annihilating with the valence electrons. This can be produced when positrons annihilate in vacancies or pores and/or o-Ps annihilate in free volumes with low-kinetic momentum electrons of the outer orbital of the neighboring atoms present at the wall of the pores or the free volumes inside the materials. The estimated S- and W-parameters' values as a function of sonication time are shown in Figure 6.

The results show that the S-parameter values decreased. However, there is a directly proportional relationship between the W-parameter and sonication time at 7 g commercial silicon powder with 6 wt.% of KOH, as shown in Figure 6A. On the other hand, there are a few variations of the S– and W–parameters for the samples of 7 g commercial silicon powder with 4.5 and 3 wt.% of KOH and 5 g of commercial silicon powder with 3 wt.% of KOH at the sonication times 2, 3, and 4 h, as shown in Figure 6B–D. These variances are in agreement with the PALS information, due to the existence of the Ps; the distribution of momentum is correlated with Ps (o-Ps and p-Ps). This is thinner than that linked with the e^+ annihilation ("free" positron gain). Thence, the overall line width has to increase at the Ps intensity decreases, and vice versa, as is actually observed.

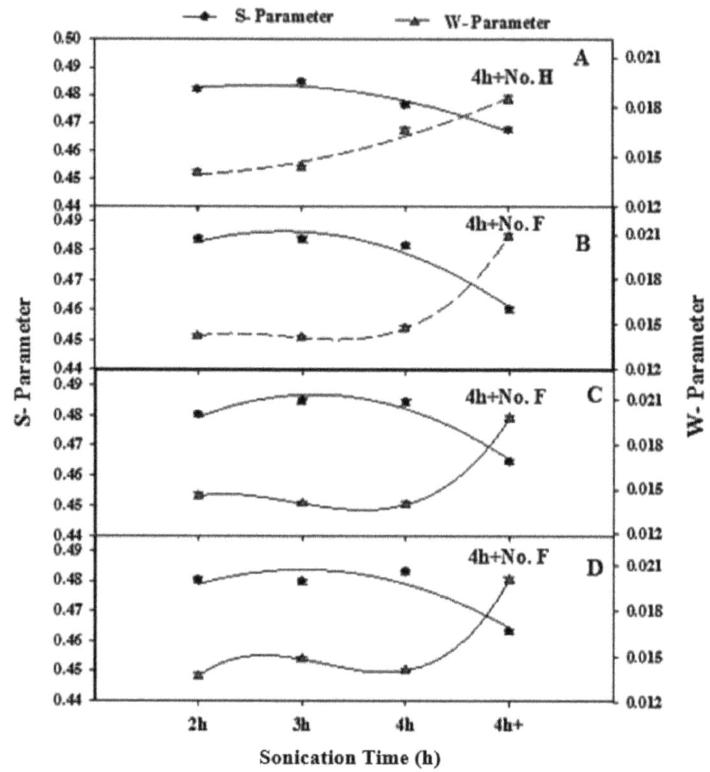

Figure 6. The variation of W–parameters at 7 g commercial silicon powder at sonication time (2, 3 and 4 h) with special conditions with different weights of KOH (wt.%): (**A**) at 6 wt.% KOH, (**B**) 4.5 wt.% KOH and (**C**) 3 wt.% KOH and (**D**) 5 g commercial silicon powder and 3 wt.% KOH.

A falling behavior compared with the steep growth of S- and W-parameters, respectively, was recorded for samples in the special conditions of 4 h without filtration, as seen in Figure 6B–D. The steep decrease is perhaps as a result of valance e⁻'s reduction, defect size, and the concentration of particles [31]. Figure 7 shows the defect type number that is obtained by plotting the S-parameter verses the W-parameter. For a sample with one kind of defect, the plot of S against W is linear. From these figures, one notices that the W-parameter is inversely proportional to the S-parameter values for all samples. Thus, only one kind of defect exists in these samples. As shown in Figure 7A,D, the only exception was found in the samples with 7 g commercial silicon powder with 6 wt.% KOH and 5 g commercial silicon powder with 3 wt.% KOH at a sonication time of 2 h.

3.4. Basic Dyes Decolourization Process onto the Synthesized NPS Powder Using a Batch Adsorption Technique

3.4.1. Effect of Contact Time

The contact time effect on the basic Congo red (CR) adsorption onto the NPS powder surface is presented (Figure 8). The experiments are performed at an initial dye concentration of 10 ppm, with 10 g/L of NPS as an adsorbent, and with a 600 rpm agitation speed at several time interval ranging from 0 to 60 min. It is stated that the CR adsorbed amount is directly proportional to the contact time, and at 5 min reaches its maximum value. The equilibrium time can be considered at 15 min for ensuring the full dye sorption atop the prepared NPS. Therefore, the maximum dye removal above the synthesized NPS powder occurred within 5 min, and subsequently, the system reaches an equilibrium point.

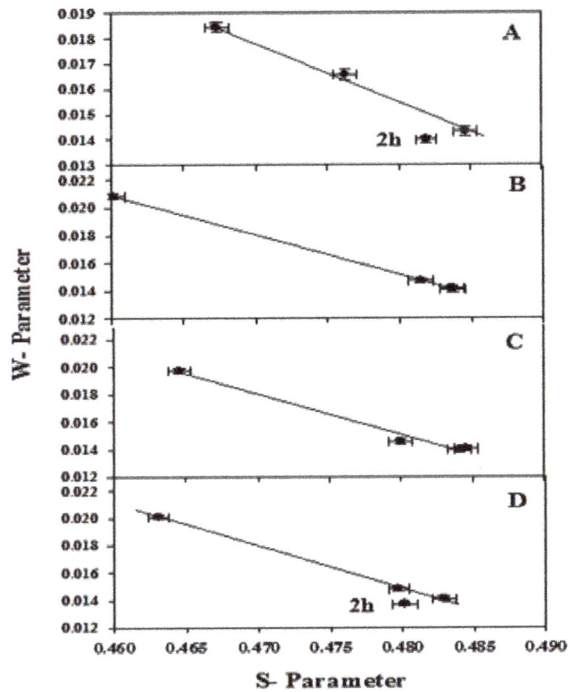

Figure 7. The S versus W plot for 7 g commercial silicon powder at sonication times of 2, 3 and 4 h with special conditions in different weights of KOH (wt.%): (**A**) at 6 wt.% KOH, (**B**) 4.5 wt.% KOH and (**C**) 3 wt.% KOH and (**D**) 5 g commercial silicon powder and 3 wt.% KOH. The solid line represents a linear fitting of the experimental data.

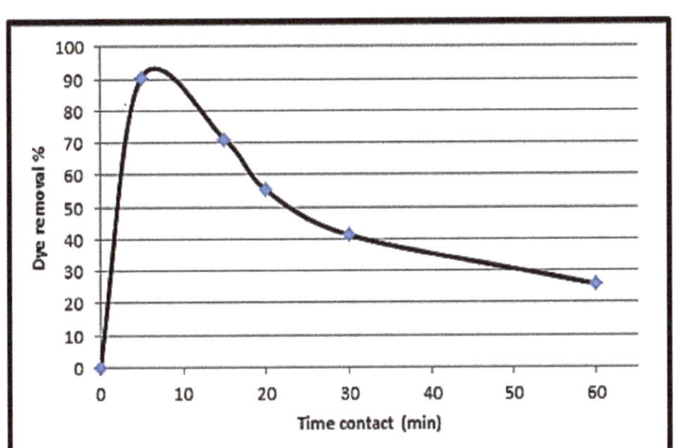

Figure 8. The contact time impact for Congo red removal using NPS (pH = 7; adsorbent dose = 10 g/L; initial CR concentration = 10 mg/L).

3.4.2. Effect of NPS Powder Dosage

The NPS dosage is an important factor that sets the NPS adsorption capacity at an initial CR concentration of 50 ppm. The CR removal percentage via various NPS dosages and the equilibrium sorption capacity is illustrated in Figure 9a,b. From this figure, the direct proportionality between the NPS dosage value and the removal percentage of CR dye is noticeable. Furthermore, the amount of CR removed per gram of NPS powder tends to reduce with the enhancement of its amount. When raising the NPS dosage at the CR dye concentration of 50 ppm, it supplies a more exposed area for dye adsorption, and thus leads to the enhancement of the extent of CR removal. Otherwise, the amount CR dye removed per gram of NPS reduces; essentially, this is due to the presence of NPS sites, and the rest being unreacted due to the dye's adsorption. Furthermore, regarding the prepared NPS dosage (over 10 g/L), a trivial increase was recorded with the increase in NPS dosage up to 33.3 g/L. Thus, 10 g/L of the NPS is chosen as the optimum adsorbent material dosage for CR dye removal.

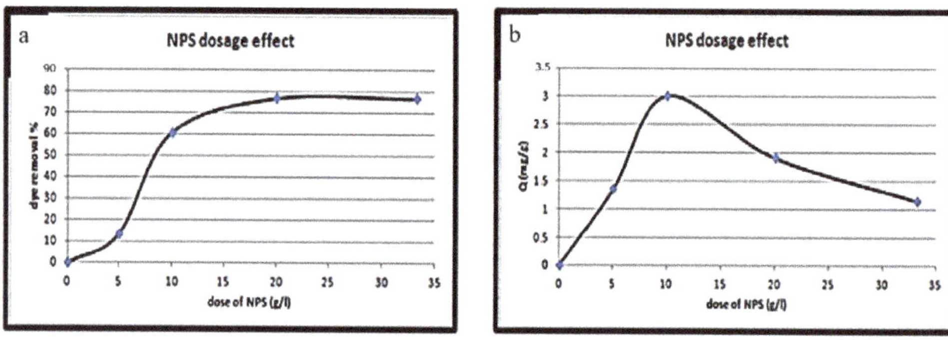

Figure 9. Effect of NPS dosage on: (**a**) CR dye removal, (**b**) the quality of the adsorption process, at initial dye concentration = 50 ppm, temperature = 298 K, pH = 7, contact time = 15 min).

3.4.3. Initial Dye Concentration Impact

The quantitative analysis of the CR removal percentage at equilibrium on the NPS surface at various initial dye concentrations is presented in Figure 10. It clarifies the inversely proportional relationship between the dye adsorption percentages and the initial dye concentration. Moreover, the adsorbed amount of dye per adsorbent unit mass is affected by raising the concentration of the initial dye.

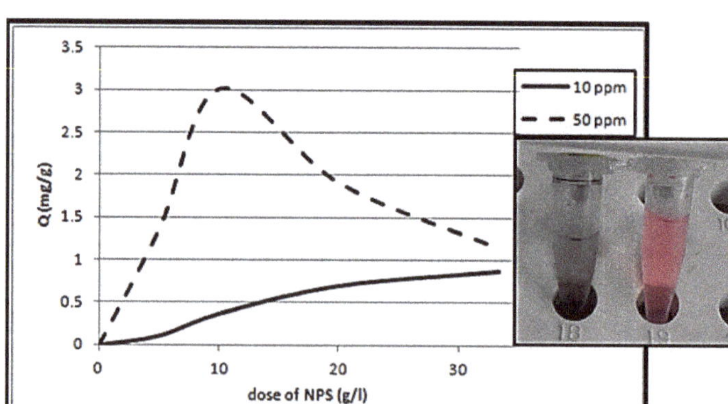

Figure 10. Impact of CR initial concentration on the adsorption values (at different adsorbent doses; T = 298 K; pH = 7; t = 15 min).

At high dye concentrations, the dye adsorption onto the prepared NPS reduced. This is as a result of the ratio value; the initial mole number of the dye to the adsorbent material's surface area. Hence, the fractional factor between the adsorbent and the adsorbate is dependent on the adsorption process. Then, the dye concentration initially supplies a significant driving force to overcome the resistance of the dye mass transfer between both aqueous and solid phases. So, at the highest initial dye concentration, the ion number for the available sites on the NPS surface is high too, ameliorating the basic CR adsorption capacity [40–42].

Furthermore, the impacts of the temperature and acidity on the value of CR adsorption are studied. When changing the pH and temperature values, no obvious impact on the CR adsorption values is observed. At room temperature and pH = 7, no change in the removal percentage can be seen compared to the previous cases. Then, all the prepared samples (as previously mentioned in Table 1) are tested in the batch technique for dye adsorption. However, the result was negative, except for the samples which were prepared at the preparation conditions of 7 g commercial silicon powder, with a sonication time of 4 h with a slow drying process (without heat treatment (4 h + No. H)) and without a separation process (without filtration (4 h + No. F)). This agrees with the results of the free volume values (I_3), which agree with the PALS measurement values. The PALS measurements and the TEM images also agree with the final results of the CR adsorption process.

4. Conclusions

The morphological and crystalline description of the NPS product records the high purity state with good crystallinity. The NPS powder is prepared using the combination of two techniques. The PALS results show that positron annihilation can be a useful technique to characterize the NPS product. The results indicate an enhancement of the I_3% values at a sonication time of 4 h; without heat treatment (No. H) and without a filtration process (No. F). The produced results are in line with the TEM measurements. DBAR measurements show an inversely linear relationship between S and W for all samples. This suggests that only one type of defect is present in these samples.

The NPS product is used effectively for CR dye adsorption from aqueous solutions. The dye removal percentage is reinforced at the increased contact time value. The best dye removal occurs at 5 min, and afterwards the equilibrium point is reached by the system. The best CR adsorption capacity of the NPS product is 2665.3 mg/g, at an initial CR concentration of 100 mg/L and an adsorption time of 30 min, with no pH and temperature effect. Therefore, 5 min is sufficient for removing 82.12% of CR at an initial concentration of 30 mg/L.

Author Contributions: Conceptualization, M.N. and K.R.M.; methodology, M.N., K.R.M. and R.N.; formal analysis, M.N., K.R.M. and R.N.; investigation, M.N., K.R.M. and R.N.; writing—original draft preparation, M.N., K.R.M. and R.N.; writing—review and editing, E.-M.E.-M. and H.M.; visualization, E.-M.E.-M. and H.M.; supervision, E.-M.E.-M. and H.M. All authors have read and agreed to the published version of the manuscript.

Funding: This research received no external funding.

Institutional Review Board Statement: Not applicable.

Informed Consent Statement: Not applicable.

Data Availability Statement: The data presented in this study are available on request from the corresponding author.

Conflicts of Interest: The authors declare no conflict of interest.

References

1. Jiménez, J.J.; Algarra, M.; Guimarães, V.; Bobos, I.; Rodríguez-Castellón, E. The application of functionalized pillared porous phosphate heterostructures for the removal of textile dyes from wastewater. *Materials* **2017**, *10*, 1111. [CrossRef]
2. Anjaneyulu, Y.; Chary, N.S.; Raj, D.S.S. Decolourization of industrial effluents—Available methods and emerging technologies—A review. *Rev. Environ. Sci. Bio/Technol.* **2005**, *4*, 245–273. [CrossRef]
3. Grégorio, C. Non-conventional low-cost adsorbents for dye removal: A review. *Bioresour. Technol.* **2006**, *97*, 1061–1085.
4. Abd El-Latif, M.M.; Ibrahim, A.M.; El-Kady, M.F. Adsorption equilibrium, kinetics and thermodynamics of methylene blue from aqueous solutions using biopolymer oak sawdust composite. *J. Am. Sci.* **2010**, *6*, 267–283.
5. Kay, G.M. The adsorption of dyestuffs from aqueous solution using activated carbon: Analytical solution for batch adsorption based on external mass transfer. *Chem. Eng. J.* **1983**, *27*, 187–196.

6. Kay, G.M. Analytical solution using a pore diffusion model for a pseudo irreversible isotherm for the adsorption of basic dye on silica. *AIChE. J.* **1984**, *30*, 692–697.
7. Salehi, R.; Arami, M.; Mahmoodi, N.M.; Bahrami, H.; Khorramfar, S. Novel biocompatible composite (Chitosan zinc oxide nanoparticle): Preparation, characterization and dye adsorption properties. *Colloids. Surf. B Biointerfaces* **2010**, *80*, 86–93. [CrossRef]
8. El-Sheikh, A.H.; Newman, A.P.; Al-Daffaee, H.; Phull, S.; Cresswell, N.; York, S. Deposition of anatase on the surface of activated carbon. *Surf. Coat. Technol.* **2004**, *187*, 284–292. [CrossRef]
9. Kashyout, A.H.; Soliman, H.M.A.; Nabil, M.; Bishara, A.A. Fabrication of congo red/oxidized porous silicon (CR/OPS) pH-sensors. *Mater. Sci. Appl.* **2013**, *4*, 79–87. [CrossRef]
10. Pastor, E.; Balaguer, M.; Bychto, L.; Salonen, J.; Lehto, V.P.; Matveeva, E.; Chirvony, V. Porous silicon for photosensitized formation of singlet oxygen in water and in simulated body fluid: Two methods of modification by undecylenic acid. *J. Nanosci. Nanotechnol.* **2008**, *8*, 1–7. [CrossRef]
11. Gidley, D.W.; Peng, H.G.; Vallery, R.S. Positron annihilation as a method to characterize porous materials. *Ann. Rev. Mater. Res.* **2006**, *36*, 49–79. [CrossRef]
12. Eldrup, M.; Lightbody, D.; Sherwood, J.N. The temperature dependence of positron lifetimes in solid pivalic acid. *Chem. Phys.* **1981**, *63*, 51–58. [CrossRef]
13. Tao, S.J.J. Positronium annihilation in molecular substances. *Chem. Phys.* **1972**, *56*, 5499–5510. [CrossRef]
14. Nakanishi, H.; Ujihira, Y. Application of positron annihilation to the characterization of zeolites. *J. Phys. Chem.* **1982**, *86*, 4446–4450. [CrossRef]
15. Ito, K.; Nakanishi, H.; Ujihira, Y. Extension of the equation for the annihilation lifetime of ortho-positronium at a cavity larger than 1 nm in radius. *J. Phys. Chem. B* **1999**, *103*, 4555–4558. [CrossRef]
16. Dutta, D.; Ganguly, B.; Gangopadhyay, D.; Mukherjee, T.; Dutta-Roy, B.J. General trends of positronium pick-off annihilation in molecular substances. *Phys. Condens. Matter.* **2002**, *14*, 7539–7549. [CrossRef]
17. Mackenzie, I.K.; Eady, I.A.; Gingerich, R.R. The interaction between positrons and dislocations in copper and in an aluminum alloy. *Phys. Lett. A* **1970**, *33*, 279–280. [CrossRef]
18. Dutta, D.; Pujari, P.K.; Sudarshan, K.; Sharma, S.K. Effect of confinement on the phase transition of benzene in nanoporous silica: A positron annihilation study. *J. Phys. Chem. C* **2008**, *112*, 19055–19060. [CrossRef]
19. Grafutin, V.I.; Zaluzhnyi, A.G.; Kalugin, V.V.; Ilyukhina, O.V.; Myasishcheva, G.G.; Prokop'ev, E.P.; Timoshenkov, S.P.; Funtikov, Y.V.; Khmelevskii, N.O. On the feasibility of investigation of some defect and porous systems by means of positron annihilation spectroscopy. *High Energy Chem.* **2008**, *42*, 478–484. [CrossRef]
20. Biasini, M.; Ferro, G.; Monge, M.A.; Franci, G.D.; Ferrara, V.L. Study of the structure of porous silicon via positron annihilation experiments. *J. Phys. Condens. Matter.* **2000**, *12*, 5961–5970. [CrossRef]
21. Dannefaer, S.; Kerr, D.; Craigen, D.; Bretagnon, T.; Taliercio, T.; Foucaran, A. A positron annihilation investigation of porous silicon. *J. Appl. Phys.* **1996**, *79*, 9110–9117. [CrossRef]
22. Itoh, Y.; Murakami, H.; Kinoshita, A. Positron annihilation in porous silicon. *Appl. Phys. Lett.* **1993**, *63*, 2798–2799. [CrossRef]
23. Surowiec, Z.; Wiertel, M.; Zaleski, R.; Budzyński, M.; Goworek, J. Positron annihilation study of iron oxide nanoparticles in mesoporous silica MCM-41 template. *Nukleonika* **2010**, *55*, 91–96.
24. Williams, J.F.; Guagliardo, P.; Sudarshan, K.; Ranganathaiah, C.; Koutsantonis, G.; Hondow, N.; Samarin, S. Positron annihilation studies of mesoporous silica MCM-41. *J. Phys Conf. Ser.* **2013**, *443*, 012063–012066. [CrossRef]
25. Ismail, A.M.; Mahmoud, K.R.; Abd-El Salam, M.H. Electrical conductivity and positron annihilation characteristics of ternary silicone rubber/carbon black/TiB_2 nanocomposites. *Polym. Test.* **2015**, *48*, 37–43. [CrossRef]
26. El-meniawi, M.A.H.; Mahmoud, K.R.; Megahed, M. Positron annihilation spectroscopy and mechanical properties studies for epoxy matrices reinforced with different nanoparticles. *J. Polym. Res.* **2016**, *23*, 181–192. [CrossRef]
27. Mahmoud, K.R.; Khodair, A.I.; Shaban, S.Y. Positron annihilation lifetime studies of changes in free volume on some biorelevant nitrogen heterocyclic compounds and their S-glycosylation. *Appl. Radiat. Isotop.* **2015**, *105*, 303–307. [CrossRef] [PubMed]
28. Mahmoud, K.R.; Refat, M.S.; Sharshar, T.; Adam, A.M.A.; Manaaa, E.S.A. Synthesis of amino acid iodine charge transfer complexes in situ methanolic medium: Chemical and physical investigations. *J. Mol. Liq.* **2016**, *222*, 1061–1067. [CrossRef]
29. Shaban, S.Y.; Mahmoud, K.R.; Sharshar, T. Positron annihilation studies of bio-related N_2S_2-tetradentate ligands and their zinc complexes. *Rad. Phys. Chem.* **2013**, *83*, 12–15. [CrossRef]
30. Mahmoud, K.R.; Al-Sigeny, S.; Sharshar, T.; El-Hamshary, H. Positron annihilation study on free volume of amino acid modified, starch-grafted acrylamide copolymer. *Radiat. Phys. Chem.* **2006**, *75*, 590–595. [CrossRef]
31. Nabil, M.; Mahmoud, K.R.; Nomeir, R.; El-Maghraby, E.M.; Motaweh, H.A. 3D Porous Silicon (Nanorods Array, Nanosheets, and Nanoclusters) Production. *Egypt. J. Chem.* **2020**, *63*, 1269–1278. [CrossRef]
32. Kansy, J. Microcomputer program for analysis of positron annihilation lifetime spectra. *Nucl. Instrum. Meth. A* **1996**, *374*, 235–244. [CrossRef]
33. Mc-Gonigle, E.A.; Liggat, J.J.; Pethrick, R.A.; Jenkins, S.D.; Daly, G.H.; Hayward, D. Permeability of N_2, Ar, He, O_2 and CO_2 through biaxially oriented polyester films dependence on free volume. *Polymer* **2001**, *42*, 2413–2426. [CrossRef]
34. Porto, A.O.; Silva, G.G.; Magalhães, W.F. Free volume-size dependence on temperature and average molecular-weight in poly (ethyleneoxide) determined by positron annihilation life time spectroscopy. *J. Polym. Sci.* **1999**, *37*, 219–226. [CrossRef]

35. Jean, Y.C.; Hong, X.; Liu, J.; Huang, C.M.; Cao, H.; Chung, C.Y.; Dai, G.H.; Cheng, K.L.; Hsinjin, Y. High sensitivity of positron annihilation lifetime to time and pressure effects in gas-exposed polymers. *J. Radioanal. Nucl. Chem.* **2005**, *210*, 513–518. [CrossRef]
36. Rafatullah, M.; Sulaiman, O.; Hashim, R.; Ahmad, A. Adsorption of copper (II), chromium (III), nickel (II) and lead (II) ions from aqueous solutions by meranti sawdust. *J. Hazard. Mater.* **2009**, *170*, 969–977. [CrossRef]
37. Nabil, M.; Motaweh, H.A. Enhanced thermal stability of promising nano-porous silicon powder. *Adv. Nanopart.* **2016**, *5*, 199–205. [CrossRef]
38. Kashyout, A.H.; Soliman, H.M.A.; Nabil, M.; Bishara, A.A. Fabrication of nano-porous silicon using alkali etching process. *Mater. Lett.* **2013**, *100*, 184–187. [CrossRef]
39. Dannefaer, S.; Wiebe, C.; Kerr, D. Positron annihilation investigation of porous silicon heat treated to 1000 °C. *J. Appl. Phys.* **1998**, *84*, 6559–6564. [CrossRef]
40. Itoh, Y.; Murakami, H.; Kinoshita, A. Characterization of porous silicon by positron annihilation. *J. Phys. IV Coll.* **1993**, *3*, 193–195. [CrossRef]
41. Idris, M.N.; Ahmad, Z.A.; Ahmad, M.A. Adsorption equilibrium of malachite green dye onto rubber seed coat based activated carbon. *Int. J. Basic Appl. Sci.* **2011**, *11*, 38–43.
42. Mezenner, N.Y.; Bensmaili, A. Biosorption behavior of basic red 46 and violet 3 by dead pleurotus mutilus from single- and multicomponent systems. *Chem. Eng. J.* **2009**, *147*, 87–98. [CrossRef]

materials

MDPI

Review

Improving Gas-Sensing Performance Based on MOS Nanomaterials: A Review

Shirui Xue, Sicheng Cao, Zhaoling Huang *, Daoguo Yang * and Guoqi Zhang

School of Mechanical and Electrical Engineering, Guilin University of Electronic Technology, Guilin 541000, China; shiruixue4268@163.com (S.X.); cao_sicheng@163.com (S.C.); G.Q.Zhang@tudelft.nl (G.Z.)
* Correspondence: zhaoling_huang@guet.edu.cn (Z.H.); d.g.yang@guet.edu.cn (D.Y.); Tel.: +86-0773-2316270 (D.Y.)

Abstract: In order to solve issues of air pollution, to monitor human health, and to promote agricultural production, gas sensors have been used widely. Metal oxide semiconductor (MOS) gas sensors have become an important area of research in the field of gas sensing due to their high sensitivity, quick response time, and short recovery time for NO_2, CO_2, acetone, etc. In our article, we mainly focus on the gas-sensing properties of MOS gas sensors and summarize the methods that are based on the interface effect of MOS materials and micro–nanostructures to improve their performance. These methods include noble metal modification, doping, and core-shell (C-S) nanostructure. Moreover, we also describe the mechanism of these methods to analyze the advantages and disadvantages of energy barrier modulation and electron transfer for gas adsorption. Finally, we put forward a variety of research ideas based on the above methods to improve the gas-sensing properties. Some perspectives for the development of MOS gas sensors are also discussed.

Keywords: MOS gas sensors; gas-sensing properties; improvement methods; gas-sensing mechanism; research ideas

Citation: Xue, S.; Cao, S.; Huang, Z.; Yang, D.; Zhang, G. Improving Gas-Sensing Performance Based on MOS Nanomaterials: A Review. *Materials* **2021**, *14*, 4263. https://doi.org/10.3390/ma14154263

Academic Editor: Avelino Núñez-Delgado

Received: 20 June 2021
Accepted: 19 July 2021
Published: 30 July 2021

1. Introduction

In daily life, gas sensors have been used in various areas, including environmental monitoring, medical diagnosis, and agriculture [1–5]. In 1953, Brattain et al. [6] found the properties of semiconductors were affected by the change in the components of surrounding gases. In 1962, Seiyama et al. [7] manufactured the first metal oxide semiconductor-based gas sensor, which solved the problem of toxic gas adsorption and detection. With the development of advanced manufacturing technology and new materials, high-performance gas sensors based on different principles and structures have been widely developed [8,9]. Multiwalled carbon nanotubes and a graphene gas sensor have been successfully developed and manufactured by Dilonardo et al. [10] and Hayasaka et al. [11]. However, traditional nanomaterials sensitize the adsorption of toxic gases, accompanied by a decline in performance and the generation of by-products. In order to develop a stable and efficient gas sensor, the metal oxide semiconductor (MOS) has attracted researchers' attention due to its excellent properties in gas sensing.

Gas sensors based on MOS materials have many advantages compared to others such as the fast response, low cost, and easy operation [12]. Shendage et al. [13] reported a WO_3 thin-film sensor whose response was about 10 towards 5 ppm NO_2 and about 131.75 towards 100 ppm NO_2. Choi et al. [14] fabricated a SnO_2 nanowire gas sensor. When the NO_2 concentration was 0.5 and 5 ppm, its responses were 18 and 180 in 200 °C, respectively. However, there are some factors limiting its performance. The operation temperature of pristine MOS gas sensors ranges from 150 to 400 °C in general, which can cause high power consumption [8,15]. It is also harmful to the reliability of integrated sensors. Ordered mesoporous materials may be a solution as they can improve selectivity in high-temperature environments [16–18]. Wang et al. [19] synthesized hierarchical Cr-doped WO_3 microspheres and achieved a significant improvement towards H_2S in 80 °C. Some researchers tried to use hierarchical metal oxides and binary metal oxides to solve this problem. Joshi et al. [20] prepared hierarchical $NiCo_2O_4$ structures and improved the response to O_3 gas. Additionally, they also researched the feasibility of binary metal oxides in gas sensing. The yolk-shelled $ZnCo_2O_4$ micro–nanostructure was proved to have a fast response and shorter recovery time to 80 ppb O_3 gas [21].

In order to demonstrate the energy dependence of the dynamical barrier and grasp the key points for MOS gas sensors, we pay great attention to the interfacial properties of gas-sensing materials and nanostructures for enhancing sensitivity, responsivity, and recovery time. Many methods have been developed to enhance the property of MOS gas sensors such as noble metal modification [22], doping [23], and core–shell (C-S) nanostructure [24]. Although these methods have been mentioned in some articles, the concepts and application are introduced briefly [4,15,18]. Differently, we summarize the gas sensing performance of various interface structures based on the sensing mechanism of material interface. Meanwhile, we have classified these methods according to the mechanisms to help readers further understand the types of gas sensors. For example, C-S nanostructure gas sensors could be summarized as heterojunction gas sensors. We summarize these gas-sensing methods and adsorption mechanisms in this review article. In addition, the main properties of MOS gas sensors based on micro–nanomaterials are discussed. Finally, some perspectives for the development of MOS gas sensors are proposed in this article.

2. The Properties of MOS Gas Sensors

The gas-sensing properties of MOS gas sensors are evaluated by the response [15], selectivity [17], and stability [25]. Generally, response represents the ability of gas sensors to detect target gas concentrations [26]. The resistance in air is named Ra, while the resistance exposed to the target gas is named Rg. Ia is the current in the air, and Ig is the current exposed to the target gas [27]. "a" is short for air, and "g" is short for target gas [28]. Response is expressed as the ratio of Ra and Rg, or the change in Ia and Ig [29]. Similarly, the response is described as the change in currents in the target gas to air for FET [30]. Selectivity is the ability of the gas sensors to detect one or more target gases in a mixture of gases [25,31]. Stability is the ability of a gas sensor to reproduce the results for a certain period [32]. Stability is one of the key properties of sensor devices, which is related to whether the device can effectively detect toxic gases over a long time in the detection process. Moreover, there is also recovery time, response time, and LOD (limit of detection, which expresses the smallest concentration of the target gas).

3. The Methods to Improve the Properties

It is essential to improve the properties on account of the extensive research for MOS gas sensors. The methods to improve the properties of materials can be divided into six aspects on the basis of our research. They consist of a change in nanostructure morphology, noble metal decorating, doping, C-S nanostructures, carbon nanomaterials, conducting polymers, 2D metal dichalcogenides, temperature modulating, heating and ultraviolet irradiation (UV irradiation).

3.1. Changing the Morphology of Nanostructures

This method is used to improve response or selectivity of gas sensors by changing the surface-to-volume ratios. The preparation process is usually used the hydrothermal method, CVD, ALD technique, etc. [33–35]. The morphology of nanostructures has been classified into four kinds: zero-dimensional nanostructures [33], one-dimensional nanostructures [36,37], two-dimensional nanostructures [38], and three-dimensional nanostructures [39]. Next, we will focus on several typical types of nanostructures.

3.1.1. Nanoparticles

Nanoparticles have higher surface-to-volume ratios, which is the cause of the deep research of nanoparticles in nanostructures. Li et al. [33] synthesized α-Fe_2O_3 nanoparticles via a hydrothermal reaction and calcination treatment (Figure 1a). The α-Fe_2O_3 nanoparticles could detect H_2S gas whose concentration is 0.05 ppm at 300 °C.

Figure 1. Typical nanostructures. (**a**) Nanoparticles. Adapted from [33] copyright (2015), with permission from Elsevier. (**b**) Nanowires. Adapted from [34] copyright (2008), with permission from Elsevier. (**c**) Nanorods. Adapted from [36] copyright (2014), with permission from Elsevier. (**d**) Nanofibers. Adapted from [37] copyright (2009), with permission from Elsevier. (**e**) Nanosheets. Adapted from [38] copyright (2010), with permission from Taylor & Franics. (**f**) Nanoflowers. Reprinted from [39].

3.1.2. Nanowires

Nanowires are representative of one-dimensional nanostructured materials. Liu et al. [34] used a chemical thermal evaporation method to manufacture Ga_2O_3 nanowire gas sensors (Figure 1b). The experimental results indicated that the response to 5 ppm O_2 was 10 at 300 °C, and the response to 500 ppm CO was 5 at 100 °C.

Networked nanowires are also an effective method to enhance sensing properties. Park et al. [40] succeeded in synthesizing ZnO networked nanowires using thermal oxidation of ZnSe nanowires. Single-crystal ZnO nanowire gas sensors were compared with networked nanowire gas sensors at 300 °C and 10 ppm NO_2; the responsivity of the latter was 237, and that of the former was only 6.5. When the concentration of NO_2 was 10 ppm, the recovery time of multinetworked ZnO nanowire gas sensors was shorter (about 180 s), and that of single-crystal ZnO nanowire gas sensors was 510 s.

Some researchers have used UV irradiation to improve the properties of nanowire gas sensors. A ZnO nanowire gas sensor was synthesized on a plastic substrate to detect ethanol gas by Lin et al. [41]. Under UV irradiation, it detected ethanol gas at 60 °C and achieved the purpose of reducing power consumption. The principle can be explained as follows: UV irradiation provided the power required for oxygen ions to reduce the operating temperature to room temperature (RT). On the one hand, the absence of nooks and crannies in nanowire-based devices contributes to the direct adsorption/desorption of gas molecules from the surface of 1D nanomaterials structures [41]. On the other hand, the bent morphology of nanowires is suitable for manufacturing flexible gas sensors. In one-dimensional nanostructures, nanowires are a research hotspot. This is due to the morphology advantages of nanowires.

3.1.3. Nanorods

Nanorods are typical one-dimensional nanostructures. They usually exhibit the form of a nanorod array. In contrast to networked nanowires, nanorod arrays have a higher longitudinal orientation and better field electron emission properties [42].

Lim et al. [35] reported a vertical ZnO nanorod array on the Nb electrode by a two-step method. First, Al film and Nb films were thermally evaporated on a Si substrate. Then, they fabricated an AAO (deblock copolymers, polycarbonates, and anodic aluminum oxides) nanotemplate with several vertical pores using the chemical etching method. Finally, they used ALD techniques (atomic layer deposition techniques) to deposit ZnO film and finish the vertical ZnO nanorod array (Figure 2). It had a higher response to H2 at 350 °C. The response to 5 ppm H2 was 21 and to 500 ppm was 162 at 350 °C.

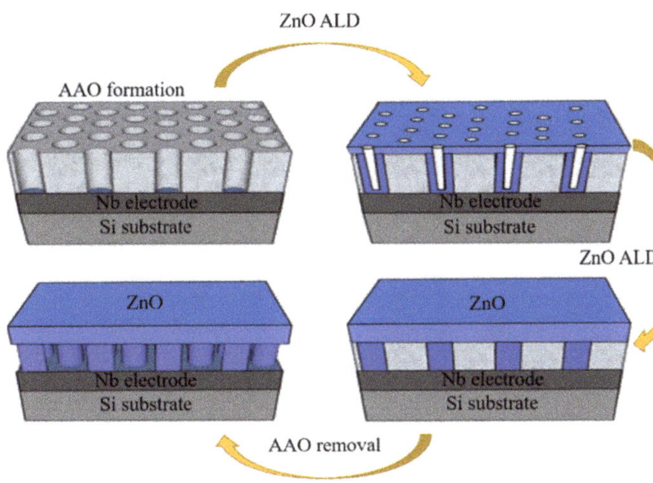

Figure 2. Schematic diagram of ZnO array fabricated by ALD technology. Reprinted from [35] copyright (2013), with permission from Elsevier.

Aside from ALD techniques, Zhang et al. [36] used a ZnO nanorod array fabricated by post-annealing treatment to realize the detection of H_2 (Figure 1c). At 425 °C, the gas sensing response was 3.56 corresponding to the H_2 concentration of 25×10^{-6}. They also proved that post-annealing treatment improved the crystal quality and enhanced the H_2 gas sensing properties [36].

3.1.4. Nanofibers

Nanofibers are another group of important one-dimensional nanostructures. Zheng et al. [37] used electrospinning to synthesize In_2O_3 nanofiber gas sensors for ethanol gas. At 300 °C and 10–500 ppm ethanol, the response was fast (1 s) and the recovery time was short (5 s). Their morphology was characterized by SEM and TEM (Figure 1d). The nanofiber structure was beneficial for ethanol molecule conduction and improved the rate at which carriers passed through the barriers [37]. Katoch et al. [43] manufactured SnO_2 and ZnO nanofiber gas sensors by electrospinning. The response of ZnO nanofibers was higher (the response of ZnO nanofibers was 63.8, and the response of SnO_2 nanofibers was 5.9.) in the experiment of detecting up to 10 ppm H_2. The surface metallization of ZnO nanograins induced by H_2 may be the reason for the enhancement of their properties.

3.1.5. Nanosheets

Recently, two-dimensional nanostructures such as nanosheets have entered the view of researchers. Nanosheets can provide more adsorption sites and strong connections allowing more channels for electron transfer [44]. Hexagonal ZnO nanosheets, whose thickness was 17 nm, were synthesized by Guo et al. [45]. At 350 °C, this sensor, synthesized the using hydrothermal method, had a short response time (9 s) and recovery time (11 s). The response to up to 50 ppm formaldehyde gas was 37.8. Jia et al. [38] also used the hydrothermal method to synthesize monodisperse and stable CuO nanosheets (Figure 1e). The response to ethanol was about 3.

3.1.6. Nanoflowers

Nanoflowers are layered, three-dimensional nanostructures that can effectively increase the contact area to enlarge the reaction between the target gas and sensor. This is helpful in promoting the property of gas sensors. Song et al. [39] synthesized a SnO_2 nanoflower gas sensor to detect methanol gas via the hydrothermal method and calcination method (Figure 1f). The response of the sensor to methanol gas with a concentration of 100 ppm was about 58 at 200 °C. The response time and recovery time were 4 and 8 s, respectively, at the same temperature. The sensing mechanism can be summarized as follows: Oxygen or air seized free electrons from it and turned them into oxygen ions, and the electron depletion layer was generated with it. The electron depletion layer caused resistance to rising. When the sensor was exposed to methanol gas, oxygen ions with methanol gas reacted and released electrons into the layered SnO_2 nanoflower. This process reduced the thickness and resistance of the electron depletion layer. The change in resistance usually expresses the responsivity of a sensor. The delamination, adsorption site, and contact area of nanostructures can greatly promote the reaction between oxygen and methanol gas. The significant change in the electron depletion layer was due to the increase in adsorption position and contact area.

In this section, we briefly introduce several nanostructures. Compared with the traditional structure of gas sensors, they can increase the absorption part of gas and the surface-to-volume ratio, so as to improve the performance of gas sensors. However, the thermal stability of special nanostructures is a significant problem. When the characteristic size is on a nanometer scale or smaller, the melting temperature of MOS will decrease [46]. In this case, nanostructures can be deformed or damaged. Moreover, zero-dimensional nanostructures have the largest ratio of surface to volume; the nanostructure stability is the worst owing to the smallest characteristic size. The feature size should be increased appropriately to prevent damage in the application process. The characteristic sizes of two-dimensional and three-dimensional nanostructures are larger than those of zero-dimensional nanostructures. Their structure stability is better than that of zero-dimensional nanostructures. When these nanostructures are close to each other, the adsorption sites could be sheltered owing to their complex morphologies [39,44,45]. Thus, we need to prevent them from forming clusters and hindering the gas adsorption in the manufacturing process.

However, the main limitation of nanoflower gas sensors is stability over their long-term operation of bending and stretching [47]. Thus, we should improve the synthetic technology of nanoflowers and add some other materials to enhance structural stability. In summary, the structural and thermal stability of nanostructures are important in the fabrication of gas sensors. In addition, the nanoflowers' structure may have the most potential for the application of gas sensors due to the huge surface area.

3.2. Noble Metal Decorating

This method mainly depends on the electron sensitization and chemical catalysis of noble metals on the interface of materials [48,49]. Noble metal decorating can validly enhance the responsivity and selectivity of MOS sensors [50–54]. Some noble metal particles also increase recovery time [55–57]. When the surface of the material is decorated with a noble metal, some chemical reactions often occur at the micro level, while the change of resistance structure is observed at the macro level [58–60]. According to these changes, we often classify sensors based on this method as chemical resistance sensors [60,61]. We demonstrate some types of noble metal nanoparticles and some target gases in Table 1. In the next sections, we use the example of NO_2 gas to describe the effect of noble metal nanoparticles [56,59,60].

Table 1. Nanostructures decorated and loaded with noble metal catalysts.

Noble Metal Catalysts	Technique	Content	Material Structure	Operation Temperature (°C)	Target Gas	Gas Concentration (ppm)	Response (Ra/Rg)	Reference
Pd	Decorating	0.5 mol%	SnO_2 Films	150	NO	0.5	542.8	[48]
	Decorating	/	SnO_2 Nanowires	300	H_2	1	8.02	[50]
	Loading	10 wt%	Co_3O_4 Membranes	150	H_2	100	2.95	[51]
	Loading	/	Fe_2O_3 Nanocubes	139	Acetone	100	25.7	[49]
Pt	Decorating	6 %	ZnO Nanosheets	240	CH_4	50	63.45	[52]
	Decorating	0.5 mol%	SnO_2 Films	300	CO	150	406.2	[48]
	Loading	0.5 wt%	WO_3 Mesoporous	125	CO	100	10 ± 1	[53]
	Loading	2 wt%	WO_3 Nanosheets	300	Acetone	1.5	5.1	[54]
Au	Decorating	4 wt%	SnO_2 Nanosheets	260	Ethanol	100	70.2	[55]
	Decorating	10%	VO_2 Nanowires	25	NO_2	5	3.22	[56]
	Decorating	1.5 wt%	SnO_2 Nanoflowers	120	CH_4	100	4.973	[57]
Ag	Decorating	1 wt%	ZnO Nanorods	360	Ethanol	50	21.5	[58]
	Decorating	0.5%	WO_3 Films	200	NO_2	3	12.22	[59]
	Loading	0.5%	WO_3 Mesoporous	75	NO_2	1	44	[60]
Rh	Decorating	2 deposition cycles	WO_3 Films	350	CH_4	5	63.1	[61]

Decorating (loading) with Au and Ag nanoparticles could improve response and selectivity. For example, Liang et al. [56] and Zhang et al. [62] used Au nanoparticles to decorate a VO_2 nanowire sensor and bilayer WO_3 nanoporous thin-film sensor, respectively. This was useful to improve response and selectivity for NO_2 gas. Moreover, decoration with Ag nanoparticles had the same effect. Kamble et al. [59] improved the performance of WO_3 film in sensors by modification with silver nanoparticles. The response speed was increased by 6 times. Xiao et al. [26] manufactured a Ag-In_2O_3 nanosphere sensor. The best response was 58 toward NO_2 gas with a concentration of 10 ppb, while the pristine In_2O_3 nanospheres' response was 25.5 at 120 °C. The selectivity of the sensor to NO_2 was more significant than that to some volatile organic compounds (VOCs). Additionally, this sensor exhibited several responses under different concentrations of NO_2 and proved that decoration with Ag nanoparticles can reduce recovery time.

The performance of MOS gas sensors can be improved by modification with metal materials through electron sensitization and chemical sensitization. The electronic sensitization mechanism improves the response of MOS gas sensors in two ways. On the one hand, when noble metal nanoparticles contact MOS nanomaterial, their Fermi levels will be aligned together [58]. Due to the different work functions, electrons flow in their energy bands, causing their energy bends to be bent. When the Fermi level arrives at a new balance, the depletion region and the Schottky barrier will be created in the interfaces (Figure 3) [56]. They influence the concentration of carriers or improve the mobility of carriers [58,59]. In these circumstances, the baseline resistance changes, and the sensor response is improved. On the other hand, when noble metal nanoparticles were decorated on the surface of the gas sensor nanomaterial, adsorption sites were increased, and the rate of gas adsorption was accelerated. In contrast to the electronic sensitization mechanism, the chemical sensitization mechanism can be defined as the catalysis of noble metals [63]. The activation energy of the reaction between iron oxide and the target gas can be reduced by using gold nanoparticles as a catalyst [64]. The chemical sensitization mechanism is also known as the spillover effect [64]. Furthermore, Pt and Pd nanoparticles can be used to decorate the surface of a MOS to improve the properties. More details can be found in [53] and [65]. To reduce costs, we can decorate with transition metal oxide nanoparticles instead of the noble metal nanoparticles to enhance the properties. Na et al. [66] decorated ZnO nanowires with Co_3O_4 nanoparticles. This improved response and selectivity to NO_2 and C_2H_5OH. Ko et al. [67] synthesized SnO_2 nanowires decorated with V_2O_5 nanoparticles with a better response to NO_2. The mechanism of the transition metal oxide nanoparticles is characterized by the form of heterojunctions. With the help of heterojunctions, the electron depletion layer and the mobility of carriers obtain modulation. Then, baseline resistance can be altered. Based on this, response and selectivity were improved.

In summary, improving the performance of the gas sensors via the electronic sensitization and chemical sensitization of the noble metal is very effective. The sensitization mechanism with the catalytic effects has been pointed out in our article. Furthermore, the impact on the selectivity of gas sensors depends on the type of noble metal. For instance, Au nanoparticles show good selectivity for NO_2 or CO [23,31], while Ag nanoparticles are sensitive to NO_2 or ethanol [26,58]. A noble metal with sensitive materials has the possibility of forming a cluster and hindering gas adsorption. Therefore, we should uniformly disperse the noble metal nanoparticles on the surface of host-sensitive materials [67–69]. Considering the structural stability, the noble metal is more easily destroyed than the host-sensitive nanostructure owing to the size of nanoparticles. Thus, we should pay attention to the preparation process of the noble metal decorating, the operation temperature, and the thermal stability of gas sensors [49,53,56,60].

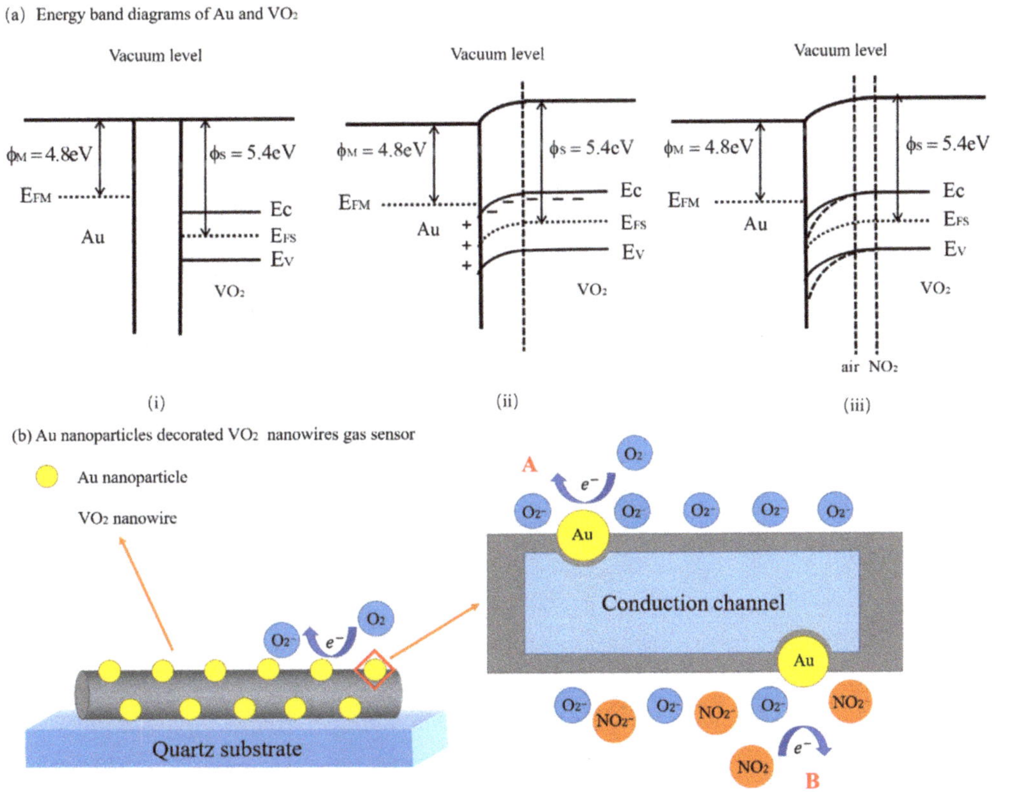

Figure 3. Schematic diagram of the mechanism of Au nanoparticles [56]. Reprinted from [56] copyright (2018), with permission from Elsevier. (**a**) The energy band diagrams of Au and VO_2; (**b**) The gas sensing mechanism of Au nanoparticles.

3.3. Doping

In addition to noble metal decorating, doping can also be used to improve the properties of chemical resistance gas sensors. Some researchers have used doping to increase MOS sensor properties, such as response, response time, and recovery time [27,70]. Metal oxide [27], metal [28,71], nonmetallic elements [70,72], and so on can be used as dopants.

Han et al. [27] reported a self-doped nanocolumnar vanadium oxide gas sensor. Due to the effect of self-doping, response and selectivity were both enhanced to NO_2 gas. Bayata et al. [28] synthesized an Al-doped titania gas sensor. The best response to hydrogen was acquired under 300 °C. The response time and recovery time were shortened to different degrees. Moreover, more adsorption sites occur due to doping. Yu et al. [71] produced a 2% Al-doped ZnO nanovase gas sensor. Compared to the pristine ZnO nanovase gas sensor, the produced sensor had higher response and selectivity to CO, and its response time was shortened. Basu et al. [70] fabricated an F-doped SnO_2 film gas sensor. Its response time and recovery time were shortened to 22 and 52 s, respectively. The mechanism can be described in that dopants can modulate the concentration of carriers or expand the width of the electron depletion layer to change baseline resistance or conductivity.

Compared with noble metal decorating, dopants cannot form a cluster on the surface of host-sensitive nanostructures. Moreover, dopants can not only decrease the activation energy and control the specific exposed facets but also lead to a catalysis effect [69]. However, excessive doping may cause poor electron mobility [72]. Therefore, we need to monitor the number of dopants to avoid the adverse influence of excessive doping.

3.4. Core-Shell (C-S) Nanostructures

3.4.1. Overview

The Definition of C-S Nanostructure

C-S nanostructure is a special nanocomposite and plays an important role in gas sensing [31]. It is usually composed of a core nanomaterial and a shell nanomaterial covering the core. Compared with non-C-S structures, C-S nanostructure provides a way to maximize the interfacial area between two or more materials [73]. In addition, C-S nanostructures can protect the core nanomaterial from the surrounding environment, so as to improve physical and chemical properties [74].

Due to above, C-S nanostructure has been applied in zero-dimensional nanomaterials (nanoparticles [75]), one-dimensional nanomaterials (nanowires [56,76–78], nanorods [79–81], nanofibers [82–85]), two-dimensional nanomaterials (nanosheets [86]), and three-dimensional nanomaterials (microcubes [87]). Figure 4 shows these C-S nanostructures. Because different C-S nanostructures have diverse applications, their material combinations are also different.

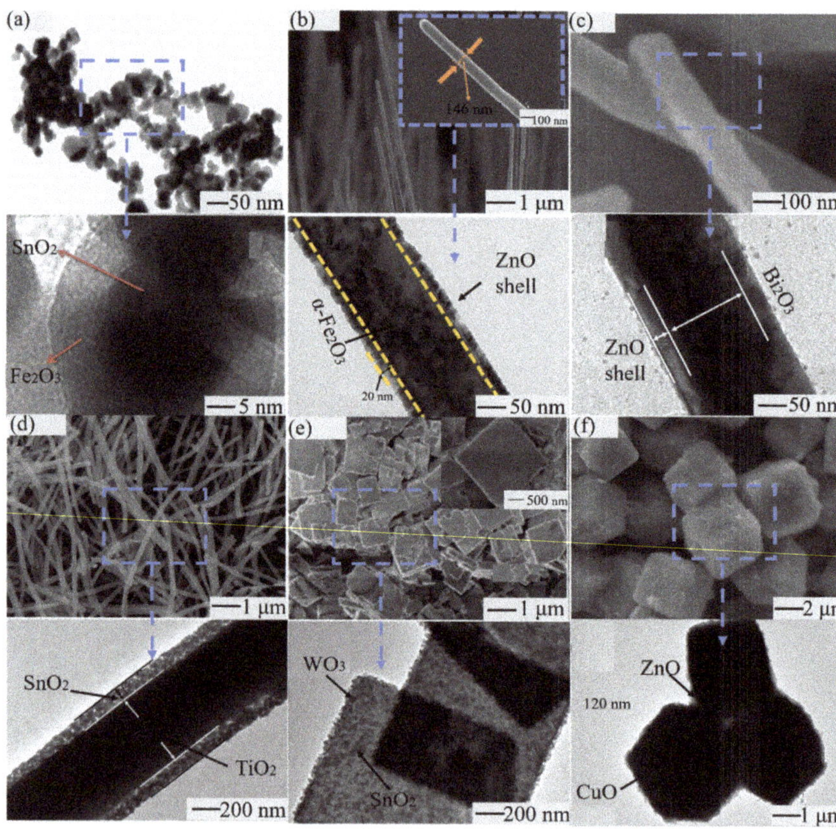

Figure 4. Typical C-S nanostructures. (**a**) C-S nanoparticles. Adapted from [75] copyright (2019), with permission from Elsevier. (**b**) C-S nanowires. Adapted from [76] copyright (2020), with permission from Elsevier. (**c**) C-S nanorods. Adapted from [81] copyright (2017), with permission from Elsevier. (**d**) C-S nanofibers. Adapted from [82] copyright (2017), with permission from Elsevier. (**e**) C-S nanosheets. Adapted from [86] copyright (2018), with permission from Elsevier. (**f**) C-S microcubes. Adapted from [87] copyright (2016), with permission from Elsevier.

The Composition of C-S Nanostructures

The composition of C-S nanostructure materials can be roughly divided into the following categories: noble metal/noble metal [85,88], metal oxide/metal oxide [31,89–91], metal oxide/metal sulfide [92,93], and metal oxide/noble metal (including core shell exchange) [94,95]. Next, we introduce some typical examples.

(1) Metal oxide/metal oxide

Metal oxides are usually applied in metal oxide semiconductors (MOSs) as important functional materials in gas sensing. MOSs are typically divided into n-type MOSs with the electron as the carrier and p-type MOSs with the hole as the carrier. When oxidizing gas acquires electrons contacting an n-type MOS, the concentration of electrons in the n-type MOS will decrease and conductivity will weaken. When reducing gas releases electrons contacting an n-type MOS, the concentration of electrons in the n-type MOS will increase and conductivity will be enhanced. However, the conductivity of p-type MOS is contrary to that of the n-type MOSs, as shown in Table 2. MOS conductivity will have different changes in different gas environments. Thus, we can choose relevant a MOS to form the heterojunction at the interface of C-S nanostructures such as p–n heterojunction [96,97], n–n heterojunction [80], or p–p heterojunction [98]. The mechanism and cases will be demonstrated in the section on applications of the C-S nanostructure.

Table 2. The conductivity of different MOSs in different gas conditions.

Semiconductor Type	Majority Carrier	Target Gas	Conductivity Performance
n-type	Free Electron	Oxidizing Gas	Reduce
		Reducing Gas	Increase
p-type	Hole	Oxidizing Gas	Increase
		Reducing Gas	Reduce

(2) Metal oxide/metal sulfide

The metal oxide/metal sulfide combinations are similar to the metal oxide semiconductor material combinations. Different types of semiconductor (n-type, p-type) materials are used to form the corresponding heterostructure (such as p–p heterojunction) to improve the properties of the sensor. The mechanism and cases are demonstrated in the section on applications of the C-S nanostructure.

(3) Metal oxide/noble metal

The catalytic effect and the carrier mobility are amplified due to the participation of noble metal nanoparticles. The mechanism and cases are demonstrated in the section on applications of the C-S nanostructure.

3.4.2. The Thickness of Shell Layer

In the C-S nanostructure, the types of MOS (p-type and n-type) that constitute the core and shell layer can be chosen according to the type of target gas (oxidizing gas or reducing gas). However, the thickness of the shell layer will affect the response of the sensor.

Kim et al. [77] investigated the effect of the thickness of the shell layer on sensor response under different gas concentrations. First, they measured the dynamic resistance changes of ZnO-SnO_2 C-S nanowire gas sensors with different thicknesses of the SnO_2 shell layer. The dynamic curve of resistance changes showed that the sensor has the properties of an n-type semiconductor for detecting the C_6H_6, C_7H_8, and CO gas with concentrations of 1, 5, and 10 ppm respectively. Then, they explored how the thickness of the SnO_2 shell layer affects the ZnO-SnO_2 C-S nanowire gas sensor's properties. The shell thickness of 40 nm is an important parameter for the gas sensing performance of ZnO-SnO_2 materials, which can be seen in the bell curve in Figure 5. In short, though the mechanisms of all kinds of C-S structures are different, there is an optimum shell thickness where the response will arrive at the peak [77]. The influencing factor of the optimum shell thickness is the Debye length (λD) of the shell layer. When the shell layer's thickness is close to the Debye length (λD) of the shell layer, an ideal response will be acquired. Besides, many articles relate the shell layer's thickness with the Debye length [81,99,100].

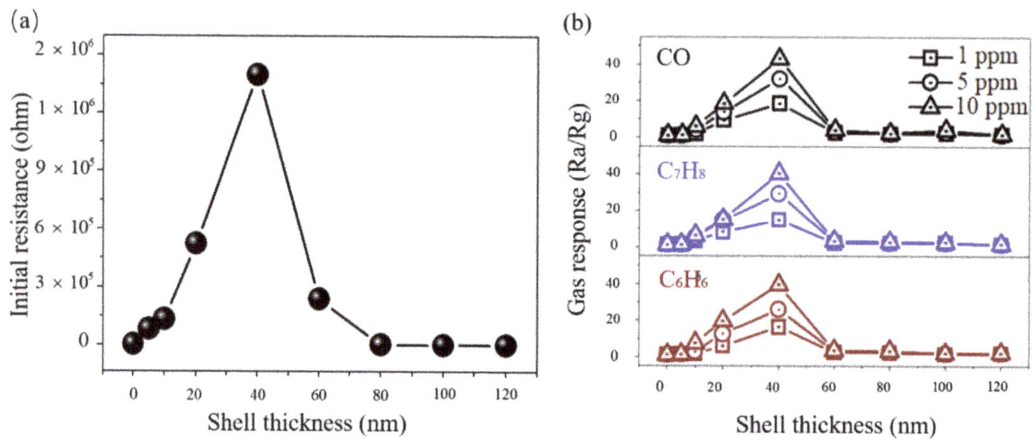

Figure 5. The influence of shell thickness of sensor response. Reprinted from [77] copyright (2020), with permission from Elsevier. (**a**) Relationship between the Initial resistance and the shell thickness; (**b**) The dynamic curve of gas response with shell thickness.

The Debye length, also known as the Debye radius, is a typical length describing the action scale of charge in plasma and is an important parameter of plasma. The mechanism of a C-S nanostructure is relevant to the changes in the electron depletion layer. The change of the electron depletion layer is also related to the Debye length. Taking reducing gas as an example, when reducing gas reacts with adsorbed oxygen ions and the shell material, several electrons will be released to the electron depletion layer. If the shell layer's thickness is smaller or equal to the Debye length, the electron depletion layer can be changed from the whole electron depletion layer to the part electron depletion layer. It will cause a significant change in resistance, allowing a high response to be acquired [77]. If the shell layer thickness is greater than the Debye length, the initial electron depletion layer will not be wholly depleted. When released electrons contact the shell material, the resistance change will not be evident, and the response will not be higher [99]. In short, when the shell layer's thickness is close to the Debye length, a significant response can be acquired.

3.4.3. The Manufacture of C-S Nanostructures

ALD techniques [99,101] and coaxial electrospinning [102–105] are used to produce C-S nanostructures. Atomic layer deposition techniques are abbreviated as ALD techniques. The process is usually composed of several cycles. Every cycle includes precursor pulse, reactant pulse, and purification [31]. Users can change the number of cycles to control the shell layer's thickness. There are numerous advantages to ALD techniques, especially the precise control of the thickness and the uniform coverage ability [101]. Coaxial electrospinning is another method to produce C-S nanostructure gas sensors. The coaxial electrospinning device is mainly composed of a high-voltage DC power supply, liquid supply system, composite nozzle, and collecting plate [103].

The process is as follows: First, the required solution is mixed, and magnetic stirring is used to finish the manufacture of the precursor solution. After that, the precursor solution is put into the electrospinning syringe and then ejected through the composite nozzle. At last, the finished C-S nanostructure is collected [102,104]. The main advantages of coaxial electrospinning are simple synthesis and reliable structure [105]. Besides, there are numerous methods such as the hydrothermal method [106], coprecipitation, and sol–gel processes that can be used to prepare a core–shell nanostructure [75,78].

3.4.4. The Application of C-S Nanostructures

C-S nanostructures have been researched to detect inorganic gas and VOCs (Table 3). To further introduce the application of C-S nanostructures, we will summarize the research progress of C-S nanostructure gas sensors of three types: metal oxide/metal oxide [31,89], metal oxide/metal sulfide [92,93], and metal oxide/noble metal [94,107].

Metal Oxide/Metal Oxide

The p–n heterostructures, which are composed of n-type and p-type MOSs, are widely researched C-S nanostructures in gas sensing.

Liang et al. [108] produced ZnO-$NiCo_2O_4$ C-S nanofibers via a chemical deposition method. Their response time and recovery time towards methanol were 37 and 175 s, respectively. Those of pristine ZnO nanofibers were 123 and 338 s. The responses to methanol of 5, 10, 20, 50, and 100 ppm were 1.96, 3.02, 3.97, 4.88, and 6.77, respectively, which were much higher than those of pristine ZnO at the same concentrations. The improvement of the property was due to the p–n heterojunction and the unique C-S porous structures. Li et al. [89] synthesized a ZnO-Co_3O_4 C-S nanostructure. At the best temperature (200 °C), the response to ethanol (100 ppm) gas was 38.87. Compared with single shell ZnO-Co_3O_4 nanostructure, the best temperature had decreased 40 °C and the response had increased 25.07. Majhi et al. [109] produced a PdO-ZnO C-S nanostructure gas sensor to detect acetaldehyde gas. Due to p–n heterostructure, at 350 °C, the best response was 76 to acetaldehyde gas with 100 ppm, while that of the pristine ZnO nanostructure was 18. Moreover, PdO nanoparticle catalytic behavior was not ignored. That is why the response time was shortened to 20 s. Xu et al. [110] anchored NiO porous nanosheets on α-MoO_3 nanobelts to synthesize α-MoO_3-NiO-2 C-S nanobelts and α-MoO_3-NiO-1 C-S nanobelts. The α-MoO3-NiO-2 and α-MoO_3-NiO-1 are differentiated by the content of anchored NiO. Their responses to acetone were 17.2 times and 16 times greater than those of the pristine α-MoO_3 structure. Compared with pristine NiO structure, α-MoO_3-NiO-1 C-S nanobelts' response was 6.6 times higher. According to their experiment, though pristine NiO nanosheets had much higher surface-to-volume ratios, the improvement of properties depended on the heterostructure. Kim et al. [31] also studied the effect of p–n heterostructure on NO_2. They used ALD techniques to produce SnO_2-Cu_2O C-S nanofibers. When the thickness of Cu_2O was 30 nm and the concentration of NO_2 was 10 ppm, SnO_2-Cu_2O C-S nanofibers' response time and recovery time had been shortened by 137 and 46 s compared to the pristine SnO_2 nanofibers.

Table 3. Summary of some MOS C-S nanostructure applications.

C-S Heterostructure	Shell Deposition Technique	Nanostructure	Operation Temperature (°C)	Target Gas	Gas Concentration (ppm)	Response (Ra/Rg)	Reference
CuO-SnO_2	ALD	Nanowires	250	HCHO	50	2.42	[101]
SnO_2-NiO	ALD	Nanowires	500	H_2	500	114	[97]
α-Fe_2O_3-ZnO	ALD	Nanowires	250	H_2S	5	5.98	[76]
WO_3-SnO_2	ALD	Nanosheets	200	NH_3	15	1.55	[99]
Ag-TiO_2	Sol–gel	Nanowires	240	NH_3	100	9	[78]
SnO_2-Fe_2O_3	Sol–gel	Nanoparticles	Room Temperature	2-methoxyethanol	100	2080	[75]
Au-In_2O_3	Hydrothermal	Nanoparticles	300	H_2	100	34.4	[107]
TiO_2-NiO	Hydrothermal	Nanorods	400	Acetone	200	9.81	[79]
WO_3-SnO_2	Hydrothermal	Nanosheets	260	Acetone	50	32.1	[106]
In_2O_3-SnO_2	Coaxial Electrospinning	Nanofibers	280	TMA	10	7.11	[102]
$ZnO@In_2O_3$	Coaxial Electrospinning	Nanofibers	225	Ethanol	100	31.87	[83]
Co_3O_4-α-Fe_2O_3	Coaxial Electrospinning	Nanofibers	240	Acetone	50	11.7	[103]
ZnO-CeO_2	Coaxial Electrospinning	Nanofibers	370	Acetone	1	8.2	[90]

Besides, n–n heterostructure and p–p heterostructure have also been produced with C-S nanostructures. Jayababu et al. [111] synthesized CeO_2-Fe_2O_3 C-S nanoparticles via the sol–gel method. Their response time and recovery time towards 100 ppm of ethanol at RT were 3 and 7 s. Compared with pristine CeO_2 nanoparticles and pristine Fe_2O_3 nanoparticles, the C-S nanostructure's properties were enhanced. The main cause of enhanced properties is the n–n heterostructure. The catalytic behavior of Fe_2O_3 nanoparticles also helped in the shortening of response/recovery time. In Yin et al.'s work [106], a WO_3-SnO_2 nanosheet was synthesized by the hydrothermal method, and its properties and sensing mechanism were investigated. Taking the WO_3-SnO_2 nanosheet whose particle density of SnO_2 was 0.5% as an example, the working temperature decreased by 80 °C and response time shortened by 3 s. On the one hand, the n–n heterostructure and the modulation of barrier height improve properties. On the other hand, the interface between WO_3 nanosheets and SnO_2 nanoparticles improved, which accelerates the reaction of target gas and sensor [106]. Wan et al. [112] used the hydrothermal method and electrospinning to produce In_2O_3-SnO_2 C-S nanofibers. At 120 °C, their response to formaldehyde gas with a concentration of 100 ppm reached 180.1. Compared with pristine In_2O_3 nanofibers and SnO_2 nanofibers, this represents an increase by 9 times and 5 times, respectively. Diao et al. [90] synthesized ZnO-CeO_2 nanofibers to detect acetone gas. The best operation temperature was 370 °C. At this temperature, the response arriving at the peak was 8.2. Moreover, the nanostructure can detect target gas at a lower concentration. At 0.2 ppm acetone, the response was 3.8. In Wang et al.'s work [98], p–p heterostructure was introduced to detect H_2S. The response of the CuO-NiO C-S microspheres designed by them was 47.6 at 260 °C. It almost was 3 times greater than that of pristine CuO microspheres. The mechanism can be summarized as follows: p–p heterostructure, wrinkles on the surface of the NiO shell layer, and the catalysis of the NiO shell layer.

The C-S nanostructure is a typical heterostructure. The mechanism of this kind of C-S nanostructure gas sensor is mainly attributed to the formation of the electron depletion layer or the hole depletion layer and the modulation of barrier height [101,106,111]. When two nanomaterials contact each other, the heterostructure will form at the interface between them [113]. When p-type MOS and n-type MOS contact with each other, p–n heterostructure, p–p heterostructure, or n–n heterostructure will be formed. Usually, the compositions of the core layer and shell layer are different, so the work functions are different. The work function is determined by the composition of the Fermi level and the electron depletion layer or the hole depletion layer at the interface of the heterostructure [113]. To balance the Fermi Level, when MOS materials contact each other, charges will transfer at their interior. With the transfer of charges, the barrier height will change, and the electron depletion layer or the hole depletion layer will arise on the side of the output electrons.

Generally, the process by which the electron depletion layer or the hole depletion layer and the barrier height affect the response of the sensor is divided into three parts. Taking the PdO-ZnO C-S nanostructure gas sensor made by Majhi et al. [109] as an example, we will demonstrate this process. First, the work function of PdO (7.9 eV) is higher than the work function of ZnO (5.3 eV). As an effect, the electrons in the conduction band of ZnO move to the conduction band of PdO, and the hole moves to ZnO. When the Fermi level of this system is balanced, an electron depletion layer will be formed near ZnO, and a hole depletion layer will be formed near PdO at the interface of the PdO-ZnO heterojunction. This also can lead to band bending (Figure 6). Then, if this sensor is exposed to oxygen or air, the oxygen will be absorbed by the surface of the sensor, and oxygen molecules will change into oxygen atoms. Some of the oxygen atoms will trap electrons from the conduction band of ZnO and form oxygen ions. This process is described by Equations (1) and (2).

$$O_2(\text{gas}) \rightarrow 2O(\text{ads}), \quad (1)$$

$$O(\text{ads}) + e^-(\text{from ZnO}) \rightarrow O^-(\text{ads}), \quad (2)$$

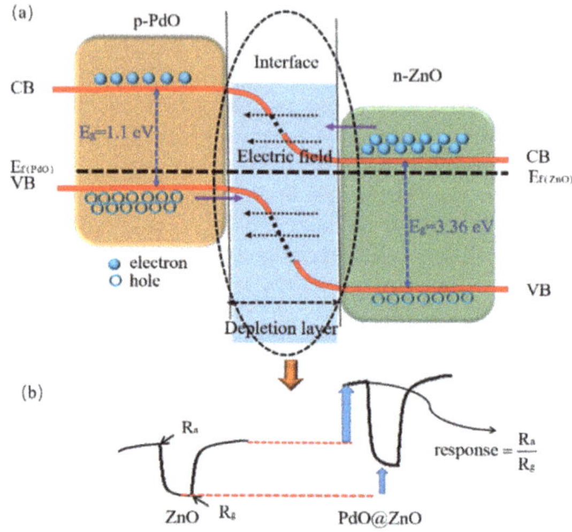

Figure 6. Schematic diagram of the mechanism of core-shell structure. Reprinted from [109]. (**a**) The energy band diagrams of PdO-ZnO heterojunction; (**b**) Response signals of the prisitine ZnO nanostructure and PdO-ZnO heterojunction.

This process will produce a new electron depletion layer at the surface of ZnO nanomaterials. With the effect of oxygen ions, the width of the electron depletion layer at the PdO-ZnO heterojunction could be increased. As the electron depletion layer changes, high potential barriers hinder electron transfer and produce high resistance between ZnO and PdO nanoparticles. In this process, Ra denotes the pristine resistance of this sensor. At last, when the sensor is exposed to acetaldehyde gas, those absorbed oxygen ions will react with acetaldehyde gas and release trapped electrons to the conduction bands of PdO and ZnO. The barriers between ZnO nanoparticles and PdO nanoparticles will have decreased, and the width of electron depletion layers will also have decreased. Owing to the change of electron depletion layers and the modulation of barrier height, resistance will have been reduced. Rg describes the ultimate resistance of this sensor. We can obtain the final response by the radio of Ra and Rg. In this kind of C-S nanostructure gas sensor mechanism, the depletion layers and barrier height have a variety of different changes in oxygen or air. Table 4 summarizes the change of depletion layer thickness for oxidizing gas or reducing gas.

Table 4. Variation of depletion layer thickness in oxidizing gas and reducing gas.

Heterojunction Type	Main Carrier	Target Gas	Main Depletion Layer Types	Layer Thickness
p–n Type	Free Electrons and Holes	Oxidizing Gas	Electron Depletion Layer	Increase
			Hole Depletion Layer	Reduce
		Reducing Gas	Electron Depletion Layer	Reduce
			Hole Depletion Layer	Increase
n–n Type	Free Electrons	Oxidizing Gas	Electron Depletion Layer	Increase
		Reducing Gas		Reduce
p–p Type	Holes	Oxidizing Gas	Hole Depletion Layer	Reduce
		Reducing Gas		Increase

Metal Oxide/Metal Sulfide

ZnO is one of the popular materials in gas sensing. Pristine ZnO displays n-type properties because of oxygen vacancies and Zn interstitial atoms. However, researchers doping other elements such as nitrogen change n-type ZnO to p-type ZnO. Aside from metal oxide, parts of metal sulfide can be applied to gas sensing.

Chang et al. [92] used p-type ZnO and MoS_2 to produce a p-ZnO-MoS_2 C-S nanosheet gas sensor by the hydrothermal method. Firstly, they tested this sensor's response to 500 ppb acetone gas at 350 °C. It was 18 times that of pristine p-ZnO. Secondly, in a low concentration acetone experiment (100 ppb), it had a near 80 times increase in response compared to pristine p-ZnO and pristine MoS_2. Additionally, the recovery time and response time were shortened. They succeeded in detecting ultra-low concentration acetone gas. This sensor not only detected 5 ppb acetone but also possessed a fast response time and recovery time (60 and 40 s). Its mechanism is based on the change of depletion layers. When exposed to air, oxygen will trap electrons from the surface of p-type ZnO. In this process, the hole concentration will increase, and the hole depletion layer will be reduced at the interface of p-ZnO-MoS_2 heterojunction. This decreases the pristine resistance (Ra). If the acetone gas contacts it, oxygen ions will react with acetone gas and release electrons to the conduction band. With the generation of electrons, the hole concentration will decrease, and the hole depletion layer will expand. The sensor's ultimate resistance (Rg) will increase. According to Chang et al. [92], the response is described by Equation (3).

$$\text{Response} = (\text{Rg} - \text{Ra})/\text{Ra} \times 100\% \tag{3}$$

In their latest article, UV irradiation was introduced to hollow p-ZnO-MoS_2 C-S nanosheets [93]. Compared with hollow p-ZnO-MoS_2 C-S nanosheets, UV-irradiated hollow p-ZnO-MoS_2 C-S nanosheets' response to 20 ppm acetone had increased 2.32 times at 100 °C. The UV-irradiation mechanism can be summarized as follows: it can induce more electron–hole pairs, which is helpful to oxygen adsorption and the rate of oxygen reacting with electrons.

Metal Oxide/Noble Metal

In addition to the combination of the two types mentioned earlier, the metal oxide–noble material combination is also promising for core-shell nanostructures.

Majhi et al. [94] synthesized Au-NiO C-S nanoparticles by wet chemical methods at 85 °C. Compared with pristine NiO nanoparticles, operation temperature dropped from 300 to 200 °C, the response to 100 ppm ethanol increased from 1.68 to 2.54 at 200 °C, the response time decreased from 400 to 250 s, and the recovery time decreased from 540 to 420 s. The sensing mechanism can be attributed to the formation of the Schottky junction and the catalysis of metal particles. Yang et al. [78] prepared a kind of Ag-TiO_2 C-S nanowires. The response of these nanowires to ammonia was 200 at 240 °C. Compared with pristine TiO_2 nanowires, the operation temperature decreased by 20 °C. Besides, the response time of pristine TiO_2 nanowires in ammonia with concentrations of 20, 50, 100, 300, and 500 ppm was 29, 30, 31, 33, and 35 s, respectively. Due to the catalysis of Au particles and the formation of the Schottky barrier, the response of this structure can be effectively shortened to 26, 28, 27, 28, and 30 s, respectively. Zhao et al. [95] prepared a NO_2 sensor based on Au-WO_3 C-S nanospheres. The response of the nanosphere to 5 ppm NO_2 was 136 at 100 °C, which is 5 times the response of pristine WO_3 nanospheres under the same conditions. The response time of the structure is 4 s, while that of the pristine WO_3 nanosphere is 218 s, and their recovery times are 59 and 2649 s. Moreover, they verified the sensor could maintain good NO_2 sensing performance at high humidity of 75% RH.

This combination is different from the previous methods, and the sensing mechanism is mainly attributed to the Schottky junction and the catalytic behavior of metal particles. We take the Ag-TiO_2 C-S nanowire gas sensor designed by Yang et al. [78] as an example to summarize. First of all, it will create the Schottky junction on the contact interface because of the existence of Ag nanowires when Ag nanowires contact the TiO_2 shell layer.

Because the work function (4.3 eV) of TiO_2 is smaller than that of Ag (4.6 eV), free electrons begin to flow. The electron depletion layer (side of the TiO_2 shell) is generated on the contact interface between Ag nanowires and the TiO_2 shell layer. Then, when the sensor is exposed to oxygen, the ionized oxygen ions capture electrons from the TiO_2 layer and form adsorbed oxygen. During the process, the Ag nanowires continuously provide electrons to the surface of TiO_2. The electron depletion layer continues to widen, and the Schottky barrier height increases, leading to an improvement in the resistance of the sensor in air (Ra). When the sensor is in contact with the reducing gas NH_3, the ionized oxygen ions react with NH_3. The trapped electrons are released from the surface of TiO_2 to the Ag nanowires. Therefore, the widened electron depletion layer is gradually reduced, and the Schottky barrier height decreases. Finally, the resistance of the sensor in ammonia (Rg) decreases. The response is described by Equation (1). Besides the Schottky junction, the chemical catalytic behavior of Ag nanoparticles is also an important factor in enhancing response. Ag nanoparticles can reduce the reaction barrier between the target gas and oxygen ions and promote the surface reaction. The chemical catalytic behavior of noble metals has been discussed in detail in Section 3.2.

3.4.5. C-S Nanostructure and Noble Metal Decorating/Doping

Ju et al. [114] synthesized Au-Loaded ZnO-SnO_2 C-S nanorods by pulsed laser deposition (PLD) and DC sputtering. The response to 50 ppm of triethylamine (TEA) is about 12.4 and the response time is 1.2 s at 40 °C. The performance of core-shell nanorods is much better than that of pristine ZnO nanorods. The improvement of properties mainly depends on the Schottky junction between the Au nanoparticles and SnO_2 shell and the n–n heterojunction between the ZnO layer and SnO_2 layer. Kim et al. [1] loaded Au nanoparticles on the SnO_2-ZnO C-S nanowires. When the working voltage was 5 V, the response to 0.1, 1, 10, and 50 ppm CO was about 1, 1, 1.16, and 1.25. When the working voltage was 20 V, the response reached about 1, 1.25, 1.40, and 1.62. Besides, they also verified that the formation of the Schottky barrier and the catalysis of Au nanoparticles can enhance the response to CO. Gong et al. [115] used heterogeneous precipitation and sintering treatment to prepare Ga-doped (1 mol%) Pt-ZnO C-S nanoparticles. Compared with the Pt-ZnO C-S nanoparticles without Ga doping, the Ga-doped C-S nanostructure achieved the ability to detect acetone at 10 ppb and 20 ppb. The response of this structure to 1 ppm acetone increased from 2.4 to 13.8, and the optimum operating temperature dropped to 275 °C from 300 °C. Bonyani et al. [81] decorated Bi_2O_3-ZnO C-S nanorods with Pd nanoparticles. The special structure of the sensor effectively shortened the response and recovery times. The response to 200 ppm benzene was 28 at 300 °C, which is higher than that for the pristine Bi_2O_3 (1.7) and pristine ZnO (6.8) nanorods. It was found that decorating (loading) or doping the C-S nanostructure with noble metal can improve the performance of the sensor effectively. Kim et al. [116] decorated SnO_2-ZnO C-S nanowires with CuO; combined with the self-heating effect, this could also improve the selectivity to H_2S.

The mechanism of this combination is mainly attributed to the synergetic effect including the Schottky junction, heterojunction, and the catalytic behavior of metal and some metal oxide particles. We take the Au-loaded ZnO-SnO_2 C-S nanorod gas sensor to detect triethylamine (TEA) in the experimental work of Ref. [114].

Due to the influence of different work functions, the electron depletion layer is formed on the contact interface of ZnO and SnO_2 (side of the SnO_2 shell) when Au nanoparticles are loaded on the surface of SnO_2-ZnO C-S nanorods. The electrons will flow from the SnO_2 shell to Au nanoparticles, which leads to the Schottky junctions forming on the contact interface of Au and SnO_2. Then, the flow of electrons widens the depletion layer on the SnO_2. Besides, due to the existence of Schottky junctions and heterojunction, the electron depletion layer of the SnO_2 shell further expands and leads to resistance improvement when the sensor is exposed to oxygen. By contrast, the trapped electrons are released from the reaction of TEA and O ions when the sensor is exposed to TEA. The electron depletion layer of the SnO_2 shell layer is reduced, resulting in a decrease in resistance. The sensor response is described by Equation (2), where Ra and Rg are the resistances of the sensors in air and target gas, respectively. Its mechanism is shown in Figure 7. To sum up, the C-S nanostructure can effectively improve the properties of MOS gas sensors.

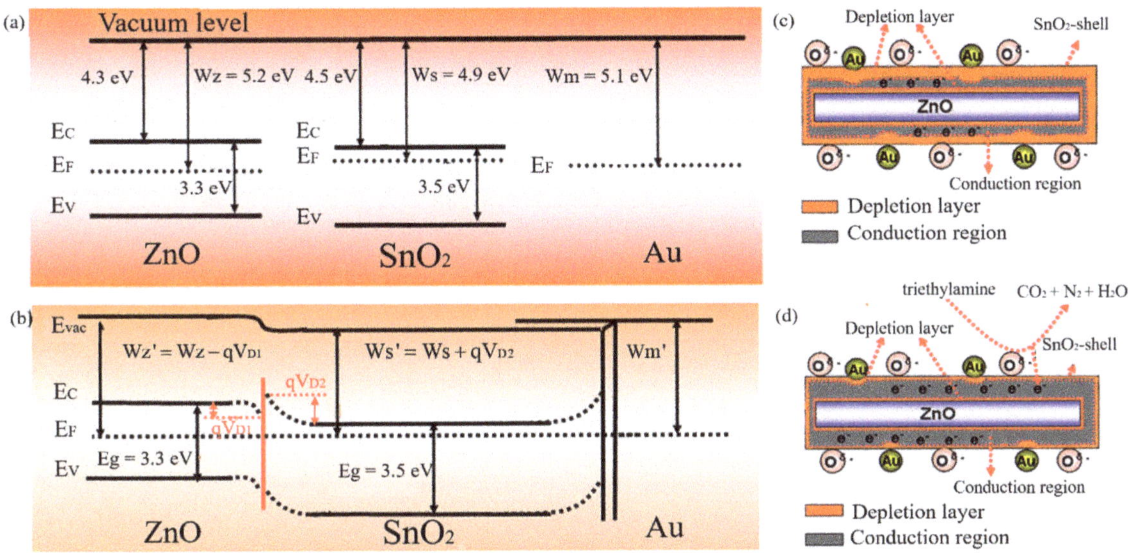

Figure 7. Schematic diagram of the mechanism of interaction between core–shell structure and noble metal nanoparticles. Reprinted from [114] copyright (2015), with permission from American Chemical Society. (**a**) Schematic diagram of energy bands for ZnO, SnO_2, and Au; (**b**) Schematic diagram of energy bands for Au-SnO_2/ZnO heterojunction; (**c**) Schematic diagram of Au-SnO_2/ZnO sensor exposed to air; (**d**) Schematic diagram of Au-SnO_2/ZnO sensor exposed to TEA.

To sum up, the C-S nanostructure is composed of the host-sensitive nanomaterials (core layer nanostructure) and the external-sensitive nanomaterials (shell layer nanostructure). Generally speaking, the thickness of the external-sensitive layer affects C-S nanostructure properties, owing to the Debye length. Because the sensing mechanism of C-S nanostructure usually involves the heterojunction, we can roughly classify C-S nanostructure gas sensors according to the different materials of the heterojunction, namely p–n heterojunction, n–n heterojunction, and p–p heterojunction gas sensors. For MOS/MOS and MOS/metal sulfide C-S nanostructure sensors, the sensing mechanism can be summarized as the formation of heterojunction and the modulation of barrier height. As for MOS/noble metal C-S nanostructure, the mechanism also includes the catalytic behavior of metal particles. More details about the catalytic behavior of metal particles can be found in Section 3.2. Obvious advantages of C-S nanostructure gas sensor can be summarized as follows: (1) higher response due to the synergistic effect of multiple sensing mechanisms [78,79,101,106]; (2) reducing the interference of unnecessary other gases [117]; (3) protecting the host-sensitive nanostructure [87]; (4) making full use of MOS advantages. For example, the α-MoO_3-NiO-2 C-S nanobelt gas sensor manufactured by Xu et al. displayed a high response to acetone gas and good thermal stability [107]. This is due to the p-type MOS having better thermal stability and the n-type MOS having better carrier mobility. Moreover, the critical points to consider in achieving better properties in C-S nanostructures are mainly as follows: (1) the modulation of energy barrier; (2) the catalytic behavior of other materials; (3) the mechanism of carrier mobility [68]. However, the synthesis process for C-S nanostructures is more tedious and the cost is more expensive [82]. Therefore, we should research new preparation processes to solve these problems. Besides, the external additives influence the adsorption capacity and chemical reactivity of the host-sensitive material's surface [118], which has often been disregarded in gas sensor studies. So, when we select suitable materials to make the C-S nanostructure, we should pay attention to the interaction of the C-S interface in the context of density functional theory (DFT), molecular dynamics, or other theories.

3.5. Carbon Nanomaterials

In recent years, carbon nanomaterials have been widely used in the field of gas sensing due to their excellent conductivity and mechanical and thermal properties. Among the most representative are graphene and its derivatives, which have a high surface-to-volume ratio and active functional groups on the surface [119–121], and carbon nanotubes (CNTs) and their products with excellent electrical properties and high flexibility [122–124].

There is often a limitation due to poor selectivity and high operation temperature when using conventional metal oxide semiconductor sensors. Due to its unique properties, graphene can effectively improve the selectivity and carrier mobility of the composites with the synergy of metal oxide. Wang et al. [125] synthesized a kind of nanocomposite with ZnO nanosheets and graphene oxide, effectively increasing the contact area of the target gas as it has a high surface-to-volume ratio and more gas molecular adsorption sites. Moreover, the response–recovery ability was improved compared with pristine ZnO nanosheets by modulating the barrier at the materials' interface. Feng et al. [126] used electrospinning technology to prepare a kind of rGO-encapsulated Co_3O_4 composite nanofiber that can monitor ammonia at room temperature. The response of the composite nanofibers to ammonia gas was significantly higher than that of the composite without rGO, and the response to 50 ppm ammonia was above that of the other interfering gases by 10 times. The improvement of the selectivity of the nanofibers is probably attributed to two reasons: On the one hand, these polarized bonds of rGO and Co^{3+} centers had an even stronger interaction with ammonia that has one lone pair of electrons. On the other hand, the capacity of the pore walls to adsorb different gases is different when gas diffuses in the mesopores of carbon nanofibers. As a derivative of graphene, the defects in the preparation process and electrical properties of the residual oxygen components contributing to the main carrier in rGO are holes. Generally, rGO under ambient conditions exhibits p-type behavior due to the electron-withdrawing nature of defects [127]. Li et al. [128] synthesized rGO-decorated TiO_2 microspheres by the hydrothermal method, and the p–n heterojunction formed between n-type TiO_2 and p-type rGO enhanced the selectivity to ammonia. The heterojunction also suppressed the response to other alcohol gases. They also measured the effect of humidity on the sensor performance in the experiment: the capacity of rGO to adsorb H_2O molecules was stronger than its capacity to adsorb ammonia molecules. With the increase in relative humidity, more water molecules covering the rGO membrane providing electrons led to the decrease in the ammonia recognition ability. The content of carbon nanomaterials in the composite materials had a significant influence on the overall sensing performance [129]. The operation temperature of NO_2 gas was tested by adjusting the content of rGO in the composite [130]. In the composite material, the active adsorption sites increased with the increase in rGO content, which led to the increased response performance of their sensor (Figure 8). Then, the operation temperature gradually dropped to room temperature.

Carbon nanotubes and their products have excellent electrical conductivity and mechanical properties, but they cannot detect specific gases [131]. However, the composites of carbon nanotubes with metal and metal oxide not only inherit the unique properties of carbon nanotubes but also gain the ability to recognize some gases. Schutt et al. [132] obtained a mixed sensing material (ZnO-T-CNT) by attaching carbon nanotubes to the surface of a ZnO-T (tetrapodal ZnO) network, which had a high response to NH_3 (Figure 9). ZnO-T-CNT had a porosity of up to 93%, which greatly enhanced the adsorption and desorption capacity of the mixed materials. Due to CNTs' high conductivity, electrons can effectively transfer from ammonia molecules attached by CNTs to ZnO-T, enhancing the sensing performance of the network effectively. Bhat et al. [133] synthesized a ZnO-MWCNT nanocomposite. Although the sensitivity of the composite was less than that of pure ZnO, the response–recovery property was improved. In addition to the heterojunction interface between ZnO-MWCNT, another possible reason for the property improvement is MWCNTs acting as a catalyst in the experimental process. This leads to the change in the reaction rate by influencing various reaction sites on the surface and accelerates the overall response.

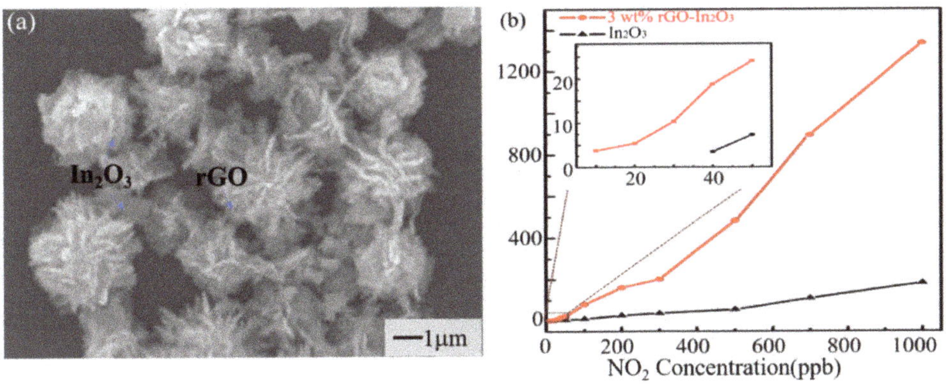

Figure 8. rGO-In_2O_3 gas sensor. Adapted from [130] copyright (2017), with permission from Elsevier. (**a**) The morphology of rGO-In_2O_3 composite nanostructure; (**b**) the response of rGO-In_2O_3 composite nanostructure to NO_2.

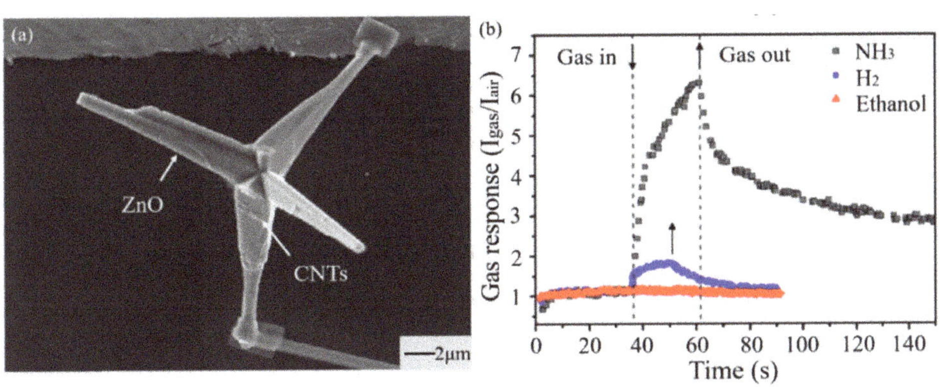

Figure 9. ZnO-T-CNT gas sensor. Adapted from [132] copyright (2017), with permission from American Chemical Society. (**a**) The morphology of CNTs and ZnO-T composite nanostructure; (**b**) the response of CNTs and ZnO-T composite nanostructure to NH_3.

In summary, this method can effectively decrease the operation temperature and increase the selectivity of MOS gas sensors. The mechanism is mainly attributed to the increase in absorption sites and the formation of heterojunctions. More details about heterojunctions can be found in Section 3.4. This is also a factor, in that the carbon nanomaterials could effectively enlarge the channel of carrier transfer and accelerate the transfer of carriers to improve the properties [68]. Moreover, combining carbon nanomaterials with MOS nanomaterials can prove an effective method to improve the electron transfer in the interface. It provides an idea for the fabrication of heterojunction gas sensors. Additionally, owing to the fragile structure of carbon nanomaterials, we should prevent gas sensors from being destroyed in processing.

3.6. Conducting Polymers

Due to conducting polymers' sensing mechanism, their selectivity is higher than that of MOSs in general and the response is lower than that of MOSs [134]. Therefore, some researchers tried to use the conducting polymers and MOSs to overcome the shortcomings of MOS gas sensors. Wang and coworkers [135] synthetized a nanocomposite of polyaniline (PANI)-CeO_2 C-S nanoparticles to detect NH_3 (Figure 10). At room temperature, the response to 65 ppm of NH_3 was 6.5. According to their experiment, this sensor can maintain this response for 15 days. Jiang et al. [136] also manufactured this nanocomposite of SnO_2 and polypyrrole (Ppy) in their research, and it could detect 20 ppb of H_2.

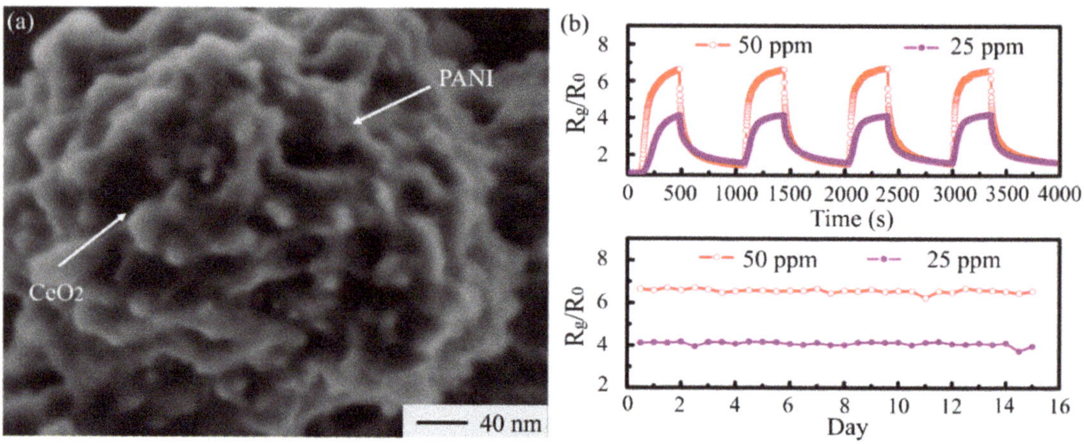

Figure 10. PANI-CeO_2 gas sensor polymers. Adapted from [135] copyright (2014), with permission from American Chemical Society. [135]. (**a**) The morphology of PANI-CeO_2 nanoparticles; (**b**) the response of PANI-CeO_2 nanoparticles to NH_3.

The sensing mechanism is characterized by the effect of heterojunctions and conducting polymers. The incorporation of conducting polymers with MOSs could increase the concentration of carriers and reaction sites, which is beneficial for the target gas adsorption. More details about heterojunction can be found in Section 3.4.

This method not only makes full use of the advantages of heterojunctions but also efficiently decreases the operation temperate of MOS gas sensors [135]. The disadvantage of this method is the high affinity of conductive polymers toward volatile organic compounds (VOCs) and humidity in the atmosphere [68]. It may be more suitable for making heterojunction sensors to detect inorganic gases. MOSs with conductive polymers could cause a response drop, which may increase the zero-drift of gas sensors. Therefore, we tried to use a function correction to decrease the zero-drift.

3.7. 2D Metal Dichalcogenides

Inspired by the appealing properties of graphene, researchers have made great efforts in exploring other 2D nanomaterials for gas sensing such as 2D metal dichalcogenides [17]. Han et al. [137] made a MoS_2-SnO_2 heterostructure gas sensor. It was able to detect NO_2 at room temperature, and the response to 5 ppm of NO_2 was 18.7. Compared with other gases, the selectivity was increased (Figure 11). Furthermore, this sensor had reliable long-term stability. Kim et al. [138] synthetized WS_2-SnO_2 C-S nanosheets by ALD. At the optimum shell thickness, this sensor indicated a good selectivity to CO. $MoSe_2$ was used to manufacture the gas sensor. Abun et al. [139] designed a $MoSe_2$-ZnO heterostructure gas sensor to detect H_2. Compared with pristine ZnO and $MoSe_2$, the selectivity and response were greatly increased.

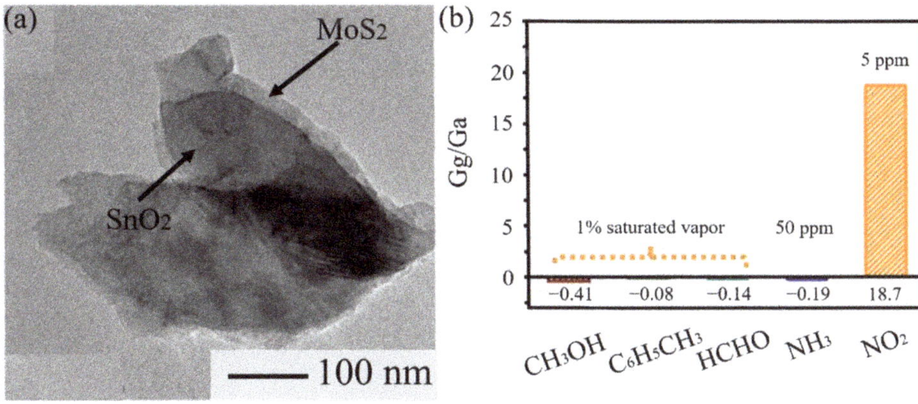

Figure 11. MoS_2-SnO_2 heterostructure gas sensor. Adapted from [137] copyright (2019), with permission from Elsevier. (**a**) The morphology of MoS_2-SnO_2 heterostructure; (**b**) the response of MoS_2-SnO_2 heterostructure to NO_2.

The mechanism is due to the formation of heterojunctions. Under the action of heterojunction, response, selectivity, and operation temperature have been improved in varying degrees [137–139]. More details about heterojunctions can be found in Section 3.4. The combination of one-dimensional MOS material and two-dimensional metal material is a novel method owing to the large surface area and high surface-to-volume ratio. It is helpful to fabricate this type of heterojunction MOS gas sensor to detect inorganic gases. However, it is difficult to fit them firmly [46]. Therefore, it is necessary to improve the preparation process. Moreover, the unique layered structure of 2D metal dichalcogenides has provided the possibility of fabricating a flexible sensor substrate.

3.8. Temperature Modulating

Usually, gas sensors are used to monitor the target gas in a complex gas environment. There is the inevitable problem of cross-sensitivity. It could be understood that sensors have a response to multiple gases at the same time. Therefore, improving the selectivity of gas sensors is essential. Temperature modulating is considered a beneficial method to solve this problem. Yuan et al. [140] synthetized ZnO gas sensors detecting VOCs (volatile organic compounds) via the thermal method. To solve the cross-sensitivity, they used the trapezoidal wave temperature modulation improved by a rectangle to detect target gases and the GRNN to recognize gas species. The rates at which different gases combine with oxygen ions are different, so the optimal temperatures of reactions are different. With the trapezoidal change in temperature, the ZnO sensor showed a good response to every target gas at different temperatures in their experiment. Likewise, Yuan et al. [141] also discussed the feasibility of the application of temperature modulation to improve the rose-like MoO_3/MoS_2/rGO gas sensor selectivity to multiple gases (including acetone, methanol, ethanol, benzene, toluene, and ammonia). When exposed to the condition of temperature changing in the form of a sine wave, it showed the best response to ammonia and the lowest response to acetone. According to different responses to multiple gases, we could ensure it had a good selectivity to ammonia in multiple gases (including acetone, methanol, ethanol, benzene, toluene, and ammonia). Moreover, Krivetskiy et al. [142] proposed successfully using a Temperature modulation combined with statistical shape analysis to modify the SnO_2/Au-SnO_2/Au and Pd-SnO_2 nanocrystalline gas sensor selectivity (Figure 12).

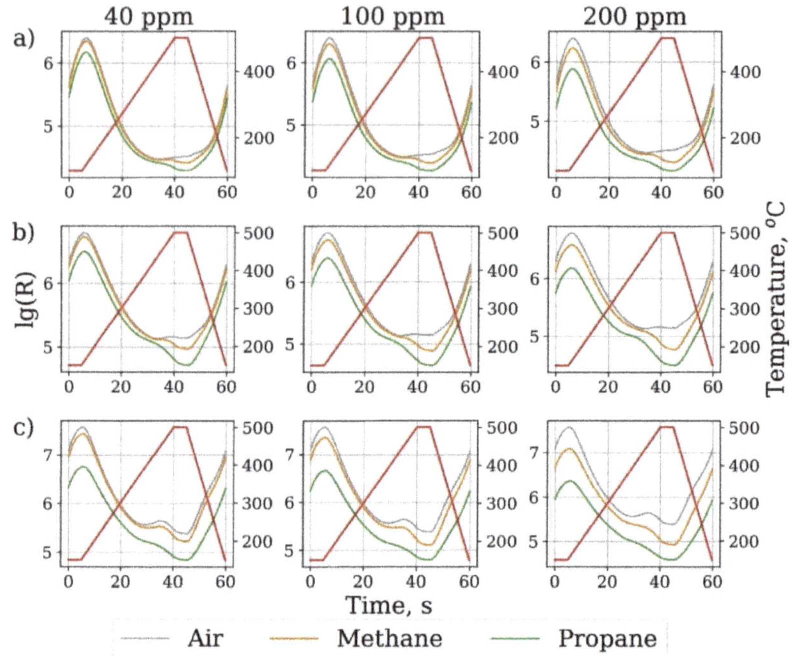

Figure 12. Schematic diagram of temperature modulation. Reprinted from [142] copyright (2021), with permission from Elsevier. (**a**) SnO_2 sensors towards air, methane, and propane at different concentrations during a temperature modulation cycle; (**b**) Au-SnO_2 sensors towards air, methane, and propane at different concentrations during a temperature modulation cycle; (**c**) Au/Pd-SnO_2 sensors towards air, methane, and propane at different concentrations during a temperature modulation cycle.

The mechanism can be summarized as follows: The reaction temperatures of different target gases and oxygen ions are different for a kind of nanomaterial-based gas sensor. When the ambient temperature changes with time for a certain range, each target gas has an optional reaction temperature at a certain point. Then, we can obtain a dynamic response curve, which can reveal the target gases and improve the selectivity of gas sensing (Figure 12). Based on the above, this method may provide a feasible way to detect both inorganic gases and VOCs. The advantage of this method is that it reduces the temperature drift effect by using the response information in static detection and eliminating the interference [141]. The category and concentration of mixed gases are distinguished by the errors in the distinction of adjacent concentrations [140]. Therefore, it is necessary to improve the recognition rate of the algorithm and the accuracy of the data processing algorithm. Moreover, the thermal stability of gas sensors should be improved to avoid their destruction by the temperature cycle.

3.9. Heating

Heating is a method used to promote the property of gas sensors by assembling an extra heater [143–146] or the self-heating effect [147–149] to maintain the temperature condition that the gas reaction needs. An integrated heater is shown in Figure 13. Moon et al. [145] designed a NO_2 gas sensor based on a microheater, which significantly reduced power consumption and recovery time and improved response. Compared with assembling an extra heater, the self-heating effect not only reduces the power consumption of the sensor but also facilitates the fabrication and miniaturization of sensor arrays. Tan et al. [149] designed a gas sensor based on SnO_2 nanowires that can detect a NO_2

concentration from 25.6 to 2.5 ppm, and they proved the feasibility of the self-heating effect (Figure 14). Moreover, they also demonstrated that the application of gas sensors with the self-heating effect could further reduce power consumption by narrowing the size of the sensor in the experiment. In general, the main function of heating is providing the operation temperature of the experiment and reducing humidity impact. Moreover, heating also contributes to the requirement of electron exchange between chemically adsorbed oxygen and MOS and between target gas and active oxygen species on the MOS surface [150]. Based on this information, we can infer that the electron is activated by heating to increase the possibility of electron transition.

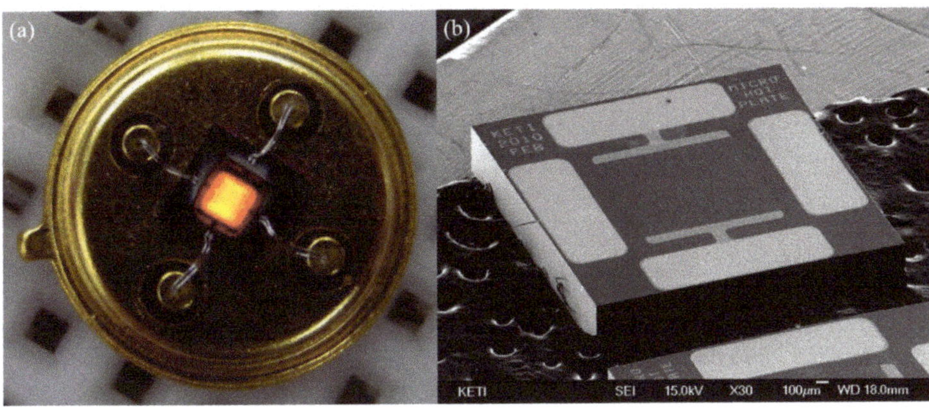

Figure 13. Schematic diagram of external micro-heater. Reprinted from [143]. (**a**) Packaged the micro-heater using TO39 package; (**b**) the SEM image of the micro-heater.

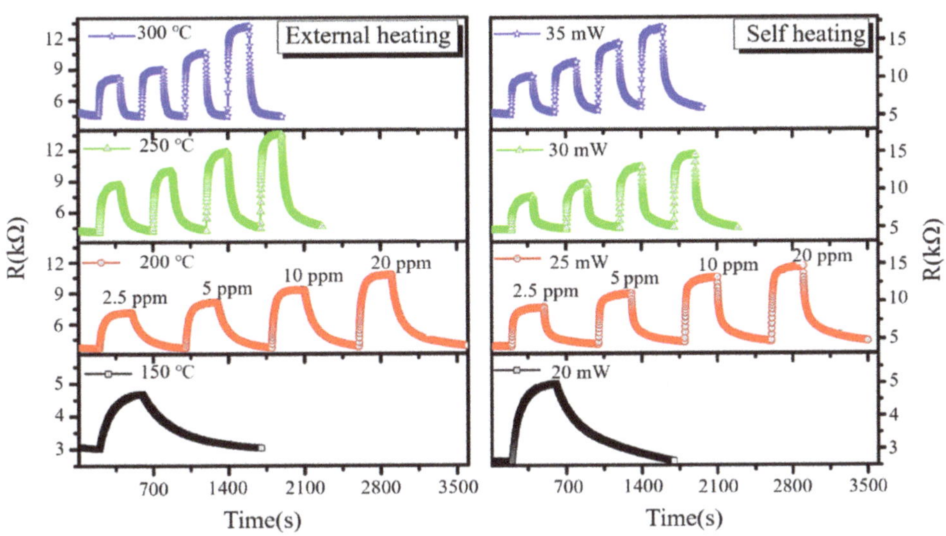

Figure 14. Performance of a sensor under external heating and self-heating effect. Adapted from [149] copyright (2017), with permission from American Chemical Society.

As the gas reaction temperature is several hundred degrees Celsius, heating is an essential method for the improvement of gas sensor properties. However, the higher temperature could cause a decline in MOS gas sensors' reliability. For example, with a reduction in the characteristic size of nanostructures, their melting temperatures decreased significantly [46]. Nanostructures in MOS sensors may be destroyed in a high-temperature environment. This is obviously harmful for their long-term application. Microheaters increase the volume of the whole sensor, and the circuit's thermal consumption is inevitable [144,145]. If we integrate two kinds of gas sensors whose operation temperatures are similar, using an extra heater to provide a suitable temperature condition may be a good choice. Self-heating is helpful for the integration of other components and provides a new idea for wearable gas sensors [147–149]. Compared with extra heaters, the actual heating effect is difficult to control. We must consider the thermal loss by the substrate, the surrounding gases, and the contact pads [146]. Moreover, exploring a suitable MOS material to use in self-heating is also important. In the long term, both an extra heater and self-heating provide a suitable operation temperature to detect gas, which is harmful to the thermal stability of MOS gas sensors. Therefore, UV irradiation is a potential method for MOS gas sensing.

3.10. UV Irradiation

Although ultraviolet irradiation (UV irradiation) can also improve the performance from the perspective of energy, different from temperature modulation and heating, UV irradiation makes more use of photocatalysis, while the former is the change of thermal energy. To replace conventional high-temperature gas sensors, UV-LED can be used to enhance MOS gas sensors [151]. Karaduman et al. [152] manufactured NO_2 gas sensors based on Al-Al_2O_3-p-Si and Al-TiO_2-Al_2O_3-p-Si. The sensor response of two different materials was analyzed in a UV contrast test. The results showed that UV light can significantly improve the selectivity of a TiO_2 sensor, and the response and recovery times were shortened to 6 and 12 s. Fan et al. [153] studied the properties of sensors based on zinc oxide film and nanowires in hydrogen; both sensors were able to detect hydrogen at the ppm level. The response of film and nanowire sensors with a width of 400 nm was increased to 9% and 19%, respectively (as shown in Figure 15). This method is beneficial for improving the response of MOS gas sensors [154]. Under the effect of UV irradiation, MOS material could absorb more energy, generate more charge carriers, and increase the density of free electron–hole pairs. It is helpful in promoting the electrical conductivity and properties of the sensor.

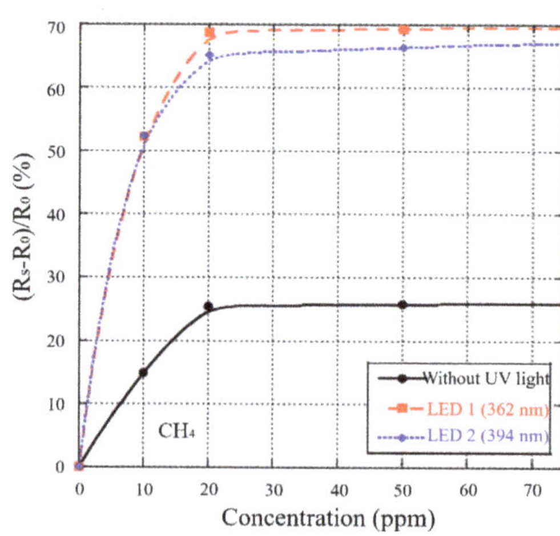

Figure 15. Performance of a gas sensor in the absence and presence of UV irradiation. Reprinted from [154] copyright (2016), with permission from Elsevier.

UV irradiation may be beneficial to achieve room-temperature gas sensing. It can effectively avoid high consumption and decrease thermal deformation by photocatalysis [155]. Moreover, UV irradiation also eliminates the influence of humidity. Water molecules are dissociated by photogenerated electrons and holes under different UV light intensities [156]. However, the disadvantages should not be ignored. For example, conventional UV lamps usually are large and make it difficult to integrate other sensors. The selectivity for reducing gases and VOCs has not obviously increased [155]. This is due to the interference of oxygen and other gases. Therefore, we can try to manufacture UV-LEDs and integrate them with MOS gas sensors to decrease the volume of the whole device, as well as decorating MOS gas sensors with some noble metals (Au, Ag). With the effect of noble metal, the selectivity will be improved. The synergistic effect of noble metal and UV irradiation is helpful to achieve gas sensing at room temperature. According to Trawka et al. [154], these gas sensing results mainly depend on the wavelength of ultraviolet light. Thus, determining how to precisely control the wavelength of ultraviolet light is a significant problem.

4. Conclusions

In this review, we summarized the advantages and disadvantages of MOS nanomaterials and introduced the main gas-sensing properties of MOS gas sensors. Then, we focused on the fabrication methods of interface micro–nanostructures to improve the gas-sensing properties.

MOS nanomaterials have become important nanomaterials in the fabrication of gas sensors owing to their high sensitivity, easy operation, and low cost. We can change the morphology of nanostructures by enlarging the surface-to-volume ratio in order to improve the gas-sensing properties. C-S nanostructures can improve the electron orientation through the formation of heterojunctions and the modulation of energy barriers. Carbon materials, conducting polymers, and 2D metal dichalcogenides improve the gas-sensing properties by forming heterojunctions. The carbon materials and conducting polymers increase the gas adsorption sites, while 2D metal dichalcogenides enlarge the surface-to-volume ratio. Although temperature modulation, self-heating effects, and UV irradiation methods can effectively enhance the MOS gas-sensing properties, the main limitations of the above methods mainly concern two aspects: On the one hand, MOS nanostructures are more fragile than conventional structures. MOS nanostructures are easily destroyed in the preparation of MOS gas sensors. On the other hand, MOS gas sensors can be deformed by thermal stress during the gas reaction.

In summary, thermal stability and structural stability are two important improvement directions for MOS gas sensors in the future. Additionally, UV irradiation provides a novel way to achieve room-temperature gas sensing. Thus, the C-S nanostructure can improve structural stability. These methods could provide some ideas to improve gas-sensing properties and point out development directions for MOS gas sensors. The advances in knowledge in all our endeavors can be a foundation and useful experience for sensing technology, surface science, catalysis, fluidic mechanics, and microelectronics.

Author Contributions: Conceptualization, Z.H. and S.X.; writing—review and editing, S.X.; funding acquisition, Z.H. and D.Y.; supervision, G.Z. and D.Y.; investigation, S.X. and S.C.; writing—original draft preparation, Z.H.; resources, Z.H. and S.C. All authors have read and agreed to the published version of the manuscript.

Funding: This research was funded by the Natural Science Foundation of Guangxi Province (grant number 2017GXNSFDA198006), the Guangxi Postdoctoral Special Foundation (C21RSC90YX03), and the Guangxi Young and Middle-Aged Teachers Promotion Project (2021KY0210).

Institutional Review Board Statement: Not applicable.

Informed Consent Statement: Not applicable.

Data Availability Statement: No new data were created in this study.

Conflicts of Interest: The authors declare no conflict of interest.

References

1. Kim, J.-H.; Mirzaei, A.; Kim, H.W.; Kim, S.S. Low power-consumption CO gas sensors based on Au-functionalized SnO_2-ZnO core-shell nanowires. *Sens. Actuators B* **2018**, *267*, 597–607. [CrossRef]
2. Feng, S.; Farha, F.; Li, Q.; Wan, Y.; Xu, Y.; Zhang, T.; Ning, H. Review on Smart Gas Sensing Technology. *Sensors* **2019**, *19*, 3760. [CrossRef]
3. Park, S. Acetone gas detection using TiO_2 nanoparticles functionalized In_2O_3 nanowires for diagnosis of diabetes. *J. Alloy. Compd.* **2017**, *696*, 655–662. [CrossRef]
4. Alrammouz, R.; Podlecki, J.; Abboud, P.; Sorli, B.; Habchi, R. A review on flexible gas sensors: From materials to devices. *Sens. Actuators A* **2018**, *284*, 209–231. [CrossRef]
5. Kim, J.H.; Mirzaei, A.; Kim, H.W.; Kim, S.S. Low-Voltage-Driven Sensors Based on ZnO Nanowires for Room-Temperature Detection of NO_2 and CO Gases. *ACS Appl. Mater. Interfaces* **2019**, *11*, 24172–24183. [CrossRef]
6. Brattain, W.H.; Bardeen, J. Surface Properties of Germanium. *Bell Syst. Tech. J.* **1953**, *32*, 1–41. [CrossRef]
7. Seiyama, T.; Kato, A.; Fukiishi, K.S.; Nagatini, M. A new detector for gaseous components using semiconductive thin films. *Anal. Chem.* **1962**, *34*, 1502–1503. [CrossRef]
8. Nazemi, H.; Joseph, A.; Park, J.; Emadi, A. Advanced Micro- and Nano-Gas Sensor Technology: A Review. *Sensors* **2019**, *19*, 1285. [CrossRef]
9. Huang, Z.L.; Zeng, Q.; Hui, Y.; Alhai, M.E.E.; Qin, S.J.; Wu, T.Z. Fast Polymerization of Polydopamine Based on Titanium Dioxide for High-Performance Flexible Electrodes. *ACS Appl. Mater. Interfaces* **2020**, *12*, 14495–14506. [CrossRef] [PubMed]
10. Dilonardo, E.; Penza, M.; Alvisi, M.; Rossi, R.; Cassano, G.; Di Franco, C.; Palmisano, F.; Torsi, L.; Cioffi, N. Gas sensing properties of MWCNT layers electrochemically decorated with Au and Pd nanoparticles. *Beilstein J. Nanotechnol.* **2017**, *8*, 592–603. [CrossRef]
11. Hayasaka, T.; Kubota, Y.; Liu, Y.; Lin, L. The influences of temperature, humidity, and O_2 on electrical properties of graphene FETs. *Sens. Actuators B* **2019**, *285*, 116–122. [CrossRef]
12. Nikolic, M.V.; Milovanovic, V.; Vasiljevic, Z.Z.; Stamenkovic, Z. Semiconductor Gas Sensors: Materials, Technology, Design, and Application. *Sensors* **2020**, *20*, 6694. [CrossRef] [PubMed]
13. Shendage, S.S.; Patil, V.L.; Vanalakar, S.A.; Patil, S.P.; Harale, N.S.; Bhosale, J.L.; Kim, J.H.; Patil, P.S. Sensitive and selective NO_2 gas sensor based on WO_3 nanoplates. *Sens. Actuators B* **2017**, *240*, 426–433. [CrossRef]
14. Choi, Y.-J.; Hwang, I.-S.; Park, J.-G.; Choi, K.-J.; Park, J.-H.; Lee, J.-H. Novel fabrication of a SnO_2 nanowire gas sensor with high gas response. *Nanotechnology* **2008**, *19*, 095508. [CrossRef]
15. Li, W.; Zhao, D. An overview of the synthesis of ordered mesoporous materials. *Chem. Commun.* **2013**, *49*, 943–946. [CrossRef]
16. Mirzaei, A.; Hashemi, B.; Janghorban, K. α-Fe_2O_3 based nanomaterials as gas sensors. *J. Mater. Sci. Mater. Electron.* **2016**, *27*, 3109–3144. [CrossRef]
17. Liu, X.H.; Ma, T.T.; Pinna, N.; Zhang, J. Two-Dimensional Nanostructured Materials for Gas Sensing. *Adv. Funct. Mater.* **2017**, *27*, 1702168. [CrossRef]
18. Moseley, P.T. Progress in the development of semiconducting metal oxide gas sensors: A review. *Meas. Sci. Technol.* **2017**, *28*, 082001. [CrossRef]
19. Wang, Y.R.; Liu, B.; Xiao, S.H.; Wang, X.H.; Sun, L.M.; Li, H.; Xie, W.Y.; Li, Q.H.; Zhang, Q.; Wang, T.H. Low-Temperature H_2S Detection with Hierarchical Cr-Doped WO_3 Microspheres. *ACS Appl. Mater. Interfaces* **2016**, *8*, 9674–9683. [CrossRef]
20. Joshi, N.; Silva, L.F.D.; Jadhav, H.; M'Peko, J.C.; Torres, B.B.M.; Aguir, K.; Mastelaro, V.R.; Oliveira Jr, O.N. One-step approach for preparing ozone gas sensors based on hierarchical $NiCo_2O_4$ structures. *RSC Adv.* **2016**, *6*, 92655–92662. [CrossRef]
21. Joshi, N.; Silva, L.F.D.; Jadhav, H.S.; Shimizu, F.M.; Suman, P.H.; M'Peko, J.C.; Orlandi, M.O.; Seo, J.G.; Mastelaro, V.R.; Oliveira, O.N., Jr. Yolk-shelled $ZnCo_2O_4$ microspheres: Surface properties and gas sensing application. *Sens. Actuators B* **2018**, *257*, 906–915. [CrossRef]
22. Yousefi, H.R.; Hashemi, B.; Mirzaei, A.; Roshan, H.; Sheikhi, M.H. Effect of Ag on the ZnO nanoparticles properties as an ethanol vapor sensor. *Mater. Sci. Semicond. Process.* **2020**, *117*, 105172. [CrossRef]
23. Manasa, M.V.; Sarala Devi, G.; Prasada Reddy, P.S.; Sreedhar, B. High performance CO_2 gas sensor based on noble metal functionalized semiconductor nanomaterials for health and environmental safety. *Mater. Res. Express.* **2019**, *6*, 125041. [CrossRef]
24. Liu, C.; Kuang, Q.; Xie, Z.; Zheng, L. The effect of noble metal (Au, Pd and Pt) nanoparticles on the gas sensing performance of SnO_2-based sensors: A case study on the {221} high-index faceted SnO_2 octahedra. *Cryst. Eng. Comm.* **2015**, *17*, 6308–6313. [CrossRef]
25. Kim, J.H.; Wu, P.; Kim, H.W.; Kim, S.S. Highly Selective Sensing of CO, C_6H_6, and C_7H_8 Gases by Catalytic Functionalization with Metal Nanoparticles. *ACS Appl. Mater. Interfaces* **2016**, *8*, 7173–7183. [CrossRef] [PubMed]
26. Xiao, B.; Song, S.; Wang, P.; Zhao, Q.; Chuai, M.; Zhang, M. Promoting effects of Ag on In_2O_3 nanospheres of sub-ppb NO_2 detection. *Sens. Actuators B* **2017**, *241*, 489–497. [CrossRef]
27. Han, S.D.; Moon, H.G.; Noh, M.-S.; Pyeon, J.J.; Shim, Y.-S.; Nahm, S.; Kim, J.-S.; Yo, K.S.; Kang, C.-Y. Self-doped nanocolumnar vanadium oxides thin films for highly selective NO_2 gas sensing at low temperature. *Sens. Actuators B* **2017**, *241*, 40–47. [CrossRef]
28. Bayata, F.; Saruhan-Brings, B.; Urgen, M. Hydrogen gas sensing properties of nanoporous Al-doped titania. *Sens. Actuators B* **2014**, *204*, 109–118. [CrossRef]
29. Kim, J.-H.; Mirzaei, A.; Kim, H.W.; Kim, S.S. Extremely sensitive and selective sub-ppm CO detection by the synergistic effect of Au nanoparticles and core–shell nanowires. *Sens. Actuators B* **2017**, *249*, 177–188. [CrossRef]

30. Sharma, B.; Sharma, A.; Kim, J.-S. Recent advances on H_2 sensor technologies based on MOX and FET devices: A review. *Sens. Actuators B* **2018**, *262*, 758–770. [CrossRef]
31. Kim, J.H.; Lee, J.H.; Kim, J.Y.; Mirzaei, A.; Kim, H.W.; Kim, S.S. Enhancement of CO and NO_2 sensing in n-SnO_2-p-Cu_2O core-shell nanofibers by shell optimization. *J. Hazard. Mater.* **2019**, *376*, 68–82. [CrossRef]
32. Dey, A. Semiconductor metal oxide gas sensors: A review. *Mater. Sci. Eng. B Adv.* **2018**, *229*, 206–217. [CrossRef]
33. Li, Z.; Huang, Y.; Zhang, S.; Chen, W.; Kuang, Z.; Ao, D.; Liu, W.; Fu, Y. A fast response & recovery H_2S gas sensor based on alpha-Fe_2O_3 nanoparticles with ppb level detection limit. *J. Hazard. Mater.* **2015**, *300*, 167–174. [PubMed]
34. Liu, Z.; Yamazaki, T.; Shen, Y.; Kikuta, T.; Nakatani, N.; Li, Y. O_2 and CO sensing of Ga_2O_3 multiple nanowire gas sensors. *Sens. Actuators B* **2008**, *129*, 666–670. [CrossRef]
35. Lim, Y.T.; Son, J.Y.; Rhee, J.S. Vertical ZnO nanorod array as an effective hydrogen gas sensor. *Ceram. Int.* **2013**, *39*, 887–890. [CrossRef]
36. Zhang, J.-j.; Guo, E.-j.; Wang, L.-p.; Yue, H.-y.; Cao, G.-j.; Song, L. Effect of annealing treatment on morphologies and gas sensing properties of ZnO nanorods. *Trans. Nonferrous Met. Soc.* **2014**, *24*, 736–742. [CrossRef]
37. Zheng, W.; Lu, X.; Wang, W.; Li, Z.; Zhang, H.; Wang, Y.; Wang, Z.; Wang, C. A highly sensitive and fast-responding sensor based on electrospun In_2O_3 nanofibers. *Sens. Actuators B* **2009**, *142*, 61–65. [CrossRef]
38. Jia, X.; Fan, H.; Yang, W. Hydrothermal Synthesis and Primary Gas Sensing Properties of CuO Nanosheets. *J. Dispers. Sci. Technol.* **2010**, *31*, 866–869. [CrossRef]
39. Song, L.; Lukianov, A.; Butenko, D.; Li, H.; Zhang, J.; Feng, M.; Liu, L.; Chen, D.; Klyui, N.I. Facile Synthesis of Hierarchical Tin Oxide Nanoflowers with Ultra-High Methanol Gas Sensing at Low Working Temperature. *Nanoscale Res. Lett.* **2019**, *14*, 84. [CrossRef]
40. Park, S.; An, S.; Ko, H.; Jin, C.; Lee, C. Synthesis of nanograined ZnO nanowires and their enhanced gas sensing properties. *ACS Appl. Mater. Interfaces* **2012**, *4*, 3650–3656. [CrossRef]
41. Lin, C.-H.; Chang, S.-J.; Hsueh, T.-J. A low-temperature ZnO nanowire ethanol gas sensor prepared on plastic substrate. *Mater. Res. Express.* **2016**, *3*, 095002. [CrossRef]
42. Jung, S.H.; Oh, E.; Lee, K.H.; Park, W.; Jeong, S.H. A Sonochemical Method for Fabricating Aligned ZnO Nanorods. *Adv. Mater.* **2007**, *19*, 749–753. [CrossRef]
43. Katoch, A.; Choi, S.W.; Kim, H.W.; Kim, S.S. Highly sensitive and selective H_2 sensing by ZnO nanofibers and the underlying sensing mechanism. *J. Hazard. Mater.* **2015**, *286*, 229–235. [CrossRef]
44. Choi, P.G.; Izu, N.; Shirahata, N.; Masuda, Y. Fabrication and H_2-Sensing Properties of SnO_2 Nanosheet Gas Sensors. *ACS Omega* **2018**, *3*, 14592–14596. [CrossRef]
45. Guo, W.; Fu, M.; Zhai, C.; Wang, Z. Hydrothermal synthesis and gas-sensing properties of ultrathin hexagonal ZnO nanosheets. *Ceram. Int.* **2014**, *40*, 2295–2298. [CrossRef]
46. Korotcenkov, G. Current Trends in Nanomaterials for Metal Oxide-Based Conductometric Gas Sensors: Advantages and Limitations. Part 1: 1D and 2D Nanostructures. *Nanomaterials* **2020**, *10*, 1392. [CrossRef] [PubMed]
47. Wang, Y.; Duan, L.; Deng, Z.; Liao, J.H. Electrically Transduced Gas Sensors Based on Semiconducting Metal Oxide Nanowires. *Sensors* **2020**, *20*, 6781. [CrossRef]
48. Xie, F.; Li, W.H.; Zhang, Q.Y.; Zhang, S.P. Highly Sensitive and Selective $CO/NO/H_2/NO_2$ Gas Sensors Using Noble Metal (Pt, Pd) Decorated MOx (M = Sn, W) Combined with SiO_2 Membrane. *IEEE Sens. J.* **2019**, *19*, 10674–10679. [CrossRef]
49. Zhang, S.D.; Yange, M.J.; Liang, K.Y.; Turak, A.; Zhang, B.X.; Meng, D.; Wang, C.X.; Qu, F.D.; Cheng, W.L.; Yang, M.H. An acetone gas sensor based on nanosized Pt-loaded Fe_2O_3 nanocubes. *Sens. Actuators B* **2019**, *290*, 59–67. [CrossRef]
50. Kim, J.H.; Mirzaei, A.; Kim, H.W.; Kim, S.S. Improving the hydrogen sensing properties of SnO_2 nanowire-based conduc-tometric sensors by Pd-decoration. *Sens. Actuators B* **2019**, *285*, 358–367. [CrossRef]
51. Akamatsu, T.; Itoh, T.; Izu, N.; Shin, W. Effect of Noble Metal Addition on Co_3O_4-Based Gas Sensors for Selective NO Detection. *Sens. Mater.* **2016**, *28*, 1191–1201.
52. Zhou, Q.; Hong, C.X.; Li, Z.G.; Peng, S.D.; Wu, G.L.; Wang, Q.; Zhang, Q.Y.; Xu, L.N. Facile Hydrothermal Synthesis and Enhanced Methane Sensing Properties of Pt-Decorated ZnO Nanosheets. *J. Nanosci. Nanotechol.* **2018**, *18*, 3335–3340. [CrossRef]
53. Ma, J.H.; Ren, Y.; Zhou, X.R.; Liu, L.L.; Zhu, Y.H.; Cheng, X.W.; Xu, P.C.; Li, X.X.; Deng, Y.H.; Zhao, D.Y. Pt Nanoparticles Sensitized Ordered Mesoporous WO_3 Semiconductor: Gas Sensing Performance and Mechanism Study. *Adv. Funct. Mater.* **2018**, *28*, 1705268.
54. Li, Y.; Hua, Z.Q.; Wu, Y.; Zeng, Y.; Qiu, Z.L.; Tian, X.M.; Wang, M.J. Surface Modification of Pt-loaded WO_3 Nanosheets for Acetone Sensing Application. *Chem. Lett.* **2018**, *47*, 167–170. [CrossRef]
55. Liu, L.; Song, P.; Wei, Q.; Zhong, X.; Yang, Z.X.; Wang, Q. Synthesis of porous SnO_2 hexagon nanosheets loaded with Au na-noparticles for high performance gas sensors. *Mater. Lett.* **2017**, *201*, 211–215. [CrossRef]
56. Liang, J.R.; Zhu, K.L.; Yang, R.; Hu, M. Room temperature NO_2 sensing properties of Au-decorated vanadium oxide nanowires sensor. *Ceram. Int.* **2018**, *44*, 2261–2268. [CrossRef]
57. Xue, D.P.; Zhang, Z.Y.; Wang, Y. Enhanced methane sensing performance of SnO2 nanoflowers based sensors decorated with Au nanoparticles. *Mater. Chem. Phys.* **2019**, *237*, 121864. [CrossRef]

58. Wei, Y.; Wang, X.D.; Yi, G.Y.; Zhou, L.X.; Cao, J.L.; Sun, G.; Chen, Z.H.; Bala, H.; Zhang, Z.Y. Hydrothermal synthesis of Ag modified ZnO nanorods and their enhanced ethanol-sensing properties. *Mater. Sci. Semicond. Process.* **2018**, *75*, 327–333. [CrossRef]
59. Kamble, C.; Panse, M.; Nimbalkar, A. Ag decorated WO_3 sensor for the detection of sub-ppm level NO_2 concentration in air. *Mater. Sci. Semicond. Process.* **2019**, *103*, 104613. [CrossRef]
60. Wang, Y.L.; Cui, X.B.A.; Yang, Q.Y.; Liu, J.; Gao, Y.; Sun, P.; Lu, G.Y. Preparation of Ag-loaded mesoporous WO_3 and its en-hanced NO_2 sensing performance. *Sens. Actuators B* **2016**, *255*, 544–552. [CrossRef]
61. Tan, Y.; Lei, Y. Atomic layer deposition of Rh nanoparticles on WO_3 thin film for CH_4 gas sensing with enhanced detection characteristics. *Ceram. Int.* **2020**, *46*, 9936–9942. [CrossRef]
62. Zhang, H.W.; Wang, Y.Y.; Zhu, X.G.; Li, Y.; Cai, W.P. Bilayer Au nanoparticle-decorated WO_3 porous thin films: On-chip fabrication and enhanced NO_2 gas sensing performances with high selectivity. *Sens. Actuators B* **2019**, *280*, 192–200. [CrossRef]
63. Yuan, Z.Y.; Zhang, J.J.; Meng, F.L.; Li, Y.; Li, R.; Chang, Y.L.; Zhao, J.P.; Han, E.C.; Wang, S.Y. Highly Sensitive Ammonia Sensors Based on Ag-Decorated WO_3 Nanorods. *IEEE Trans. Nanotechnol.* **2018**, *17*, 1252–1258. [CrossRef]
64. Meng, F.L.; Zheng, H.X.; Chang, Y.L.; Zhao, Y.; Li, M.Q.; Wang, C.; Sun, Y.F.; Liu, J.H. One-Step Synthesis of Au/SnO_2/RGO Nanocomposites and Their VOC Sensing Properties. *IEEE Trans. Nanotechnol.* **2018**, *17*, 212–219. [CrossRef]
65. Park, S.; Kim, S.; Lee, S.; Lee, C. Acetone Gas-sensing Properties of Multiple-networked Pd-decorated Bi_2O_3 Nanorod Sensors. *Bull. Korean Chem. Soc.* **2015**, *36*, 468–472.
66. Na, C.W.; Woo, H.S.; Kim, I.D.; Lee, J.H. Selective detection of NO_2, C_2H_5OH using a Co_3O_4-decorated ZnO nanowire network sensor. *Chem. Commun.* **2011**, *47*, 5148–5150. [CrossRef] [PubMed]
67. Ko, W.C.; Kim, K.M.; Kwon, Y.J.; Choi, H.; Park, J.K.; Jeong, Y.K. ALD-assisted synthesis of V_2O_5 nanoislands on SnO_2 nanowires for improving NO_2 sensing performance. *Appl. Surf. Sci.* **2020**, *509*, 144821. [CrossRef]
68. Jian, Y.Y.; Hu, W.W.; Zhao, Z.H.; Cheng, P.F.; Haick, H.; Yao, M.S.; Wu, W.W. Gas Sensors Based on Chemi Resistive Hybrid Functional Nanomaterials. *Nano-Micro Lett.* **2020**, *12*, 71. [CrossRef]
69. Qiu, Z.L.; Hua, Z.Q.; Li, Y.; Wang, M.J.; Huang, D.; Tian, C.; Zhang, C.S.; Tian, X.M.; Li, E.P. Acetone Sensing Properties and Mechanism of Rh-Loaded WO_3 Nanosheets. *Front. Chem.* **2018**, *6*, 385. [CrossRef] [PubMed]
70. Basu, S.; Wang, Y.H.; Ghanshyam, C.; Kapur, P. Fast response time alcohol gas sensor using nanocrystalline F-doped SnO_2 films derived via sol-gel method. *Bull. Mater. Sci.* **2013**, *36*, 521–533. [CrossRef]
71. Yu, L.M.; Fan, X.H.; Cao, L.; Qi, L.J.; Yan, W. Gas sensing enhancement of aluminum-doped ZnO nanovase structure with many gas facile diffusivity paths. *Appl. Surf. Sci.* **2013**, *265*, 108–113.
72. Du, W.; Si, W.; Du, W.; Ouyang, T.; Wang, F.; Gao, M.; Wu, L.; Liu, J.; Qian, Z.; Liu, W. Unraveling the promoted nitrogen dioxide detection performance of N-doped SnO_2 microspheres at low temperature. *J. Alloy. Compd.* **2020**, *834*, 155209. [CrossRef]
73. Miller, D.R.; Akbar, S.A.; Morris, P.A. Nanoscale metal oxide-based heterojunctions for gas sensing: A review. *Sens. Actuators B* **2014**, *204*, 250–272. [CrossRef]
74. Chaudhuri, R.G.; Paria, S. Core/shell nanoparticles: Classes, properties, synthesis mechanisms, characterization, and applications. *Chem. Rev.* **2012**, *112*, 2373–2433. [CrossRef] [PubMed]
75. Jayababu, N.; Poloju, M.; Ramana Reddy, M.V. Facile synthesis of SnO_2-Fe_2O_3 core-shell nanostructures and their 2-methoxyethanol gas sensing characteristics. *J. Alloy. Compd.* **2019**, *780*, 523–533. [CrossRef]
76. Yang, J.-H.; Yuan, K.-P.; Zhu, L.-Y.; Hang, C.-Z.; Li, X.-X.; Tao, J.-J.; Ma, H.-P.; Jiang, A.-Q.; Lu, H.-L. Facile synthesis of α-Fe_2O_3/ZnO core-shell nanowires for enhanced H_2S sensing. *Sens. Actuators B* **2020**, *307*, 127617. [CrossRef]
77. Kim, J.-H.; Mirzaei, A.; Kim, H.W.; Kim, S.S. Variation of shell thickness in ZnO-SnO_2 core-shell nanowires for optimizing sensing behaviors to CO, C_6H_6, and C_7H_8 gases. *Sens. Actuators B* **2020**, *302*, 127150. [CrossRef]
78. Yang, X.; Fu, H.; Zhang, L.; An, X.; Xiong, S.; Jiang, X.; Yu, A. Enhanced gas sensing performance based on the fabrication of polycrystalline Ag@TiO_2 core-shell nanowires. *Sens. Actuators B* **2019**, *286*, 483–492. [CrossRef]
79. Sun, G.-J.; Kheel, H.; Choi, S.; Hyun, S.K.; Lee, C. Prominent Gas Sensing Performance of TiO_2-Core/NiO-Shell Nanorod Sensors. *J. Nanosci. Nanotechol.* **2017**, *17*, 4099–4102. [CrossRef]
80. Park, B.G.; Reddeppa, M.; Kim, Y.H.; Kim, S.G.; Kim, M.D. Hydrogenation-produced In_2O_3/InN core-shell nanorod and its effect on NO_2 gas sensing behavior. *Nanotechnology* **2020**, *31*, 335503. [CrossRef]
81. Bonyani, M.; Lee, J.K.; Sun, G.-J.; Lee, S.; Ko, T.; Lee, C. Benzene sensing properties and sensing mechanism of Pd-decorated Bi_2O_3-core/ZnO-shell nanorods. *Thin Solid Film.* **2017**, *636*, 257–266. [CrossRef]
82. Li, F.; Gao, X.; Wang, R.; Zhang, T.; Lu, G. Study on TiO_2-SnO_2 core-shell heterostructure nanofibers with different work function and its application in gas sensor. *Sens. Actuators B* **2017**, *248*, 812–819. [CrossRef]
83. Huang, B.Y.; Zhang, Z.X.; Zhao, C.H.; Cairang, L.M.; Bai, J.L.; Zhang, Y.X.; Mu, X.M.; Du, J.W.; Wang, H.; Pan, X.J.; et al. Enhanced gas-sensing performance of ZnO@In_2O_3 core@shell nanofibers prepared by coaxial electrospinning. *Sens. Actuators B* **2018**, *255*, 2248–2257. [CrossRef]
84. Li, F.; Zhang, T.; Gao, X.; Wang, R.; Li, B. Coaxial electrospinning heterojunction SnO_2/Au-doped In_2O_3 core-shell nanofibers for acetone gas sensor. *Sens. Actuators B* **2017**, *252*, 822–830. [CrossRef]
85. Katoch, A.; Choi, S.W.; Sun, G.J.; Kim, H.W.; Kim, S.S. Mechanism and prominent enhancement of sensing ability to reducing gases in p/n core-shell nanofiber. *Nanotechnology* **2014**, *25*, 175501. [CrossRef] [PubMed]

86. Kim, J.-H.; Mirzaei, A.; Kim, H.W.; Kim, S.S. Realization of Au-decorated WS_2 nanosheets as low power-consumption and selective gas sensors. *Sens. Actuators B* **2019**, *296*, 126659. [CrossRef]
87. Yin, M.; Wang, F.; Fan, H.; Xu, L.; Liu, S. Heterojunction CuO@ZnO microcubes for superior p-type gas sensor application. *J. Alloy. Compd.* **2016**, *672*, 374–379. [CrossRef]
88. Song, H.; Luo, Z.; Liu, M.; Zhang, G.; Peng, W.; Wang, B.; Zhu, Y. Centrifugal Deposited Au-Pd Core-Shell Nanoparticle Film for Room-Temperature Optical Detection of Hydrogen Gas. *Sensors* **2018**, *18*, 1448. [CrossRef] [PubMed]
89. Li, B.; Liu, J.; Liu, Q.; Chen, R.; Zhang, H.; Yu, J.; Song, D.; Li, J.; Zhang, M.; Wang, J. Core-shell structure of ZnO/Co_3O_4 composites derived from bimetallic-organic frameworks with superior sensing performance for ethanol gas. *Appl. Surf. Sci.* **2019**, *475*, 700–709. [CrossRef]
90. Diao, Q.; Yin, Y.; Zhang, X.; Li, J.; Jiao, M.; Cao, J.; Qin, Q.; Yang, K.; Zhu, G.; Xu, X. Fabrication of $ZnO@CeO_2$ core–shell hetero-structural nanofibers and enhanced gas sensing performance for acetone. *Funct. Mater. Lett.* **2020**, *13*, 2050013. [CrossRef]
91. Diao, K.; Xiao, J.; Zheng, Z.; Cui, X. Enhanced sensing performance and mechanism of CuO nanoparticle-loaded ZnO nanowires: Comparison with ZnO-CuO core-shell nanowires. *Appl. Surf. Sci.* **2018**, *459*, 630–638. [CrossRef]
92. Chang, X.; Li, X.; Qiao, X.; Li, K.; Xiong, Y.; Li, X.; Guo, T.; Zhu, L.; Xue, Q. Metal-organic frameworks derived $ZnO@MoS_2$ nanosheets core/shell heterojunctions for ppb-level acetone detection: Ultra-fast response and recovery. *Sens. Actuators B* **2020**, *304*, 127430. [CrossRef]
93. Chang, X.; Qiao, X.; Li, K.; Wang, P.; Xiong, Y.; Li, X.; Xia, F.; Xue, Q. UV assisted ppb-level acetone detection based on hollow ZnO/MoS_2 nanosheets core/shell heterostructures at low temperature. *Sens. Actuators B* **2020**, *317*, 128208. [CrossRef]
94. Majhi, S.M.; Naik, G.K.; Lee, H.-J.; Song, H.-G.; Lee, C.-R.; Lee, I.-H.; Yu, Y.-T. Au@NiO core-shell nanoparticles as a p-type gas sensor: Novel synthesis, characterization, and their gas sensing properties with sensing mechanism. *Sens. Actuators B* **2018**, *268*, 223–231. [CrossRef]
95. Zhao, S.; Shen, Y.; Zhou, P.; Zhong, X.; Han, C.; Zhao, Q.; Wei, D. Design of $Au@WO_3$ core−shell structured nanospheres for ppb-level NO_2 sensing. *Sens. Actuators B* **2019**, *282*, 917–926. [CrossRef]
96. Nakate, U.T.; Ahmad, R.; Patil, P.; Wang, Y.S.; Bhat, K.S.; Mahmoudi, T.; Yu, Y.T.; Suh, E.K.; Hahn, Y.B. Improved selectivity and low concentration hydrogen gas sensor application of Pd sensitized heterojunction n-ZnO/p-NiO nanostructures. *J. Alloy. Compd.* **2019**, *797*, 456–464. [CrossRef]
97. Raza, M.H.; Kaur, N.; Comini, E.; Pinna, N. Toward Optimized Radial Modulation of the Space-Charge Region in One-Dimensional SnO_2-NiO Core-Shell Nanowires for Hydrogen Sensing. *ACS Appl. Mater. Interfaces* **2020**, *12*, 4594–4606. [CrossRef] [PubMed]
98. Wang, Y.; Qu, F.; Liu, J.; Wang, Y.; Zhou, J.; Ruan, S. Enhanced H_2S sensing characteristics of CuO-NiO core-shell microspheres sensors. *Sens. Actuators B* **2015**, *209*, 515–523. [CrossRef]
99. Yuan, K.P.; Zhu, L.Y.; Yang, J.H.; Hang, C.Z.; Tao, J.J.; Ma, H.P.; Jiang, A.Q.; Zhang, D.W.; Lu, H.L. Precise preparation of $WO_3@SnO_2$ core shell nanosheets for efficient NH_3 gas sensing. *J. Colloid Interf. Sci.* **2020**, *568*, 81–88. [CrossRef]
100. Kim, W.; Baek, M.; Yong, K. Fabrication of ZnO/CdS, ZnO/CdO core/shell nanorod arrays and investigation of their ethanol gas sensing properties. *Sens. Actuators B* **2016**, *223*, 599–605. [CrossRef]
101. Zhu, L.-Y.; Yuan, K.; Yang, J.-G.; Ma, H.-P.; Wang, T.; Ji, X.-M.; Feng, J.-J.; Devi, A.; Lu, H.-L. Fabrication of heterostructured p-CuO/n-SnO_2 core-shell nanowires for enhanced sensitive and selective formaldehyde detection. *Sens. Actuators B* **2019**, *290*, 233–241. [CrossRef]
102. Li, F.; Gao, X.; Wang, R.; Zhang, T.; Lu, G.; Barsan, N. Design of Core-Shell Heterostructure Nanofibers with Different Work Function and Their Sensing Properties to Trimethylamine. *ACS Appl. Mater. Interfaces* **2016**, *8*, 19799–19806. [CrossRef]
103. Cao, J.; Wang, Z.; Wang, R.; Liu, S.; Fei, T.; Wang, L.; Zhang, T. Core–shell Co_3O_4/α-Fe_2O_3 heterostructure nanofibers with enhanced gas sensing properties. *RSC Adv.* **2015**, *5*, 36340–36346. [CrossRef]
104. Zeng, X.; Li, S.; He, Y.; Zhao, B.; Ju, X.; Chen, W.; Lu, B.; Li, H.; Li, Y.; Liu, L.; et al. Gas sensors based on pearl-necklace-shaped In_2O_3 nanotubes with highly enhanced formaldehyde-sensing performance. *J. Mater. Sci. Mater. Electron.* **2019**, *30*, 18362–18373. [CrossRef]
105. Peng, X.; Santulli, A.C.; Sutter, E.; Wong, S.S. Fabrication and enhanced photocatalytic activity of inorganic core–shell nanofibers produced by coaxial electrospinning. *Chem. Sci.* **2012**, *3*, 1262–1272. [CrossRef]
106. Yin, M.; Yao, Y.; Fan, H.; Liu, S. WO_3-SnO_2 nanosheet composites: Hydrothermal synthesis and gas sensing mechanism. *J. Alloy. Compd.* **2018**, *736*, 322–331. [CrossRef]
107. Yu, Y.-T.; Majhi, S.M.; Song, H.-G. Synthesis and Gas Sensing Properties of $Au@In_2O_3$ Core-shell Nanoparticles. *Proc. Eng.* **2016**, *168*, 227–230. [CrossRef]
108. Liang, Y.; Liu, W.; Hu, W.; Zhou, Q.; He, K.; Xu, K.; Yang, Y.; Yu, T.; Yuan, C. Synthesis and gas-sensing properties of $ZnO@NiCo_2O_4$ core@shell nanofibers. *Mater. Res. Bull.* **2019**, *114*, 1–9. [CrossRef]
109. Majhi, S.M.; Lee, H.-J.; Choi, H.-N.; Cho, H.-Y.; Kim, J.-S.; Lee, C.-R.; Yu, Y.-T. Construction of novel hybrid PdO–ZnO p–n heterojunction nanostructures as a high-response sensor for acetaldehyde gas. *Cryst. Eng. Comm.* **2019**, *21*, 5084–5094. [CrossRef]
110. Xu, K.; Duan, S.; Tang, Q.; Zhu, Q.; Zhao, W.; Yu, X.; Yang, Y.; Yu, T.; Yuan, C. P-N heterointerface-determined acetone sensing characteristics of α-MoO_3@NiO core@shell nanobelts. *Cryst. Eng. Comm.* **2019**, *21*, 5834–5844. [CrossRef]
111. Jayababu, N.; Poloju, M.; Shruthi, J.; Reddy, M.V.R. Ultrasensitive resistivity-based ethanol sensor based on the use of CeO_2-Fe_2O_3 core-shell microclusters. *Mikrochim. Acta* **2019**, *186*, 712. [CrossRef] [PubMed]

112. Wan, K.; Wang, D.; Wang, F.; Li, H.; Xu, J.; Wang, X.; Yang, J. Hierarchical In_2O_3@SnO_2 Core-Shell Nanofiber for High Efficiency Formaldehyde Detection. *ACS Appl. Mater. Interfaces* **2019**, *11*, 45214–45225. [CrossRef]
113. Karnati, P.; Akbar, S.; Morris, P.A. Conduction mechanisms in one dimensional core-shell nanostructures for gas sensing: A review. *Sens. Actuators B* **2019**, *295*, 127–143. [CrossRef]
114. Ju, D.X.; Xu, H.Y.; Qiu, Z.W.; Zhang, Z.C.; Xu, Q.; Zhang, J.; Wang, J.Q.; Cao, B.Q. Near Room Temperature, Fast-Response, and Highly Sensitive Triethylamine Sensor Assembled with Au-Loaded ZnO/SnO_2 Core-Shell Nanorods on Flat Alumina Substrates. *ACS Appl. Mater. Interfaces* **2015**, *7*, 19163–19171. [CrossRef] [PubMed]
115. Gong, Y.; Wu, X.; Zhou, X.; Li, X.; Han, N.; Chen, Y. High acetone sensitive and reversible P- to N-type switching NO_2 sensing properties of Pt@Ga-ZnO core-shell nanoparticles. *Sens. Actuators B* **2019**, *289*, 114–123. [CrossRef]
116. Kim, J.-H.; Mirzaei, A.; Bang, J.H.; Kim, H.W.; Kim, S.S. Selective H_2S sensing without external heat by a synergy effect in self-heated CuO-functionalized SnO_2-ZnO core-shell nanowires. *Sens. Actuators B* **2019**, *300*, 126981. [CrossRef]
117. Korotcenkov, G.; Brinzari, V.; Cho, B.K. In_2O_3 and SnO_2-based ozone sensors: Design and characterization. *Crit. Rev. Sol. St. Mater. Sci.* **2017**, *43*, 83–132. [CrossRef]
118. Marikutsa, A.; Rumyantseva, M.; Konstantinova, E.A.; Gaskov, A. The Key Role of Active Sites in the Development of Selective Metal Oxide Sensor Materials. *Sensors* **2021**, *21*, 2554. [CrossRef]
119. Chen, Z.; Wang, J.; Umar, A.; Wang, Y.; Li, H.; Zhou, G. Three-Dimensional Crumpled Graphene-Based Nanosheets with Ultrahigh NO_2 Gas Sensibility. *ACS Appl. Mater. Interfaces* **2017**, *9*, 11819–11827. [CrossRef]
120. Karaduman, I.; Er, E.; Çelikkan, H.; Erk, N.; Acar, S. Room-temperature ammonia gas sensor based on reduced graphene oxide nanocomposites decorated by Ag, Au and Pt nanoparticles. *J. Alloy. Compd.* **2017**, *722*, 569–578. [CrossRef]
121. Lu, X.; Song, X.; Gu, C.; Ren, H.; Sun, Y.; Huang, J. Freeze drying-assisted synthesis of Pt@reduced graphene oxide nanocomposites as excellent hydrogen sensor. *J. Phys. Chem. Solid.* **2018**, *116*, 324–330. [CrossRef]
122. Lupan, O.; Schutt, F.; Postica, V.; Smazna, D.; Mishra, Y.K.; Adelung, R. Sensing performances of pure and hybridized carbon nanotubes-ZnO nanowire networks: A detailed study. *Sci. Rep.* **2017**, *7*, 14715. [CrossRef] [PubMed]
123. Shi, J.; Li, X.; Cheng, H.; Liu, Z.; Zhao, L.; Yang, T.; Dai, Z.; Cheng, Z.; Shi, E.; Yang, L.; et al. Graphene Reinforced Carbon Nanotube Networks for Wearable Strain Sensors. *Adv. Funct. Mater.* **2016**, *26*, 2078–2084. [CrossRef]
124. Duy, N.V.; Hoa, N.D.; Dat, N.T.; Le, D.T.T.; Hieu, N.V. Ammonia-Gas-Sensing Characteristics of WO_3/Carbon Nanotubes Nanocomposites: Effect of Nanotube Content and Sensing Mechanism. *Sci. Adv. Mater.* **2016**, *8*, 524–533.
125. Wang, P.; Wang, D.; Zhang, M.; Zhu, Y.; Xu, Y.; Ma, X.; Wang, X. ZnO nanosheets/graphene oxide nanocomposites for highly effective acetone vapor detection. *Sens. Actuators B* **2016**, *230*, 477–484. [CrossRef]
126. Feng, Q.; Li, X.; Wang, J.; Gaskov, A.M. Reduced graphene oxide (rGO) encapsulated Co_3O_4 composite nanofibers for highly selective ammonia sensors. *Sens. Actuators B* **2016**, *222*, 864–870. [CrossRef]
127. Ye, Z.; Tai, H.; Xie, T.; Yuan, Z.; Liu, C.; Jiang, Y. Room temperature formaldehyde sensor with enhanced performance based on reduced graphene oxide/titanium dioxide. *Sens. Actuators B* **2016**, *223*, 149–156. [CrossRef]
128. Li, X.; Zhao, Y.; Wang, X.; Wang, J.; Gaskov, A.M.; Akbar, S.A. Reduced graphene oxide (rGO) decorated TiO_2 microspheres for selective room-temperature gas sensors. *Sens. Actuators B* **2016**, *230*, 330–336. [CrossRef]
129. Bai, S.; Sun, X.; Han, N.; Shu, X.; Pan, J.; Guo, H.; Liu, S.; Feng, Y.; Luo, R.; Li, D.; et al. rGO modified nanoplate-assembled ZnO/CdO junction for detection of NO_2. *J. Hazard. Mater.* **2020**, *394*, 121832. [CrossRef]
130. Liu, J.; Li, S.; Zhang, B.; Wang, Y.; Gao, Y.; Liang, X.; Wang, Y.; Lu, G. Flower-like In_2O_3 modified by reduced graphene oxide sheets serving as a highly sensitive gas sensor for trace NO_2 detection. *J. Colloid Interf. Sci.* **2017**, *504*, 206–213. [CrossRef]
131. Guo, T.; Zhou, T.; Tan, Q.; Guo, Q.; Lu, F.; Xiong, J. A Room-Temperature CNT/Fe_3O_4 Based Passive Wireless Gas Sensor. *Sensors* **2018**, *18*, 3542. [CrossRef] [PubMed]
132. Schutt, F.; Postica, V.; Adelung, R.; Lupan, O. Single and Networked ZnO-CNT Hybrid Tetrapods for Selective Room-Temperature High-Performance Ammonia Sensors. *Acs Appl. Mater. Interfaces* **2017**, *9*, 23107–23118. [CrossRef] [PubMed]
133. Bhat, P.; Nagaraju, P. Synthesis and characterization of ZnO-MWCNT nanocomposites for 1-butanol sensing application at room temperature. *Phys. B* **2019**, *570*, 139–147. [CrossRef]
134. Zhang, J.; Liu, X.H.; Neri, G.; Pinna, N. Nanostructured Materials for Room-Temperature Gas Sensors. *Adv. Funct. Mater.* **2016**, *28*, 795–831. [CrossRef]
135. Wang, L.L.; Huang, H.; Xiao, S.H.; Cai, D.P.; Liu, Y.; Liu, B.; Wang, D.D.; Wang, C.X.; Li, H.; Wang, Y.R.; et al. Enhanced sensitivity and stability of room-temperature NH_3 sensors using core-shell CeO_2 nanoparticles@cross-linked PANI with p-n heterojunctions. *ACS Appl. Mater. Interfaces* **2014**, *6*, 14131–14140. [CrossRef]
136. Jiang, T.T.; Wang, Z.J.; Li, Z.Y.; Wang, W.; Xu, X.R.; Liu, X.C.; Wang, J.F.; Wang, C. Synergic effect within n-type inorganic–p-type organic nano-hybrids in gas sensors. *J. Mater. Chem. C* **2013**, *1*, 3017–3025. [CrossRef]
137. Han, Y.T.; Ma, Y.J.; Liu, Y.; Xu, S.S.; Chen, X.W.; Zeng, M.; Hu, N.T.; Su, Y.J.; Zhou, Z.H.; Yang, Z. Construction of MoS_2/SnO_2 heterostructures for sensitive NO_2 detection at room temperature. *Appl. Sur. Sci.* **2019**, *493*, 613–619. [CrossRef]
138. Kim, J.H.; Mirzaei, A.; Bang, J.H.; Kim, Y.W.; Kim, S.S. Achievement of self-heated sensing of hazardous gases by WS_2 (core)–SnO_2 (shell) nanosheets. *J. Hazard. Mater.* **2021**, *412*, 125196. [CrossRef]
139. Abun, A.; Huang, B.R.; Saravanan, A.; Kathiravan, D.; Hong, P.D. Exfoliated $MoSe_2$ Nanosheets Doped on the Surface of ZnO Nanorods for Hydrogen Sensing Applications. *Acs Appl. Nano Mater.* **2020**, *3*, 12139–12147. [CrossRef]

140. Yuan, Z.Y.; Han, E.C.; Meng, F.L.; Zuo, K.Y. Detection and Identification of Volatile Organic Compounds Based on Temperature-Modulated ZnO Sensors. *IEEE Trans. Instrum. Meas.* **2020**, *69*, 4533–4544. [CrossRef]
141. Yuan, Z.Y.; Liu, Y.; Zhang, J.J.; Meng, F.L.; Zhang, H. Rose-Like MoO_3/MoS_2/rGO Low-Temperature Ammonia Sensors Based on Multigas Detection Methods. *IEEE Trans. Instrum. Meas.* **2021**, *70*, 1–9.
142. Krivetskiy, V.V.; Andreev, M.D.; Efitorov, A.O.; Gaskov, A.M. Statistical shape analysis pre-processing of temperature modulated metal oxide gas sensor response for machine learning improved selectivity of gases detection in real atmospheric conditions. *Sens. Actuators B* **2021**, *329*, 129187. [CrossRef]
143. Hwang, W.J.; Shin, K.S.; Roh, J.H.; Lee, D.S.; Choa, S.H. Development of Micro-Heaters with Optimized Temperature Compensation Design for Gas Sensors. *Sensors* **2011**, *11*, 2580–2591. [CrossRef]
144. Choi, W.S.; Kim, B.J.; Lee, H.J.; Choi, J.W.; Kim, S.D.; Min, N.K. Study on the micro-heater geometry in In_2O_3 micro electro mechanical systems gas sensor platforms and effects on NO_2 gas detecting performances. *J. Nanosci. Nanotechnol.* **2012**, *12*, 1170–1173. [CrossRef]
145. Moon, S.E.; Lee, H.K.; Choi, N.J.; Lee, J.; Yang, W.S.; Kim, J.; Jong, J.J.; Yoo, D.J. Low-power-Consumption metal oxide NO_2 gas sensor based on micro-heater and screen printing technology. *J. Nanosci. Nanotechnol.* **2012**, *12*, 5543–5546. [CrossRef] [PubMed]
146. Rajput, G.Y.; Gofane, M.S.; Dhobale, S. Design of Micro-heater on 3D-SnO_2 Gas Sensor, Computing. *Commun. Signal. Process.* **2019**, 621–629.
147. Kim, J.-H.; Mirzaei, A.; Kim, H.W.; Kim, S.S. Pd-functionalized core-shell composite nanowires for self-heating, sensitive, and benzene-selective gas sensors. *Sens. Actuators A* **2020**, *308*, 112011. [CrossRef]
148. Zhu, L.F.; She, J.C.; Luo, J.Y.; Deng, S.Z.; Chen, J.; Ji, X.W.; Xu, N.S. Self-heated hydrogen gas sensors based on Pt-coated $W_{18}O_{49}$ nanowire networks with high sensitivity, good selectivity and low power consumption. *Sens. Actuators B* **2011**, *153*, 354–360. [CrossRef]
149. Tan, H.M.; Hung, C.M.; Ngoc, T.M.; Nguyen, H.; Hoa, D.; Van Duy, N.; Van Hieu, N. Novel Self-Heated Gas Sensors Using on-Chip Networked Nanowires with Ultralow Power Consumption. *ACS Appl. Mater. Interfaces* **2017**, *9*, 6153–6162. [CrossRef]
150. Sharma, S.; Madou, M. A new approach to gas sensing with Nanotechnology. *Phil. Trans. R. Soc. A* **2012**, *370*, 2448–2473. [CrossRef]
151. Espid, E.; Taghipour, F. UV-LED Photo-activated Chemical Gas Sensors: A Review. *Crit. Rev. Solid State Mat. Sci.* **2016**, *42*, 416–432. [CrossRef]
152. Karaduman, I.; Yıldız, D.E.; Sincar, M.M.; Acar, S. UV light activated gas sensor for NO_2 detection. *Mater. Sci. Semicond. Process.* **2014**, *28*, 43–47. [CrossRef]
153. Fan, S.-W.; Srivastava, A.K.; Dravid, V.P. UV-activated room-temperature gas sensing mechanism of polycrystalline ZnO. *Appl. Phys. Lett.* **2009**, *95*, 142106. [CrossRef]
154. Trawka, M.; Smulko, J.; Hasse, L.; Granqvist, C.-G.; Annanouch, F.E.; Ionescu, R. Fluctuation enhanced gas sensing with WO_3-based nanoparticle gas sensors modulated by UV light at selected wavelengths. *Sens. Actuators B* **2016**, *234*, 453–461. [CrossRef]
155. Chizhov, A.; Rumyantseva, M.; Gaskov, A. Light activation of nanocrystalline metal oxides for gas sensing: Principles, achievements, challenges. *Nanomaterials* **2021**, *11*, 892. [CrossRef]
156. Wang, J.; Shen, H.C.; Xia, Y.; Komarneni, S. Light-activated room-temperature gas sensors based on metal oxide nanostructures: A review on recent advances. *Ceram. Int.* **2021**, *47*, 7353–7368. [CrossRef]

materials

MDPI

Article

Removal Performance and Mechanism of Benzo(*b*)Fluorathene Using MnO_2 Nanoflower/Graphene Oxide Composites

Qingqing Cao [1], Siqi Lu [2], Wenjun Yin [1,2,*], Yan Kang [3], Naihao Yang [4], Yudong Hou [5] and Zizhang Guo [6,*]

1 School of Architecture and Urban Planning, Shandong Jianzhu University, Jinan 250014, China; caoqingqing18@sdjzu.edu.cn
2 College of Environmental Science and Engineering, Tongji University, Shanghai 200092, China; 1851284@tongji.edu.cn
3 College of Environment and Safety Engineering, Qingdao University of Science and Technology, Qingdao 266042, China; kangyan@qust.edu.cn
4 Jinan Engineering Consulting Institute, Jinan 250014, China; qdhfx.238@163.com
5 Majian International Architectural Design Consulting Co., Ltd., Jinan 250014, China; hou.yu.dong@163.com
6 Shandong Key Laboratory of Water Pollution Control and Resource Reuse, School of Environmental Science and Engineering, Shandong University, Qingdao 266237, China
* Correspondence: yinwenjun1991@163.com (W.Y.); guozizhang@sdu.edu.cn (Z.G.)

Abstract: High-ring polycyclic aromatic hydrocarbons (PAHs, Benzo[b]fluorathene (BbFA), etc.) are difficult to biodegrade in the water environment. To address this issue, an innovative method for the preparation of MnO_2 nanoflower/graphene oxide composite (MnO_2 NF/GO) was proposed for adsorption removal of BbFA. The physicochemical properties of MnO_2 NF/GO were characterized by SEM, TEM, XRD, and N_2 adsorption/desorption and XPS techniques. Results show that the MnO_2 NF/GO had well-developed specific surface area and functional groups. Batch adsorption experiment results showed that adsorption capacity for BbFA was 74.07 mg/g. The pseudo-second-order kinetic model and Freundlich isotherm model are fitted well to the adsorption data. These show electron-donor-acceptor interaction; especially π-π interaction and π complexation played vital roles in BbFA removal onto MnO_2 NF/GO. The study highlights the promising potential adsorbent for removal of PAHs.

Keywords: MnO_2 nanoflower; graphene oxide; PAHs; adsorption

Citation: Cao, Q.; Lu, S.; Yin, W.; Kang, Y.; Yang, N.; Hou, Y.; Guo, Z. Removal Performance and Mechanism of Benzo(*b*)Fluorathene Using MnO_2 Nanoflower/Graphene Oxide Composites. *Materials* **2021**, *14*, 4402. https://doi.org/10.3390/ma14164402

Academic Editor: Avelino Núñez-Delgado

Received: 10 June 2021
Accepted: 2 August 2021
Published: 6 August 2021

1. Introduction

Polycyclic aromatic hydrocarbons (PAHs) refer to aromatic hydrocarbons containing two or more benzene rings which are formed by the incomplete combustion or pyrolysis of fossil fuels such as coal, oil and natural gas, wood, paper, and other hydrocarbons under reduced conditions [1]. The toxicity, genotoxicity, mutagenicity and carcinogenicity of PAHs cause a variety of harms to the human body, such as damage to the respiratory system, circulation system, nervous system, liver, and kidney damage [2,3]. PAHs have recently attracted much attention in studies on water, soil and air pollution as a result of the United States Environmental Protection Agency blacklisting 16 PAHs as "priority-controlled pollutants" [4,5]. Currently, many different techniques, such as liquid-phase adsorption, photocatalytic degradation, bioremediation, and electrochemical remediation, have been extensively investigated in treating PAH-contaminated water environments in wastewater reclamation [6–9]. Among them, adsorption technology seems to be a potential method for PAH control due to its selectivity, low operating cost, affordability, simplicity, high efficiency, and the adsorbent reusability [10,11]. Kumar et al. used pyrolysis-assisted potassium hydroxide-induced palm shell activated carbon to remove PAHs from aqueous solution and the maximum adsorption capacity was 131.7 mg/g [12]. Bhadra et al. proved the adsorption capacity of MOF-derived carbons on naphthalene (237 mg/g), anthracene (284 mg/g), and pyrene (307 mg/g) [13].

Among all kinds of adsorbents, graphene has a high specific surface area, strong hydrophobicity and a unique delocalized large π bond, lending it broad application prospects in the adsorption and treatment of aromatic pollutants from wastewater [14]. Huang et al. [15] synthesized a reduced graphene oxide-hybridized polymeric high-internal phase emulsion with an open-cell structure and hydrophobicity to absorb PAH (47.5 mg/g). However, the complex preparation process, limited adsorption capacity and high cost remain significant obstacles to the large-scale application of graphene in wastewater treatment [16]. Sun et al. compared the adsorption capacity of graphene oxide (GO), reduced graphene oxide (rGO) and graphite (G) for naphthalene, anthracene and pyrene in aqueous solution, and the rGO had the optimal adsorption capacity [17]. Herein, the development of the excellent performance adsorbents had always been the research hotspot in removing PAHs from wastewater.

The previous research show that the metal cations could be the adsorption-active component which could interact with the aromatic ring [18]. Furthermore, the hybridization of metal cations to the adsorption substrate could form the π- complexation interaction which could efficiently realize the extraction of PAHs [19]. The hybridization of metal cations to the graphene materials could be an innovative technology to improve the strength of π complexation in carbon-based materials [20]. MnO_2 nanoflowers (MnO_2 NFs) are regarded as optimal nanostructures because of their rational open structure, increasing the adsorption point between the adsorbent and the contaminant [21,22]. At the same time, due to the stable properties and large specific surface area, graphene oxide (GO) can also serve as a porous backbone to support functional materials, thus leading to much mass loading of MnO_2 NFs for pollutant removal [23,24]. In addition, there is little literature on the performance and mechanism of PAHs using MnO_2 NFs/GO synthetic materials.

In this study, Benz[b]fluorathene (BbFA) was chosen as the model PAH. It is one of the carcinogenic PAHs, which was included in the list of carcinogenic 2B carcinogens by the International Agency for Research on Cancer (IARC, World Health Organization) in 2017 [25]. In this work, an innovative preparation method of MnO_2 nanoflower/graphene oxide (MnO_2 NF/GO) composites was developed to remove BbFA from wastewater. The objectives of the research are (1) to prepare MnO_2 NF using a direct reduction of $KMnO_4$ with poly-(dimethyl diallyl ammonium chloride) (PDDA) and to synthesize MnO_2 NF/GO in a hydrothermal reactor; (2) to characterize the physicochemical properties of MnO_2 NF/GO; and (3) to evaluate the BbFA adsorption performance and mechanism of the MnO_2 NF/GO composite by batch adsorption experiment.

2. Results

2.1. Structural and Morphology Characterization of MnO_2 NF/GO Composites

Based on the SEM and TEM observations and analyses of MnO_2 NF, the nano-agglomerated structure can be observed in Figure 1a,b, which is consistent with the experimental anticipation and similar research results [26]. As displayed in Figure 1c, the morphological characteristics of MnO_2 NF/GO composites showed an irregular porous structure with a distributed rippled and crumpled morphology, which may increase the surface area of the adsorbent. The HRTEM image (Figure 1d) demonstrated that the lattice distance of 0.341 nm corresponds to the (002) plane of the as-prepared composite [27]. Notably, the STEM and X-ray elemental mappings (Figure 1e) confirm that the MnO_2 NFs are homogeneously deposited and distributed into the GO.

The XRD pattern of as-synthesized composites was used to analyze the precise crystal structure and the results are demonstrated in Figure 2a. MnO_2 NF had weak diffraction peaks ascribed to the fact that MnO_2 NF possess insufficient crystalline property. The peaks at 2θ values of 37.28 and 67.94 in MnO_2 NF were observed, which could be assigned to the (−111) and (114) planes of the MnO_2 structure (JCPDS card No: 80-1098) [26]. The diffractogram peak at 2θ the value of 23.5 in MnO_2 NF/GO composites is attributed to the amorphous carbon with low graphitization, corresponding to the highly ordered laminar structure with an interlayer distance of 0.34 nm along with the (002) orientation [28]. The diffraction peak at the 2θ value of 45 in MnO_2 NF/GO composites indicates a short-range order in stacked graphene layers. The results were consistent with the previous SEM results.

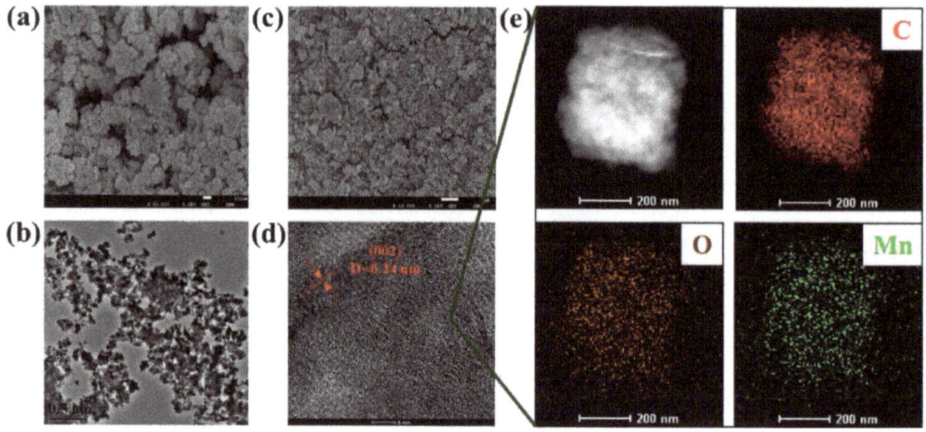

Figure 1. SEM (**a**) and TEM (**b**) images of MnO_2 NF; SEM (**c**) and HRTEM (**d**) images of MnO_2 NF/GO composites; and (**e**) HAADF-STEM image of MnO_2 NF/GO composites and corresponding elemental mapping images of C, O and Mn.

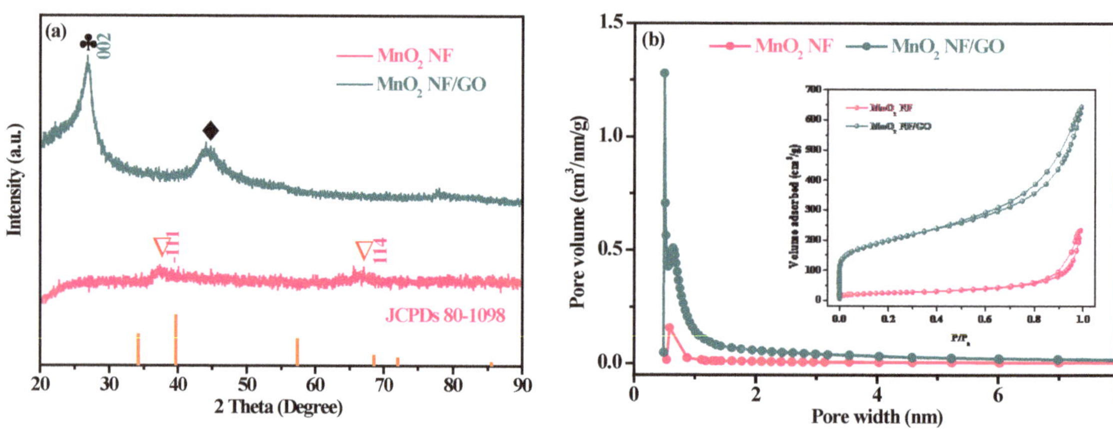

Figure 2. XRD (**a**); the pore size distributions and N_2 adsorption/desorption isotherms (**b**) of as-synthesized composites.

As displayed in Figure 2b, the adsorbent pores of MnO_2 NF and MnO_2 NF/GO composites mainly contain micro-mesoporous structures (diameter < 2 nm) according to the IUPAC classification [29]. The N_2 adsorption/desorption isotherms for MnO_2 NF and MnO_2 NF/GO composites showed a sharp increase at low relative pressure (P/P_0), consistent with the typical curve (type I and IV) with H4 hysteresis loop, which was the microporous structure characteristic [29]. The detailed textural parameters of the as-synthesized composites are shown in Table 1. The S_{BET} of MnO_2 NF/GO composites (694.30 m^2/g) was larger than that of MnO_2 NF (87.78 m^2/g), and the volume adsorbed (V_{tot}) of MnO_2 NF/GO composite is higher than MnO_2 NF at a high P/P_0 value. These results suggest that MnO_2 NF/GO composites could facilitate pollutant adsorption due to the nano-agglomerated structure and the large specific surface area and volume adsorbed of MnO_2 NF/GO composites.

Table 1. Textural parameters of as-synthesized composites.

Identification	MnO_2 NF	MnO_2 NF/GO
S_{BET} [a] (m^2/g)	87.78	694.30
S_{mic} [b] (m^2/g)	11.40	182.19
S_{ext} [c] (m^2/g)	76.38	512.11
V_{tot} [d] (cm^3/g)	0.3584	0.9606
V_{mic} [e] (cm^3/g)	0.0046	0.0778
V_{mes} [f] (cm^3/g)	0.3538	0.8828

[a] BET surface area, [b] micropore surface area, [c] external surface area, [d] total pore volume, [e] micropore volume, [f] external volume.

2.2. Chemical Characteristics of MnO_2 NF/GO Composites

The surface chemical characteristics of as-synthesized composites were further analyzed with the typical XPS spectra. As displayed in Figure 3a, C, O and Mn were the major elemental compositions on the surface of MnO_2 NF/GO composites. Moreover, the Mn $_{2p}$ spectrum for MnO_2 NF/GO illustrated the successful fabrication of MnO_2 NF on the GO surface. The high-resolution O 1s spectrum (Figure 3b) shows peaks at 530.6 and 532.2 eV, attributed to Mn-O-Mn and Mn-O-H bonding [30]. The ratio of Mn-O-Mn/Mn-O-H was 4.05 based on the peak area ratios calculation results. These findings suggest that Mn primarily exists in the oxide form (MnO_2) on the MnO_2 NF/GO composites, consistent with experimental expectations. As displayed in Figure 3c, the high-resolution wide-range Mn $2p_{1/2}$ (652.9 eV) and Mn $2p_{3/2}$ (641.1 eV) peaks using the XPS best peak fitting with Gaussian modes were caused by the overlap of Mn^{3+} and Mn^{4+} ions [31]. Additionally, the separation value (>5.9 eV) between Mn $2p_{3/2}$ and Mn $2p_{1/2}$ was consistent with published reports [32]. The presence of carboxyl group and hydroxyl group was conductive for the pollutant adsorption according to the wide-range C 1s spectrum of MnO_2 NF/GO composites in Figure 3d.

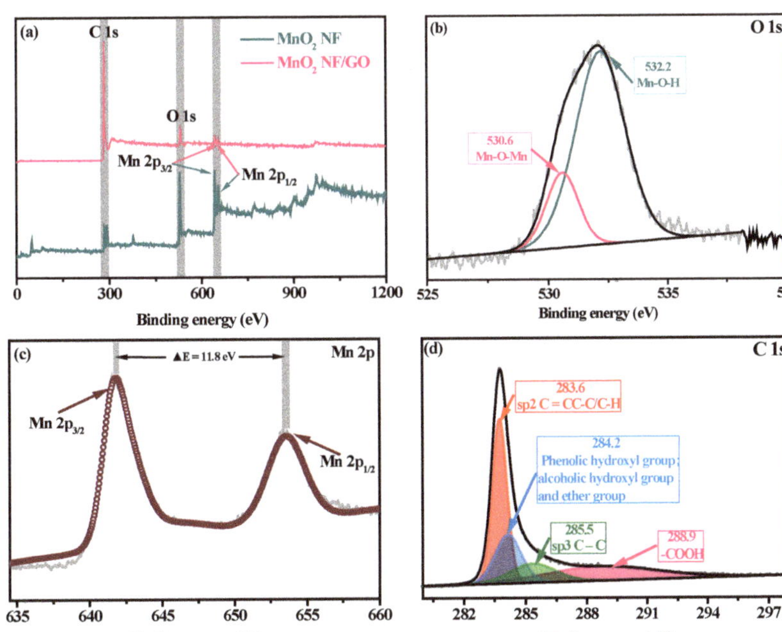

Figure 3. XPS survey spectra of MnO_2 NF and MnO_2 NF/GO composites (**a**); O 1s (**b**); Mn 2p (**c**); and C 1s (**d**) spectra of MnO_2 NF/GO composites.

2.3. Effect of Contact Time and Adsorption Kinetics

Figure 4a showed the effect of contact time for BbFA adsorption capacity on as-synthesized composites. It could be seen that BbFA was rapidly adsorbed onto adsorbents during the initial 30 min, which can be explained by the rapid diffusion speed of BbFA due to the higher initial BbFA concentration and the initial sufficient adsorption sites of adsorbents. In addition, the large number of aromatic ring structures of BbFA determine the adsorption rate. The BbFA concentration and diffusion speed decrease continuously with continuous contact reaction, while the BbFA adsorption capacity on as-synthesized composite increased. At the same time, the BbFA adsorption capacity on MnO_2 NF/GO composites was six times higher than that of MnO_2 NF due to MnO_2 NF/GO composites' larger specific surface area and volume adsorbed. To identify the possible rate-controlling steps and reaction mechanisms in the BbFA adsorption process, the pseudo-first-order model and the pseudo-second-order model were used to simulate the experimental data [33].

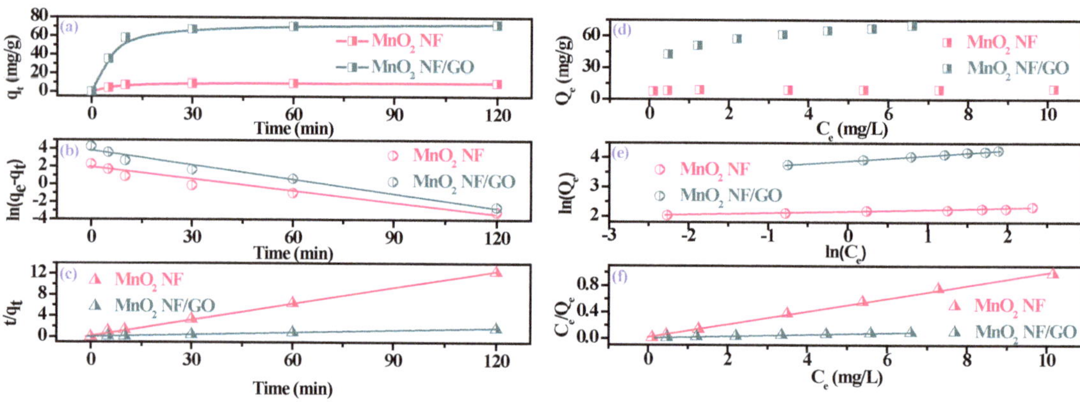

Figure 4. Effect of contact time for BbFA adsorption on as-synthesized composites (**a**); the linear plots of the pseudo-first-order model (**b**) and the pseudo-second-order model (**c**); effect of initial concentration for BbFA adsorption on the as-synthesized composites (**d**); the linear plots of Langmuir isotherm (**e**) and Freundlich isotherm (**f**). (dosage = 0.2 g/L, contact time = 2 h; temperature = 25 ± 1 °C).

The pseudo-first-order model, which is based on solid capacity, was defined as follows (Equation (1)):

$$ln(q_e - q_t) = lnq_e - k_1 t \quad (1)$$

The pseudo-second-order model, which predicts the behavior of the whole adsorption range, was defined as follows (Equation (2)):

$$\frac{t}{q_t} = \frac{1}{k_2 q_e^2} + \frac{1}{q_e} t \quad (2)$$

where q_e (mg/g) and q_t (mg/g) are adsorption capacities at equilibrium and time t, respectively. k_1 (1/h) are the rate constants of the pseudo-first-order model, and k_2 (g/(mg·min)) are the rate constants of the pseudo-second-order model, respectively.

Figure 4b and c present the plots for the BbFA adsorption of as-synthesized composites by applying the kinetic models in this study, and the slopes and intercepts of these curves were used to determine the fitting parameters. The calculated constants of the kinetics and the corresponding linear regression correlation are shown in Table 2. The high correlation coefficients value (R^2 > 99%) and the excellent agreement between the experimental (q_e) and calculated values (q_{cal}) indicate that the pseudo-second-order model resulted in a better fit than the pseudo-first-order model. Therefore, the pseudo-second-order model was more suitable for describing the adsorption of BbFA onto MnO_2 NF/GO composites, and the critical rate-controlling steps were multiple processes, especially the activated or chemisorption processes [34].

Table 2. Parameters of kinetics models for the BbFA adsorption by as-synthesized composites.

Kinetic Models	Constants	MnO_2 NF	MnO_2 NF/GO
Pseudo-first-order parameters	$Q_{e,cal}$ (mg/g)	5.272	40.62
	K_1 (1/min)	0.042	0.053
	R^2	0.9522	0.9721
Pseudo-second-order parameters	$Q_{e,cal}$ (mg/g)	9.9	74.07
	K_2 (g/mg/min 10^{-4})	299	48.73
	R^2	0.998	0.9986

2.4. Adsorption Isotherm

The adsorption isotherms were generated by changing the initial concentration of BBFA, and the mechanism is that the higher initial concentration of BBFA provides a prominent driving force to control the resistance of BBFA transfer from liquid to solid part in the adsorption system. As displayed in Figure 4d, the adsorption isotherms showed a sharp initial slope due to the fact that the amount of BbFA could not meet as-synthesized composites' abundance of available adsorption sites in low equilibrium BbFA concentration, resulting in a weakening maximum adsorption capacity. As the equilibrium BbFA concentration increased further, the maximum adsorption capacity increased gradually as its active sites were gradually occupied by BbFA.

The Langmuir isotherm model assumed monolayer coverage of the adsorbate over a homogenous adsorbent surface [35]. The Freundlich equation described the adsorption from low and medium concentrations, when the monolayer was not filled, and the parameter *n* described the heterogeneity of adsorption sites [36]. In this study, the Langmuir and Freundlich isotherms were used to describe the adsorption isotherm in detail (Figure 4a,e,f). The isotherm models were given by Equations (4) and (5):

$$Q_e = \frac{Q_0 K_L C_e}{1 + K_L C_e} \tag{3}$$

$$lnQ_e = lnK_F + \frac{1}{n} lnC_e \tag{4}$$

where Q_e (mg/g) is the maximum adsorption capacity of adsorbents; C_e (mg/L) is the equilibrium BbFA concentration; Q_0 (mg/g) is the initial adsorption capacity; K_L (L/mg) and K_F ((mg/g)/(L/mg)1/n) are the Langmuir isotherm constant and Freundlich affinity coefficient, respectively; and *n* is the adsorption intensity.

The isotherm lines, isotherm constants, and correlation coefficients of isotherm models are summarized in Figure 4b,c and Table 3. The Langmuir isotherm model exhibited a better fit to the BbFA adsorption process of the MnO_2 NF (Figure 5), which indicated that the BbFA adsorption tended to be homogeneous and showed monolayer coverage due to the strong interactions between the surface of MnO_2 NF and BbFA. Further, the Freundlich model was the best for describing the BbFA adsorption process onto the MnO_2 NF/GO composites, explaining the complex chemical and multi-layer adsorption process due to the metal oxides' hybridization in MnO_2 NF/GO composites. In addition, the Freundlich constant *1/n* values were in the range of 0–1, suggesting that the MnO_2 NF/GO composites can actively adsorb BbFA. As displayed in Table 3, the maximum adsorption capacities (Q_e) of MnO_2 NF/GO composites (74.07 mg/g) were higher than those of MnO_2 NF (9.9 mg/g), which were consistent with 2.3 results.

Table 3. Langmuir and Freundlich constants related to the adsorption isotherms of BbFA for as-synthesized composites.

Isotherm Models	Constants	MnO_2 NF	MnO_2 NF/GO
Langmuir	K_L (L/mg)	5.364	74.07
	Q_m (mg/g)	10.13	1.824
	R^2	0.9973	0.9964
Freundlich	K_F (mg/g·(L/mg)·1/n)	8.701	49.11
	$1/n$	0.0582	0.1878
	R^2	0.9647	0.9976

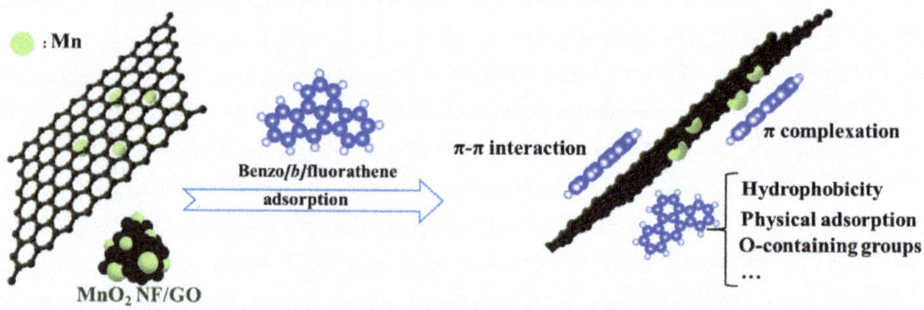

Figure 5. BbFA adsorption mechanism on MnO_2 NF/GO composites.

2.5. BbFA Adsorption Mechanism

Despite the destruction of the graphene conjugated structure during the GO oxidation process, GO still retains a unique delocalized π bond and surface hydrophobic properties. The π bond on the MnO_2 NF/GO surface could form π–π interactions with the aromatic ring of BbFA. Many studies have also shown that π–π interaction was an essential way for adsorbents to adsorb PAHs. GO could be used as a metal oxide carrier to synthesize compounds, the compounds can prevent agglomeration from taking place and form π complexing bonds due to metal oxide doping; the adsorption capacity of BbFA on MnO_2 NF/GO was improved by π complexation. The hydrophobic properties and multilayer structure characteristics of MnO_2 NF/GO provide sufficient adsorption sites for BbFA. Many oxygen-containing functional groups (carboxyl and hydroxyl) are introduced into MnO_2 NF/G during GO synthesis and metal oxide doping. At the same time, the adsorption of BbFA molecules would lead to changes in the morphology of the MnO_2 NF/G, thus generating new adsorption active sites for BbFA removal. The BbFA adsorption capacity of these active sites still needs to be further studied. In general, the results of batch adsorption experiments and model fitting showed that the adsorption of BbFA onto MnO_2 NF/G was dominated by chemisorption, and the π-π interaction, π complexation, and hydrophobicity of nanoflowers have played a role in the adsorption of BbFA.

3. Methods

3.1. Materials

Poly dimethyl diallyl ammonium chloride (PDDA, 20%) and potassium permanganate ($KMnO_4$, analytical grade) were purchased from ALADDIN Co. Ltd. (Shanghai, China). The fabrication method of graphene oxide (GO) is provided in the Supporting Information Text S1. Benzo[*b*]fluorathene (BbFA) solid (purchased from Aladdin Industrial Corporation) was of 98% purity. Benzo[*b*]fluorathene (BbFA) was analyzed with GC-MS (please refer to Table S1 for details of method parameters). Additionally, all solutions necessary for analytical procedures were prepared with distilled water, and all the chemicals used were of analytical grade.

3.2. *Fabrication of MnO_2 NF/GO Composites*

As displayed in Figure 6, the 4.5 mL PDDA was mixed with 20 mL ultrapure water and heated to 120 °C. Afterward, 4.0 g of $KMnO_4$ was added to the mixed solution while stirring at 220 rpm for 60 min until the aqueous dispersion mixture turned dark brown, which was defined as MnO_2 nanoflower aqueous dispersion. The partial MnO_2 nanoflower aqueous dispersion was centrifuged at 8000 rpm for 10 min and further washed twice with ethanol and three times with distilled water, respectively. Furthermore, the resultant product was dried at 60 °C for 12 h to obtain MnO_2 nanoflower particles, which were defined as MnO_2 NFs. The rest of the MnO_2 nanoflower aqueous dispersion was mixed with the prepared GO (2 g) under continuous stirring at room temperature for 10 min. Then, the resultant mixture was transferred into a hydrothermal reactor and heated to 120 °C for 4 h. Subsequently, the dark precipitate powders were collected and washed with distilled water several times. Finally, they were dried at 60 °C for 12 h to obtain the MnO_2 NF/GO composite.

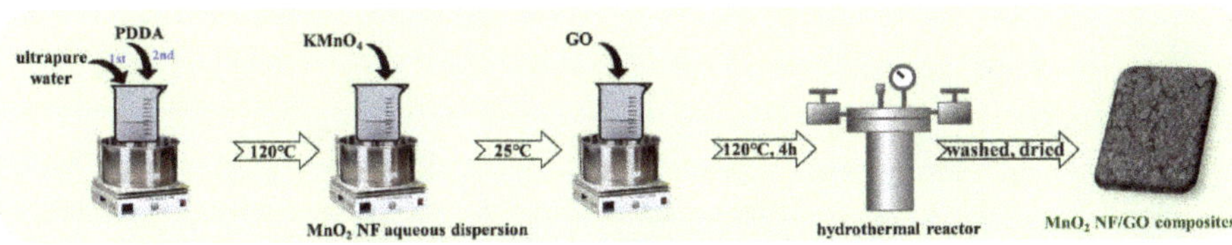

Figure 6. Fabrication of MnO_2 NF/GO composites.

3.3. *Characterization Methods*

The characterization methods are provided in the Supporting Information Text S2.

3.4. *Adsorption Experiments*

The kinetic experiments were conducted to investigate the effect of contact time and evaluate the kinetic properties. The as-synthesized composites (0.2 g) were added into 1 L of BbFA solution with initial concentrations of 0.3 mg/L. The mixture solution had natural pH, which was detected using a pH meter (Model PHS-3C, Shanghai, China). The mixture solution was agitated on the magnetic stirrers (Model 78-1) at a 250 ± 10 rpm speed with control of 25 ± 1 °C. The flasks were wrapped in aluminum foil to prevent photolysis. The 5 ± 0.5 mL samples were taken and filtered at desired adsorption duration (0–240 min), then the mixture was filtered from the liquid phase using a Millipore membrane filter (0.45 μm), and the residual BbFA concentrations were enriched into 10 mL CH_2Cl_2 through solid-phase extraction, followed by concentration determination of BbFA using GCMS.

The adsorption capacities of adsorbent were calculated using the following (Equation (5)):

$$Q = (C_0 - C_e)V/M \tag{5}$$

where Q (mg/g) represents the remove capacities; C_0 and C_e (mg/L) are the initial and equilibrium concentrations of BbFA, respectively; V (L) is the volume of the BbFA solution; and M (g) is the mass of adsorbent added.

In the batch adsorption experiments, the stock solution of BbFA (1 g/L) was prepared by dissolving 0.05 g of powder BbFA in a 500 mL CH_2Cl_2 solution, and the desired concentrations were obtained by dilution, followed by magnetic stirring to ensure the complete dissolution of BbFA in water solutions. The batch equilibrium BbFA adsorption studies were performed with a series of brown conical flasks (500 mL) containing a volume of 100 mL of the fixed initial concentration of BbFA (0-350 mg/L). Subsequently, the adsorbents (20 mg) were added to each flask, and the flasks were shaken at 200 ± 10 rpm in a shaded water bath shaker (SHZ-88) at 25 ± 1 °C for 24 h until the equilibrium achieved. The residual BbFA concentration was analyzed using the same method described above.

4. Conclusions

MnO_2 NF aqueous dispersion composed of PDDA and $KMnO_4$ was used to produce MnO_2 NF/GO in a hydrothermal reactor. MnO_2 NF/GO composites showed excellent removal performance of BbFA from wastewater. The batch adsorption experiments revealed that the adsorption isotherms agreed well with Freundlich isotherm and kinetics obeyed the pseudo-second-order kinetics model and adsorption capacity of 74.07 mg/g. The result was related to the well-developed physicochemical properties of MnO_2 NF/GO composites. The first reason is that it has a larger specific surface area and adsorption sites, and another important reason is that it has strong electron donor–acceptor interaction (EDA interaction, especially π–π interaction and π complexation). Thus, MnO_2 NF/GO composites could be cost-effective functional materials for BbFA removal. However, further studies are required to improve oxidative degradation of MnO_2 NF/GO composites.

Supplementary Materials: The following are available online at https://www.mdpi.com/article/10.3390/ma14164402/s1, Text S1: The fabrication method of graphene oxide (GO). Text S2: The characterization methods. Table S1: Shimadzu GCMS-QP 2010 method parameters for BbF concentration determination. The Supporting Information is available free of charge on the ACS Publications website at DOI.

Author Contributions: Data curation, Q.C., S.L. and N.Y.; Formal analysis, S.L. and Z.G.; Investigation, Y.K.; Methodology, W.Y.; Project administration, N.Y.; Resources, W.Y.; Supervision, Y.H.; Validation, S.L.; Writing—original draft, Q.C., Y.K. and Y.H.; Writing—review & editing, W.Y. and Z.G. All authors have read and agreed to the published version of the manuscript.

Funding: This research was funded by Guo, Z. of National Natural Science Foundation of China, grant number 51908326; was funded by Cao, Q. of Natural Science Foundation of Shandong Province, grant number ZR2020QC041; was funded by Cao, Q. of the Science Foundation of Shandong Jianzhu University, grant number X18047ZX; and was funded by Kang, Y. of Science Foundation of Qingdao University of Science and Technology, grant number 1203043003575.

Institutional Review Board Statement: Not applicable.

Informed Consent Statement: Not applicable.

Data Availability Statement: Data sharing is not applicable to this article.

Conflicts of Interest: The authors declare no conflict of interest.

References

1. Li, J.; Li, F.; Liu, Q. PAHs behavior in surface water and groundwater of the Yellow River estuary: Evidence from isotopes and hydrochemistry. *Chemosphere* **2017**, *178*, 143–153. [CrossRef]
2. Dat, N.-D.; Chang, M.B. Review on characteristics of PAHs in atmosphere, anthropogenic sources and control technologies. *Sci. Total Environ.* **2017**, *609*, 682–693. [CrossRef]
3. Han, J.; Liang, Y.; Zhao, B.; Wang, Y.; Xing, F.; Qin, L. Polycyclic aromatic hydrocarbon (PAHs) geographical distribution in China and their source, risk assessment analysis. *Environ. Pollut.* **2019**, *251*, 312–327. [CrossRef]
4. Meng, Y.; Liu, X.; Lu, S.; Zhang, T.; Jin, B.; Wang, Q.; Tang, Z.; Liu, Y.; Guo, X.; Zhou, J. A review on occurrence and risk of polycyclic aromatic hydrocarbons (PAHs) in lakes of China. *Sci. Total Environ.* **2019**, *651*, 2497–2506. [CrossRef]
5. Bandowe, B.A.M.; Meusel, H. Nitrated polycyclic aromatic hydrocarbons (nitro-PAHs) in the environment—A review. *Sci. Total Environ.* **2017**, *581*, 237–257. [CrossRef]
6. Hale, S.E.; Elmquist, M.; Braendli, R.; Hartnik, T.; Jako, L.; Henriksen, T.; Werner, D.; Comelissen, G. Activated carbon amendment to sequester PAHs in contaminated soil: A lysimeter field trial. *Chemosphere* **2012**, *87*, 177–184. [CrossRef]
7. Zhou, W.; Wang, X.; Chen, C.; Zhu, L. Removal of polycyclic aromatic hydrocarbons from surfactant solutions by selective sorption with organo-bentonite. *Chem. Eng. J.* **2013**, *233*, 251–257. [CrossRef]
8. Li, C.H.; Wong, Y.S.; Wang, H.Y.; Tam, N.F.Y. Anaerobic biodegradation of PAHs in mangrove sediment with amendment of $NaHCO_3$. *J. Environ. Sci.* **2015**, *4*, 148–156. [CrossRef]
9. Sun, K.; Liu, J.; Gao, Y.; Jin, L.; Gu, Y.; Wang, W. Isolation, plant colonization potential, and phenanthrene degradation performance of the endophytic bacterium Pseudomonas sp. Ph6-gfp. *Sci. Rep.* **2014**, *4*, 5462. [CrossRef]
10. Yin, W.; Guo, Z.; Zhao, C.; Xu, J. Removal of Cr(VI) from aqueous media by biochar derived from mixture biomass precursors of Acorus calamus Linn. and feather waste. *J. Anal. Appl. Pyrolysis* **2019**, *140*, 86–92. [CrossRef]

11. Yin, W.; Zhao, C.; Xu, J. Enhanced adsorption of Cd (II) from aqueous solution by a shrimp bran modified Typha orientalis biochar. *Environ. Sci. Pollut. Res.* **2019**, *26*, 37092–37100. [CrossRef] [PubMed]
12. Kumar, J.A.; Amarnath, D.J.; Sathish, S.; Jabasingh, S.A.; Saravanan, A.; Hemavathy, R.; Anand, K.V.; Yaashikaa, P. Enhanced PAHs removal using pyrolysis-assisted potassium hydroxide induced palm shell activated carbon: Batch and column investigation. *J. Mol. Liq.* **2019**, *279*, 77–87. [CrossRef]
13. Bhadra, B.N.; Song, J.Y.; Lee, S.-K.; Hwang, Y.K.; Jhung, S.H. Adsorptive removal of aromatic hydrocarbons from water over metal azolate framework-6-derived carbons. *J. Hazard. Mater.* **2018**, *344*, 1069–1077. [CrossRef]
14. Zhong, Y.; Wang, S.; He, Y.; Song, G. Synthesis of Magnetic/Graphene Oxide Composite and Application for High-Performance Removal of Polycyclic Aromatic Hydrocarbons from Contaminated Water. *Nano Life* **2015**, *5*, 1542006. [CrossRef]
15. Huang, Y.; Zhang, W.; Ruan, G.; Li, X.; Cong, Y.; Du, F.; Li, J. Reduced graphene oxide-hybridized polymeric high-internal phase emulsions for highly efficient removal of polycyclic aromatic hydrocarbons from water matrix. *Langmuir* **2018**, *34*, 3661–3668. [CrossRef]
16. Yan, H.; Wu, H.; Li, K.; Wang, Y.; Tao, X.; Yang, H.; Li, A.; Cheng, R. Influence of the surface structure of graphene oxide on the adsorption of aromatic organic compounds from water. *ACS Appl. Mater. Interfaces* **2015**, *7*, 6690–6697. [CrossRef]
17. Sun, Y.; Yang, S.; Zhao, G.; Wang, Q.; Wang, X. Adsorption of Polycyclic Aromatic Hydrocarbons on Graphene Oxides and Reduced Graphene Oxides. *Chem. Asian J.* **2013**, *8*, 2755–2761. [CrossRef]
18. Zhang, W.; Zheng, J.; Zheng, P.; Tsang, D.; Qiu, R. The roles of humic substances in the interactions of phenanthrene and heavy metals on the bentonite surface. *J. Soils Sediments* **2015**, *15*, 1463–1472. [CrossRef]
19. Liang, X.; Zhu, L.; Zhuang, S. Sorption of polycyclic aromatic hydrocarbons to soils enhanced by heavy metals: Perspective of molecular interactions. *J. Soil Sediments* **2016**, *16*, 1509–1518. [CrossRef]
20. Takahashi, A.; Yang, R.T. New adsorbents for purification: Selective removal of aromatics. *Aiche J.* **2002**, *48*, 1457–1468. [CrossRef]
21. Lin, M.; Chen, Z. A facile one-step synthesized epsilon-MnO2 nanoflowers for effective removal of lead ions from wastewater. *Chemosphere* **2020**, *250*, 126329. [CrossRef]
22. Ouyang, H.; Lu, Q.; Wang, W.; Song, Y.; Tu, X.; Zhu, C.; Smith, J.N.; Du, D.; Fu, Z.; Lin, Y. Dual-readout immunochromatographic assay by utilizing MnO_2 nanoflowers as the unique colorimetric/chemiluminescent probe. *Anal. Chem.* **2018**, *90*, 5147–5152. [CrossRef]
23. Ma, X.; Zhang, L.; Xia, M.; Li, S.; Zhang, X.; Zhang, Y. Mimicking the active sites of organophosphorus hydrolase on the backbone of graphene oxide to destroy nerve agent simulants. *ACS Appl. Mater. Interfaces* **2017**, *9*, 21089–21093. [CrossRef]
24. Lopez, A.; Zhao, Y.; Huang, Z.; Guo, Y.; Guan, S.; Jia, Y.; Liu, J. Poly-Cytosine Deoxyribonucleic Acid Strongly Anchoring on Graphene Oxide Due to Flexible Backbone Phosphate Interactions. *Adv. Mater. Interfaces* **2021**, *8*, 2001798. [CrossRef]
25. Ding, H.X.; Tao, X.M.; Kang-Le, L.V.; Zhang, N. Distribution Characteristics and Risk Analysis of PAHs and PCBs in Soils of Lanzhou. *Adm. Tech. Environ. Monit.* **2018**, *30*, 25–29.
26. Sadak, O.; Wang, W.; Guan, J.; Sundramoorthy, A.K.; Gunasekaran, S. MnO_2 nanoflowers deposited on graphene paper as electrode materials for supercapacitors. *ACS Appl. Nano Mater.* **2019**, *2*, 4386–4394. [CrossRef]
27. Park, W.-Y.; Wada, T.; Joo, S.-H.; Han, J.; Kato, H. Novel hierarchical nanoporous graphene nanoplatelets with excellent rate capabilities produced via self-templating liquid metal dealloying. *Mater. Today Commun.* **2020**, *24*, 101120. [CrossRef]
28. Al-Gaashani, R.; Najjar, A.; Zakaria, Y.; Mansour, S.; Atieh, M. XPS and structural studies of high quality graphene oxide and reduced graphene oxide prepared by different chemical oxidation methods. *Ceram. Int.* **2019**, *45*, 14439–14448. [CrossRef]
29. Sing, K.S.; Williams, R.T. Physisorption hysteresis loops and the characterization of nanoporous materials. *Adsorpt. Sci. Technol.* **2004**, *22*, 773–782. [CrossRef]
30. Stranick, M.A. MnO_2 by XPS. *Surf. Sci. Spectra* **1999**, *6*, 31–38. [CrossRef]
31. Stobinski, L.; Lesiak, B.; Malolepszy, A.; Mazurkiewicz, M.; Mierzwa, B.; Zemek, J.; Jiricek, P.; Bieloshapka, I. Graphene oxide and reduced graphene oxide studied by the XRD, TEM and electron spectroscopy methods. *J. Electron Spectrosc. Relat. Phenom.* **2014**, *195*, 145–154. [CrossRef]
32. Xu, H.; Hu, X.; Yang, H.; Sun, Y.; Hu, C.; Huang, Y. Flexible asymmetric micro-supercapacitors based on Bi2O3 and MnO2 nanoflowers: Larger areal mass promises higher energy density. *Adv. Energy Mater.* **2015**, *5*, 1401882. [CrossRef]
33. Simonin, J.-P. On the comparison of pseudo-first order and pseudo-second order rate laws in the modeling of adsorption kinetics. *Chem. Eng. J.* **2016**, *300*, 254–263. [CrossRef]
34. Guo, Z.; Zhang, X.; Kang, Y.; Zhang, J. Biomass-derived carbon sorbents for Cd (II) removal: Activation and adsorption mechanism. *ACS Sustain. Chem. Eng.* **2017**, *5*, 4103–4109. [CrossRef]
35. Marczewski, A.W. Analysis of kinetic Langmuir model. Part I: Integrated kinetic Langmuir equation (IKL): A new complete analytical solution of the Langmuir rate equation. *Langmuir* **2010**, *26*, 15229–15238. [CrossRef]
36. Saadi, R.; Saadi, Z.; Fazaeli, R.; Fard, N.E. Monolayer and Multilayer Adsorption Isotherm Models for Sorption from Aqueous Media. *Korean J. Chem. Eng.* **2015**, *32*, 787–799. [CrossRef]

materials

MDPI

Article

An Innovative Digestion Method: Ultrasound-Assisted Electrochemical Oxidation for the Onsite Extraction of Heavy Metal Elements in Dairy Farm Slurry

Chenyu Li [1], Bin Xue [1], Shang Wang [1], Xi Zhang [1], Chen Zhao [1], Xiaobo Yang [1], Run Zhao [2], Lin Dai [3], Shengqi Su [1], Haoqi Xu [1,2], Zhiqiang Shen [1], Zhigang Qiu [1,*] and Jingfeng Wang [1,*]

1 Department of Environment and Health, Tianjin Institude of Environmental and Operational Medicine, Tianjin 300050, China; nk_lcy710430@hotmail.com (C.L.); xue_bin04@163.com (B.X.); wsh847@163.com (S.W.); Zhangxi0820@126.com (X.Z.); zhaochen212@126.com (C.Z.); 18072712080@163.com (X.Y.); ssqsuqi@163.com (S.S.); Haoqi_xu@126.com (H.X.); tianjinszq922@sohu.com (Z.S.)

2 Agro-Environmental Protection Institute, Ministry of Agriculture and Rural Affairs, Tianjin 300191, China; 15900389657@163.com

3 Tianjin Key Laboratory of Pulp and Paper, Tianjin University of Science and Technology, Tianjin 300457, China; dailin@tust.edu.cn

* Correspondence: zhigangqiu99@gmail.com (Z.Q.); wangjingfeng0116@163.com (J.W.); Tel.: +86-22-84655052 (J.W.); Fax: +86-22-23328809 (J.W.)

Citation: Li, C.; Xue, B.; Wang, S.; Zhang, X.; Zhao, C.; Yang, X.; Zhao, R.; Dai, L.; Su, S.; Xu, H.; et al. An Innovative Digestion Method: Ultrasound-Assisted Electrochemical Oxidation for the Onsite Extraction of Heavy Metal Elements in Dairy Farm Slurry. *Materials* **2021**, *14*, 4562. https://doi.org/10.3390/ma14164562

Academic Editor: Avelino Núñez-Delgado

Received: 11 June 2021
Accepted: 11 August 2021
Published: 13 August 2021

Abstract: Dairy farm slurry is an important biomass resource that can be used as a fertilizer and in energy utilization and chemical production. This study aimed to establish an innovative ultrasound-assisted electrochemical oxidation (UAEO) digestion method for the rapid and onsite analysis of the heavy metal (HM) contamination level of dairy slurry. The effects of UAEO operating parameters on digestion efficiency were tested based on Cu and Zn concentrations in a dairy slurry sample. The results showed that Cu and Zn digestion efficiency was (96.8 ± 2.6) and (98.5 ± 2.9)%, respectively, with the optimal UAEO operating parameters (digestion time: 45 min; ultrasonic power: 400 W; NaCl concentration: 10 g/L). The digestion recovery rate experiments were then operated with spiked samples to verify the digestion effect on broad-spectrum HMs. When the digestion time reached 45 min, all digestion recovery rates exceeded 90%. Meanwhile, free chlorine concentration, particle size distribution, and micromorphology were investigated to demonstrate the digestion mechanism. It was found that 414 mg/L free chorine had theoretically enough oxidative ability, and the ultrasound intervention could deal with the blocky undissolved particles attributed to its crushing capacity. The results of particle size distribution showed that the total volume and bulky particle proportion had an obvious decline. The micromorphology demonstrated that the ultrasound intervention fragmented the bulky particles, and electrochemical oxidation made irregular blocky structures form arc edge and cellular structures. The aforementioned results indicated that UAEO was a novel and efficient method. It was fast and convenient. Additionally, it ensured digestion efficiency and thus had a good application prospect.

Keywords: biomass resource; dairy slurry; digestion efficient; heavy metals; ultrasound-assisted electrochemical oxidation

1. Introduction

Dairy farm slurry is an important, cheap biomass resource [1–3] rich in mineral nutrients and lignocellulose. With the rapid increase in the number of large-scale livestock farms in the last few years in China, plenty of livestock feces and slurry are discharged into the nearby natural environment, resulting in ecological pressure [4,5]. The resource utilization methods of dairy slurry include returning cropland as a fertilizer [6,7], energy engineering [8], and chemical production [9,10]. Practical experience indicates that returning cropland is the most practical and common treatment approach for farm slurry because it is rich in nutrients (N, P, and K) [11]. Unfortunately, the high abundance of heavy metals (HMs) may result in a serious ecological hazard. The HMs in farm slurry can be fixed in the soil along with returning cropland, keeping an average concentration of several mg/kg to tens of mg/kg for a long time [12]. HM pollution consistently threatens environmental ecology due to its toxic, accumulative, and persistent nature in the environment [5]. More seriously, the transfer of HMs from soil to cereals and plants is a major HM intake route for humans. Excessive HMs may accumulate in specific human organs and interact with proteins and enzymes, making them damaged or inactive [13].

Although the Chinese government has developed many management policies and treatment technologies to reduce the HM pollution from livestock slurry to environment, the lack of the onsite digestion-detection method is still the bottleneck to realize the effective control of the ecological risk due to HMs during returning cropland. Some novel detection methods have been developed for the onsite quantitative analysis of HMs in the last few years. Wen et al. established a portable tungsten coil electrothermal atomic absorption spectrometer for HM field analysis [14–16]. Wang et al. designed and implemented a field-based HM detection system involving electrochemical differential pulse anodic stripping voltammetry [17]. Additionally, with the advent of microfluidics technology, colorimetric sensors for the rapid detection of HMs can even work on a handheld device [18].

Compared with the detection methods, these recently reported digestion methods [19–23] were still challenging to implement in the field. In theory, the digestion procedure should ensure that treated samples are completely dissolved and HMs are released in a positive ion form compatible with the analytical method [24,25]. The conventional wet digestion procedures were relatively complicated and hazardous. Heavy use of strong acids ($HClO_4$, HF, H_2SO_4, HCl, etc.) and heating devices restricted the digestion procedures to be operated in a normative chemical laboratory. Although some modified digestion methods have been recently established to reduce operation difficulty, the inevitable use of acids and laboratory devices (such as a microwave) still cannot meet the onsite digestion requirement.

Therefore, more safe and simple digestion methods need to be established to analyze the HM contamination level of livestock slurry. This study reported an innovative digestion method based on ultrasound-assisted electrochemical oxidation (UAEO) theory. The UAEO method was designed as a specialized technology for the dairy farm slurries onsite digestion, which had the potential to be applied in other farm slurry and sewage samples. The digestion extraction effects for major (Cu and Zn) and trace (Cr, Cd, Pb, Ba, Co, Ni, Bi, and Ag) HMs were proved with the livestock slurry samples. Furthermore, the digestion mechanism of the reported method was analyzed and discussed to elaborate the functioning process.

2. Materials and Method

2.1. Reagents, Standards, and Samples

All reagents used in this study were at least analytical grade. All solutions were prepared in deionized water (resistivity $\geq$18.2 MΩ·cm). A commercially multi-element standard solution (Thermo Scientific Co. Ltd., Waltham, MA, USA) was used to analyze the standard digestion rate.

The dairy farm slurry samples were collected from three typical large-scale crop–animal mixed farms in July 2019 in Tianjin of China. All breeding varieties were Chinese Holstein Cattle. The cattle breeding stock of every farm was above 1000 units. The fecal slurry process flow in three large-scale dairy farms is shown in Figure 1. One of the dairy farms employed biogas engineering as a recycling treatment technique, while the other two carried out only a sedimentation liquid–solid separation. In any case, the lagoon was the unique storage facility before returning cropland. Therefore, all slurry samples were collected from the lagoon in experiments.

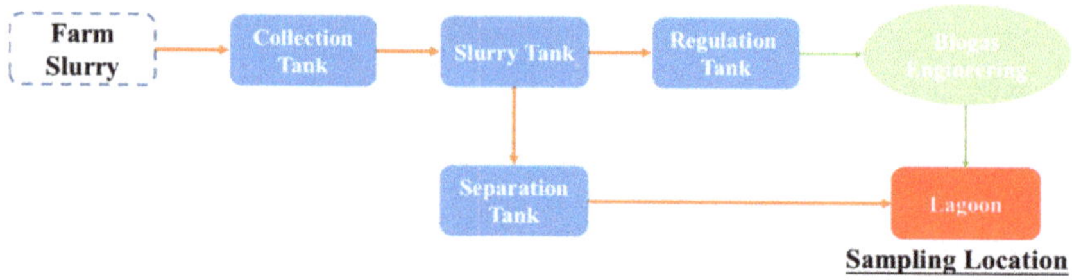

Figure 1. Fecal slurry process flow in three large-scale dairy farms.

The slurry equally sampled from three random points in a lagoon was uniformly mixed and stored in a clean-washed polyethylene plastic bottle. All samples were filtered with 10-mesh sieves to remove macroparticle impurities.

2.2. Construction of the UAEO Digestion Apparatus

As shown in Figure 2, the UAEO digestion apparatus was mainly constructed using two cuvettes connected with a NaCl salt bridge. The volume of both cuvettes was approximately 120 mL. The ingredients of the salt bridge solution were 20% (w/v) NaCl and 2% (w/v) agar. The aforementioned solution was heated to a boil, infused into the connected part, and cooled down to form a gel. The electrochemical oxidation function was implemented using a classic three-electrode system. The working and auxiliary electrodes were made using Ru–Ir-coated titanium ((Ru–Ir)@Ti) inert metal material, while the reference electrode was an Ag–AgCl electrode. The working potential was controlled at 2 V using a CHI-760D electrochemical workstation (Science Days Technology Co. Ltd., Beijing, China). The ultrasound function was implemented using a Scientz-IID ultrasonic generator (Scientz Co. Ltd., Ningbo, China), while the ultrasonic duty ratio was set to 50% (5 s/5 s). In the digestion cuvette, a stirrer was installed at the bottom to blend the solution along with digestion.

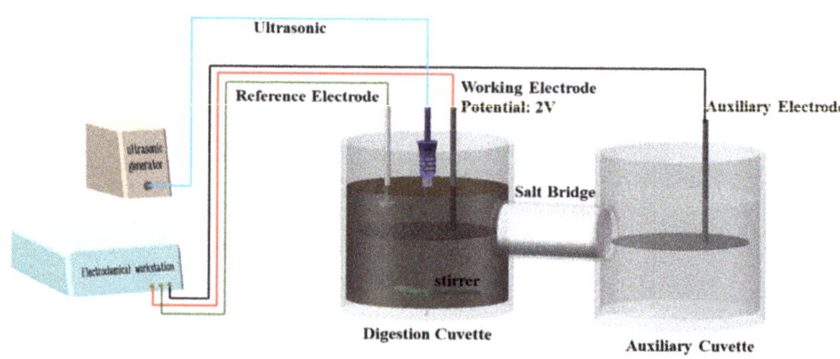

Figure 2. Schematic diagram of the UAEO digestion apparatus.

2.3. UAEO Digestion Method and Zn and Cu Digestion Efficiency Test

To determine Cu and Zn digestion efficiency in slurry samples, the aqua regia with closed-vessel microwave digestion (AD) method was applied and compared with the UAEO digestion method. The AD method was performed as previously reported [26]. Briefly, 5.0 g of accurately weighed slurry sample, 10 mL of aqua regia (3:1, v/v, HCl:HNO_3), and 5 mL of H_2O_2 (30%, v/v) were successively added in a poly tetra fluoroethylene (PTFE) vessel. The PTFE vessel was capped tightly and placed in a WX-6000 microwave apparatus (PreeKem Co. Ltd., Shanghai, China). The operating program of the microwave is shown in Table S1. After cooling down, the digestion solution was completely transferred and made up to 50 mL with deionized water.

The UAEO digestion method was operated with the following steps: (I) 10.0 g slurry sample and a certain quality (testing range: 0.1–10 g) NaCl were accurately added into deionized water to make 100 mL. The aforementioned suspension was poured into the digestion cuvette. (II) A NaCl solution with the same volume and concentration was poured into the auxiliary cuvette as in the digestion cuvette. (III) The electrochemical workstation and the ultrasonic generator were started at the same time, sampling approximately 5 mL of digestion suspension in 0, 10, 20, 30, 45, and 60 min. The ultrasonic power was set in the test range of 0–600 W.

Both AD and UAEO samples were filtered to remove undissolved solid before HM analysis. The concentrations of Cu and Zn were tested using an AA-7000 atomic absorption spectrophotometer (Shimadzu Co. Ltd., Kyoto, Japan). The digestion efficiency was calculated using Equation (1).

$$\text{Digestion efficiency} = C_{UAEO}/C_{AD} \times 100\% \quad (1)$$

where C_{UAEO} is the concentration of the tested element digested by the UAEO method, while C_{AD} is that digested by the AD method.

2.4. Digestion Recovery Rate

The digestion recovery rate experiment was performed to monitor the UAEO method digestion effect for trace HMs. The slurry sample with the standard substance was prepared before the test. Specifically, 10 g slurry sample and 10 mL of multi-element standard solution were mixed accurately and shaken on an HNYC-203T constant-temperature shaking table (Honour Co. Ltd., Tianjin, China) for 6 h to form nonionic compounds (such as coordination compounds). A NaCl solution was used to increase the volume of the aforementioned sample to 100 mL. The UAEO method was executed with the optimal NaCl concentration and ultrasonic power verified by the digestion efficiency result. The digestion suspension was sampled for 0, 20, and 45 min to analyze the standard digestion rate.

The concentrations of 10 objective HMs (Zn, Cu, Cr, Cd, Pb, Ba, Co, Ni, Bi, and Ag) were tested using a Thermo 7400 Inductively Coupled Plasma Optical Emission Spectrometer (ICP-OES) (Thermo Scientific Co. Ltd., Waltham, MA, USA). The ICP-OES instrumental parameters for the analysis are listed in Table S2. The digestion recovery rate of every HMs was calculated using Equation (2).

$$\text{Digestion recovery rate} = (C_1 - C_0)/C_{spk} \times 100\% \quad (2)$$

where C_0 is the concentration of the tested element digested by the UAEO method without standard solution, while C_1 is that by standard solution. C_{spk} represents the theoretical concentration of spiked elements.

2.5. Chlorine Quantitative Analysis

The digestion solution was taken out from the digestion cuvette in 0, 5, 10, 15, 20, 25, 30, 35, 40, and 45 min during the UAEO digestion method. Every sample was immediately placed in the ice-water bath to cool down, diluted, and filtered to be tested.

The concentration of chlorine was measured according to the *N*,*N*-diethyl-*p*-phenylenediamine (DPD) colorimetric method [27] using a UV 2600 spectrophotometer (Shimadzu, Kyoto, Japan).

2.6. Particle Size Distribution and Micromorphology

The particle size distribution of the digestion sample was measured using a Mastersizer 3000 laser particle size analyzer (Malvern Panalytical. Ltd., Cambridge, UK). The microtopographic analysis of the insoluble solid in the digestion sample was carried out using a field-emission scanning electron microscope (SEM) (JSM-IT300LV, Jeol, Japan, operating at 20 kV) and an optical microscope (OM) (BX51, Olympus, Tokyo, Japan).

2.7. Statistical Analysis

All statistical analyses were performed using PASW Statistics 18 software (SPSS Inc., Armonk, NY, USA) with analysis of variance (ANOVA) and Dunnet's test [28].

3. Results

3.1. Effects of UAEO Operating Parameters on Digestion Efficiency

As shown in Figure 3, the digestion efficiency was positively related to the digestion time, ultrasonic frequency, and NaCl concentration. Further, Cu and Zn digestion efficiency was tested for different periods at 400 W ultrasonic power with a NaCl concentration of 10 g/L (Figure 3A). The digestion efficiency increased rapidly in the first 30 min and then reached a plateau in 30–60 min. The Dunnet's test results showed no significant difference ($p > 0.05$) in digestion efficiency between 45 and 60 min. As shown in Figure 3B, the digestion efficiency increased with the increase in ultrasonic power with 45 min digestion time and 10 g/L NaCl concentration; the solute temperature was raised at the same time. The digestion efficiency had no significant difference between 400 W and 600 W ultrasonic power, although the solute temperature slightly increased. Finally, the optimal concentration of oxidation substrate NaCl was tested from 1 to 100 g/L with 45 min digestion time and 400 W ultrasonic power, as shown in Figure 3C. The results showed that an increase in NaCl concentration led to more working current and higher digestion efficiency until the NaCl concentration exceeded 10 g/L. With the optimal UAEO operating parameters (digestion time: 45 min; ultrasonic power: 400 W; NaCl concentration: 10 g/L), Cu and Zn digestion efficiency was (96.8 ± 2.6) and (98.5 ± 2.9)%, respectively.

3.2. UAEO Digestion Recovery Rate

The digestion recovery rate experiments were operated with spiked samples to evaluate the UAEO digestion effect for trace HMs. The 20 min and 45 min standard digestion rate for 10 HMs was tested, and the results are shown in Figure 4. After a 20 min UAEO digestion operation, the recovery rate of all 10 HMs exceeded 70%. When the processing duration further reached 45 min, all digestion recovery rates exceeded 90%. The specific results are listed in Table 1.

3.3. Free Chlorine

The concentration of free chlorine was tested by the DPD method to analyze the level of oxidation medium in the digestion solution. As shown in Figure 5, the ultrasound intervention would reduce the free chlorine increment speed in the reaction system. Specifically, when the ultrasound function module was turned off, the free chlorine concentration was maintained at about 1500 mg/L after digestion for 30 min. In another case, with ultrasound intervention, the free chlorine concentration could only reach the maximum (414 mg/L) at 20 min and, further, slightly decreased to 283 mg/L in 45 min instead of continuing to increase.

3.4. Effect of the UAEO Process on Particle Size Distribution and Micromorphology

As shown in Figure 6, UAEO digestion treatment indeed changed the undissolved particle size distribution. Before digestion, the particles with a diameter between 100 and 300 μm contributed to the main particle volume (1021 μm^3). Under the influence of oxidation, the total particle volume sharply declined, although the size distribution did not show an obvious change. Under the influence of ultrasound intervention, the size distribution moved toward small sizes, and the total particle volume had an obvious decline as well. When the UAEO digestion method worked normally, the effect on the change in particle size was comparable to a coupling of ultrasound and oxidation; the greatest abundance of particle size was changed to a smaller diameter range between 30 and 100 μm, which had a much smaller particle volume (193 μm^3) compared with the untreated sample.

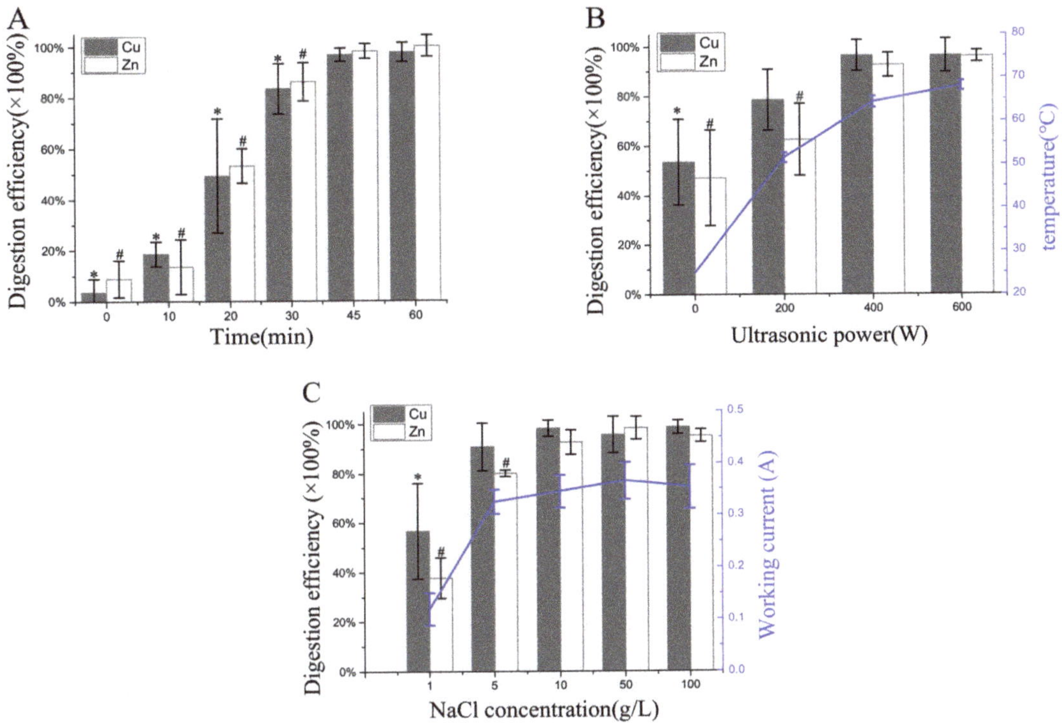

Figure 3. Variations in Cu and Zn digestion efficiency, solution temperature, and working current during UAEO digestion method. (**A**) Cu and Zn digestion efficiency was tested for different periods at 400 W ultrasonic power with a NaCl concentration of 10 g/L. Significant differences between each digestion time and 60 min group were found with the Dunnet's test; * $p < 0.05$ for Cu, # $p < 0.05$ for Zn. (**B**) Digestion efficiency was tested at 0, 200, 400, and 600 W ultrasonic power for 45 min with a NaCl concentration of 10 g/L, and the solute temperature was detected. Significant differences between each ultrasonic power and 600 W group were found with the Dunnet's test; * $p < 0.05$ for Cu, # $p < 0.05$ for Zn. (**C**) Digestion efficiency was tested with NaCl concentrations of 1, 5, 10, 50, and 100 g/L at 400 W ultrasonic power for 45 min, and the working current was tested using an electronic working station. Significant differences between each NaCl concentration and 100 g/L group were found with Dunnet's test; * $p < 0.05$ for Cu, # $p < 0.05$ for Zn.

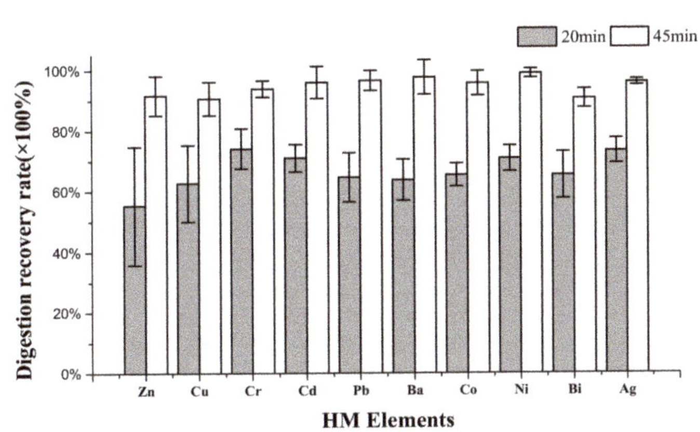

Figure 4. UAEO standard digestion rate of 10 primary HMs in dairy slurry.

Table 1. Concentrations and recovery rates (%) in the spiked samples digested by the UAEO method.

Element	Certified Values (mg/L)	Found Values (mg/L) and Recovery Rates (%)			
		20 min		45 min	
Zn	0 [a] 1 [b]	1.25 ± 0.28 1.8 ± 0.09	55.3%	2.24 ± 0.45 3.16 ± 0.44	91.7%
Cu	0 1	0.29 ± 0.05 0.92 ± 0.17	62.7%	0.54 ± 0.07 1.45 ± 0.09	90.7%
Cr	0 5	0.06 ± 0.03 3.77 ± 0.35	74.1%	0.11 ± 0.03 4.81 ± 0.16	93.9%
Cd	0 1	ND [c] 0.71 ± 0.05	71%	ND 0.96 ± 0.05	96%
Pb	0 10	ND 6.46 ± 0.81	64.6%	0.08 ± 0.04 9.74 ± 0.36	96.6%
Ba	0 1	ND 0.64 ± 0.07	63.7%	ND 0.98 ± 0.06	97.7%
Co	0 1	ND 0.65 ± 0.04	65.3%	ND 0.96 ± 0.04	95.7%
Ni	0 5	ND 3.54 ± 0.22	70.8%	ND 4.95 ± 0.08	98.9%
Bi	0 10	ND 6.53 ± 0.78	65.3%	ND 9.78 ± 0.31	97.8%
Ag	0 1	ND 0.73 ± 0.04	73.3%	ND 0.96 ± 0.01	96%

[a] Unspiked sample; [b] Sample spiked with multi-element standard solution; [c] Not detected and calculated as 0.

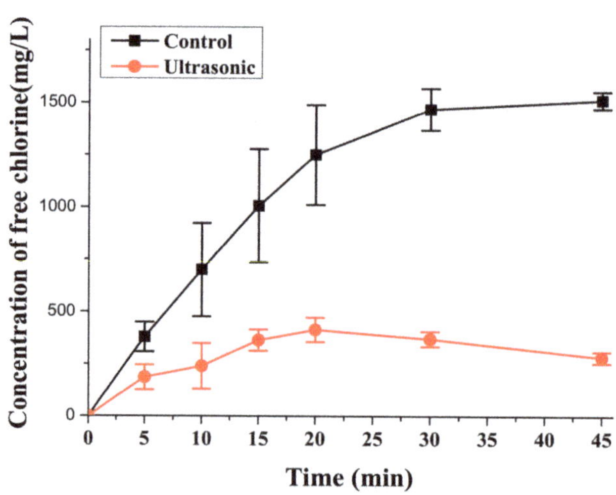

Figure 5. Variations in the concentration of free chlorine during digestion operation (45 min).

As shown in Figure 7, the micromorphology of undissolved solid matters before and after UAEO digestion was observed using FE-SEM and OM. The undissolved solid matters before digestion (Figure 7A,B) presented typical micromorphological characteristics of the dairy slurry, an irregular blocky structure. As shown in Figure 7D,E, the blocky structure appeared as arc edge and cellular structure after the UAEO digestion method. More macroscopic OM images (Figure 7C–E) showed that the digestion process not only reduced the amount of undissolved solid matters but also made them fragmented and blanched.

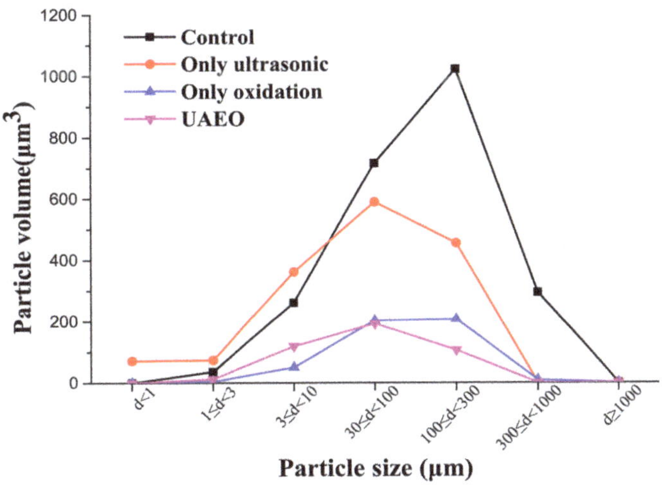

Figure 6. Particle size distribution of undissolved solid matters in the dairy slurry sample after different digestion treatments (45 min).

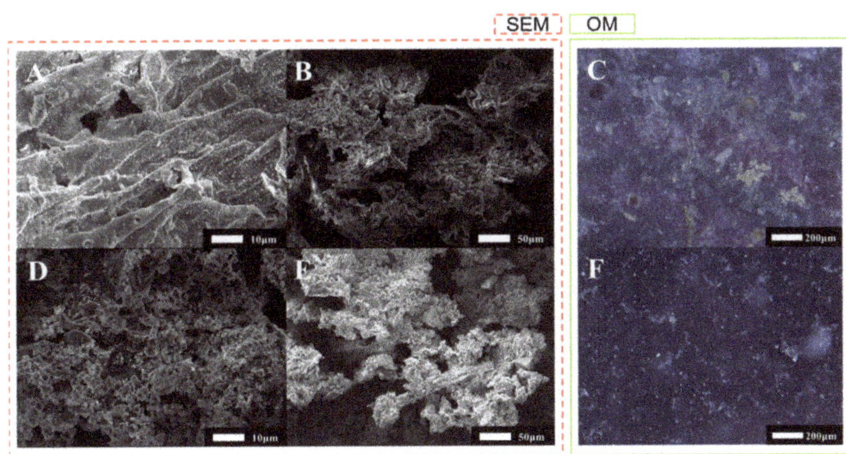

Figure 7. FE-SEM and OM images of undissolved solid matters before (**A–C**) and after (**D–F**) UAEO digestion.

4. Discussion

4.1. Digestion Theory of the UAEO Method

The constitution of UAEO technology is shown in Figure 1. The digestion device comprised two open cuvettes of NaCl solution, which was connected with a NaCl salt bridge. Under the influence of electric potential difference, negative ions moved toward the working electrode and positive ions moved toward the auxiliary electrode. The reactions in the digestion cuvette were expressed by Equations (3)–(6). Hydroxyl radicals and free chlorine were generated in the digestion process, and redundant H^+ provided an acidic environment for easy distribution of HMs. In addition, the ultrasound function module produced ultrasonic waves, generated shock waves to break large particles and solids, accelerated the redox reaction, and then improved the digestion effect [19,29]. By setting a classic three-electrode system and ultrasound function module, the synergy between ultrasonic effect and electrochemical digestion was ensured in the digestion process to improve the digestion effect.

$$(Ru\text{-}Ir)@Ti + H_2O \rightarrow (Ru\text{-}Ir)@Ti(OH) + H^+ + e^- \tag{3}$$

$$Cl^- + OH \rightarrow \cdot ClO^- + H^+ + e^- \tag{4}$$

$$2Cl^- \rightarrow Cl_2 + 2e^- \tag{5}$$

$$Cl_2 + 2H_2O \rightarrow HClO/ClO^- + Cl^- + H^+/2H^+ \tag{6}$$

4.2. Digestion Efficiency of the UAEO Method

A previous investigation [12] indicated that the concentrations of Zn and Cu were significantly higher than those of the other HMs in the dairy farm slurry. Therefore, a series of digestion efficiency experiments were designed to explore the optimal operating parameters of the UAEO apparatus based on the results of Zn and Cu digestion efficiency. Figure 3A shows the effect of digestion time on digestion efficiency at 400 W ultrasonic power with a NaCl concentration of 10 g/L. As expected, the digestion efficiency increased continually in the first 45 min due to sustaining electrochemical oxidation and ultrasound function. As shown in Figure 3B, the digestion efficiency was positively associated with ultrasonic power. The ultrasound intervention could not only increase the temperature of the digestion solution to accelerate the redox reaction but generate shock waves to break large particles and solids. When the ultrasonic power was kept at 400 W or above, the solution temperature reached 60 °C and the digestion efficiency reached the maximum. NaCl was used as an oxidation substrate to generate an oxidizing agent and working current. Figure 3C shows that a low concentration of NaCl restricted the oxidation reaction rate. However, excessively high salinity possibly had a negative impact on the follow-up analysis. Taken together, 10 g/L NaCl was the ideal concentration for the UAEO digestion method. The foregoing conclusions not only proved that the UAEO digestion method could sufficiently release Zn and Cu, which were the main HMs in the dairy slurry, but also found the optimal UAEO operating parameters (digestion time: 45 min; ultrasonic power: 400 W; NaCl concentration: 10 g/L).

An HM recovery rate test was carried out to verify the UAEO digestion effect on broad-spectrum HMs. The element form in spiked digestion samples was different from that in normal samples because the spiked process could form nonionic compounds only in solution and not in insoluble particulate matter. The results of the recovery rate could still partly prove the applicability of UAEO digestion for different HMs. As shown in Figure 4, the UAEO digestion method demonstrated adequate ability for 10 primary HM elements, illustrating that the UAEO method had indiscriminate digestion function for dairy slurry HMs.

4.3. Digestion Mechanism of the UAEO Method

Free chlorine was the main oxidation medium in the digestion solution, according to Equations (3)–(6). Therefore, the variations in the concentration of free chorine were tested to discuss the key mechanism of the UAEO method. As shown in Figure 5, the variation curves demonstrated a high concentration of free chorine without ultrasound intervention. Due to the great oxidizability of free chlorine, it was easy to degrade organics and release HMs oxidatively. Although the ultrasound intervention decreased the maximum concentration of free chorine due to vibration and temperature effect, 414 mg/L free chorine still theoretically had sufficient oxidative ability. On the contrary, the inhalation toxicology of excessive chlorine might influence the health of operators [30]. Because the dairy slurry had many undissolved particles, single electrochemical oxidation digestion could not process them. The ultrasound intervention could exactly constitute this function attributed to its crushing capacity. These hypotheses were verified by the results of particle size distribution analysis. As shown in Figure 6, the particle size distribution obviously moved toward small size, although the ultrasound intervention could not sharply decrease the total volume of particles. Moreover, under the oxidation effect, the total volume and the proportion of bulky particles obviously declined. To intuitively demonstrate the change in particle form during the digestion process, the micromorphology results are shown in Figure 7. The ultrasound intervention led to the fragmentation of the bulky particles, and electrochemical oxidation made an irregular blocky structure form arc edges and a cellular structure.

4.4. Advantages of the UAEO Digestion Method

The conventional digestion methods usually required strong acids as digestion reagents, and the digestion process was completed under high-temperature and high-pressure conditions, which depended on large-scale digestion equipment. They were difficult to meet the requirements of the onsite test. UAEO was a novel and efficient method for dairy farm slurry digestion. The electrochemical oxidation process provided a sufficient oxidation agent. The ultrasound intervention not only physically broke the particles and solids in the dairy slurry but also catalyzed the improvement in oxidation digestion efficiency. On the other hand, the UAEO digestion method is an economical approach owing to its low-cost reagent and simple apparatus design. The UAEO digestion method was fast and convenient and ensured digestion efficiency, thus having a good application prospect.

5. Conclusions

Dairy farm slurry is an important cheap biomass resource rich in mineral nutrients and lignocellulose. However, the abuse of feed and the lack of harmless treatment lead to HM pollution. An innovative digestion method was established in this study, which combined digestion efficiency and process convenience. According to the experimental results, the UAEO method can sufficiently digest all main HMs (Zn and Cu) or trace HMs in slurry. This might be further beneficial to agriculture biomass resource quality control and effective transformation.

Supplementary Materials: The following are available online at https://www.mdpi.com/article/10.3390/ma14164562/s1. Table S1, Operating program of the microwave apparatus. Table S2, ICP-OES instrumental parameters for the analysis of 10 objective HMs.

Author Contributions: Conceptualization, C.L., B.X., Z.Q. and J.W.; formal analysis, C.L., S.W. and C.Z.; investigation, X.Z., S.S. and H.X.; writing—original draft preparation, C.L. and B.X.; writing—review and editing, L.D.; visualization, X.Y.; project administration, R.Z. and Z.S.; funding acquisition, Z.Q. and J.W. All authors have read and agreed to the published version of the manuscript.

Funding: This study was funded by the National Key Research and Development Program of China (grant number 2018YFD0800104), the Special Fund (grant numbers AWS18J004 and AWS16J004), and the Natural Science Foundation of Tianjin, China (grant number 19JCYBJC23800).

Institutional Review Board Statement: Not applicable.

Informed Consent Statement: Not applicable.

Data Availability Statement: The raw/processed data required to reproduce these findings cannot be shared at this time as the data also forms part of an ongoing study.

Conflicts of Interest: The authors declare no conflict of interest.

References

1. Lu, Y.Z.; Zhuo, C.; Li, Y.J.; Li, H.S.; Yang, M.Y.; Xu, D.N.; He, H.Z. Evaluation of filamentous heterocystous cyanobacteria for integrated pig-farm biogas slurry treatment and bioenergy production. *Bioresour. Technol.* **2020**, *297*, 9. [CrossRef]
2. Muhammad, L.; Tareq, A. A novel solution towards zero waste in dairy farms: A thermodynamic study of an integrated polygeneration approach. *Energy Convers. Manag.* **2021**, *230*, 113753.
3. Tethi, B.; Shashi, B.; Kumar, P.S.; Shaon, R.C. An eco-friendly strategy for dairy wastewater remediation with high lipid microalgae-bacterial biomass production. *J. Environ. Manag.* **2021**, *286*, 112196.
4. Wang, X.; Ledgard, S.; Luo, J.; Guo, Y.; Zhao, Z.; Guo, L.; Liu, S.; Zhang, N.; Duan, X.; Ma, L. Environmental impacts and resource use of milk production on the North China Plain, based on life cycle assessment. *Sci. Total Environ.* **2018**, *625*, 486–495. [CrossRef] [PubMed]
5. Zhu, D.; Wei, Y.; Zhao, Y.; Wang, Q.; Han, J. Heavy Metal Pollution and Ecological Risk Assessment of the Agriculture Soil in Xunyang Mining Area, Shaanxi Province, Northwestern China. *Bull. Environ. Contam. Toxicol.* **2018**, *101*, 178–184. [CrossRef] [PubMed]

6. Jean, H.; Eléonore, L.; Carole, S.; Arnaud, H. Identifying the resource use and circularity in farm systems: Focus on the energy analysis of agroecosystems. *Resour. Conserv. Recycl.* **2021**, *169*, 105502.
7. Li, J.; Akdeniz, N.; Kim, H.H.M.; Gates, R.S.; Wang, X.; Wang, K. Optimal manure utilization chain for distributed animal farms: Model development and a case study from Hangzhou, China. *Agric. Syst.* **2021**, *187*, 102996. [CrossRef]
8. Phitsanu, T.; Alongkot, B.; Suwicha, K.; Srisamai, W.; Juree, P.; Suwanna, T.; Hathairad, H.; Ratchaneekorn, M.; Ramnaree, N.; Sutha, K. Comparative study of heavy metal and pathogenic bacterial contamination in sludge and manure in biogas and non-biogas swine farms. *J. Environ. Sci.* **2011**, *23*, 991–997.
9. Roa Engel Carol, A.; Van Gulik Walter, M.; Leonie, M.; Van Der Wielen Luuk, A.M.; Straathof Adrie, J.J. Development of a low pH fermentation strategy for fumaric acid production by Rhizopus oryzae. *Enzym. Microb. Technol.* **2011**, *48*, 39–47. [CrossRef]
10. Wen, Z.; Liao, W.; Chen, S. Production of cellulase by Trichoderma reesei from dairy manure. *Bioresour. Technol.* **2005**, *96*, 491–499. [CrossRef] [PubMed]
11. Li, H.; Dai, M.; Dai, S.; Dong, X. Current status and environment impact of direct straw return in China's cropland–A review. *Ecotoxicol. Environ. Saf.* **2018**, *159*, 293–300. [CrossRef]
12. Liu, W.-R.; Zeng, D.; She, L.; Su, W.-X.; He, D.-C.; Wu, G.-Y.; Ma, X.-R.; Jiang, S.; Jiang, C.-H.; Ying, G.-G. Comparisons of pollution characteristics, emission situations, and mass loads for heavy metals in the manures of different livestock and poultry in China. *Sci. Total Environ.* **2020**, *734*, 139023. [CrossRef]
13. Lin, Z.; Chen, X.; Xi, Z.; Lin, S.; Sun, X.; Jiang, X.; Tian, H. Individual heavy metal exposure and birth outcomes in Shenqiu county along the Huai River Basin in China. *Toxicol. Res.* **2018**, *7*, 444–453. [CrossRef]
14. Xiaodong, W.; Yu, Z.; Qingwen, D.; Shoulian, J.; Xia, Z.; Jie, G. Investigation of novel rapidly synergistic cloud point extraction pattern for bismuth in water and geological samples coupling with flame atomic absorption spectrometry determination. *Spectrochim. Acta Part A Mol. Biomol. Spectrosc.* **2012**, *89*, 1–6.
15. Chen, L.; Lei, Z.; Yang, S.; Wen, X. Application of portable tungsten coil electrothermal atomic absorption spectrometer for the determination of trace cobalt after ultrasound-assisted rapidly synergistic cloud point extraction. *Microchem. J.* **2017**, *130*, 452–457. [CrossRef]
16. Wen, X.; Yang, S.; Zhang, H.; Deng, Q. Combination of knotted reactor with portable tungsten coil electrothermal atomic absorption spectrometer for on-line determination of trace cadmium. *Microchem. J.* **2016**, *124*, 60–64. [CrossRef]
17. Wang, Z.; Sun, X.; Li, C.; He, X.; Liu, G. On-site detection of heavy metals in agriculture land by a disposable sensor based virtual instrument. *Comput. Electron. Agric.* **2016**, *123*, 176–183. [CrossRef]
18. Ajay, P.V.S.; Printo, J.; Kiruba, D.S.C.G.; Susithra, L.; Takatoshi, K.; Sivakumar, M. Colorimetric sensors for rapid detection of various analytes. *Mater. Sci. Eng. C* **2017**, *78*, 1231–1245.
19. Santos Daniele, C.M.B.; Carvalho Larissa, S.B.; Lima Daniel, C.; Leão Danilo, J.; Teixeira Leonardo, S.G.; Korn Maria Graças, A. Determination of micronutrient minerals in coconut milk by ICP OES after ultrasound-assisted extraction procedure. *J. Food Compos. Anal.* **2014**, *34*, 75–80. [CrossRef]
20. Luisa, A.M.; Elisabetta, M.; Carmela, P.; Matteo, V.; Elisa, S.; Paola, M.; Silvia, C. Optimization and validation of a fast digestion method for the determination of major and trace elements in breast milk by ICP-MS. *Anal. Chim. Acta* **2018**, *1040*, 49–62.
21. Fernandes, D.O.A.; Santos, C.; Rossana, B.S.; Araujo, N.A.R. The use of diluted formic acid in sample preparation for macro- and microelements determination in foodstuff samples using ICP OES. *J. Food Compos. Anal.* **2018**, *66*, 7–12. [CrossRef]
22. Krishna, M.V.B.; Chandrasekaran, K.; Venkateswarlu, G.; Karunasagar, D. Development of a simple and rapid microwave-assisted extraction method using very dilute solutions of perchloric acid and hydrogen peroxide for the multi-elemental analysis of food materials by ICP-OES: A green analytical method. *Microchem. J.* **2019**, *146*, 807–817.
23. Kuznetsova, O.V.; Burmii, Z.P.; Orlova, T.V.; Sevastyanov, V.S.; Timerbaev, A.R. Quantification of the diagenesis-designating metals in sediments by ICP-MS: Comparison of different sample preparation methods. *Talanta* **2019**, *200*, 468–471. [CrossRef] [PubMed]
24. Rovasi, A.F.; Cícero, D.N.P.; Camera, L.G.; Denise, B.; Carine, V.; Machado, L. Simultaneous determination of Fe and Ni in guarana (Paullinia cupana Kunth) by HR-CS GF AAS: Comparison of direct solid analysis and wet acid digestion procedures. *J. Food Compos. Anal.* **2020**, *88*, 103459.
25. Da Silva, I.J.S.; Lavorante André, F.; Paim, A.P.S.; Da Silva, M.J. Microwave-assisted digestion employing diluted nitric acid for mineral determination in rice by ICP OES. *Food Chem.* **2020**, *319*, 126435. [CrossRef]
26. Vimlesh, C.; Surendra, P. ICP-OES assessment of heavy metal contamination in tropical marine sediments: A comparative study of two digestion techniques. *Microchem. J.* **2013**, *111*, 53–61.
27. Rice, E.W.; Baird, R.B.; Eaton, A.D.; Clesceri, L.S. *Standard Methods for the Examination of Water and Wastewater*; American Water Works Association: Washington, DC, USA, 2012.
28. Ge, B.; Hunter, J.; Hunter, W. *Satistics for Experimenters: Design, Innovation, and Discovery*, 2nd ed.; Wiley: Hoboken, NJ, USA, 2005.
29. Tadeo, J.L.; Sánchez-Brunete, C.; Albero, B.; García-Valcárcel, A.I. Application of ultrasound-assisted extraction to the determination of contaminants in food and soil samples. *J. Chromatogr. A* **2010**, *1217*, 2415–2440. [CrossRef]
30. Musah, S.; Schlueter, C.F.; Humphrey, D.M.; Powell, K.S.; Roberts, A.M.; Hoyle, G.W. Acute lung injury and persistent small airway disease in a rabbit model of chlorine inhalation. *Toxicol. Appl. Pharmacol.* **2017**, *315*, 1–11. [CrossRef] [PubMed]

sustainability

MDPI

Review

Remediation of Petroleum-Contaminated Soils with Microbial and Microbial Combined Methods: Advances, Mechanisms, and Challenges

Xin Sui [1], Xuemei Wang [1], Yuhuan Li [1] and Hongbing Ji [1,2,*]

[1] Beijing Key Laboratory of Resource-Oriented Treatment of Industrial Pollution, School of Energy and Environmental Engineering, University of Science and Technology Beijing, Beijing 100083, China; sx4426@163.com (X.S.); wangxuemei0000@126.com (X.W.); hyulily@163.com (Y.L.)

[2] Beijing Municipal Key Laboratory of Resource Environment and GIS, College of Resource Environment and Tourism, Capital Normal University, Beijing 100048, China

* Correspondence: ji.hongbing@hotmail.com; Tel.: +86-1831-1312-619

Abstract: The petroleum industry's development has been supported by the demand for petroleum and its by-products. During extraction and transportation, however, oil will leak into the soil, destroying the structure and quality of the soil and even harming the health of plants and humans. Scientists are researching and developing remediation techniques to repair and re-control the afflicted environment due to the health risks and social implications of petroleum hydrocarbon contamination. Remediation of soil contamination produced by petroleum hydrocarbons, on the other hand, is a difficult and time-consuming job. Microbial remediation is a focus for soil remediation because of its convenience of use, lack of secondary contamination, and low cost. This review lists the types and capacities of microorganisms that have been investigated to degrade petroleum hydrocarbons. However, investigations have revealed that a single microbial remediation faces difficulties, such as inconsistent remediation effects and substantial environmental consequences. It is necessary to understand the composition and source of pollutants, the metabolic genes and pathways of microbial degradation of petroleum pollutants, and the internal and external aspects that influence remediation in order to select the optimal remediation treatment strategy. This review compares the degradation abilities of microbial–physical, chemical, and other combination remediation methods, and highlights the degradation capabilities and processes of the greatest microbe-biochar, microbe–nutrition, and microbe–plant technologies. This helps in evaluating and forecasting the chemical behavior of contaminants with both short- and long-term consequences. Although there are integrated remediation strategies for the removal of petroleum hydrocarbons, practical remediation remains difficult. The sources and quantities of petroleum pollutants, as well as their impacts on soil, plants, and humans, are discussed in this article. Following that, the focus shifted to the microbiological technique of degrading petroleum pollutants and the mechanism of the combined microbial method. Finally, the limitations of existing integrated microbiological techniques are highlighted.

Keywords: petroleum contaminated soil; composition of petroleum; harm of petroleum; microbial remediation; combined microbial methods; phytoremediation; biochar

Citation: Sui, X.; Wang, X.; Li, Y.; Ji, H. Remediation of Petroleum-Contaminated Soils with Microbial and Microbial Combined Methods: Advances, Mechanisms, and Challenges. *Sustainability* **2021**, *13*, 9267. https://doi.org/10.3390/su13169267

Academic Editor: Avelino Núñez-Delgado

Received: 13 July 2021
Accepted: 12 August 2021
Published: 18 August 2021

1. Introduction

Extraction, processing, and transportation (pipe rupture) all contribute to the entry of petroleum into the soil environment [1,2]. The primary contaminants in petroleum-contaminated soil are toxic and hazardous aliphatic, cycloaliphatic, and aromatic hydrocarbons [3]. They decrease the diversity of plants and microbes in the soil, deplete soil fertility, disrupt soil ecological balance, and even put human health at risk [4]. Crop germination is delayed, the chlorophyll content is poor, and some crops perish when grown in high petroleum-contaminated soil [5]. Furthermore, pollutants can enter the human body by breathing, skin contact, or eating petroleum-contaminated food, causing contact dermatitis, visual and auditory hallucinations, and gastrointestinal disorders, as well as substantially raising the risk of leukemia in children. Although certain low-molecular-weight hydrocarbon pollutants will weather and decay over time, high-molecular-weight hydrocarbon pollutants will remain in the soil for a long period due to their hydrophobicity, causing secondary contamination in the ecosystem [6,7]. According to statistics, Chevron Texaco's oilfields in Ecuador's Amazon region have harmed human health, the water supply, and the ecosystems in the area. The plaintiff (30,000 individuals of mixed races and indigenous peoples) was awarded USD 9.5 billion by the Cuban Supreme Court in 2013 [8]. This demonstrates that petroleum pollutants have a negative influence on society in addition to destroying the environment. As a result, the issue of restoring petroleum-contaminated soil has become a hot topic.

Incineration, landfill, leaching, chemical oxidation, and microbiological treatment are now used to remediate petroleum-contaminated soil. These technologies can extract, remove, transform, or mineralize petroleum pollutants in a contaminated environment, transforming them into a less damaging, harmless, and stable form [9]. Although incineration and chemical oxidation can remove 99.0% and 92.3% of total petroleum hydrocarbons, respectively, both restoration procedures have disadvantages [10,11]. Toxic substances such as dioxins, furans, polychlorinated biphenyls, and volatile heavy metals will be released into the atmosphere as a result of incomplete petroleum burning [12]. The carbon in the soil is reduced by 49–98% as the incineration temperature rises from 200 °C to 1050 °C, and the organic matter and carbonate in the soil are decomposed into light hydrocarbons (C_2H_2, C_2H_4, and CH_4) and carbon dioxide separately [13,14]. The total number of soil microorganisms decrease from 10^4 CFU/g to 10^3 CFU/g to 10^2 CFU/g after chemically oxidizing petroleum pollutants in the soil with 5 percent hydrogen peroxide and persulfate for 10 days. The bacteria will continue to develop slowly over the following 10 days [15]. The incomplete combustion of petroleum increases the hidden dangers of environmental safety, while the loss of carbon and organic matter limits the recovery ability of the soil ecosystem. The addition of oxidants will inhibit the growth of soil microorganisms. Therefore, while reducing the concentration of soil petroleum pollutants, it will not cause secondary pollution to the soil and the surrounding environment, which has become the main consideration for selecting remediation technologies.

Microbial remediation is inexpensive, and it can completely mineralize organic pollutants into carbon dioxide, water, inorganic compounds and cell proteins, or convert complex organic pollutants into other simpler organics [16]. Microorganisms can utilise organic pollutants as their only source of carbon, allowing them to degrade organic pollutants in the soil [17,18]. Microorganisms destroyed 62–75% of petroleum hydrocarbons in the soil in 150 to 270 days [19,20]. Free microorganisms destroyed 2.3–6.8% of petroleum hydrocarbons in 60 days, however when biochar was employed as a carrier, 7.2–30.3% of petroleum hydrocarbons were degraded in 60 days [21]. Petroleum degraded at a rate of 29.8% in the immobilized system (sodium alginate-diatomite beads), whereas free cells degraded at a rate of 21.2% in 20 days [22]. At 4 °C and 10 °C, microbial mineralization of hexadecane generates 45% CO_2, while at 25 °C, 68% CO_2 is generated in 50 days, indicating that microorganisms can better digest hexadecane [23]. When the soil salinity is higher than 8%, and the pH value is lower than 4 and higher than 9, the activity of *Acinetobacter baylyi ZJ2* is affected, and a certain amount of lipopeptide surfactant cannot be produced, thereby reducing the degradation of petroleum by microorganisms [24].

In conclusion, extreme environmental conditions (soil temperature below 10 °C, pH below 4 and more than 9) decrease microbial activity, which diminishes the removal impact of petroleum pollutants. Furthermore, a pH 5.5–8.8, temperature 15–45 °C, oxygen content 10%, low clay or silt content soil type, and C/N/P ratio of 100:10:1 are the optimum conditions for microbial remediation of oily soil, according to current research [25,26]. Long remediation times and low remediation efficacy of free microorganisms are issues with microbial remediation. The microbial combination technique is used to increase the bio-degradation efficiency of microorganisms in order to overcome the challenge of microbial remediation of petroleum in the soil.

The source, categorization, and content of hydrocarbon contamination in soil, as well as its influence on the environment and human health, are discussed in this article. Following that, the different forms of combination microbial repair methods are explained, as well as their benefits. The microbial combination method focuses on the microbial remediation of petroleum pollutants and the interaction of microbial–biochar/nutrients/plants. Finally, the benefits and limitations of contemporary microbial mixed approach repair technologies are discussed.

2. Petroleum-Contaminated Soil

2.1. Sources of Petroleum Pollutants

Figure 1 depicts the major pathways through which petroleum contaminants permeate the soil [27]. Petroleum spills are a major cause of hydrocarbon contamination in the soil. The global leakage of natural petroleum is reported to be 600,000 metric tons per year [28]. Petroleum contamination is estimated to have polluted 3.5 million sites in Europe [29]. In China, about 4.8 million hectares of soil petroleum content may exceed the safe limit [30]. Distinct nations and areas have varied sampling and transportation techniques, as well as different sources and degrees of petroleum contamination. Furthermore, contaminants are leached into the surrounding and deep soil in horizontal and vertical orientations, as well as into the groundwater system, as a result of rainfall washing and leaching.

Figure 1. Major sources of hydrocarbons in the soils.

Low-molecular-weight hydrocarbons are more volatile and more easily penetrate into groundwater than high-molecular-weight hydrocarbons, although volatilization and permeability are influenced by the physical and chemical characteristics of the soil, climate, and vegetation [29]. The natural decay half-life of petroleum hydrocarbons grows as the concentration of petroleum hydrocarbons increases (when the petroleum concentration is 250 mg/L, the half-life is 217 days) [31]. The natural half-life of alkane and aromatic pollutants rises with increasing molecular weight. Under normal conditions, the half-life of the three-ring molecule phenanthrene is 16 to 126 days, but the half-life of the five-ring molecule benzo[α]pyrene is 229 to 1400 days [32]. Though some specific bacteria in polluted soil may biodegrade and bio-transform these hydrocarbons, absorbing them into biomass in the soil [33,34], small quantities of hydrocarbons (such as long chain and high molecular weight hydrocarbons) are still challenging to handle in the environment due to the non-polarity and chemical inertness of pollutants [35].

2.2. *Composition of Petroleum Pollutants*

Petroleum-contaminated soil often contains petroleum, water, and solid particles. Petroleum pollutants are often shown as water-in-petroleum (W/O). Petroleum is made up of a variety of hydrocarbons, composed of carbon (83–87%), hydrogen (11–14%), and sulfur (0.06–0.8%), nitrogen (0.02–1.7%), oxygen (0.08–1.82%), and trace metal components (nickel, vanadium, iron, antimony, etc.) [36]. Hydrocarbons formed by the combination of carbon and hydrogen constitute the main component of petroleum, accounting for about 95% to 99%. Various hydrocarbons are classified according to their structure: alkanes, cycloalkanes, and aromatic hydrocarbons.

Alkanes are the main components of gasoline, diesel, and jet fuel [37,38]. The molecular structure is linear, branched, and cyclic. The general formula of linear-alkanes is CnH2n+2, the general formula of branched alkanes is CnH2n+2 ($n > 2$), and the general formula of cycloalkanes is CnH2n ($n > 3$). Aromatics are found in gasoline, diesel, lubricants, kerosene, tar, and asphalt [39]. They have a similar molecular structure to cycloalkanes, but they have at least one benzene ring [40]. Aromatics have the general formula CnH2n-6.

Petroleum is derived from bitumen, and the heaviest and most polar molecules in asphaltene are firmly adsorbed on the source rock, making discharge into the reservoir problematic. As a result, the most frequent are saturated hydrocarbons with the lowest polarity, followed by aromatics [41]. The molecular weight of hydrocarbons influences their degradability. Low-molecular-weight hydrocarbons have better bioavailability than high-molecular-weight hydrocarbons [42,43]. As a result, hydrocarbon susceptibility to microbial breakdown is generally: linear alkanes > branched alkanes > aromatics with low molecular weight > cyclic alkanes [16,44].

2.3. *Toxic Effects of Petroleum on the Environment*

Saturates, aromatics, and other poisonous and hazardous hydrocarbons are mostly found in petroleum [30]. Highly hazardous petroleum pollutants (PAHs, BTEX) will have an adverse effect on soil, plants, and humans. High levels of polycyclic aromatic hydrocarbons (PAHs) in the soil can induce tumors, reproductive, development, and immunological problems in terrestrial invertebrates [45]. BTEX (benzene, toluene, ethylbenzene, and xylene) can harm a person's personal neurological system, liver, kidneys, and respiratory system [46]. Pollutants obstruct soil pores, alter the content and structure of soil organic matter, diminish the activity and variety of soil microbes and plants, and, as a result, endanger human health via the food chain [47]. Deuterated PAH(dPAH) was utilized by Jose L. Gomez-Eyles et al. to evaluate the bioavailability of PAH in soil [48]. According to research, the dPAH:PAH ratio of benzo(a)pyrene in earthworm tissues is greater than the dPAH:PAH ratio obtained by normal chemical methods. The ratio of additional dPAH accumulated by earthworms is increasing as the size of PAH rises. This indicates that the toxicity of petroleum pollution on animals is much worse than previously thought. The petroleum in the soil also pollutes the groundwater environment through diffusion and migration, putting a strain on a variety of elements of human life.

2.3.1. Toxic Effects of Petroleum on Soil

Petroleum degrades the ecological structure and function of soils [6], affecting soil moisture, pH, total organic carbon, total nitrogen, exchangeable potassium, and enzyme activity substantially (urease, catalase and dehydrogenase) [49–52]. As pollutant concentrations rise, the clay content in contaminated soil rises [53], soil porosity declines, and impermeability and hydrophobicity rise [54], inhibiting the growth of plant roots and the number of bacteria in the soil. The root length of *Lepidium sativum*, *Sinapis alba*, and *Sorghum saccharatum* was reduced by 65.1%, 42.3%, and 47.3%, respectively, when the petroleum hydrocarbon concentration in the soil was 7791 mg/kg [55]. Straight-chain alkanes have the greatest influence on the number of bacteria species. The following is the order of influence: 320.5 ± 5.5 (in the control soil) > 289.1 ± 4.7 (in the aromatic hydrocarbon-contaminated soil) > 258.6 ± 2.5 (in the branched-chain-alkane-contaminated soil) > 229.7 ± 2.0 (in straight-chain- and cyclic-alkanes-hydrocarbons-contaminated soil) [56]. According to studies, the major contaminant that causes soil salinization and acidification is benzo[a]pyrene, which is present in petroleum [57].

2.3.2. Toxic Effects of Petroleum on Plants

Petroleum pollutants have the ability to permeate plant surfaces and move via the intracellular space and vascular system. Plant roots may collect petroleum pollutants in the soil, transport them to leaves and fruits, and store them, as well as transmit pollutants from leaves to roots. Corn germination rate, plant height, leaf area, and dry matter yield were all drastically reduced as a result of petroleum contamination [58]. Plant growth is slowed, stem length and diameter are shortened, aboveground tissue length is reduced, and the root length and plant leaf area are altered due to a lack of oxygen and nutrients in the polluted soil (depending on the plant species) [59]. Low concentrations of petroleum hydrocarbon (10 g/kg) have been found to increase plant root vitality, but medium concentrations (30 g/kg) and high concentrations (50 g/kg) have been shown to decrease plant root vitality. Simultaneously, the chlorophyll content of 50 g/kg petroleum-contaminated soil is almost 60% lower than that of non-contaminated soil [60].

2.3.3. Toxic Effects of Petroleum on Human Health

Exposure to petroleum and petroleum products, whether direct (breathing polluted air and direct contact with skin) or indirect (bathing in contaminated water and eating contaminated food), can cause significant health issues in people [61]. Many petroleum pollutants, such as benzene and polycyclic aromatic hydrocarbons, are toxic, mutagenic, and carcinogenic. Some aromatics have a negative impact on human liver and kidney functioning, even causing cancer [62]. Furthermore, because PAHs are extremely lipophilic, they are easily absorbed by animals through the digestive tract [45]. Long-term exposure to polluted areas can cause tiredness, respiratory problems, eye irritation, and headaches, and women are more likely to have spontaneous abortions [8]. Oil extraction in residential areas, particularly in low- and middle-income nations, has been shown to affect the health of a huge number of non-occupational contacts, according to studies. It is estimated that 638 million people in low- and middle-income countries live in rural areas close to conventional oil reservoirs [8]. Individuals who are more exposed to oil-related pollution and are not typically exposed to occupational areas, such as infants, children, pregnant women, the elderly, or people with prior health conditions, will use daily activities (such as bathing, agricultural activities, and so on) that will be affected. Simultaneously, natural gas burning in oil wells can produce volatile organic compounds (VOCs), nitrogen dioxide (NO_2), sulfur dioxide (SO_2), polycyclic aromatic hydrocarbons, and benzo[a]pyrene, all of which are harmful to non-occupationally exposed individuals.

3. Advances in the Utilization of Microorganisms in Petroleum Remediation

Articles were searched for in "web of science" databases. Databases contain the Core Collection (WOS), the Derwent Innovations Index (DII), the Korean Journal Database (KJD), MEDLINE, the Russian Science Citation Index (RSCI), and Scientific Electronic Library Online (SCIELO) six databases. The items that were retrieved were only published between 1950 and 2020. "Microbial degradation petroleum" is the result of a specific search phrase. The deadline for the search is 17 September 2020, and the findings will be analyzed statistically.

Figure 2 depicts the unprecedented number of research findings on microbial petroleum pollution cleanup from 1950 to 2020. The number of published study findings has risen year after year, suggesting that petroleum microbial remediation technology has attracted the interest of academics both at home and abroad in recent years. Figure 3 shows the statistics on different sorts of research outputs and the percentage of countries/regions that were re-searched. The data shows that the article is the most common kind of research output. The majority of research on microbial remediation of petroleum pollution takes place in Asia (34%) and Europe (34%).

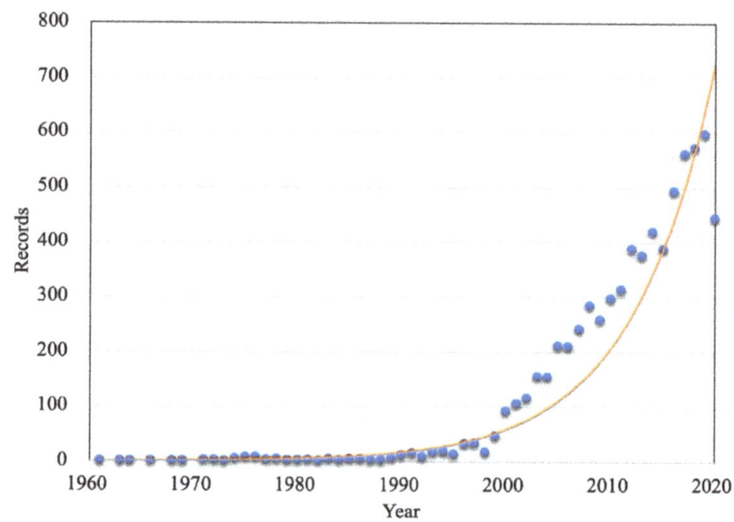

Figure 2. The record number of research results of microbial remediation of petroleum pollution.

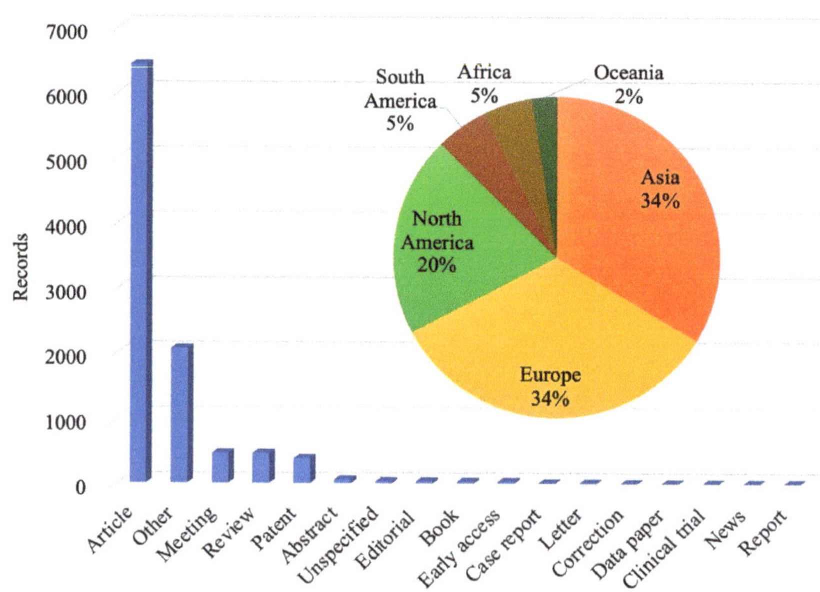

Figure 3. Statistics of research output types and the percentage of countries/regions researched.

4. Microbial Remediation

Bioaugmentation has a high practicality and economic application when compared to physical and chemical remediation techniques [63,64]. By adding lipophilic bacteria, bioaugmentation can be accomplished [65]. Oleophilic bacteria may be found in a wide range of petroleum-contaminated environments, including saltwater, coastlines, sludge, and soil [6]. They may thrive only on hydrocarbons while decomposing or mineralizing harmful and hazardous petroleum contaminants [66,67]. Different types of degrading bacteria can be found in different sorts of polluted environments. To determine the kinds and activities of soil organisms, DNA-based stable isotope probing (DNA-SIP) technology is used [68]. In soil contaminated by polycyclic aromatic hydrocarbons, actinomycetes are a common phylum. *Acidovorax, Rhodoferax, Hydrogenophaga* and *Polaromonas* were found in the soil contaminated in the Philippines. *Acidobacteria* exists in the soil contaminated with petroleum, phenanthrene, pyrene, and fluoranthene.

Studies have demonstrated that a number of bacteria, including *Rhodococcus* sp., *Pseudomonas* sp., and *Scedosporium boydii*, can degrade petroleum contaminants [69–71]. Hydrocarbons are mostly degraded by bacteria via aerobic pathways [72]. When oxygen serves as an electron acceptor, hydrocarbon catabolism is often accelerated [73]. In aerobic mode, the processes of oxidation, reduction, hydroxylation, and dehydrogenation mediate degradation. The biodegradation of hydrocarbons is assisted by enzymes such as monooxygenase, dioxygenase, cytochrome P450, peroxidase, hydroxylase, and dehydrogenase [72,74–77].

Microorganisms that degrade alkanes and PAHs in an inorganic salt liquid media have been effectively isolated for the time term (as shown in Table 1). *Pseudomonas* sp., *Acinetobacter* sp., and *Rhodococcus* sp. are currently the bacteria that have the most effect on the degradation of petroleum pollutants. Short-chain and medium-chain alkanes (C5-C16) can be oxidized by the integral membrane non-heme iron oxygenase (AlkB) or cytochrome P450 enzyme (CYP153) in the strain, according to studies. Putative flavin-binding monooxygenase (AlmA) and long-chain alkane monooxygenase (LadA) is involved in the oxidation of long-chain alkanes [78]. Several degradation genes can coexist in a single bacterium. There are at least two AlkB-type genes (AlkMa and AlkMb) and one AlmA-type gene (AlmA) in *Acinetobacter strain* DSM17874 that are responsible for degrading alkanes of varying chain lengths [79]. *Pseudomonas* sp. also contained nahAc, catechol dioxygenase (C12O and C23O), AlkB, and cytochrome P450, which are important for the degradation of alkanes and polycyclic aromatic hydrocarbons [80–84].

The major pathways for alkane and PAH metabolism in microorganisms include terminal oxidation, subterminal oxidation, ω-oxidation, and β-oxidation. The terminal oxidation pathway is the most common mechanism for alkanes to be destroyed. Alkane hydroxylase introduces molecular oxygen into hydrocarbons to oxidize terminal methyl to form alcohols, which are next oxidized to aldehydes and fatty acids, and eventually, carbon dioxide and water are produced by the β-oxidation pathway [85–87]. PAHs, on the other hand, are resistant to biodegradation due to their structural stability. PAHs are metabolized primarily through a mixed functional oxidase system mediated by the cytochrome P450 enzyme, with oxidation or hydroxylation as the initial step and the production of diols as intermediate products. These intermediates are converted to catechol intermediates via ortho- or meta-cleavage pathways, which are then integrated into the tricarboxylic acid cycle (TCA) [73,86].

Table 1. Common microorganisms that degrade alkanes and polycyclic aromatic hydrocarbons.

Substrates	Microorganisms		Source of Strain	Main Findings			Reference
				Substrate Concentration	Incubation Conditions	Degradation Rate	
PAHs	bacteria	*Achromobacter* sp. HZ01	Petroleum-contaminated seawater, China.	100 mg/kg anthracene, phenanthrene and pyrene.	10^9 cells mL^{-1}/ 28 °C/150 rpm/30 days.	Strain remove anthracene, phenanthrene and pyrene about 29.8%, 50.6%, and 38.4%, respectively.	[88]
		Acinetobacter sp. WSD	Petroleum-contaminated groundwater, Shanxi province of northern China.	1 mg/kg phenanthrene, 2 mg/kg fluorine, and 0.14 mg/kg pyrene.	5% cells suspension/ 33 °C/150 rpm/6 days.	Approximately 90% of fluorine, 90% of phenanthrene, and 50% of pyrene were degraded.	[89]
		Bacillus subtilis BMT4i (MTCC 9447)	Automobile contaminated soil, Uttarakhand, India.	50 g/mL Benzo[a]Pyrene.	1×10^8 cells mL^{-1}/ 37 °C/120 rpm/28 days.	Strain started degrading Benzo[a]Pyrene achieving maximum degradation of approximately 84.66%.	[90]
		Caulobacter sp. (T2A12002)	From King Fahd University of Petroleum and Minerals Department of Life Sciences laboratory.	100 ppm pyrene.	2% cells suspension/ 37 °C and 25 °C/ 120 rpm/18 days/pH 5.0 and pH 9.0.	Strain degraded 35% and 36% of pyrene at 25 °C and 37 °C, respecitvely.	[91]
		Enterobacter sp. (MM087)	Engine-oil-contaminated soil, Puchong and Seri Kembangan, Selangor Malaysia.	500 mg/L phenanthrene and 250 mg/L pyrene.	5% cells suspension and 1×10^6cells mL^{-1}/ 37 ± 0.5 °C/200 rpm/24 h.	Strain with 80.2% degradations for phenanthrene and 59.7% degradations for pyrene.	[92]
		Klebsiella pneumoniae AWD5	Automobile-contaminated soil, Silchar, Assam.	0.005% PAH (Pyrene, Chrysene, Benzo(a)pyrene).	Cells(OD600 = 0.4)/ 30 °C/140 rpm/9 days.	Strain degraded pyrene (56.9%), chrysene (36.5%) and benzo(a)pyrene (50.5%), respectively.	[93]
		Mycobacterium vanbaalenii PYR-1	Petroleum-contaminated sediment and water, the watershed of Redfish Bay near Port Aransas, Tex.	0.5 ug/mL pyrene.	1.5×10^6 cells mL^{-1}/ 24 °C /150 rpm/48 to 96 h.	After incubation, 47.3 to 52.4% of pyrene was mineralized to CO_2.	[94]

Table 1. *Cont.*

Substrates	Microorganisms	Source of Strain	Main Findings			Reference
			Substrate Concentration	Incubation Conditions	Degradation Rate	
	Raoultella planticola	Near a car repair station, Hangzhou, China.	20 mg L^{-1} pyrene and 10 mg L^{-1} benzo[a]pyrene.	2.0 × 10^8 cells mL^{-1}/30 °C /180 rpm/10 days.	Strain degraded 52.0% of pyrene and 50.8% of benzo[a]pyrene.	[95]
	Rhodococcus sp. P14	Petroleum-contaminated sediments, Xiamen, China.	50 mg/L phenanthrene, pyrene and benzo[a]pyrene.	1% cells suspension/30 °C/150 rpm/30 days.	Strain degraded 34% of the pyrene, about 43% of the phenanthrene and 30% of the Benzo[a]pyrene.	[96]
	Pseudomonas sp. MPDS	PAH- and petrochemical-contaminated soil and mud, Tianjin.	1 mg/mL naphthalene, 0.1 mg/mL dibenzofuran, 0.1 mg/mL dibenzothiophene, 0.1 mg/mL fluorene.	Cells(OD600 = 5.0)/25°C/200 rpm/84 h, 96 h, and 72 h.	Strain could completely degrade naphthalene in 84 h. A total of 65.7% dibenzofuran and 32.1% dibenzothiophene could be degraded in 96 h and 40.3% fluorene could be degraded in 72 h.	[97]
	Pseudoxanthomonas sp. DMVP2	Petroleum-contaminated sediment, Gujarat, India.	300 ppm phenanthrene	4% cells suspension/37 °C/150 rpm/72 h.	Strain was able to degrade 86% phenanthrene.	[98]
	Sphinogmonas sp.	Typical mangrove swamp(surface sediment (0–2 cm)), Ho Chung, Hong Kong.	5000 mg L^{-1} phenanthrene.	180 rpm/7 days.	Strain was obtained to degrade 99.4% phenanthrene at the end of 7 days.	[99]
	Stenotrophomonas sp. IITR87	—[1]	Phenanthrene(10 ppm), pyrene(10 ppm), and benzo-α-pyrene(10 ppm).	0.8% cells suspension/30 °C/175 rpm/15 days.	Strain showed >99, 98, and <50% degradation of phenanthrene, pyrene, and benzo-α-pyrene respectively.	[100]
	Streptomyces sp. (ERI-CPDA-1)	Petroleum-contaminated soil, Chennai, India.	Naphthalene(0.1%), phenanthrene(0.1%).	3% cells suspension/30 °C/200 rpm/7 days.	Strain could remove 99.14% naphthalene and 17.5% phenanthrene.	[101]

Table 1. *Cont.*

Substrates	Microorganisms	Source of Strain	Main Findings			Reference
			Substrate Concentration	Incubation Conditions	Degradation Rate	
fungus	*Aspergillus* sp. RFC-1	Rumaila oilfield(surface polluted sludge (1–10 cm)), Basra, Iraq.	50 mg/L naphthalene, 20 mg/L phenanthrene, 20 mg/L pyrene.	10% cells suspension/30 °C/ 120 rpm/7 days.	Biodegradation efficiencies of crude oil, naphthalene, phenanthrene, and pyrene were 84.6%, 50.3%, and 55.1%, respectively.	[102]
	Nocardia sp. H17-1	Petroleum-contaminated soil	Aliphatic and aromatic (1%, *w*/*v*).	30 °C/6 days.	The aliphatic and aromatic fractions were degraded 99.0 ± 0.1% and 23.8 ± 0.8%, respectively.	[103]
	Penicillium sp. CHY-2	Soil, Antarctic.	100 mg L^{-1} butylbenzene, naphthalene, acenaphthene, ethylbenzene, and benzo[a]pyrene.	20 °C/110 rpm/28 days.	Strain showed the level of degradation for butylbenzene (42.0%), naphthalene (15.0%), acenaphthene (10.0%), ethylbenzene (4.0%), and benzo[a]pyrene (2.0%).	[104]
	Trichoderma sp.	—[1]	100 mg kg^{-1} pyrene and benzo(a)pyrene.	240 h	Strain degraded 63% of pyrene (100 mg kg^{-1}) and 34% of benzo(a)pyrene (100 mg kg^{-1}) after 240 h of incubation.	[105]
	Fusarium sp.	—[1]	100 mg kg^{-1} pyrene and benzo(a)pyrene.	240 h	Strain degraded 69% of pyrene (100 mg kg^{-1}) and 37% of benzo(a)pyrene (100 mg kg^{-1}) after 240 h of incubation.	[105]

Table 1. *Cont.*

Substrates	Microorganisms		Source of Strain	Main Findings			Reference
				Substrate Concentration	Incubation Conditions	Degradation Rate	
alkanes	bacteria	*Achromobacter* sp. HZ01	Petroleum-contaminated seawater, China.	2% (*w*/*v*) diesel oil	28 °C/150 rpm/10 days.	Strain degraded the total n-alkanes reached up to 96.6%.	[88]
		Acinetobacter sp. (*KC211013)*	Coal chemical industry wastewater treatment plant, northeast China.	700 mg/L alkanes.	35 °C	The degradation rate reached 58.7%.	[106]
		Bacillus subtilis	Petroleum-polluted soil, Shengli Oilfield, China.	0.3% (*w*/*v*) crude oil.	6% cells suspension/ 30 °C/150 rpm /5 days.	The results indicated that 30–80% of the n-alkanes (C13–C30) were degraded by strain.	[107]
		Pseudomonas sp. WJ6	Xinjiang oilfield, China.	0.5% (*w*/*v*) n-alkanes.	1010 CFU mL^{-1}/ 37 °C/180 rpm/20 days.	N-dodecane (C12) was degraded by 46.65%. 42.62%, 31.69%, and 23.62% of C22, C32, and C40 were degraded, respectively.	[108]
		Rhodococcus sp.	Bay of Quinte, Ontario, Canada.	0.1% (*v*/*v*) diesel fuel.	Cells(OD600 = 0.025)/ 0 °C/150 rpm/102 days.	After 102 days of incubation at 0 °C, strain mineralized C12 (8%), C16 (6.1%), C28 (1.6%), and C32 (4.3%).	[109]
	fungus	*Cladosporium Resinae*	Soil, Australian.	12.5%(*v*/*v*) n-alkanes.	0.75–1.25% cells suspension/35 °C/ 35 days.	All higher n-alkanes from n-nonane to n-octadecane were assimilated by the fungus.	[110]
		Penicillium sp. CHY-2	Soil, Antarctic.	100 mg L^{-1} decane, dodecane and octane.	20 °C/110 rpm/28 days.	Strain was degraded decane (49.0%), dodecane (33.0%), and octane (8.0%).	[104]

Table 1. *Cont.*

Substrates	Microorganisms		Source of Strain	Main Findings			Reference
				Substrate Concentration	Incubation Conditions	Degradation Rate	
	actinomycetes	*Gordonia* sp.	Hydrocarbon-contaminated Mediterranean shoreline, west coast of Sicily, Italy.	1 g L^{-1} eicosane and octacosane.	30 °C /28 days.	Eicosane and octacosane were degraded from 53% to 99% in 28 days.	[111]
		Tsukamurella sp. MH1	Petroleum-contaminated soil, Pitești, Romania.	0.5% (*v*/*v*) liquid alkanes.	30 °C	Strain capable to use a wide range of n-alkanes as the only carbon source for growth.	[112]

[1] There is no clear description in the article.

Low-molecular-weight saturated hydrocarbons and aromatic hydrocarbons are easily degraded by microorganisms, while petroleum hydrocarbons with higher-molecular-weight have strong resistance to microbial degradation [113]. The sequence of microbial degradation is as follows: N-alkanes > branched-chain alkanes > branched alkenes > low-molecular-weight n-alkyl aromatics > monoaromatics > cyclic alkanes > polynuclear aromatics > asphaltenes [114]. The methylene concentration in asphaltenes dropped by 14% and 8%, respectively, after 45 days of degradation by *Bacillus subtilis* and *Pseudomonas aeruginosa* [115]. *Pseudomonas aeruginosa* can degrade 63.8% of n-hexadecane within 60 days [116].

The most critical issue affecting the globe is the elimination of persistent environmental contaminants. PAHs have emerged as one of the most significant environmental contaminants, due to their hydrophobicity [117]. The following is a ranking of PAHs based on the order of the mineralization rate and the estimated half-life (in weeks): naphthalene (2.4–4.4), hexadecane (2.2–4.2), phenanthrene (4–18), 2-methyl Base naphthalene (14–20), pyrene (34–>90), 3-methylcholanthrene (87–>200), and benzo[a] pyrene (200–>300) [118]. Long-term exposure to low-level petroleum hydrocarbons lasts two to four times longer than PAHs surviving in the original environment. Despite the discovery of microbes able to degrade naphthalene, phenanthrene, and pyrene, the biodegradation of polycyclic aromatic hydrocarbons with large molecular weight remains a challenge.

The majority of microbial degradation of petroleum pollutants research are conducted in the laboratory using a mineral basal medium (liquid) (as indicated in Table 1) and have not been applied to actual petroleum-contaminated soil. Although some studies have shown that a single strain may degrade petroleum-contaminated soil, there are still issues with a single bioremediation technique, such as lengthy repair times, unstable microbial activity, and inadequate destruction of free microorganisms. Within 30 days, *S. changbaiensis* and *P. stutzeri* may decompose 39.2 ± 1.9% and 47.2 ± 1.2% of TPH in soil, respectively (the initial oil concentration is 1026 ± 50 mg/kg) [119]. *T. versicolor* can degrade 50% of TPH within 280 days (the initial oil content of the soil is 1727 mg/kg) [120]. Therefore, to increase degradation impact and practical application, combined microbial methods (synergistic repair incorporating microorganisms in the degradation process) are utilized.

5. Combined Microbial Methods Remediation

Microorganism–physical, microorganism–chemical, and microorganism–biology are the three primary types of microbial combination methods. In the microbial combined method of decomposing petroleum-contaminated soil, a variety of materials and procedures have been employed (Table 2). Most remediation combination methods are designed to enhance the microbial activity and aeration of polluted soil because of the hydrophobicity and fluidity of petroleum. To increase the system's degradation rate, an electric field, fertilizers, biocarrier, biochar, biosurfactants, and plants were applied to the petroleum-contaminated soil [121–123]. As shown in Table 2, the combination of microbes and physical or chemical technologies can improve the efficiency of microbial degradation of petroleum pollutants. In high-concentration petroleum-contaminated soil (≥10,000 mg/kg), the addition of biochar, electric fields, nutrients, and biosurfactants can all make the removal rate of petroleum pollutants reach more than 60%. The combination of ryegrass and mixed microbial strains had the best degradation effect within 162 days, with a degradation rate of 58% (the initial oil content was 6.19%). The combination of alfalfa and microorganisms can degrade 63% of petroleum hydrocarbons within 60 days (the initial oil content is 12%).

Table 2. The microbial combined materials and methods were used for the degradation of petroleum-contaminated soil.

Methods	Materials	Main Findings			Reference
		Substrate Concentration	Incubation Conditions	Degradation Rate	
Microorganism–physical	Biochar (walnut shell biochar (900 °C)/pinewood biochar (900 °C))	24,000, 16,000 and 21,000 mg/kg total petroleum hydrocarbons (TPH).	50 g soil/5% pinewood biochar/C:N:P at 800:13.3:1/25 °C/60 days.	The combined remediation of biochar and fertilizer reduces the TPH in the soil to 10,000 mg/kg (the US EPA clean up standard).	[121]
	Biochar (rice straw (500 °C))	16,300 mg kg^{-1} TPH (saturated hydrocarbons, 8260 mg kg^{-1}; aromatic hydrocarbons, 5130 mg kg^{-1}; polar components, 2910 mg kg^{-1}).	1000 g soil/2% (*w*/*w*) biochar/60% water holding capacity/C:N:P ratio 100:10:5/80 days.	TPH removal rate was 84.8%.	[124]
	electrokinetics	12,500 mg/kg TPH.	600 g soil/C:N:P 100:10:1/30 days.	The degradation rate of TPH was 88.3%.	[122]
	β-cyclodextrin	1000 mg/kg PAHs	1.5, 3.0, 5.0 mmol kg^{-1} β-cyclodextrin/25 °C.	Compared with the co-metabolism of glucose, the addition of β-cyclodextrin more strongly enhanced oil remediation in soil.	[125]
	bulking agents (chopped bermudagrass-hay/sawdust/vermiculite) aeration (tillage/forced aeration).	10% TPH	C:N:P 1000:10:1/15–35 °C/ 12 weeks.	Tillage and adding bulking agents enhanced remediation of oil-contaminated soil. The most rapid rate of remediation occurred during the first 12 weeks, where the TPH decreased 82% and the initial concentration of TPH was 10%.	[126]
	Biocarrier (activated carbon/zeolite)	49.81 mg g^{-1} TPH	800 g soil/50 g biocarrier + 150 mL planktonic bacterial culture/C:N:P 100:10:1/ 30 °C/33 days.	Biocarrier enhanced the biodegradation of TPH, with 48.89% removal, compared to natural attenuation with 13.0% removal.	[127]

Table 2. *Cont.*

Methods	Materials	Main Findings			Reference
		Substrate Concentration	Incubation Conditions	Degradation Rate	
	biostimulation	19.8 ± 0.38 g kg^{-1} TPH	0.8 kg soil/10^8 cfu g^{-1} petroleum degrading flora/15% soil moisture/C:N:P 100:10:1/ 24 °C/12 weeks.	Biostimulation achieved 60% oil hydrocarbon degradation.	[26]
	biosurfactants(rhamnolipids)	47.5 g kg^{-1} TPH	500 g soil/7 g of rhamnolipids (dissolved in 1 L deionized water)/500 mL bacterial consortium (in sterile 0.9% NaCl solution)/20% (*w*/*w*) moisture content/C:N:P 100:10:1/30 days.	TPH degradation of 77.6% was observed in the soil inoculated with hydrocarbon-degrading bacteria supplemented with rhamnolipids and nutrients.	[123]
	permanganate/activated persulfate/modified-Fenton/Fenton	263.6 ± 73.3 and 385.2 ± 39.6 mg·kg^{-1}Σ16 PAHs.	50 g soil/the final volume of the Milli-Q water and oxidant was 100 mL/150 rpm/ 15 days.	The removal efficiency of PAHs was ordered: permanganate (90.0–92.4%) > activated persulfate (81.5–86.54%) > modified Fenton (81.5–85.4%) > Fenton (54.1–60.0%).	[11]
	activator (low ammonia and acetic acid)	29,500 mg kg^{-1} TPH.	18–20% moisture content/ 12 weeks.	Macro-alkanes in soils were efficiently degraded.	[128]
Microorganism–biology	Lolium perenne	6.19% TPH	750 g soil/20–30% moisture content/162 days.	The results show that the combination of ryegrass with mixed microbial strains gave the best result with a degradation rate of 58%.	[129]
	Medicaga sativa	30% (40% TPH oily sludge)+70% non-pollution soil.	1 kg soil/N:P 10:1/75–80% moisture content/60 days.	Consortium degraded more than 63% TPH.	[130]

Table 2. *Cont.*

Methods	Materials	Main Findings			Reference
		Substrate Concentration	Incubation Conditions	Degradation Rate	
	Medicaga sativa/vicia faba/Lolium perenne	1.13% TPH.	2 kg soil/18 months.	The TPH degradation in the soil cultivated with broad beans and alfalfa was 36.6% and 35.8%, respectively, compared with 24% degradation in case of ryegrass.	[131]
	biopiles (bark chips)	700 mg kg^{-1} TPH	soil to bulking agent was approximately 1:3/ 15–20 °C/5 months.	The TPH content in the pile with oil-contaminated soil decreased with 71%.	[132]
	biopiles (peanut hull powder)	29,500 mg kg^{-1} TPH	5 kg of soil/15% *w*/*w* peanut hull powder/18–20% moisture content/C:N:P 100:10:1/25–30 °C/12 weeks.	Biodegradation was enhanced with free-living bacterial culture and biocarrier with a TPH removal ranging from 26% to 61%.	[133]
	biopiles (food waste)	2% diesel oil	soil [77% (*w*/*w*)] and food waste [23% (*w*/*w*)]/C:N 11:1/13 days.	84% of the TPH was degraded, compared with 48% of removal ratio in control reactor without inoculum.	[134]
	earthworms (*Eisenia fetida*/*Allolobophora chlorotica*/*Lumbricus terrestris*)	10,000 mg kg^{-1} TPH	1000 g soil/ten adult worms per container/28 days.	The TPH concentration decreased by 30–42% in samples with *L. terrestris*, by 31–37% in samples with *E. fetida*, and by 17–18% in samples with *A. chlorotica*.	[135]

The time period is confined to 2016–2020 according to six databases in the "web of research." "Microbial biochar (electrokinetics/bulking agents/aeration/biocarrier/biostimulation/biosurfactants/permanganate/activated persulfate/fenton/activator/plant/biopiles/earthworms) remediation of petroleum polluted soil," according to the results of the particular search keywords. The limit for the search is 17 September 2020, and the findings will be examined statistically. From 2016 to 2020, the papers employing the three combined microbial approaches to treat petroleum-polluted soil had the highest citation frequency (Figure 4). The microorganism–biochar, microorganism–nutrients, and microorganism–plant combined microbiological techniques have been extensively used for hydrocarbon degradation in current study, according to the gathered data.

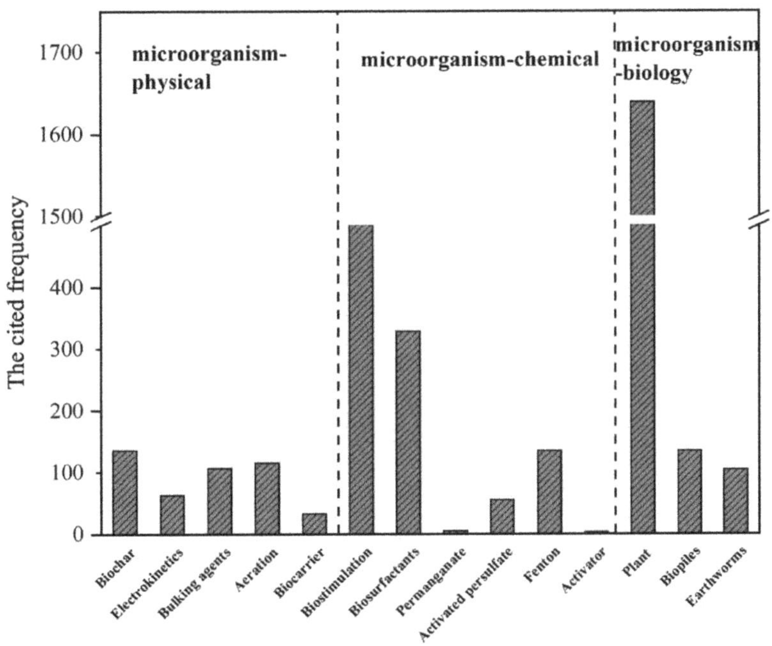

Figure 4. The frequency of citations of the article with three microbial combined methods for remediation of petroleum contaminated soil.

5.1. Microorganism–Biochar Interactions in Remediation of Hydrocarbons

Biochar has a high carbon content, excellent adsorption capacity, good stability, and the best bacteria and nutrient immobilization capacity. Biochar's porous structure can provide attachment sites and appropriate habitats for microorganisms to survive. The addition of various types of biochar to the soil promotes the enrichment of particular functional groups of microorganisms as well as an increase in biological activity [48,136]. The functional groups on the surface of biochar, as well as the easily decomposable carbon source and nitrogen source, assist to increase microbial activity and influence their growth, development, and metabolism. The use of biochar to immobilize microorganisms with various functional properties can enhance the release of particular nutrients in the soil and the efficiency with which pollutants are degraded. Biochar has been found in studies to absorb contaminants in petroleum, decreasing soil toxicity while having no discernible detrimental influence on soil microbes [124]. Furthermore, combining biochar with petroleum-degrading bacteria enhances the variety of microbial populations as well as the hydrocarbon bioavailability [137].

The basic interactions that occur in the microorganism–biochar remediation of pollutants are illustrated in Figure 5. The three modes of microorganism–biochar remediation include adsorption, biodegradation, and mineralization, or a combination of these three. Because of the huge specific surface area and rough surface structure of biochar, associated microorganisms produce biofilm, which improves the adsorption and degradation rate of hydrocarbons while also increasing the quantity of soil and active microorganisms. Simultaneously, studies have demonstrated that fixed bacteria may employ carbon chains more broadly than free bacteria, and the clearance rate of hydrocarbons has risen by around 21% to 49% [137]

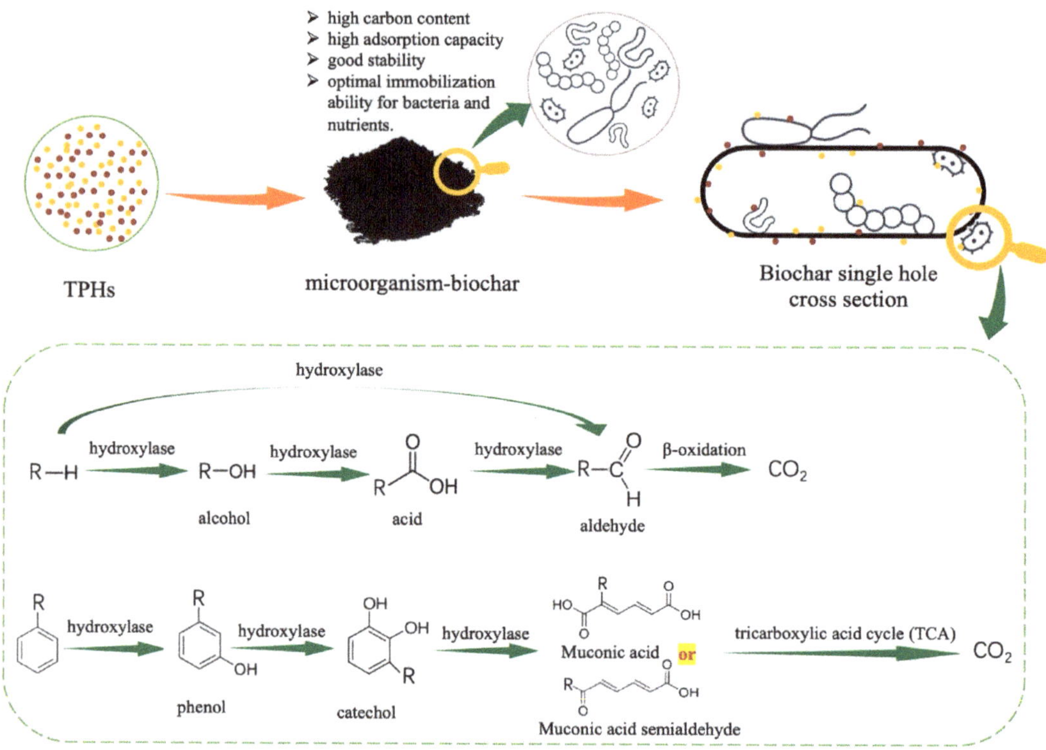

Figure 5. Proposed mechanism for the microbial metabolization of alkanes and aromatic hydrocarbons.

5.2. Microorganism–Nutrients Interactions in Remediation of Hydrocarbons

The input of a large amount of carbon sources (petroleum pollutants) frequently results in the rapid depletion of the available pools of the main inorganic nutrients (such as nitrogen (N) and phosphorus (P)) in the soil, whereas the essential nutrients (such as N, P, and terminal electron acceptors (TEA), etc.) are key factors in reducing the rate of microbial metabolism [138]. Although soil microorganisms have apparent pollution remediation potential, a shortage of necessary nutrients or activation of the degradation metabolic pathways inhibits or delays microbial repair. As a result, additional nutrients must be added to stimulate the biodegradation of inorganic pollutants [139].

If the soil environment is anaerobic for an extended period of time and the pollutant has a high carbon content, the metabolism of denitrifying bacteria in the soil will lower the total nitrogen level of the soil, therefore restricting this nutrient [140]. According to research, the amount of ammonium nitrogen (NH_4^+-N) and phosphorus (PO_4^{3-}-P) in soil decreases quickly 15 days after restoration [141]. Nitrate has a major benefit in enhancing the capacity for organic pollutant biodegradation in soil. Adding N to nutrient-deficient samples rich in hydrocarbons can accelerate cell growth and hydrocarbon degradation. Because nitrate has thermodynamic benefits over TEA, it participates in the absorption and/or dissimilatory reduction process under oxygen limitation and anaerobic circumstances, promoting heterotrophic or autotrophic denitrification while oxidizing organic matter (especially alkanes) [142]. At the same time, the phosphorous concentration of the terrestrial subsurface environment is quite low. Although apatite is common in some locations, it cannot be utilized by life. Several inorganic and organic forms of phosphate have been effectively utilized to stimulate pollution in the environment [143]. As a result, the addition of nutrients nitrogen and phosphorus promotes the efficient oxidation of carbon substrates while also accelerating bacterial growth and hydrocarbon catabolism [138]. Currently, the optimum C:N:P ratio for effective biodegradation of petroleum hydrocarbons has been observed to be 100:10:1 [144].

5.3. Microorganism–Plant Interactions in Remediation of Hydrocarbons

The most popular technique for in situ remediation is the microorganism–plant combination method. Organic contaminants are mostly metabolized by plant-related microorganisms in phytoremediation, according research. It has also been reported that the re-mediation capacity of plants is influenced by the quantity of bacteria in their surroundings [145]. As a result, in the process of pollution remediation, the synergy between plants and microbes increases pollutant degradation and mineralization. Special enzymes and other chemicals found in plants and microbes can transform many hazardous and complicated chemical molecules into simpler and less poisonous ones. Under polluted environments, this mechanism promotes their development. Plant rhizospheres can offer microorganisms with nutrition, oxygen, and area for growth and development [146,147]. These bacteria expand the surface area of plant roots, allowing them to make contact with the soil and acquire more nutrients required for plant growth. As a result, the inoculation bacteria are more concentrated in the soil near the vegetation's roots [148]. Simultaneously, plant root exudates can promote the destruction of microorganisms by altering the composition of the microbial community and increasing microbial activity [149].

Plants such as *Merr.,Setaria viridis Beauv., Plantago asiatica L., Phragmites communis, Medicago sativa, Festuca elata Keng ex E.Alexeev*, and *Lolium perenne L.* have been shown in studies to be suitable for the climate and environment in China and are candidates for the phytoremediation of petroleum-contaminated soil in China [150]. The petroleum removal rate after 90 days of restoring petroleum-contaminated soil by *Festuca elata Keng ex E.Alexee* is around 64% [151]. *Festuca elata Keng ex E.Alexeev* not only successfully removes benzopyrene from soil [152], but its development also improves soil biological activity in saline-alkali regions contaminated with petroleum [54]. The basic interactions that occur in the microorganism–plant remediation of pollutants are depicted in Figure 6.

The mechanisms of microorganism–plant remediation can be classified as degradation, extraction, inhibition, or a combination of the three. Roots not only give oxygen to microorganisms in the rhizosphere through respiration, but they also stimulate the release of root exudates and the degradation of rhizosphere contaminants [153]. Plants and microorganisms then degrade hydrocarbons into simpler organic molecules by expressing specific enzymes such as nitroreductase, dehalogenase, laccase, and peroxidase, among others [154]. Some pollutants are adsorbed on the root surface and accumulate in the root via the hemicellulose of the plant cell wall and the lipid bilayer of the plasma membrane [155]. A part of the pollutants are absorbed via phytoextraction/plant transfer to the upper section of plants (stems and leaves) [156]. Finally, phytovolatilization releases certain contaminants into the atmosphere [157]. Some plants, as a self-protection strategy, limit the transfer of hydrocarbons from the roots to the ground, retaining more hydrocarbons in the root tissues. This limitation preserves the chlorophyll and other nutrient synthesis mechanisms of plants and ensures that photosynthesis continues to function normally [156]. This is to guarantee that photosynthetic processes of the plants are regular, allowing them to produce more energy for survival and repair.

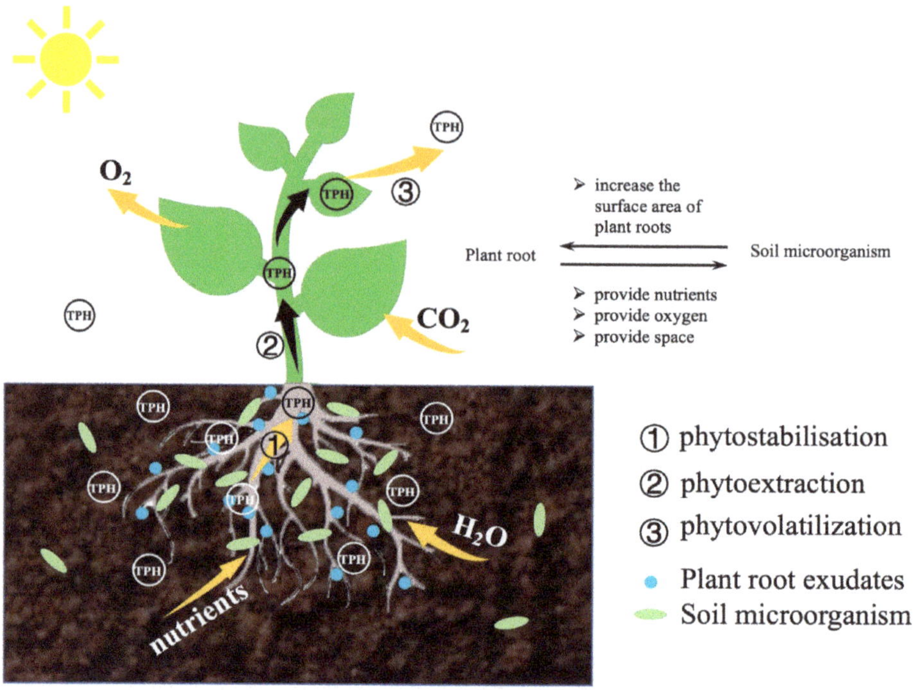

Figure 6. The common interactions that occur during pollutant remediation by microorganisms and plants.

6. Advantages and Challenges in Combined Microbial Methods Application for Hydrocarbon Removal

The discharge of hazardous contaminants into the soil environment has increased substantially as a result of petroleum extraction. Bioremediation offers the advantages of ease of use, economic feasibility, and no secondary contamination, among other things, and is currently a research hotspot for oily soil remediation [64]. The addition of biochar, nutrients, and plants to microorganisms not only enhances their biological stability and activity, but it also improves their capacity to degrade petroleum pollutants. The benefits of three combined microbial methods are as follows. These methods will not harm the soil ecosystem, physical, chemical, or biological characteristics, and will actually improve them following restoration. They may also degrade organic pollutants into entirely non-polluting inorganic molecules (carbon dioxide and water) without causing secondary contamination. The study found that after integrating oily soil remediation with microbial biomass and the number of PAH-degrading bacteria, soil enzyme activity, microbial biomass, and the number of polycyclic-aromatic-hydrocarbon-degrading bacteria were significantly higher than in other treatments [158,159]. The diversity, richness, and homogeneity of soil microbial communities have altered following restoration, according to Biolog analysis [160]. Joint restoration has enhanced the genetic variety of soil microbial communities, according to a DNA sequencing study of soil microbial diversity [161].

The three repair approaches are currently only at the laboratory stage, and few strains are utilized in engineering repair. Many contributing variables and degradation processes are yet unknown, necessitating more investigation. Figure 7 summarizes some of the problems of the three integrated microbiological techniques. The long-term stability and tolerance of biochar is one of the challenges with microbial–biochar composite repair. The most essential feature influencing the thermal decomposition of biomass and the characteristics of biochar is the pyrolysis temperature. The physicochemical characteristics and structure of biochar, such as element composition, pore structure, surface area, and functional groups, are affected by the pyrolysis temperature [162]. Biochar is rich in oxygen-containing functional groups when the pyrolysis temperature is 300 to 500 °C. There are less oxygen-containing functional groups, a higher mineral concentration, and more micropores when the pyrolysis temperature is 500 to 700 °C [163]. Their environmental activities are determined by these features. Furthermore, the pyrolysis temperature affects carbon retention throughout the pyrolysis process as well as biochar carbon stability [164]. According to research, the higher the temperature, the lower the H/C ratio, the greater the electron donor–acceptor interaction, the higher the quantity of non-decomposable carbon, and the higher the adsorption effectiveness of biochar [165]. However, investigations have indicated that at a moderate temperature of around 500 °C, the residual carbon in biochar is only around 50% on average [164]. Soil microorganisms will mineralize a portion of the biochar after joint remediation. As a result, certain techniques for adjusting the pyrolysis process should be presented in order to maximize biochar's overall carbon sequestration capability while taking carbon retention and carbon persistence into account.

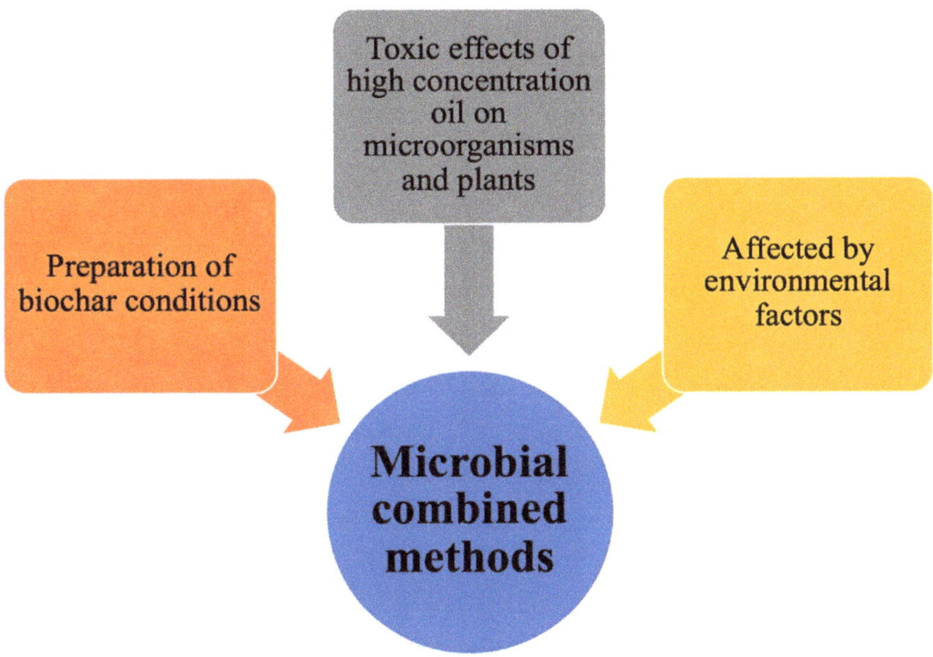

Figure 7. Some challenges of three microbial combined methods.

Most microorganisms and plants are more suitable to soil remediation with a petroleum pollution concentration of less than 5%, according to previous studies [26,151,166]. The remediation potential of microorganisms and plants is rapidly negatively affected when the concentration of petroleum pollutants in the soil increases (5%). The original oil content was 1.21%, and the removal rate of *Testuca arundinacea* for TPH was 64.0 ± 1.6% after 90 days of repair, and the removal rate of biological flora was 54.6 ± 1.3% [151]. After 90 days of repair, the stem and root biomass of ryegrass is lower than the control group when the soil oil concentration is 3% [167]. Tall fescue can remove 48.4% of oil pollution after 70 days of restoration when the soil oil concentration is 5% [168]. Microorganisms could remove 15% of petroleum pollutants after 70 days of remediation when the soil oil content is 5.6% [169]. When soil oil levels are too high, it is hazardous to plants and microorganisms, reducing their capacity to degrade petroleum contaminants and potentially causing deaths in microorganisms and plants. Extreme climatic circumstances (soil temperature below 10 °C, pH value below than 4 and higher than 9), on the other hand, will limit the activity of microorganisms and plants, lowering the removal of petroleum pollutants. Changes in soil pH and abiotic or biodegradation of biochar, on either hand, will increase the desorption of PAH from biochar into sediments. In conclusion, despite the benefits of minimal secondary contamination and low cost, microbial remediation still confronts significant challenges.

7. Conclusions

This article explains the use of a combination of microbiological methods to remediate petroleum-contaminated soil. Although a combination of microorganisms–biochar/nutr ients/plants can be utilized to remediate petroleum-contaminated soil to solve the issues of a unique remediation, no one method is best for all types of pollutants or all unique site circumstances that occur in the impacted area. As a result, an efficient combined remediation method based on the physical and chemical characteristics of soil at various polluted sites as well as the kinds of contaminants is required. In addition, scientists are working at the movement, distribution, and degradation mechanisms of contaminants in the combined system, as well as their interactions and relationships with microorganisms. Clarify the internal and external elements that impact the restoration before selecting particular therapeutic treatments. Therefore, to find out the key factors and mechanisms that increase the degradation rate of microbial joint remediation, and to design a microbial joint remediation technology with high degradation efficiency, sustainability, and environmental friendliness is a problem that needs to be solved urgently.

Author Contributions: Writing—original draft preparation, X.S.; supervision, X.W.; data curation, Y.L.; conceptualization and supervision, review and editing, H.J. All authors have read and agreed to the published version of the manuscript.

Funding: This research received no external funding.

Institutional Review Board Statement: Not applicable.

Informed Consent Statement: Not applicable.

Data Availability Statement: The data presented in this study are openly available in reference.

Acknowledgments: This study was supported by the Beijing Key Laboratory of Resource-oriented Treatment of Industrial Pollutants, International Science and Technology Cooperation Base for Environmental and Energy Technology of MOST.

Conflicts of Interest: The authors declare no conflict of interest.

References

1. Abbaspour, A.; Zohrabi, F.; Dorostkar, V.; Faz, A.; Acosta, J.A. Remediation of an oil-contaminated soil by two native plants treated with biochar and mycorrhizae. *J. Environ. Manag.* **2020**, *254*, 109755. [CrossRef]
2. Freedman, B. Oil pollution. In *Environmental Ecology*, 2nd ed.; Academic Press: San Diego, CA, USA, 1995; pp. 159–188.
3. Suganthi, S.H.; Murshid, S.; Sriram, S. Enhanced biodegradation of hydrocarbons in petroleum tank bottom oil sludge and characterization of biocatalysts and biosurfactants. *J. Environ. Manag.* **2018**, *220*, 87–95. [CrossRef]
4. Ansari, N.; Hassanshahian, M.; Ravan, H. Study the microbial communities' changes in desert and farmland soil after crude oil pollution. *Int. J. Environ. Res.* **2018**, *12*, 391–398. [CrossRef]
5. Ekundayo, E.O.; Emede, T.O.; Osayande, D.I. Effects of crude oil spillage on growth and yield of maize (Zea mays L.) in soils of midwestern Nigeria. *Plant Foods Hum. Nutr.* **2001**, *56*, 313–324. [CrossRef] [PubMed]
6. Kumari, S.; Regar, R.K.; Manickam, N. Improved polycyclic aromatic hydrocarbon degradation in a crude oil by individual and a consortium of bacteria. *Bioresour. Technol.* **2018**, *254*, 174–179. [CrossRef] [PubMed]
7. Ozigis, M.S.; Kaduk, J.D.; Jarvis, C.H. Detection of oil pollution impacts on vegetation using multifrequency SAR, multispectral images with fuzzy forest and random forest methods. *Environ. Pollut.* **2020**, *256*, 113360. [CrossRef] [PubMed]
8. O'Callaghan-Gordo, C.; Orta-Martínez, M.; Kogevinas, M. Health effects of non-occupational exposure to oil extraction. *Environ. Health* **2016**, *15*, 1–4. [CrossRef]
9. Maletić, S.; Dalmacija, B.; Rončević, S. Petroleum hydrocarbon biodegradability in soil–implications for bioremediation. *Intech* **2013**, *43*, 43–64.

10. Sankaran, S.; Pandey, S.; Sumathy, K. Experimental investigation on waste heat recovery by refinery oil sludge incineration using fluidised-bed technique. *Environ. Lett.* **1998**, *33*, 829–845. [CrossRef]
11. Liao, X.; Wu, Z.; Li, Y.; Cao, H.; Su, C. Effect of various chemical oxidation reagents on soil indigenous microbial diversity in remediation of soil contaminated by PAHs. *Chemosphere* **2019**, *226*, 483–491. [CrossRef]
12. Giannopoulos, D.; Kolaitis, D.I.; Togkalidou, A.; Skevis, G.; Founti, M.A. Quantification of emissions from the co-incineration of cutting oil emulsions in cement plants—Part II: Trace species. *Fuel* **2007**, *86*, 2491–2501. [CrossRef]
13. Vidonish, J.E.; Zygourakis, K.; Masiello, C.A.; Sabadell, G.; Alvarez, P.J.J. Thermal treatment of hydrocarbon-impacted soils: A Review of technology innovation for sustainable remediation. *Engineering* **2016**, *2*, 426–437. [CrossRef]
14. Kang, C.-U.; Kim, D.-H.; Khan, M.A.; Kumar, R.; Ji, S.-E.; Choi, K.-W.; Paeng, K.-J.; Park, S.; Jeon, B.-H. Pyrolytic remediation of crude oil-contaminated soil. *Sci. Total Environ.* **2020**, *713*, 136498. [CrossRef]
15. Ku-Fan, C.; Yu-Chen, C.; Wan-Ting, C. Remediation of diesel-contaminated soil using in situ chemical oxidation (ISCO) and the effects of common oxidants on the indigenous microbial community: A comparison study. *J. Chem. Technol. Biotechnol.* **2016**, *91*, 1877–1888.
16. Das, N.; Chandran, P. Microbial degradation of petroleum hydrocarbon contaminants: An overview. *Biotechnol. Res. Int.* **2011**, *10*, 1–13. [CrossRef] [PubMed]
17. Senko, J.M.; Campbell, B.S.; Henriksen, J.R.; Elshahed, M.S.; Dewers, T.A.; Krumholz, L.R. Barite deposition resulting from phototrophic sulfide-oxidizing bacterial activity. *Geochim. Cosmochim. Acta* **2004**, *68*, 773–780. [CrossRef]
18. Velacano, M.; Castellanohinojosa, A.; Vivas, A.F.; Toledo, M.V.M. Effect of Heavy Metals on the Growth of Bacteria Isolated from Sewage Sludge Compost Tea. *Adv. Microbiol.* **2014**, *4*, 644–655. [CrossRef]
19. Chaîneau, C.H.; Rougeux, G.; Yéprémian, C.J. Oudot. Effects of nutrient concentration on the biodegradation of crude oil and associated microbial populations in the soil. *Soil Biol. Biochem.* **2005**, *37*, 1490–1497. [CrossRef]
20. Chalneau, C.-H.; Morel, J.-L.; Oudot, J. Microbial degradation in soil microcosms of fuel oil hydrocarbons from drilling cuttings. *Environ. Sci. Technol.* **1995**, *29*, 1615–1621. [CrossRef]
21. Pino, N.J.; Muñera, L.M.; Peñuela, G.A. Bioaugmentation with immobilized microorganisms to enhance phytoremediation of PCB-contaminated soil. *Soil Sediment Contam. Int. J.* **2016**, *25*, 419–430. [CrossRef]
22. Wang, Z.-Y.; Xu, Y.; Wang, H.-Y.; Zhao, J.; Gao, D.-M.; Li, F.-M.; Xing, B. Biodegradation of crude oil in contaminated soils by free and immobilized microorganisms. *Pedosphere* **2012**, *22*, 717–725. [CrossRef]
23. Hamamura, N.; Fukui, M.; Ward, D.M.; Inskeep, W.P. Assessing soil microbial populations responding to crude-oil amendment at different temperatures using phylogenetic, functional gene (alkB) and physiological analyses. *Environ. Sci. Technol.* **2008**, *42*, 7580–7586. [CrossRef] [PubMed]
24. Zou, C.; Wang, M.; Xing, Y.; Lan, G.; Ge, T.; Yan, X.; Gu, T. Characterization and optimization of biosurfactants produced by Acinetobacter baylyi ZJ2 isolated from crude oil-contaminated soil sample toward microbial enhanced oil recovery applications. *Biochem. Eng. J.* **2014**, *90*, 49–58. [CrossRef]
25. Das, M.; Adholeya, A. Role of Microorganisms in Remediation of Contaminated Soil. In *Microorganisms in Environmental Management*; Springer: Berlin/Heidelberg, Germany, 2012.
26. Wu, M.; Wu, J.; Zhang, X.; Ye, X. Effect of bioaugmentation and biostimulation on hydrocarbon degradation and microbial community composition in petroleum-contaminated loessal soil. *Chemosphere* **2019**, *237*, 124456. [CrossRef] [PubMed]
27. Fanaei, F.; Moussavi, G.; Shekoohiyan, S. Enhanced treatment of the oil-contaminated soil using biosurfactant-assisted washing operation combined with H2O2-stimulated biotreatment of the effluent. *J. Environ. Manag.* **2020**, *271*, 110941. [CrossRef] [PubMed]
28. Abioye, O.P. Biological remediation of hydrocarbon and heavy metals contaminated soil. *Soil Contam.* **2011**, *7*, 127–142.
29. Adipah, S. Introduction of petroleum hydrocarbons contaminants and its human effects. *J. Environ. Sci. Public Health* **2018**, *3*, 001–009. [CrossRef]
30. Wang, S.; Xu, Y.; Lin, Z.; Zhang, J.; Norbu, N.; Liu, W. The harm of petroleum-polluted soil and its remediation research. *Aip Conf. Proc.* **2017**, *1864*, 020222.
31. Kachieng'a, L.; Momba, M.N.B. Kinetics of petroleum oil biodegradation by a consortium of three protozoan isolates (*Aspidisca* sp., *Trachelophyllum* sp. and *Peranema* sp.). *Biotechnol. Rep.* **2017**, *15*, 125–131. [CrossRef]
32. Sarma, H.; Nava, A.R.; Prasad, M.N.V. Mechanistic understanding and future prospect of microbe-enhanced phytoremediation of polycyclic aromatic hydrocarbons in soil. *Environ. Technol. Innov.* **2019**, *13*, 318–330. [CrossRef]
33. Brakstad, O.G.; Ribicic, D.; Winkler, A.; Netzer, R. Biodegradation of dispersed oil in seawater is not inhibited by a commercial oil spill dispersant. *Mar. Pollut. Bull.* **2018**, *129*, 555–561. [CrossRef]
34. Włodarczyk-Makuła, M. Half-Life of carcinogenic polycyclic aromatic hydrocarbons in stored sewage sludge. *Arch. Environ. Prot.* **2012**, *38*, 33–44. [CrossRef]
35. Elumalai, P.; Parthipan, P.; Karthikeyan, O.P.; Rajasekarcorresponding, A. Enzyme-mediated biodegradation of long-chain n-alkanes (C32 and C40) by thermophilic bacteria. *3 Biotech* **2017**, *7*, 1–10. [CrossRef] [PubMed]
36. Aguelmous, A.; Zegzouti, Y.; Khadra, A.; Fels, L.E.; Souabi, S.; Hafidi, M. Landfilling and composting efficiency to reduce genotoxic effect of petroleum sludge. *Environ. Technol. Innov.* **2020**, *20*, 101047. [CrossRef]
37. Schirmer, A.; Rude, M.A.; Li, X.; Popova, E.; Cardayre, S.B.D. Microbial Biosynthesis of Alkanes. *Science* **2010**, *329*, 559–562. [CrossRef]

38. Ryckaert, J.P.; Bellemans, A. Molecular dynamics of liquid alkanes. *Faraday Discuss. Chem. Soc.* **1978**, *66*, 95–106. [CrossRef]
39. Zakaria, M.P.; Bong, C.-W.; Vaezzadeh, V. Fingerprinting of petroleum hydrocarbons in malaysia using environmental forensic techniques: A 20-year field data review. In *Oil Spill Environmental Forensics Case Studies*; Butterworth-Heinemann: Oxford, UK, 2018; Chapter 16, pp. 345–372.
40. Sadeghbeigi, R. Chapter 3—FCC Feed Characterization. In *Fluid Catalytic Cracking Handbook*, 3rd ed.; Butterworth-Heinemann: Oxford, UK, 2012.
41. Tissot, B.P.; Welte, D.H. Composition of crude oils. In *Petroleum Formation and Occurrence*; Springer: Berlin/Heidelberg, Germany, 1984; Chapter 1, pp. 333–368.
42. Vasconcelos, I.U.; FrançaI, F.P.D.; OliveiraII, F.J.S. Removal of high-molecular weight polycyclic aromatic hydrocarbons. *Química Nova* **2011**, *34*, 218–221. [CrossRef]
43. Haines, J.R.; Alexander, M. Microbial degradation of high-molecular-weight alkanes. *Appl. Microbiol. Biotechnol.* **1974**, *28*, 1084–1085. [CrossRef]
44. Imam, A.; Suman, S.K.; Ghosh, D.K. Kanaujia, P. Analytical approaches used in monitoring the bioremediation of hydrocarbons in petroleum-contaminated soil and sludge. *TrAC Trends Anal. Chem.* **2019**, *118*, 50–64. [CrossRef]
45. Abdel-Shafy, H.I.; Mansour, M.S.M. A review on polycyclic aromatic hydrocarbons: Source, environmental impact, effect on human health and remediation. *Egypt. J. Pet.* **2016**, *25*, 107–123. [CrossRef]
46. Fooladi, M.; Moogouei, R.; Jozi, S.A.; Golbabaei, F.; Tajadod, G. Phytoremediation of BTEX from indoor air by Hyrcanian plants. *Environ. Health Eng. Manag. J.* **2019**, *6*, 233–240. [CrossRef]
47. Wei, Z.; Wang, J.J.; Meng, Y.; Li, J.; Gaston, L.A.; Fultz, L.M.; DeLaune, R.D. Potential use of biochar and rhamnolipid biosurfactant for remediation of crude oil-contaminated coastal wetland soil: Ecotoxicity assessment. *Chemosphere* **2020**, *253*, 126617. [CrossRef] [PubMed]
48. Gomez-Eyles, J.L.; Collins, C.D.; Hodson, M.E. Using deuterated PAH amendments to validate chemical extraction methods to predict PAH bioavailability in soils. *Environ. Pollut.* **2011**, *159*, 918–923. [CrossRef] [PubMed]
49. Achuba, F.I.; Peretiemo-Clarke, B.O. Effect of spent engine oil on soil catalase and dehydrogenase activities. *Int. Agrophys.* **2008**, *22*, 1–4.
50. Polyak, Y.; Bakina, L.G.; Bakina, L.G.; Chugunova, M.V.; Bure, V. Effect of remediation strategies on biological activity of oil-contaminated soil—A field study. *Int. Biodeterior. Biodegrad.* **2018**, *126*, 57–68. [CrossRef]
51. Wei, Y.; Li, G. Effect of oil pollution on water characteristics of loessial soil. *IOP Conf. Ser. Earth Environ. Sci.* **2018**, *170*, 032154. [CrossRef]
52. Barua, D.; Buragohain, J.; Sarma, S.K. Certain physico-chemical changes in the soil brought about by contamination of crude oil in two oil fields of Assam. *NE India Pelagia Res. Libr.* **2011**, *1*, 154–161.
53. Osuji, L.C.; Idung, I.D.; Ojinnaka, C.M. Preliminary investigation on Mgbede-20 oil-polluted site in Niger Delta, Nigeria. *Chem. Biodivers.* **2006**, *3*, 568–577. [CrossRef]
54. Xin, L.; Hui-hui, Z.; Bing-bing, Y.; Nan, X.; Wen-xu, Z.; Ju-wei, H.; Guang-yu, S. Effects of Festuca arundinacea on the microbial community in crude oil-contaminated saline-alkaline soil. *Chin. J. Appl. Ecol.* **2012**, *23*, 3414–3420.
55. Steliga, T.; Kluk, D. Application of Festuca arundinacea in phytoremediation of soils contaminated with Pb, Ni, Cd and petroleum hydrocarbons. *Ecotoxicol. Environ. Saf.* **2020**, *194*, 110409. [CrossRef] [PubMed]
56. Mangse, G.; Werner, D.; Meynet, P.; Ogbaga, C.C. Microbial community responses to different volatile petroleum hydrocarbon class mixtures in an aerobic sandy soil. *Environ. Pollut.* **2020**, *264*, 114738. [CrossRef]
57. Buzmakov, S.A.; Khotyanovskaya, Y.V. Degradation and pollution of lands under the influence of oil resources exploitation. *Appl. Geochem.* **2020**, *113*, 104443. [CrossRef]
58. Uzoho, B.; Oti, N.; Onweremadu, E. Effect of crude oil pollution on maize growth and soil properties in Ihiagwa, Imo State, Nigeria. *Int. J. Agric. Rural Dev.* **2006**, *5*, 91–100. [CrossRef]
59. Lorestani, B.; Kolahchi, N.; Ghasemi, M.; Cheraghi, M.; Yousefi, N. Survey the effect of oil pollution on morphological characteristics in Faba Vulgaris and Vicia Ervilia. *J. Chem. Health Risks* **2012**, *2*, 2251–6727.
60. Zhen, M.; Chen, H.; Liu, Q.; Song, B.; Wang, Y.; Tang, J. Combination of rhamnolipid and biochar in assisting phytoremediation of petroleum hydrocarbon contaminated soil using Spartina anglica. *J. Environ. Sci.* **2019**, *85*, 107–118. [CrossRef]
61. Kuppusamy, S.; Raju, M.N.; Mallavarapu, M.; Kadiyala, V. Impact of Total Petroleum Hydrocarbons on Human Health. In *Total Petroleum Hydrocarbons*; Springer: Cham, Germany, 2020; pp. 139–165.
62. Haller, H.; Jonsson, A. Growing food in polluted soils: A review of risks and opportunities associated with combined phytoremediation and food production (CPFP). *Chemosphere* **2020**, *254*, 126826. [CrossRef] [PubMed]
63. Hussain, I.; Puschenreiter, M.; Gerhard, S.; Sani, S.G.A.S.; Khan, W.-u.-d.; Reichenauer, T.G. Differentiation between physical and chemical effects of oil presence in freshly spiked soil during rhizoremediation trial. *Environ. Sci. Pollut. Res.* **2019**, *26*, 18451–18464. [CrossRef] [PubMed]
64. Tao, K.; Liu, X.; Chen, X.; Hu, X.; Cao, L.; Yuan, X. Biodegradation of crude oil by a defined co-culture of indigenous bacterial consortium and exogenous Bacillus subtilis. *Bioresour. Technol.* **2016**, *224*, 327–332. [CrossRef]
65. Abena, M.T.B.; Li, T.; Shah, M.N.; Zhong, W. Biodegradation of total petroleum hydrocarbons (TPH) in highly contaminated soils by natural attenuation and bioaugmentation. *Chemosphere* **2019**, *234*, 864–874. [CrossRef] [PubMed]

66. Bacosa, H.P.; Suto, K.; Inoue, C. Bacterial community dynamics during the preferential degradation of aromatic hydrocarbons by a microbial consortium. *Int. Biodeterior. Biodegrad.* **2012**, *74*, 109–115. [CrossRef]
67. Lee, D.W.; Lee, H.; Kwon, B.-O.; Khim, J.S.; Yim, U.H.; Kim, B.S.; Kim, J.-J. Biosurfactant-assisted bioremediation of crude oil by indigenous bacteria isolated from Taean beach sediment. *Environ. Pollut.* **2018**, *241*, 254–264. [CrossRef] [PubMed]
68. Wang, B.; Teng, Y.; Yao, H.; Christie, P. Detection of functional microorganisms in benzene [a] pyrene-contaminated soils using DNA-SIP technology. *J. Hazard. Mater.* **2021**, *407*, 124788. [CrossRef]
69. Pi, Y.; Chen, B.; Bao, M.; Fan, F.; Cai, Q.; Ze, L.; Zhang, B. Microbial degradation of four crude oil by biosurfactant producing strain *Rhodococcus* sp. *Bioresour. Technol.* **2018**, *232*, 263–269. [CrossRef] [PubMed]
70. Yuan, X.; Zhang, X.; Chen, X.; Kong, D.; Liu, X.; Shen, S. Synergistic degradation of crude oil by indigenous bacterial consortium and T exogenous fungus Scedosporium boydii. *Bioresour. Technol.* **2018**, *264*, 190–197. [CrossRef] [PubMed]
71. Varjani, S.J.; Upasani, V.N. Biodegradation of petroleum hydrocarbons by oleophilic strain of Pseudomonas aeruginosa NCIM 5514. *Bioresour. Technol.* **2016**, *222*, 195–201. [CrossRef]
72. Wang, L.; Wang, W.; Lai, Q.; Shao, Z. Gene diversity of CYP153A and AlkB alkane hydroxylases in oil-degrading bacteria isolated from the Atlantic Ocean. *Environ. Microbiol.* **2010**, *12*, 1230–1242. [CrossRef]
73. Cao, B.; Nagarajan, K.; Loh, K.C. Biodegradation of aromatic compounds: Current status and opportunities for biomolecular approaches. *Appl. Microbiol. Biotechnol.* **2009**, *85*, 207–228. [CrossRef]
74. Margesin, R.; Zimmerbauer, A.; Schinner, F. Soil lipase activity–A useful indicator of oil biodegradation. *Biotechnol. Tech.* **1999**, *13*, 859–863. [CrossRef]
75. Sirajuddin, S.; Rosenzweig, A.C. Enzymatic Oxidation of Methane. *Biochemistry* **2015**, *54*, 2283–2294. [CrossRef] [PubMed]
76. Liu, Y.-C.; Li, L.-Z.; Wu, Y.; Tian, W.; Zhang, L.-P.; Xu, L.; Shen, Q.-R.; Shen, B. Isolation of an alkane-degrading *Alcanivorax* sp. strain 2B5 and cloning of the alkB gene. *Bioresour. Technol.* **2010**, *101*, 310–316.
77. Huan, L.; Jing, X.; Rubing, L.; Jianhua, L.; ONE, D.A.J.P. Characterization of the Medium- and Long-Chain n-Alkanes Degrading Pseudomonas aeruginosa Strain SJTD-1 and Its Alkane Hydroxylase Genes. *PLoS ONE* **2014**, *9*, e105506.
78. Liu, J.; Zhao, B.; Lan, Y.; Ma, T. Enhanced degradation of different crude oils by defined engineered consortia of Acinetobacter venetianus RAG-1 mutants based on their alkane metabolism. *Bioresour. Technol.* **2021**, *327*, 124787. [CrossRef]
79. Mimmi, T.-H.; Alexander, W.; Trond, E.E.; Hans-Kristian, K.; Sergey, B.Z. Identification of novel genes involved in long-chain n-alkane degradation by *Acinetobacter* sp. strain DSM 17874. *Appl. Environ. Microbiol.* **2007**, *73*, 3327–3332.
80. Abouseoud, M.; Yataghene, A.; Amrane, A.; Maachi, R. Biosurfactant production by free and alginate entrapped cells of Pseudomonas fluorescens. *J. Ind. Microbiol. Biotechnol.* **2008**, *35*, 1303–1308. [CrossRef] [PubMed]
81. Hesham, E.L.; Mawad, A.M.M.; Mostafa, Y.M.; Shoreit, A.J.M. Study of enhancement and inhibition phenomena and genes relating to degradation of petroleum polycyclic aromatic hydrocarbons in isolated bacteria. *Microbiology* **2014**, *83*, 599–607. [CrossRef]
82. Varjani, S.J. Microbial degradation of petroleum hydrocarbons. *Bioresour. Technol.* **2016**, *223*, 277–286. [CrossRef]
83. Champion, P.M.; Gunsalus, I.C.; Wagner, G.C. Resonance Raman investigations of cytochrome P450CAM from Pseudomonas putida. *J. Am. Chem. Soc.* **1978**, *100*, 3743–3751. [CrossRef]
84. Noble, M.E.M.; Cleasby, A.; Johnson, L.N.; Egmond, M.R.; Frenken, L.G.J. The crystal structure of triacylglycerol lipase from Pseudomonas glumae reveals a partially redundant catalytic aspartate. *FEBS Lett.* **1993**, *331*, 123–128. [CrossRef]
85. Rojo, F. Degradation of alkanes by bacteria. *Environ. Microbiol.* **2009**, *11*, 2477–2490. [CrossRef]
86. Xia, Y.; Min, H.; Rao, G.; Lv, Z.; Liu, J.; Ye, Y.; Duan, X. Isolation and character- ization of phenanthrene-degrading Sphingomonas paucimobilis strain ZX4. *Biodegradation* **2005**, *16*, 393–402. [CrossRef]
87. Oyetibo, G.O.; Chien, M.F.; Ikeda-Ohtsubo, W.; Suzuki, H.; Obayori, O.S.; Adebusoye, S.A.; Ilori, M.O.; Amund, O.O.; Endo, G. Biodegradation of crude oil and phenanthrene by heavy metal resistant Bacillus subtilis isolated from a multi-polluted industrial wastewater creek. *Int. Biodeterior. Biodegrad.* **2017**, *120*, 143–151. [CrossRef]
88. Deng, M.-C.; Li, J.; Liang, F.-R.; Wang, J.-H. Isolation and characterization of a novel hydrocarbon-degrading bacterium Achromobacter sp HZ01 from the crude oil-contaminated seawater at the Daya Bay, southern China. *Mar. Pollut. Bull.* **2014**, *83*, 79–86. [CrossRef]
89. Shao, Y.; Wang, Y.; Wu, X.; Li, B. Biodegradation of PAHs by Acinetobacter isolated from karst groundwater in a coal-mining area. *Environ. Earth Sci.* **2015**, *73*, 7479–7488. [CrossRef]
90. Lily, M.K.; Bahuguna, A.; Dangwal, K.; Garg, V. Degradation of Benzo [a] Pyrene by a novel strain Bacillus subtilis BMT4i (MTCC 9447). *Braz. J. Microbiol.* **2009**, *40*, 884–892. [CrossRef] [PubMed]
91. Al-Thukair, A.A.; Malik, K. Pyrene metabolism by the novel bacterial strains Burkholderia fungorum (T3A13001) and Caulobacter sp (T2A12002) isolated from an oil-polluted site in the Arabian Gulf. *Int. Biodeterior. Biodegrad.* **2016**, *110*, 32–37. [CrossRef]
92. Darma, U.Z.; Aziz, N.A.A.; Zulkefli, S.Z.; Mustafa, M. Identification of Phenanthrene and Pyrene degrading bacteria from used engine oil contaminated soil. *Int. J. Sci. Eng. Res.* **2016**, *7*, 680–686.
93. Rajkumari, J.; Singha, L.P.; Pandey, P. Genomic insights of aromatic hydrocarbon degrading Klebsiella pneumoniae AWD5 with plant growth promoting attributes: A paradigm of soil isolate with elements of biodegradation. *3 Biotech* **2018**, *8*, 1–22. [CrossRef]
94. Heitkamp, M.A.; Freeman, J.P.; Miller, D.W.; Cerniglia, C.E. Pyrene degradation by a *Mycobacterium* sp.: Identification of ring oxidation and ring fission products. *Appl. Environ. Microbiol.* **1988**, *54*, 2556–2565. [CrossRef] [PubMed]

95. Ping, L.; Guo, Q.; Chen, X.; Yuan, X.; Zhang, C.; Zhao, H. Biodegradation of pyrene and benzo[a]pyrene in the liquid matrix and soil by a newly identified Raoultella planticola strain. *3 Biotech* **2017**, *7*, 56. [CrossRef]
96. Song, X.; Xu, Y.; Li, G.; Zhang, Y.; Huang, T.; Hu, Z. Isolation, characterization of Rhodococcus sp. P14 capable of degrading high-molecular-weight polycyclic aromatic hydrocarbons and aliphatic hydrocarbons. *Mar. Pollut. Bull.* **2011**, *62*, 2122–2128. [CrossRef]
97. Liu, Y.; Hu, H.; Zanaroli, G.; Xu, P.; Tang, H. A Pseudomonas sp. strain uniquely degrades PAHs and heterocyclic derivatives via lateral dioxygenation pathways. *J. Hazard. Mater.* **2021**, *403*, 123956. [CrossRef]
98. Patel, V.; Cheturvedula, S.; Madamwar, D. Phenanthrene degradation by Pseudoxanthomonas sp. DMVP2 isolated from hydrocarbon contaminated sediment of Amlakhadi canal, Gujarat, India. *J. Hazard. Mater.* **2012**, *201*, 43–51. [CrossRef]
99. Chen, J.; Wong, M.H.; Wong, Y.S.; Tam, N.F.Y. Multi-factors on biodegradation kinetics of polycyclic aromatic hydrocarbons (PAHs) by Sphingomonas sp. a bacterial strain isolated from mangrove sediment. *Mar. Pollut. Bull.* **2008**, *57*, 695–702. [CrossRef] [PubMed]
100. Tiwari, B.; Manickam, N.; Kumari, S.; Tiwari, A. Biodegradation and dissolution of polyaromatic hydrocarbons by *Stenotrophomonas* sp. *Bioresour. Technol.* **2016**, *216*, 1102–1105. [CrossRef] [PubMed]
101. Balachandran, C.; Duraipandiyan, V.; Balakrishna, K.; Ignacimuthu, S. Petroleum and polycyclic aromatic hydrocarbons (PAHs) degradation and naphthalene metabolism in *Streptomyces* sp. (ERI-CPDA-1) isolated from oil contaminated soil. *Bioresour. Technol.* **2012**, *112*, 83–90. [CrossRef]
102. Al-Hawash, A.B.; Zhang, X.; Ma, F. Removal and biodegradation of different petroleum hydrocarbons using the filamentous fungus Aspergillus sp. RFC-1. *Microbiologyopen* **2019**, *8*, e00619. [CrossRef] [PubMed]
103. Baek, K.-H.; Yoon, B.-D.; Oh, H.-M.; Kim, H.-S.; Lee, I.-S. Biodegradation of aliphatic and aromatic hydrocarbons by *Nocardia* sp. H17-1. *Geomicrobiol. J.* **2007**, *23*, 253–259. [CrossRef]
104. Govarthanan, M.; Fuzisawa, S.; Hosogai, T.; Chang, Y.-C. Biodegradation of aliphatic and aromatichydrocarbons using the filamentous fungusPenicillium sp. CHY-2 and characterization of itsmanganese peroxidase activity. *RSC Adv.* **2017**, *7*, 20716–20723. [CrossRef]
105. Wang, X.; Gong, Z.; Li, P.; Zhang, L.; Hu, X. Degradation of pyrene and benzo(a)pyrene in contaminated soil by immobilized fungi. *Environ. Eng. Sci.* **2008**, *25*, 677–684. [CrossRef]
106. Liu, Y.; Han, H.; Fang, F. Degradation of long-chain n-alkanes by *Acinetobacter* sp. *Adv. Mater. Res.* **2013**, *726–731*, 2151–2155. [CrossRef]
107. Wang, D.; Lin, J.; Lin, J.; Wang, W.; Li, S. Biodegradation of petroleum hydrocarbons by bacillus subtilis BL-27, a strain with weak hydrophobicity. *Molecules* **2019**, *24*, 3021. [CrossRef] [PubMed]
108. Xia, W.; Du, Z.; Cui, Q.; Dong, H.; Wang, F.; He, P.; Tang, Y. Biosurfactant produced by novel Pseudomonas sp. WJ6 with biodegradation of n-alkanes and polycyclic aromatic hydrocarbons. *J. Hazard. Mater.* **2014**, *276*, 489–498. [CrossRef] [PubMed]
109. Whyte, L.G.; Hawari, J.; Zhou, E.; Bourbonnière, L.; Inniss, W.E.; Greer, C.W. Biodegradation of Variable-Chain-Length Alkanes at Low Temperatures by a Psychrotrophic Rhodococcussp. *Appl. Environ. Microbiol.* **1998**, *64*, 2578–2584. [CrossRef] [PubMed]
110. Teh, J.S.; Lee, K.H. Utilization of n-Alkanes by Cladosporium resinae. *Appl. Microbiol.* **1973**, *25*, 454–457. [CrossRef] [PubMed]
111. Quatrini, P.; Scaglione, G.; Pasquale, C.D.; Riela, S.; Puglia, A.M. Isolation of Gram-positive n-alkane degraders from a hydrocarbon-contaminated Mediterranean shoreline. *J. Appl. Microbiol.* **2007**, *104*, 251–259. [CrossRef] [PubMed]
112. Chiciudean, I.; Nie, Y.; Tănase, A.-M.; Stoica, I.; Wu, X.-L. Complete genome sequence of *Tsukamurella* sp. MH1: A wide-chain length alkane-degrading actinomycete. *J. Biotechnol.* **2018**, *268*, 1–5. [CrossRef]
113. Atlas, R.M. Microbial degradation of petroleum hydrocarbons: An environmental perspective. *Microbiol. Rev.* **1981**, *45*, 180–209. [CrossRef]
114. Chen, Y.-A.; Liu, P.-W.G.; Whang, L.-M.; Wu, Y.-J.; Cheng, S.-S. Effect of soil organic matter on petroleum hydrocarbon degradation in diesel/fuel oil-contaminated soil. *J. Biosci. Bioeng.* **2020**, *129*, 603–612. [CrossRef]
115. Zhang, H.; Hu, Z.; Hou, S.; Xu, T. Effects of microbial degradation on morphology, chemical compositions and microstructures of bitumen. *Constr. Build. Mater.* **2020**, *248*, 118569. [CrossRef]
116. Hajieghrari, M.; Hejazi, P. Enhanced biodegradation of n-Hexadecane in solid-phase of soil by employing immobilized Pseudomonas Aeruginosa on size-optimized coconut fibers. *J. Hazard. Mater.* **2020**, *389*, 122134. [CrossRef]
117. Premnath, N.; Mohanrasu, K.; Rao, R.G.R.; Dinesh, G.H.; Prakash, G.S.; Ananthi, V.; Ponnuchamy, K.; Muthusamy, G.; Arun, A. A crucial review on polycyclic aromatic Hydrocarbons–Environmental occurrence and strategies for microbial degradation. *Chemosphere* **2021**, *280*, 130608. [CrossRef]
118. Heitkamp, M.A.; Cerniglia, C.E. Effects of chemical structure and exposure on the microbial degradation of polycyclic aromatic hydrocarbons in freshwater and estuarine ecosystems. *Environ. Toxicol. Chem.* **1987**, *6*, 535–546. [CrossRef]
119. Li, Q.; Huang, Y.; Wen, D.; Fu, R.; Feng, L. Application of alkyl polyglycosides for enhanced bioremediation of petroleum hydrocarbon-contaminated soil using Sphingomonas changbaiensis and Pseudomonas stutzeri. *Sci. Total Environ.* **2020**, *719*, 137456. [CrossRef]
120. Lladó, S.; Solanas, A.M.; Lapuente, J.D.; Borràs, M.; Viñas, M. A diversified approach to evaluate biostimulation and bioaugmentation strategies for heavy-oil-contaminated soil. *Sci. Total Environ.* **2012**, *435*, 262–269. [CrossRef]

121. Mukome, F.N.D.; Buelowa, M.C.; Shang, u.; Peng, J.; Rodriguez, M.; Mackay, D.M.; Pignatello, J.J.; Sihota, N.; Hoelen, T.P.; Parikh, S.J. Biochar amendment as a remediation strategy for surface soils impacted by crude oil. *Environ. Pollut.* **2020**, *265*, 115006. [CrossRef] [PubMed]
122. Dong, Z.-Y.; Huang, W.-H.; Xing, D.-F.; Zhang, H.-F. Remediation of soil co-contaminated with petroleum and heavy metals by the integration of electrokinetics and biostimulation. *J. Hazard. Mater.* **2013**, *260*, 399–408. [CrossRef]
123. Tahseen, R.; Afzal, M.; Iqbal, S.; Shabir, G.; Khan, Q.M.; Khalid, Z.M.; Banat, I.M. Rhamnolipids and nutrients boost remediation of crude oil-contaminated soil by enhancing bacterial colonization and metabolic activities. *Int. Biodeterior. Biodegrad.* **2016**, *115*, 192–198. [CrossRef]
124. Qin, G.; Gong, D.; Fan, M.-Y. Bioremediation of petroleum-contaminated soil by biostimulation amended with biochar. *Int. Biodeterior. Biodegrad.* **2013**, *85*, 150–155. [CrossRef]
125. Zhao, L.; Deng, J.; Hou, H.; Li, J.; Yang, Y. Investigation of PAH and oil degradation along with electricity generation in soil using an enhanced plant-microbial fuel cell. *J. Clean. Prod.* **2019**, *221*, 678–683. [CrossRef]
126. Rhykerd, R.L.; Crews, B.; Mclnnes, K.J.; Weaver, R.W. Impact of bulking agents, forced aeration, and tillage on remediation of oil-contaminated soil. *Bioresour. Technol.* **1999**, *67*, 279–285. [CrossRef]
127. Liang, Y.; Zhang, X.; Dai, D.; Li, G. Porous biocarrier-enhanced biodegradation of crude oil contaminated soil. *Int. Biodeterior. Biodegrad.* **2009**, *63*, 80–87. [CrossRef]
128. Xu, J.; Zheng, Y.; Fan, P.; Xu, L. Oil-addicted biodegradation of macro-alkanes in soils with activator. *Biochem. Eng. J.* **2020**, *159*, 107578. [CrossRef]
129. Tang, J.; Wang, R.; Niu, X.; Zhou, Q. Enhancement of soil petroleum remediation by using a combination of ryegrass (Lolium perenne) and different microorganisms. *Soil Tillage Res.* **2010**, *110*, 87–93. [CrossRef]
130. Shahzad, A.; Siddiqui, S.; Bano, A.; Sattar, S.; Hashmi, M.Z.; Qin, M.; Shakoor, A. Hydrocarbon degradation in oily sludge by bacterial consortium assisted with alfalfa (*Medicago sativa* L.) and maize (*Zea mays* L.). *Arab. J. Geosci.* **2020**, *13*, 1–12. [CrossRef]
131. Yateem, A.; Balba, M.T.; El-Nawawy, A.S.; Al-Awadhi, N. Plants-associated microflora and the remediation of oil-contaminated soil. *Int. J. Phytoremediat.* **2008**, *2*, 183–191. [CrossRef]
132. Jørgensen, K.S.; Puustinen, J.; Suortti, A.-M. Bioremediation of petroleum hydrocarbon-contaminated soil by composting in biopiles. *Environ. Pollut.* **2000**, *107*, 245–254. [CrossRef]
133. Xu, Y.; Lu, M. Bioremediation of crude oil-contaminated soil: Comparison of different biostimulation and bioaugmentation treatments. *J. Hazard. Mater.* **2010**, *183*, 395–401. [CrossRef]
134. Joo, H.-S.; Ndegwa, P.M.; Shoda, M.; Phae, C.-G. Bioremediation of oil-contaminated soil using Candida catenulata and food waste. *Environ. Pollut.* **2008**, *156*, 891–896. [CrossRef]
135. Schaefer, M.; Petersen, S.O.; Filser, J. Effects of Lumbricus terrestris, Allolobophora chlorotica and Eisenia fetida on microbial community dynamics in oil-contaminated soil. *Soil Biol. Biochem.* **2005**, *37*, 2065–2076. [CrossRef]
136. Luo, C.; Lü, F.; Shao, L.; He, P. Application of eco-compatible biochar in anaerobic digestion to relieve acid stress and promote the selective colonization of functional microbes. *Water Res.* **2014**, *70*, 710–718. [CrossRef]
137. Zhang, B.; Zhang, L.; Zhang, X. Bioremediation of petroleum hydrocarbon-contaminated soil by petroleum-degrading bacteria immobilized on biochar. *RSC Adv.* **2019**, *9*, 35304–35311. [CrossRef]
138. Margesin, R.; Schinner, F. Bioremediation (natural attenuation and biostimulation) of diesel-oil-contaminated soil in an alpine glacier skiing area. *Appl. Environ. Microbiol.* **2001**, *67*, 3127–3133. [CrossRef]
139. Kanissery, R.G.; Sims, G.K. Biostimulation for the enhanced degradation of herbicides in soil. *Appl. Environ. Soil Sci.* **2011**, *2011*, 988–1027. [CrossRef]
140. Hazen, T.C. Bioremediation. In *Handbook of Hydrocarbon and Lipid Microbiology*; Springer: Berlin/Heidelberg, Germany, 2010. [CrossRef]
141. Sun, Y.; Chen, W.; Wang, Y.; Guo, J.; Hu, X. Nutrient depletion is the main limiting factor in the crude oil bioaugmentation process. *J. Environ. Sci.* **2021**, *100*, 317–327. [CrossRef]
142. Sarkar, J.; Kazy, S.K.; Gupta, A.; Dutta, A.; Mohapatra, B.; Roy, A.; Bera, P.; Mitra, A.; Sar, P. Biostimulation of Indigenous Microbial Community for Bioremediation of Petroleum Refinery Sludge. *Front. Microbiol.* **2016**, *7*, 1407. [CrossRef]
143. Bento, F.M.; Camargo, F.A.O.; Okeke, B.C.; Frankenberger, W.T. Comparative bioremediation of soils contaminated with diesel oil by natural attenuation, biostimulation and bioaugmentation. *Bioresour. Technol.* **2005**, *96*, 1049–1055. [CrossRef]
144. Varjani, S.; Upasani, V.N. Influence of abiotic factors, natural attenuation, bioaugmentation and nutrient supplementation on bioremediation of petroleum crude contaminated agricultural soil. *J. Environ. Manag.* **2019**, *245*, 358–366. [CrossRef] [PubMed]
145. Rehman, K.; Imran, A.; Amin, I.; Afzal, M. Inoculation with bacteria in floating treatment wetlands positively modulates the phytoremediation of oil field wastewater. *J. Hazard. Mater.* **2018**, *349*, 242–251. [CrossRef] [PubMed]
146. Weyens, N.; Lelie, D.V.D.; Taghavi, S.; Newman, L.; Vangronsveld, J. Exploiting plant–microbe partnerships to improve biomass production and remediation. *Trends Biotechnol.* **2009**, *27*, 591–598. [CrossRef] [PubMed]
147. Muhammad, A.; Sohail, Y.; Thomas, G.R.; Angela, S. The inoculation method affects colonization and performance of bacterial inoculant strains in the phytoremediation of soil contaminated with diesel oil. *Int. J. Phytoremediat.* **2018**, *14*, 35–47.
148. Shehzadi, M.; Afzal, M.; Khan, M.U.; Islam, E.; Mobin, A.; Anwar, S.; Khan, Q.M. Enhanced degradation of textile effluent in constructed wetland system using Typha domingensis and textile effluent-degrading endophytic bacteria. *Water Res.* **2014**, *58*, 152–159. [CrossRef] [PubMed]

149. Kadam, S.K.; Watharkar, A.D.; Chandanshive, V.V.; Khandare, R.V.; Jeon, B.-H.; Jadhav, J.P.; Govindwar, S.P. Co-planted floating phyto-bed along with microbial fuel cell for enhanced textile effluent treatment. *J. Clean. Prod.* **2018**, *203*, 788–798. [CrossRef]
150. Wang, B.; Xie, H.-L.; Ren, H.-Y.; Li, X.; Chen, L.; Wu, B.-C. Application of AHP, TOPSIS, and TFNs to plant selection for phytoremediation of petroleum-contaminated soils in shale gas and oil fields. *J. Clean. Prod.* **2019**, *233*, 13–22. [CrossRef]
151. Cai, B.; Ma, J.; Yan, G.; Dai, X.; Li, M.; Guo, S. Comparison of phytoremediation, bioaugmentation and natural attenuation for remediating saline soil contaminated by heavy crude oil. *Biochem. Eng. J.* **2016**, *112*, 170–177. [CrossRef]
152. Mang, L.; Zhong-Zhi, Z.; Jing-Xiu, W.; Min, Z.; Yu-Xin, X.; Xue-Jiao, W. Interaction of heavy metals and pyrene on their fates in soil and tall fescue *(Festuca arundinacea)*. *Environ. Sci. Technol.* **2014**, *48*, 1158–1165.
153. Wilhantoro, Y.; Lowe, A. Phytoremediation of crude oil contaminated soil: The effect of growth of Glycine max on the physico-chemistry and crude oil contents of soil. *Molecules* **2015**, *20*, 22–30.
154. Cherian, S.; Oliveira, M.M. Transgenic plants in phytoremediation: recent advances and new possibilities. *Environ. Sci. Technol.* **2005**, *39*, 9377–9390. [CrossRef]
155. Abdullah, S.R.S.; Al-Baldawi, I.A.; Almansoory, A.F.; Purwanti, I.F.; Al-Sbani, N.H.; Sharuddin, S.S.N. Plant-assisted remediation of hydrocarbons in water and soil: Application, mechanisms, challenges and opportunities. *Chemosphere* **2020**, *247*, 125932. [CrossRef]
156. Kiamarsi, Z.; Kafi, M.; Soleimani, M.; Nezami, A.; Lutts, S. Conjunction of Vetiveria zizanioides L. and oil-degrading bacteria as a promising technique for remediation of crude oil-contaminated soils. *J. Clean. Prod.* **2020**, *253*, 119719. [CrossRef]
157. Rodríguez, M.D.F.; Gómez, M.C.G.; Blazquez, N.A.; Tarazona, J.V. Soil pollution remediation. *Encycl. Toxicol.* **2014**, *4*, 344–355.
158. Li, X.; Li, J.; Qu, C.; Yu, T.; Du, M. Bioremediation of clay with high oil content and biological response after restoration. *Sci. Rep.* **2021**, *11*, 9725. [CrossRef] [PubMed]
159. Chakravarty, P.; Deka, H. Enzymatic defense of Cyperus brevifolius in hydrocarbons stress environment and changes in soil properties. *Sci. Rep.* **2021**, *11*, 718. [CrossRef] [PubMed]
160. Dekang, K.; Hongqi, W.; Zili, L.; Jie, X.; Ying, X. Remediation of Petroleum Hydrocarbon Contaminated Soil by Plant-Microbe and the Change of Rhizosphere Microenvironment. *Asian J. Ecotoxicol.* **2017**, *12*, 644–651.
161. Iffis, B.; St-Arnaud, M.; Hijri, M. Petroleum Contamination and Plant Identity Influence Soil and Root Microbial Communities While AMF Spores Retrieved from the Same Plants Possess Markedly Different Communities. *Front. Plant Sci.* **2017**, *8*, 1381. [CrossRef]
162. Yaashikaaa, P.R.; Senthil Kumar, P.; Varjanic, S.; Saravanan, A. A critical review on the biochar production techniques, characterization, stability and applications for circular bioeconomy. *Biotechnol. Rep.* **2020**, *28*, e00570. [CrossRef]
163. Zhang, Y.; Xu, X.; Zhang, P.; Zhao, L.; Qiu, H.; Cao, X. Pyrolysis-temperature depended quinone and carbonyl groups as the electron accepting sites in barley grass derived biochar. *Chemosphere* **2019**, *232*, 273–280. [CrossRef] [PubMed]
164. Nan, H.; Yin, J.; Yang, F.; Luo, Y.; Zhao, L.; Cao, X. Pyrolysis temperature-dependent carbon retention and stability of biochar with participation of calcium: Implications to carbon sequestration. *Environ. Pollut.* **2021**, *287*, 117566. [CrossRef]
165. Bianco, F.; Race, M.; Papirio, S.; Oleszczuk, P.; Esposito, G. The addition of biochar as a sustainable strategy for the remediation of PAH–contaminated sediments. *Chemosphere* **2021**, *263*, 128274. [CrossRef]
166. Iqbal, A.; Mukherjee, M.; Rashid, J.; Khan, S.A.; Ali, M.A.; Arshad, M. Development of plant-microbe phytoremediation system for petroleum hydrocarbon degradation: An insight from alkb gene expression and phytotoxicity analysis. *Sci. Total Environ.* **2019**, *671*, 696–704. [CrossRef]
167. Abedi-Koupai, J.; Vossoughi-Shavari, M.; Yaghmaei, S.; Borghei, M.; Ezzatian, A.R. The effects of microbial population on phytoremediation of petroleum contaminated soils using tall fescue. *Int. J. Agric. Biol.* **2007**, *9*, 242–246.
168. Tang, J.; Wang, R.; Niu, X.; Wang, M.; Zhou, Q. Characterization on the rhizoremediation of petroleum contaminated soil as affected by different influencing factors. *Biogeosci. Discuss.* **2010**, *7*, 4665–4688.
169. Zhang, M.; Guo, P.; Wu, B.; Guo, S. Change in soil ion content and soil water-holding capacity during electro-bioremediation of petroleum contaminated saline soil. *J. Hazard. Mater.* **2020**, *387*, 122003. [CrossRef] [PubMed]

processes MDPI

Review

Anti-Smog Building and Civil Engineering Structures

Elżbieta Stanaszek-Tomal

Faculty of Civil Engineering, PK Cracow University of Technology, 24 Warszawska Street, 31-155 Cracow, Poland; estanaszek-tomal@pk.edu.pl

Abstract: Currently, people worldwide, in the period from September to April, observe with their own eyes and feel the pollution of the air, called smog, in their own breath. The biggest cause of smog and the source of air pollution is burning rubbish in stoves. Other causes include exhaust fumes from large factories, burning coal in furnaces, and car exhaust fumes. Smog is an unnatural phenomenon, directly related to human activity. The weather is becoming worse. On no-wind, foggy days, the smog phenomenon is the most troublesome for city dwellers. Smog persists in European countries from November to April, during the heating season. The harmful effect of smog affects almost the entire human body. Every year, air pollution causes the death of approximately 26,000–48,000 people. At the same time, poor air quality reduces life expectancy by up to a year. The purpose of this article is to present buildings and finishing elements that can help in the fight against air pollution.

Keywords: smog; pollutions; sustainable development

Citation: Stanaszek-Tomal, E. Anti-Smog Building and Civil Engineering Structures. *Processes* **2021**, *9*, 1446. https://doi.org/10.3390/pr9081446

Academic Editor: Elza Bontempi

Received: 20 July 2021
Accepted: 17 August 2021
Published: 19 August 2021

1. Introduction

Pollution of the natural environment is a state of the environment resulting from the introduction into the air, water, or land or from the accumulation on the surface of the earth of solid, liquid, or gaseous substances or energy in such quantities or in such composition that it may adversely affect human health, living nature, climate, soil, or water, or cause other adverse changes, e.g., corrosion of materials [1].

Sometimes environmental pollution is understood as exceeding environmental quality standards or acceptable emission factors, i.e., the actual occurrence of an unacceptable level of environmental pollution. According to UNESCO experts, currently the most dangerous polluting factors are:

- carbon dioxide (CO_2) one of the causes of the greenhouse effect;
- carbon monoxide (CO);
- sulphur dioxide and nitrogen dioxide (SO_2 and NO_2), causing acidification, phosphorus, and eutrophication;
- mercury and lead, bioaccumulating;
- crude oil;
- DDT and other pesticides;
- radiation [2].

At the same time, many risks arise from the contamination of people's immediate environment, including indoor air (presence of CO_2 and CO, NO_x, volatile organic compounds, radon, cigarette smoke, and oxygen deficiency), drinking water, and food.

Air pollution called smog has increased in recent years. It usually arises in large cities where levels of exhaust emissions and energy consumption are very high [3]. Road transport has a significant impact on the formation of photochemical smog.

Often, due to comfort or lack of other options, people choose to use a personal car over public transport. This results in heavy traffic on the streets. This increases the emission of exhaust fumes from tailpipes and the escape of dust from worn tyres and asphalt into the atmosphere. Passing cars also stir up the pollutants lying on the streets. The problem is the condition of cars. Quite often they do not meet the standards. Rapidly developing countries have a huge problem with air pollution and thus the environment as such [4]. This also has a negative impact on human health [5]. Each component of smog adversely affects human health, but just as the composition of smog can vary and is constantly changing, so too are its effects. The components it contains are very dangerous for humans.

As shown by researchers at the Health Effects Institute (HEI), smog is the fourth most serious risk factor, just behind hypertension, smoking, and obesity. It affects almost all organ systems in the human body, causing many different diseases:

- nervous system (headache, central nervous system disorders);
- the respiratory system (chronic lung disease, lung cancer, asthma);
- the cardiovascular system (ischaemic heart disease, heart attack);
- the digestive system (liver disorders);
- reproductive system (disorders of internal organs);
- the immune system (allergies).

Even small and short-lasting exceedance of nitrogen oxide concentration in the air may worsen the feeling of well-being. At the concentration of only 1.5 mg/m^3, it may cause cough or irritation of the nasal mucosa. When the concentration rises, this leads to sore throat and eyes, a drop in blood pressure, dizziness, and an increase in methemoglobin levels in the blood; prolonged breathing in such air results in shortness of breath and swelling of the lungs. This can lead to inflammation of the airways [6]

2. Smog

The word "smog" is a combination of the words "smoke" and "fog". This newly coined word or word cluster has spread around the world. It seems to define the phenomenon very well. Smog is an artificial fog, an unnatural atmospheric phenomenon consisting of the coexistence of air pollution caused by human activity and natural phenomena: considerable haze and windless weather [7,8]. There are two main types of smog, which are shown in Figure 1 [9].

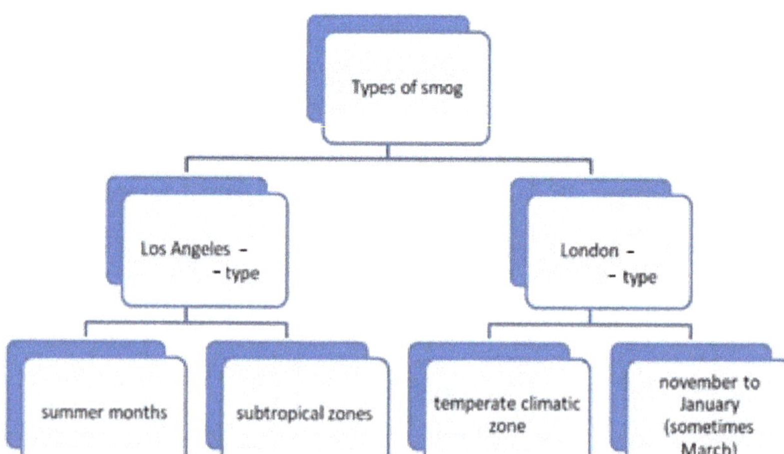

Figure 1. Diagram: types of smog (developed by the author).

Los Angeles-type smog [9,10], otherwise known as photochemical smog, occurs during the summer months of the year and usually in subtropical areas. Under such conditions, under the influence of strong sunlight, the chemical compounds contained in car exhausts are strongly transformed and oxidants are formed. It consists of gases including carbon oxides, nitrogen oxides, and hydrocarbons. It also forms a very noxious brown haze. The London type [9,11], on the other hand, is formed in temperate climate zones and occurs during winter periods from November to January and even into March. It consists of different dusts, enriched with sulphur, nitrogen, and carbon oxides, as well as soot. As the smoke from chimneys rises higher, its temperature drops. In normal weather conditions, the air temperature decreases with height and the smoke rises freely into the upper layers of the atmosphere. In autumn and winter, however, a so-called temperature inversion often occurs. This creates the opposite situation. The higher the altitude, the higher the temperature. Under such conditions, the temperatures-of the pollutants from the chimney and the air - quickly equalise. As a result, the pollution stays closer to the ground because it cannot escape into the higher layers of the atmosphere. If, in addition, there is no wind and high humidity, the smog accumulates more and more [12,13].

On the one hand, smog is caused by the air mixing with pollutants and exhaust fumes produced by human activity. Factories, the increasing number of cars, burning coal, and wood and other solid fuels in cookers are responsible for this. On the other hand, weather, climate, and general conditions of the area influence its formation. Pollutants lingering over a city located in a basin, in windless weather, cannot spread and dilute. This causes them to hover over the town in question [13,14].

There is also another type of smog, the so-called influx smog. This involves pollution spreading to other areas. The air clears in one place but becomes more polluted in another [15].

Smog has a negative impact on human health. It has direct and indirect effects. The direct ones are allergies, heart failure, cardiovascular and heart diseases, and lowering of the general immunity of the body. Indirect ones are related to eating smog-contaminated animal or plant meat. Toxins entering the respiratory system also irritate its mucous membranes, lead to inflammation, and cause symptoms similar to those of the common cold—cough, runny nose, and throat irritation. In addition, smog contributes to the formation of acid rain [4].

In Poland and many European countries, there is London-type smog, but also Los Angeles-type smog due to traffic pollution. In Poland, however, particulate smog has yet to be distinguished [15]. Smog comes from different areas of life, as shown in Figure 2.

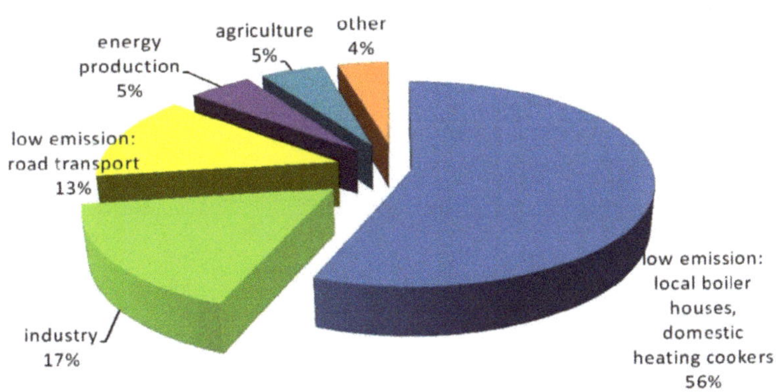

Figure 2. Areas of life with an impact on smog formation (developed by the author).

The largest emissions of smog, including PM10 and PM2.5 particles, are caused by burning low-quality coal in old and often poorly regulated boilers and household furnaces, as well as by various types of waste [16]. PM2.5 are atmospheric aerosols that are less than 2.5 micrometres in diameter. This type of particulate matter is considered to be the most dangerous to human health. PM10 is a mixture of airborne particles whose diameter does not exceed 10 micrometres. It is harmful due to its content of elements such as benzopyrenes, furans, and dioxins, i.e., carcinogenic heavy metals.

Low emission is the emission of harmful substances and dusts from local boiler houses, domestic heating cookers, and car traffic [17]. Combustion products contributing to low emissions include the following gases: carbon dioxide (CO_2), carbon monoxide (CO), sulphur dioxide (SO_2), nitrogen oxides (NO_x; NO_2, NO), polycyclic aromatic hydrocarbons such as benzo(α)pyrene and dioxins, and heavy metals (lead, arsenic, nickel, cadmium) and particulate matter (PM10 and PM2.5). These are point air pollutants. They may occur throughout the year.

"Low emissions" of particulate matter are also caused by industry, especially the power, chemical, mining, and metallurgical industries [18], but due to the height of the emitters and the legislation in force regulating the permissible values of emissions, these sources usually have a much smaller impact on air quality.

Three main air pollutants can be distinguished in European countries: PM10, PM 2.5, and polycyclic aromatic hydrocarbons (PAHs), including benzo(α)pyrene and, to a lesser extent, PM1. PM 2.5 contains particulate matter with a very small diameter of only up to 2.5 m. Due to their size, once they enter the respiratory system, they travel very easily and poison the bloodstream. PM10, on the other hand, consists of grains up to 10 m in diameter. This size allows it to penetrate deep into the lungs. Pollutants from home heating (so-called "low emissions") account for the largest percentage (over 83%) of PM10 concentrations in the air. The situation is very similar for benzo(α)pyran (polycyclic aromatic hydrocarbons). It is emitted in the highest amount during individual heating of buildings, followed by the operation of coking plants and road transport [6].

The building industry is committed to sustainable development. It is defined as a way of farming in which meeting the needs of the present generation will not reduce the chances of meeting the needs of future generations. The breakdown in sustainability is shown in Figure 3.

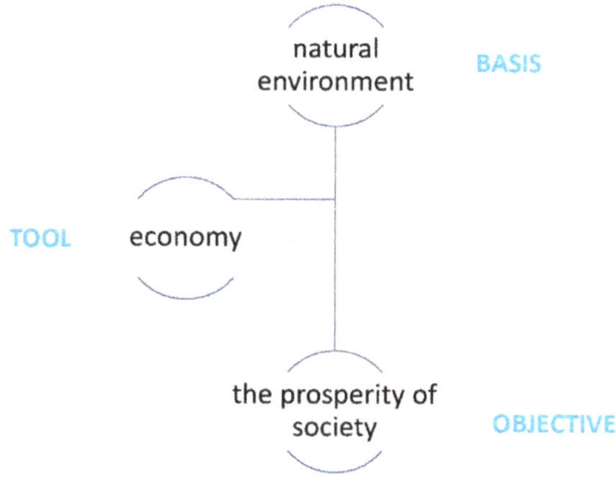

Figure 3. Division in sustainable development (developed by the author).

In order to achieve sustainable development, coherence is needed between three key elements: economic growth, social inclusion, and environmental protection. These are interlinked and all are critical to achieving well-being for individuals and societies as a whole [19].

Sustainable development is implemented in many ways. One of them is the gradual elimination of hazardous and toxic substances from economic processes and keeping emissions within the limits set by the assimilative capacity of the environment. Such actions entail striving to ensure the sense of security and well-being of citizens, understood as creating conditions conducive to their physical, mental, and social health. Sustainable development is implemented through

- restricting the use of renewable resources to the limit of their recovery;
- reducing the consumption of non-renewable resources to a level that allows them to be gradually replaced by suitable substitutes;
- the progressive elimination of hazardous and toxic substances from economic processes;
- keeping emissions of pollutants within the limits set by the assimilative capacity of the environment;
- restoration and permanent protection of biological diversity at landscape, ecosystem, species, and gene levels;
- socialisation of decision-making processes concerning the local natural environment [19].

In the analysis of the above, efforts should be made to minimise the amount of smog. At the same time, the causes of smog should be eliminated, i.e., low-emission cookers and reducing the number of old cars, especially diesels, which emit the greatest amount of pollution. In addition, the effects must be fought. In the circles of city halls, architects, town planners, and builders, the topic of smog removal has been frequently discussed in recent years.

In the following part of this article, different ways of fighting smog in architectural and construction aspects are presented.

3. Building with Nano Additives

Nanoparticles are compounds with at least one dimension below 100 nm. Nanoparticles can be used as nanomodifiers for various materials. The most commonly used nanoparticles include nanosilver, titanium dioxide, nanogold, nanopalladium, nanocopper, nanoplatin, zirconium dioxide, zinc dioxide, graphene, nanotubes, and fullerenes, as well as carbon nanofibers [20].

One of the most commonly used nano additives is nano-TiO_2, which has specific physical and chemical properties. These include photocatalytic activity, hydrophilicity, and strong UV absorption. One solution to clean up the surrounding environment is to apply coatings with nano-TiO_2 to building materials. This additive, after absorbing UV radiation, is able to effectively decompose organic compounds such as VOCs (volatile organic compounds), microorganisms, and NO_x pollutants. The photocatalytic activity of the nano additive requires access to sunlight, and thus thin layers (coatings) of a few millimetres are sufficient to achieve the desired properties [21]. Various applications of nano-TiO_2 are presented in Figure 4.

Titanium dioxide exists in three crystal structures: anatase (distorted tetragonal crystal structure), rutile (also tetragonal), and brucite (rhombic crystal structure). Of these, only rutile and anatase can be of practical use. This is due to the fact that they have a wide semiconductor bandgap [22]. Anatase is more efficient in degrading organic as well as inorganic contaminants in the gas and liquid phases [23]. Rutile and brucite phases are more applicable for selective oxidation of organic syntheses [24]. The simultaneous use of anatase and rutile phase increases the photocatalytic activity compared to each individual component [25]. The high photocatalytic activity of anatase has led to its widespread use as a photocatalytic coating on various substrates under low intensity, i.e., near UV light [26]. Consequently, visible light is not energetic enough to induce photocatalytic activity in anatase [27].

The photocatalytic degradation reactions of pollutants are shown in Figure 5.

The mechanism is based on the generation of radicals as a result of irradiation of the photocatalyst substances, and then transforming the pollutants into harmless compounds [28]. Figure 2 shows the photocatalytic reaction to eliminate NO_x pollution by photocatalysis of concrete pavements.

Figure 4. The use of nano-TiO_2 products (developed by the author).

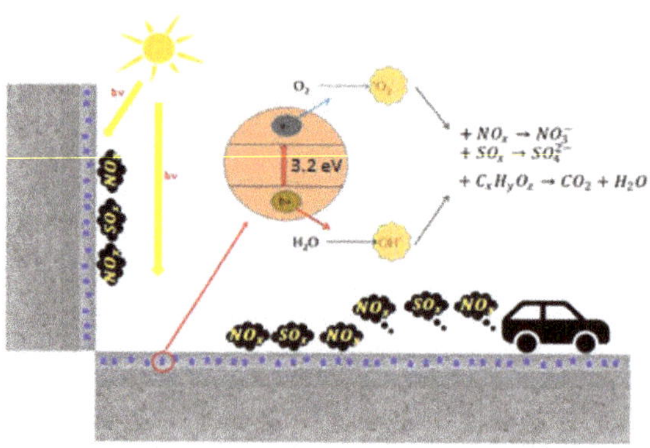

Figure 5. Photocatalytic degradation reactions of pollutants (developed by the author).

NO is considered a primary pollutant that is mainly introduced into the atmosphere directly from high-temperature combustion in transport and industrial activities, while NO_2 is considered a secondary pollutant as it is mainly formed in the atmosphere due to the interaction of NO with O_2 or O_3 and/or sunlight [29].

Concrete pavements and building exteriors are ideal for photocatalytic application of materials because their flat configurations would facilitate exposure of the photocatalyst to sunlight [30].

In photocatalytic cements, the resulting NO_3^- reacts with the calcium in the cement to form a water-soluble salt, calcium nitrate, which is easily removed by rainwater. Effective elimination of air pollutants at concentrations in the range of 0.1–10 ppm is possible with such photocatalytic cement materials [31].

3.1. Construction Works

Portland cement modified with nano-TiO_2 has become one of the elements for reducing air pollution. Thanks to its self-purifying properties, i.e., the conversion of organic compounds, sulphur, and nitrogen oxides into harmless substances, i.e., water and CO_2, titanium dioxide makes it possible to clean the air of pollutants [32]. Its use in cement or concrete is already well known. Due to the fact that only a certain thickness of the material with this nano-additive has the possibility to be activated by UV radiation, it is only used on top surfaces. Therefore, these materials are used in the form of façades, coatings, or paints [21].

3.1.1. Façades

There are façade panels on the market from Berlin-based architecture firm Elegant Embellishments that absorb pollutants from the air. The technology is called Prosolve370e [33]. The individual modules are coated with very fine titanium oxide particles. The panels are designed to absorb harmful substances and light over as large a surface area as possible. According to research, the panel system cleans the air of pollution from a thousand cars a day. The technology has been used in many cities around the world, including those in Mexico, the United Arab Emirates, and Australia.

One of the first buildings to be built with a purifying façade is the Jubilee Church in Rome dedicated to the Merciful God the Father, designed by Richard Meier. Self-cleaning and smog-absorbing cast concrete blocks with titanium oxide and white architectural concrete were used in its construction. It was completed in 2003 [33].

Another example is the Hotel de Police building in Bordeaux, France. It was built in 2003, designed by Claude Marty and Lacrouts Massicaults SA Architects. The building is located in the city centre, where there is a lot of traffic and therefore also a lot of pollution. The façade of the Hotel de Police is made of prefabricated concrete panels. There are 750 panels, 700 of which are white, covering a total area of 5400 m^2. In addition to cleaning the air, the technology also has a self-cleaning function so that, even after many years, the original appearance and especially the colour of the building will not change [34,35]. TX Cement Active® technology was also used by Luc Declercq in the creation of The Commodore building in Ostend. In collaboration with E&L Projects, he designed an apartment complex to purify the air in the Belgian town. The first six floors of the building are covered in polished concrete with titanium dioxide to help purify the air. Due to its coastal location, the building is also exposed to organic pollutants, which develop more quickly in a humid environment. The building was constructed in 2005, and tests carried out several years later confirmed the material's self-cleaning properties. This shows that TX Active® technology allows the aesthetic appearance to be maintained for many years, regardless of the location of the building and the pollutants affecting it [33].

Another facility that uses titanium nano-dioxide façades to reduce smog is the Manuel Gea Gonzalez Hospital in Mexico City. The façade has a decorative appearance, i.e., the white openwork panels resemble honeycombs, a spider web, or a coral reef. This increases the photocatalytic surface. The unusual shapes exhibit larger active surfaces, which allow for increased removal of contaminants. A 2500 m^2 façade can reduce pollution from 1000 cars per day. In 1992, the city had only nine smog-free days per year. Today, it is one of the leaders in the fight against smog [34].

The Palazzo Italia designed by Nemesi and Partners for Expo 2015 [36] in Milan is a building that also uses nano modifications. The façade used in this building is designed to resemble growing trees, in which the tree crowns are made of nanocement. The effectiveness of this surface (9000 m^2) measured by the reduction of air pollution has been found to be between 20 and 70%, depending on weather conditions. This refers to the purifying properties of plants, which clean the air of pollutants, as does the façade of the pavilion. The decorative element is made of cement with titanium [35].

3.1.2. Plasters

In buildings with complex shapes, cement-based plasters or paints are used. This is because the low thickness of the coating provides the same performance on concrete surfaces while reducing the consumption of titanium dioxide. An interesting example of paint application is a historical building undergoing renovation is Matrice Church in Italy. A coating product based on natural hydraulic lime with low cement content was applied [2].

Another example is the Umberto I Tunnel in Rome, where plaster was applied to the walls over a length of about 400 m and a width of 8 m. Due to the lack of access to UV rays, dual-function lamps were used, providing normal light and UV radiation. The efficiency of pollution reduction in the tunnel reaches 20 to 70% [37].

Cement paints doped with TiO_2 can be used to create a mural. In Poznań, Poland, a mural depicting Poznań of the future was created on the wall of the "Wiktoria" cultural centre in the Winograd Zwycięstwa housing estate. The painting is 77 sq m in size.

A mural with similar features can be found in Italy in Rome of the Yourban 2030 project called "Pollution Hunt" with an area of 1000 m^2. It depicts the tricoloured heron, an endangered species that is struggling to survive. The mural absorbs as much smog as 30 trees. The creators of the technology ensure that the paint is able to eliminate up to 88.8% of air pollutants and 99.9% of bacteria. Many murals have been created by the Converse City Forests project. They can be admired in Rio de Janeiro, Lima, Belgrade, Sydney, Santiago, Bangkok, Johannesburg, Saigon, Manila, Mexico City, Sao Paulo, Jakarta, Ratchaburi, and Chiang Mai, among others [38].

3.1.3. Paints

Artists have also taken an interest in the use of nano-TiO_2 paints. They create murals to decorate the walls of buildings and neutralise nitrogen dioxide. They can be found in many cities around the world, e.g., in England and Italy, while in Poland they can be seen in Warsaw and Poznań, among others [39].

This is a cheap and simple solution with double benefits. Apart from cleaning the air, the paints also have a self-cleaning function. They remove unpleasant odour and have an antibacterial effect.

3.1.4. Organic Coatings

Nano additives can also be added to organic coatings such as acrylic, fluorinated ethylenetetrafluoroethylene (abbreviated ETFE), and silane polymers. They are also used for hybrid coatings such as silica and polydimethylsiloxane.

3.2. Road Surfaces

Cementitious materials with titanium dioxide additives can also be used in road construction. The first anti-smog road construction projects were carried out at the beginning of the 21st century. The City Park of Antwerp in Belgium built a pavement of about 10,000 m^2 with concrete bricks containing nano-TiO_2 [40]. Along a road in Antwerp, parking lanes have been created with two-layer photocatalytic concrete pavers. Such solutions are still in use, among others in the Netherlands, where some roads are made of air-purifying cobblestones.

3.2.1. Concrete Asphalt

Researchers at the Technical University of Eindhoves in the Netherlands tested concrete asphalt with photocatalytic properties. They obtained efficiencies of between 19 and 45% reduction in airborne pollutants with this material [41]. Very similar studies were carried out in Chicago in the United States. They obtained a slightly higher efficiency, ranging from 20 to 70% reduction of pollutants. Of course, the effectiveness depended on the type of road in question and the weather conditions. However, the implementation of such a system showed that it works better on local roads than on busy roads, i.e., motorways. The reason behind this is cars. When there is a lot of traffic on the road, it is difficult for particles to fall onto the asphalt, and thus the cleaning processes is prevented [42].

3.2.2. Paving Blocks

Paving blocks consist of two layers: the base layer and the surface layer. Only to the surface layer (texture) is nano titanium dioxide added. They are used to make pavements, as well as bicycle paths. This type of solution is used all over the world. Examples include Chicago in the USA, Bergamo in Italy, Nowa Sól in Poland, in the surroundings of a kindergarten in Bietigheim-Bissingen (Germany), pavements in Tatton Park in Knutsford (Great Britain), and pavements in Warsaw and Krakow in Poland. In Bergamo, both the pavement and the roadway are lined with these paving stones. Traffic routes clean the air up to a height of 2–2.5 m. The effect decreases with distance. They are effective in reducing NO_2 by about 30% under optimal atmospheric conditions [40].

Another air-purifying solution for roads are pavements. A frequently used solution, similarly to roads, is paving blocks made of "green concrete". In Poland, such projects are carried out, among others, by Skanska. The solution is able to reduce the concentration of carbon dioxide by 30%, and in laboratory conditions even by 70% [43]. Photovoltaic tiles are also used for the construction of pavements and cycle paths.

3.3. Acoustic Screens

Acoustic screens are photocatalytic coatings that break down nitrogen oxides into harmless compounds when exposed to sunlight. These are then flushed from the facilities by precipitation. Even on a cloudy day, sufficient sunlight is provided for the screen to work effectively. Many factors influence the effectiveness of the photocatalysis process—porosity, particle size, irradiation time, atmosphere exposure, and pollutant concentration. Such a solution has been applied, for example, in the Philippines. In the capital city, concrete noise barriers along a 15 km city ring road have been painted. It absorbs the pollution produced by tens of thousands of cars every year. Such screens can reduce road air pollution by 15–25% [44].

3.4. Roof Tiles

Cement or ceramic roof tiles are also used as anti-smog materials. However, the nano-additive is not applied to the tile itself, but they are coated with an impregnation with photocatalytic properties. The effectiveness of the tiles: a roof of 2000 m^2/foot can oxidise the nitrogen oxides released by driving a car 10,000 miles a year.

Students at the University of California, Riverside, have developed roof tiles that clean the air of harmful nitrogen oxides. They coated ordinary tiles with a layer of anatase (a special variety of titanium dioxide), which has the ability to actively clean the air. Then, by placing them in a special chamber, they tested their photocatalytic properties and air purification function depending on the thickness of the coating. The result showed that the number of TiO_2 layers had no significant effect on the effectiveness, and that the anatase layer, working together with the sun, is able to remove up to 97% of nitrogen oxides from the air. The stronger the UV radiation, the better the effect. A roof covered with such tiles is able to clean the air of 21 tons of harmful oxides per day [45,46].

4. Building with Vegetation

In addition to the nano-additive and its pollutant-reducing properties through the photocatalysis process, there are other methods used in building structures. Plants can help combat pollution by producing oxygen. Large cities suffer from a lack of space to create new squares, parks, and gardens. These include green roofs, green façades, and artificial trees.

Green roofs, green façades, and living walls are elements of sustainable construction [47]. The use of green roofs is used worldwide [48]. Vertical green systems (VGS), i.e., green façades and living walls, are very rarely used. This is due to the fact that the number of plants possible to use is small compared to the number of plants used in green roofs

Living walls differ from green façades in that the plants are rooted in a structural support. This is attached to the wall itself. The plants receive water and nutrients from the vertical support rather than from the ground [48].

4.1. Pollution-Reducing Façades

The growing popularity of green façades has influenced the number of solutions available for placing vegetation on a building. They differ in terms of:

- construction;
- the degree of independence from the façade plane.

Vegetated vertical systems include two main categories: green façades and living walls [49]. These two categories are characterised by different locations of growing media for plant roots. Green façades use plants placed directly on the ground at the base of the building or in pots at different façade heights. Self-supporting plants can grow directly stuck to the wall in the case of direct green façades; climbing or hanging plants can climb onto support structures placed a short distance from the wall in the case of intermediate green façades. In the latter solution, an air gap is created between the wall and the vegetation layer. Some climbing plant species can cause damage to the building surface if cracks are already present [50]. The presence of an air gap in intermediate green façades and the appropriate choice of plants mitigate these negative effects. Living walls are based on the use of a growing medium that is attached vertically to the building envelope or to the frame. Vertical green systems can reduce the frequency of maintenance interventions on the building envelope due to limited temperature fluctuations of the wall surface [51].

Several green façade systems can be distinguished. The first is to create wall space for climbing plants, guided along a special structure made of stainless steel or impregnated wood. The plants are then planted at the base of the wall. This eliminates the need to install additional irrigation systems. This is the most commonly used system.

The second one consists in creating a wall on the basis of a system of pots, which are fixed to a special construction made of stainless steel. The advantage of this solution is the possibility to use various types of greenery and integrate it into the appearance of the façade. This method is slightly more advanced than the first system [52].

The third one consists in keeping the vegetation in good condition thanks to an automatic irrigation system, on the basis of the principle of horizontally placed pipes through which water is pumped.

The last solution, the so-called living wall, is the most technologically advanced method. It makes it possible to reduce the construction elements to a minimum. It consists in using ready-made system inserts made of a special material resembling horticultural foam. Plants that are part of the façade are rooted in the inserts and fixed to the frame. This solution, like the previous ones, makes it possible to water the greenery using irrigation systems [52].

The issue of fauna and flora on building façades and roofs does not only relate to how to build a green roof or living wall, but first and foremost to what role they are to play in the natural and social aspect and to what extent they correspond to current trends in shaping urban space.

The environmental performance of green vertical systems can be influenced by different green systems, weather conditions, building types, selected plant species, building orientation, and materials, etc. [53]. A green façade reduces nitrogen oxide by up to 20%, PM10 by 35%, and the temperature drop around the wall by up to 16 °C. The green wall also stores 750 L of water (100 m^2/24 h) and produces oxygen through photosynthesis (155 m^2 of green wall provides enough oxygen for one person for a whole day). It absorbs carbon dioxide and works with great efficiency. Only 100 m^2 will absorb 250 kg CO_2 per year, a figure comparable to a row of street trees [54].

The façade greening of building walls, known as vertical green systems (VGS), requires the use of climate-appropriate and non-allergenic vegetation. Trees that have a positive and significant impact on air quality include maples (*Acer* spp.). In addition to native species such as Norway maple, sycamore, and field maple, North American sugar maple (*A. saccharum*), silver maple (*A. saccharinum*), and red maple (*A. rubrum*) are particularly valuable for cities. Some authors also mention that the most effective pollutant-absorbing plants include common yew (*Taxus baccata*), lime trees (*Tilia* spp.), Sabine juniper (*Juniperus sabina*), Meyer lilac (*Syringa meyeri*), and Siberian microbiota (*Microbiota decussata*), and from American reports, in addition to some already mentioned are black walnut (*Juglans nigra*), ash (*Fraxinus excelsior*), common beech (*Fagus sylvatica*), Lawson's cypress (*Chamecyparis lawsoniana*), and Chinese metasequoia (*Metasequoia glyptostroboides*) [52].

Some species that retain their leaves for a long period of time in autumn, such as the shrub Ligustrum vulgare and the climbing Akebia quinata or the evergreen Hedera helix, which starts flowering in autumn, are also important for improving air quality. Ivy can be planted in roadside screens, but not on the street side, because unlike akebia it is much more sensitive to salt spray. In Prof. B. Wolverton's research, ivy was also found to be one of the more effective plants in purifying indoor air.

Evergreen trees are more effective at removing pollutants because the leaf activity period is longer. The size of the tree is important, as it determines the amount of CO_2 absorbed, retained, and stored in the biomass, as well as the total leaf area ready to absorb pollutants from the air, including particulate matter [51,52].

An example of a building with air pollution removal properties is the Bosco Verticale skyscraper designed by Boeri Studio in Milan [54,55]. Part of the building is shown in Figure 6. Its façade is formed by a vertically planted forest. The 9000 m^2 of terraces are planted with trees, 11,000 selected plants such as perennials and ground-cover plants, and 5000 shrubs. Such façades have another function, i.e., they regulate the indoor temperature [56].

Figure 6. This is a figure. Schemes follow the same formatting. Photograph of part of the Bosco Verticale building. Polina Chistyakova z Pexels (common creative).

Another example of a similar façade is the Tao Zhu Yin Yuan Tower building (Taipei, Taiwan), designed by Vincent Callebaut Architectures based in Paris. The skyscraper is in the form of a DNA double helix. This signals the unity of people with nature. A total of 23,000 trees are planned to be planted on balconies, terraces, and in the immediate vicinity of the skyscraper. The efficiency of this facility is 130 Mg of carbon dioxide per year. Construction of the Tao Zhu Yin Yuan Tower in Taipei, Taiwan, was completed in 2018. It was designed and built by Paris-based Vincent Callebaut Architectures. The building contains 42 flats, spread over 21 floors. Its design is inspired by the structure of the double helix of DNA, twisted from base to top by 90°, the source of life, the harmony of man with nature. The main idea behind the project is to fight global warming by absorbing carbon and saving energy. The building has wide balconies and is to be planted with a total of 23,000 trees and shrubs, whose light is to be facilitated by the unusual shape of the building. They are to absorb 130 tonnes of carbon dioxide emissions annually. It has many other green features: systems for recycling organic waste and used water, a 4.5° horizontal rotation of the floors provides natural lighting, and a photovoltaic solar roof to provide additional power. The construction materials and equipment used to make Tao Zhu Yin Yuan Tower were recycled and/or fully recyclable. The design combines the concepts of energy conservation, carbon reduction, and human–ecosystem harmony [57].

According to *Business Insider*, Boeri Studio is currently in the process of building a similar project, but on a much larger scale. In China, a "forest city" is being built along the Liujiang River, on an area of 175 hectares. It is to be a district that will contain 40,000 trees and almost 1 million plants of over 100 different species.

In addition to housing, there will be hotels, schools, and hospitals, and thus it will be fully self-sufficient. It will be connected to the city of Liuzhou by an electric railway. The district will annually absorb 10 thousand tonnes of carbon dioxide and 57 tonnes of pollution and produce about 900 tonnes of oxygen. The project is to incorporate many other green features, including the use of geothermal energy and solar panels. When completed, it will be able to house up to 30,000 people [58].

Citicape House in London is a project to be completed in 2024. It includes, among other things, a five-star hotel, offices, a spa, and a restaurant. The façade of the corner 11-storey building in central London will be largely covered with vegetation. Its surface area is 24,500 m^2. It was designed by the British design studio Sheppard Robson. On the roof of the building, apart from the bar, there will be a terrace with a meadow, decorated with native endangered flower species. In addition to improving air quality, the designers want to raise public awareness of this issue. The building is located in a busy area and is full of hustle and bustle. An estimated 400,000 plants will cover the façade, producing 6 tonnes of oxygen per year. The flowers will be placed in special frames and then fixed to the façade. They will be watered by rainwater collected in special tanks. According to estimated calculations, the amount of pollution will be reduced by 8 tonnes per year and about 500 kg of particulate matter will be absorbed. It will also lower the temperature by 3–5 °C. It will be the largest green wall in Europe [59].

4.2. Green Roofs/Inverted Green Roofs

Green roofs are a common nature-based solution [60]. They carry numerous benefits such as reducing building energy consumption, rainwater management, mitigating the urban heat island effect, improving air pollution, and enhancing the green aesthetics of buildings [61].

Green roofs are most commonly constructed on inverted roofs. This is due to the fact that they are more durable and less prone to mechanical damage. The plants present on them are permanently attached to the structure. Such a roof may be designed as a roof over an underground garage, or a terrace on top of a skyscraper.

The division of green roofs is related to the way they are used. There are

- extensive roofs;
- intensive roofs.

The first group of roofs is non-utilised. It has a decorative function and is used to reduce the amount of rainwater discharged into the sewage system. Plants used in this system must have a flat root system and low vegetative requirements. This means that they do not need to be nurtured. They can maintain and grow themselves, such as grass, herbs, sedums, and succulents. Systems for extensive green roofs are lightweight roof structures that are ideal for bus shelters and buildings that cannot take a heavy load.

The second group of roofs is suitable for use and can play the role of a private garden in the city centre. They are comparable to green areas on the ground, such as gardens and parks. They are planted with lawns, shrubs, and trees, and are used in the building of a gazebo and in the making of ponds. Such compositions are representative. Plants planted on them require more care, and in some cases additional irrigation is also needed. Intensive green roofs can be installed on buildings such as detached houses, blocks of flats, or public buildings, as well as on underground structures such as car parks [62].

Green roofs involve higher production and maintenance costs. This is because a more complex structure must be designed to simultaneously store water and drain excess water.

Extensive and intensive green roof systems, which can be used to roof almost any structure, give it a green appearance while reducing harmful substances in the air. Green roofs on bus shelters provide shade for passengers on hot days, while on the roofs of buildings, they replace traditional gardens for residents. When used on underground structures, they allow a piece of developed space to be reclaimed. In addition, they provide bus stops and buildings a green and positive appearance, improving the living environment for city dwellers [63].

However, whether it is an extensive or intensive type of green roof, they are more expensive than the classic ones, being more complicated to install and possibly to repair.

Pollution in the park can be 20–40% lower than in other areas of the city. An area of about 1.5 m^2, covered with uncut grass, produces as much oxygen per year as the annual demand of one person. Therefore, the use of green roofs offers great opportunities.

Another solution is the use of green roofs. They are planted with flower meadows as well as low trees and shrubs. The roofs can be used in an inverted system.

However, green roofs are also arranged on residential roofs. Of course, there must be sufficient moisture insulation and a layer in which plants can grow. A thickness of 5 to 10 cm is suitable for grasses and lower plants, such as shrubs, whereas a thickness of 30 to 50 cm allows for the growth of taller, more strongly rooted trees. Of course, the roof structures in these cases must be reinforced [63].

4.3. Anti-Smog Towers

Anti-smog towers mostly work by using ionisation. They suck in polluted air from above and release it in a purified form at the bottom.

The hexagonal tower called Smog Free Tower, an idea by Daan Roosegaarde [64], is 7 m long. It is made of aluminium and has two storeys. It filters 30,000 m^3 of air per hour. The operation of the tower is based on ionisation technology, i.e., capturing harmful particles. It is powered by solar panels. It is completely environmentally friendly. The Smog Free Tower was first presented on 4 September 2015 in front of the Roosegaarde design studio in Rotterdam. However, it has no fixed location. In accordance with the vision of the originator, it travels the world, cleaning the air in the most polluted cities. In 2018, a second tower was constructed, which could be seen for two months in Krakow [38]. However, research by scientists from the AGH University of Science and Technology did not confirm the possibilities declared by the designer. The concentration of suspended dust at a distance of 10 m from the facility decreased by 12%, but no improvement was recorded within a radius of 50 m. This was confirmed by tests conducted in China [38].

Another tower was developed in India. India's Symbiosis Studio has developed a system of two types of pollution-absorbing towers. Larger, 60 m high towers are to be set up along the city boundary to absorb pollution coming from the suburbs. Each tower will clean the air in an area of about 2.5 km^2 and operate within a radius of 900 m. Smaller,

18 m high towers will be placed in different parts of New Delhi. They are to clean the air during atmospheric silence, i.e., when there is no air circulation. One is able to purify 30 million m^3 of air. Additionally, 60–70% of the surface of each tower is to be covered with vegetation, which will further reduce harmful substances, and inside them there will be docks for drones to monitor the state of pollution in the city [65,66].

Another country with a major smog problem is China. The Chinese city of Xi'an is home to the world's largest anti-smog tower [67]. There is also a coal mine there with low prices, which makes the air very polluted. The biggest problem arises in winter when the heating season starts. The tower is 100 m high and is able to clean 10 million m^3 of air per day from an area of 10 km^2. It works by drawing air into greenhouses of about 3.5 thousand m^2 on its surface. The air is then heated by UV rays, causing it to rise. This is how it gets inside, where it is filtered several times and, after being cleaned, is released. After confirming that the tower meets the claimed properties, Chinese engineers plan to build other, even larger towers [65].

5. Concrete with Activated Carbon

A major challenge is the use of anti-smog materials based on concrete in tunnels or underground car parks [68]. Currently, research is being conducted in Poland and worldwide on the effectiveness of concrete with activated carbon in absorbing and cleaning air of NO_x oxides. Researchers have looked into activated carbon. Activated charcoal is an adsorbent of NO_x compounds [69]. Worldwide literature [70] reports that it shows significant chemical affinity to nitrogen oxides in terms of adsorption and reactivity. The material owes its strong adsorbent capacity to its very high specific surface area per unit mass. It is almost 3300 m^2/g. This surface area is so large due to the highly porous internal structure of this material, which consists of micropores and mesopores [71]. Activated carbon starts its action when it comes into contact with pollutants, acting as a particle-catching filter. By means of chemical reactions, the pollutants are converted into harmless nitrate ions.

NO_2 can react in alkaline aqueous solutions to produce nitrite and nitrate ions [72]. Therefore, NO_2 reduction may be related to calcium hydroxide content. The authors of [73] hypothesised that NO_2 first dissolves in the adsorbed aqueous layer that covers alkali cement-based hydrates at 60% relative humidity [74]. According to the authors of [73], the course of reaction for activated carbon modified concrete is as follows:

$$2\,Ca(OH)_2 + 4\,NO_2 \Rightarrow (Ca^{2+}, 2\,NO_3^-) + (Ca^{2+}, 2\,NO_2^-) + 2\,H_2O$$

Due to the low weight of the activated carbon particles, it flows to the surface of the slurry and settles right at the mashing surface. In order to increase the effectiveness, the expansion surface of the activated carbon particles is sought. This is done by increasing the roughness of the surface and texturing the face layer [68].

The authors of [68] created a garage from concrete containing AC (activated carbon). As a result of experiments, the authors found that a garage made of concrete modified with activated carbon had a NO_2 absorption of 20–25%.

The performance of activated carbons towards NO_2 has been reported by several research groups [75–77]. It is known that the major part of NO_x from power plants and car engines is NO. Removal of NO_2 by activated carbons has become a potential technique.

6. Results

The solutions outlined above will not immediately clean the air of harmful substances. However, one has to start somewhere, according to the proverb/sentence "Rome was not built in a day". Material/plant solutions are the way forward for designing sustainable buildings that will have a positive impact on the environment.

Biophilic design [78] helps to reduce urban heat islands, improving outdoor air quality. It provides a better indoor climate through shading. It reduces the need for air conditioning, which helps to create more sustainable cities by reducing energy consumption. There is a reduction in greenhouse gas emissions and gaseous reactive nitrogen compounds.

To conclude the above examples, it is important to look at them in the right way. Usually, the amount of CO_2 and NO_x absorbed in the case of buildings with vegetated façades or green roofs are given as conversion figures, i.e., a given amount of plants will absorb a certain amount of harmful compounds. There are no measurements or tests to check how far the designers' predictions agree with reality.

In the case of concretes with nano-TiO_2 and activated carbon, tests were performed. In the case of nano-TiO_2, field studies show that the absorption of NO_x was around 65–70% [38]. Laboratory conditions showed 80–90%. For activated carbon, field conditions showed 20–25% NO_x absorption capacity.

Solutions using vegetation or nano-added TiO_2 or activated carbon can be considered in terms of advantages and disadvantages. As any solution will have both. The advantages are presented in the text. The biggest disadvantage by far will be the cost of producing or constructing and planting the vegetation. If these types of buildings were built only to absorb pollutants (NO_x), the profitability of such investments would be low. However, they do have other advantages that outweigh the disadvantages that may still arise. The solutions presented in the article are definitely better than using an anti-smog tower of one of the companies, which worked on the principle of air ionisation (capturing harmful particles). It is powered by solar panels. However, the designer's research was not confirmed. The concentration of particulate matter at a distance of 10 m from the facility reduced by 12%, but there was no improvement within a 50 m radius.

Facilities with vegetation should be looked at through the prism of large, built-up cities, where there is no space for vegetation and green zones. These types of buildings (with vegetated façades or green roofs) are the answer to fast-growing cities and cramped buildings.

Of course, the buildings themselves will not help in the fight against smog, but they can be a good start to the changes that large cities must undergo in order to be able to breathe fresh air in a few years' time.

7. Conclusions

This article presents an overview of solutions that can be applied to buildings and infrastructure to reduce smog and other pollutants. Several solutions are given. The first is the use of plants as a biophilic solution in the form of green façades and green roofs. The second is the use of anti-smog towers with different mechanisms of action. The next two are solutions related to concrete and the use of surface-modifying admixtures. These include nano additives in the form of nano titanium dioxide and activated carbon.

The awareness of states, cities, and even the public allows for the introduction of new, feasible, and usable solutions. Of course, buildings, facilities, and infrastructure alone will not bring about complete clean-up. Changes in other areas are also needed. It is not enough to build anti-smog buildings to reduce smog. Nor is it enough to change heating systems. It is also necessary to phase out cars that produce a large amount of pollution and switch to hybrid and electric cars. Renewable energy sources should be used, waste should be reused, and filters should be installed on chimneys. We should begin to eliminate the causes of pollution, which to a large extent include low emissions. Then, it will be much easier to fight their effects, including smog.

The examples mentioned in the article and those that are still a vision of the future, such as the appearance of Paris in 2050, where buildings consisting of traditional buildings will be complemented by green towers are technological innovations or solutions that already exist, but gain new functions. However, they show the direction for designing new buildings for the future. They set a new direction for researchers, designers, and architects to create new materials or use existing solutions to create new architectural designs that preserve fresh air.

Funding: This research received no external funding.

Institutional Review Board Statement: Not applicable.

Informed Consent Statement: Not applicable.

Data Availability Statement: Not applicable.

Conflicts of Interest: The authors declare no conflict of interest.

References

1. Ukaogo, P.O.; Ewuzie, U.; Onwuka, C.V. Environmental pollution: Causes, effects, and the remedies. In *Microorganisms for Sustainable Environment and Health*; Chowdhary, A., Raj, A., Verma, D., Akhter, Y., Eds.; Elsevier: London, UK, 2020.
2. Cheremisinoff, N.P. *Pollution Control Handbook for Oil and Gas Engineering—General Description*; John Wiley & Sons: New York, NJ, USA, 2016.
3. Saraf, A.K.; Bora, A.K.; Das, J.; Rawat, V.; Sharma, K.; Jain, S.K. Winter fog over the Indo-Gangetic Plains: Mapping and modelling using remote sensing and GIS. *Nat. Hazards* **2011**, *58*, 199–220. [CrossRef]
4. Manisalidis, I.; Stavropoulou, E.; Stavropoulos, A.; Bezirtzoglou, E. Environmental and Health Impacts of Air Pollution: A Review. *Front. Pub. Health* **2020**, *8*. [CrossRef] [PubMed]
5. Yue, R.P.H.; Lee, H.F.; Hart, M.A. The human dimension of visibility degradation in a compact city. *Nat. Hazards* **2016**, *82*, 1683–1702. [CrossRef]
6. Lotko, W.; Lisowska, A.; Łodygowski, K. Koncepcja zastosowania dodatku węglowodorów syntetycznych do paliw celem ograniczenia emisji wybranych toksycznych składników spalin w aglomeracji wrocławskiej. *Autobusy Tech. Eksploat. Syst. Transp.* **2016**, *6*, 255–259.
7. Smog—Definicja i Informacje o Zjawisku. Available online: https://airly.org/pl/smog-definicja-skutki-i-przyczyny/ (accessed on 19 July 2021).
8. Li, H.C. Smog and Air Pollution: Journalistic Criticism and Environmental Accountability in China. *J. Rural Stud.* **2019**, in press. [CrossRef]
9. Kumar, S.; Narwal, D.; Sethi, A. Smog: Anthropogenic pollution. *Int. J. Adv. Res. Sci. Eng.* **2017**, *6*, 1.
10. Haagen-Smit, A.J. Chemistry and Physiology of Los Angeles Smog. *Ind. Eng. Chem.* **1952**, *44*, 1342–1346. [CrossRef]
11. Mishra, S. Is smog innocuous? Air pollution and cardiovascular disease. *Indian Heart J.* **2017**, *69*, 425–429. [CrossRef]
12. Science behind Smog and Its Ominous Implications. Available online: https://www.downtoearth.org.in/blog/air/science-behind-smog-and-its-ominous-implications-60080 (accessed on 19 July 2021).
13. Smog. Available online: https://smogtok.com/indexOld.html#!/smog (accessed on 19 July 2021).
14. Rani, B.; Singh, U.; Chuhan, A.; Sharma, D.; Maheshwari, R.K. Photochemical Smog Pollution and Its Mitigation Measures. *J. Adv. Sci. Res.* **2011**, *2*, 28–33.
15. Czerwińska, J.; Wielgosiński, G.; Szymańska, O. Is the Polish smog a new type of smog? *Ecol. Chem. Eng. S* **2019**, *26*, 465–474. [CrossRef]
16. Tayanç, M.; Göçmen, G. Measurement and Analysis of Photochemical Smog over İstanbul, Turkey. In *Air Pollution Modeling and Its Application*, 13th ed.; Gryning, S.E., Batchvarova, E., Eds.; Springer: Boston, MA, USA, 2000; pp. 737–738. [CrossRef]
17. Tucki, K.; Orynycz, O.; Wasiak, A.; Swięc, A.; Mieszkalski, L.; Wichłacz, J. Low Emissions Resulting from Combustion of Forest Biomass in a Small Scale Heating Device. *Energies* **2020**, *13*, 5495. [CrossRef]
18. Munsif, R.; Zubair, M.; Aziz, A.; Zafar, M.N. Industrial Air Emission Pollution: Potential Sources and Sustainable Mitigation. In *Environmental Emissions*; Viskup, R., Ed.; IntechOpen: London, UK, 2021.
19. Ruggerio, C.A. Sustainability and sustainable development: A review of principles and definitions. *Sci. Total Environ.* **2021**, *786*. [CrossRef] [PubMed]
20. Bhatia, S. Nanoparticles Types, Classification, Characterization, Fabrication Methods and Drug Delivery Applications. In *Natural Polymer Drug Delivery Systems*; Springer: Cham, Switzerland, 2016. [CrossRef]
21. Stanaszek-Tomal, E.; Kozak, A. Mineral and organic coatings modified nano-tio2 addition as elements of sustainable building. In *Energy Efficient, Sustainable Building Materials and Products*, 1st ed.; Hager, I., Ed.; Cracow University of Technology: Kraków, Poland, 2019.
22. Bellardita, M.; Di Paola, A.; Megna, B.; Palmisano, L. Determination of the crystallinity of TiO_2 photocatalysts. *J. Photochem. Photobiol. A Chem.* **2018**, *367*, 312–320. [CrossRef]
23. Li, Z.; Ding, S.; Yu, X.; Han, B.; Ou, J. Multifunctional cementitious composites modified with nano-titanium dioxide: A review. *Compos. Part A Appl. Sci. Manuf.* **2018**, *111*, 115–137. [CrossRef]
24. Palmisano, G.; Yurdakal, S.; Augugliaro, V.; Loddo, V.; Palmisano, L. Photocatalytic Selective Oxidation of 4-Methoxybenzyl Alcohol to Aldehyde in Aqueous Suspension of Home-Prepared Titanium Dioxide. *Adv. Synth. Catal.* **2007**, *349*, 964–970. [CrossRef]
25. Hwang, J.Y.; Moon, G.-h.; Kim, B.; Tachikawa, T.; Majima, T.; Hong, S.; Cho, K.; Kim, W.; Choi, W. Crystal phase-dependent generation of mobile OH radicals on TiO_2: Revisiting the photocatalytic oxidation mechanism of anatase and rutile. *Appl. Catal. B Environ.* **2021**, *286*. [CrossRef]

26. Fujishima, A.; Rao, T.N.; Tryk, D.A. Titanium dioxide photocatalysis. *J. Photochem. Photobiol. C Photochem. Rev.* **2000**, *1*, 1–21. [CrossRef]
27. Cassar, L.; Beeldens, A.; Pimpinelli, N.; Guerrini, G. Photocatalysis of cementitious materials. In Proceedings of the International RILEM Symposium on Photocatalysis, Environment and Construction Materials, Rome, Italy, 8 October 2007; pp. 131–145.
28. Zhong, L.; Haghighat, F. Photocatalytic air cleaners and materials technologies–Abilities and limitations. *Build. Environ.* **2015**, *91*, 191–203. [CrossRef]
29. Seinfeld, J.H.; Pandis, S.N. *Atmospheric Chemistry and Physics: From Air Pollution to Climate Change*; JohnWiley & Sons: Hoboken, NJ, USA, 2012.
30. Sikkema, J.K. Photocatalytic Degradation of NO_x by Concrete Pavement Containing TiO_2. Ph.D. Dissertation, Graduate Theses and Dissertations. Iowa State University, Ames, IA, USA, 2013.
31. Hamidi, F.; Aslani, F. TiO_2-based photocatalytic cementitious composites: Materials, properties, influential parameters, and assessment techniques. *Nanomaterials* **2019**, *9*, 1444. [CrossRef]
32. Jayapalan, A.R.; Lee, B.-Y.; Kurtis, K.E. Can nanotechnology be 'green'? Comparing efficacy of nano and microparticles in cementitious materials. *Cem. Concr. Compos.* **2013**, *36*, 16–24. [CrossRef]
33. Andaloro, A.; Mazzucchelli, E.S.; Lucchini, A.; Pedeferri, M.P. Photocatalytic self-cleaning coatings for building facade maintenance. Performance analysis through a case-study application. *J. Facade Des. Eng.* **2016**, *4*, 1–15. [CrossRef]
34. Topçu, I.B.; Akkan, E.; Uygunoğlu, T.; Çalişkan, K. Self-Cleaning Concretes: An Overview. *J. Cem. Based Compos.* **2020**, *2*, 6–12.
35. Jackiewicz-Rek, W. Betony inne niż wszystkie. *Przegląd Bud.* **2020**, *2*, 22–31.
36. Guranowska-Gruszecka, K.; Chudzińska, A. Urban theories of the emergence, development and endurance of commercial streets. *Space* **2020**, *41*, 101–140. [CrossRef]
37. Chilmon, K.; Jackiewicz-Rek, W. Beton fotokatalityczny a możliwość oczyszczania powietrza. *Bud. Technol. Archit.* **2019**, *2*, 66–69.
38. Broniewicz, P. Architektoniczne metody walki z zanieczyszczeniem powietrza. *Środowisko Mieszk.* **2018**, *23*, 141–148. [CrossRef]
39. La Russa, M.F.; Rovella, N.; Alvarez de Buergo, M.; Belfiore, C.M.; Pezzino, A.; Crisci, G.M.; Ruffolo, S.A. Nano-TiO_2 coatings for cultural heritage protection: The role of the binder on hydrophobic and self-cleaning efficacy. *Prog. Org. Coat.* **2016**, *91*, 1–8. [CrossRef]
40. Staub de Melo, J.; Trichęs, G.; Gleize, P.; Villena, J. Development and evaluation of the efficiency of photocatalytic pavement blocks in the laboratory and after one year in the field. *Constr. Build. Mater.* **2012**, *37*, 310–319. [CrossRef]
41. Mexico City's Manuel Gea Gonzalez Hospital Has an Ornate Double Skin That Filters Air Pollution. Available online: https://inhabitat.com/mexico-citys-manuel-gea-gonzalez-hospital-has-an-ornate-double-skin-that-filters-air-pollution/prosolve-torre-de-especialidaes1/ (accessed on 15 July 2021).
42. Toro, C.; Jobson, B.; Haselbach, L.; Shen, S.; Chung, S.; Chung, S. Photoactive roadways: Determination of CO, NO and VOC uptake coefficients and photolabile side product yields on TiO_2 treated asphalt and concrete. *Atmos. Environ.* **2016**, *139*, 37–45. [CrossRef]
43. Witkowski, H.; Tryfon-Bojarska, A.; Jackiewicz-Rek, W.; Chilmon, K.; Szerszeń, K.; Jarosławski, J.; Gąsiewski, A. Wykorzystanie betonu fotokatalitycznego z cementem Tiocem® do redukcji stężenia Nox na projekcie Generation Park w Warszawie. In Proceedings of the XVIII Konferencja Naukowo-Techniczna Reologia w Technologii Betonu, Warszawa, Polska, 14 May 2019.
44. Projekt Realizowany w Ramach Wspólnego Przedsięwzięcia RID, Finansowany ze środków Narodowego Centrum Badań i Rozwoju oraz Generalnej Dyrekcji Dróg Krajowych i Autostrad, Ochrona Przed hałasem Drogowym Zadanie 8. Innowacyjne Metody i środki w Kompleksowej Ochronie Otoczenia Drogi Przed hałasem z Oceną ich skuteczności i Uwarunkowań Stosowania, Projekt RID–I/76. 2018. Available online: https://www.gov.pl/web/gddkia/rid (accessed on 19 August 2021).
45. Hadnadjev, K.M.; Ranogajec, J.; Snezana, P.; Markov, S.; Ducman, V.; Marinkovic-Neducin, R. Design of self-cleaning TiO_2 coating on clay roofing tiles. *Philos. Mag.* **2010**, *90*, 2989–3002. [CrossRef]
46. Ranogajec, J.; Radeka, M. Self-Cleaning Surface of Clay Roofing Tiles. In *Self-Cleaning Materials and Surfaces*; Daoud, W.A., Ed.; Wiley John and Sons: London, UK, 2013. [CrossRef]
47. Hopkins, G.; Goodwin, C. *Living Architecture: Green Roofs and Walls*, 1st ed.; CSIRO Publishing: Melbourne, Australia, 2011; ISBN 9780643103078.
48. Brković Dodig, M.; Radic, M.; Auer, T. Green Facades and Living Walls—A Review Establishing the Classification of Construction Types and Mapping the Benefits. *Sustainability* **2019**, *11*, 4579. [CrossRef]
49. Blanco, I.; Schettini, E.; Scarascia, G.; Vox, G. Thermal behaviour of green façades in summer. *J. Agric. Eng.* **2018**, *49*, 183–190. [CrossRef]
50. Manso, M.; Castro-Gomes, J.P. Green wall systems: A review of their characteristics. *Renew. Sustain. Energy Rev.* **2015**, *41*, 863–871. [CrossRef]
51. Wong, N.H.; Kwang Tan, A.Y.; Chen, Y.; Wong, N.C. Thermal evaluation of vertical greenery systems for building walls. *Build. Environ.* **2010**, *45*, 663–672. [CrossRef]
52. Perini, K.; Ottelé, M.; Haas, E.M.; Raiteri, R. Vertical greening systems, a process tree for green façades and living walls. *Urban Ecosyst.* **2013**, *16*, 265–277. [CrossRef]
53. Feng, H.; Hewage, K. Lifecycle assessment of living walls: Air purification and energy performance. *J. Clean. Prod.* **2014**, *69*, 91–99. [CrossRef]

54. Chakre, O. Choice of eco-friendly trees in urban environment to mitigate airborne particulate pollution. *J. Hum. Ecol.* **2006**, *20*, 135–138. [CrossRef]
55. Flannery, J.A.; Smith, K.M. Bosco Verticale. In *Eco-Landscape Design*; Springer International Publishing: Cham, Switzerland, 2015. [CrossRef]
56. High-Rise Forests in Italy Are Fighting Air Pollution. Available online: https://www.theverge.com/2017/8/9/16112758/milan-vertical-forest-stefano-boeri-video (accessed on 7 July 2021).
57. Tao Zhu Yin Yuan Apartment Building/Vincent Callebaut Architectures. Available online: https://www.archdaily.com/955926/tao-zhu-yin-yuan-vincent-callebaut-architectures (accessed on 7 July 2021).
58. Is Building a 'Forest City' to Fight Air Pollution/Mental Floss. Available online: https://www.mentalfloss.com/article/502326/China (accessed on 15 July 2021).
59. Europe's Largest Green Wall "Will Absorb Eight Tonnes of Pollution Annually" in London. Available online: https://www.dezeen.com/2019/11/11/citicape-house-green-wall-architecture-sheppard-robson/ (accessed on 15 July 2021).
60. Thuring, C.E.; Dunnett, N.P. Persistence, loss and gain: Characterizing mature green roof vegetation by functional composition. *Landsc. Urban Plan.* **2019**, *185*, 228–236. [CrossRef]
61. Tabatabaee, S.; Mahdiyar, A.; Durdyev, S.; Ismail, S. An Assessment Model of Benefits, Opportunities, Costs, and Risks of Green Roof Installation: A Multi Criteria Decision Making Approach. *J. Clean. Prod.* **2019**, *238*, 117956. [CrossRef]
62. Liu, H.; Kong, F.; Yin, H.; Middel, A.; Zheng, X.; Huang, J.; Xu, H.; Wang, D.; Wen, Z. Impacts of green roofs on water, temperature, and air quality. *Build. Environ.* **2021**, *196*. [CrossRef]
63. Drozd, W. Problems and benefits of using green roofs in Poland. *IOP Conf. Ser. Earth Environ. Sci.* **2019**, *214*, 012076. [CrossRef]
64. Zheng, T. Smog Free Tower: Studio Roosegaarde, Beijing, September 2016—Present. *Technol. Archit. Des.* **2017**, *1*, 253–254. [CrossRef]
65. Guttikunda, S.; Jawahar, J. Can We Vacuum Our Air Pollution Problem Using Smog Towers? *Atmosphere* **2020**, *11*, 922. [CrossRef]
66. Breathing Lungs for Delhi: Aũra Towers and Drones by Studio Symbiosis. Available online: https://www.stirworld.com/see-features-breathing-lungs-for-delhi-aura-towers-and-drones-by-studio-symbiosis (accessed on 15 July 2021).
67. Beware China's 'Anti-Smog Tower' and other Plans to Pull Pollution from the Air. Available online: https://theconversation.com/beware-chinas-anti-smog-tower-and-other-plans-to-pull-pollution-from-the-air-90596 (accessed on 15 July 2021).
68. Horgnies, M.; Florence, S.; Dubois-Brugger, I.; Gartner, E.M. NO_x de-pollution using activated charcoal concrete -from laboratory experiments to tests with prototype garages. In Proceedings of the 4th International Conference on Environmental Pollution and Remediation, Prague, Czeh Republic, 11–13 August 2014.
69. Zhang, W.J.; Bagreev, A.; Rasouli, F. Reaction of NO_2 with activated charcoal at ambient temperature. *Ind. Eng. Chem. Res.* **2008**, *47*, 4358–4362. [CrossRef]
70. Mejdi, J.; Meriem, B.; Limousy, L.; Bennici, S. Adsorption/Reduction of Nitrogen Dioxide on Activated Carbons: Textural Properties versus Surface Chemistry—A Review. *Chem. Eng. J.* **2018**, *347*. [CrossRef]
71. Horgnies, M.; Dubois-Brugger, I.; Gartner, E.M. NO_x de-pollution by hardened concrete and the influence ofactivated charcoal additions. *Cem. Concr. Res.* **2012**, *4*. [CrossRef]
72. Ignarro, L.J.; Fukuto, J.M.; Griscavage, J.M.; Rogers, N.E.; Byrns, R.E. Oxidation of nitric oxide in aqueous solution to nitrite but not nitrate: Comparison with enzymatically formed nitric oxide from L-arginine. *Proc. Nat. Acad. Sci. USA* **1993**, *90*, 8103–8107. [CrossRef]
73. Horgnies, M.; Dubois-Brugger, I.; Stora, E. An innovative de-polluting concrete doped with activated carbon to enhance air quality. In Proceedings of the 10th International Concrete Sustainability Conference, NRMCA0, Miami, FL, USA, 4–7 October 2015.
74. Baroghel-Bouny, V. Water vapour sorption experiments on hardened cementitious materials Part I: Essential tool for analysis of hygral behaviour and its relation to pore structure. *Cem. Concr. Res.* **2007**, *37*, 414–437. [CrossRef]
75. Ridnour, L.A.; Sim, J.E.; Hayward, M.A.; Wink, D.A.; Martin, S.M.; Buettner, G.R.; Spitz, D.R. A Spectrophotometric Method for the Direct Detection and Quantitation of Nitric Oxide, Nitrite, and Nitrate in Cell Culture Media. *Anal. Biochem.* **2000**, *281*, 223–229. [CrossRef] [PubMed]
76. Sager, U.; Schmidt, W.; Schmidt, F.; Suhartiningsih, S. Catalytic reduction of nitrogen oxides via nanoscopic oxide catalysts within activated carbons at room temperature. *Adsorption* **2013**, *19*, 1027–1033. [CrossRef]
77. Levasseur, B.; Ebrahim Amani, M.; Burress, J.; Bandosz, T. Interactions of NO_2 at ambient temperature with cerium–zirconium mixed oxides supported on SBA-15. *J. Hazard. Mater.* **2011**, *197*, 294–303. [CrossRef] [PubMed]
78. Kellert, S.R.; Heerwagen, J.; Mador, M. *Biophilic Design. The Theory, Science and Practice of Bringing Buildings to Life*; John Wiley & Sons: New York, NY, USA, 2011.

sustainability

MDPI

Article

A Unique Unified Wind Speed Approach to Decision-Making for Dispersed Locations

Ayman M. Mansour [1], Abdulaziz Almutairi [2], Saeed Alyami [2,*], Mohammad A. Obeidat [3], Dhafer Almkahles [4] and Jagabar Sathik [4]

[1] Department of Communication, Electronics and Computer Engineering, Tafila Technical University, Tafila 66110, Jordan; Mansour@ttu.edu.jo
[2] Department of Electrical Engineering, College of Engineering, Majmaah University, Al-Majmaah 11952, Saudi Arabia; adalmutiri@mu.edu.sa
[3] Department of Electrical Power and Mechatronics Engineering, Tafila Technical University, Tafilah 66110, Jordan; maobaidat76@ttu.edu.jo
[4] Renewable Energy Lab, Prince Sultan University, Riyadh 12435, Saudi Arabia; dalmakhles@psu.edu.sa (D.A.); mjsathik@ieee.org (J.S.)
* Correspondence: s.alyami@mu.edu.sa

Abstract: The repercussions of high levels of environmental pollution coupled with the low reserves and increased costs of traditional energy sources have led to the widespread adaptation of wind energy worldwide. However, the expanded use of wind energy is accompanied by major challenges for electric grid operators due to the difficulty of controlling and forecasting the production of wind energy. The development of methods for addressing these problems has therefore attracted the interest of numerous researchers. This paper presents an innovative method for assessing wind speed in different and widely spaced locations. The new method uses wind speed data from multiple sites as a single package that preserves the characteristics of the correlations among those sites. Powerful Waikato Environment for Knowledge Analysis (Weka) machine learning software has been employed for supporting data preprocessing, clustering, classification, visualization, and feature selection and for using a standard algorithm to construct decision trees according to a training set. The resultant arrangement of the sites according to likely wind energy productivity facilitates enhanced decisions related to the potential for the effective operation of wind energy farms at the sites. The proposed method is anticipated to provide network operators with an understanding of the possible productivity of each site, thus facilitating their optimal management of network operations. The results are also expected to benefit investors interested in establishing profitable projects at those locations.

Keywords: data mining; decision tree; wind speed; renewable energy; system modeling; machine learning

Citation: Mansour, A.M.; Almutairi, A.; Alyami, S.; Obeidat, M.A.; Almkahles, D.; Sathik, J. A Unique Unified Wind Speed Approach to Decision-Making for Dispersed Locations. *Sustainability* **2021**, *13*, 9340. https://doi.org/10.3390/su13169340

Academic Editors: Avelino Núñez-Delgado and Pablo García Triviño

Received: 6 June 2021
Accepted: 10 August 2021
Published: 20 August 2021

1. Introduction

Growing global interest in reducing the environmental pollution created by heavy reliance on oil derivatives for the production of electric power has motivated governments to take significant steps toward the implementation of renewable energy. One of the most important renewable energy sources is wind, with the 2019 total world capacity of wind energy estimated to be 650 gigawatts [1] and the annual global increase in wind energy calculated at 20% [2]. This expansion has resulted in wind energy technology becoming a principal source of energy in terms of sales and technical development. In spite of these advances, this energy resource remains unreliable at high rates, and increasing dependence on this technology is associated with the emergence of numerous problems for electrical system operators. Examples of these challenges are the substantial changes in wind production arising from the random behavior of wind speeds, as well as the difficulty of accurately forecasting wind production, which gives rise to many issues during the operation of the electric grid. Any decision to increase the use of wind energy hence requires careful planning along with highly reliable methods of making rational and informed decisions [3,4].

To analyze and evaluate the effects of inconsistent wind behavior on the reliability and stability of an electrical grid as well as on short-term operation and long-term planning, several researchers have applied and reported probability-based methods. For example, in [5,6] a sequential Monte Carlo simulation (SMCS) method was used for representing the probability distribution and time-series characteristics of wind speed. Another efficient method is a Monte Carlo Markov chain (MCMC) method [7,8], which is based on the dependence of the wind speed at a given point in time on the speed during the previous moment. This feature makes this method effective for preserving the chronological characteristics of wind speed. Some studies [9–11] have also dealt with correlations between the output levels of wind turbines installed in separate geographical areas or between those of multiple wind farms in adjacent areas. These studies led to the conclusion that a determination of the type of correlation (positive, negative, or zero) is related to several factors, including the way the turbines are arranged on the site and the method employed for connecting the turbines with one another as well as with the electrical network.

An examination of the correlation between the output levels of distant wind energy sites is not usually of interest to researchers because the relationship is often a zero or an inverse correlation. However, we believe that reconsidering this factor is very important, especially with respect to the correlations among multiple wind energy sites in different regions of the same country or in different countries, which might be interconnected in an electrical network. Conducting such studies would offer several advantages: (1) Knowing the diversity of and variations in wind energy production from different sites would be beneficial for grid operation in terms of power quantity and time of supply. (2) Prior knowledge of the amount of variation and the type of correlation, even if negative, would help network operators achieve effective management of grid operations, such as load flow and network stability. (3) Identification of the potential of wind energy in each region of a country is crucial information for investors or decision makers.

The results of such a study would be very important for countries that feature large areas and substantial regional diversity. With an area of 2.25 million square meters encompassing regions that exhibit varied environmental characteristics, the Kingdom of Saudi Arabia (KSA) is one such country. The KSA is also one of the largest countries in the Middle East, and most of the nearby countries rely on the KSA for resilient grid interconnections for ensuring power security and economic benefits. The Saudi government is taking rapid steps toward diversification of energy sources and is investing heavily in sustainable energy. This trend is one of the main priorities and objectives of the KSA's Vision 2030. One of the most important of these subsidized projects is wind energy, since it is expected that wind energy capacity will reach 9.5 gigawatts by 2030 [12]. In 2018, the Renewable Energy Development Office (REDO) nominated about 50 companies to begin implementing the planned renewable projects [12], which include solar power stations with a capacity of 300 megawatts and wind farms with a capacity of 400 megawatts. These stations are to be operating and connected to the electric grid before the end of 2021. The Saudi government recently announced new renewable energy projects estimated at $50 billion, with implementation expected to be completed in 2023. Establishing such projects requires accurate technical and economic studies so that suitable construction locations can be determined.

In the past few years, numerous studies dealing with wind energy in Saudi Arabia have appeared in the literature. As reported in [13–15], several studies involved the analysis of statistical parameters associated with different wind farm sites and the extraction of Weibull distribution parameters for each individual site. The limitation of these studies is that their findings with respect to site productivity were dependent on the overall assessment of the available wind speed data for each site. Based on this method, the evaluation might indicate that a site is currently unsuitable for a wind project, but that site might in fact be considered a good choice for a specific period. These studies also relied on the assumption that an appropriate distribution for all sites is a Weibull distribution. Since such an assumption is neither accurate nor valid for all sites, the results could be over-approximations, according to [16–18]. To the best of our knowledge, no study has taken into account either wind speed data collected for different, distanced locations or the processing of those data as a single package to maintain the characteristics of the correlations among locations and thus to provide more accurate and detailed standard measures of wind speed productivity at those locations.

Addressing this point represents the core contribution of the work presented in this paper. Data mining techniques have recently been used in numerous applications because of the benefits these techniques offer with respect to developing models and making decisions.

Several studies have employed artificial intelligence techniques for renewable systems. For example, artificial neural networks are used in [19] to characterize PV modules. Application of data mining procedures that include support vector machines and fuzzy logic is also applied in several studies. In [20], a new methodology combining both Gaussian-kernel support vector machine and adaptive fuzzy inference system is developed. This methodology extracts the fuzzy rules directly from the training data to be used in the testing stage. In [21,22], EEG signals are analyzed using SVM, ANN, Naïve Bayes, and decision trees for epilepsy detection. In [23], authors have used the decision tree technique to detect adverse drug reactions and the system was optimized using a genetic algorithm. An efficient feature selection method was developed in [24] for enhancing Arabic text classification. In [25–27], texture classification techniques are developed based on independent component analysis and naïve base classifier.

In this study, a decision tree algorithm is used and the major contributions of this study in comparison to existing studies are as follows:

1. A unique and unified method for predicting wind speeds at diversified locations in the KSA is proposed. The proposed model enables the examination of deviations and correlations of wind speeds at different locations.
2. A model is developed that deals and examines an extensive range of data for a variety of sites. In addition, conclusions about the characteristics of these data using the least possible number of classifications can also be drawn to facilitate the understanding of the data and to expedite their use. The goal was to help decision makers arrive at quick, accurate, and informed decisions.
3. Finally, the capability of the assessed locations can be ranked to enable system operators to ascertain in advance the monthly productivity of each site so that they can implement appropriate planning and operating actions.

2. System Design and Methodology

This section provides details about the developed prediction system, which is based on a decision tree algorithm. Numerous decision tree algorithms are currently available, including random forests, random trees, the J48, and classification and regression trees (CART). A decision tree algorithm employs training data to build a tree model that is used for classification purposes. The developed classification algorithm involves three phases: data gathering, data preprocessing, and learning and classification. In the data-gathering phase, the training and test set is collected from wind station databases. The second phase involves the preprocessing of the data, including outlier detection and elimination, missing data treatment, and averaging. In the learning and classification phase, the goal is to develop an intelligent decision mechanism. A test set is then applied for determining the accuracy of the developed model.

2.1. Data-Gathering Phase

The five locations whose wind speed data were examined in this study were carefully selected to include all regions of the KSA [28]. Five sites were chosen to be representative of each region: center, east, west, south, and north. The selection corresponds to the operational divisions of the Saudi Arabian electrical system. Figure 1 shows the sites where the data were collected.

Figure 1. Regional map of Saudi Arabia.

Table 1 provides a statistical summary of the data collected for each site. These statistics are a collection of indices that provide meaningful information regarding the location and variability of the data. To facilitate their interpretation, brief definitions of some of the statistics are given here [29]. The most common indicator of the central tendency of a random variable is the mean, which represents the average number of data points. For the selected sites, it can be noted that the means are about 3 m/s to 4 m/s, with the exception of the east region, where 1.9 m/s is the recorded mean. The standard error (SE) is the measure that indicates how close the mean of the sampled data is to the true population mean. An SE of 0.05 or less implies that the sample data are quite similar to those for the whole population, with a confidence level of 95%. As can be observed from a review of the results, the SE values for all sites are less than 5%, so the data sample for each site is thus large enough to represent the true population. The median is another measure of central tendency, and the mode refers to the most frequently or commonly occurring number in the data. Standard deviation and variance denote the spread of the data distribution. Kurtosis identifies whether the tails of a given distribution contain extreme values. Skewness is the measure of the symmetry of distribution, and it differentiates extreme values in one versus the other tail. The minimum is the smallest value in the data set while the maximum is the largest value in the data set. The sum shows the summation of the wind speeds of all data sets. The count shows how many items the data have. The results listed in Table 1 reveal noticeable differences among the statistical values associated with different sites. These discrepancies were expected due to the divergent distances between the sites and the diverse nature of the local weather.

Table 1. Data set statistics.

Statistics	Center	East	West	North	South
Mean	3.935	1.923	3.623	3.270	3.001
Standard Error	0.008	0.006	0.011	0.013	0.014
Median	3.800	1.800	3.300	3.000	2.600
Mode	3.300	1.700	2.900	2.700	0.000
Standard Deviation	1.572	0.999	1.939	1.688	1.891
Sample Variance	2.470	0.998	3.758	2.849	3.577
Kurtosis	0.548	0.711	0.563	1.257	−0.338
Skewness	0.541	0.713	0.813	0.901	0.516
Minimum	0.000	0.000	0.000	0.000	0.000
Maximum	12.20	7.600	13.70	15.20	11.10
Sum	97,878	39,455	77,928	57,344	57,031
Count	24,871	20,523	21,511	17,539	19,007

The data is a part of the Renewable Resource Monitoring and Mapping (RRMM) program prepared by King Abdullah City for Atomic and Renewable Energy (KACARE). KACARE monitored and recorded the wind speed data at different installed stations in the Kingdom of Saudi Arabia at 3 m height. Table 2 provides an example of data for one of the five sites. The size of the sample is associated with the amount of information provided and the determination of the precision or level of confidence about the desired estimate. Wind speed estimate always has an associated level of uncertainty, which depends upon the underlying variability of the data as well as the sample size: the smaller the sample size, the greater the uncertainty in the estimate. Similarly, a larger sample size can provide more information, thus the uncertainty is reduced. In this study, the sample size in all selected sites ranges from 19,000 to 25,000 data points. We tried to collect this large sample size to reduce the amount of uncertainty associated with the estimate and achieve reasonable results. The steps involved in the proposed model through the Weka tool consider different concepts of data mining, which are as follows. First, the Weka software allows preprocessing step for raw data to detect the outliers and irrelevant data by cleaning and clustering the data using the k- means technique. In addition, the data mining techniques cater to the uncertainty. This is noticed in the used decision tree methodology when applying the Gini impurity measure to decide the optimal split from a root node and subsequent splits. The Gini impurity measures the frequency at which any element of the dataset will be mislabeled when it is randomly labeled. The entropy is another way of measuring that is based on the selection of the optimum split for the features with less entropy.

Table 2. Sample from the south site database.

Site	Latitude	Longitude	Date	Wind Speed (m/s)	Irradiance (Wh/m^2)
Jazan University	16.96035	42.545865	14/01/2015 07:00:00	1.5	2.1
Jazan University	16.96035	42.545865	14/01/2015 08:00:00	2.2	83.4
Jazan University	16.96035	42.545865	14/01/2015 09:00:00	3.1	249.5
Jazan University	16.96035	42.545865	14/01/2015 10:00:00	3.9	452.7
Jazan University	16.96035	42.545865	14/01/2015 11:00:00	4	366.6
Jazan University	16.96035	42.545865	14/01/2015 12:00:00	4.9	519

A subset of the combined database is shown in Table 3. The data were collected from 9 January 2013, to 31 December 2016. The subset consists of 34,872 records. The information in Table 3 is only a small subset of the available database. Zero irradiances for the north region in this table were recorded at 4 and 5 am; this is normal at sunset time when the sun disappears.

Table 3. A subset from the combined database.

Region	Latitude	Longitude	Date	Wind Speed at 3 m (m/s)	Irradiance (Wh/m^2)
West	21.49604	39.24492	29/05/2013 10:00:00	2.6	674.3
West	21.49604	39.24492	29/05/2013 11:00:00	4	840.5
North	27.39	41.42	1/1/2015 4:00	2.9	**0**
North	27.39	41.42	1/1/2015 5:00	2.5	**0**
East	25.34616	49.5956	29/05/2013 08:00:00	3	471.6
East	25.34616	49.5956	29/05/2013 09:00:00	3.5	671.7
Center	24.52958	46.43635	9/1/2013 11:00	4.9	611.3
Center	24.52958	46.43635	9/1/2013 12:00	3.8	697.9
South	16.96035	42.545865	5/11/2014 8:00	0.6	206.9
South	16.96035	42.545865	5/11/2014 9:00	1.5	411.5

2.2. Data Preprocessing Phase

Data preprocessing includes data cleaning and missing data treatment. In this phase, information not needed for the wind speed model, such as the irradiance and the latitude and longitude, are removed from the database. Wind speed data missing for a specific date are then replaced by the average value of the wind speeds for that day [30–33]. That date is eliminated and simply replaced by the corresponding month; i.e., 29/05/2013 10:00:00 is replaced by May, as shown in Table 4.

Table 4. Database following preprocessing.

Region.	Date	Wind Speed at 3 m (m/s)
West	May	2.6
West	May	4
West	May	2.5
North	January	2.9
North	January	2.5
North	January	2.8
East	May	3
East	May	3.4
East	May	3.7
Center	January	4.9
Center	January	3.8
Center	January	3.7
South	November	0.6
South	November	1.5
South	November	2.6

The combined database is then rearranged to add an output label to a new set of input attributes. The new set of input attributes are defined as indicated in Table 5: month, center wind speed, south wind speed, east wind speed, north wind speed, and west wind speed. The output attribute consists of multi-labeled data: case 1 to case 120. Since the number of locations is five, the resultant possible number of output cases is 5! = 120 possibilities.

Table 5. Attribute list sample for developing the decision tree model.

Month	Center	South	East	North	West	Output
	Wind Speed (m/s)					
Jan	4.00	3.36	1.79	3.30	2.77	case 1
Feb	3.96	3.04	0.91	3.83	2.98	case 2
Mar	4.64	3.21	2.25	3.67	3.51	case 2
Apr	4.17	3.18	2.09	4.08	3.45	case 3
May	3.77	3.15	2.03	3.62	3.97	case 4
Jun	3.95	3.44	2.36	3.04	4.84	case 5
Jul	3.64	3.28	1.99	2.61	3.55	case 6
Aug	3.70	3.38	2.11	2.96	4.08	case 5
Sep	3.73	2.84	1.75	2.87	4.60	case 4
Oct	3.95	2.17	1.76	3.17	3.64	case 7
Nov	3.89	2.81	1.79	3.85	3.01	case 3
Dec	3.48	2.77	1.66	3.34	2.89	case 3

With the use of an association rule algorithm [34–37], the number of possible cases can be decreased to eight. The association rule algorithm caters for the correlation between wind speeds in different areas.

The association algorithm can be summarized in the following steps:

Step 1: Generate all association rules in the form if {A,B,C,D,...} then {E,F,G,...}, where A, B, C, D, E, F, G,... are items.

Step 2: Calculate confidence of the generated rules, i.e., if A then B using:

$$\text{Confidence} = \frac{\text{number of records containing both A and B}}{\text{number of records containing A}}$$

Step 3: Calculate support of the generated rules, i.e., if A then B using:

$$\text{Support} = \frac{\text{number of records containing both A and B}}{\text{total number of records}}$$

Step 4: Check if support is less than a pre-defined threshold, i.e., minsup.

Step 5: Check if confidence is less than a pre-defined threshold, i.e., minconf

Step 6: Prune rules that fail the minsup and minconf thresholds.

The wind speed of each location is labeled using a rank-based system. The developed ranking system distributes wind speeds evenly, measuring them only relative to a given location, but not according to the real value of any given speed. The developed ranking-based system includes five labels that identify the level of the wind speed: very high (VH), high (H), medium (M), low (L), and very low (VL). The database resulting after the labels have been assigned based on the wind speed ranking is shown in Table 6.

Table 6. Attribute list with assigned labels.

Record	Center	South	East	North	West	Output
1	VH	H	VL	M	L	case 1
2	VH	M	VL	H	L	case 2
3	VH	M	VL	H	L	case 2
4	VH	L	VL	H	M	case 3
5	H	L	VL	M	VH	case 4
6	H	M	VL	L	VH	case 5
7	VH	M	VL	L	H	case 6
8	H	M	VL	L	VH	case 5
9	H	L	VL	M	VH	case 4
10	VH	L	VL	M	H	case 7
11	VH	L	VL	H	M	case 3
12	VH	L	VL	H	M	case 3

To minimize the number of output attributes, an association rule algorithm is applied for analyzing all of the relations between the cases. Table 7 shows the resulting cases and the corresponding locations of the rules that produce support and confidence levels greater than a given minimal support threshold (minsup = 0.01) and a given minimal confidence threshold (minconf = 0.5).

Table 7. Cases ordered according to location preferences.

	Region 1	Region 2	Region 3	Region 4	Region 5
case 1	Center	South	North	West	East
case 2	Center	North	South	West	East
case 3	Center	North	West	South	East
case 4	West	Center	North	South	East
case 5	West	Center	South	North	East
case 6	Center	West	South	North	East
case 7	Center	West	North	South	East
case 8	West	Center	North	East	South

Table 8 provides a sample of association rules with their support and confidence levels. The table shows the minimum number of cases that can be achieved using the association algorithm with a unity confidence level.

Table 8. Support and confidence levels of sample rules.

Rule	Support	Confidence
Center = VH, South = H, East = VL, North = M, West = L→ Case 1	1/12 = 0.083	1
Center = VH, South = H, East = VL, North = M, West = L→ Case 2	2/12 = 0.16	1
Center = VH, South = H, East = VL, North = M, West = L→ Case 3	3/12 = 0.25	1
Center = VH, South = H, East = VL, North = M, West = L→ Case 4	2/12 = 0.16	1
Center = VH, South = H, East = VL, North = M, West = L→ Case 5	2/12 = 0.16	1

Figure 2 summarizes all of the steps described above for the data-gathering and preprocessing stages.

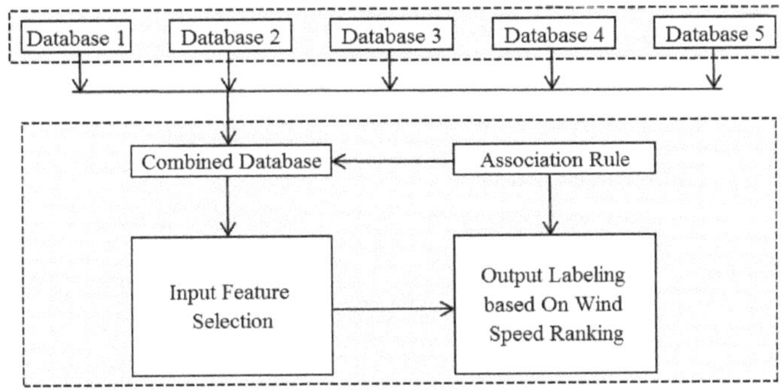

Figure 2. Data-gathering and preprocessing stages.

2.3. Learning and Classification Phase

Figure 3 displays a flowchart of the developed classification algorithm, which governs the processing of the data through three stages: training, testing, and validation. First, the training data are applied to the decision tree algorithm to obtain the initial model. For each iteration, the accuracy and precision are then calculated as a means of achieving the optimal model; the test data are applied so that the performance and efficiency of the model can be verified; and in the final step, the remaining verification data are employed to ensure that the results produced by the model have a high degree of accuracy and precision.

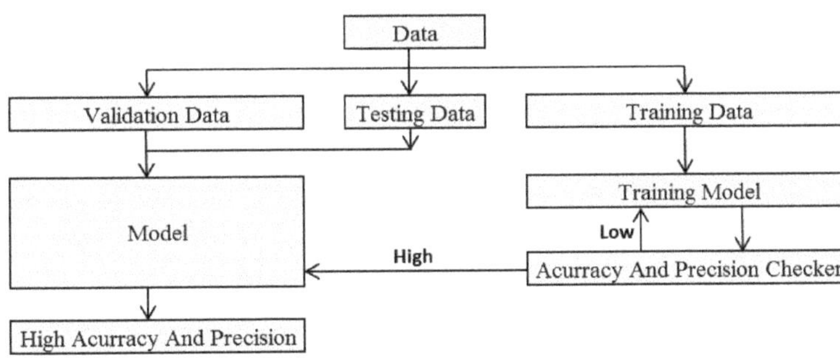

Figure 3. Developed classification algorithm.

A decision tree partitions the input space of the dataset into mutually exclusive regions by assigning each region a label. The decision tree begins with a root node and ends with a leaf node [23]. Multiple branches are formed between the root and the leaf nodes. The decision tree algorithm is performed based on splitting data into multiple regions and each region is divided into small parts. Furthermore, splitting continues until the terminal node reaches leaf nodes. The splitting is formed based on an impurity measure. Two common measures are used to obtain impurity values, Gini index, and entropy. In this paper, entropy is used as impurity measure that evaluates the homogeneity of the partition nodes too. The following steps summarizes the decision tree algorithm.

Step 1: the entropy of the root node with n branches is calculated as

$$E(\text{root}) = -\sum_{i=1}^{n} p_i \log_2 p_i \quad (1)$$

where p is the fraction of records that belongs to class i at the node.

Step 2: the entropy of each partition with J sub classes is calculated as

$$E(\text{partition}) = -\sum_{i=1}^{J} p_i \log_2 p_i \quad (2)$$

Step 3: The branch entropy is calculated using the individual k partition entropies as

$$E(\text{branch}) = \sum_{i=1}^{k} \frac{n_i}{n} E(\text{partition } i) \tag{3}$$

where n_i is the number of records at partition i,
n is number of records at branch, and
E is the entropy.

Step 4: The $GAIN_{Split}$ which is used to decide the best partition is the best. The partition that produces the most reduction is chosen The $GAIN_{Split}$ is shown below

$$GAIN_{Split} = E(\text{root}) - E(\text{branch}) \tag{4}$$

where E is the entropy.

If all input attributes are used, the algorithm for decision tree induction is as shown in Figure 4.

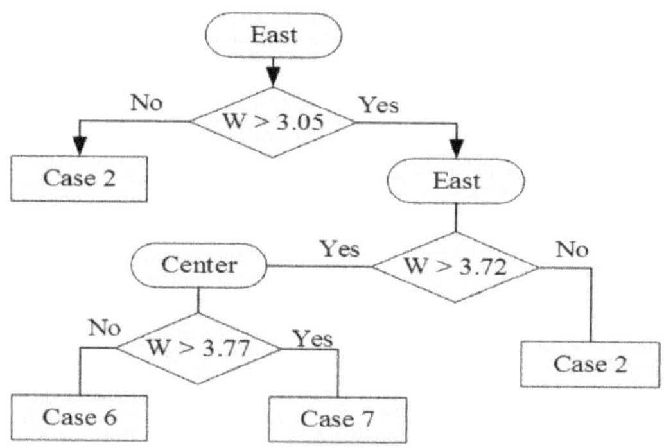

Figure 4. Decision tree model based on each location and its wind speeds.

If the prediction order is requested for a specific month and the wind speeds are unavailable at that moment, the decision tree induction model shown in Figure 5 is used. This model is based on a single input attribute: "month".

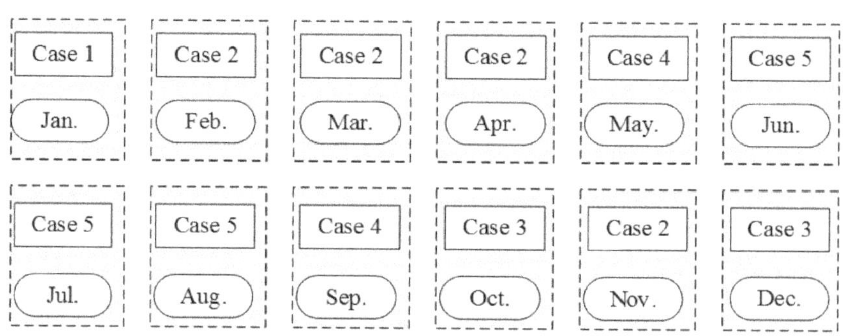

Figure 5. Decision tree model based on the month.

A new model, Model 2, is implemented based on the output of the previous model, Model 1, as shown in Figure 6. The implementation involves a comparison of the output for the five cases generated from the first model with that of the eight cases from the original training data. The output from these five cases along with the output from the original cases is then used as input to a similarity algorithm.

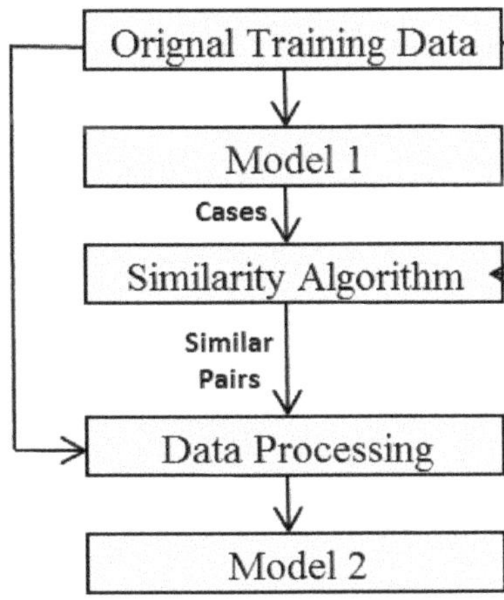

Figure 6. Development phases of Model 2.

Next, the similarity algorithm measures the similarity score between the five cases and each case from the original data, i.e., Case 8 is similar to Case 4, Case 7 is similar to Case 3, and Case 6 is similar to Case 5. The algorithm relies on edit distance, which is a technique for quantifying how dissimilar two strings (e.g., words) are to one another based on a count of the minimum number of operations required to transform the first string into the second. The edit distance between two cases for the five locations is the minimum number of operations required for transforming one case into another case. For example, the edit distance between "case 1 case 2 case 3 case 4 case 5" and "case 1 case 3 case 2 case 4 case 5" is two. A flowchart of the similarity algorithm is shown in Figure 7.

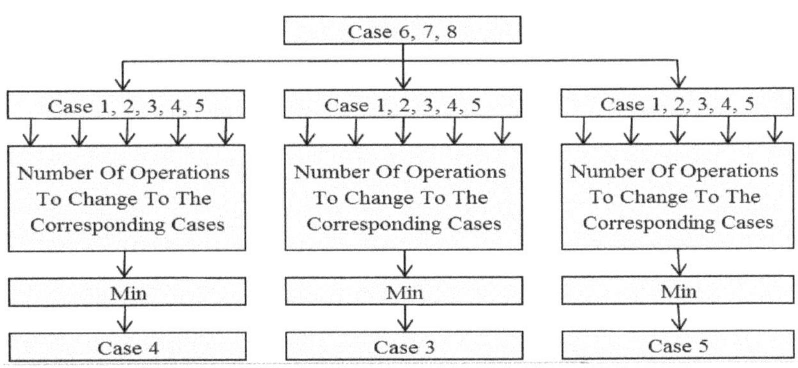

Figure 7. Similarity algorithm.

The resulting similarity pairs are employed for reprocessing the original training data through the replacement of the original cases with the similar cases, as shown in Table 9 compared with Table 6. The final step is that the resulting training data are applied for teaching Model 2, with the use of the decision tree as previously performed for developing Model 1. The degree of accuracy of Model 2 is then increased to 100%.

Table 9. The resulted database after similarity algorithm.

Record	Center	South	East	North	West	Output
1	VH	H	VL	M	L	case 1
2	VH	M	VL	H	L	case 2
3	VH	M	VL	H	L	case 2
4	VH	L	VL	H	M	case 3
5	H	L	VL	M	VH	case 4
6	H	M	VL	L	VH	case 5
7	VH	M	VL	L	H	case 5
8	H	M	VL	L	VH	case 5
9	H	L	VL	M	VH	case 4
10	VH	L	VL	M	H	case 3
11	VH	L	VL	H	M	case 3
12	VH	L	VL	H	M	case 3

3. Experiments and Results

For this study, Waikato Environment for Knowledge Analysis (Weka) software was employed [38] for constructing decision trees according to the training set, using the standard J48 algorithm [39–42]. This algorithm has been selected as one of the top 10 algorithms in data mining [43]. Java was used as the development language with J2SDK version 1.6.0_22. Weka version 3.8.4 was employed for the experimental component of the model development.

The first use of Weka software is to do data pre-processing before applying machine learning algorithms on it. The wind speed data for selected sites are recalled from CSV files. This can be done by clicking the "Open file" button and loading the data file. The loaded dataset is then processed to Cross-validation to randomly partition the data into k subsamples for training and testing. The number entered in the Fold section is used to divide the dataset into the number of Folds specified. Then classifier J48 is used as a decision tree to create a pruned tree. The Classifier Model part illustrates the model as a tree and gives some information about the tree, like the number of leaves, size of the tree, etc. Next is the stratified cross-validation part and it shows the error rates. It shows how successful the model is. By right-clicking "Visualize tree", the developed model's tree can be visualized.

The performance measurements for this work were recall, precision, the classifier F1-score, and accuracy. Examining the data for accuracy and precision establishes the credibility of the results. Accuracy refers to how closely the measurements match the desired "true" value. Precision indicates how well repeated measurements agree with and are approximate to one another. As with the order of decisions about wind speed location, it is important that the values be close, i.e., a high level of precision, and at the same time, that the decisions be correct, i.e., a high degree of accuracy. The accuracy and the precision is defined in (5) and (6)

$$\text{Accuracy} = \frac{T_P + T_N}{T_P + T_N + F_P + F_N} \tag{5}$$

where T_P is true positive, T_N is true negative, F_P is false positive and F_N is false negative. The true positive and true negative is the outcomes where the developed model correctly predicts the cases. By contrast, a false positive and a false negative are the outcomes for which the model incorrectly predicts the cases.

$$\text{Precision}(P) = \frac{T_P}{T_P + F_P} \tag{6}$$

Recall (R) is the ratio of the accurate data to the total relevant data. Its formula is shown in (7).

$$R = \frac{T_P}{T_P + F_N} \tag{7}$$

where T_P is true positive and F_N is false negative.

The classifier F1-score is calculated based on the harmonic mean. It is given as

$$F_1 = \frac{2{*}P * R}{P + R} \tag{8}$$

where P is the precision and R is the recall.

The performance measurement results are listed in Table 10.

Table 10. Overall performance results (training and validation set).

Model	Model 1	Model 2
Total number of instances	11.43	11.43
Correctly classified instances	95.26%	100%
Kappa statistic	0.93	1
Mean absolute error	0.027	0.07
Root mean squared error	0.11	0.12
Relative absolute error	5.64%	25.03%
Root relative squared error	24.36%	31.60%

Measurements from another performance indicator established with the use of a confusion matrix are presented in Table 11. The confusion matrix was built based on the data testing, and a confusion matrix was constructed for each class in the form shown in Table 12.

Table 11. Recall, precision, and F1-score measurements for each class.

Recall		Precision		F1-Score		Class
Model 1	**Model 2**	**Model 1**	**Model 2**	**Model 1**	**Model 2**	
96.89%	100%	98.19%	100%	0.97541	1	**case1**
92.84%	100%	90.96%	100%	0.91895	1	**case2**
92.36%	100%	88.80%	100%	0.90546	1	**case3**
93.80%	100%	95.58%	100%	0.94681	1	**case4**
97.43%	100%	98.45%	100%	0.97941	1	**case5**

Because of the limited number of training cases, exercising care when minimizing and reserving the number of training samples for testing purposes is extremely important. Cross-validation was employed for testing, checking, and verifying the generalizability of the model. In training any model, a frequent tendency is to overfit, and cross-validation was applied as a means of avoiding this effect. The best way to improve the performance of a system is to reserve a small portion of the training data itself for use in validating the model since this approach provides an idea of the ability of the model to predict the previously unseen reserved data. K-fold cross-validation is a technique commonly used for this purpose. In a 10-fold version of k-fold cross-validation, the training set is randomly split into groups of 10 that have approximately the same size. The classifier is then trained using eight subsets. One of the two remaining subsets is used for validation and the last, for testing. This process is repeated until all folds, one by one, have an opportunity to be the assigned test version. This technique establishes the generalizability of the model, especially when limited data makes it difficult to break the data down into test data and training data. Table 13 shows the average degree of accuracy for 2-fold, 4-fold, 6-fold, and 8-fold cross-validation and for the 10-fold cross-validation used in this paper.

Table 12. Confusion matrix (training and validation set).

		Real System				
		case 1	**case 2**	**case 3**	**case 4**	**case 5**
Model 1	**case 1**	2500	28	13	0	5
	case 2	18	1803	74	13	74
	case 3	2	91	1475	89	4
	case 4	16	20	33	1665	8
	case 5	44	0	2	8	3450
Model 2	**case 1**	2578	0	0	0	0
	case 2	0	1942	0	0	0
	case 3	0	0	1597	0	0
	case 4	0	0	0	1775	0
	case 5	0	0	0	0	3541

Table 13. Degrees of accuracy for 2-fold, 4-fold, 6-fold, 8-fold, and 10-fold cross-validation.

K-fold	**Accuracy (%)**	
	Model 1	**Model 2**
2-fold	68.654	69.38
4-fold	65.145	88.62
6-fold	78.224	95.64
8-fold	87.325	98.32
10-fold	95.26	100.00

In this research, a unique system was developed to arrange places according to wind speed. The process was carried out through three stages, i.e., the data collection stage, the processing stage, and the design stage. In the first stage, data are collected from different places, for example in the center, north, south, east, and west of the region. These data contain wind speed and other additional information such as location data from longitude and latitude and the date of collected samples. The data are collected in a central database and this database contains all the information deduced from the databases spread in different places. In the second stage (data processing stage), the information that is not useful in this research, such as longitude and latitude, is discarded and the date is replaced by the month. Then the central database is rearranged and the number of cases is reduced by using the association rules (a famous method of finding relationships) and this is done by studying all cases and their relationship to each other. This developed theory can be used for other places and other databases, and the developed method does not exist before in the literature. Machine learning methods depend on a set of algorithms, and these algorithms are applied to a set of data to build models that help in making decisions. This model is not limited to these data. This model can be used as a solid foundation to address similar problems in different areas. Other factors such as the direction of the wind, the maximum and minimum wind speed per day are important and might serve different applications. In this paper, however, the focus was on the wind speed to achieve a specific goal of providing the network operators with an understanding of the possible productivity of each wind site location, thus facilitating the optimal management and installation of wind plants and network operations. Such other factors open the door for great future work. The wind direction especially will play an important role in determining the place of the wind plants and the layout of wind turbines.

The proposed model shows great promise, so that two locations are sufficient for obtaining the order of preference of the locations. For example, if it is known only that the wind speed in the east region is below 3.05 m/s, then this scenario follows Case 2. Once the cause is known, the order of the wind speed values at all locations can be determined. If the wind speed in the east region is greater than 3.5 m/s but less than 3.72 m/s, the status of the wind speed at the other locations can be extracted from the Case 4 scenario. If the wind speed in the east region is greater than 3.77 m/s, the status of the wind speed at the center location and whether it follows Case 6 or Case 7 can be determined. Indeed, this feature of the proposed model saves the time and effort that would otherwise be required for predicting the wind speed at multiple locations. This model can thus be very helpful to system operators who desire an easy, quick, and accurate method of determining the status of the wind speeds at different locations.

4. Conclusions

This paper has presented a machine learning-based decision-making method for the assessment of potential wind speed productivity in different locations. To preserve the characteristics of the correlations among these sites, the new method employs wind speed data from multiple sites as a single package. Machine learning using Weka software is then employed to test the correlations among the sites to rank the sites into different cases. Wind speed becomes the primary classification factor for prioritizing the sites in order. The implementation of training tests for big data sets improves the prediction of appropriate locations for wind farms. Using real data, the decision model has been constructed, tested, and verified. The data is a part of the Renewable Resource Monitoring and Mapping (RRMM) Program prepared by King Abdullah City for Atomic and Renewable Energy (KACARE). KACARE monitored and recorded the wind speed data at different installed stations in the Kingdom of Saudi Arabia at 3 m height. 10-fold cross-validation was used in the experimental part. The proposed model shows great results, so that the information about two locations is sufficient for obtaining the order of the remaining locations. The developed model shows high accuracy (up to 95.26%) in the test data. The final performance of Model 1 has been improved by developing Model 2, where the accuracy has increased to 100%. Electric network planners could use the proposed model as a means of enhancing their ability to conduct feasibility studies for any plans for establishing wind farm projects at different distanced locations. A system operator could also use this method for assessing likely wind power productivity at each site so that network operational activities can be managed effectively. The results of this study also offer electricity market investors helpful input for making appropriate investment decisions.

Author Contributions: The authors' conribations are as follows: data mining, data analysis and software, A.M.M.; Concepulazation and methodology, A.A. and S.A.; investigation, M.A.O.; writing—original draft preparation, D.A. and J.S. All authors have read and agreed to the published version of the manuscript.

Funding: This research received funding from the Deputyship for Research and Innovation, Ministry of Education, Saudi Arabia under project number (IFP-2020-02).

Institutional Review Board Statement: Not Applicable.

Informed Consent Statement: Not Applicable.

Data Availability Statement: The data that support the findings of this study are available from the corresponding author.

Acknowledgments: The authors extends their appreceiations to Deputyship for Research and Innovation, Ministry of Education, Saudi Arabia, for funding this research work through project number (IFP-2020-02).

Conflicts of Interest: The authors declare no conflict of interest.

References

1. Statista. Installed Wind Power Capacity—Worldwide, 2001–2019. Available online: https://www.statista.com/statistics/268363/installed-wind-power-capacity-worldwide (accessed on 28 April 2020).
2. Renewables Information 2019—Analysis. Comprehensive Historical Review and Current Market Trends in Renewable Energy, IEA. Available online: https://www.iea.org/reports/renewables-information-overview (accessed on 1 July 2020).
3. Georgilakis, P.S. Technical challenges associated with the integration of wind power into power systems. *Renew. Sustain. Energy Rev.* **2008**, *12*, 852–863. [CrossRef]
4. Albadi, M.H.; El-Saadany, E.F. Overview of wind power intermittency impacts on power systems. *Electr. Power Syst. Res.* **2010**, *80*, 627–632. [CrossRef]
5. Billinton, R.; Wangdee, W. Reliability-based transmission reinforcement planning associated with large-scale wind farms. *IEEE Trans. Power Syst.* **2007**, *22*, 34–41. [CrossRef]
6. Billinton, R.; Karki, R.; Gao, Y.; Huang, D.; Hu, P.; Wangdee, W. Adequacy assessment considerations in wind integrated power systems. *IEEE Trans. Power Syst.* **2012**, *27*, 2297–2305.
7. Chao, H.; Hu, B.; Xie, K.; Tai, H.-M.; Yan, J.; Li, Y. A Sequential MCMC Model for Reliability Evaluation of Offshore Wind Farms Considering Severe Weather Conditions. *IEEE Access* **2019**, *7*, 132552–132562. [CrossRef]
8. Almutairi, A.; Ahmed, M.H.; Salama MM, A. Use of MCMC to incorporate a wind power model for the evaluation of generating capacity adequacy. *Electr. Power Syst. Res.* **2016**, *133*, 63–70. [CrossRef]
9. Gao, Y.; Billinton, R. Adequacy assessment of generating systems containing wind power considering wind speed correlation. *IET Renew. Power Gener.* **2009**, *3*, 217–226. [CrossRef]
10. Chen, F.; Li, F.; Wei, Z.; Sun, G.; Li, J. Reliability models of wind farms considering wind speed correlation and WTG outage. *Electr. Power Syst. Res.* **2015**, *119*, 385–392. [CrossRef]
11. Sun, M.; Feng, C.; Zhang, J. Conditional aggregated probabilistic wind power forecasting based on spatio-temporal correlation. *Appl. Energy* **2019**, *256*, 113842. [CrossRef]
12. Hasan, S.; Al-Aqeel, T.; Peerbocus, N. Saudi Arabia's Unfolding Power Sector Reform: Features, Challenges and Opportunities for Market Integration. *ResearchGate* **2020**. [CrossRef]
13. Baseer, M.; Meyer, J.; Rehman, S.; Alam, M. Wind power characteristics of seven data collection sites in Jubail, Saudi Arabia using Weibull parameters. *Renew. Energy* **2017**, *102*, 35–49. [CrossRef]
14. Rehman, S.; Halawani, T.; Mohandes, M. Wind power cost assessment at twenty locations in the Kingdom of Saudi Arabia. *Renew. Energy* **2003**, *28*, 573–583. [CrossRef]
15. Bassyouni, M.; Gutub, S.A.; Javaid, U.; Awais, M.; Rehman, S.; Hamid, S.M.-S.A.; Abdel-Aziz, M.H.; Abouel-Kasem, A.; Shafeek, H. Assessment and analysis of wind power resource using weibull parameters. *Energy Explor. Exploit.* **2015**, *33*, 105–122. [CrossRef]
16. Qin, Z.; Li, W.; Xiong, X. Generation system reliability evaluation incorporating correlations of wind speeds with different distributions. *IEEE Trans. Power Syst.* **2013**, *28*, 551–558. [CrossRef]
17. Almutairi, A.; Nassar, M.E.; Salama, M.M.A. Statistical evaluation study for different wind speed distribution functions using goodness of fit tests. In Proceedings of the IEEE Electrical Power and Energy Conference (EPEC) 2016, Ottawa, ON, Canada, 12–14 October 2016.
18. Ouarda, T.; Charron, C.; Shin, J.-Y.; Marpu, P.R.; Al-Mandoos, A.; Al-Tamimi, M.; Ghedira, H.; Al Hosary, T. Probability distributions of wind speed in the UAE. *Energy Convers. Manag.* **2015**, *93*, 414–434. [CrossRef]
19. Almonacid, F.J.M.F.; Rus, C.; Hontoria, L.; Munoz, F.J. Characterisation of PV CIS module by artificial neural networks. *A comparative study with other methods. Renew. Energy* **2010**, *35*, 973–980.
20. Khait, J.A.; Mansour, A.M.; Obeidat, M. Classification based on Gaussian-kernel Support Vector Machine with Adaptive Fuzzy Inference System. *Margin* **2018**, *5*, 16–24.
21. Mansour, A.M.; Alaqtash, M.M.; Obeidat, M. Intelligent Classifiers of EEG Signals for Epilepsy Detection. *WSEAS Trans. Signal Process.* **2019**, *15*, 2224–3488.
22. Obeidat, M.A.; Mansour, A.M. EEG Based Epilepsy Diagnosis System using Reconstruction Phase Space and Naïve Bayes Classifier. *WSEAS Trans. Circuits Syst.* **2018**, *17*, 2224–2266.
23. Mansour, A.M. Decision Tree-Based Expert System for Adverse Drug Reaction Detection using Fuzzy Logic and Genetic Algorithm. *Int. J. Adv. Comput. Res.* **2018**, *8*, 110–128. [CrossRef]
24. Hawashin, B.; Mansour, A.; Aljawarneh, S. An Efficient Feature Selection Method for Arabic Text Classification. *Int. J. Comput. Appl.* **2013**, *83*, 17. [CrossRef]
25. Ayman, M.M. Texture Classification using Naïve Bayes Classifier. *Int. J. Comput. Sci. Netw. Secur.* **2018**, *18*, 112–120.
26. Al Nadi, D.A.; Mansour, A.M. Independent Component Analysis (ICA) for texture classification. In Proceedings of the 5th International Multi-Conference on Signals and Devices, IEEE SSD, Amman, Jordan, 20–23 July 2008.
27. Hawashin, B.; Mansour, A.; Abukhait, J.; Khazalah, F.; Alzubi, S.; Kanan, T.; Obaidat, M.; Elbes, M. Efficient Texture Classification Using Independent Component Analysis. In Proceedings of the IEEE Jordan International Joint Conference on Electrical Engineering and Information Technology (JEEIT), Amman, Jordan, 9–11 April 2019; pp. 544–547.
28. Renewable Resource Atlas. King Abdullah City for Atomic and Renewable Energy. Available online: http://rratlas.kacare.gov.sa (accessed on 10 June 2020).

29. Navida, W. *Statistics for Engineers and Scientists*, 3rd ed.; McGraw-Hill: New York, NY, USA, 2011; ISBN 978-0-07-337633-2.
30. Zhang, Z. Missing data imputation: Focusing on single imputation. *Ann. Transl. Med.* **2016**, *4*, 1.
31. Curley, C.; Krause, R.M.; Feiock, R.; Hawkins, C.V. Dealing with Missing Data: A Comparative Exploration of Approaches Using the Integrated City Sustainability Database. *Urban Aff. Rev.* **2019**, *55*, 591–615. [CrossRef]
32. Ordiano, J.; Ángel, G.; Waczowicz, S.; Reischl, M.; Mikut, R.; Hagenmeyer, V. Photovoltaic power forecasting using simple data-driven models without weather data. *Comput. Sci. Res. Dev.* **2017**, *32*, 237–246.
33. Khan, S.I.; Hoque, A.S.M.L. SICE: An improved missing data imputation technique. *J. Big Data* **2020**, *7*, 3. [CrossRef]
34. Angulakshmi, M.; Deepa, M.; Sudha, S.; Brindha, K. Association Rule Modeling using UML and Apriori Algorithm. In Proceedings of the International Conference on Emerging Trends in Information Technology and Engineering (ic-ETITE), Vellore Institute of Technology, Vellore, India, 24–25 February 2020; pp. 1–5.
35. Agapito, G.; Milano, M.; Guzzi, P.H.; Cannataro, M. Mining Association Rules from Disease Ontology. In Proceedings of the 2019 IEEE International Conference on Bioinformatics and Biomedicine (BIBM), San Diego, CA, USA, 18–21 November 2019; pp. 2239–2243.
36. Kharya, S.; Soni, S.; Swarnkar, T. Weighted Bayesian Association Rule Mining Algorithm to Construct Bayesian Belief Network. In Proceedings of the 2019 International Conference on Applied Machine Learning (ICAML), Bhubaneswar, India, 25–26 May 2019; pp. 27–33.
37. Chen, C.; Chou, H.; Hong, T.; Nojima, Y. Cluster-Based Membership Function Acquisition Approaches for Mining Fuzzy Temporal Association Rules. *IEEE Access* **2020**, *8*, 123996–124006. [CrossRef]
38. Hall, M.; Frank, E.; Holmes, G.; Pfahringer, B.; Reutemann, P.; Witten, I.H. The WEKA data mining software: An update. *ACM SIGKDD Explor. Newsl.* **2009**, *11*, 10–18. [CrossRef]
39. Bhargava, N.; Sharma, G.; Bhargava, R.; Mathuria, M. Decision tree analysis on j48 algorithm for data mining. *Int. J. Adv. Res. Comput. Sci. Softw. Eng.* **2013**, 3, 1114–1119.
40. Kaur, G.; Chhabra, A. Improved J48 classification algorithm for the prediction of diabetes. *Int. J. Comput. Appl.* **2014**, *98*, 13–17. [CrossRef]
41. Ruggieri, S. Efficient C4.5 [classification algorithm]. *IEEE Trans. Knowl. Data Eng.* **2002**, *14*, 438–444. [CrossRef]
42. Hssina, B.; Merbouha, A.; Ezzikouri, H.; Erritali, M. A comparative study of decision tree ID3 and C4.5. *Int. J. Adv. Comput. Sci. Appl.* **2014**, *4*, 13–19. [CrossRef]
43. Wu, X.; Kumar, V.; Quinlan, J.R.; Ghosh, J.; Yang, Q.; Motoda, H.; McLachlan, G.J.; Ng, A.; Liu, B.; Yu, P.S.; et al. Top 10 algorithms in data mining. *Knowl. Inf. Syst.* **2008**, *14*, 1–37. [CrossRef]

applied sciences

MDPI

Article

Application of an Electrical Low Pressure Impactor (ELPI) for Residual Particle Measurement in an Epitaxial Growth Reactor

Seungjae Lee [1,†], Dongbin Kim [2,3,†], Yujin Cho [4], Eunmi Kim [1], Pengzhan Liu [1], Dong-Bin Kwak [2], Seungho Keum [5], Hongkang Lim [5] and Taesung Kim [1,6,*]

1 School of Mechanical Engineering, Sungkyunkwan University, 2066 Seobu-ro, Jangan-gu, Suwon 16419, Korea; nount30@skku.edu (S.L.); kem920819@naver.com (E.K.); pengzhan@skku.edu (P.L.)
2 Particle Technology Laboratory, Mechanical Engineering, University of Minnesota, 111 Church St., S.E., Minneapolis, MN 55455, USA; tierraprimavera@gmail.com (D.K.); kwak0068@umn.edu (D.-B.K.)
3 Department of Land, Water and Environmental Research, Korea Institute of Civil Engineering and Building Technology (KICT), Goyang 10223, Korea
4 Convergence Research Center for Energy and Environmental Sciences, Sungkyunkwan University, 2066 Seobu-ro, Jangan-gu, Suwon 16419, Korea; gluck4u@skku.edu
5 SK Siltron Co. Ltd., 53 Imsu-ro, Gyeongsangbuk-do, Gumi 39386, Korea; seungho.keum@sk.com (S.K.); hongkang.lim@sk.com (H.L.)
6 SKKU Advanced Institute of Nanotechnology, Sungkyunkwan University, 2066 Seobu-ro, Jangan-gu, Suwon 16419, Korea
* Correspondence: tkim@skku.edu
† These authors contributed equally to this work.

Featured Application: The application considered in this study is the real-time monitoring of chamber purge performance after semiconductor processing for preventive maintenance.

Citation: Lee, S.; Kim, D.; Cho, Y.; Kim, E.; Liu, P.; Kwak, D.-B.; Keum, S.; Lim, H.; Kim, T. Application of an Electrical Low Pressure Impactor (ELPI) for Residual Particle Measurement in an Epitaxial Growth Reactor. *Appl. Sci.* **2021**, *11*, 7680. https://doi.org/10.3390/app11167680

Academic Editor: Avelino Núñez-Delgado

Received: 18 June 2021
Accepted: 17 August 2021
Published: 20 August 2021

Abstract: The purpose of this study was to determine the feasibility of using an electrical low pressure impactor (ELPI) for analyzing residual particles in a Si epitaxial growth process chamber and establish an application technique. Prior to experimental measurements, some preliminary works were conducted, including an inlet improvement of a cascade impactor, vacuum fitting fastening and flow rate adjustment, and a vacuum leak test. After that, residual particles in the process chamber were measured during N_2 gas purge using an ELPI due to its advantages including the real-time measurement of particles and the ability to separate and collect particles by their diameters. In addition, ELPI could be used to obtain particle size distribution and see the distribution trend for both number and mass concentration. The results of the real-time analysis of the total particle count revealed that the concentration at the endpoint compared to that at the beginning of the measurement by decreased 36.9%. Scanning electron microscopy/energy-dispersive X-ray spectroscopy (SEM–EDS) analysis of collected particles was performed using two types of substrates: Al foil and a Si wafer. The results showed that most particles were Si particles, while few particles had Si and Cl components. ELPI has the clear advantages of real-time particle concentration measurement and simultaneous collection. Thus, we believe that it can be more actively used for particle measurement and analysis in the semiconductor industry, which has many critical micro/nanoparticle issues.

Keywords: semiconductor process; process particle; epitaxial growth; electrical low pressure impactor (ELPI); particle size distribution (PSD)

1. Introduction

Due to recent increase in demand for personal mobile devices, the scale of the semiconductor industry is expanding and competition among related companies is intensifying [1]. Accordingly, highly integrated chip production techniques on large-area wafers are rapidly being developed and applied as methods to increase the yield and profits of semiconductor companies. In the production of such high-performance semiconductor products, since the supply of high-quality silicon wafers is the most important requirement, an increasingly higher-performance wafer manufacturing process is required. For the fabrication of a high-quality wafer, the epitaxial growth process is used, which grows a desired single-crystal thin film on a wafer for application to a device [2]. This process requires a high level of chamber cleanliness because the influx of contaminated particles can interfere with single-crystal growth, leading to defects such as dislocation [3]. For this purpose, it is essential to identify and remove contaminant particles. Prior to that, various in-depth analyses are required through the measurement and collection of particles in the process facility [4].

Research and development of various equipment technologies have been conducted for the purpose of identifying particulate contaminants in semiconductor process facilities [5,6]. However, due to extreme process characteristics, such as a high temperature and high vacuum environment and the use of corrosive and explosive gases, the actual implementation and application of real-time measurement and collection of particulate contaminants is difficult [7]. Currently, the best method is to collect and analyze particles detected on the substrate after the process or through the detection of particles remaining on the inner wall of the chamber by costly and inefficient dismantling of equipment [8]. Therefore, we attempted to introduce new measurement equipment based on the aerosol engineering principle not yet used for particle measurement or collection in semiconductor facilities. Another purpose of this study was to establish a corresponding method of application.

A scanning mobility particle sizer (SMPS) classifies particles based on their electrical mobility and optically measures their numbers. It is a representative measurement device used for the measurement of particulate pollutants in the atmosphere [9–14]. According to its operational principles, it has the advantage of obtaining an accurate particle size distribution (PSD). However, it is not possible to obtain PSD in every second because it has time resolution of 1–2 min for each scan/each measurement. For a PSD measurement at every second, a fast mobility particle sizer (FMPS) has been developed that can simultaneously measure the number of all particles using multiple electrometers [15]. However, since SMPS has been used and studied for a longer period of time with high reliability of measurement results, it is used more frequently even now [16]. Although these electromobility-based particle measuring instruments can measure particles with a size of at least several nm, they cannot measure particles of micron size due to their upper detection limits of several hundred nm. Using a different principle, an aerodynamic particle sizer (APS) and an electrical low pressure impactor (ELPI) that can measure PSD based on aerodynamic diameters of particles [17,18] are also widely used. Of these two, ELPI has the advantage of categorizing particles by size based on the multistage impactor principle, thus enabling real-time PSD measurement. It also has a wide measurement range from several nm to several μm particles. In addition, it can facilitate additional analysis after collecting particles by size. Another excellent feature of ELPI is its robustness, which makes it easier to use when collecting samples in very dusty atmospheres, which is not possible with either FMPS or SMPS [19]. Due to this advantage, ELPI has typically been used to analyze atmospheric particulate contaminants in a living environment [20–22]. The application of ELPI has been expanded to the analysis of generated particles and the evaluation of cleanliness in various industrial environments [23,24], and various experiments in the laboratory [25,26]. Despite these advantages, no study has reported the use of ELPI in the semiconductor industry where measurement and control of particles are important because this industry has more micro/nanoparticle issues than any other industry.

Based on these aforementioned advantages, the objective of this study was to determine whether ELPI could be used for particle analysis in a Si epitaxial growth process equipment for concentration measurement. Additional analysis was performed after collection of particles to establish and present its utilization.

2. Materials and Methods

2.1. Test Process Equipment and Experimental Setup

This study used a 300 mm ATM Si Epi-reactor of Applied Materials, Inc., of SK Siltron's silicon wafer manufacturing line located in Gumi, Gyeongsangbuk-do. Chemical reaction of the Si epitaxy process for production of single-crystal thin film on a wafer is shown as follows:

$$SiHCl_3(gas) + H_2(gas) \rightarrow Si(solid) + 3HCl(gas) \quad (1)$$

Figure 1 shows the setup schematic of residual particle measurement during N_2 gas purge. The exhaust of the scrubber in the process chamber was used as the particle sampling port. Due to extreme conditions of the in-process environment using toxic, ignitable, and high-temperature gas, there were difficulties in real-time measurement during the process. Therefore, residual particles were measured and collected for further analysis during chamber inactivation and N_2 gas purge before preventive maintenance. The purge flowrate of the N_2 gas inside the chamber was 25 LPM. The purge was performed every 1 min and 30 s at 20 s intervals. When using the ELPI, the flowrate incoming to the inlet of the cascade impactor should be fixed to be around 10 LPM for correct measurement and collection after separating particles by their sizes. The flowrate adjustment was performed using an orifice gasket for a 10 LPM flow when fastening the VCR fitting connection between the chamber exhaust and the ELPI inlet. When using an orifice, particles may inertially collide with the front surface, resulting in particle loss. Although this could not be prevented, all measurements were carried out by checking that the sampling was conducted under constant velocity conditions.

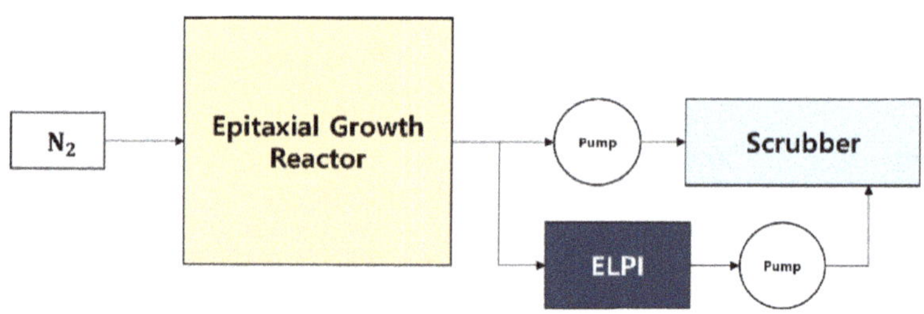

Figure 1. Schematic of the residual particle measurement setup during N_2 gas purge.

2.2. Instrument for Concentration Measurement of Particles in the Process Chamber: Electrical Low Pressure Impactor (ELPI)

Measurements of number concentration and mass concentration by particle size were carried out using an ELPI+ (Dekati Ltd., Tampere, Finland). The instrument is capable of real-time measurement of aerosol-phase particles with particle sizes of about 5 nm to 10 μm [27]. Figure 2 shows a schematic and cut-off diameters (D50%) of the cascade impactor employed in ELPI. D50% was the collection efficiency of an impactor, meaning a particle diameter that resulted in 50% collection efficiency.

Detailed principles of measurement and collection are as follows. First, particles introduced to the inlet were charged with a large number of cations generated in the corona glow region when high voltage was applied to the needle in the middle. These charged particles were then categorized and collected in 15 stages depending on the aerodynamic diameter. The topmost stage, with a D50% value of 10 μm, simply acted as a pre-separator to prevent inflow of large particles. At this time, the generated current at the substrate was measured through electrodes connected to the impactor. The current value was then converted into a number concentration through formulas [28,29]. If the principal component of the incoming particle is known, the number concentration can be converted to a mass concentration by entering the density value into the own program for data processing.

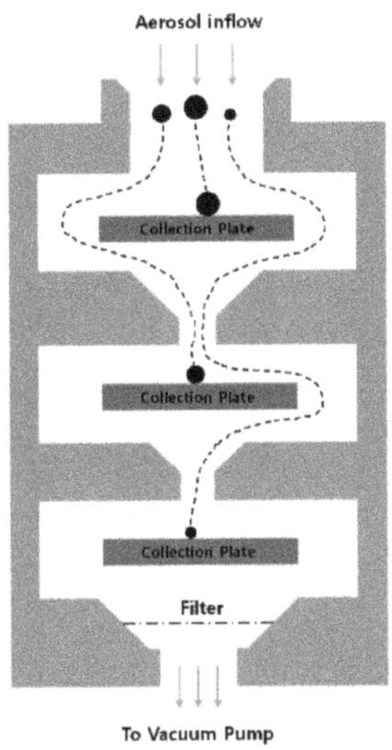

Impactor Stage No.	D_{50}% [μm]	Di [μm]
15	10	10
14	6.69	8.24
13	3.97	5.15
12	2.46	3.13
11	1.61	1.99
10	0.99	1.26
9	0.637	0.794
8	0.393	0.5
7	0.255	0.317
6	0.165	0.205
5	0.101	0.129
4	0.057	0.076
3	0.029	0.041
2	0.017	0.022
1	0.006	0.01

Figure 2. A schematic of the cascade impactor and cut-off diameters of each stage in the ELPI cascade impactor.

Even within a clean room, the leakage of particles from the chamber and instrument to the outside might affect the human body during measurement. Thus, prior to particle measurement in the epitaxial growth chamber, the ELPI was improved to check and prevent gas leakage to the outside during measurement. First, the inlet of the impactor was improved to use international organization for standardization (ISO) quick flanges (aka QF, KF, or NW) for vacuum fitting fastening. It was then confirmed that there was no gas leakage factor between the equipment and the outside using Heliot 900 model (ULVAC, Kanagawa, Japan), a pressurization-type vacuum leak tester. After that, the measurement of particles in the epitaxial growth chamber was conducted for about 22 min and averaged.

2.3. Additional Analysis after Collecting Particles from the Epitaxial Chamber

The collection of particles was carried out using the cascade impactor in the ELPI as described above at the same time as the measurement. Particles were separated by their sizes and collected on each plate of the cascade impactor. Aluminum foil is generally used as a particle collection substrate with an ELPI. However, varied types of substrates can be used to ease additional analysis, distinguishing them from surface shapes and components of the substrate for a more diverse and discriminative analysis of collected particles [30,31]. As shown in Figure 3, a 10 × 10 mm^2 coupon Si wafer was used as the collection substrate in addition to aluminum foil, the typical substrate of ELPI. If sizes of incoming particles were too large for each stage, or if the incoming velocity was too high, particle bouncing could occur. When bouncing occurs, recoiled particles can flow into a lower stage and become deposited and unmeasured. By applying grease on the surface of the substrate, this bouncing phenomenon can be somewhat prevented [19,32,33]. Thus, grease was applied to the surface using a vacuum chamber cleaning wiper. It was then wiped off with another clean wiper to leave a thin layer. The collection of particles in the epitaxial growth chamber was carried out for about one hour during and after the measurement to improve the ease of detection of particles on the substrate during further analysis.

Energy dispersive spectroscopy (EDS) and mounted scanning electron microscopy (SEM) (JSM 7600F, JEOL, Akishima, Japan), (Quanta Inspect F; FEI Co., Hillsboro, OR, USA) analyses were performed to check compositions, particle sizes, and shape of particles collected from the chamber.

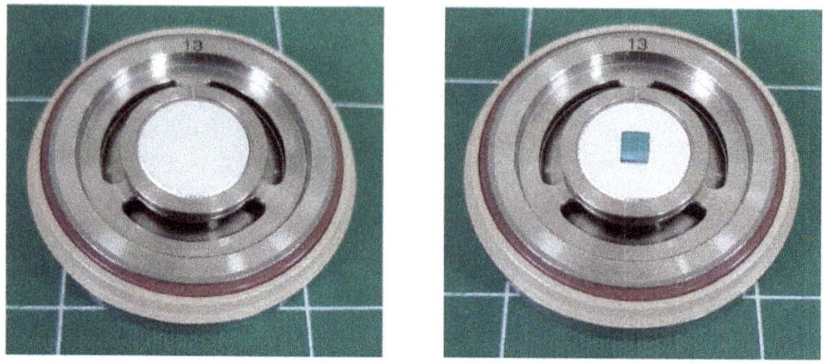

(a) **(b)**

Figure 3. Images of an impactor with different collection substrates. (**a**) Aluminum foil, (**b**) Si wafer.

3. Results and Discussion

3.1. Number and Mass Concentration

Figure 4 is a graph showing the ratio of the total number of particles by averaging measurement results of residual particles in the epitaxial growth chamber after the epitaxial process. First, looking at the number concentration measurement results in (a), about 86% of particles were measured in the first stage and 11% of particles were measured in the second stage. The total amount of particles measured in these two stages was 97%, confirming that most particles had a size of less than 20 nm. Looking at the mass concentration measurement results in (b), the percentage of detected particles was found to be 79% at the 15th stage and 14% at the 14th stage. Thus, a total of 93% of particles were detected at these two stages. Contrary to the number concentration measurement results, the number of detected particles was mainly measured at top stages, where micrometer-sized particles were mainly collected. The reason is that the cascade impactor uses an aerodynamic diameter, which detects a virtual spherical particle having the same sedimentation rate and unit density when separating each particle size and the volume of the sphere has a cubic value of the radius. Thus, the difference according to particle size significantly changed. This is a natural phenomenon. It is a limitation of mass-based particle measurement techniques. Therefore, for the accurate analysis of nano-sized particles, techniques for measuring number concentration are required.

Figure 5 shows the particle number concentration graph by time. The y-axis is the total number concentration of all stages divided by the initial total concentration. The valleys that occur regularly in the graph correspond to the purge interval. It was confirmed that the number of measured particles gradually decreased with the passage of time for about 22 min, during which the measurement was possible. At the end of the measurement, the particle number concentration decreased to 36.9% of the initial measurement value. The measurement results confirmed that the number of residual particles was reduced well by the purging operation in the chamber. The removal from the chamber of particles that can directly adversely affect subsequent process results is essential. In order to confirm that the particle amount was reduced to an appropriate amount and to find an appropriate purge time for cost and time efficiency, such real-time monitoring technique was required. When using an ELPI to measure particle concentration during the purging of the chamber after the process and to confirm that the particle number falls to a certain level before proceeding to the next process, it is important to prevent the effect that residual particles may have on the process.

3.2. Image and Element Analysis of Particles Collected by ELPI

Figure 6 shows SEM–EDS analysis results of particles from the epitaxial growth chamber collected on aluminum foil, the basic substrate of ELPI. The cutoff diameter (D50%) was 0.101 μm for the 5th stage and 6.69 μm for the 14th stage. However, due to the high surface roughness of the aluminum foil used, it was difficult to identify the shapes of the collected particles through the detection and separation of the particles collected from the substrate during image analysis. In area-specific EDS analysis results, in addition to substrate components and components arising from atmospheric exposure during sample transfer, Si components were detected in all samples in significant amounts. This was expected because component Si particles would inevitably occur during the Si epitaxy process and remain in the chamber.

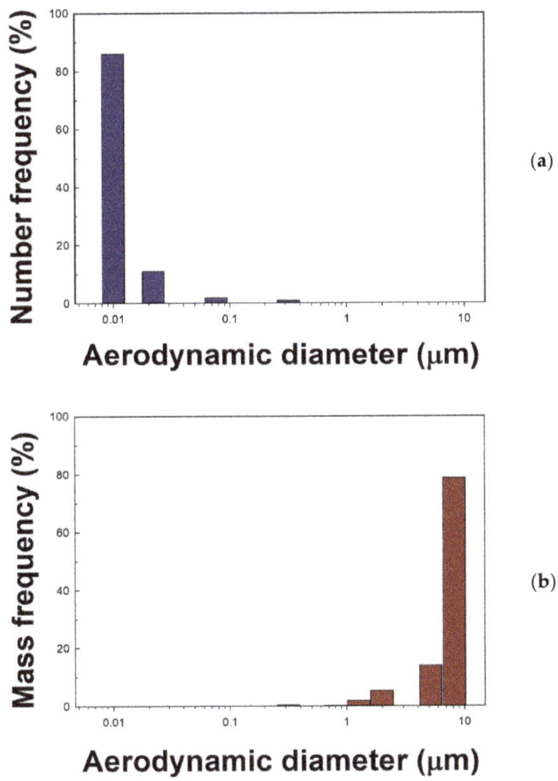

Figure 4. Particle size distributions in the epitaxial growth chamber. (**a**) Number concentration, and (**b**) Mass concentration.

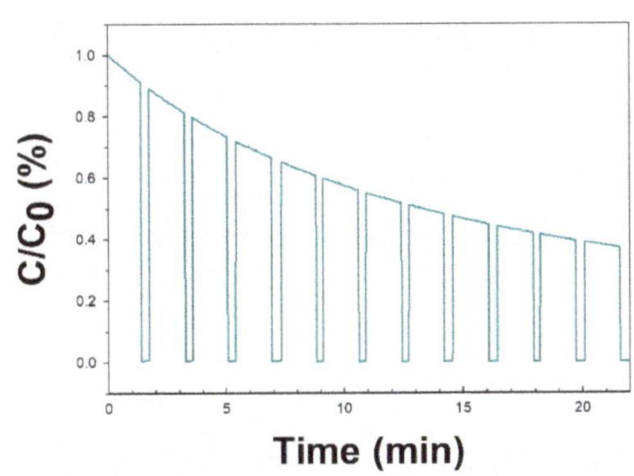

Figure 5. Particle number concentration graph by time. X-axis: time (min); Y-axis: the number proportion compared to the initial total concentration (C/C_0).

Figure 7 shows the SEM–EDS analysis results of particles on the Si wafer substrate collected from the epitaxial growth chamber. The cutoff diameter (D50%) was 0.637 μm for the 9th stage and 3.97 μm for the 13th stage. To solve the difficulty of detecting particles when using aluminum foil, a typical substrate, an alternative substrate with low surface roughness was used to collect particles and conduct SEM analysis. By changing the substrate, particle detection was facilitated and images were successfully obtained. In the EDS analysis results of particles detected on the substrate on the 9th stage, it was confirmed that particles had Si components, as in the analysis results of the particles on the aluminum foil substrate. Likewise, the analysis results of particles detected at other stages were mostly particles having only the Si element. In the EDS analysis results of particles detected at the 13th stage, significant amounts of Cl components were detected, in addition to Si. Particles containing these Si and Cl components at the same time were rarely found across several stages. These detections were due to the use of SiHCl3 gas in the epitaxial growth process. It is presumed that it was adsorbed into the inner wall of the chamber through chemisorption of the SiCl2 formed through gas phase decomposition [34]. If sufficient time is secured and various analyses are performed, such as quantifying components and crystal structures after particle measurement and collection, it is possible to estimate and control the generation/inflow path.

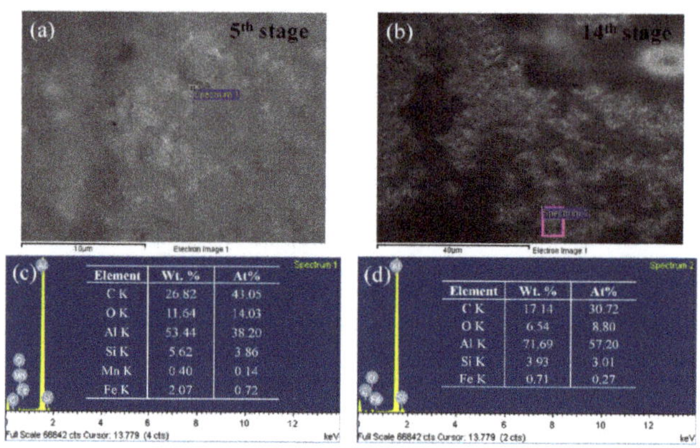

Figure 6. SEM–EDS analysis results of particles collected on (**a**,**c**) the aluminum foil substrate by ELPI substrate of the 5th stage and the (**b**,**d**) substrate of 14th stage.

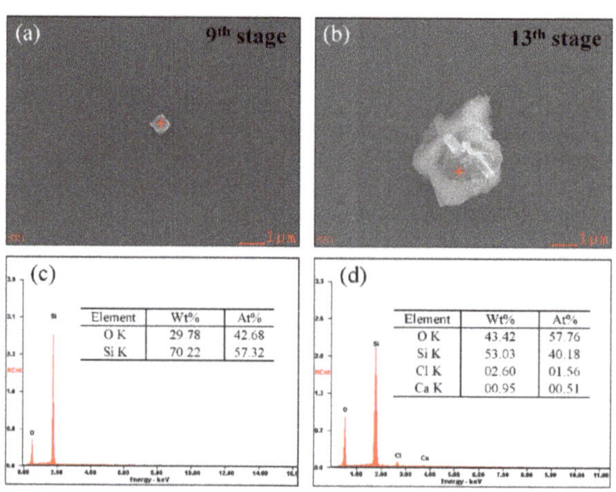

Figure 7. SEM–EDS analysis results of particles collected on the Si wafer substrate with ELPI. (**a**,**c**) Particles from the 9th stage; (**b**,**d**) particles from the 13th stage.

4. Conclusions

In order to analyze the particles remaining in the epitaxial growth chamber after the process, an ELPI was applied. As a result of measuring the concentrations of residual particles in the epitaxial growth chamber during the N_2 purge using ELPI, it was possible to obtain particle size distribution and see the distribution trend for both the number concentration and mass concentration. Using the advantage of real-time measurement, it was confirmed that N_2 purge operation was well performed after the process. The particle shape and composition were confirmed by SEM analysis of particles collected at each stage of the cascade impactor. Since various substrates could be mounted on each stage of the cascade impactor of the ELPI, a substrate with a low surface roughness is recommended when detecting particle shape images. To accurately identify the occurrence and inflow paths of each component, after collecting a large number of particles using a sufficient collection time, various in-depth additional analyses are required.

We performed a cornerstone study of the application of ELPI for particle analysis for process equipment in the semiconductor industry and other industries. The first application point found in this study is the real-time monitoring of chamber purge performance after the process for preventive maintenance using the advantage of real-time measurement. After confirming that the number of particles has fallen to a certain level before proceeding to the next process, it is possible to prevent the possible influence of residual particles on the process. This novel measurement mechanism provides critical information allowing us to reduce impurities and increase the rate of crystal growth. In addition, as this is a process that does not use explosive gas and which is conducted at normal pressure and room temperature, the concentration of the particles generated during the process can be monitored in real time. In addition, the ELPI method has a clear advantage of real-time particle concentration measurement and simultaneous collection. Thus, it can be more actively used for particle measurement and analysis in the semiconductor industry, which is plagued by many micro/nanoparticle issues.

Author Contributions: Conceptualization, S.L., D.K.; Methodology, S.L., D.K., S.K., H.L.; Formal analysis, S.L., E.K., P.L.; Investigation, S.L., D.K., E.K., D.-B.K.; Resources, S.K., H.L.; Data curation, S.L., Y.C.; Writing—original draft preparation, S.L.; Writing—review and editing, S.L., D.K., Y.C., T.K., D.-B.K.; Visualization, S.L., P.L., Y.C.; Supervision, T.K.; Project administration, S.L., T.K., S.K., H.L.; Funding acquisition, D.K., T.K., H.L. All authors have read and agreed to the published version of the manuscript.

Funding: This research was funded by Korean government (MSIT), grant number 2020M3H1A1077095, MOTIE (Ministry of Trade, Industry, and Energy), grant number P0008746, SK Siltron Co. Ltd. of funder, grant number 201807059.

Institutional Review Board Statement: Not applicable.

Informed Consent Statement: Not applicable.

Data Availability Statement: Not applicable.

Acknowledgments: This research was supported by Korea Initiative for fostering University of Research and Innovation Program of the National Research Foundation (NRF) funded by the Korean government (MSIT) (No.2020M3H1A1077095). This research was also supported by the MOTIE (Ministry of Trade, Industry, and Energy) of Korea under the Fostering Global Talents for Innovative Growth Program (P0008746) supervised by the Korea Institute for Advancement of Technology (KIAT). We also give our deepest appreciation to SK Siltron Co., Ltd. for their support and cooperation.

Conflicts of Interest: The authors have no conflict of interest relevant to this study to disclose.

References

1. Varas, A.; Varadarajan, R.; Goodrich, J.; Yinug, F. Government incentives and US competitiveness in semiconductor manufacturing. *BCG Glob. Sept.* **2020**, *9*, 2.
2. Müller, S.G.; Sanchez, E.; Hansen, D.; Drachev, R.; Chung, G.; Thomas, B.; Zhang, J.; Loboda, M.; Dudley, M.; Wang, H. Volume production of high quality SiC substrates and epitaxial layers: Defect trends and device applications. *J. Cryst. Growth* **2012**, *352*, 39–42. [CrossRef]
3. Jacobson, H.; Birch, J.; Yakimova, R.; Syväjärvi, M.; Bergman, J.; Ellison, A.; Tuomi, T.; Janzén, E. Dislocation evolution in 4H-SiC epitaxial layers. *J. Appl. Phys.* **2002**, *91*, 6354–6360. [CrossRef]
4. Jean-Luc, B.; Bruno, D. Contamination monitoring and analysis in semiconductor manufacturing. *Semicond. Technol.* **1999**, *1999*, 57–78.
5. Ziemann, P.J.; Liu, P.; Rao, N.P.; Kittelson, D.B.; McMurry, P.H. Particle beam mass spectrometry of submicron particles charged to saturation in an electron beam. *J. Aerosol Sci.* **1995**, *26*, 745–756. [CrossRef]
6. Miyashita, H.; Kikuchi, T.; Kawasaki, Y.; Katakura, Y.; Ohsako, N. Particle measurements in vacuum tools by in situ particle monitor. *J. Vac. Sci. Technol. A Vac. Surf. Film.* **1999**, *17*, 1066–1070. [CrossRef]
7. Takahashi, K.M.; Daugherty, J.E. Current capabilities and limitations of in situ particle monitors in silicon processing equipment. *J. Vac. Sci. Technol. A Vac. Surf. Film.* **1996**, *14*, 2983–2993. [CrossRef]
8. Brundle, C.; Uritsky, Y. Full wafer particle defect characterization. In Proceedings of the AIP Conference, Gaithersburg, MD, USA, 26–29 June 2000; pp. 285–291.
9. Shen, S.; Jaques, P.A.; Zhu, Y.; Geller, M.D.; Sioutas, C. Evaluation of the SMPS–APS system as a continuous monitor for measuring PM2.5, PM10 and coarse (PM2.5–10) concentrations. *Atmos. Environ.* **2002**, *36*, 3939–3950. [CrossRef]
10. Xiao, S.; Wang, M.; Yao, L.; Kulmala, M.; Zhou, B.; Yang, X.; Chen, J.; Wang, D.; Fu, Q.; Worsnop, D. Strong atmospheric new particle formation in winter in urban Shanghai, China. *Atmos. Chem. Phys.* **2015**, *15*, 1769–1781. [CrossRef]
11. Harrison, R.M.; Jones, M.; Collins, G. Measurements of the physical properties of particles in the urban atmosphere. *Atmos. Environ.* **1999**, *33*, 309–321. [CrossRef]
12. Wiedensohler, A.; Birmili, W.; Nowak, A.; Sonntag, A.; Weinhold, K.; Merkel, M.; Wehner, B.; Tuch, T.; Pfeifer, S.; Fiebig, M. Mobility particle size spectrometers: Harmonization of technical standards and data structure to facilitate high quality long-term observations of atmospheric particle number size distributions. *Atmos. Meas. Tech.* **2012**, *5*, 657–685. [CrossRef]
13. Hogrefe, O.; Lala, G.G.; Frank, B.P.; Schwab, J.J.; Demerjian, K.L. Field evaluation of a TSI model 3034 scanning mobility particle sizer in New York City: Winter 2004 intensive campaign. *Aerosol Sci. Technol.* **2006**, *40*, 753–762. [CrossRef]
14. Kheirkhah, P.; Baldelli, A.; Kirchen, P.; Rogak, S. Development and validation of a multi-angle light scattering method for fast engine soot mass and size measurements. *Aerosol Sci. Technol.* **2020**, *54*, 1083–1101. [CrossRef]
15. Jeong, C.-H.; Evans, G.J. Inter-comparison of a fast mobility particle sizer and a scanning mobility particle sizer incorporating an ultrafine water-based condensation particle counter. *Aerosol Sci. Technol.* **2009**, *43*, 364–373. [CrossRef]
16. Jørgensen, R.B. Comparison of four nanoparticle monitoring instruments relevant for occupational hygiene applications. *J. Occup. Med. Toxicol.* **2019**, *14*, 28. [CrossRef] [PubMed]
17. Keskinen, J.; Pietarinen, K.; Lehtimäki, M. Electrical low pressure impactor. *J. Aerosol Sci.* **1992**, *23*, 353–360. [CrossRef]
18. Baron, P.A. Calibration and use of the aerodynamic particle sizer (APS 3300). *Aerosol Sci. Technol.* **1986**, *5*, 55–67. [CrossRef]
19. Jørgensen, R.B.; Kero, I.T. Real-time measurements and characterization of airborne particulate matter from a primary silicon carbide production plant. *Int. J. Environ. Res. Public Health* **2017**, *14*, 1611. [CrossRef]
20. Held, A.; Zerrath, A.; McKeon, U.; Fehrenbach, T.; Niessner, R.; Plass-Dülmer, C.; Kaminski, U.; Berresheim, H.; Pöschl, U. Aerosol size distributions measured in urban, rural and high-alpine air with an electrical low pressure impactor (ELPI). *Atmos. Environ.* **2008**, *42*, 8502–8512. [CrossRef]
21. Hsieh, Y.-K.; Chen, L.-K.; Hsieh, H.-F.; Huang, C.-H.; Wang, C.-F. Elemental analysis of airborne particulate matter using an electrical low-pressure impactor and laser ablation/inductively coupled plasma mass spectrometry. *J. Anal. At. Spectrom.* **2011**, *26*, 1502–1508. [CrossRef]
22. Liu, Z.; Ge, Y.; Johnson, K.C.; Shah, A.N.; Tan, J.; Wang, C.; Yu, L. Real-world operation conditions and on-road emissions of Beijing diesel buses measured by using portable emission measurement system and electric low-pressure impactor. *Sci. Total Environ.* **2011**, *409*, 1476–1480. [CrossRef]
23. Kero, I.; Naess, M.K.; Tranell, G. Particle size distributions of particulate emissions from the ferroalloy industry evaluated by electrical low pressure impactor (ELPI). *J. Occup. Environ. Hyg.* **2015**, *12*, 37–44. [CrossRef] [PubMed]
24. Kero, I.T.; Jørgensen, R.B. Comparison of three real-time measurement methods for airborne ultrafine particles in the silicon alloy industry. *Int. J. Environ. Res. Public Health* **2016**, *13*, 871. [CrossRef] [PubMed]
25. Lee, M.-H.; Yang, W.; Chae, N.; Choi, S. Aerodynamic diameter distribution of aerosols from plasma arc cutting for steels at different cutting power levels. *J. Radioanal. Nucl. Chem.* **2020**, *323*, 613–624. [CrossRef]
26. Glover, W.; Chan, H.-K. Electrostatic charge characterization of pharmaceutical aerosols using electrical low-pressure impaction (ELPI). *J. Aerosol Sci.* **2004**, *35*, 755–764. [CrossRef]
27. Järvinen, A.; Aitomaa, M.; Rostedt, A.; Keskinen, J.; Yli-Ojanperä, J. Calibration of the new electrical low pressure impactor (ELPI+). *J. Aerosol Sci.* **2014**, *69*, 150–159. [CrossRef]

28. Pagels, J.; Gudmundsson, A.; Gustavsson, E.; Asking, L.; Bohgard, M. Evaluation of aerodynamic particle sizer and electrical low-pressure impactor for unimodal and bimodal mass-weighted size distributions. *Aerosol Sci. Technol.* **2005**, *39*, 871–887. [CrossRef]
29. Marjamäki, M.; Keskinen, J.; Chen, D.-R.; Pui, D.Y. Performance evaluation of the electrical low-pressure impactor (ELPI). *J. Sci.* **2000**, *31*, 249–261. [CrossRef]
30. Kim, B.; Lee, J.S.; Choi, B.-S.; Park, S.-Y.; Yoon, J.-H.; Kim, H. Ultrafine particle characteristics in a rubber manufacturing factory. *Ann. Occup. Hyg.* **2013**, *57*, 728–739.
31. Baldelli, A.; Rogak, S.N. Morphology and Raman spectra of aerodynamically classified soot samples. *Atmos. Meas. Tech.* **2019**, *12*, 4339–4346. [CrossRef]
32. Dzubay, T.; Hines, L.; Stevens, R. Particle bounce errors in cascade impactors. *Atmos. Environ.* **1976**, *10*, 229–234. [CrossRef]
33. Lawson, D.R. Impaction surface coatings intercomparison and measurements with cascade impactors. *Atmos. Environ.* **1980**, *14*, 195–199. [CrossRef]
34. Habuka, H.; Aoyama, Y.; Akiyama, S.; Otsuka, T.; Qu, W.-F.; Shimada, M.; Okuyama, K. Chemical process of silicon epitaxial growth in a SiHCl3–H2 system. *J. Cryst. Growth* **1999**, *207*, 77–86. [CrossRef]

processes

MDPI

Article

Antibiotic Resistance Gene Transformation and Ultrastructural Alterations of Lettuce (*Lactuca sativa* L.) Resulting from Sulfadiazine Accumulation in Culture Solution

Liang Mei, Ying-Xin Chen, Chao Wang, Jia-Hua Chen, Zhi-Jin Zhang, Min-Yao Zhou, Jin-Tao Feng and Yan Wang *

College of Animal Science, South China Agricultural University, Guangzhou 510642, China; meiliang@stu.scau.edu.cn (L.M.); 19927537760@163.com (Y.-X.C.); WC20202122009@163.com (C.W.); 19927538670@163.com (J.-H.C.); z19927535779@163.com (Z.-J.Z.); zhouminyao00@163.com (M.-Y.Z.); fengjintaott@163.com (J.-T.F.)

* Correspondence: ywang@scau.edu.cn

Abstract: The research herein explored the possible mechanism of toxicity of the antibiotic sulfadiazine (SD) and the related antibiotic resistance gene transformation in lettuce by systematically investigating its growth responses, ultrastructural changes, and antibiotic resistance gene transformation via solution culture experiments. The results showed that SD mainly accumulated in the roots of lettuce at concentrations ranging from 6.48 to 120.87 μg/kg, which were significantly higher than those in leaves (3.90 to 16.74 μg/kg). Lower concentrations of SD (0.5 and 2.0 mg/L) in the culture nutrient solution exerted little effect on lettuce growth, while at SD concentrations higher than 10 mg/L, the growth of lettuce was significantly inhibited, manifesting as shorter root length and lower dry matter yield of whole lettuce plants. Compared with that for the control group, the absolute abundance of bacteria in the root endophyte, rhizosphere, and phyllosphere communities under different concentrations of SD treatment decreased significantly. *sul1* and *sul2* mainly accumulated in the root endophyte community, at levels significantly higher than those in the leaf endophyte community. Studies of electrolyte leakage and ultrastructural characteristics of root and leaf cells indicated that lettuce grown in culture solutions with high SD concentrations suffered severe damage and disintegration of the cell walls of organs, especially chloroplasts, in leaves. Furthermore, the possible mechanism of SD toxicity in lettuce was confirmed to start with the roots, followed by a free flow of SD into the leaves to destroy the chloroplasts in the leaf cells, which ultimately reduced photosynthesis and decreased plant growth. Studies have shown that antibiotic residues have negative effects on the growth of lettuce and highlight a potential risk of the development and spread of antibiotic resistance in vegetable endophyte systems.

Keywords: sulfadiazine; lettuce; phytotoxic; hydroponic; antibiotic resistance gene

Citation: Mei, L.; Chen, Y.-X.; Wang, C.; Chen, J.-H.; Zhang, Z.-J.; Zhou, M.-Y.; Feng, J.-T.; Wang, Y. Antibiotic Resistance Gene Transformation and Ultrastructural Alterations of Lettuce (*Lactuca sativa* L.) Resulting from Sulfadiazine Accumulation in Culture Solution. *Processes* **2021**, *9*, 1451. https://doi.org/10.3390/pr9081451

Academic Editors: Avelino Núñez-Delgado, Zhien Zhang, Elza Bontempi, Mario Coccia, Marco Race and Yaoyu Zhou

Received: 21 July 2021
Accepted: 17 August 2021
Published: 20 August 2021

1. Introduction

Intensive animal farming is often associated with the use of considerable amounts of drugs, including antibiotics for disease prevention and animal growth [1,2]. Sulfadiazine [SD, 4-amino-N-(2-pyrimidinyl) benzene sulfonamide] is a potent antibacterial agent belonging to a large group of structurally related antibiotics: the sulfonamides. Attributed to its high efficacy and affordable price, SD has been long and widely used in intensive livestock production. Although the amounts of antibiotics excreted from animals vary with the type and dosage level of antibiotics, as well as the species and age of the animals [3], the overall majority (25–75%) of the administered drugs are excreted via feces and urine [4].

Due to its high nutrient content, animal manure is often applied as an organic fertilizer to arable lands for crop production, which has resulted in the large-scale introduction of antibiotics into the terrestrial environment. Furthermore, antibiotics could exert pressure on microorganisms, inducing antibiotic resistance, in the receiving environment. To date, various kinds of antibiotics, including SD, have been widely detected at high concentrations ranging from 1.93 to 760 μg/kg in soils [5–7]. In recent years, the ecotoxicity of residual antibiotics and antibiotic resistance genes in soils have received increasing attention, and antibiotics have been confirmed to cause toxicity to a wide range of organisms, including soil microorganisms, earthworms, insects, and other endpoints [8–11]. With antibiotic contamination in soils, many planted edible crops have been reported to accumulate antibiotics in their organs, which could ultimately enter the human body through the food chain [12]. The occurrence of antibiotic resistance genes in soil, water, and livestock manure has been studied extensively [13–15]. In contrast, only limited studies are available regarding plant microbial resistance [16]. Lettuce may be consumed with little processing, and this may spread antibiotic resistance to humans through the food chain [16], posing health risks to humans.

Lettuce (*Lactuca sativa* L.) is a widely consumed vegetable worldwide, and it has a high germination rate, sensitivity to contaminants, and low genetic variability, making it economically and ecologically relevant for toxicity studies [17,18]. For example, Song reported changes in the antioxidative system of lettuce in response to arsenic toxicity [19]; Yadzi evaluated cadmium accumulation in lettuce cultivated in contaminated environments and found that several phytochelatin synthase genes contribute to the accumulation of cadmium [17]. Recently, antibiotics as pollutants and their toxicity in plants have received particular attention. Xu et al. reported that the phytotoxicity of sulfanilamide antibiotics on *Arabidopsis thaliana* generates both toxic effects and hormesis related to plant drug uptake [20]. Nasri et al. studied the impact of five antibiotics on phytotoxicity and genotoxicity in rice [9]. Antibiotic resistance has become a global problem; antibiotics are considered emerging environmental pollutants and have thus attracted extensive attention worldwide [21]. Recently, antibiotic resistance in various manure-fertilized vegetables, such as celery, pakchoi, and cucumber, has been reported [22]. As mentioned above, although research has been conducted on the effects of antibiotics on several plant species [23], the possible growth response mechanisms and ultrastructural alterations of leafy vegetables in response to antibiotics and whether the antibiotic resistance of endophytic bacteria can be directly impacted by antibiotic pollution in the environment are still unclear.

Therefore, in this study, the responses of *Lactuca sativa* L. organs, including roots and leaves, to SD and antibiotic resistance gene accumulation during the growth processes under SD stress in culture solutions were systematically studied to explore the possible mechanism of lettuce toxicity and antibiotic resistance gene selection induced by antibiotics. To achieve this goal, experiments were conducted to (i) investigate the quantitative responses of roots and leaves of *Lactuca sativa* L. to SD stress, (ii) determine SD and antibiotic resistance gene accumulation levels in different organs with different SD concentrations in cultures, and (iii) explore the extent of organ damage and ultrastructure alterations during the growth processes of *Lactuca sativa* L. under SD stress via the culture substrate. Our findings will facilitate a more accurate assessment of the potential risks posed by antibiotic contamination to food quality and environmental health.

2. Materials and Methods

Commercial seeds of lettuce (*Lactuca sativa* L.) were used as the source of plant materials in this study. Seeds were sterilized with a 6% sodium hypochlorite solution for 30 min, followed by eight rounds of vigorous washing in distilled water. Then, seeds were transferred into a temporary plastic germination box with sandy soil under continuous artificial lighting at 20 °C. The germinated seedlings were carefully dug up from the germination box. After being washed to remove sediments and soils from roots, the seeds were immediately transferred to containers filled with nutrient solution for a 10-day adaptation period to hydroponic conditions. The nutrient solution contained 2 mM $Ca(NO_3)_2$, 0.1 mM KH_2PO_4, 0.5 mM $MgSO_4$, 0.1 mM KCl, 0.7 mM K_2SO_4, 10 μM H_3BO_3, 0.5 μM $MnSO_4$, 0.5 μM $ZnSO_4$, 0.2 μM $CuSO_4$, 0.01 μM Na_2MoO_4, and 100 μM Fe(II)-EDTA, with pH values ranging from 5.5 to 6.0. On Day 11, the lettuce seedlings were carefully transferred and allocated to their respective treatments (plastic containers 10 cm wide × 21 cm long × 10 cm deep). The SD (Sigma-Aldrich, St. Louis, MO, USA) concentrations in the culture nutrient solutions were set at 0.01, 0.05, 0.5, 2, 10, and 50 mg/L, with conditions without SD as the control. All experiments were conducted in triplicate, and all plants were maintained in an incubator at 20 °C.

The primary roots of the 0, 0.01, 0.05, 0.5, 2, 10, and 50 mg/L SD-treated plants were collected on Days 0, 7 and 14 and after the last measurement on Day 14. After first measuring the root length, all plants were carefully washed in distilled water and then individually separated into roots and leaves. Three roots from each treatment were randomly selected for the membrane permeability study. Plant samples from the control, 2 mg/L SD, and 10 mg/L SD treatments were selected to examine the cell structures via transmission electron microscopy (TEM). The lettuces in all the different treatments were harvested and weighed to study the dry matter yields and SD concentrations in lettuces after they were freeze-dried for 24 h.

A previously described sequential extraction procedure [24] was used for the extraction of SD with a minor modification. Briefly, 250 mg of leaf or root sample (in dry weight) was first extracted using 2 mL 0.01 M $CaCl_2$ as the mobile fraction for the total dissolved SD, followed by a second extraction using 10 mL methanol to extract the bioavailable SD fraction. The total SD concentrations in the organs were the sum of both of the above two extracted fractions. For each extraction step, samples were first sonicated for 30 min and then centrifuged for 30 min at 1700× *g*. The supernatants were mixed, evaporated, and dried by nitrogen stream to 5 mL.

After evaporation by nitrogen stream flow, a high-performance liquid chromatography (HPLC) system (Waters 2487, MILFORD, MA, USA) was used for the quantitative determination of SD concentrations during its transformation reactions. The HPLC system was equipped with a Waters 1525 binary pump and a dual λ absorbance UV/Vis detector (Waters 2487, UV–vis SPD-10AVP), with the wavelength used for detecting SD set at 254 nm. Samples of 20 μL were injected into the column through the sampling loop for analysis. SD separation was carried out using a Symmetry C18 column (5 μm beads, 250 × 4.6 mm I.D.) A mobile phase consisting of 75% methanol, 24% water, and 1% acetic acid (*v*/*v*/*v*) was applied at a flow rate of 1.0 mL/min.

Fresh leaf and root samples of lettuce plants collected from the control, 0.01, 0.05, 0.5, and 10 mg/L SD treatment groups were used to quantify antibiotic resistance genes. DNA extraction using the FastDNA™ SPIN Kit for Soil for root endophytes and leaf endophytes was performed using the FastPrep®-24 Instrument (both MP Biomedicals, Santa Ana, CA, USA) according to the manufacturer's instructions with minor modifications as described elsewhere [25]: Bead-beating was run twice for 45 s at a speed of 6.5 m s^{-1}. The samples were then centrifuged for 5 min at 14,000× *g* and room temperature. Approximately 2 g of lettuce leaves or roots from each replicate was transferred into a 50 mL centrifuge tube, shaken at 200 rpm for 2 h after addition to 50 mL autoclaved 1 × phosphate buffer, and sonicated for 15 min. The washing solution was filtered with a 0.22 μm sterilized nylon net to collect the phyllosphere and rhizosphere microbiomes for DNA extraction using a DNeasy PowerWater Kit, Qiagen USA, Valencia, CA, USA. Primers developed for total bacteria, *sul1*, *sul2*, and *tnpA* genes and validated in previous studies were used in this study. Quantitative real-time PCR was performed on triplicate DNA extracts in independent runs for total bacteria, *sul1*, *sul2*, and *tnpA*. Each qPCR reaction was conducted using a Bio-Rad CFX96 Real-Time PCR Detection System (Bio-Rad, Herculers, CA, USA). The total reaction volume of 20 μL contained 1.0 μL DNA, 10.0 μL Talent qPCR mix (SYBR Green), and 0.5 μL of each primer (forward and reverse). The qPCR conditions for all of the genes consisted of an initial denaturation of 95 °C for 2 min, followed by 40 cycles of 15 s of denaturation at 95 °C, 15 s of annealing at the temperature specified in Table 1, and 40 s at 72 °C. A melt curve was run following each plate for primer specificity. The abundance of each gene in each sample was calculated by multiplying the number of copies per well by the total volume of DNA per well (1.0 μL). DNA standards were prepared from *E. coli* strains carrying plasmids with total bacteria, *sul1*, *sul2*, and *tnpA* gene fragment inserts. Amplified DNA from SYBR Green assays was subjected to the melting curve analysis and gel electrophoresis to assure primer specificity. Samples of DNA were also selected from soil matrices for PCR product sequencing. DNA extracted from the soil was amplified with both forward and reverse primers (without SYBR green to prevent interference with the sequencing process), and the reaction products were purified using the E.Z.N.A.TM Gel Extraction Kit. The purified product subsamples were then submitted to the Biotechnology Company for sequencing.

Table 1. Primers employed in the present study for real-time quantitative PCR.

Target Gene	Primer Pair a	Sequence (5′→3′)	Annealing Temp. (°C)	Reference
16s rRNA	16s rRNA-F 16s rRNA-R	GTGSTGCAYGGYTGTCGTCA ACGTCRTCCMCACCTTCCTC	60	[26]
sul1	*sul1-F* *sul1-R*	CGCACCGGAAACATCGCTGCAC TGAAGTTCCGCCGCAAGGCTCG	62	[27]
sul2	*sul2-F* *sul2-R*	CTCCGATGGAGGCCGGTAT GGGAATGCCATCTGCCTTGA	60	[28]
tnpA	*tnpA-F* *tnpA-R*	CCGATCACGGAAAGCTCAAG GGCTCGCATGACTTCGAATC	60	[29]

The membrane permeability of roots was studied through electrolyte leakage (EL) measurements based on a procedure reported in previous studies. Briefly, 0.2 g of fresh roots was rinsed thoroughly with distilled water to remove surface contamination; then, the rinsed roots were cut into 1 cm segments and placed in individual vials containing 10 mL of distilled water. After exposure to vacuum treatment at 25 °C for 3 h, the electrical conductivity (EC) of the solution (EC_1) was measured using an electrical conductivity meter (SY-2, Institute of Soil Science, Chinese Academy Sciences, Nanjing, China). Samples were then placed in a thermostatic water bath at 100 °C for 15 min, and a second reading (EC_2) was determined after the solutions were cooled to room temperature. The EL values were calculated using the equation $EL = 100 \times EC_2/EC_1$.

Fresh leaf and root samples of lettuce collected from the control, 2 mg/L, and 10 mg/L SD treatment groups were used for the cell structure study via transmission electron microscopy, which was carried out on a Zeiss E.M.95 at 60 kV. Plant tissues were cut into 1 mm^2 slices, which were prefixed with 4% glutaraldehyde solution and then stored at 4 °C for 24 h after washing with 0.1 M sodium cacodylate buffer 3 times for 10 min each time. The samples were fixed in 1% osmium tetroxide at 4 °C for 2 h and then washed again with 0.1 M sodium cacodylate buffer 3 times for 10 min each. The samples were then dehydrated in acetone at 35, 50, 70, 95% (30 min each), and 100% (45 min each), 3 times. The dehydrated samples were infiltrated with 1/1 (*v*/*v*) resin and acetone for 2 h, 1/3 (*v*/*v*) resin and acetone overnight, and 100% resin overnight; then, each specimen was placed into beam capsules, filled with resin, and polymerized in an oven at 60 °C for 24 h. Then, 60- to 90-nanometer-thick sections were cut with a diamond knife on a Reichert ultramicrotome, stained for 10 min with uranyl acetate, and washed with 50% filtered alcohol and distilled water 2 times each. Sections were then stained with lead for 10 min and washed with double-distilled water.

The root length, dry matter yield, electrolyte leakage, and SD concentrations in plant organs were analyzed for deviations using Statistical Product and Service Solutions (SPSS 22.0). Duncan's multiple range test was used to compare the treatment means, and a 0.05 probability level was used to identify differences.

3. Results

3.1. Effect of SD Accumulation on Lettuce Growth

At all the studied SD concentrations, *Lactuca sativa* L. survived cultivation for 14 d, which indicated that this vegetable can grow under high SD concentrations, even up to 50 mg/L, in culture solutions. However, obvious morphological differences were found among the lettuces under treatments with different concentrations of SD. At 0.01, 0.05, and 0.5 mg/L SD, both the roots and leaves exhibited little difference from those under the control treatments without SD in the culture solutions. With increasing SD concentration, phenomena of inhibited root and leaf growth were found, especially with SD concentrations higher than 10 mg/L, at which the lettuces showed an obvious decrease in biomass.

Accordingly, with increasing SD concentration in the culture solution, the dry matter yields in both the leaves and roots of the lettuce plants were reduced (Figure 1). There were differences in the dry mass among the SD treatments in the control, 0.01, 0.05, 0.5, and 2.0 mg/L groups, but the differences were nonsignificant. However, the total dry matter (roots + leaves) yields were significantly decreased in the treatments with SD concentrations of 10 and 50 mg/L. The mean dry mass values in each lettuce were 0.221, 0.22, 0.214, 0.22, and 0.19 g under the 0.01, 0.05, 0.5 and 2.0 mg/L SD treatments and the control, respectively. At higher SD concentrations, the dry mass decreased to 0.14 and 0.125 g under the SD treatments at concentrations of 10 and 50 mg/L, respectively.

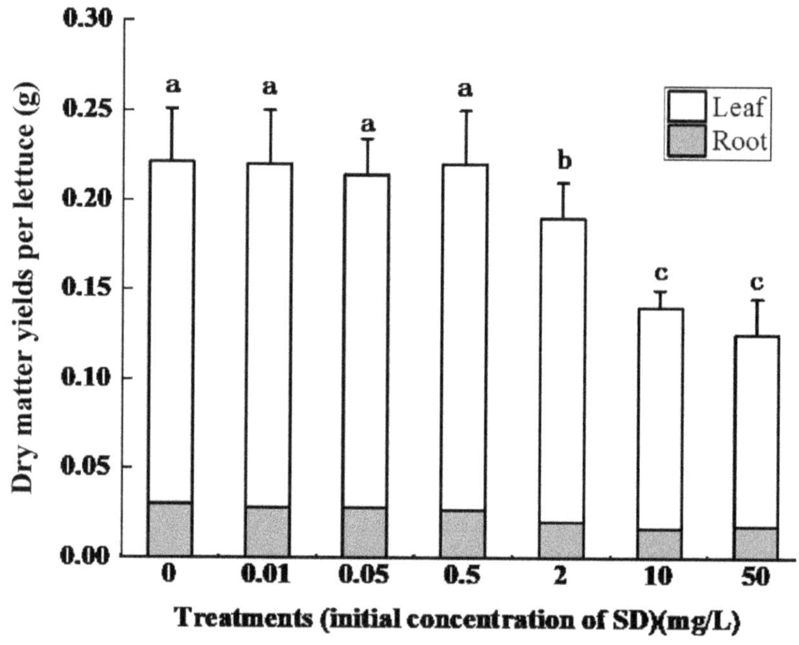

Figure 1. Effect of SD concentration on the dry matter yields per lettuce ($n = 10$). Error bars represent standard deviations of the means. a, b, and c on the top of the error bars represent significant differences ($p < 0.05$) among different initial concentrations of treatments.

The growth of roots is most sensitive to toxic effects and can act as a highly indicative parameter for evaluating the growth rate under SD stress environments; roots are the first and the primary organs through which plants contact the soil/aquatic environment, and they are also known to be the storage organs of nutrients, including toxic compounds [30,31]. The effects of different SD concentrations on the length of primary roots of lettuce on Days 0, 7, and 14 were studied (Figure 2). Without SD in the culture solution, the mean length of roots increased from 14.86 cm to 22.88 cm over the 14-day period, an increase of 1.5-fold. However, in the presence of SD, root growth was inhibited, and the inhibition rate was dependent on the SD concentration and exposure duration, as indicated by Figure 2. Up to Day 7, the root lengths grew to 3.63 cm in the control but only to 3.41, 3.24, 2.58, 2.38, 1.71, and 0.14 cm in SD stress environments with concentrations of 0.01, 0.05, 0.5, 2, 10, and 50 mg/L, respectively. Onward to Day 14, the root lengths under all SD treatments ($p < 0.05$) continuously increased, but a large increase of 8.02 cm was observed in the control treatment, while under the 0.01, 0.05, 0.5, 2, 10, and 50 mg/L SD treatments, the increases in values compared with those on Day 7 were 7.78, 8.38, 5.37, 2.47, 1.87, and 0.03 cm, respectively. The above results clearly indicate the highly toxic effect of SD accumulation on the growth of lettuce roots and the whole dry mass, integrating the results of Figure 1.

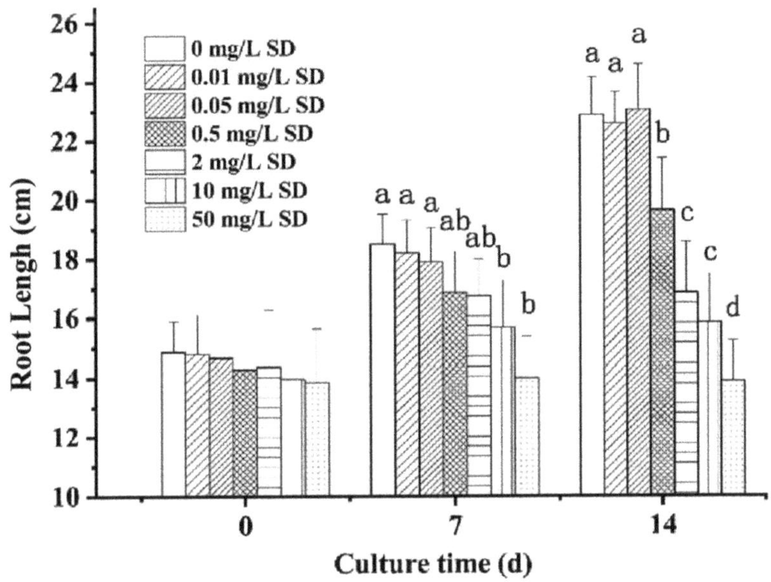

Figure 2. Effect of SD concentration on the root length of lettuce. Error bars represent the standard deviations of the means ($n = 15$), and a–d on top of the error bars represent significant differences ($p < 0.05$) among different initial concentrations of treatments.

3.2. Extent of SD Accumulation in Lettuce Organs

The morphological responses of lettuce to SD stress were attributed to the accumulation of SD in the organs of lettuce, which ultimately caused toxic effects on the plant. The accumulated concentrations of SD in roots and leaves of lettuce under different treatments are shown in Table 2. In the control experiments, no SD was detected in either the leaves or the roots. With SD in the culture nutrient solution, SD accumulated in the lettuce, including the roots and leaves; the accumulated SD concentrations in roots were higher than those in leaves under the four highest treatment concentrations (0.5, 2, 10, 50 mg/L) but not under the two lowest concentrations (0.01, 0.05 mg/L). In both roots and leaves, the accumulated SD concentrations significantly increased with increasing SD concentration in the culture nutrient solution. After 14 d of growth in the culture solution with 0.5 mg/L SD, the accumulated SD concentrations were 3.90 ± 0.91 and 6.48 ± 1.62 μg/kg in the leaves and roots, respectively. When the SD concentration in the solution was 2.0 mg/L, the accumulated SD concentrations increased by 2.4 and 3.5 times compared with those under 0.5 mg/L SD for leaves and roots, respectively. The increased accumulation rates of SD in roots were also found to be significantly higher than the corresponding increased rates in leaves under the same increased SD concentrations in the culture solutions.

Table 2. Concentrations (μg/kg in dry mass) of SD (±standard deviation) in the roots and leaves in different treatments.

Treatment	Leaf	Root
Control	Not detectable	Not detectable
0.01 mg/L SD	Not detectable	Not detectable
0.05 mg/L SD	Not detectable	Not detectable
0.5 mg/L SD	3.90 ± 0.91 c	6.48 ± 1.62 b
2 mg/LSD	9.55 ± 1.73 b	22.86 ± 2.41 c
10 mg/L SD	13.06 ± 1.76 a	39.80 ± 2.34 b
50 mg/L SD	16.74 ± 1.88 a	120.87 ± 17.33 a

a, b, and c within the same column represent significant differences ($p < 0.05$) among different initial concentrations of treatments ($n = 10$).

3.3. Antibiotic Resistomes in Plant Microbiomes

The roots and leaves of lettuce samples after 14 d of growth from the control and 0.01, 0.05, 0.5, and 10 mg/L SD treatment groups were collected to study the antibiotic resistance genes transformed in an SD stress culture environment. The absolute abundance levels of 16S rRNA in the lettuce samples after growth for 14 d under different treatments are shown in Figure 3. The absolute abundance levels of 16S rRNA in the phyllosphere and rhizosphere were significantly lower than those in the leaf endophytes and root endophytes, respectively. When SD was added to the culture solutions, the absolute abundance of 16S rRNA in all of the lettuce samples was reduced. The absolute abundance levels of 16S rRNA in the rhizosphere, root endophytes, and phyllosphere of lettuce under all SD treatments were significantly lower than those in the control group. In particular, the absolute abundance of 16S rRNA in the treatments with an SD concentration of 10 mg/L was significantly lower than that under the other SD concentrations.

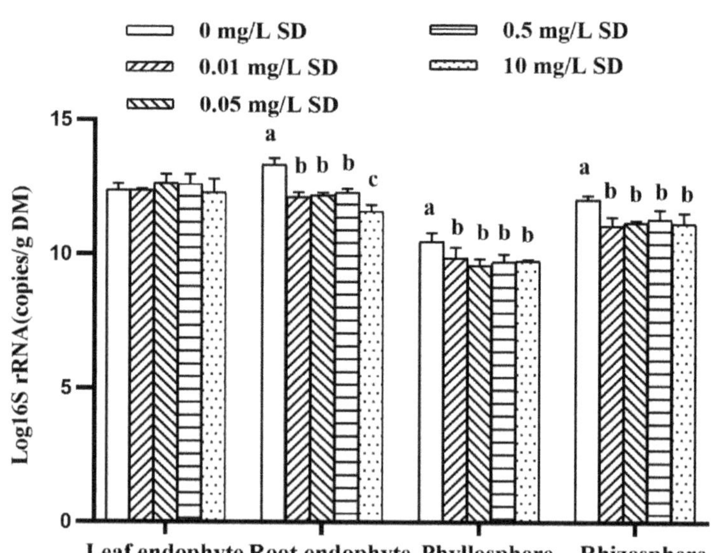

Figure 3. Effects of SD concentration on the log 16S rRNA absolute abundance in different parts of lettuce. Error bars represent standard deviations of the means. a, b, and c on top of the error bars represent significant differences ($p < 0.05$) among different initial concentrations of treatments.

Two common sulfonamide resistance genes, *sul*1 and *sul*2, and a transposon related to horizontal transfer, *tnpA*, were detected, and their relative abundance levels are shown in Figure 4. For *sul*1 and *sul*2, the relative abundance levels of root endophytes and leaf endophytes were significantly lower than those of the rhizosphere and phyllosphere, and the leaf endophytes had the lowest abundance among all the samples. For both *sul*1 and *sul*2, the relative abundance of leaf endophytes was significantly lower than that of root endophytes. For the relative abundance levels of *sul*1 and *sul*2 in leaf endophytes, the abundance levels were significantly higher under SD concentrations of 10 mg/L than under other SD concentrations, while the *sul*2 relative abundance in the phyllosphere was similar.

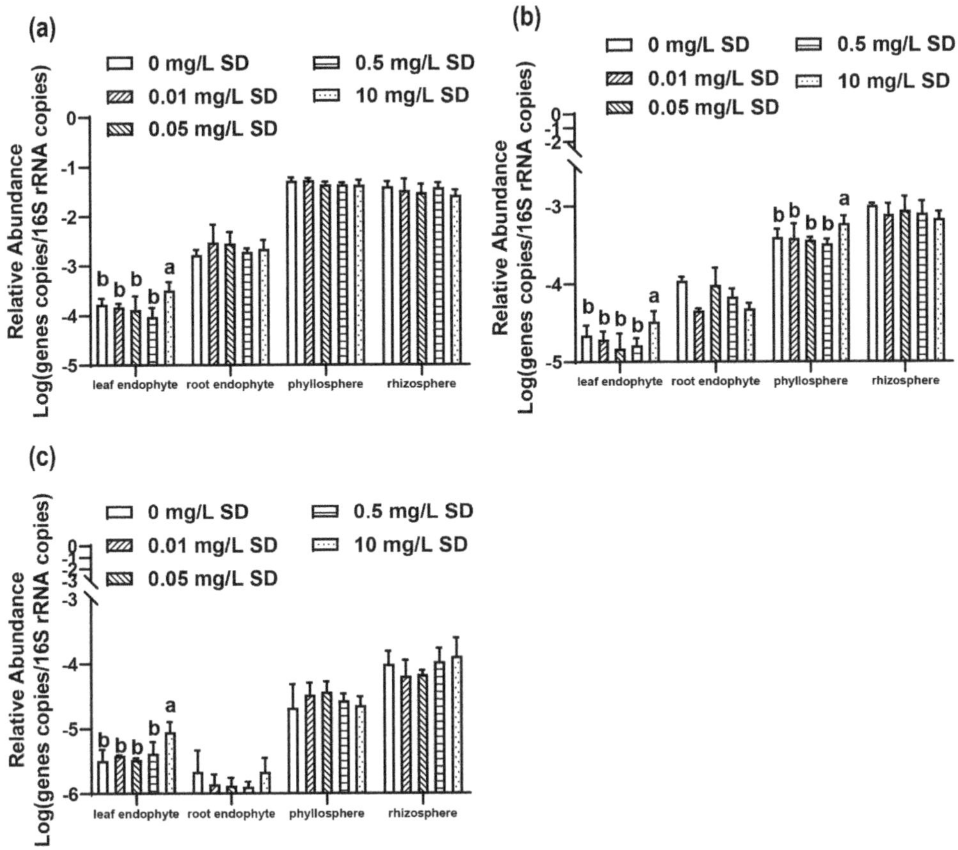

Figure 4. Effects of SD concentration on the relative abundance of antibiotic resistance genes (*sul*1, *sul*2) and *tnpA* in different parts of lettuce. Error bars represent standard deviations of the means. a and b on top of the error bars represent significant differences ($p < 0.05$) among different initial concentrations of treatments. (**a**) *sul*1; (**b**) *sul*2; (**c**) *tnpA*.

3.4. Membrane Permeability of the Roots

Electrolyte leakage (EL) has been suggested to be an indicator of membrane permeability and the degree of damage in roots [32]. The results of EL analyses of the lettuce roots after growth for 14 d under different treatments are shown in Figure 5. With the accumulation of SD in the roots during the lettuce growth processes in an SD stress environment, the leakage rates of roots were significantly increased. In the control group, the EL value of the roots was 20.0. Under SD stress environments at 0.5 and 2.0 mg/L SD, there were no significant differences compared with the control. When the SD concentration increased to 10 mg/L in the culture solution, the EL increased greatly, with an increase of 3.4-fold. When the SD concentration was further increased to 50 mg/L, the EL value increased to 79.0, with a significant difference obtained compared with that under 10 mg/L SD. The results also indicate that an SD concentration of 10 mg/L is the threshold for SD toxicity in lettuce and can result in obvious physiological damage to the roots, ultimately leading to the inhibition of plant growth.

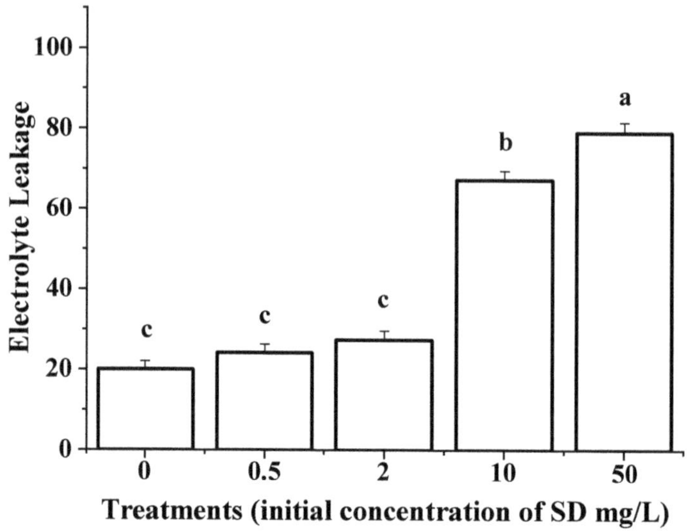

Figure 5. Effects of SD concentration on the electrolyte leakage of excised roots. Error bars represent standard deviations of the means. a, b, and c on top of the error bars represent significant differences ($p < 0.05$) among different initial concentrations of treatments.

3.5. The Ultrastructural Responses of Lettuce to the SD Stress Culture Environment

The roots and leaves of lettuce samples from the control, 2, and 10 mg/L SD treatment groups after 14 d of growth were collected to study their ultrastructural responses to the SD stress culture environment; the TEM results of root cells are shown in Figure 6a–c. The root cells of lettuce grown in the solution without SD showed a well-defined cell wall and regular cell contents for a normal lettuce root cell (Figure 6a), indicating no damage to lettuce under this condition. When the root cells of lettuce were grown with 2.0 mg/L SD, ultrastructural alterations, including thickening, were observed in the cell walls (Figure 6b). With a higher SD concentration of 10 mg/L, obvious damage and disintegration of cell walls were observed, including irregular cell walls and detached particles within the cell contents (Figure 6c).

Ultrastructural alterations were also observed in the leaf cells through TEM photographs of lettuce leaves collected from different treatment groups (Figure 6d–f). From the results, we observed that the chloroplasts of leaves from the control group were well elongated, having clear stoma and thylakoids with a typical grana structure and starch grains (Figure 6d). However, the chloroplasts in lettuce leaves under SD stress conditions at 2.0 mg/L showed damage and disintegration, although to a small extent, as indicated by Figure 6e. A further increase in SD concentrations in the culture nutrient solutions resulted in more obvious alterations of the chloroplasts. As shown in Figure 6f, the chloroplast structures in lettuce leaves under treatment with 10 mg/L SD, which was also observed as the threshold for SD toxicity in the study of membrane permeability, were severely altered in terms of the grana structure and starch grains, which appeared to be badly dissolved.

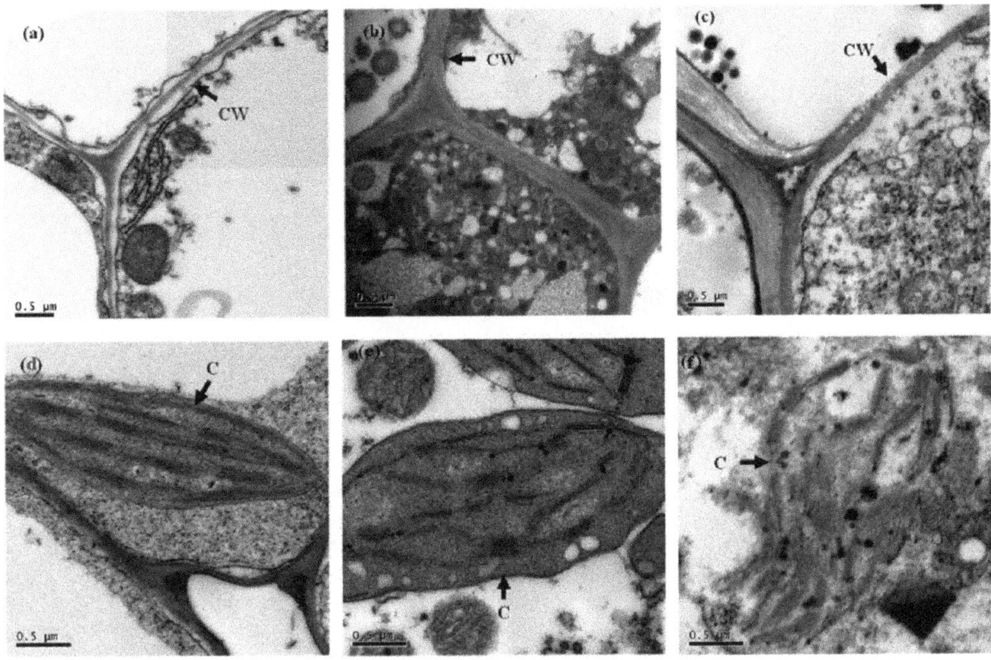

Figure 6. Transmission electron micrographs of lettuce root and leaf cells after 14 days of treatment under different SD concentrations: lettuce root under (**a**) control, (**b**) 2 mg/L, and (**c**) 10 mg/L; lettuce leaf under (**d**) control, (**e**) 2 mg/L, and (**f**) 10 mg/L. C denotes cell wall and CW denotes chloroplasts.

4. Discussion

Vegetable plants such as lettuce can absorb and accumulate residual SD antibiotics from the environment, which are ultimately enriched in the human body through direct consumption via the food chain. The USEPA reported that after being cultured in nutrient solution with 100 mg/L SD, the dry mass of *Lupinus albus* was reduced by 13% compared with that in the control [33]. In this study, *Lactuca sativa* L. was found to be more sensitive to SD than *Lupinus albus*: the dry mass yield reduction rates reached 36% with 10 mg/L SD in the culture solution. Our results also indicated that SD concentrations lower than 2.0 mg/L exerted no significant effect on the growth of lettuce, although the accumulation of SD and damage to the organs were still observed. Only at higher concentrations of SD, e.g., higher than 10 mg/L, did the morphology of lettuce, including the roots and leaves, exhibit significant changes in response to SD toxicity. Attributed to their direct and prolonged contact with SD residues, roots may be the main organs for accumulation and deposition in plants. As our results showed, the accumulated SD was mainly concentrated in the roots, with levels significantly higher than those in the leaves under all the different SD concentrations. On average, the SD concentrations in roots were two- to eightfold higher than those in leaves. Ahmed et al. also reported the bioaccumulation of six antibiotics in several plant species with obviously higher concentrations in the roots than in the stems [34]. The SD concentrations in *Lactuca sativa* L. are more related to the health of the human body since lettuces are usually directly eaten rather than processed.

Plant microbiota harbor intrinsic or acquired (from the environment) resistance genes that may be transmitted to humans through the food chain [16]. Environmental bacteria, especially rhizosphere bacteria, are an important type of plant endophytic bacteria [35], which can enter endophytic systems through plant tissues. Our results showed that the root and leaf endophyte microbiota were more abundant than the rhizosphere and phyllosphere microbiota. With the increase in SD concentration, the microbial abundance levels in the root endophyte, rhizosphere, and leaf endophyte communities were significantly lower than those in the control group. This indicated that the addition of SD reduced the plant microbiota, which was also in good accordance with SD being a potent antibacterial agent. However, there was no significant change in the abundance of microbes within the leaf endophyte community because SD in lettuce is transported from roots to leaves and the amount of SD accumulation within the leaves is low.

*sul*1 and *sul*2 are the most common resistance genes to sulfonamides, so we selected them for this study. The abundance levels of these genes in both the rhizosphere and phyllosphere were significantly higher than those in both the root endophytes and leaf endophytes, which is consistent with previous reports on the transfer of antibiotic resistance from manure-amended soil to vegetables [36]. There are two potential explanations for the lower abundance of *sul*1 and *sul*2 in lettuce endophytes: (1) the endophyte community diversity is low [37]; and (2) there was no direct contact between endophytes and SD [38], which served as a selective pressure [36]. The relative abundance levels of *sul*1 in the leaf endophytes and in the rhizosphere and *sul*2 in the leaf endophytes were only significantly higher under the 10 mg/L SD treatments than those under the control treatments, and those under all other concentrations were not significantly different from the control levels. *sul*1 and *sul*2 were mainly concentrated in the roots and were significantly more abundant in roots than in the leaves under all the different SD concentrations, which is consistent with previous reports [36]. The relative abundance of *tnpA* was similar to that of *sul*1 and *sul*2 at different SD concentrations. In our study, SD was added to the nutrient solution as a pure chemical. The situation may differ from the conditions under actual agricultural practices, as they likely represent a mixed input of antibiotics, antibiotic-resistant bacteria, and antibiotic resistance genes.

The toxicity mechanisms of antibiotics accumulated in plants have been researched for a long time but remain unclear and, at times, even controversial. It has been reported that the phytotoxicity of enrofloxacin in *Lactuca sativa* plants generates both toxic effects and hormesis, which manifest as quickly accelerated growth; the toxic effect is reported to be induced by high concentrations of antibiotics (e.g., 5000 μg/L), while the hormetic effect occurs at relatively low concentrations (e.g., 50 and 100 μg/L) [39]. Liu et al. found that in Welsh onion leaves, chloroplasts are sensitive to antibiotics by promoting ROS accumulation [40], while other researchers suggested that antibiotics potentially inhibit plant growth because of hydrophobicity [41]. In the present study, our results from a cell structure study provide direct evidence to confirm that the chloroplasts in lettuce leaves are the structures most sensitive to SD accumulation that affects growth. The accumulation of a sufficiently high level of SD antibiotics in the leaf cell could damage the structure and function of chloroplasts, thereby affecting photosynthesis.

Based on the above results, the proposed possible mechanism of SD toxicity in lettuce is presented in Figure 7. The toxic effect of lettuce was expected to first occur in the roots due to the direct contact of this organ with SD during the growth process. This was also evidenced by the thickening of root cell walls in the antibiotic-stressed lettuce. The thickening phenomenon of the cell wall is believed to be a natural self-protection mechanism of the root cells to prevent further accumulation of the antibiotic from the rhizosphere and/or to prevent the enlargement of the cells due to the increased pressure built up within the cell contents [42]. However, under high SD stress with extremely high concentrations in the culture solutions, the root cells of the lettuce were still physiologically damaged (Figure 6c), which was expected to lead to an equilibrium in the SD distribution between the external solution and the sap of the xylem. The physiological damage to the root cells was also reflected by the drastic increase in the EL rates obtained in the roots from lettuce grown under treatments with high SD concentrations.

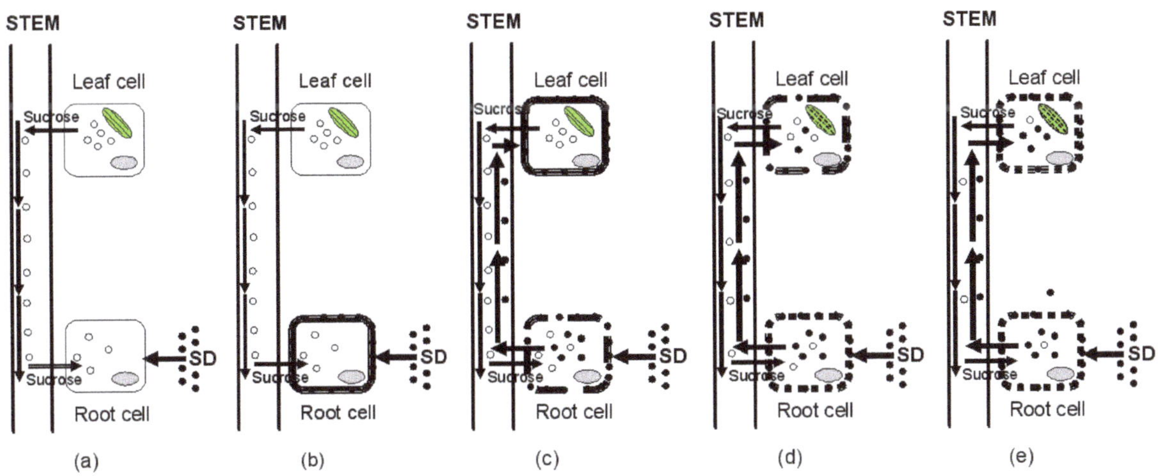

Figure 7. Proposed toxicity mechanism of accumulated SD in lettuces grown in SD stressed culture solutions. (**a**) lettuce at first contact with SD; (**b**) thickening of the lettuce root cell in the SD-stressed; (**c**) lettuce root cell damage and SD translocated from the roots to the leaves; (**d**) lettuce leaf cell damage and SD damage to the chloroplasts; (**e**) lettuce leaf cell broken, thus ultimately inhibiting plant growth.

After entering the xylem, the antibiotic can be easily translocated from the roots to the leaves along with the transpiration stream, resulting in accumulation of the antibiotic in the leaves and leading to further damage to the leaf cells, especially the chloroplasts. Chloroplasts are the organelles directly responsible for photosynthesis, and after being destroyed by the toxic effects of antibiotics, they lose the important ability to produce chlorophyll for the conversion of solar energy into chemical/food energy in the plant, thus ultimately inhibiting plant growth.

5. Conclusions

In conclusion, this study presents a possible mechanism of SD accumulation in lettuces grown in SD stressed culture solutions and antibiotic resistance gene transformation. The mechanism we have proposed is unverified, so SD may inhibit lettuce growth via other ways. Our study provides valuable information on the uptake and accumulation of antibiotics and antibiotic resistance genes in plants, which is useful for environmental fate and risk assessments of antibiotics for human exposure.

Author Contributions: Conceptualization, Y.W.; project administration, Y.W., L.M.; data curation, L.M., Y.-X.C.; performed the experiments, Y.W., L.M., Y.-X.C., C.W., J.-H.C., Z.-J.Z., M.-Y.Z., J.-T.F.; writing—review and editing, L.M., Y.W. All authors have read and agreed to the published version of the manuscript.

Funding: This study was supported by This study was supported by Guangdong agricultural research projects (YUECAINONG202137).

Institutional Review Board Statement: Not applicable.

Informed Consent Statement: Not applicable.

Data Availability Statement: Data are contained within the article.

Acknowledgments: We acknowledge the support in terms of time and facilities from South China Agricultural University for this study.

Conflicts of Interest: The authors declare no conflict of interest.

References

1. Mathew, A.G.; Cissell, R.; Liamthong, S. Antibiotic resistance in bacteria associated with food animals: A United States perspective of livestock production. *Foodborne Pathog. Dis.* **2007**, *4*, 115–133. [CrossRef]
2. Micinski, J.; Pogorzelska, J.; Slyamowa, A.; Kobzhassarov, T.; Bermagambetova, N.; Dzik, S.; Kowalski, P.M.; Zaborowska-Sapeta, K.; Kowalski, I.M. Hazards to Humans and Animals Associated with Antibiotic Misuse. *J. Elementol.* **2015**, *20*, 1077–1086. [CrossRef]
3. Sarmah, A.K.; Meyer, M.T.; Boxall, A.B. A global perspective on the use, sales, exposure pathways, occurrence, fate and effects of veterinary antibiotics (VAs) in the environment. *Chemosphere* **2006**, *65*, 725–759. [CrossRef] [PubMed]
4. Zhou, X.; Wang, J.; Lu, C.; Liao, Q.H.; Gudda, F.O.; Ling, W.T. Antibiotics in animal manure and manure-based fertilizers: Occurrence and ecological risk assessment. *Chemosphere* **2020**, *255*, 127006. [CrossRef] [PubMed]
5. Menz, J.; Olsson, O.; Kummerer, K. Antibiotic residues in livestock manure: Does the EU risk assessment sufficiently protect against microbial toxicity and selection of resistant bacteria in the environment? *J. Hazard. Mater.* **2019**, *379*, 120807. [CrossRef] [PubMed]
6. Lyu, J.; Yang, L.S.; Zhang, L.; Ye, B.X.; Wang, L. Antibiotics in soil and water in China-a systematic review and source analysis. *Environ. Pollut.* **2020**, *266*, 115147. [CrossRef]
7. Li, Y.W.; Wu, X.L.; Mo, C.H.; Tai, Y.P.; Huang, X.P.; Xiang, L. Investigation of Sulfonamide, Tetracycline, and Quinolone Antibiotics in Vegetable Farmland Soil in the Pearl River Delta Area, Southern China. *J. Agric. Food Chem.* **2011**, *59*, 7268–7276. [CrossRef] [PubMed]
8. Peltzer, P.M.; Lajmanovich, R.C.; Attademo, A.M.; Junges, C.M.; Teglia, C.M.; Martinuzzi, C.; Curi, L.; Culzoni, M.J.; Goicoechea, H.C. Ecotoxicity of veterinary enrofloxacin and ciprofloxacin antibiotics on anuran amphibian larvae. *Environ. Toxicol. Pharmacol.* **2017**, *51*, 114–123. [CrossRef]
9. Nasri, A.; Hannachi, A.; Allouche, M.; Barhoumi, B.; Saidi, I.; Dallali, M.; Harrath, A.H.; Mansour, L.; Mahmoudi, E.; Beyrem, H.; et al. Chronic ecotoxicity of ciprofloxacin exposure on taxonomic diversity of a meiobenthic nematode community in microcosm experiments. *J. King Saud Univ. Sci.* **2020**, *32*, 1470–1475. [CrossRef]
10. Xie, Z.X.; Tang, J.; Wu, X.W.; Fan, S.S.; Cheng, H.M.; Li, X.D.; Hua, R.M. Bioconcentration and ecotoxicity of sulfadiazine in the aquatic midge Chironomus riparius. *Environ. Toxicol. Pharmacol.* **2019**, *66*, 69–74. [CrossRef]
11. Pino, M.R.; Val, J.; Mainar, A.M.; Zuriaga, E.; Espanol, C.; Langa, E. Acute toxicological effects on the earthworm Eisenia fetida of 18 common pharmaceuticals in artificial soil. *Sci. Total Environ.* **2015**, *518*, 225–237. [CrossRef]
12. Pan, M.; Chu, L.M. Transfer of antibiotics from wastewater or animal manure to soil and edible crops. *Environ. Pollut.* **2017**, *231*, 829–836. [CrossRef] [PubMed]
13. Zhang, Y.J.; Hu, H.W.; Gou, M.; Wang, J.T.; Chen, D.; He, J.Z. Temporal succession of soil antibiotic resistance genes following application of swine, cattle and poultry manures spiked with or without antibiotics. *Environ. Pollut.* **2017**, *231*, 1621–1632. [CrossRef] [PubMed]
14. Xie, W.Y.; McGrath, S.P.; Su, J.Q.; Hirsch, P.R.; Clark, I.M.; Shen, Q.; Zhu, Y.G.; Zhao, F.J. Long-Term Impact of Field Applications of Sewage Sludge on Soil Antibiotic Resistome. *Environ. Sci. Technol.* **2016**, *50*, 12602–12611. [CrossRef]
15. Chen, Y.; Su, J.Q.; Zhang, J.; Li, P.; Chen, H.; Zhang, B.; Gin, K.Y.; He, Y. High-throughput profiling of antibiotic resistance gene dynamic in a drinking water river-reservoir system. *Water Res.* **2019**, *149*, 179–189. [CrossRef] [PubMed]
16. Chen, Q.L.; Cui, H.L.; Su, J.Q.; Penuelas, J.; Zhu, Y.G. Antibiotic Resistomes in Plant Microbiomes. *Trends Plant Sci.* **2019**, *24*, 530–541. [CrossRef]
17. Yazdi, M.; Kolahi, M.; Kazemi, E.M.; Barnaby, A.G. Study of the contamination rate and change in growth features of lettuce (Lactuca sativa Linn.) in response to cadmium and a survey of its phytochelatin synthase gene. *Ecotoxicol. Environ. Saf.* **2019**, *180*, 295–308. [CrossRef]
18. Hayes, K.L.; Mui, J.; Song, B.; Sani, E.S.; Eisenman, S.W.; Sheffield, J.B.; Kim, B. Effects, uptake, and translocation of aluminum oxide nanoparticles in lettuce: A comparison study to phytotoxic aluminum ions. *Sci. Total Environ.* **2020**, *719*, 137393. [CrossRef]
19. Song, Y.; Zhang, F.L.; Li, H.P.; Qiu, B.; Gao, Y.; Cui, D.; Yang, Z.G. Antioxidant defense system in lettuces tissues upon various As species exposure. *J. Hazard. Mater.* **2020**, *399*, 123003. [CrossRef]
20. Xu, D.M.; Pan, H.; Yao, J.C.; Feng, Y.X.; Wu, P.P.; Shao, K. Stress responses and biological residues of sulfanilamide antibiotics in Arabidopsis thaliana. *Ecotoxicol. Environ. Saf.* **2020**, *199*, 110727. [CrossRef]
21. Berendonk, T.U.; Manaia, C.M.; Merlin, C.; Fatta-Kassinos, D.; Cytryn, E.; Walsh, F.; Burgmann, H.; Sorum, H.; Norstrom, M.; Pons, M.N.; et al. Tackling antibiotic resistance: The environmental framework. *Nat. Rev. Microbiol.* **2015**, *13*, 310–317. [CrossRef] [PubMed]
22. Yang, Q.X.; Ren, S.W.; Niu, T.Q.; Guo, Y.H.; Qi, S.Y.; Han, X.K.; Liu, D.; Pan, F. Distribution of antibiotic-resistant bacteria in chicken manure and manure-fertilized vegetables. *Environ. Sci. Pollut. Res.* **2014**, *21*, 1231–1241. [CrossRef]
23. Bellino, A.; Lofrano, G.; Carotenuto, M.; Libralato, G.; Baldantoni, D. Antibiotic effects on seed germination and root development of tomato (Solanum lycopersicum L.). *Ecotoxicol. Environ. Saf.* **2018**, *148*, 135–141. [CrossRef] [PubMed]
24. Xu, Y.G.; Yu, W.T.; Ma, Q.; Wang, J.; Zhou, H.; Jiang, C.M. The combined effect of sulfadiazine and copper on soil microbial activity and community structure. *Ecotoxicol. Environ. Saf.* **2016**, *134*, 43–52. [CrossRef] [PubMed]
25. Hemkemeyer, M.; Christensen, B.T.; Martens, R.; Tebbe, C.C. Soil particle size fractions harbour distinct microbial communities and differ in potential for microbial mineralisation of organic pollutants. *Soil Biol. Biochem.* **2015**, *90*, 255–265. [CrossRef]

26. Rafraf, I.D.; Lekunberri, I.; Sanchez-Melsio, A.; Aouni, M.; Borrego, C.M.; Balcazar, J.L. Abundance of antibiotic resistance genes in five municipal wastewater treatment plants in the Monastir Governorate, Tunisia. *Environ. Pollut.* **2016**, *219*, 353–358. [CrossRef] [PubMed]
27. Pei, R.T.; Kim, S.C.; Carlson, K.H.; Pruden, A. Effect of River Landscape on the sediment concentrations of antibiotics and corresponding antibiotic resistance genes (ARG). *Water Res.* **2006**, *40*, 2427–2435. [CrossRef]
28. Liao, H.; Lu, X.; Rensing, C.; Friman, V.P.; Geisen, S.; Chen, Z.; Yu, Z.; Wei, Z.; Zhou, S.; Zhu, Y. Hyperthermophilic Composting Accelerates the Removal of Antibiotic Resistance Genes and Mobile Genetic Elements in Sewage Sludge. *Environ. Sci. Technol.* **2018**, *52*, 266–276. [CrossRef] [PubMed]
29. Zhu, Y.G.; Johnson, T.A.; Su, J.Q.; Qiao, M.; Guo, G.X.; Stedtfeld, R.D.; Hashsham, S.A.; Tiedje, J.M. Diverse and abundant antibiotic resistance genes in Chinese swine farms. *Proc. Natl. Acad. Sci. USA* **2013**, *110*, 3435–3440. [CrossRef] [PubMed]
30. Lv, Y.; Li, Y.Y.; Liu, X.H.; Xu, K. The tolerance mechanism and accumulation characteristics of Phragmites australis to sulfamethoxazole and ofloxacin. *Chemosphere* **2020**, *253*, 126695. [CrossRef]
31. Yang, L.; Feng, Y.X.; Zhang, H.; Yu, X.Z. Estimating the synergistic and antagonistic effects of dual antibiotics on plants through root elongation test. *Ecotoxicology* **2020**, 1–12. [CrossRef]
32. Eraslan, F.; Inal, A.; Savasturk, O.; Gunes, A. Changes in antioxidative system and membrane damage of lettuce in response to salinity and boron toxicity. *Sci. Hortic Amst.* **2007**, *114*, 5–10. [CrossRef]
33. United States Environmental Protection Agency. *Development Document for the Proposed Revisions to the National Pollutant Discharge Elimination System Regulation and the Effluent Guidelines for Concentrated Animal Feeding Operations by Engineering and Analysis Division, Office of Science and Technology*; United States Environmental Protection Agency: Washington, DC, USA, 2001.
34. Ahmed, M.B.M.; Rajapaksha, A.U.; Lim, J.E.; Vu, N.T.; Kim, I.S.; King, H.M.; Lee, S.S.; Ok, Y.S. Distribution and Accumulative Pattern of Tetracyclines and Sulfonamides in Edible Vegetables of Cucumber, Tomato, and Lettuce. *J. Agric. Food Chem.* **2015**, *63*, 398–405. [CrossRef]
35. Compant, S.; Clement, C.; Sessitsch, A. Plant growth-promoting bacteria in the rhizo- and endosphere of plants: Their role, colonization, mechanisms involved and prospects for utilization. *Soil Biol. Biochem.* **2010**, *42*, 669–678. [CrossRef]
36. Zhang, Y.J.; Hu, H.W.; Chen, Q.L.; Singh, B.K.; Yan, H.; Chen, D.L.; He, J.Z. Transfer of antibiotic resistance from manure-amended soils to vegetable microbiomes. *Environ. Int.* **2019**, *130*, 104912. [CrossRef]
37. Wang, F.H.; Qiao, M.; Chen, Z.; Su, J.Q.; Zhu, Y.G. Antibiotic resistance genes in manure-amended soil and vegetables at harvest. *J. Hazard. Mater.* **2015**, *299*, 215–221. [CrossRef] [PubMed]
38. Zhu, B.; Chen, Q.; Chen, S.; Zhu, Y.G. Does organically produced lettuce harbor higher abundance of antibiotic resistance genes than conventionally produced? *Environ. Int.* **2017**, *98*, 152–159. [CrossRef]
39. Migliore, L.; Cozzolino, S.; Fiori, M. Phytotoxicity to and uptake of enrofloxacin in crop plants. *Chemosphere* **2003**, *52*, 1233–1244. [CrossRef]
40. Liu, X.N.; Lv, Y.; Gao, S.; Xu, K. Ofloxacin induces etiolation in Welsh onion leaves. *Chemosphere* **2021**, *267*, 128918. [CrossRef]
41. Fu, L.; Huang, T.; Wang, S.; Wang, X.H.; Su, L.M.; Li, C.; Zhao, Y.H. Toxicity of 13 different antibiotics towards freshwater green algae Pseudokirchneriella subcapitata and their modes of action. *Chemosphere* **2017**, *168*, 217–222. [CrossRef]
42. Davi, V.; Tanimoto, H.; Ershov, D.; Haupt, A.; De Belly, H.; Le Borgne, R.; Couturier, E.; Boudaoud, A.; Minc, N. Mechanosensation Dynamically Coordinates Polar Growth and Cell Wall Assembly to Promote Cell Survival. *Dev. Cell* **2018**, *45*, 170–182. [CrossRef] [PubMed]

processes MDPI

Article

Influence of Hydrothermal Pretreatment Temperature on the Hydration Properties and Direct Carbonation Efficiency of Al-Rich Ladle Furnace Refining Slag

Yi Huang [1,*] and Guo Xiong [2]

[1] School of Materials and Chemical Engineering, Hunan City University, Yiyang 413002, China
[2] Hunan Hualin Xiangtan Iron and Steel Co., Ltd., Xiangtan 411101, China; xgsz50@163.com
* Correspondence: Huangyi@hncu.edu.cn

Abstract: The influence of hydrothermal pretreatment temperature on the hydration products and carbonation efficiency of Al-rich LF slag was investigated. The results showed that the carbonation efficiency was strongly dependent on the morphology of hydration products and the hydration extent of the raw slag. Hydrothermal pretreatment at 20 °C or 80 °C favored the formation of flake-shaped products with a higher specific surface area and therefore resulted in a higher CO_2 uptake of 20 °C and 80 °C-pretreated slags (13.66 wt% and 10.82 wt%, respectively). However, hydrothermal pretreatment at 40 °C, 60 °C or 100 °C led to the rhombohedral-shaped calcite layer surrounding the unreacted core of the raw slag and the formation of fewer flake-shaped products, resulting in a lower CO_2 uptake of 40 °C, 60 °C and 100 °C-pretreated slags (9.21 wt%, 9.83 wt%, and 6.84 wt%, respectively).

Keywords: LF slag; hydrothermal pretreatment; temperature; hydration products; CO_2 uptake

Citation: Huang, Y.; Xiong, G. Influence of Hydrothermal Pretreatment Temperature on the Hydration Properties and Direct Carbonation Efficiency of Al-Rich Ladle Furnace Refining Slag. *Processes* **2021**, *9*, 1458. https://doi.org/10.3390/pr9081458

Academic Editor: Federica Raganati

Received: 29 July 2021
Accepted: 19 August 2021
Published: 21 August 2021

1. Introduction

The growth of global greenhouse gas emissions was 2.0% in 2018 and there is no sign that any of these emissions are peaking yet. The six largest emitters of greenhouse gases, together accounting for 62% globally, are China (26%), the United States (13%), the European Union (more than 8%), India (7%), the Russian Federation (5%), and Japan (almost 3%) [1]. China's carbon emission peak is a matter of international focus. Recently, China made a solemn promise to peak its carbon dioxide emission by 2030 and achieve carbon neutrality by 2060. Some of the main measures China will use to reduce CO_2 emissions over next 10 years are: changing energy and industrial structures, transforming the development mode, promoting clean energy, and appropriately increasing carbon sequestration ability. Among the current CO_2 sequestration routes, mineral carbonation is regarded as a potential technology because of its advantages; it is environmentally benign, it enables the permanent trapping of CO_2 in the form of carbonate, and it does not require post-storage surveillance for CO_2 leakage [2]. In general, mineral carbonation can be divided into two categories, namely direct carbonation and indirect carbonation. Direct mineral carbonation is accomplished through the reaction of a solid alkaline mineral with CO_2, either in gaseous or in aqueous phase [3].

Alkaline solid wastes such as red mud, steel slag, blast furnace slag, fly ash, etc., are used for direct mineral carbonation as efficient and economically available capturers of CO_2 [4–9]. For the direct mineral carbonation of steel slag, the formation of an increasingly thick and dense carbonate layer surrounding the unreacted core of the solid particle hinders further carbonation and results in the lower CO_2 capture capacity [10]. In our previous study [11], the improvement in the direct carbonation efficiency of Al-rich ladle furnace refining slag (LF slag) by hydrothermal pretreatment was investigated. The results showed that after hydrothermal pretreatment at 80 °C, the morphology of $Ca_{12}Al_{14}O_{33}(C_{12}A_7)$ in the slag transformed from separated particles to the flake-shaped $Ca_3Al_2O_6{\cdot}xH_2O(C_3AH_x)$, resulting in an increased reaction surface area and carbonation efficiency. However, this study did not discuss the effect of hydrothermal temperature on the carbonation efficiency. In fact, the hydration product of $C_{12}A_7$ is dependent on the hydration temperature. Koplík et al. [12] reported that at 20 °C the major hydration products of $C_{12}A_7$ were $Ca_2Al_2O_5{\cdot}8H_2O(C_2AH_8)$ and $CaAl_2O_4{\cdot}10H_2O(CAH_{10})$; at 30 °C CAH_{10} disappeared and only C_2AH_8 remained; at 60 °C the only stable hydrates–$Ca_3Al_2O_6{\cdot}6H_2O(C_3AH_6)$ and $Al(OH)_3(AH)$ were formed. Edmonds et al. [13] stated that both C_2AH_8 and CAH_{10} can be produced during the hydration of $C_{12}A_7$ at 4 °C, while no trace of CAH_{10} was spotted when $C_{12}A_7$ was hydrated at 20 or 40 °C. Given that the morphology of the hydration product of C12A7 has a significant effect on the carbonation efficiency of LF slag and the type of hydration product produced is related to temperature, the aim of this study was to investigate the influence of hydrothermal pre-treatment temperature on the hydration properties and the carbonation efficiency of Al-rich LF slag at ambient temperature and pressure. Moreover, the relation between the morphology of the hydration product and the carbonation efficiency was clarified in this work.

2. Materials and Methods

2.1. Materials

The Al-rich LF slag used in this study was collected from the Xiangtan steel plant in Hunan province, China. The chemical composition of the slag determined by X-ray Fluorescence (XRF) is listed in Table 1. Before pretreatment, the raw slag was crushed and ground into a powder <20 mesh particle size. Distilled water was used in this study for slag suspension preparation.

Table 1. Chemical composition of raw slag as determined by XRF analysis.

Component	CaO	Al_2O_3	SiO_2	MgO	TiO_2	SO_3	Fe_2O_3	Others
wt%	52.0056	23.66	15.83	4.05	0.90	2.45	0.54	0.57

2.2. Hydrothermal Pretreatment of LF Slag at Different Temperatures

At first, the raw slag was fully mixed with water at a solid/water (S/W) ratio of 1:10 in a beaker. Then, the suspension was stirred for 30 min at 20, 40, 60, 80 and 100 °C (designated as 20H, 40H, 60H, 80H and 100H-slag, respectively). Next, the suspension was filtered and the obtained solid was sufficiently washed and dried to a constant weight at 105 °C for further characterization and for the following carbonation experiment.

2.3. Direct Aqueous Carbonation Process

The schematic diagram of the aqueous carbonation experimental system is shown in Figure 1. The pretreated slag suspension with a solid/water ratio of 1:10 was placed in a conical flask into an electric-heated thermostatic water bath equipped with a mechanical stirrer. The temperature of the water bath was kept at 40 °C. Then, 99.99% pure CO_2 from the CO_2 cylinder was injected into the suspension at a flow rate of 5 L/min controlled by a flowmeter, and was simultaneously stirred for carbonation. The suspension underwent the carbonation process for 60 min and was then filtered. The obtained solid was dried to a constant weight at 105 °C for further characterization.

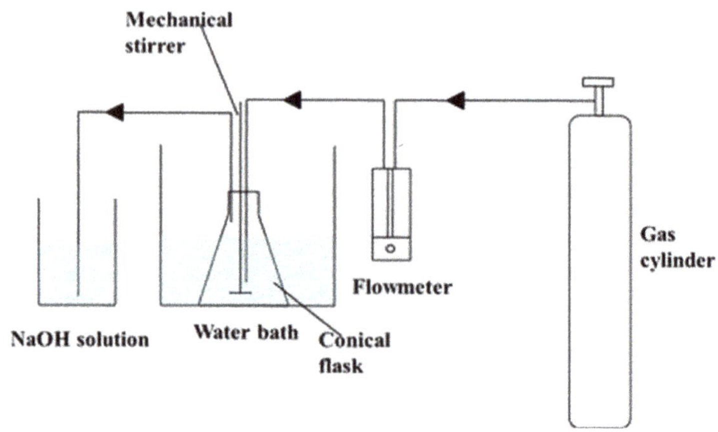

Figure 1. Schematic diagram of the aqueous carbonation experimental system.

2.4. Characterization of Slag

X-ray diffraction (XRD, Bruker AXS company D8 Advance, Karlsruhe, Germany) was conducted on the slags to identify their main mineral phases. The scanning range was from 5° to 70° 2θ at 2°/min. TG-DSC analysis was performed using a METTLER TOLEDO 1600 LF thermal gravimetric analyzer. A field-emission scanning electron microscope (FESEM, Hitachi company SU8010, Tokyo, Japan) was used to characterize the morphology of the slag. The specific surface area of the slags was measured by the N_2 gas adsorption Brunauer–Emmet–Teller (BET) method (ASAP 2020, Micromeritics, Norcross, GA, USA). The pH value of the slag suspension was determined by a PHS-3C pH meter.

2.5. Analysis of Carbonation Efficiency

In order to compare the carbonation efficiency of Al-rich LF slags under the hydrothermal pretreatment at different temperatures, the CO_2 uptake of the slags was measured based on the weight fraction of the TG curve ($\Delta m_{600-800\ ^\circ C}$) and the dry weight (m) [14] expressed in terms of CO_2 (wt%), Equation (1):

$$CO_2(wt\%) = \frac{\Delta m_{600-800\ ^\circ C}}{m} \times 100 \quad (1)$$

3. Results and Discussion

3.1. Influence of Hydrothermal Temperature on Hydration Properties of Al-Rich LF Slag

Figure 2 shows the XRD patterns of the raw slag and the pretreated slags. The main mineral phases of raw slag were $C_{12}A_7$ and $Ca_2SiO_4(CS_2)$. For all the slags with pretreatment, the peaks of $C_{12}A_7$ were reduced, indicating its hydration. The 40H-slag and 100H-slag presented relatively more intense residual $C_{12}A_7$ peaks than the other slags, suggesting the lower hydration extent of these two slags. For the 20H-slag, $3CaO{\cdot}Al_2O_3{\cdot}CaCO_3{\cdot}11H_2O(C_4A\overline{C}H_{11})$ was the dominant hydration product. A small amount of $(C_4A\overline{C}H_{11})$ also appeared in the 40H-slag and C_3AH_6 was the other main product for this slag. Hydrocalumite $(Ca_4Al_2(OH)_{12}CO_3{\cdot}5H_2O)$ was only present in the 100H-slag, and this slag had the most intense C_3AH_6 peaks. With respect to the 60H-slag, only C_3AH_6 crystal was found. In general, C_3AHx, $C_4A\overline{C}H_{11}$, and C_3AH_6 were the main hydration products, while the other products mentioned in the "Introduction" (such as CAH_{10} and C_2AH_8) were not observed. This could be explained by the fact that CAH_{10} and C_2AH_8 are the transition phases and can be converted to the ultimate stable products (e.g., C_3AH_6 and C_3AH_x), described by Equations (2)–(4), respectively [15].

$$2CAH_{10} \rightarrow C_2AH_8 + AH_3 + 9H \quad (2)$$

$$3C_2AH_8 \rightarrow 2C_3AH_6 + AH_3 + 9H \quad (3)$$

$$3C_2AH_8 \rightarrow 2C_3AH_x + AH_3 + (21 - 2x)H \quad (4)$$

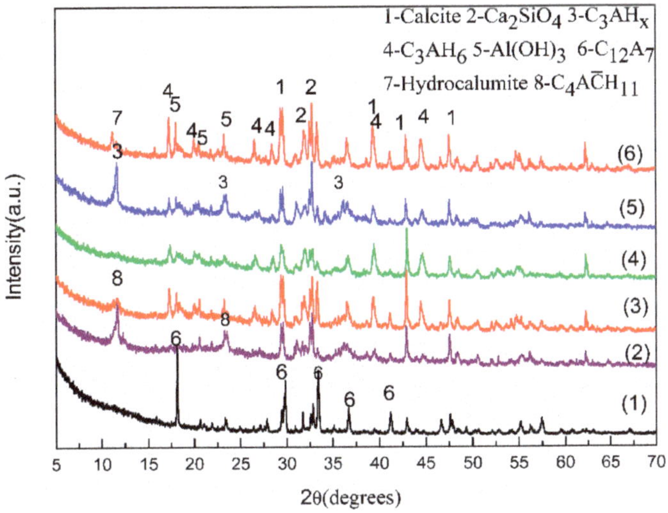

Figure 2. XRD patterns of (1) raw slag, (2) 20H-slag, (3) 40H-slag, (4) 60H-slag, (5) 80H-slag, and (6) 100H-slag.

Figure 3 displays the FESEM pictures of the raw slag and the pretreated slags. The raw slag appeared as irregular-shaped particles with dense and coarse surfaces (Figure 3a). After hydration at 20 °C, the slag surface became smooth due to the formation of flake-shaped $C_4A\overline{C}H_{11}$ (Figure 3b). In addition, metastable hydrates in the form of hexagonal platelets [14] were observed (Figure 3b). The microstructure of the 40H-slag presented as a mixture of $C_4A\overline{C}H_{11}$, metastable hydrated hexagonal-shaped platelets [16], and unhydrated slag particles (Figure 3c). The edge of the unhydrated particles in the 40H-slag was covered by rhombohedral-shaped $CaCO_3$ (calcite) particles and AH gel with a grain size of 0.5 μm, which may hinder the further hydration of $C_{12}A_7$ (Figure 3d). Once again, this verified that the thick and dense $CaCO_3$ layer surrounding the unreacted core of the solid particle was the main cause of the low carbonation efficiency of the slag without hydrothermal pretreatment, as illustrated in our previous studies [9]. $CaCO_3$ and AH gel should be generated by the indirect carbonation reaction between $C_{12}A_7$ and CO_2 in the air, which can be described by Equations (5) and (6) [17]. The occurrence of Equation (5) resulted in the alkalinity of the slag suspensions (the final pH values were 10.42, 10.13, 10.05, 10.56, and 10.02 for the slag suspensions treated at 20 °C, 40 °C, 60 °C, 80 °C, and 100 °C, respectively).

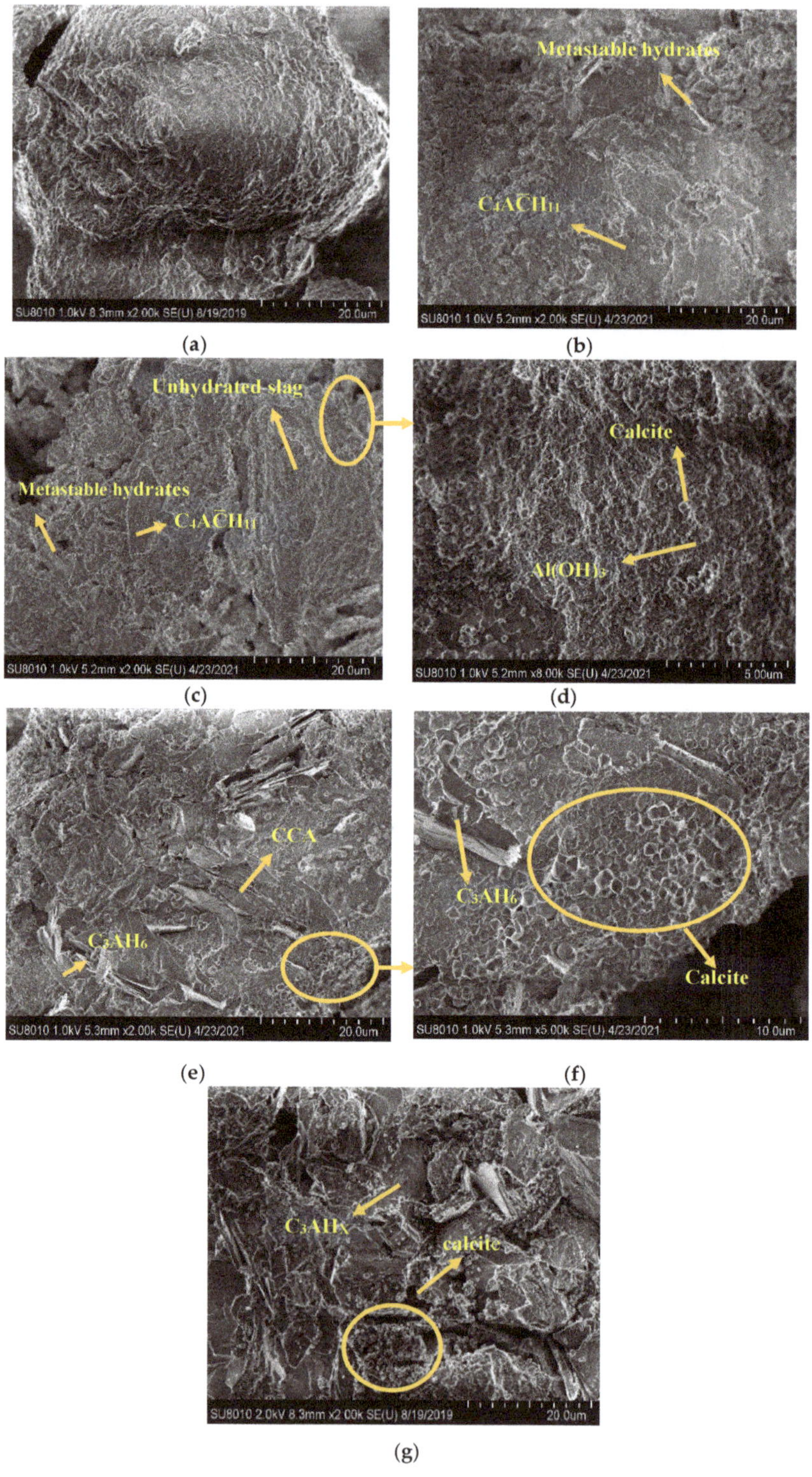

Figure 3. *Cont.*

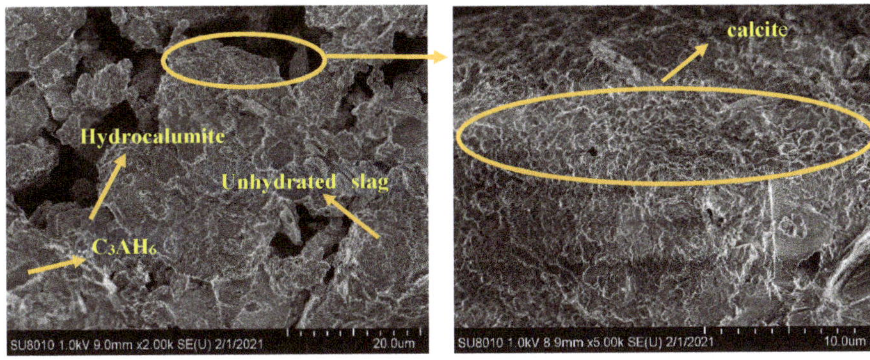

(**h**) (**i**)

Figure 3. FESEM pictures of (**a**) raw slag 2000×, (**b**) 20H-slag 2000×, (**c**) 40H-slag 2000×, (**d**) 40H-slag 8000×, (**e**) 60H-slag 2000×, (**f**) 60H-slag 5000×, (**g**) 80H slag 2000×, (**h**) 100H-slag 2000× and (**i**) 100H-slag 5000×.

The FESEM image of the 60H-slag (Figure 3e) indicates the formation of flake-shaped and cubic hydrates that should be amorphous calcium carboaluminate (CCA) and C_3AH_6, respectively. In addition, a dense rhombohedral-shaped calcite layer covered part of the slag surface (Figure 3f). Therefore, it is assumed that CCA was formed through the reaction between C_3AH_6 and calcite [18,19]. This also provides an explanation for why the 20H-slag contained large amounts of $C_4A\overline{C}H_{11}$ but small amounts of calcite formation (the reaction between C_3AH_6 and calcite can be described by Equation (7) [19]). After hydration at 80 °C, the morphology of the slag changed from separated particles to continuous gel (Figure 3g). The flake-shaped gel should be C_3AH_x, and calcite particles were scattered on the surface of C_3AH_X in the 80H-slag (Figure 3g). For the 100H-slag, flake-shaped hydrocalumite and cubic C_3AH_6 particles were embedded in the unhydrated slag particles (Figure 3h), and a rhombohedral-shaped calcite layer deposited on the edge of the slag particles (Figure 3i), similarly to the 40H and 60H slag.

$$Ca_{12}Al_{14}O_{33} + 33H_2O \rightarrow 12Ca^{2+} + 14Al(OH)_4^{-} + 10OH^{-} \quad (5)$$

$$Ca^{2+} + CO_2 + 2Al(OH)_4^{-} \rightarrow CaCO_3\downarrow + 2Al(OH)_3\downarrow + H_2O \quad (6)$$

$$CaCO_3 + Ca_3Al_2O_6{\cdot}6H_2O + 5H_2O \rightarrow 3CaO{\cdot}Al_2O_3{\cdot}CaCO_3{\cdot}11H_2O \quad (7)$$

The BET specific surface area (S_{BET}) of the slags were listed in Table 2. The S_{BET} of the 20H-slag and the 80H-slag was more than two times that of the raw slag, while other pretreated slags demonstrated only a slight S_{BET} increase compared with the raw slag. This should be attributed to the larger amount of flake-shaped hydrates in the 20H-slag and the 80H-slag [20].

Table 2. The BET specific surface area of the pretreated slags.

	Raw Slag	20H-Slag	40H-Slag	60H-Slag	80H-Slag	100H-Slag
SBET (m^2/g)	4.68	8.32	5.89	6.15	9.40	5.15

Figure 4 shows the TG-DSC analysis results of the pretreated slags. The endothermic peak of around 260–270 °C denoted the decomposition of $Al(OH)_3$ and appeared in all slags [19]. This peak was overlapped by the endothermic peak between 280 °C and 325 °C, which was attributed to the dehydration of C_3AH_6 in the 40H-slag, 60H-slag, and 80H-slag [18]. These results corresponded well with the XRD and FESEM analysis. The endothermic peak at 157 °C in the 20H-salg indicated the dehydration of $C_4A\overline{C}H_{11}$ [16], while this peak became broad for the 40H-slag due to the low crystallinity of $C_4A\overline{C}H_{11}$ [16], as was also reflected in the broad peak of XRD patterns (Figure 1). The absence of $C_4A\overline{C}H_{11}$ in the other slags can be explained by its instability in temperatures above 40 °C [19]. A very broad endothermic peak between 80 and 200 °C was observed in the 60H-slag, generated by the dehydration of amorphous CCA [14]. The endothermic peak at 155 °C in the 80H-slag represented the dehydration of C_3AH_x which is close to the dehydration temperatures of CAH_{10} and C_2AH_8 [15]. With respect to the 100H-slag, the dehydration of hydrocalumite was reflected in the endothermic peak around 146 °C. In general, the dehydration of hydrates mainly occurred over the temperature range of 105–325 °C, resulting in significant weight loss. The other significant weight loss region was between 600 and 800 °C, which was ascribed to the $CaCO_3$ decomposition. During the hydrothermal process, $C_{12}A_7$ was transformed into calcium aluminates hydrate (CAH), CAC, AH, and $CaCO_3$; therefore, the mass loss ratio of the slags (See Table 3) above the temperature range of 105–800 °C should be an indicator of $C_{12}A_7$ hydration extent. It may be concluded from the results in Table 2 that the 20H-slag and the 80H-slag had significantly higher hydration extent than other three slags, in good agreement with the XRD and FESEM analysis.

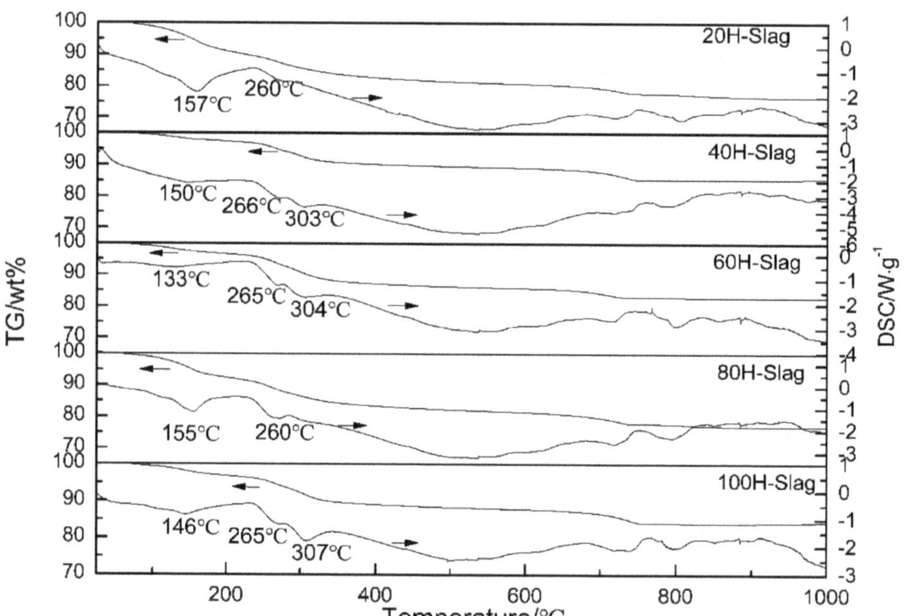

Figure 4. TG-DSC curves of the pretreated slags.

Table 3. The mass loss ratio of the pretreated slags between 105 and 800 °C.

Sample	Mass Loss Ratio between 105 and 800 °C (wt%)
20H-slag	20.58
40H-slag	13.80
60H-slag	15.95
80H-slag	20.70
100H-slag	15.07

In conclusion, cubic C_3AH_6 was a main hydration product for the 40H-slag, 60H-slag, and 100H-slag. Part of C_3AH_6 could react with $CaCO_3$ to generate CCA while the unreacted rhombohedral-shaped $CaCO_3$ layer covered the slag surface, resulting in the hindrance of further hydration for these three slags. By contrast, flake-shaped $C_4A\overline{C}H_{11}$ and C_3AH_x were the main hydration products for the 20H-slag and the 80H-slag, respectively, and their higher specific surface area may accelerate the carbonation reaction.

3.2. Carbonation Efficiency of Slags Pretreated at Different Temperatures

The XRD patterns of the slags after carbonation are shown in Figure 5. After carbonation, the peaks of C_3AH_x, $C_4A\overline{C}H_{11}$, and hydrocalumite disappeared or showed a significant decrease in intensity while the calcite peaks increased in intensity. This suggests the carbonation of these hydrates. On the contrary, C_3AH_6 appeared less active in terms of its carbonation, which may be the main cause of the less intense calcite peaks in the carbonated100H-slag with C_3AH_6 as the main hydration product (see Figure 1). In addition, residual $C_4A\overline{C}H_{11}$ peaks were observed, indicating the incomplete carbonation of $C_4A\overline{C}H_{11}$ in the 20H-slag during carbonation. This resulted in the less intense calcite peaks in the carbonated 20H-slag compared with the carbonated 80H-slag.

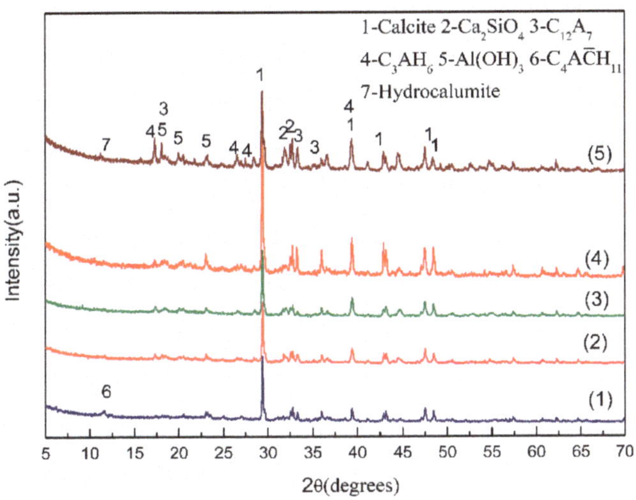

Figure 5. XRD patterns of (1) carbonated 20H-slag, (2) carbonated 40H-slag, (3) carbonated 60H-slag, (4) carbonated 80H-slag, and (5) carbonated 100H-slag.

Figure 6 exhibits the FESEM pictures of the slags after carbonation. For the carbonated 20H-slag, flake-shaped $C_4A\overline{C}H_{11}$ was decomposed and cubic calcite crystals were observable (Figure 6a). In the carbonated 40H-slag and 60H-slag, some of the cubic calcite crystals were surrounded by unreacted hydrates (Figure 6b,c). Larger amounts of cubic calcite crystals appeared in the carbonated 80H-slag than in the other slags, leading to the breakdown of continuous C_3AH_x gel (Figure 6d). Moreover, the carbonation products were covered by a small amount of unreacted C_3AH_x debris (Figure 6d). The microstructure of the carbonated 100H-slag (Figure 6e) was similar to the carbonated 40H-slag and 60H-slag; it was composed of unreacted hydration products, unhydrated slag, and some cubic calcite crystals. In each of the slags, the cubic calcite was generated by the direct reaction of CO_2 with the hydrates and amorphous AH, as the other reaction product surrounded the cubic calcite crystals [11]. In each of the carbonated slags, calcite appeared as non-uniform aggregated crystal particles, which indicate direct carbonation [21]. Direct carbonation of alkaline slag involved two stages: CO_2 dissolution and carbonation reaction [21,22]. The simplified direct carbonation mechanism of slags with hydrothermal pretreatment in this study was summarized in Table 4.

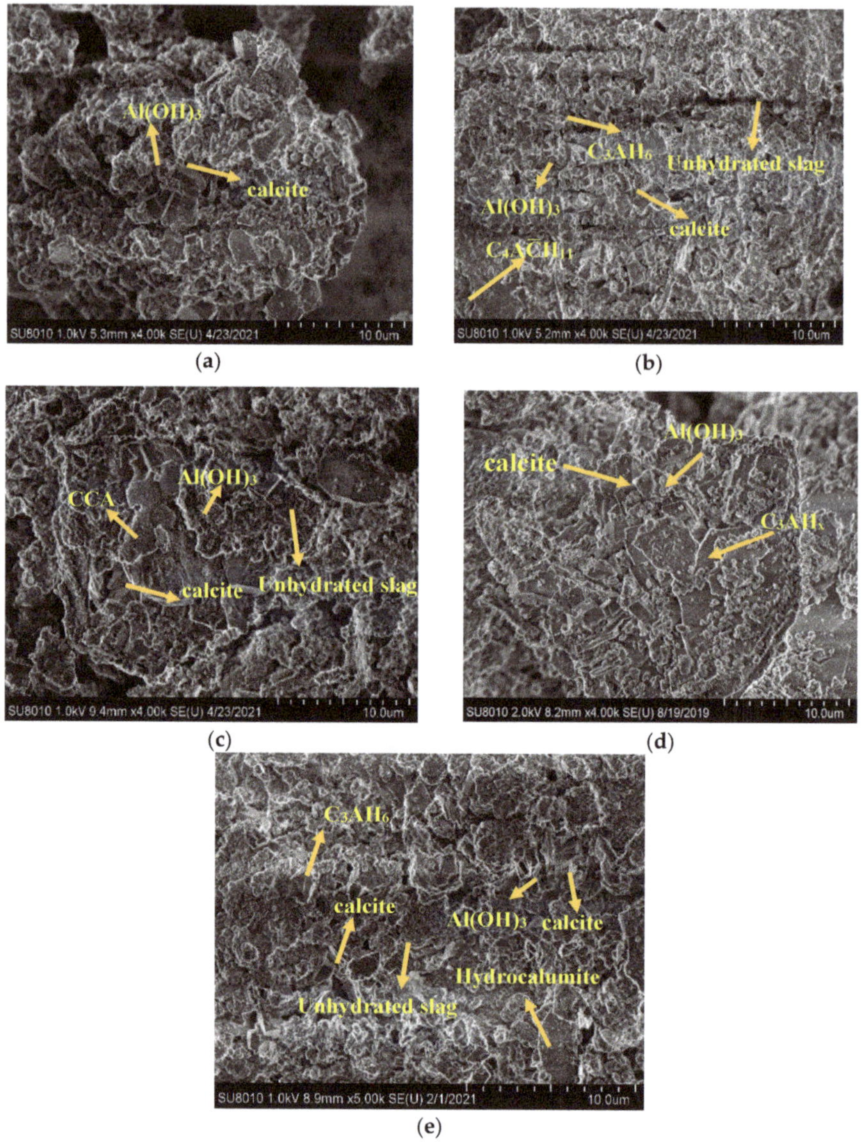

Figure 6. FESEM pictures of (**a**) carbonated 20H-slag, (**b**) carbonated 40H-slag, (**c**) carbonated 60H-slag, (**d**) carbonated 80H-slag, and (**e**) carbonated 100H-slag.

Table 4. Chemical reaction equations of the direct carbonation of the pretreated slags in this study.

Stage	Chemical Reaction Equation
CO_2 dissolution	$CO_{2(g)} \rightarrow CO_{2(aq)}$ $CO_{2(aq)} + 2OH^-_{(aq)} \rightarrow CO_3{}^{2-}{}_{(aq)} + H_2O_{(l)}$
Carbonation	$Ca_3Al_2O_6 \cdot 6H_2O_{(s)} + 3CO_3{}^{2-}{}_{(aq)} \rightarrow 3CaCO_{3(s)} + 2Al(OH)_{3(s)} + 6OH^-_{(aq)}$ $Ca_3Al_2O_6 \cdot xH_2O_{(s)} + 3CO_3{}^{2-}{}_{(aq)} \rightarrow 3CaCO_{3(s)} + 2Al(OH)_{3(s)} + 6OH^-_{(aq)} + (x-6)H_2O(l)$ $3CaO \cdot Al_2O_3 \cdot CaCO_3 \cdot 11H_2O_{(s)} + 3CO_3{}^{2-}{}_{(aq)} \rightarrow 4CaCO_{3(s)} + 2Al(OH)_{3(s)} + 6OH^-_{(aq)} + 5H_2O(l)$

Based on the TG curves of the carbonated slags (Figure 7), CO_2 uptake (wt%) was calculated with Equation (1) where m was the dry weight at 325 °C (at this temperature, free water and chemically bound water evaporated). The results of CO_2 uptake were listed in Table 5. The CO_2 uptake of slags followed this order: 80H-slag (13.66 wt%) > 20H-slag (10.82 wt%) > 60H-slag (9.83 wt%) > 40H-slag (9.21 wt%) > 100H-slag (6.84 wt%), which corresponded well with the XRD and FESEM results. This is attributed to the following reasons: (1) a dense $CaCO_3$ or AH gel layer covered the unhydrated slag surface in the 40H-slag and the 60H-slag, therefore resulting in the hindrance of further hydration and carbonation; (2) flake-shaped hydrates such as $C_4A\overline{C}H_{11}$ in the 20H-slag or C_3AH_x in the 80H-slag provided a larger reaction surface aera than the cubic C_3AH_6 and raw slag particles, avoiding calcite and AH gel layer formation on the unreacted hydrates surface. In short, the carbonation efficiency was strongly dependent on the type and morphology of the hydrates of LF slag.

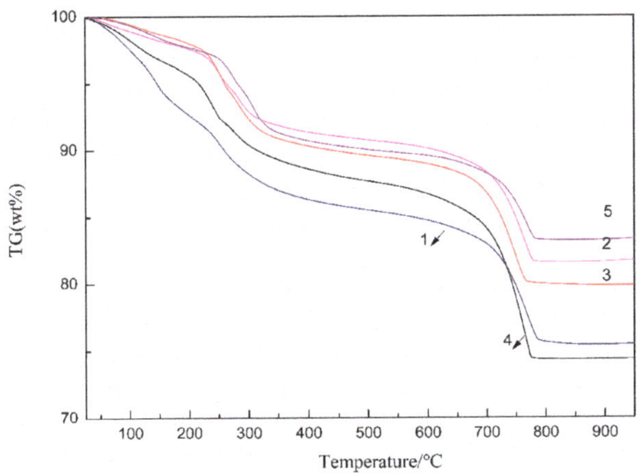

Figure 7. TG curves of (1) carbonated 20H-slag, (2) carbonated 40H-slag, (3) carbonated 60H-slag, (4) carbonated 80H-slag, and (5) carbonated 100H-slag.

Table 5. CO_2 uptake of pretreated slags.

	20H-Slag	40H-Slag	60H-Slag	80H-Slag	100H-Slag
CO_2 Uptake(wt%)	10.82	9.21	9.83	13.66	6.84

The maximum CO_2 uptake among these five pretreated slags was 13.66 wt% (namely 136.6 g of CO_2/1 kg of slag) which was attained by the 80H-slag. Compared with previous studies about wet direct carbonation of steelmaking slags conducted under ambient temperature and pressure (See Table 6), the CO_2 uptake of the 80H-slag in this study was considerable and the process was attractive. With an annual (2019–2020) output of LF slag of about 5 MT in China, this waste could, under 80 °C hydrothermal pretreatment, capture about 0.67 MT CO_2 if the mineral carbonation process is applied in steel plants.

Table 6. Comparison of current with previous steelmaking slag carbonation studies conducted under ambient temperature and pressure.

Slag Type	CO_2 Uptake (g of CO_2/1 kg of Slag)	Conditions	Reference
Al-rich LF slag	136.6	Temperature: 40 °C S/W ratio: 1:10 Reaction time: 1 h	This study

Table 6. *Cont.*

Slag Type	CO_2 Uptake (g of CO_2/1 kg of Slag)	Conditions	Reference
EAF steel slag	87	Temperature: 20 °C S/W ratio: 1:10 Reaction time: 37 min	[23]
BOF steel slag	116.4	Temperature: 70 °C S/W ratio: 1:2 Reaction time: 2 h	[24]
LF slag	67	Temperature: 20 °C S/W ratio: 1:20 Reaction time: 1 h	[25]
BOF steel slag	168.32	Temperature: 60 °C S/W ratio: 1:30 Reaction time: 10 h	[26]
LF slag	56	Temperature: 20 °C S/W ratio: 1:5 Reaction time: 70 min	[27]
BOF steel slag	215	Temperature: 60 °C S/W ratio: 1:5 Reaction time: 3 h	[28]

4. Conclusions

This study investigated the temperature of hydrothermal pretreatment on the hydrate formation and carbonation efficiency of Al-rich LF slag at ambient temperature and pressure. The main results are as follows:

During hydrothermal pretreatment, cubic C_3AH_6 was a main hydration product for 40 °C, 80 °C, and 100 °C-pretreated slags while $C_4A\overline{C}H_{11}$ and C_3AH_x with flaked shapes were the main hydrates for 20 °C and 80 °C-pretreated slags, respectively. Rhombohedral-shaped $CaCO_3$ was generated by the reaction between $C_{12}A_7$ in the slag and CO_2 in the air; and then $CaCO_3$ reacted with C_3AH_6 to form flake-shaped CCA. Flake-shaped products presented higher BET specific surface area. In 40 °C, 60 °C, and 100 °C-pretreated slags, a dense $CaCO_3$ layer surrounded the unreacted core of the slag particle, resulting in the hindrance of further $C_{12}A_7$ hydration.

Flake-shaped products could provide a lager reaction surface area and avoid the calcite and AH gel layer formation on the surface of the unreacted hydrates. Therefore, 80 °C and 20 °C-pretreated slags containing a larger number of flake-shaped hydrates had larger CO_2 uptake (13.66 wt% and 10.82 wt%, respectively). Cubic C_3AH_6 crystal and unhydrated raw slag particles were less inactive for carbonation, resulting in the smaller CO_2 uptake for 40 °C, 60 °C, and 100 °C-pretreated slags (9.21 wt%, 9.83 wt% and 6.84 wt%, respectively). In short, the carbonation efficiency of the pretreated slag was strongly associated with the morphology of the hydration products and the hydration extent of LF slag.

Author Contributions: Investigation, Y.H.; resources, G.X.; writing—original draft preparation, G.X.; writing—review and editing, Y.H.; supervision, G.X.; funding acquisition, Y.H. Both authors have read and agreed to the published version of the manuscript.

Funding: This work was funded by the Natural Science Foundation of Hunan Province, China (Grant No. 2020JJ4157).

Data Availability Statement: The data presented in this study are available on request from the corresponding author. The data are not publicly available due to patent issues.

Acknowledgments: The authors thank the support from the Research Foundation of the Natural Science Foundation of Hunan Province, China (Grant No. 2020JJ4157).

Conflicts of Interest: The authors declare no conflict of interest.

References

1. International Energy Agency. *Global Energy & CO_2 Status Report: The Latest Trends in 2018*; International Energy Agency: Pairs, France, 2019.
2. Sun, Y.; Yao, M.S.; Zhang, J.P.; Yang, G. Indirect CO_2 mineral sequestration by steelmaking slag with NH_4Cl as leaching solution. *Chem. Eng. J.* **2011**, *173*, 437–445. [CrossRef]
3. Rushendra Revathy, T.D.; Palanivelu, K. Direct mineral carbonation of steelmaking slag for CO_2 sequestration at room temperature. *Environ. Sci. Pollut. Res.* **2016**, *23*, 7349–7359. [CrossRef] [PubMed]
4. Liu, W.Z.; Yin, S.; Luo, D.M.; Zhang, G.Q.; Yue, H.R.; Liang, B.; Wang, L.M.; Li, C. Optimizing the recovery of high-value-added ammonium alum during mineral carbonation of blast furnace slag. *J. Alloys Compd.* **2019**, *774*, 1151–1159. [CrossRef]
5. Rushendra Revathy, T.D.; Ramachandran, A.; Palanivelu, K. Sequestration of CO_2 by red mud with flue gas using response surface methodology. *Carbon Manag.* **2021**, *12*, 139–151. [CrossRef]
6. Wang, Y.J.; Zeng, Y.N.; Li, J.G.; Zhang, Y.Z.; Zhao, Q.Z. Carbonation of argon oxygen decarburization stainless steel slag and its effect on chromium leachability. *J. Clean. Prod.* **2020**, *256*, 120377. [CrossRef]
7. Ho, H.J.; Lizuka, A.; Shibata, E. Utilization of low-calcium fly ash via direct aqueous carbonation with a low-energy input: Determination of carbonation reaction and evaluation of the potential for CO_2 sequestration and utilization. *J. Waste Manag.* **2021**, *288*, 112411.
8. Liu, W.; Teng, L.; Rohani, S.; Qin, Z.; Zhao, B.; Xu, C.C.; Ren, S.; Liu, Q.; Liang, B. CO_2 mineral carbonation using industrial solid wastes: A review of recent developments. *J. Chem. Eng.* **2021**, *416*, 129093. [CrossRef]
9. Ibrahim, M.H.; El-Naas, M.H.; Benamor, A.; Al-Sobhi, S.S.; Zhang, Z. Carbon mineralization by reaction with steel-making waste: A review. *Processes* **2019**, *7*, 115. [CrossRef]
10. Santos, R.M.; François, D.; Mertrn, G.; Elsen, J.; Gerven, T.V. Ultrasound-enhanced mineral carbonation. *Appl. Therm. Eng.* **2013**, *157*, 109–116.
11. Huang, Y.; Zeng, Z. Improvement of desulfurization efficiency of Al-rich ladle furnace refining slag with an aqueous carbonation method by hydrothermal or ultrasound pretreatment. *Environ. Sci. Pollut. Res.* **2021**, *28*, 27703–27711. [CrossRef]
12. Koplík, J.; Tomala, L.; Novotný, R. Hydration of calcium aluminate phases at different temperatures. *Adv. Mater. Res.* **2014**, *1000*, 24–27. [CrossRef]
13. Edmonds, R.N.; Majumdar, A.J. The hydration of $12CaO.7Al_2O_3$ at different temperatures. *Cem. Concr. Res.* **1988**, *18*, 473–478. [CrossRef]
14. El-Naas, M.H.; Maisa, E.G.; Hameedi, S.; Mohamed, A.-M.O. CO_2 sequestration using accelerated gas-solid carbonation of pre-treated EAF steel-making bag house dust. *J. Environ. Manag.* **2015**, *156*, 218–224.
15. Wang, Z.P.; Yang, H.Y.; Zhao, Y.T.; Lu, L.L.; Jia, D.L. Delaying Effect of NaCl on the conversion of hydrates of calcium aluminate cement. In Proceedings of the 11th International Conference on High-Performance Ceramics, Kunming, China, 25–29 May 2019.
16. Ewa, L.; Dominika, M. Structure, microstructure and thermal stability characterizations of C_3AH_6 synthesized from different precursors through hydration. *J. Therm. Anal. Calorim.* **2020**, *139*, 1693–1706.
17. Liu, W.; Zhang, H.; Zhou, Q.S.; Peng, Z.H.; Qi, T.G.; Li, X.B.; Liu, G.H. Reaction tricalcium aluminate hexahydrate(C_3AH_6) with carbon dioxide. *J. Cent. South Univ.* **2011**, *42*, 595–598.
18. Xiao, J.; Gou, F.; Jin, Y.G.; Wang, Y.H. Effect of $CaCO_3$ on hydration characteristics of C_3A. *J. Cent. South Univ. Technol.* **2010**, *17*, 918–923. [CrossRef]
19. Luz, A.P.; Pandolfelli, V.C. $CaCO_3$ addition effect on the hydration and mechanical strength evolution of calcium aluminate cement for endodontic applications. *Ceram. Int.* **2012**, *38*, 1417–1425. [CrossRef]
20. Feng, Y.B.; Tang, C.M.; Qiu, T. Effect of ball milling and moderate surface oxidization on the microwave absorption properties of FeSiAl composites. *Mater. Sci. Eng. B* **2013**, *178*, 1005–1011. [CrossRef]
21. Sun, Y.; Yang, G.; Li, K.; Zhang, L.C.; Zhang, L. CO_2 mineralization using basic oxygen furnace slag: Process optimization by response surface methodology. *Environ. Earth. Sci.* **2016**, *75*, 1335. [CrossRef]
22. Yi, Y.R.; Han, M.F. The reactivity of carbon dioxide capture with calcium-based waste solid by wet process. *J. China Coal Soc.* **2012**, *37*, 1205–1210.
23. Uibu, M.; Kuusik, R.; Andreas, L.; Kirsimäe, K. The CO_2-binding by Ca–Mgsilicates in direct aqueous carbonation of oil shale ash and steel slag. *Energy Procedia* **2011**, *4*, 925–932. [CrossRef]
24. Wang, C.Y.; Bao, W.J.; Guo, Z.C.; Li, H.Q. Carbon Dioxide Sequestration via Steelmaking Slag Carbonation in Alkali Solutions: Experimental Investigation and Process Evaluation. *Acta Metall. Sinica* **2018**, *31*, 771–784. [CrossRef]
25. Paris, K.A.; Hills, C.D.; Maries, A.; Gunning, P.J.; Wray, D.S. Enhancement of accelerated carbonation of alkaline waste residues.by ultrasound. *Waste Manag.* **2016**, *50*, 121–129.
26. Chen, Z.M.; Li, R.; Zheng, X.M.; Liu, J.X. Carbon sequestration of steel slag and carbonation for activating RO phase. *Cem. Concr. Res.* **2021**, *139*, 106271. [CrossRef]
27. Yi, Y.R.; Lin, Y.; Du, Y.C.; Bai, S.Q.; Ma, Z.L.; Chen, Y.G. Accelerated carbonation of ladle furnace slag and characterization of its mineral phase. *Cem. Concr. Res.* **2021**, *276*, 122235.
28. Chen, K.W.; Pan, S.Y.; Chen, C.T.; Chen, Y.H.; Qiang, R.C. High-gravity carbonation of basic oxygen furnace slag for CO_2 fixation and utilization in blended cement. *J. Cleaner Prod.* **2016**, *124*, 350–360. [CrossRef]

processes

MDPI

Review

Analysis of the Status and Improvement of Microalgal Phosphorus Removal from Municipal Wastewater

Yilin Mao [1], Rongwei Xiong [1], Xiufang Gao [1,2,*], Li Jiang [1], Yancong Peng [1] and Yan Xue [1]

[1] College of Resources and Environment, Yangtze University, Wuhan 430100, China; 13995992011@163.com (Y.M.); xiongrw1225@163.com (R.X.); 15038742477@163.com (L.J.); 15392973742@163.com (Y.P.); XY15122677013@163.com (Y.X.)

[2] Engineering Research Center of Ecology and Agricultural Use of Wetland, Ministry of Education, Jingzhou 434025, China

* Correspondence: gxf20210703@163.com; Tel.: +86-132-1271-9979

Abstract: Phosphorus, as one of the main pollutants in municipal sewage, has received increasing attention recently. Phosphorus recovery also increases the sustainable development of municipal wastewater. Since algae have the ability to effectively redirect nutrients, including phosphorus, from municipal sewage to algae biomass, municipal sewage treatments involving microalgae have piqued the interest of many researchers. The phosphorus removal depends on the potential of the microalgae to absorb, preserve, or degrade phosphorus in municipal wastewater. It is, therefore, of great interest to study the mechanisms underlying the absorption, storage, and degradation of phosphorus by microalgae to ensure the viability of this phosphorus removal process in wastewater. The objectives of this review were to summarize phosphorus metabolism in microalgae, examine key external and internal factors impacting phosphorous removal by microalgae from wastewater, and examine the status of phosphorous-metabolism-related research to improve our understanding of microalgae-based municipal wastewater treatments. In addition, the methods of recovery of microalgae after phosphorous removal were summarized to ensure the sustainability of municipal wastewater treatment. Finally, a potential approach using nanomaterials was proposed to enhance the overall phosphorous removal performance in municipal wastewater through the addition of nanoparticles such as magnesium and iron.

Keywords: microalgae; municipal wastewater; phosphorus removal; immobilization and recycle technology

Citation: Mao, Y.; Xiong, R.; Gao, X.; Jiang, L.; Peng, Y.; Xue, Y. Analysis of the Status and Improvement of Microalgal Phosphorus Removal from Municipal Wastewater. *Processes* **2021**, *9*, 1486. https://doi.org/10.3390/pr9091486

Academic Editor: Avelino Núñez-Delgado

Received: 8 June 2021
Accepted: 18 August 2021
Published: 24 August 2021

1. Introduction

Cities are generally densely populated and have great demands for water. For some cities, the efficient treatment of urban sewage to achieve water recycling has become a strategic solution to water shortages [1]. In China, the amount of wastewater treated has increased by more than 50% in the last decade, and the increase in the volume of wastewater has significantly endangered human and environmental welfare [2]. The fecal sludge collected by the sewage systems in some industrial cities is untreated, causing significant damage to water supplies across such cities [3]. While urban sewage treatment in developed European countries strongly reduces pollution in wastewater, the amount of sludge generated is still growing year on year. Finding methods to efficiently treat sludge is an urgent challenge [4]. In comparative studies and simulation forecasts regarding the statuses of cities in Nanjing (domestic), South Asia, and Southeast Asia, the results indicated that by 2030 the water quality in samples sites in Manila and Jakarta will have deteriorated further [5].

Phosphorus is an essential basic element in organisms that is also commonly present in water bodies and primarily occurs in a dissolved form and in association with other particles via chemical adsorption; however, regardless of whether it originates from the natural world or via human activities, phosphorus significantly impacts the environment [6]. Excessive phosphorus in a water body, for example, accelerates the growth of algae and other microorganisms, subsequently creating nutrient imbalance in the water body and accelerating eutrophication [6]. Some phosphorous-containing compounds in water can be converted into minerals and may block water supply lines and sewage treatment facilities [7,8]. In municipal drainage systems, phosphorus originates primarily from domestic sewage, with 85% of the total phosphorus entering the wastewater from domestic sewage containing human excrement and detergents [6]. Phosphorus-containing sewage compounds in urban life exist mainly in the form of orthophosphoric acid, tripolyphosphate, and pyrophosphate [9,10]. Modern technologies used for the elimination of phosphorus are primarily categorized into biological, chemical, and physical technologies [11]. The removal and recovery of phosphorus from municipal sewage not only provides conditions for the utilization of phosphorus resources, but also reduces the eutrophication and increases the sustainability of municipal sewage; however, many conventional phosphorous reduction processes used for urban sewage entail high running and maintenance costs and are not sustainable, and may even cause other pollution to the water body [12,13].

Phosphorus also has numerous constructive uses. For example, certain phosphates can be applied to water supply pipes, where they combine with heavy metals in rainwater, lowering the concentration of heavy metals in the water while still acting as a corrosion inhibitor [14,15]. To avoid direct interactions between the metal and the food in metal food containers, phosphides are often added to form an inert coating [16]. Furthermore, household detergents, toothpaste, and shampoos are often also incorporated phosphorous compounds to improve their washing efficiency, with sodium tripolyphosphate being the most popular phosphorous-containing compound in detergents [17]. The overall amount of phosphorus in the world is small and nearly 40 million tons of phosphorus is absorbed worldwide per year, making the recycling of phosphorus critically significant [18].

CiteSpace is a tool used for the analysis of scientific literature. It can help an author to explore research hotspots and research frontiers in a certain research field and to predict future development trends. The Web of Science database was used to perform a comprehensive study of the literature, using the search terms "microalgae" and "phosphorous removal". In total, 678 manuscripts published in the past five years were retrieved and a visual study of the keywords in the manuscripts was carried out using CiteSpace (version 5.7.R2), with the results shown in Figure 1. Based on the findings of the search, CiteSpace was used again to analyze clusters, with the results shown in Figure 2. Figures 1 and 2 demonstrate that the study hotspots regarding the reduction of phosphorus by microalgae focus primarily on the treatment of municipal sewage, recovery, and resource utilization. The results of this analysis provided theoretical significance for this review of relevant research frontiers regarding the treatment of phosphorus in urban sewage and the prediction of future development trends for microalgal phosphorus removal technology.

This review summarizes phosphorus elimination from municipal wastewater by microalgae and analyzes the factors influencing this phosphorous removal and the associated methods. The aim of this review is to further understand the process of phosphorus removal by microalgae in municipal wastewater, so as to optimize the process and sustainability. In this paper, the problems related to microalgae dephosphorization are introduced for researchers who are interested in municipal wastewater dephosphorization technology.

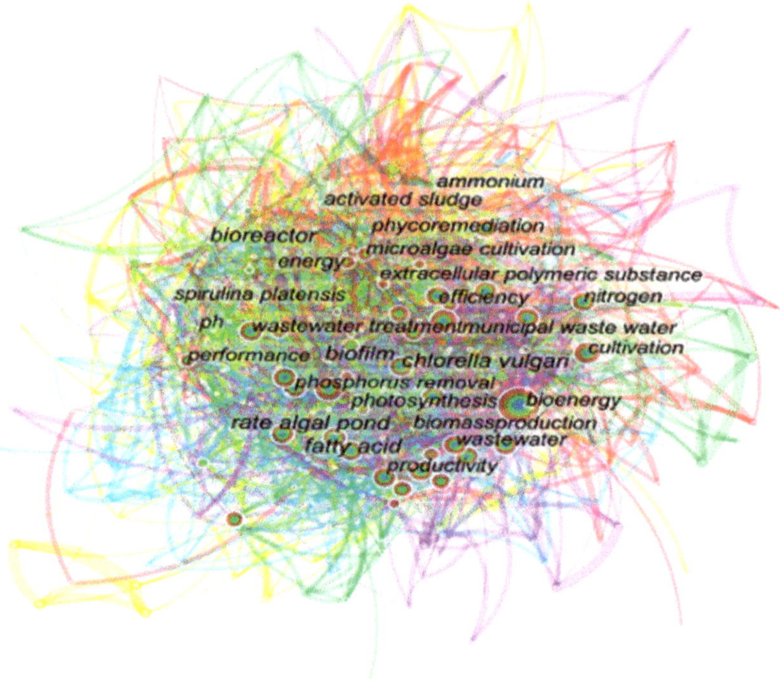

Figure 1. CiteSpace visual analysis diagram.

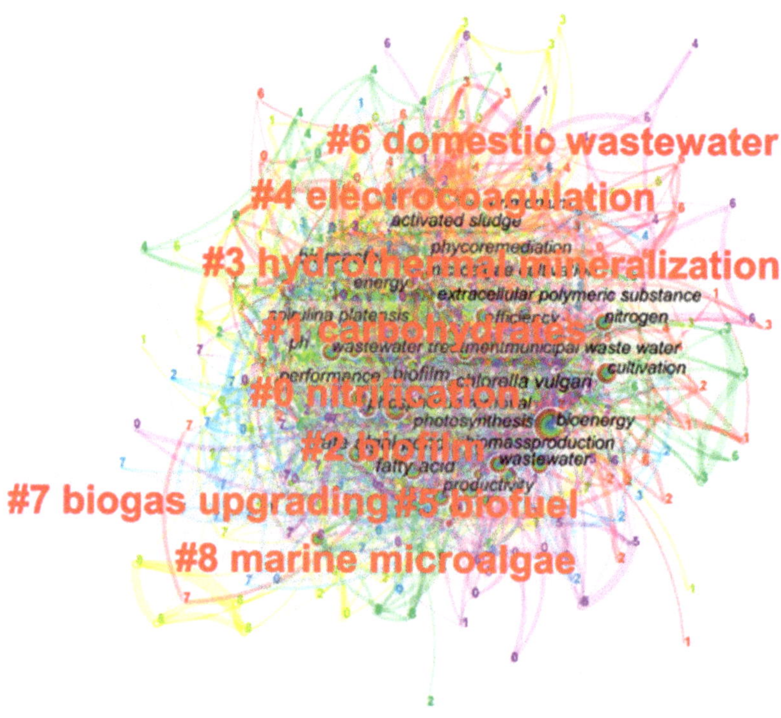

Figure 2. CiteSpace cluster analysis diagram.

2. Various Systems and Strategies for the Reduction of Phosphorus in Municipal Wastewater

Phosphorus removal from municipal sewage is mainly categorized into three categories of processes: physical, chemical, and biological. The respective technological methods and respective advantages and drawbacks of each process are shown in Table 1. Physical phosphorus removal technologies can eliminate all forms of particulate phosphorus compounds [19]. Membrane technology not only extracts phosphorus from complete suspended solids but also removes dissolved phosphorus. Membrane bioreactors (MBRs) and reverse osmosis (RO) devices with good phosphorous removal capacity have been widely used in full-scale sewage treatment plants [20]. The fundamental theory behind the elimination of chemical phosphorus is to crystallize or condense phosphorous compounds by adding chemical agents or by modifying certain reaction conditions. For example, in the coagulation and flocculation process, phosphorous-containing materials are flocculated by the addition of polymers or metal ions. This approach is effective for removing larger molecules, and its quality is determined by the charge of the salt ions [21]. Since biological phosphorus removal is used to treat municipal sewage, anaerobic or anoxic treatment is usually required first, followed by aerobic treatment and other procedures to remove phosphorus from activated sludge in municipal sewage. Microorganisms have been shown in studies to have the largest reduction effect on total phosphorus under anaerobic conditions, with removal rates exceeding 80–90% [21]. Compared to the large operational and repair costs of conventional physical and chemical phosphorous removal systems and the complexities of certain biological treatment processes, the use of microalgae has proved to be a cost-effective and long-term alternative for biological phosphorus removal, which is now commonly utilized [22]. The following is an introduction to the microalgae dephosphorization technology mechanism.

2.1. Microalgae Culture Methods

There are two types of microalgae culture structures: open and closed systems. The term "open systems" refers to growing systems of outdoor waters, such as lakes and reservoirs. In order to provide adequate light for the microalgae, the system's water depth is usually no greater than 0.5 m [23]. While the open system layout is simple and easy to manage, nutrients can become diluted by pollution when exposed to the open environment over a long period [24]. Closed devices are segregated from the external environment, thereby shielding the system from the harmful effects of the external environment. Photobioreactors (PBRs) are widely used as closed structures for microalgae cultivation. PBRs are usually classified into stubs and flat plates, and are generally made from glass or plastic, air, carbon dioxide (CO_2), and other gases may be fed into the PBRs [23,25]. On the basis of being highly controlled, the closed method can be used to evaluate the characteristics of microalgae and the effects on the purification of wastewater under different sets of conditions. It also provides a culture system for improving the conditions for the absorption of phosphorus by microalgae in urban wastewater.

Microalgae culture modes can be separated into continuous and semi-continuous batch modes [24]. A lot of the management costs are avoided in closed batches because the culture material does not always need to be replaced. The growth of the microalgae, however, will be inhibited if the nutrients in the batch system are depleted or if certain factors occur, such as cell self-shading, pH variations, and contamination, inhibiting the growth. In addition, the device must ensure a successful exchange of gas [23]. Compared to the batch model, the semi-continuous model can achieve higher biomass despite the need for periodic substitution of culture material and the continuous removal of wastewater [26].

2.2. Phosphorus Uptake and the Metabolism Mechanism of Microalgae

The absorption and metabolism of phosphorus by microalgae are often distinct for different types of phosphorus or under different environmental conditions. Microalgae can induce phosphatase to absorb external organophosphorus and synthesize high-affinity inorganic phosphorus transporters to assist inorganic phosphorus absorption [27]. The absorption of inorganic phosphorus also relies on the inorganic phosphorus charge and the pH of the microalgae cell membrane [28–30]. In general, the lower the molecular charge, the higher the bioavailability of inorganic phosphorus for the microalgae [28]. Most microalgae assimilate inorganic salts such as HPO^-, HPO^{2-}, and $PO_4{}^{3-}$ [31].

Polyphosphates include acid-soluble and acid-insoluble polyphosphates. Although certain microalgae do not use polyphosphates as their primary supply of phosphorus [32], in the absence of phosphorus, microalgae can assimilate and metabolize polyphosphate [31]. In addition, under the condition of excess inorganic phosphorus, microalgae can take up excess phosphorus and deposit it in the form of insoluble polyphosphate acid, where it can be used for cell metabolism when inorganic phosphorus is lacking [33]. Excessive phosphorus and high light intensity in municipal wastewater tend to facilitate the removal of phosphorus by microalgae [31,34]; however, some studies have shown that excessive phosphorus can impede the growth of some microalgae due to excessive accumulation of polyphosphate in the cells [35].

Photosynthesis is the basis of the metabolism of microalgae. Over the entire photosynthesis process, phosphorus is required in the reaction that produces the energy substance ATP. The equation for this reaction is as follows:

$$2\,H_2O + 2\,NADP^+ + 3\,ADP + 3\,P_i + 8\text{-}10\text{ Photons (light)} \rightarrow 2\,NADPH + 2\,H^+ + 3\,ATP + O_2$$

To summarize, the electrons in the water are transferred to $NADP^+$ after the absorption of light energy by the microalgae. H^+ formed by water allows ADP and inorganic phosphorus to form ATP on the thylakoid membrane in the cell [31]. Phosphorus is also important for the synthesis of DNA, RNA, and cell membranes [36].

Table 1. Different physical, biological, and chemical phosphorus removal technologies used in wastewater treatment processes [20,37–39].

Methods	Technologies	Advantages	Disadvantages
physical methods	physical absorption	widely used for phosphorus removal	not yet perfect for phosphorus adsorption
	sand filtration	removes all P compounds	only for the primary stage
	the membrane purification	simple and efficient	high operation and maintenance costs
	ion exchange	can treat hazardous waste and higher concentrations of phosphorus	lack of selectivity for specific ions and complex process
chemical methods	by precipitation of metal salts and lime	high phosphorus removal efficiency and economical	may cause secondary contamination
	crystal	reusable, little environmental harm	need to add chemicals and low stability
	Coagulation and flocculation	can be used for reaction by adding metal ions such as polymers or aluminum	need high charge for salt ions
biological methods	artificial aeration	mainly used for dephosphorization of lakes	no significant effect in shallow lakes
	enhanced biological phosphorus removal	no chemicals need to be added	low stability and biological population competition
	photosynthetic microorganisms immobilized on cellulose, ceramic, or gel carriers	can effectively immobilize and remove more than one type of microorganism or contaminant	not easily removed for most phototrophs
	phosphoric acid binds proteins	can work in low phosphorus environments	the use of this protein is limited

3. Factors Impacting the Elimination of Phosphorus from Municipal Wastewater by Microalgae

There are several internal and external factors influencing the removal of phosphorus by microalgae, such as the temperature, strength and period of illumination, and pH [31,40]. These common factors have been extensively studied and summarized, although a review regarding their impacts on the treatment of urban sewage by microalgae is lacking. In this chapter, the impacts of microalgae on the removal of phosphorus in municipal wastewater will be examined in terms of the hydraulic retention time, ratio of nitrogen to phosphorus, carbon dioxide concentration, species of microalgae, and different types of wastewater.

3.1. Hydraulic Retention Time

The hydraulic retention time (HRT) for microalgae in bioreactors influences their growth and phosphorous removal performance. An appropriate HRT not only improves the efficiency of the microalgae wastewater treatment but also reduces the operating and maintenance costs of the system. In the batch regime pilot scale photobioreactor system, activated microalgae sludge is used for the treatment of municipal sewage. Findings have shown that when the hydraulic retention time is 2–6 days, the phosphorous removal efficiency of the microalgae is improved by around 30% to 90% [41]; however, owing to the deterioration of the switch virtual interface (SVI) and other factors in the later stage, the phosphorous removal performance of the microalgae does not improve with the increase in hydraulic retention time [41]. Similarly, the optimum hydraulic retention time in the high-level algae pond also occurs on the sixth day when the bacteria and algae system is used to treat sewage [42].

The short HRT means that ammonia nitrogen ions in steam pools used for wastewater treatment cannot be fully nitrified [43–45]. By separating HRT and sludge retention times (SRT), a next-generation anaerobic–aerobic algal bioreactor was developed to solve the problem of inadequate HRT [44–46]. Toledo-Cervantes et al. [45] used a new form of photobioreactor involving hypoxia–aerobic algae to investigate the removal of phosphorus in water under varying hydraulic retention periods. Their findings revealed that when the HRT of the bioreactor decreased from 4 to 2 days, the removal rate of P-PO_4^{3-} decreased from about 22% to approximately 11%. A schematic diagram of the photobioreactor involving anoxic–aerobic algae–bacteria is shown in Figure 3. Similarly, when a mixed-culture microalgae membrane bioreactor was used to treat secondary wastewater from municipal wastewater, the optimal retention time was decreased to 1 day [39]. In general, the optimum hydraulic retention time for the removal of phosphorus by microalgae is approximately 6 days. By enhancing the treatment technology, such as the selection of a suitable bioreactor, the hydraulic retention time can be reduced, along with the running and repair costs of the treatment.

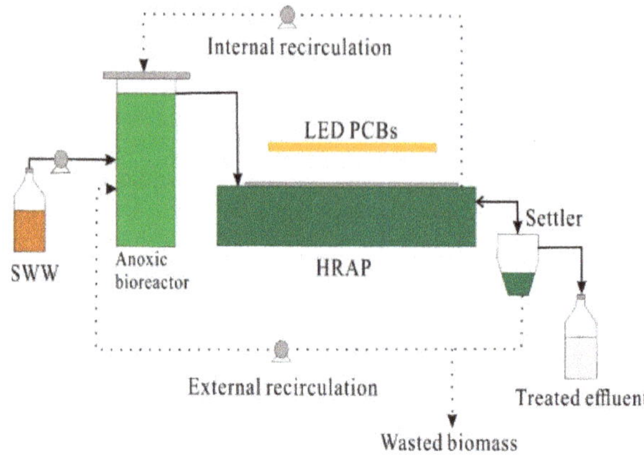

Figure 3. Schematic diagram of anoxic–aerobic algae–bacteria photobioreactor [45]. SWW: secondary wastewater; HRAP: open photobioreactor.

3.2. *The Ratio of Nitrogen to Phosphorus*

Nitrogen and phosphorus are important urban wastewater elimination indices, as well as important microalgae nutrient sources. An appropriate N/P ratio provides a good growth environment for microalgae and increases the phosphorous removal performance of the microalgae. The optimal N/P ratio range for microalgae development in freshwater ranges is 6.8 to 10 [47]. In the photobioreactor method, when microalgae are used to treat urban wastewater, the optimum N/P ratio range for the elimination of total phosphorus is 5–30 [48].

Molazadeh et al. [49] performed post-screening, biological treatment, and disinfection treatment of wastewater drained from sedimentation tanks of urban wastewater treatment plants. They controlled the N/P ratio and CO_2 concentration through the injection of potassium dihydrogen phosphate and used a high CO_2 concentration. The analyses showed that *Chlorella vulgaris* demonstrated strong potential to remove phosphorus under all concentrations of CO_2 and N/P, exhibiting a removal range of 70.0–96.0%. Under conditions with 16% CO_2 and a ratio of 10:1 N:P, algae biomass was the highest, with an increase in lipid productivity, which makes a powerful contribution to the eventual recovery of microalgae for biofuel [49]. To summarize, the reduction of total phosphorus from urban wastewater by microalgae is not only related to the required N/P ratio but is also directly proportional to the biomass of the microalgae [48]; however, as microalgae process urban waste, the feedbacks between the N/P ratio, the concentration of CO_2 in the water body, and the regulation of the optimum concentration remain unclear [49].

3.3. *Carbon Dioxide Concentration*

Carbon is the most fundamental component of living things. When microalgae are used to treat urban waste, the source of carbon comes not only from sewage but also from CO_2 in the air. Shanshan Ma [50] added 10% CO_2 mixed gas to unsterilized sewage to support *Tetradesmus obliquus* PF3 for the treatment of sewage and nutrient recovery. Compared to the addition of air, sewage with 10% CO_2 added shows greater TP (99 $\pm$ 0%) removal performance under unsterilized conditions. This is due to the increased supply of carbon and the high concentration of CO_2 changing the pH to an optimal growth range (6.8–7.8) [50]. Chaudhary et al. [51] used Chlorella ATCC13482 for the treatment of urban wastewater in bubble column photobioreactors at a volume of 7 liters. The findings revealed that the rate of microalgae orthophosphate elimination with 5% CO_2 air was as high as 92.8%.

Increasing CO_2 concentrations not only enhances phosphorous reduction by the microalgae but also increases the biomass of the microalgae. The higher the biomass of the microalgae, the higher the phosphorous removal performance [48]. Studies have shown that in the case of *Nannochloropsis* sp., where the concentration of CO_2 was 15%, the biomass of the microalgae and intracellular lipids was dramatically increased [52]. In general, high CO_2 concentrations not only boost the phosphorous removal performance of microalgae but also improve the lipid content of microalgae cells and increase the recovery value of the microalgae.

3.4. Species of Microalgae

The optimal growth conditions for each microalgae species are different, such that suitable algae species are chosen for different initial concentrations of urban waste in order to achieve the maximum benefit of phosphorous elimination; however, in experimental or practical applications, a single type of algae is rarely used to treat municipal wastewater.

By using mixed microalgae for the treatment of municipal wastewater, dominant algae species can be chosen on the basis of the sewage characteristics. For example, Toledo-Cervantes et al. [45] increased the rate of phosphate elimination from a water body from about 10% to around 50% by reducing the C/N ratio from 9 to 7. At the same time, *Chlorella vulgaris*, the dominant species of algae, was eventually replaced by *Phormidium* sp. [45]. This type of research approach can reliably and efficiently find appropriate microalgae for certain sewage treatment plants by screening the dominant algae species and adding mixed microalgae for realistic conditions.

More studies have shown that relative to single algae species in wastewater treatment, there is a cooperative or competitive partnership between mixed algae species, resulting in biodiversity and making the treatment system more stable and efficient. Paches et al. [53] performed batch and mixed cultures for four types of microalgae using anaerobic membrane bioreactors. Their findings showed that the mixed microalgae culture could increase the rate of phosphorous removal and the productivity of water by letting the species compete with each other [53]. Devi et al. [54] also revealed through their research that mixed microalgae showed a high degree of phosphorous elimination in wastewater and concluded that using mixed cultures was one of the better methods to handle municipal wastewater and other low-toxicity wastewater.

3.5. Different Municipal Wastewater Treatment Technologies

In a city sewage treatment facility, the municipal wastewater can be separated into three levels based on the extent of treatment, ranging from primary wastewater treatment, where the wastewater has not yet been deeply treated, up to tertiary wastewater treatment, in which the wastewater is at the final cleaning process. Although primary wastewater exhibits several negative factors, such as high optical density (OD) and bacterial contamination, the concentration of nutrient species in primary wastewater is much higher than in other wastewater treatment levels, making it more favorable for microalgae development [50,55]. Secondary wastewater is partially treated and most of the nitrogen at this stage is available as nitrate due to nitrification. This is a negative factor for microalgae, as microalgae preferentially absorb nitrogen in the form of ammonia [50]. Bellucci et al. [56] used microalgae to treat secondary wastewater and evaluated the combined function of microalgae as a disinfectant and nutrient remover. Their findings revealed that the microalgae contributed to an E. coli count equal to that of standard ultraviolet therapy in the batch disinfection test, and that the count was smaller than that of light experiments without microalgae. The E. coli population decreased by an order of magnitude in subsequent continuous studies. The rate of elimination of total phosphorus in the secondary wastewater was 100%. In addition to the level of municipal wastewater treatment, the forms of municipal wastewater can also be categorized according to the special new wastewater created by the treatment process. Various processes and techniques are used to treat different forms of wastewater with microalgae, as shown in Table 2.

Anaerobic digestion is a mechanism that converts polluted waste to energy materials; however, this method volatilizes harmful gasses such as high-viscosity, high-moderation, and highly volatile fatty acids. In addition, these reactive compounds are also poisonous to plants and microalgae. Products from untreated anaerobic digests cannot, therefore, be released immediately into nature [57]. The immediate discharge of untreated anaerobic digestive fluid into bodies of water can cause eutrophication [57]. The integrated technology of using microalgae to treat digested products is a technology that can offer economic gains, while also being environmentally sustainable [58]. In regards to the uptake of nitrogen and phosphate from wastewater, algae have demonstrated higher removal efficiency than other microorganisms [59]. Ermis et al. [60] used an experimental batch sequencing device to investigate the use of mixed microalgae in the treatment of anaerobic liquor digestion. The digestive juice was diluted to 2%, 5%, 7%, and 10%, so that the original concentrations of ammonia nitrogen and phosphate in the digestive juice were regulated at 18.6–87.1 mg L^{-1} and 1.85–6.88 mg L^{-1}, respectively. It was found that the absorption of nitrogen by mixed microalgae was 10 times greater than that of phosphorus. Based on a biokinetic coefficient of the phosphorus measurements, the reaction rate coefficient was 0.21 mg PO_4-P mg^{-1} chl a day^{-1} and the saturation constant was 2.94 mg L^{-1}, with a yield coefficient of 5.03 mg chl a mg^{-1} PO_4-P.

The main goal of treating eutrophic water bodies is to remove organic and inorganic compounds from the wastewater. Nitrogen and phosphorus, however, are not readily eliminated [61]. The utilization of photosynthetic–autotrophic digestion by microalgae means that CO_2 or inorganic carbon in the water or air can be used as a carbon source and source of energy. Autotrophic microalgae release extracellular organic matter (EOM) that converts inorganic carbon to organic carbon, increasing the concentration of organic carbon in the water [62]. While microalgae do not specifically remove organic matter from eutrophic water sources, they can be mixed with bacteria and other heterotrophic microorganisms to treat bodies of water with high amounts of organic matter; this specific topic will be detailed in the next chapter.

Sludge ozone technology can not only degrade several refractory organic compounds so that the production of the sludge can be decreased to 50–100%, but no harmful by-products are generated during the application of this technology [63,64]; however, owing to the high concentrations of nitrogen, ammonia, COD, and heavy metals in the excess ozone sludge, an additional burden is placed on the sewage treatment plant, which decreases the effectiveness of the sewage treatment [64–68]. Lei et al. verified the possibility of growing algae in sludge-concentrated wastewater ozone. While generating biomass, the microalgae can also extract nutrients from the water [64]. Their findings revealed that the bacteria–algae system had greater elimination effects in terms of total phosphorus removal than the pure microalgae system, with the systems showing 93% and 53.9% elimination effectiveness, respectively.

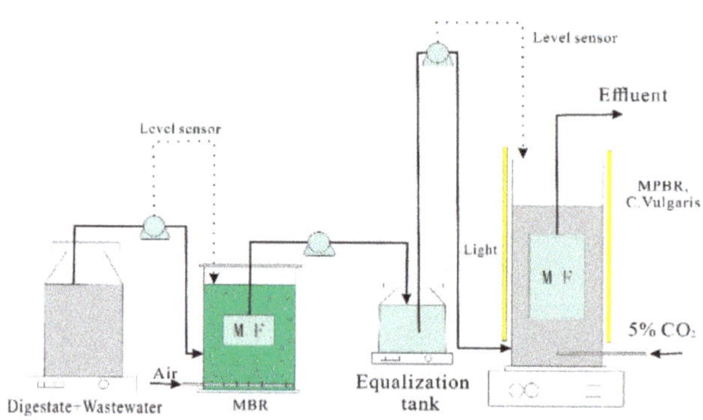

Figure 4. Schematic diagram of the MBR-MPBR experimental setup [72].

Table 2. Phosphorus removal effects of microalgae on municipal wastewater samples treated using different technologies.

Sewage Source	Microalgal Species	Initial Conditions	Experimental Conditions	Results	Cites	Notes
General municipal sewage	Mutant *Chlorella*	After 121 °C autoclave treatment COD_{Cr}: 190–230 mg/L TP: 4.5–5.6 mg/L TN: 40–60 mg/L NH_3-N: 20–35 mg/L pH: 6.6–7.6	a symbiotic system of PAOs and bacillariophyta	absorb 3.05 mg/L phosphorus, keep TP below 0.46 mg/L	[69]	
Synthetic domestic wastewater	*Chlorella vulgaris* and *Phormidium* sp.	COD: 632 ± 45 mg/L TOC: 196 ± 9 mg/L IC: 195 ± 12 mg/L TN: 43 ± 3 mg/L N-NH_4^+: 24 ± 3 mg/L P-PO_4^{3-}: 13.1 ± 0.8 mg/L	Anoxic–aerobic algal–bacterial photobioreactor structure	the maximum removal rate of P-PO_4^{3-} was 47 ± 5%	[45]	low C/N ratio, Chlorella is the main algae, otherwise Phormidium SP will be dominant
	Chlorella vulgaris	COD: 300 mg/L TN: 30 mg/L TP: 10 mg/L	the new MAIFAS SBR	more than 51% phosphorus was removed without mechanical aeration	[70]	
Aerobic wastewater	Mixed microalgae collected in lakes	pH: 7.7 ± 0.2 TN: 99.5 mg/L TP: 5.5 mg/L COD: 475 mg/L TOC: 245.6 mg/L pH: 7.2	Photoperiod:12 h/d, immobilized microalgae, operated at 5 different HRTS for 2–10 days	the removal rate of phosphorus was 93%	[71]	collected in an aeration tank of a distributed domestic sewage treatment plant based on ASP
Unsterilized sewage	*Tetradesmus obliquus*	N-NH_4^+: 28 mg/L	the mixed gas containing 10% CO_2 was added to the unsterilized sewage	The removal rate of TP was 99.0%	[50]	
Anaerobic digester	*Chlorella* sp. and *Scenedesmus* sp.	COD: 12600 ± 300 mg/L TKN: 1692 ± 256 mg/L NH_3-N: 900 ± 62 mg/L NO_3-N: 0.13 ± 0.02 mg/L TP: 105 ± 7.5 mg/L PO_4-P: 64 ± 6 mg/L TSS: 15880 ± 932 mg/L pH: 9.00–9.15	in an adaptive room with continuous illumination: 150 mol photon M^{-2} S^{-1}, 25 ± 2 °C. cultured at a dilution ratio of 2%, 5%, 7% and 10%	reaction rate coefficient: 0.21 mg PO_4-Pmg^{-1} CHl a day^{-1}, saturation constant:2.94 mg L^{-1}, yield coefficient: 5.03 mg CHL A mg^{-1} PO_4-P	[60]	

Table 2. *Cont.*

Sewage Source	Microalgal Species	Initial Conditions	Experimental Conditions	Results	Cites	Notes
	Chlorella	activated sludge: COD: 500 mg/L; NH_4^{+1}-N: 40 mg/L; NO_3^--N: 2 mg/L; PO_4^{3-}-P: 8 mg/L anaerobic digester: COD: 5–10 g/L; NH_4^{+1}-N: 0.7–1.2 g/L; NO_3^--N: 90–300 mg/L; PO_4^{3-}-P: 60–190 mg/L	treated in a membrane photobioreactor (MPBR) in a continuous mode	the removal rate of orthophosphate exceeded 99%	[72]	schematic diagram is shown in Figure 4
Ozonation sludge wastewater	*Scendesmus* sp. is the dominant species	MLSS: 1500 mg/L algae: sludge = 1:3 (*w*/*w*).	run for 10 days, under 2500 lx on the inner wall of the reactor, photoperiod: 12 h/d (from 5:00–17:00), magnetic stirring rod (80 RPM)	the removal rate of TP was 53.9 ± 1.4%, higher than microalgae alone	[64]	sludge is obtained from secondary sedimentation tanks
Secondary wastewater from sewage treatment plants	Natural algal bloom (*Chlorella* mainly)	TP: 0.43 mg/L TN: 7 mg/L Mg: 0.45 mM Ca: 1.12 mM	continuous bubbling tower photobioreactor (BCPBR), flocculation–precipitation method	the removal rate of total dissolved phosphorus was greater than 99% under continuous operation	[73]	
Secondary wastewater from sewage treatment plants	*Chlorella*	COD: 111 mg/L pH: 7.9 ± 0.9 NH_3-N: 22 ± 2.6 mg/L NO_3-N: 0.30 ± 0.42 mg/L PO_4^{-3}-P: 3.2 ± 1.3 mg/L Turbidity: 184 ± 23 FAU *E. coli*: $4.7 \times 10^6 \pm 3 \times 10^6$ CFU 100 m/L	laboratory-scale photobioreactor, 10% of the effluent mixed with secondary effluent from a large municipal wastewater treatment plant, tertiary disinfection by ultraviolet treatment	The removal rate of TP was 100%	[56]	

Table 2. *Cont.*

Sewage Source	Microalgal Species	Initial Conditions	Experimental Conditions	Results	Cites	Notes
Sewage discharged from sedimentation tanks of municipal wastewater treatment plants	Common *Chlorella*	NH_4^{+1}-N: 64.84 mg/L NO_3^{-1}-N: 4.21 mg/L PO_4^{-3}-P: 3.78 mg/L COD: 82.00 mg O_2/L pH: 8.52 Alkalinity: 91.80 mg $CaCO_3$/L	Through the different concentrations of CO_2 and different N/P ratios	Absorbance of 95.00% phosphorus for the medium supplemented under 16% CO_2 and N:P ratio of 10	[49]	the wastewater was screened, biotreated, and disinfected.
Synthetic wastewater from municipal wastewater and laterite nickel mine	*Chlorella*	The two types of sewage were mixed in different proportions	temperature: 25 °C, light intensity: 4000 lux, Photoperiod:14 h/d, sterilized before experiment, added after sampling high-pressure deionized water of the same volume 6 times	The removal rate of TP was 39.3%	[74]	
Primary sedimentation tank wastewater	*Chlorella*	NH_4^{+1}-N: 25 ± 1.24 mg/L TKN: 42.0.47 mg/L NO_3-N: 2.5 ± 0.39 mg/L sCOD: 156 ± 2.6 mg O_2/L pH: 6.7 ± 0.05 DO: 3.5 ± 0.08 mg/L sBOD: 65 ± 3.4 mg/L TOC: 45.3 ± 1.12 mg/L TIC: 1.24 ± 0.07 mg/L TN: 46 ± 1.25 mg/L	carried out in a 7 L bubbling photobioreactor, temperature: 25 ± 2 °C, Photoperiod:14 h/d, pumped into the air with different concentrations of CO_2	the removal rate of orthophosphate was 92.8% under 5% CO_2 (*v/v*) for 7 days	[51]	
Settlement of sewage	Mixed algae	/	Wastewater Treatment and Resource Recovery (STaRR) system	the phosphorus recovery content was 71.6%	[75]	

4. Research Status Analysis of Phosphorus Removal from Municipal Wastewater by Microalgae

4.1. Symbiotic Systems of Bacteria and Algae

The treatment of wastewater by pure microalgae is usually limited to laboratory conditions, while sterile water is difficult to find in sewage treatment plants. Microalgae often work with endophytic bacteria to purify the wastewater [76]. There are high concentrations of activated sludge in some urban sewage treatment plants. Urban sewage provides a culture substrate for microalgae, which can reduce the high costs associated with microalgal artificial culture medium. Cultures of microalgae can also be mixed with heterotrophic microorganisms in activated sludge to meet the sustainability requirements for urban sewage purification [77]. Bacteria and algae can form a good symbiotic relationship [78], enhancing the effects of the microalgae in the purification of urban sewage. For the treatment of municipal wastewater and industrial wastewater, there is a trend of combining selected algal and bacterial species [79,80]. For example, a culture made up of *C. vulgaris* and *P. putida* can remove organic matter and other nutrients and shows good performance in synthetic municipal wastewater [81,82]. Lananan et al. [83] co-cultured Chlorella and effective microorganisms (EM-1), and their findings revealed that this mix could extract 99.15% of the total phosphorus from domestic sewage. Qing et al. [69] screened Klebsiella from activated sludge and treated municipal wastewater with *C. pyrenoidosis*. Their findings revealed that the phosphorous microbe not only boosted the phosphorous absorption performance of the microalgae (up to 3.05 mg/L), helping to regulate the total phosphorous concentration in the water to 0.46 mg/L, but also increased the lipid yield and the average productivity of the microalgae (90.1% and 13.6%, respectively).

4.2. Adding Metal Compounds

Magnesium ions (Mg^{2+}) are some of the most essential components for microalgae photosynthesis. P in wastewater can be removed by trimagnesium diphosphate ($Mg_3(PO_4)_2$) and $MgNH_4PO_4$ precipitation with other ions such as NH_4^+-N and Mg^{2+}. The assimilation of PO_4^{3-}-P could be hindered to some degree under Mg^{2+} deficiency [54,84–87]. Studies have shown that the development of *C. vulgaris* is inhibited in media without Mg^{2+}, whereas microalgae grown in media with Mg^{2+} are four times more productive than the blank group [88,89]. The concentration of Mg^{2+} has a significant influence on the metabolism of microalgae in urban wastewater treatment [59]. Nickel laterite ore wastewater (NLOWW) provided by the hydrometallurgical recovery of the nickel contains high concentrations of Mg^{2+} in the range of 20–40 g L^{-1} [74]. Conventional NLOWW treatment for recovery of Mg^{2+} consists of a series of integrated chemical–physical processes requiring investment in equipment and chemicals that are energy intensive and produce solid waste requiring further treatment [90]. Chen et al. [74] mixed urban and lateral nickel ore wastewater to cultivate *C. sorokiniana*. Their findings showed that the growth of microalgae cells in a culture without nickel laterite ore wastewater was slower and had a low biomass yield, whereas the microalgae biomass production rate in mixed wastewater containing nickel laterite ore increased by 1.89 times, the photosynthetic activity (Fv/Fm value) increased by 3.77 times, and the phosphorus removal rate increased by 39.3%; however, for 100% nickel laterite ore wastewater, excess Mg^{2+} can contain high amounts of reactive oxygen species, which inhibit the growth of microalgae.

As an essential micronutrient for the growth of algae, iron ions also play an important role in the physiological synthesis and enzymatic reactions of algae. Iron can coordinate active oxygen in algal cells and take part in electron transport, enzyme reactions, photosynthesis and respiration, and the synthesis of proteins and nucleic acids, and can promote the metabolism and absorption of nutrients [91,92]. As mentioned earlier, phosphorus is an important nutrient for synthesizing cell proteins and nucleic acids. Qiu et al. [93] compared the effects of various forms of iron on the growth of *Anabaena flos-aquae*, and the results revealed that ferric ammonium citrate, EDTA-Fe, iron ions, and ferric oxalate are the forms of iron that can stimulate the development of microalgae. When the iron concentration was regulated in the range of 0.1 mg L^{-1} to 0.8 mg L^{-1}, the impact of the iron type on microalgae growth was still greater than that of the iron concentration.

4.3. Biofilm Technology

In the 21st century, several wastewater treatment plants in the United States found that combining mobile bed bioreactor and fixed-membrane-activated sludge technologies not only enhanced the wastewater nitrification technology but also reduced the footprint of the facilities [70,94]. The integrated fixed-membrane-activated sludge process is an innovative biological wastewater treatment process that incorporates biofilm carriers into conventional activated sludge to eliminate nitrifiers, resulting in an improved retention time for the heterotrophic bacteria [70]. The nitrifying bacteria can be applied to the biofilm without being affected by the washing of the nitrifying agent, while the biological nitrogen can be eliminated by the nitrifying reaction [70]. Compared to the moving bed bioreactors (MBBR) system, the Integrated Fixed-Film Activated Sludge (IFAS) system can decouple the SRT of nitrifiers and polyphosphate bioaccumulators (PAO) by maximizing the elimination performance of biological nitrogen and phosphorus [70].

Jared Church et al. [70] incorporated microalgae into an optimized fixed-membrane-activated sludge configuration for photooxygenation and examined the symbiotic reactions of microalgae and bacteria to suspended matter and IFAS biofilms. In sequential batch mode, the microalgae were combined with the IFAS method to remove 51% of the phosphorus without mechanical aeration. This study also showed that the addition of microalgae to the IFAS system modified the metabolic function of multiple bacterial populations. This study was not only desirable for the reduction of phosphorus in water sources, but it also offers new research ideas for the improvement of various water bodies, the use of microalgae–IFAS technologies to modify the behavior of bacterial species, and the evolution of water quality.

Abeysiriwardana-Arachchige et al. [75] suggested a research approach for the treatment and recovery of wastewater based on algae (STaRR). Their findings revealed that the STARR device had a recovery output of 71.6% of nutrient phosphorus and that the removal of phosphorus per unit of energy consumption was calculated to be 0.1 g/kJ. This indicates that the STaRR system could be a green alternative for water treatment and nutrient recovery.

Anaerobic membrane bioreactor technology has the benefits of absorbing less energy and producing less sludge relative to more conventional aerobic systems, while still producing biomethane.; however, inorganic contaminants such as nitrogen and phosphorus cannot be removed from anaerobic reactors. Other nitrogen-containing compounds convert ammonia, which increases the concentration of ammonia in the water, one of the major microalgae nutrients [53,95,96]. Microalgae, thus, play an important role in the production of this system. Microalgae demonstrate good growth conditions in 5–30% nitrified digestion solutions combined with municipal wastewater. For example, the addition of a 10% nitrification solution for digestion in a two-stage bacterial-microalgal phase can eliminate 77% of the phosphate. Under the same conditions, the continuous use of a microalgae-based photobioreactor (MPBR) membrane will extract more than 99% of the phosphate [72].

4.4. Recovery Technology

From an economic and sustainable development point of view, the recovery of microalgae is a significant link in the treatment of wastewater by microalgae [97,98]. Currently, more mainstream approaches for the recovery of microalgae include centrifugation, filtration, sedimentation by gravity, and flocculation [98,99]. In addition, flocculant recovery technology and immobilized recovery technology are used.

4.4.1. Flocculant Recovery Technology

Microalgae can also be flocculated and retrieved by inorganic coagulants, such as aluminum sulfate or ferric chloride, polymeric flocculants, mixtures of the two components, or by using automated chemical flocculation methods such as pH modification [100–102]. Mennaa et al. [73] researched the continuous activity of the BCPBR (bubble column photobioreactor) using a flocculation–precipitation system to examine the effects of phosphorous removal and recovery of natural microalgae plants in urban wastewater. Their findings showed that continuous-mode experiments extract up to 99% of the total dissolved phosphorus without controlling the volume of CO_2 or regulating pH. PAC, $Fe_2(SO_4)_3$, and $Al_2(SO_4)_3$ were found to have positive effects on the recovery of microalgae; however, the addition of a flocculant not only increases the cost, but also may cause other pollution to the water body, reducing the sustainability of urban sewage.

4.4.2. Immobilized Recovery Technology

Compared with flocculant recovery technology, immobilized recovery technology can avoid secondary water pollution and increase the possibility of the sustainable development of urban sewage. The elimination of sewage from suspended algae systems can result in a low concentration of algae in the reactors, resulting in a reduction in the treatment rate [33,103,104]. The use of alginate beads to immobilize microalgae cells helps to retain a high concentration of microalgae in the reactor, which can easily remove nutrients from the water body, while the hydraulic retention period is less than 12 h [80,105]. Immobilized microalgae beads can settle rapidly to promote screening and regeneration, and these beads can be used directly as fertilizer or biomethane production after digestion [106–108]. In addition, beads produced by immobilized microalgae can also shield the culture from harmful contaminants in wastewater [109]. Kube [110] demonstrated that different concentrations of nitrogen and phosphorus can influence the absorption of phosphorus by Chlorella and that the immobilization of microalgae does not hinder the rate or ratio of nitrogen and phosphorous absorption. In addition, cell immobilization and co-cultivation of *Bacillus vulgaris* and *P. brasiliensis* can boost the removal rates of ammonia and phosphorus [111].

Katam et al. [71] used mixed microalgae to investigate the removal efficiency of carbon and nutrients in the treatment of real wastewater in an activated sludge reactor and set up two separate treatment systems for the simultaneous treatment of domestic wastewater. Their findings revealed that the total phosphorus removal performance of the immobilized microalgae system was as high as 93%, which was higher than the suspended activated sludge system. In addition, the microalgae developed higher lipid and carbon contents than the suspended activated sludge solution in the immobilized microalgae system.

4.5. *Other Improved Technologies*

Photobioreactors used to grow microalgae also have a major effect on the treatment of urban wastewater by microalgae. For example, relative to other reactors, BPBR has the advantages of high heat and mass transfer speeds, compact construction, and low operational and maintenance costs. Since the reactor has a higher surface-to-volume ratio, good mixability, lower shear stress, high scalability potential, simple sterilization, low emissions, and decreased photoinhibition, it can better monitor the growth parameters (such as temperature) of photooxidation [112].

The microalgal elimination of phosphorus can also be improved via genome building. Guerra-Renteria et al. [113] developed a genome-scale biochemical reaction network for the co-cultivation of *Chlorella* spp. and *Pseudomonas aeruginosa* bacteria using a metabolic pathway analysis (MPA). This analysis considers the metabolic ability of co-cultivation and determines the best conditions for the removal of nutrients. The theoretical phosphorous removal yield under photoheterotrophic conditions was determined as follows: 0.042 mmol of PO_4^{3-} per g DW of *C. vulgaris*, 19.43 mmol of phosphorus (Pi) per g DW of *C. vulgaris*, and 4.90 mmol of phosphorus (Pi) per g DW of *P. aeruginosa*. These theoretical yields are important because they can help in the design of biological systems and in the understanding of the theoretical requirements of oxygen and carbon dioxide in order to achieve maximum nutrient absorption. In this system, other by-products containing nitrogen or phosphorus may not even be formed, and all nutrient absorption is directed toward the growth of microalgae and bacteria [113].

5. Conclusions and Perspectives

While microalgae-based technologies provide a sustainable alternative for the removal of phosphorus from urban wastewater, the substitution of conventional water treatment technologies remains a major challenge. In this paper, we present the microalgae culture methods and the microalgae dephosphorization process. In addition, considerations influencing the elimination of microalgae phosphorus in urban water include traditional factors, but also the species characteristics of the microalgae and urban sewage types. This study, however, cannot completely summarize all of the factors influencing the dephosphorization of microalgae. It is necessary to further study the mechanisms and factors impacting microalgae dephosphorization from a microscopic perspective.

The study of phosphorous removal by microalgae showed that microalgae were often combined with other municipal wastewater treatment systems. Symbiotic relationships between bacteria and microalgae are common in municipal wastewater treatment plants. Biofilm has commonly been used in the treatment of sewage in the 21st century, and its combination with microalgae has encouraged the elimination of phosphorus from wastewater. While the microalgae biomass may be improved and phosphorous removal efficiency may be increased by increasing the concentrations of magnesium and iron ions in water, the mechanism for the removal of phosphorus by the inclusion of certain metals is less studied. In addition, certain metal nanomaterials have good adsorption and other characteristics, although the study of metal nanomaterials on microalgae is still lacking. This study also shows that immobilized microalgae technology can not only solve the problems of microalgae recycling and urban sewage sustainability, but can also improve the efficiency of phosphorus removal. This immobilization technology also offers a research concept to solve secondary contamination caused by the addition of metal ions to support the growth of microalgae.

Funding: This research was funded by National Natural Science Foundation of China, grant number 41371464. The APC was funded by Xiufang Gao.

Institutional Review Board Statement: Not applicable.

Informed Consent Statement: Not applicable.

Data Availability Statement: Not applicable.

Conflicts of Interest: The authors declare no conflict of interest.

References

1. Alkhudhiri, A.; Bin Darwish, N.; Hilal, N. Analytical and forecasting study for wastewater treatment and water resources in Saudi Arabia. *J. Water Process. Eng.* **2019**, *32*, 100915. [CrossRef]
2. Deng, Y.; Wheatley, A. Wastewater Treatment in Chinese Rural Areas. *Asian J. Water, Environ. Pollut.* **2016**, *13*, 1–11. [CrossRef]
3. Ingallinella, A.M.; Sanguinetti, G.; Koottatep, T.; Montangero, A.; Strauss, M. The challenge of faecal sludge management in urban areas—Strategies, regulations and treatment options. *Water Sci. Technol.* **2002**, *46*, 285–294. [CrossRef]
4. Wurz, A.; Kuchta, K.; Onay, T.T. Review on municipal sewage sludge management in Turkey and Europe. *Int. J. Glob. Warm.* **2011**, 3, 116. [CrossRef]
5. Kumar, P. Numerical quantification of current status quo and future prediction of water quality in eight Asian megacities: Challenges and opportunities for sustainable water management. *Environ. Monit. Assess.* **2019**, *191*, 319. [CrossRef]
6. Azam, H.M.; Alam, S.T.; Hasan, M.; Yameogo, D.D.S.; Kannan, A.D.; Rahman, A.; Kwon, M.J. Phosphorous in the environment: Characteristics with distribution and effects, removal mechanisms, treatment technologies, and factors affecting recovery as minerals in natural and engineered systems. *Environ. Sci. Pollut. Res.* **2019**, *26*, 20183–20207. [CrossRef] [PubMed]
7. Heinzmann, B.; Betriebe, B.W. Phosphorus recovery in wastewater treatment plants. In Proceedings of the Second International Conference, Moscow, Russia, 29 November–3 December 1999.

8. Fattah, K. Finding Nutrient-Related Problems in Wastewater Treatment Plants. In Proceedings of the 2nd International Conference on Environmental, Bio-Medical and Biotechnology, Dubai, United Arab Emirates, 4–5 August 2012.
9. Jenkins, D.; Ferguson, J.F.; Menar, A.B. Chemical processes for phosphate removal. *Water Res.* **1971**, *5*, 369–389. [CrossRef]
10. Stensel, H.D. *Phosphorous and Nitrogen Removal from Municipal Wastewater: Principles and Practice*; Routledge: Boca Raton, FL, USA, 1991.
11. Zhang, M.; Lawlor, P.G.; Hu, Z.; Zhan, X. Nutrient removal from separated pig manure digestate liquid using hybrid biofilters. *Environ. Technol.* **2013**, *34*, 645–651. [CrossRef] [PubMed]
12. Grégorio, C.; Eric, L. Advantages and disadvantages of techniques used for wastewater treatment. *Environ. Chem. Lett.* **2018**, *17*, 145–155.
13. Bunce, J.T.; Ndam, E.; Ofiteru, I.D.; Moore, A.; Graham, D.W. A Review of Phosphorus Removal Technologies and Their Applicability to Small-Scale Domestic Wastewater Treatment Systems. *Front. Environ. Sci.* **2018**, *6*, 8. [CrossRef]
14. Cardew, P. Measuring the benefit of orthophosphate treatment on lead in drinking water. *Water Health* **2009**, *7*, 123–131. [CrossRef]
15. Volk, C.; Dundore, E.; Schiermann, J.; Lechevallier, M. Practical evaluation of iron corrosion control in a drinking water distribution system. *Water Res.* **2000**, *34*, 1967–1974. [CrossRef]
16. House, J.E.; House, K.A. *Descriptive Inorganic Chemistry*; Academic Press: Cambridge, MA, USA, 2001.
17. Wind, T. *The Role of Detergents in the Phosphate-Balance of European Surface Waters*; European Water Management Online: Hennef, Germany, 2007; pp. 1–19.
18. Shu, L.; Schneider, P.; Jegatheesan, V. An economic evaluation of phosphorus recovery as struvite from digester supernatant. *Bioresour. Technol.* **2006**, *97*, 2211–2216. [CrossRef]
19. Ehama, M.; Hashihama, F.; Kinouchi, S.; Kanda, J.; Saito, H. Sensitive determination of total particulate phosphorus and particulate inorganic phosphorus in seawater using liquid waveguide spectrophotometry. *Talanta* **2016**, *153*, 66–70. [CrossRef]
20. Panasiuk, O. Phosphorus Removal and Recovery from Wastewater Using Magnetite. Master's Thesis, Industrial Ecology, Royal Institute of Technology, Stockholm, Sweden, 2010.
21. Morse, G.; Brett, S.; Guy, J.; Lester, J. Review: Phosphorus removal and recovery technologies. *Sci. Total. Environ.* **1998**, *212*, 69–81. [CrossRef]
22. Mohsenpour, S.F.; Hennige, S.; Willoughby, N.; Adeloye, A.; Gutierrez, T. Integrating micro-algae into wastewater treatment: A review. *Sci. Total. Environ.* **2021**, *752*, 142168. [CrossRef] [PubMed]
23. Lutzu, G.A.; Ciurli, A.; Chiellini, C.; di Caprio, F.; Concas, A.; Dunford, N.T. Latest developments in wastewater treatment and biopolymer production by microalgae. *J. Environ. Chem. Eng.* **2021**, *9*, 104926. [CrossRef]
24. Borowitzka, M.A.; Moheimani, N.R. *Open Pond Culture Systems*; Springer Science and Business Media: Cham, Switzerland, 2013; pp. 133–152.
25. Acién, F.F.G.; Fernández, S.J.M.; Molina, G.E. Photobioreactors for the production of microalgae. *Rev. Environ. Sci. Biotechnol.* **2013**, *12*, 131–151. [CrossRef]
26. Lee, Y.-K.; Shen, H. Basic Culturing Techniques. *Handb. Microalgal Cult.* **2007**, 40–56. [CrossRef]
27. Donald, K.M.; Scanlan, D.J.; Carr, N.G.; Mann, N.H.; Joint, I. Comparative phosphorus nutrition of the marine cyanobacterium *Synechococcus* WH7803 and the marine diatom *Thalassiosira weissflogii*. *J. Plankton Res.* **1997**, *19*, 1793–1813. [CrossRef]
28. Solovchenko, A.; Khozin-Goldberg, I.; Selyakh, I.; Semenova, L.; Ismagulova, T.; Lukyanov, A.; Mamedov, I.; Vinogradova, E.; Karpova, O.; Konyukhov, I.; et al. Phosphorus starvation and luxury uptake in green microalgae revisited. *Algal Res.* **2019**, *43*, 101651. [CrossRef]
29. Cembella, A.D.; Antia, N.J.; Harrison, P.J. The Utilization of Inorganic and Organic Phosphorous Compounds as Nutrients by Eukaryotic Microalgae: A Multidisciplinary Perspective: Part I. *CRC Crit. Rev. Microbiol.* **1982**, *10*, 317–391. [CrossRef] [PubMed]
30. Cembella, A.D.; Antia, N.J.; Harrison, P.J.; Rhee, G.Y. The Utilization of Inorganic and Organic Phosphorous Compounds as Nutrients by Eukaryotic Microalgae: A Multidisciplinary Perspective: Part II. *CRC Crit. Rev. Microbiol.* **1984**, *11*, 13–81. [CrossRef] [PubMed]
31. Su, Y. Revisiting carbon, nitrogen, and phosphorus metabolisms in microalgae for wastewater treatment. *Sci. Total. Environ.* **2021**, *762*, 144590. [CrossRef]
32. Diaz, J.M.; Björkman, K.M.; Haley, S.T.; Ingall, E.; Karl, D.; Longo, A.F.; Dyhrman, S.T. Polyphosphate dynamics at Station ALOHA, North Pacific subtropical gyre. *Limnol. Oceanogr.* **2015**, *61*, 227–239. [CrossRef]
33. Whitton, R.; Ometto, F.; Pidou, M.; Jarvis, P.; Villa, R.; Jefferson, B. Microalgae for municipal wastewater nutrient remediation: Mechanisms, reactors and outlook for tertiary treatment. *Environ. Technol. Rev.* **2015**, *4*, 133–148. [CrossRef]
34. Sforza, E.; Calvaruso, C.; la Rocca, N.; Bertucco, A. Luxury uptake of phosphorus in *Nannochloropsis salina*: Effect of P concentration and light on P uptake in batch and continuous cultures. *Biochem. Eng. J.* **2018**, *134*, 69–79. [CrossRef]
35. Li, Q.; Fu, L.; Wang, Y.; Zhou, D.; Rittmann, B.E. Excessive phosphorus caused inhibition and cell damage during heterotrophic growth of *Chlorella regularis*. *Bioresour. Technol.* **2018**, *268*, 266–270. [CrossRef]
36. Dyhrman, S.T. Nutrients and their acquisition: Phosphorus physiology in microalgae. In *The Physiology of Microalgae*; Borowitzka, M., Beardall, J., Raven, J., Eds.; Springer: Dordrecht, The Netherlands, 2016; pp. 155–183. [CrossRef]
37. Montgomery, J.M. *Water Treatment Principles and Design*; Wiley: New York, NY, USA, 1985.
38. Rittmann, B.E.; Mayer, B.; Westerhoff, P.; Edwards, M. Capturing the lost phosphorus. *Chemosphere* **2011**, *84*, 846–853. [CrossRef]

39. Wang, C.; Jiang, H.L. Chemicals used for in situ immobilization to reduce the internal phosphorus loading from lake sediments for eutrophication control. *Crit. Rev. Environ. Sci. Technol.* **2016**, *46*, 947–997. [CrossRef]
40. Larsdotter, K. WasteWater treatment with microalgae—A literature review. *Vatten* **2006**, *62*, 31–38.
41. Solmaz, A.; Işık, M. Optimization of membrane photobioreactor; the effect of hydraulic retention time on biomass production and nutrient removal by mixed microalgae culture. *Biomass Bioenergy* **2020**, *142*, 105809. [CrossRef]
42. Arcila, J.S.; Buitrón, G. Microalgae-bacteria aggregates: Effect of the hydraulic retention time on the municipal wastewater treatment, biomass settleability and methane potential. *J. Chem. Technol. Biotechnol.* **2016**, *91*, 2862–2870. [CrossRef]
43. Metcalf, E. *Wastewater Engineering and Reuse, 4th ed*; Mc. GrawHill: New York, NY, USA, 2003.
44. García, D.; Alcántara, C.; Blanco, S.; Pérez, R.; Bolado, S.; Muñoz, R. Enhanced carbon, nitrogen and phosphorus removal from domestic wastewater in a novel anoxic-aerobic photobioreactor coupled with biogas upgrading. *Chem. Eng. J.* **2017**, *313*, 424–434. [CrossRef]
45. Toledo-Cervantes, A.; Posadas, E.; Bertol, I.; Turiel, S.; Alcoceba, A.; Muñoz, R. Assessing the influence of the hydraulic retention time and carbon/nitrogen ratio on urban wastewater treatment in a new anoxic-aerobic algal-bacterial photobioreactor configuration. *Algal Res.* **2019**, *44*, 101672. [CrossRef]
46. Alcántara, C.; Domínguez, J.M.; García, D.; Blanco, S.; Pérez, R.; Garcia-Encina, P.A.; Muñoz, R. Evaluation of wastewater treatment in a novel anoxic–aerobic algal–bacterial photobioreactor with biomass recycling through carbon and nitrogen mass balances. *Bioresour. Technol.* **2015**, *191*, 173–186. [CrossRef] [PubMed]
47. Wang, L.; Min, M.; Li, Y.; Chen, P.; Chen, Y.; Liu, Y.; Ruan, R. Cultivation of green *Algae Chlorella* sp. in different wastewaters from municipal wastewater treatment plant. *Appl. Biochem. Biotechnol.* **2010**, *162*, 1174–1186. [CrossRef]
48. Choi, H.J.; Lee, S.M. Effect of the N/P ratio on biomass productivity and nutrient removal from municipal wastewater. *Bioprocess Biosyst. Eng.* **2015**, *38*, 761–766. [CrossRef]
49. Molazadeh, M.; Danesh, S.; Ahmadzadeh, H.; Pourianfar, H.R. Influence of CO_2 concentration and N:P ratio on Chlorella vulgaris-assisted nutrient bioremediation, CO_2 biofixation and biomass production in a lagoon treatment plant. *J. Taiwan Inst. Chem. Eng.* **2019**, *96*, 114–120. [CrossRef]
50. Ma, S.; Yu, Y.; Cui, H.; Yadav, R.S.; Li, J.; Feng, Y. Unsterilized sewage treatment and carbohydrate accumulation in Tetradesmus obliquus PF3 with CO_2 supplementation. *Algal Res.* **2020**, *45*, 101741. [CrossRef]
51. Chaudhary, R.; Tong, Y.W.; Dikshit, A.K. Kinetic study of nutrients removal from municipal wastewater by Chlorella vulgaris in photobioreactor supplied with CO_2-enriched air. *Environ. Technol.* **2018**, *41*, 617–626. [CrossRef]
52. Jiang, L.; Luo, S.; Fan, X.; Yang, Z.; Guo, R. Biomass and lipid production of marine microalgae using municipal wastewater and high concentration of CO_2. *Appl. Energy* **2011**, *88*, 3336–3341. [CrossRef]
53. Pachés, M.; Martínez-Guijarro, R.; González-Camejo, J.; Seco, A.; Barat, R. Selecting the most suitable microalgae species to treat the effluent from an anaerobic membrane bioreactor. *Environ. Technol.* **2018**, *41*, 267–276. [CrossRef]
54. Devi, M.P.; Subhash, G.V.; Mohan, S.V. Heterotrophic cultivation of mixed microalgae for lipid accumulation and wastewater treatment during sequential growth and starvation phases: Effect of nutrient supplementation. *Renew. Energy* **2012**, *43*, 276–283. [CrossRef]
55. Bohutskyi, P.; Kligerman, D.C.; Byers, N.; Nasr, L.K.; Cua, C.; Chow, S.; Su, C.; Tang, Y.; Betenbaugh, M.J.; Bouwer, E.J. Effects of inoculum size, light intensity, and dose of anaerobic digestion centrate on growth and productivity of Chlorella and Scenedesmus microalgae and their poly-culture in primary and secondary wastewater. *Algal Res.* **2016**, *19*, 278–290. [CrossRef]
56. Bellucci, M.; Marazzi, F.; Naddeo, L.S.; Piergiacomo, F.; Beneduce, L.; Ficara, E.; Mezzanotte, V. Disinfection and nutrient removal in laboratory-scale photobioreactors for wastewater tertiary treatment. *J. Chem. Technol. Biotechnol.* **2019**. [CrossRef]
57. Abdullahi, Y.; Akunna, J.; White, N.; Hallett, P.; Wheatley, R. Investigating the effects of anaerobic and aerobic post-treatment on quality and stability of organic fraction of municipal solid waste as soil amendment. *Bioresour. Technol.* **2008**, *99*, 8631–8636. [CrossRef] [PubMed]
58. Bjornsson, W.J.; Nicol, R.W.; Dickinson, K.E.; McGinn, P.J. Anaerobic digestates are useful nutrient sources for microalgae cultivation: Functional coupling of energy and biomass production. *J. Appl. Phycol.* **2013**, *25*, 1523–1528. [CrossRef]
59. Salama, E.S.; Kurade, M.B.; Abou-Shanab, R.; El-Dalatony, M.M.; Yang, I.S.; Min, B.; Jeon, B.H. Recent progress in microalgal biomass production coupled with wastewater treatment for biofuel generation. *Renew. Sust. Energy Rev.* **2017**, *79*, 1189–1211. [CrossRef]
60. Ermis, H.; Altınbaş, M. Determination of biokinetic coefficients for nutrient removal from anaerobic liquid digestate by mixed microalgae. *J. Appl. Phycol.* **2018**, *31*, 1773–1781. [CrossRef]
61. Mujtaba, G.; Lee, K. Treatment of real wastewater using co-culture of immobilized *Chlorella vulgaris* and suspended activated sludge. *Water Res.* **2017**, *120*, 174–184. [CrossRef]
62. Vandamme, D.; Foubert, I.; Fraeye, I.; Muylaert, K. Influence of organic matter generated by *Chlorella vulgaris* on five different modes of flocculation. *Bioresour. Technol.* **2012**, *124*, 508–511. [CrossRef] [PubMed]
63. Yan, S.-T.; Chu, L.-B.; Xing, X.-H.; Yu, A.-F.; Sun, X.-L.; Jurcik, B. Analysis of the mechanism of sludge ozonation by a combination of biological and chemical approaches. *Water Res.* **2009**, *43*, 195–203. [CrossRef] [PubMed]
64. Lei, Y.J.; Tian, Y.; Zhang, J.; Sun, L.; Kong, X.W.; Zuo, W.; Kong, L.C. Microalgae cultivation and nutrients removal from sewage sludge after ozonizing in algal-bacteria system. *Ecotoxicol. Environ. Saf.* **2018**, *165*, 107–114. [CrossRef]

65. Chu, L.; Yan, S.; Xing, X.-H.; Sun, X.; Jurcik, B. Progress and perspectives of sludge ozonation as a powerful pretreatment method for minimization of excess sludge production. *Water Res.* **2009**, *43*, 1811–1822. [CrossRef] [PubMed]
66. Yan, S.-T.; Zheng, H.; Li, A.; Zhang, X.; Xing, X.-H.; Chu, L.-B.; Ding, G.; Sun, X.-L.; Jurcik, B. Systematic analysis of biochemical performance and the microbial community of an activated sludge process using ozone-treated sludge for sludge reduction. *Bioresour. Technol.* **2009**, *100*, 5002–5009. [CrossRef] [PubMed]
67. Wang, J.-H.; Zhang, T.-Y.; Dao, G.-H.; Xu, X.-Q.; Wang, X.-X.; Hu, H.-Y. Microalgae-based advanced municipal wastewater treatment for reuse in water bodies. *Appl. Microbiol. Biotechnol.* **2017**, *101*, 2659–2675. [CrossRef]
68. Semblante, G.U.; Hai, F.I.; Dionysiou, D.; Fukushi, K.; Price, W.E.; Nghiem, L. Holistic sludge management through ozonation: A critical review. *J. Environ. Manag.* **2017**, *185*, 79–95. [CrossRef]
69. Wang, Q.; Jin, W.; Zhou, X.; Guo, S.; Gao, S.-H.; Chen, C.; Tu, R.; Han, S.-F.; Jiang, J.; Feng, X. Growth enhancement of biodiesel-promising microalga *Chlorella pyrenoidosa* in municipal wastewater by polyphosphate-accumulating organisms. *J. Clean. Prod.* **2019**, *240*, 118148. [CrossRef]
70. Church, J.; Ryu, H.; Sadmani, A.H.A.; Randall, A.A.; Domingo, J.S.; Lee, W.H. Multiscale investigation of a symbiotic microalgal-integrated fixed film activated sludge (MAIFAS) process for nutrient removal and photo-oxygenation. *Bioresour. Technol.* **2018**, *268*, 128–138. [CrossRef]
71. Katam, K.; Bhattacharyya, D. Simultaneous treatment of domestic wastewater and bio-lipid synthesis using immobilized and suspended cultures of microalgae and activated sludge. *J. Ind. Eng. Chem.* **2019**, *69*, 295–303. [CrossRef]
72. Praveen, P.; Guo, Y.; Kang, H.; Lefebvre, C.; Loh, K.-C. Enhancing microalgae cultivation in anaerobic digestate through nitrification. *Chem. Eng. J.* **2018**, *354*, 905–912. [CrossRef]
73. Mennaa, F.Z.; Arbib, Z.; Perales, J.A. Urban wastewater photobiotreatment with microalgae in a continuously operated photobioreactor: Growth, nutrient removal kinetics and biomass coagulation–flocculation. *Environ. Technol.* **2019**, *40*, 342–355. [CrossRef] [PubMed]
74. Chen, Z.; Qiu, S.; Amadu, A.A.; Shen, Y.; Wang, L.; Wu, Z.; Ge, S. Simultaneous improvements on nutrient and Mg recoveries of microalgal bioremediation for municipal wastewater and nickel laterite ore wastewater. *Bioresour. Technol.* **2020**, *297*, 122517. [CrossRef] [PubMed]
75. Arachchige, I.S.A.A.; Munasinghe-Arachchige, S.; Delanka-Pedige, H.M.K.; Nirmalakhandan, N. Removal and recovery of nutrients from municipal sewage: Algal vs. conventional approaches. *Water Res.* **2020**, *175*, 115709. [CrossRef] [PubMed]
76. Kouzuma, A.; Watanabe, K. Exploring the potential of algae/bacteria interactions. *Curr. Opin. Biotechnol.* **2015**, *33*, 125–129. [CrossRef] [PubMed]
77. Ferreira, A.; Ribeiro, B.; Marques, P.A.; Ferreira, A.F.; Dias, A.P.; Pinheiro, H.M.; Reis, A.; Gouveia, L. *Scenedesmus obliquus* mediated brewery wastewater remediation and CO2 bio-fixation for green energy purposes. *J. Clean. Prod.* **2017**, *165*, 1316–1327. [CrossRef]
78. He, P.; Mao, B.; Lü, F.; Shao, L.; Lee, D.; Chang, J. The combined effect of bacteria and *Chlorella vulgaris* on the treatment of municipal wastewaters. *Bioresour. Technol.* **2013**, *146*, 562–568. [CrossRef]
79. Alsan, S.; Kapdan, I.K. Batch kinetics of nitrogen and phosphorus removal from synthetic wastewater by algae. *Ecol. Eng.* **2006**, *28*, 64–70.
80. Pires, J.C.M.; Alvim-Ferraz, M.D.C.; Martins, F.; Simões, M. Wastewater treatment to enhance the economic viability of microalgae culture. *Environ. Sci. Pollut. Res.* **2013**, *20*, 5096–5105. [CrossRef] [PubMed]
81. Mujtaba, G.; Rizwan, M.; Lee, K. Simultaneous removal of inorganic nutrients and organic carbon by symbiotic co-culture of Chlorella vulgaris and Pseudomonas putida. *Biotechnol. Bioprocess Eng.* **2015**, *20*, 1114–1122. [CrossRef]
82. Gómez-Guzmán, A.; Jiménez-Magaña, S.; Guerra-Rentería, A.S.; Gómez-Hermosillo, C.; Parra-Rodríguez, F.J.; Velázquez, S.; Aguilar-Uscanga, B.R.; Solis-Pacheco, J.; González-Reynoso, O. Evaluation of nutrients removal (NO3-N, NH3-N and PO4-P) with *Chlorella vulgaris*, *Pseudomonas putida*, *Bacillus cereus* and a consortium of these microorganisms in the treatment of wastewater effluents. *Water Sci. Technol.* **2017**, *76*, 49–56. [CrossRef]
83. Lananan, F.; Hamid, S.H.A.; Din, W.N.S.; Ali, N.; Khatoon, H.; Jusoh, A.; Endut, A. Symbiotic bioremediation of aquaculture wastewater in reducing ammonia and phosphorus utilizing Effective Microorganism (EM-1) and microalgae (*Chlorella* sp.). *Int. Biodeterior. Biodegradation* **2014**, *95*, 127–134. [CrossRef]
84. Andersson, I.; Backlund, A. Structure and function of Rubisco. *Plant Physiol. Biochem.* **2008**, *46*, 275–291. [CrossRef] [PubMed]
85. Sasaki, Y.; Nagano, Y. Plant Acetyl-CoA Carboxylase: Structure, Biosynthesis, Regulation, and Gene Manipulation for Plant Breeding. *Biosci. Biotechnol. Biochem.* **2004**, *68*, 1175–1184. [CrossRef]
86. Pasternak, K.; Kocot, J.; Horecka, A. Biochemistry of magnesium. *J. Elementol.* **2010**, *15*, 601–616. [CrossRef]
87. Sydney, E.; Sturm, W.; de Carvalho, J.; Soccol, V.T.; Larroche, C.; Pandey, A.; Soccol, C.R. Potential carbon dioxide fixation by industrially important microalgae. *Bioresour. Technol.* **2010**, *101*, 5892–5896. [CrossRef] [PubMed]
88. Ayed, H.B.A.-B.; Taidi, B.; Ayadi, H.; Pareau, D.; Stambouli, M. Effect of magnesium ion concentration in autotrophic cultures of *Chlorella vulgaris*. *Algal Res.* **2015**, *9*, 291–296. [CrossRef]
89. Ren, H.; Liu, B.; Kong, F.; Zhao, L.; Xie, G.; Ren, N. Enhanced lipid accumulation of green microalga *Scenedesmus* sp. by metal ions and EDTA addition. *Bioresour. Technol.* **2014**, *169*, 763–767. [CrossRef]
90. Mu, W.N.; Shi, S.Z.; Zhai, Y.C. Magnesium Recovery from Desiliconization Slag of Nickel Laterite Ores by Carbonization. *Adv. Mater. Res.* **2013**, *813*, 255–258. [CrossRef]

91. Alexova, R.; Fujii, M.; Birch, D.; Cheng, J.; Waite, T.D.; Ferrari, B.C.; Neilan, B.A. Iron uptake and toxin synthesis in the bloom-forming Microcystis aeruginosa under iron limitation. *Environ. Microbiol.* **2011**, *13*, 1064–1077. [CrossRef]
92. Boyd, P.; Ellwood, M. The biogeochemical cycle of iron in the ocean. *Nat. Geosci.* **2010**, *3*, 675–682. [CrossRef]
93. Qiu, Y.; Wang, Z.; Liu, F.; Liu, J.; Zhou, T. Effect of different kinds of complex iron on the growth of *Anabaena flosaquae*. *Environ. Technol.* **2018**, *40*, 2889–2896. [CrossRef]
94. Onnis-Hayden, A.; Majed, N.; Schramm, A.; Gu, A.Z. Process optimization by decoupled control of key microbial populations: Distribution of activity and abundance of polyphosphate-accumulating organisms and nitrifying populations in a full-scale IFAS-EBPR plant. *Water Res.* **2011**, *45*, 3845–3854. [CrossRef] [PubMed]
95. Ho, J.; Sung, S. Methanogenic activities in anaerobic membrane bioreactors (AnMBR) treating synthetic municipal wastewater. *Bioresource Technol.* **2010**, *101*, 2191–2196. [CrossRef] [PubMed]
96. Podevin, M.; de Francisci, D.; Holdt, S.L.; Angelidaki, I. Effect of nitrogen source and acclimatization on specific growth rates of microalgae determined by a high-throughput in vivo microplate autofluorescence method. *Environ. Boil. Fishes* **2015**, *27*, 1415–1423. [CrossRef]
97. Sukenik, A.; Bilanovic, D.; Shelef, G. Flocculation of microalgae in brackish and sea waters. *Biomass Bioenergy* **1988**, *15*, 187–199. [CrossRef]
98. Grima, E.M.; Belarbi, E.-H.; Fernández, F.G.A.; Medina, A.R.; Chisti, Y. Recovery of microalgal biomass and metabolites: Process options and economics. *Biotechnol. Adv.* **2003**, *20*, 491–515. [CrossRef]
99. Lee, S.; Kim, S.B.; Kim, J.E.; Kwon, G.S.; Yoon, B.D.; Oh, H.M. Effects of harvesting method and growth stage on the flocculation of the green alga *Botryococcus braunii*. *Lett. Appl. Microbiol.* **1998**, *27*, 14–18. [CrossRef]
100. Siew Moi, P. Handbook of microalgal culture. Biotechnology and applied phycology. *J. Appl. Phycology.* **2004**, *16*, 159–160.
101. Shelef, G.; Sukenik, A.; Green, M. Microalgae Harvesting and Processing: A Literature Review. Available online: https://www.osti.gov/biblio/6204677 (accessed on 1 August 2021).
102. Chen, C.-Y.; Yeh, K.-L.; Aisyah, R.; Lee, D.-J.; Chang, J.-S. Cultivation, photobioreactor design and harvesting of microalgae for biodiesel production: A critical review. *Bioresour. Technol.* **2011**, *102*, 71–81. [CrossRef]
103. Park, J.; Craggs, R.; Shilton, A. Wastewater treatment high rate algal ponds for biofuel production. *Bioresour. Technol.* **2011**, *102*, 35–42. [CrossRef] [PubMed]
104. Filippino, K.C.; Mulholland, M.R.; Bott, C.B. Phycoremediation strategies for rapid tertiary nutrient removal in a waste stream. *Algal Res.* **2015**, *11*, 125–133. [CrossRef]
105. Whitton, R.; Santinelli, M.; Pidou, M.; Ometto, F.; Henderson, R.; Roddick, F.; Jarvis, P.; Villa, R.; Jefferson, B. Tertiary nutrient removal from wastewater by immobilised microalgae: Impact of wastewater nutrient characteristics and hydraulic retention time (HRT). *H2Open J.* **2018**, *1*, 12–25. [CrossRef]
106. De-Bashan, L.E.; Bashan, Y. Immobilized microalgae for removing pollutants: Review of practical aspects. *Bioresour. Technol.* **2010**, *101*, 1611–1627. [CrossRef]
107. Lam, M.K.; Lee, K.T. Immobilization as a feasible method to simplify the separation of microalgae from water for biodiesel production. *Chem. Eng. J.* **2012**, *191*, 263–268. [CrossRef]
108. Yadavalli, R.; Heggers, G.R.V.N. Two stage treatment of dairy effluent using immobilized *Chlorella pyrenoidosa*. *J. Environ. Health Sci. Eng.* **2013**, *11*, 36. [CrossRef]
109. Covarrubias, S.A.; De-Bashan, L.E.; Moreno, M.; Bashan, Y. Alginate beads provide a beneficial physical barrier against native microorganisms in wastewater treated with immobilized bacteria and microalgae. *Appl. Microbiol. Biotechnol.* **2012**, *93*, 2669–2680. [CrossRef]
110. Kube, M.; Spedding, B.; Gao, L.; Fan, L.; Roddick, F. Nutrient removal by alginate-immobilized *Chlorella vulgaris*: Response to different wastewater matrices. *J. Chem. Technol. Biotechnol.* **2020**, *95*, 1790–1799. [CrossRef]
111. Choix, F.J.; De-Bashan, L.E.; Bashan, Y. Enhanced accumulation of starch and total carbohydrates in alginate-immobilized *Chlorella* spp. induced by *Azospirillum brasilense*: I. Autotrophic conditions. *Enzym. Microb. Technol.* **2012**, *51*, 294–299. [CrossRef]
112. Kantarci, N.; Borak, F.; Ulgen, K.O. Bubble column reactors. *Process Biochem.* **2005**, *40*, 2263–2283. [CrossRef]
113. Guerra-Renteria, A.S.; García-Ramírez, M.A.; Gómez-Hermosillo, C.; Gómez-Guzmán, A.; González-García, Y.; González-Reynoso, O. Metabolic Pathway Analysis of Nitrogen and Phosphorus Uptake by the Consortium between *C. vulgaris* and *P. aeruginosa*. *Int. J. Mol. Sci.* **2019**, *20*, 1978. [CrossRef] [PubMed]

sustainability

MDPI

Article

Bacterial Movement in Subsurface Soil during Winter Irrigation of Reclaimed Wastewater

Guannan Ding [1], Karen Mancl [2,*], Jiyoung Lee [3,4] and Olli H. Tuovinen [5]

1. Environmental Science Graduate Program, Ohio State University, Columbus, OH 43210, USA; guannan1221@gmail.com
2. Department of Food, Agriculture and Biological Engineering, Ohio State University, Columbus, OH 43210, USA
3. Division of Environmental Health Sciences, College of Public Health, Ohio State University, Columbus, OH 43210, USA; lee.3598@osu.edu
4. Department of Food Science & Technology, Ohio State University, Columbus, OH 43210, USA
5. Department of Microbiology, Ohio State University, Columbus, OH 43210, USA; tuovinen.1@osu.edu

* Correspondence: mancl.1@osu.edu

Abstract: Processes to remove and inactivate *Escherichia coli* from wastewater effluents and drainage are complex and interrelated. The objective of this study was to determine if irrigation of undisinfected wastewater effluents in the winter moves bacteria to surface water through subsurface drainage, posing a public health risk. The central Ohio study site, an open meadow constructed in the 1970s, is irrigated with lagoon effluents each summer. The irrigated area has subsurface drainage that collects for discharge in one spot. Undisinfected wastewater from a stabilization pond was irrigated for the first time in the winter of 2013/2014. *E. coli* was measured in the subsurface discharge during the irrigated winter season and compared to the non-irrigated previous winter season. Soil temperature and moisture were also monitored. *E. coli* moved to subsurface drains when the water table was above the drain. *E. coli* also moved to subsurface drains when the shallow soil temperature dropped to near freezing. With less winter sunlight and minimal evapotranspiration, the soil stayed moist near field capacity. Temperature appears to be the most important factor in limiting natural inactivation in subsurface soil and allowing the movement of *E. coli* in undisinfected wastewater effluents to the subsurface drainage systems. The results show that winter reuse of undisinfected wastewater does pose a public health risk to surface water through subsurface drainage. Therefore, disinfection of wastewater effluents used for irrigation is strongly recommended.

Keywords: *Escherichia coli*; groundwater; land application; public health; reuse; soil moisture; soil temperature; subsurface drainage; wastewater

Citation: Ding, G.; Mancl, K.; Lee, J.; Tuovinen, O.H. Bacterial Movement in Subsurface Soil during Winter Irrigation of Reclaimed Wastewater. *Sustainability* **2021**, *13*, 9594. https://doi.org/10.3390/su13179594

Academic Editors: Avelino Núñez-Delgado, Avelino Núñez-Delgado, Elza Bontempi, Mario Coccia, Marco Race and Yaoyu Zhou

Received: 30 July 2021
Accepted: 18 August 2021
Published: 26 August 2021

1. Introduction

Land application of wastewater has increased since the amendments to the Federal Water Pollution Act of 1972 set the goal to eliminate the discharge of pollutants to navigable waters. This reuse of wastewater protects surface waters from fecal, chemical, and physical pollution; enriches the soil with organic matter; and recycles inorganic nutrients. Wastewater reuse through irrigation saves water and reduces treatment costs, waste hazards, and health risk in the environment. Due to the presence of human pathogens, irrigation practices have severe limitations if reclaimed wastewater is used for food crops because of food safety and food-borne illnesses.

Pathogenic microorganisms are considered one of the highest risk factors for surface water quality and human health [1] because of direct and indirect exposures to potential pathogens in wastewater. Reclaimed water produced from domestic wastewater contains many bacterial, viral, and parasitic human pathogens. One goal in wastewater treatment is to reduce the disease-causing agents to acceptable levels. In drinking water, the number of total coliforms should be <1/100 mL. The most commonly used fecal bacterial indicator, *Escherichia coli*, indicates the presence of fecal contamination from sewage, animal manure, or wildlife.

Water from precipitation and irrigation moves into the soil where some will run off or evaporate and some is used by plants through evapotranspiration as a part of the hydrologic cycle. The remaining water can infiltrate the soil until it reaches field capacity, which is the water remaining after the soil drains freely. The excess water usually drains away into groundwater in one to two days [2]. The drainage rate is governed by the soil texture. Silt loam and silty clay loam soils are rated as moderately permeable. The movement of water through the soil slows down in cold temperatures as the viscosity of water increases [3].

The pathway for water movement is built by macropores in soil [4]. Macropore flow can significantly reduce the retention time of bacteria and viruses in soil, thus resulting in the contamination of groundwater. Smith and others [4] concluded that any soil type with macrospores that receives enough water to saturate the pores has the potential to transport bacteria vertically. Macropores can be formed by earthworms. Williams and others [5] monitored a contaminated soil subjected to earthworm digestion over 21 days. They found that anecic earthworms such as *Lumbricus terrestris* significantly aided vertical movement of *E. coli* in soil, whereas epigeic earthworms such as *Dendrobaena veneta* significantly aided the lateral movement within the soil.

To prevent groundwater from being contaminated by pathogenic microorganisms in reclaimed water, soil absorption processes are necessary. Because bacteria and viruses have an electrical charge, they can be immobilized as they are adsorbed by soil particles [6]. The retention and reduction of bacterial numbers in soil are complicated processes that involve soil entrapment, nutrients, temperature, moisture content, soil compaction, clay minerals, humic fraction, and toxic components.

Soil must be deep enough and unsaturated to filter pathogens from treated wastewater. It has been reported [7–9] that 0.6 to 1.2 m of unsaturated soil below an on-site wastewater treatment system is sufficient to remove most bacteria and viruses. The main bacterial removal mechanisms as wastewater moves through unsaturated soil are filtration and retention. Bacterial in the size range from 0.2 to 5 μm are entrapped if the soil pores are smaller than this range. The movement of *E. coli* in soil columns decreases with increasing dry bulk density. Specifically, the presence of macropores and degree of compaction impact the transportation of *E. coli* [10,11].

Temperature impacts the survival of *E. coli* in the soil environment. The highest survival of *E. coli* is in the coldest soils [12–14]. Biotic factors affect *E. coli* survival in the soil. Die-off rates of *E. coli* have been shown to be rapid in native soils as compared to pre-sterilized soils. Predation and competition in native soil and wastewater influence *E. coli* survival [12,15]. Indigenous soil microbes can have adverse effects on *E. coli* due to predation and more favorable growth conditions. Sunlight facilitates coliform inactivation on soil surfaces more in the summer than in the winter [13,16].

In general, *E. coli* can potentially move through the soil to groundwater or subsurface drains under wet, cold, and dark winter conditions. Thus, the goal of this study was to examine *E. coli* movement in reuse of reclaimed wastewater scenarios during winter conditions. The specific objective was to determine if winter reuse of undisinfected wastewater moves bacteria to groundwater or surface water through subsurface drainage, thus posing a public health risk. *E. coli* vertical movement was examined in relation to soil temperature, moisture, and depth of saturation at the study site.

2. Materials and Methods

The study site is located within the Deer Creek State Park, which lies on the eastern edge of a till plain in south central Ohio (39.6301° N, 83.2510° W). This region is mainly woodland, and the wastewater treatment plant and irrigation field are in an open meadow area. The system was designed and constructed by the U.S. Army Corps of Engineers in the 1970s.

The wastewater comes from the visitor lodge and cabins, campgrounds, and a golf course. A bar screen serves as the pre-treatment system to remove large solids before discharge into a sewage lagoon (stabilization pond) that provides secondary treatment to remove suspended solids, organic matter, and microorganisms. A holding pond follows the sewage lagoon and wastewater is pumped from there to the spray irrigation field. A chlorination facility is positioned between the treatment lagoon and the holding pond, but the chlorinator was not in use during the study period. The wastewater treatment and irrigation system are outlined in Figure 1. The sewage lagoon is designed to operate between the depths of 0.9 m to 1.5 m by a decanting device. The lagoon hydraulic retention time ranges from 73 to 61 days. The retention time of the holding pond is no more than 28 days [17].

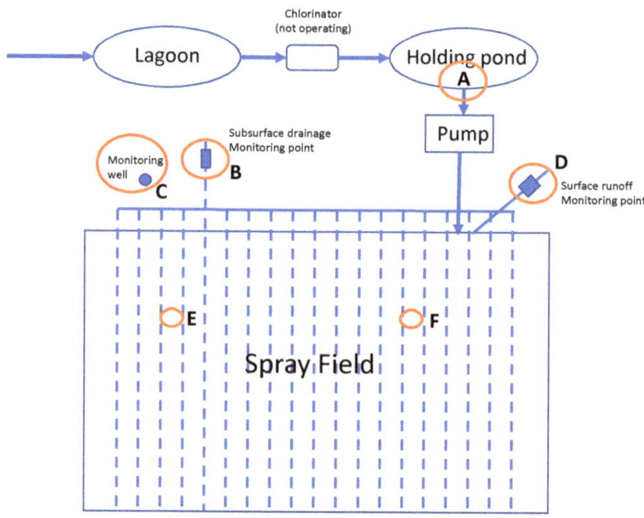

Figure 1. Wastewater treatment system and sampling locations (not to scale). The circles indicate sampling points: A is the sampling location for wastewater, B for subsurface runoff, C for groundwater depth and groundwater samples, D for surface runoff, E for soil moisture and temperature sensors. E and F are the locations of soil samples (scheme taken from [17]).

The spray field is a mowed, grass meadow to the east of the lagoon system. This 48,562 m^2 spray field is equipped with 16 lines each having 14 sprayers on 45 cm risers equipped with impact drive sprinklers (Rain Bird 29B-TNT, Rain Bird Corp., Azura, AZ, USA). The system is divided into four irrigation zones with 12,141 m^2 tracts. The irrigation changes from zone to zone at 15-min intervals.

2.1. *Sampling Locations and Sample Collection Methods*

The sampling locations are shown in Figure 1. Samples were collected at the holding pond (sampling site A) and from the subsurface drain outlet (site B). Soil moisture sensors (Decagon EC-5 sensors coupled with Em50 data loggers, ICT International, Armidale, Australia) and temperature sensors (HOBO TMC20-HD sensors coupled with HOBO U-series 4-channel dataloggers Onset, Bourne, MA, USA) were installed in the spray field.

Groundwater samples and depths were measured in a monitoring well (site C), and surface runoff was sampled at a collection weir (site D). Samples were placed in sterile Nasco Whirl-Pak® (Madison, WI, USA) sampling bags with the volume of ~800 mL and stored on ice in a dark cooler for testing within 2 h. The study site was less than a 45-min drive from the lab.

Irrigation water samples were collected directly from the holding pond (Figure 1A) and in metal pans placed in the irrigation field. Three 33 cm × 23 cm stainless pans were used to collect wastewater samples distributed in the field. Wastewater samples from the holding pond were collected with a graduated dipper with a handle length of 30 cm and 500 mL bowl. Water and wastewater samples were collected at the time of irrigation.

2.2. Sample Analysis Methods

Water samples were tested for *E. coli* by membrane filtration method (USEPA Method 1103.1). Water samples of 100 mL were filtered through sterile filter membranes (47 mm diameter and 0.45 µm pore size), and the membranes were transferred to a modified m-TEC media. The plates were incubated for 2 h at 35 °C followed by 22 h at 44.5 °C. Magenta colored colonies were counted as *E. coli* and the colony counts were converted to colony forming units (CFU) per 100 mL original sample.

The soil survey for the area was conducted by USDA and is published on the Web Soil Survey system [18]. The area includes three soil series, Miamian-Lewisburg (MIB) silt loams, Crosby (CrA) silt loam, and Kokomo (Ko) silty clay loam. Table 1 shows the series and properties of soil. The loamy soils are expected to retain at most 0.35 m^3 water/m^3 soil.

Table 1. Soil series in Deer Creek State Park, Ohio irrigation field.

Map Unit Symbol	Map Unit Name	Hectares
CrA	Crosby silt loam, 0 to 2 percent slope	2.7
Ko	Kokomo silty clay loam, 0 to 2 percent slope	0.6
MIB	Miamian-Lewisburg silt loams, 2 to 6 percent slope	4.7

2.3. Soil Conditions

Soil temperature and moisture sensors were installed in March 2013 to conduct year-long monitoring of the soil conditions. The soil sensors were placed at depths of 15, 30, 46, and 61 cm.

Daily climatology data including maximum and minimum air temperatures and precipitation were retrieved from NOAA National Climatic Data Center [19] at Circleville, OH, USA located 34.8 km (21.6 miles) southeast of the park.

To monitor performance of the irrigation system, the US Army Corp of Engineers installed groundwater monitoring wells. The monitoring well, with 15 cm diameter and 4.3 m depth, was located 15 m outside of the irrigation field. The depth of groundwater was measured by 101 P7 Water Level Meter (Solinst©, Georgetown, ON, Canada).

The subsurface drainage system under the entire field was installed with 32 PVC perforated pipes, 15 cm diameter and 186 m length, and installed a depth of 76 to 107 cm. All subsurface drainage pipes discharged to a central flume for sample collection (sampling site B).

To test for trends in soil conditions changing with time the daily means of the temperatures and soil moisture were compared using a two tailed paired *t*-test.

3. Results

Lagoon-treated wastewater was irrigated at the Deer Creek State Park on an open, flat meadow site. No wastewater was irrigated December 2012 through March 2013 winter season, and this period served as a control. Winter irrigation started on 23 October 2013 and continued through 22 March 2014. Beginning in the 1970s, wastewater was irrigated only during the summer because of concerns over equipment failure due to freezing in the winter. For this project, wastewater was irrigated during the winter of 2014 for the first time and equipment failures were not experienced.

The counts of *E. coli* in subsurface drainage water during the winters of 2013 and 2014 are shown Tables 2 and 3. When sprayed through the air and onto the plant surfaces, the *E. coli* in wastewater is exposed to sunlight as a natural inactivation process. During daytime irrigation cycles, spray irrigation effluent was collected in sterilized metal pans for 2 to 3 h and tested for *E. coli*. The wastewater in pan exposed to 2 to 3 h of sunlight experienced up to 3-log *E. coli* reductions (Table 4).

Table 2. Winter control *E. coli* counts in subsurface drainage. The drain depth ranged from 76 to 107 cm. (Gray cells show groundwater depth above the drain).

Date (m/d/y)	Depth of Precipitation 3 d before Sampling (cm) (3 d/2 d/1 d) [a]	Depth to Groundwater (m)	*E. coli* in Subsurface Drainage (CFU/100 mL)
2/7/13	1.27/0.23/Trace	0.95	10
2/12/13	0.00/0.00/0.25	1.10	10
2/20/13	Trace/0.00/0.18	1.03	<2
2/28/13	0.00/0.00/1.12	1.20	<2
3/5/13	0.03/0.03/0.00	1.30	<2
3/13/13	0.00/0.03/0.51	1.70	<2

[a] Three days, two days, and one day before the sampling.

Table 3. Winter irrigated field *E. coli* counts in subsurface drainage. The drain depth ranged from 76 to 107 cm. (Gray cells show groundwater depth above the drain).

Date (m/d/y)	Wastewater *E. coli* (CFU/100 mL)	Depth to Groundwater (m)	Soil Temperature at 15 cm (°C)	Soil Temperature at 60 cm (°C)	*E. coli* in Subsurface Drainage (CFU/100 mL)
10/23/13	55	2.3	11	14	<2
1/12/14	3	0.88	1.3	3.7	23
2/16/14	1.0×10^3	1.50	0.7	2.9	<2
2/22/14	7.9×10^3	1.60	0.9	2.4	1.1×10^2
3/5/14	7.8×10^3	0.70	0.2	2.7	3.3×10^2
3/11/14	<2	1.60	2	2.6	<2
3/22/14	9.0	1.40	4	4	<2

Table 4. *E. coli* counts in water samples collected in daytime with the metal pan exposed to sunlight.

Date (m/d/y)	*E. coli* in Wastewater (CFU/100 mL)	*E. coli* in Pan after 2 to 3 h (CFU/100 mL)
2/16/14	1.0×10^3	<2
2/22/14	7.9×10^3	40
3/5/14	7.8×10^3	1

Impact of Soil Conditions on E. coli Removal

During the non-irrigated season (December 2012 to March 2013), *E. coli* counts were detected in two out of six subsurface drainage samples. Precipitation events for the non-irrigated season are shown in Table 2. On 7 February 2013, the water table was within the depth of the subsurface drainage system and *E. coli* was detected in the drainage water. Five days later, as the groundwater dropped below the drainage system, *E. coli* was still detected. However, once the groundwater depth stayed below the subsurface drainage system, even with precipitation, *E. coli* could not be detected in the subsurface drainage water, even though the drain continued to flow throughout the winter. With no human wastewater inputs of *E. coli* during this time, the fecal source in these two samples during the non-irrigated period may be wildlife (e.g., deer, racoons, and other animals).

Soil temperature and moisture content were monitored in the irrigation field at Deer Creek State Park during the winter irrigation season. The temperature and moisture were measured at four depths before and after the irrigation practices. The data showed an increasing trend of moisture and temperature with soil depth. The moisture and soil temperature were significantly positively related (with adjusted R^2 value is at least 0.99 and p-value = 0.0016) at the same depth of soil.

The soil moisture increased in response to the application of 4.5–5 cm treated wastewater in the fall (October 23) and in the spring (March 22) (Figure 2). The shallow (15 cm) soil temperature was 10 °C in the fall and after a cold winter rose to 5 °C in the spring.

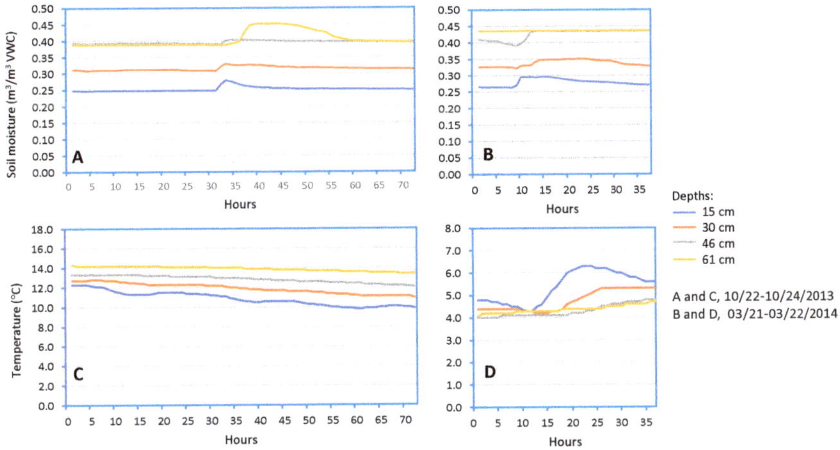

Figure 2. Soil moisture and temperature at four depths in response to irrigation in fall (October 23) and Spring (March 22). (**A**) and (**C**) show soil moisture and temperature in the fall for 10/22 to 10/24, 2013. (**B**) and (**D**) show soil moisture and temperature in the spring for 3/21 to 3/22, 2014.

Throughout the winter the shallow soil temperature dropped below 1.5 °C and was as low as 0.3 °C during January, February, and the beginning of March. During these winter months irrigation of 4.5–6 cm of treated wastewater did not change soil moisture at the site (Figure 3).

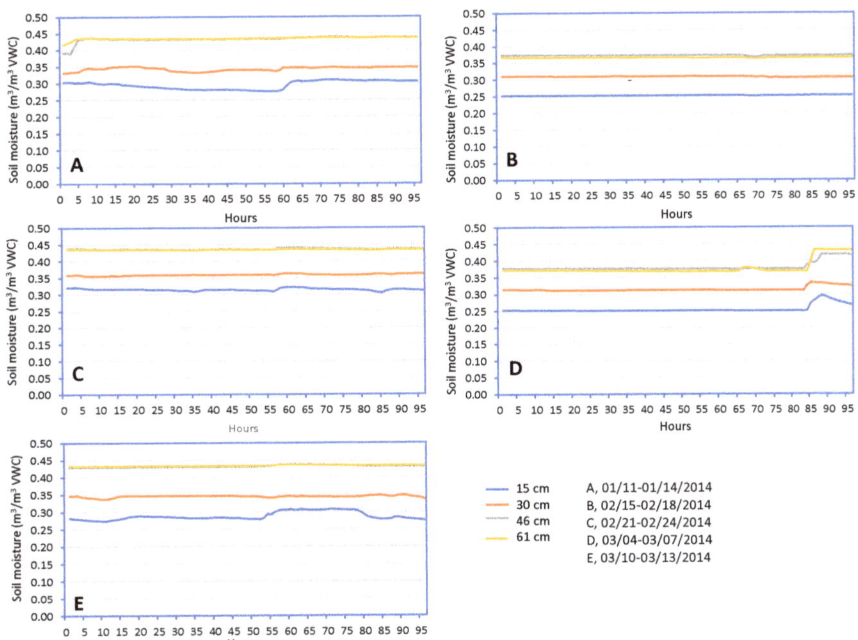

Figure 3. Soil moisture at four depths in response to irrigation during winter months (January 11 through March 13). (**A**) shows soil moisture with depth for 1/11 to 1/14, 2014. (**B**) shows soil moisture with depth for 2/15 to 2/18, 2014. (**C**) shows soil moisture with depth for 2/21 to 2/24, 2014. (**D**) shows soil moisture with depth for 3/04 to 3/07, 2014. (**E**) shows soil moisture with depth for 3/10 to 3/13, 2014.

With warmer weather and more sunlight in the fall and spring, evapotranspiration was drying out the surface soil between precipitation and irrigation events. The drier soil attenuated the movement of wastewater, allowing for natural inactivation of bacteria. When the shallow soil temperature dropped to near freezing and with less sunlight in the winter, evapotranspiration was minimal, and the soil stayed moist near field capacity. The addition of precipitation and irrigated wastewater did not change the soil moisture. With the soil at field capacity, wastewater drained through the moist soil. With cold soil temperature (less than 1.5 °C) natural inactivation was limited, and *E. coli* was moved through the soil to the drainage system.

During the winter irrigation season *E. coli* was detected three times in the subsurface drainage (Table 3). Irrigation water applied on the day of the sampling in combination with precipitation (Table 5) that occurred in the 3 days, washing bacteria to the subsurface drainage system. Even seven days later *E. coli* was detected.

Table 5. Precipitation and groundwater table depth and *E. coli* in wastewater and subsurface drainage during irrigated season. (Gray cells show groundwater depth above the drain).

Date (m/d/yr)	Depth of Precipitation 3 d before Sampling (cm) (3 d/2 d/1 d)	Depth of Irrigated Water (cm)	Depth to Groundwater (m)	Wastewater *E. coli* (CFU/100 mL)	*E. coli* in Subsurface Drain (CFU/100 mL)
10/23/13	0.00/0.00/0.30	5.0	2.3	55	<2
1/12/14	Trace/0.18/1.50	5.0	0.88	3	23
2/16/14	0.00/0.00/0.91	5.0	1.50	1.0×10^3	<2
2/22/14	0.20/0.00/0.89	6.0	1.60	7.9×10^3	1.1×10^2
3/5/14	0.03/0.61/0.00	4.5	0.70	7.8×10^3	3.3×10^2
3/11/14	0.00/0.00/0.00	5.5	1.60	<2	<2
3/22/14	0.00/0.36/Trace	4.5	1.40	9.0	<2

The wastewater irrigated in October and again in January contained very low counts of *E. coli* because natural inactivation in the lagoon facilitated die-off. Some *E. coli* was detected in the drainage system when the water table rose above the level of the drain. Starting in mid-February, *E. coli* counts in the wastewater effluent increased by 3-log (Table 5). High *E. coli* counts were initially attenuated, but within a week high numbers of *E. coli* were moving through the cold, wet soil to the subsurface drainage system.

By the end of March, soil temperatures were beginning to rise, the groundwater table had dropped below the level of the drains, and more sunlight was starting to provide natural inactivation in the wastewater lagoon. *E. coli* counts in the irrigated wastewater dropped and were not detected in the subsurface drainage system.

In this field site, *E. coli* was not detected (<2 CFU/100 mL) in the groundwater, 15 m away from the irrigation field during the irrigation and non-irrigation season. This finding indicates that the horizontal movement of bacteria was limited on the relatively flat land.

4. Discussion

In this project, the main goal was to investigate the movement of *E. coli* in the soil and infiltration through the unsaturated soil column when treated wastewater was applied on soil surface. *E. coli* counts in the reclaimed wastewater were already low due to the natural die-off processes in the lagoon system, which functioned as a stabilization pond. During daytime irrigation, *E. coli* was exposed to sunlight facilitating die-off and the movement of the remaining *E. coli* into the soil was very limited. At least a 2-log reduction in *E. coli* counts was observed when the wastewater was exposed to sunlight for 2 to 3 h, but the bacteria infiltrated into the soil relatively quickly, reducing the chance for natural removal. Standridge [16] concluded that the effect of sunlight on coliform inactivation was most noticeable in late June, when the bacteria were completely removed from samples after 1–2 h of exposure. While this experiment in the present work was conducted during late autumn and winter, when the sun irradiation is not as strong as in June, *E. coli* numbers were still somewhat reduced through natural decay.

In the field study, *E. coli* was not detected in the groundwater, 15 m away from the irrigation field both in irrigation and non-irrigation season. The Ohio Department of Health [20] recommendation is supported by this finding that all components of sewage treatment system should be at least 15 m away from any water supply source.

The irrigation practice of as much as 5 cm over two hours on the soil did not change soil moisture and temperature conditions substantially at the monitored depths. Irrigation at this rate, in combination with precipitation, did contribute to saturated soil conditions by raising the water table in the irrigation field. The irrigation rate of about 2.5 cm/hour at the Deer Creek State Park is similar to a high intensity rainstorm. Slower and more frequent irrigation would help to slow the movement of reused wastewater through wet, cold soil.

The low soil temperatures recorded in this study appeared to enhance survival and movement of *E coli*. These findings were similar to a microcosm study, where Vidovic and others [12] looked at *E. coli* survival in soils at −21, 4 and 22 °C. They found the highest survival of *E. coli* in the coldest soils. Noble and others [13] examined survival of *E. coli* in sewage and stormwater runoff. They found bacteria were inactivated more rapidly at 20 °C than at 14 °C.

During cold, wet soil conditions *E. coli* moved through 76 to 107 cm to the subsurface drain. Peterson and Ward [7] studied the movement of common wastewater bacteria in a coarse soil that was irrigated with reclaimed wastewater. They found that 60 to 120 cm of unsaturated soil below an on-site wastewater treatment system was sufficient to remove most bacteria and viruses in the test soils. Karathanasis and others [8] studied the movement of fecal coliforms through 30, 45, and 60 cm in undisturbed soil monoliths in the laboratory. They found movement of bacteria through 30 cm of coarse textured soil and that removal improved with 60 cm thickness. This study suggests that 60 cm depth is sufficient for *E. coli* removal in winter in this meadow system, and summertime irrigation should need less depth for comparable removal. Regardless of the season, supplemental disinfection should be practiced.

5. Conclusions

In summary, the outcome of this study can fill the knowledge gap about survival and movement of potential enteric bacterial pathogens in subsurface soil and ground water when wastewater is applied on soil during cold and wet conditions. Winter reuse of undisinfected wastewater does pose a public health risk to surface water through subsurface drainage. This project went on to help understand how soil conditions impact *E. coli* movement in cold, wet soil conditions.

This study investigated the removal of *E. coli* as a function of the soil temperature, moisture, and depth of soil saturation. When soil temperature was coldest in February, the soil remained wet and irrigation practices did not change the soil conditions significantly during cold weather when evapotranspiration is limited. As a result, the irrigated wastewater drained quickly through the wet and cold soil and had lower *E. coli* removal, allowing their movement to subsurface drainage.

Wastewater irrigation practices during the winter do not change the soil temperature and moisture in a short period, but high soil moisture content tends to provide a better pathway for *E. coli* movement through the soil. Greater depth to water tables shows higher *E. coli* removal efficiency as the unsaturated soil condition enhances natural inactivation processes. *E. coli* traveled downward to subsurface drainage in response to winter irrigation, and the likely driving force of this movement was the presence of macropores combined with the hydraulic force provided by irrigation water and precipitation.

Temperature appears to be the most important factor in limiting natural inactivation, allowing the movement of *E. coli* in wastewater to the subsurface drainage systems. The irrigation of undisinfected wastewater to sites with subsurface drainage is not an appropriate practice for winter as it provides the opportunity to move *E. coli* too quickly through the soil to allow for natural die-off. To protect public health, treated wastewater should be disinfected year-round and lower winter irrigation rates should be considered in an effort to control slow movement of any remaining pathogens through wet, cold soil.

Author Contributions: Conceptualization, K.M. and O.H.T.; methodology, G.D., K.M., J.L., and O.H.T.; formal analysis, G.D. and K.M.; investigation, G.D.; resources, K.M. and J.L.; writing—original draft preparation, G.D.; writing—review and editing, K.M., J.L., and O.H.T.; visualization, G.D.; supervision, K.M.; project administration, K.M.; funding acquisition, K.M. All authors have read and agreed to the published version of the manuscript.

Funding: This research was funded by the Ohio Water Development Authority. Salary and partial research support to K.M. were provided by state and federal (USDA-NIFA) funds appropriated to the Ohio Agricultural Research and Development Center. Additional research support from the Baas Memorial Endowment Fund is gratefully acknowledged.

Institutional Review Board Statement: Not applicable.

Informed Consent Statement: Not applicable.

Data Availability Statement: Data is available in Master's Thesis at http://olc1.ohiolink.edu/record=b35798607 (accessed on 29 July 2021).

Acknowledgments: We thank the operators at the Deer Creek State Park, for their support in operating the irrigation system through the winter and for their assistance in sample collection.

Conflicts of Interest: The authors declare no conflict of interest. The funders had no role in the design of the study; in the collection, analyses, or interpretation of data; in the writing of the manuscript; or in the decision to publish the results.

References

1. US EPA. 5.11 Fecal Bacteria. Water: Monitoring and Assessment. 2012. Available online: Archive.epa.gov/water/archive/web/html/vms511.html (accessed on 3 March 2021).
2. USDA. Soil Quality Indicators. Natural Resources Conservation Service. 2008. Available online: www.nrcs.usda.gov/Internet/FSE_DOCUMENTS/nrcs142p2_053288.pdf (accessed on 22 February 2021).
3. Pallardy, S.G. Absorption of water and ascent of sap. In *Physiology of Woody Plants*, 3rd ed.; Elsevier: Amsterdam, The Netherlands, 2008; pp. 287–323. [CrossRef]
4. Smith, M.S.; Thomas, G.W.; White, R.E.; Ritonga, D. Transport of *Escherichia coli* through intact and disturbed soil columns. *J. Environ. Qual.* **1985**, *14*, 87–91. [CrossRef]
5. Williams, A.P.; Roberts, P.; Avery, L.M.; Killham, K.; Jones, D.L. Earthworms as vectors of *Escherichia coli* O157:H7 in soil and vermicomposts. *FEMS Microbiol. Ecol.* **2006**, *58*, 54–64. [CrossRef] [PubMed]
6. Tyler, E.J.; Laak, R.; McCoy, E.; Sandhu, S.S. The soil as a treatment system. In Proceedings of the Second National Home Sewage Treatment Symposium, Chicago, IL, USA, 12–13 December 1977; ASAE: St. Joseph, MI, USA, 1977; pp. 22–37.
7. Peterson, T.C.; Ward, R.C. Bacterial Transport in Coarse Soils Beneath On-site Wastewater Treatment Systems. *Colo. State Univ. Tech. Bull.* **1987**, *4*, 23–25.
8. Karathanasis, A.D.; Mueller, T.G.; Boone, B.; Thompson, Y.L. Effect of soil depth and texture on fecal bacteria removal from septic effluents. *J. Water Health* **2006**, *4*, 395–404. [CrossRef] [PubMed]
9. Chen, G.S. *Typhimurium* and *E. coli* O157:H7 retention and transport in agricultural soil during irrigation practices. *Eur. J. Soil Sci.* **2012**, *63*, 239–248. [CrossRef]
10. Miles, R.J.; Rubin, R.; West, L. Fecal coliform distribution under pressure dosed onsite wastewater systems. In Proceedings of the Eleventh Individual and Small Community Sewage Systems Conference, Warwick, RI, USA, 20–24 October 2007; ASABE: St. Joseph, MI, USA, 2007; p. 6, ASABE Publication Number 701P1107.
11. Artz, R.R.E.; Townend, J.; Brown, K.; Towers, W.; Killham, K. Soil macropores and compaction control the leaching potential of *Escherichia coli* O157:H7. *Environ. Microbiol.* **2005**, *7*, 241–248. [CrossRef] [PubMed]
12. Vidovic, S.; Block, H.C.; Korber, D.R. Effect of soil composition, temperature, indigenous microflora, and environmental conditions on survival of *Escherichia coli* 0157:H7. *Can. J. Microbiol.* **2007**, *53*, 822–829. [CrossRef] [PubMed]
13. Noble, R.T.; Lee, I.M.; Schiff, K.C. Inactivation of indicator micro-organisms from various sources of faecal contamination in seawater and freshwater. *J. Appl. Microbiol.* **2004**, *96*, 464–472. [CrossRef] [PubMed]
14. Cools, D.; Merckx, R.; Vlassak, K.; Verhaegen, J. Survival of *E. coli* and *Enterococcus spp.* derived from pig slurry in soils of different texture. *Appl. Soil Ecol.* **2001**, *17*, 53–62. [CrossRef]
15. Bradford, S.A.; Segal, E. Fate of indicator microorganisms under nutrient management plan conditions. *J. Environ. Qual.* **2009**, *38*, 1728–1738. [CrossRef]

16. Standridge, J.H. Effects of sunlight on coliform in water samples. *Opflow* **1986**, *12*, 6–7. [CrossRef]
17. U.S. Corps of Engineers. *Operation and Maintenance Manual On-Land Wastewater Facilities of Deer Creek State Park*; U.S. Corps of Engineers: Washington, DC, USA, 1973.
18. USDA. Web Soil Survey, Pickaway County, Ohio. Available online: Websoilsurvey.sc.egov.usda.gov/App/HomePage.htm (accessed on 22 April 2014).
19. NOAA. National Climate Data Center, Circleville, Ohio. Available online: http://gis.ncdc.noaa.gov/map/viewer/#app=cdo&cfg=cdo&theme=daily&layers=111&node=gis (accessed on 23 May 2014).
20. Ohio Department of Health. Ohio Administrative Code Chapter 3701-29: Sewage Treatment System Rules. 23–26. 2014. Available online: https://defiancecohealth.org/wp-content/uploads/2014/02/Defiance-County-Supplement-to-3701-29-of-the-OAC.pdf (accessed on 22 May 2014).

processes

MDPI

Article

Bisphenol A Effects in Aqueous Environment on *Lemna minor*

Cristian-Emilian Pop [1,2,*], Sorin Draga [3], Roxana Măciucă [2,4], Roxana Niță [3], Nicolae Crăciun [5] and Robert Wolff [6]

1 Department of Biochemistry and Molecular Biology, Faculty of Biology, University of Bucharest, 91-95 Splaiul Independenței Str., 050095 Bucharest, Romania
2 Non-Governmental Research Organization Biologic, 14 Schitului Str., 032044 Bucharest, Romania; contact@ngobiologic.com
3 R&D Department, Biotehnos SA, 3-5 Gorunului Str., 075100 Otopeni, Romania; sorin.draga@biotehnos.com (S.D.); rnita@biotehnos.com (R.N.)
4 Department of Biology, Faculty of Biology, University of Bucharest, 91-95 Splaiul Independenței Str., 050095 Bucharest, Romania
5 Zoology Section, Department of Biochemistry and Molecular Biology, Faculty of Biology, University of Bucharest, 91-95 Splaiul Independenței Str., 050095 Bucharest, Romania; nicolai.craciun@bio.unibuc.ro
6 College of Nursing and Public Health, South University, 9 Science Ct., Columbia, SC 29203, USA; rwolff@southuniversity.edu
* Correspondence: pop.cristian-emilian@s.bio.unibuc.ro; Tel.: +40-740627944

Abstract: The link between different plastic waste pollutants and their impact on the natural aquatic environment and food chain remains a constant and growing issue. Bisphenol A (BPA), a known endocrine disruptor produced in large quantities primarily in the industry of polycarbonate plastics, can accumulate in vegetal and animal tissue, thus magnifying through trophic levels. In this study we exposed viable specimens of the aquatic plant *Lemna minor* under controlled conditions to 50, 100 and 200 ppm BPA levels in order to partially observe the toxic effects of BPA. Colonies ceased to form during the exposure and chlorosis was present especially in the 100 ppm group. Interestingly enough, a high density formation of non-fermenting bacteria as well as coliforms was also observed in the BPA exposed cultures but not in the control groups. The levels of Malondialdehyde (MDA) in the vegetal tissue indicated cellular insults and severe damage, results that were correlated with the HPLC BPA determined concentrations of 0.1%, 0.2% and 0.4%.

Keywords: Bisphenol A; BPA; aqueous; ecotoxicity; bacteria; *Lemna minor*

Citation: Pop, C.-E.; Draga, S.; Măciucă, R.; Niță, R.; Crăciun, N.; Wolff, R. Bisphenol A Effects in Aqueous Environment on *Lemna minor*. *Processes* **2021**, *9*, 1512. https://doi.org/10.3390/pr9091512

Academic Editor: Avelino Núñez-Delgado

Received: 26 July 2021
Accepted: 23 August 2021
Published: 26 August 2021

1. Introduction

Plastic pollution induces an alarming impact on aquatic ecosystems, becoming a severe issue worldwide. The unprecedented accumulation of growing plastic waste pollutants in aquatic ecosystems leads to disturbances to ecosystems' structure and functions [1]. It has been determined that there are chemical additives in the composition of some microplastics that are known as reproductive toxins, carcinogens and mutagens [2]. These micropollutants can bioaccumulate in the food chain and the resulting trophic transfer of microplastics and chemical constituents can have a serious impact on the stability of the ecosystem [1].

Bisphenol A (BPA), a compound used in the production of polycarbonate plastics and epoxy resins, has been detected in aquatic ecosystems, with about 100 tons of BPA being released into the atmosphere annually during production [3]; this does not include plastic trash or BPA from thermal papers. Evidence of BPA has been discovered in river and marine sediments in variable concentrations [4,5].

Both vegetal and animal tissues are able to absorb BPA, which has proven to be an important endocrine disruptor [6,7]. In the case of mammalian systems, BPA can interact with nuclear receptors, mimicking the action of natural hormones including progestins, estrogens, androgens, glucocorticoids, vitamin D3, thyroid and retinoid hormones. Through these interactions, BPA can affect the normal function of these receptors even in very low doses [8]. In vegetal organisms, BPA can accumulate in the tissue, interacting with lipids and therefore inducing modifications into the cell membrane which can lead to cell damage and, eventually, to apoptosis [9].

Aquatic floating plants such as *Lemna minor* have the capacity to accumulate for the removal of micropollutants. *Lemna minor* has a wide distribution around the world, a simple structure and can be easily cultivated. It can double its biomass in two days, growing faster than most other plants and proving to be an ideal test organism [10,11].

Few studies have explored the effects of BPA on macrophytic plants in aquatic environments. Adamakis et al. [12] found leaf elongation was impaired and defects in the cytoskeleton of a seagrass occurred at BPA concentrations that were environmentally relevant. In an industry sponsored study, Mihaich et al. [13] found a significant decrease in frond density and growth rate in *Lemna gibba*, but were dismissive of environmental impact. Lemna was shown to remove BPA from the water [14] due to uptake.

The objective of this paper was to identify the toxicity of BPA in the aquatic plant *Lemna minor*, known to be a sensitive test organism to pollutants. Viable specimens of *Lemna minor* were studied under controlled conditions to different BPA levels. The toxic effects of BPA were measured by lipid peroxidation, growth inhibition and the total chlorophyll content.

2. Materials and Methods

2.1. Bisphenol A Aqueous Solution

The protocol used to obtain the experimental Bisphenol A (BPA) aqueous solutions were performed after Motoyama et al. [15], in which 20, 10 and 5 mg of BPA $\geq$ 99% (Sigma-Aldrich, Darmstadt, Germany) were dissolved in 1 mL anhydrous ethanol, then 99 mL of culture water was added. The azeotrope BPA containing solutions were stirred at 600 rpm on a magnetic stirrer hotplate at 25 °C for 2 h; due to evaporation each solution was supplemented with 8 mL of distilled water to reach the total volume of 100 mL. As the maximum solubility of BPA in water is 200–300 ppm [16], a higher concentration was not used as it had no fundament.

2.2. Lemna Minor Culturing Conditions

For culturing *Lemna minor*, 500 mL of lake water was collected and vacuum filtered using qualitative filter papers grade 202 with 5–8 μm retention (Frisenette, Knebel, Denmark). The filtered water was then re-filtered with sterile 0.22 μm Polyethersulfone (PES) syringe filters (Isolab, Schweitenkirchen, Germany) in order to remove fine particles including bacteria and fungi prior to BPA being added, as stated above. Filtered water without BPA was used as the control culture media.

Mature, actively growing specimens of *Lemna minor* were selected, rinsed thoroughly with distilled water and then enclosed in sterile 400 mL capacity transparent tissue culture flasks (VWR, Philadelphia, PA, USA), each with 100 mL of the culture water containing BPA or the control. The specimens were incubated at 25 °C under controlled circadian conditions: 12 h light/12 h dark, where the light was provided by a fluorescent 4000K white light lamp with 300 lumens intensity. The experiment took place for 1 week at concentrations of 200, 100 and 50 ppm of BPA.

2.3. Macroscopic Observations

Lemna minor plants were evaluated macroscopically in order to observe budding formation and possible abnormalities. The culture water turbidity was measured at A_{600} against a blank, and Eosin Methylene Blue base (EMB base) (Himedia, Mumbai, India) plates were inoculated with culture water from the BPA experimental groups and control group. Apart from the control group, a separate group which contained only the culture water that was kept in the same experimental conditions was also assayed microbiologically. EMB base is a modification of Levine EMB Agar; without lactose, the peptone serves as a source of carbon, nitrogen, and other essential growth nutrients while eosin and methylene blue dyes differentiate between lactose fermenters and non-fermenters. The chromogenic media is used for the differentiation of *Escherichia coli* and *Enterobacter aerogenes* as well as for the rapid identification of *Candida albicans*, and is also recommended for the detection, enumeration and differentiation of members of the coliform group [17].

2.4. Total Chlorophyll Content Assay

Chlorophyll content was determined by first washing the plants with distilled water prior to their roots being removed, and then briefly dabbing them dry on paper towels before weighing [18]. Approximately 300 mg of tissue was kept in 5 mL of anhydrous ethanol in the dark at 4 °C for 24 h. The samples were gently shaken and 1:10 (v/v) dilutions in anhydrous ethanol were made using the supernatant. A clear ethanol solution served as the reference blank, and the absorbance of the plant extracts was calculated from the ratio of the response of the extract solution to the reference blank, according to the equations: concentration in units of μg/mL of chlorophyll $a = 13.70(A_{665}) - 5.76(A_{649})$; and concentration in units of μg/mL of chlorophyll $b = 25.8(A_{649}) - 7.60(A_{665})$ [19].

2.5. Lipid Peroxidation—TBARS Assay

A thiobarbituric acid reactive substances (TBARS) assay was adapted after Buege and Aust, 1978 [20]. TBARS are formed as byproducts of lipid peroxidation (i.e., as degradation products of fats) determined as malondialdehyde (MDA), a pink compound that can be spectrophotometrically read. To perform the assay, plants were weighed and 10% homogenate was made in cold 0.154M KCl. The homogenate was then centrifuged for 15 min at 10.000 rpm (Eppendorf Minispin Plus, Leipzig, Germany). The supernatant was collected in clean tubes and to each 500 μL of supernatant, 500 μL of 10% trichloroacetic acid (Fisher, New Jersey, NJ, USA) was added. The samples were then centrifuged at 3500 rpm and the supernatant was collected. To 500 μL of supernatant, 500 μL of 0.67% thiobarbituric acid (Merck, Darmstad, Germany) was added. The samples were then kept on boiling water for 15 min, left to cool and then measured at 531 nm (Shimadzu, Tokyo, Japan) against blank. Concentrations were measured using the molar extinction coefficient of 0.641×10^{-5} to obtain the MDA concentration in nM [20].

2.6. HPLC Analysis

Lemna minor samples were placed at −80 °C in an ultrafreezer (Thermofisher, Lenexa, KS, USA) for 48 h and then cryodesiccated for 5 days. The resulting dry mass was weighed and HPLC analyzed. The system used was an Agilent series 1260 Infinity (Agilent, Santa Clara, CA, USA), equipped with a binary pump and DAD detector. Data analysis was performed via OpenLAB CDS software, version C 01.02. The chromatographic separation was made on a Kinetex C18 column (100 × 2.1 mm, 2.6 μm, Phenomenex, Torrance, CA, USA) at 35 °C, using a mobile phase containing water-acetonitrile at 70:30 (v/v) with isocratic elution, at 0.3 mL/min flow rate, followed by column washing with 70% acetonitrile (Carlo Erba, Milano, Italy) and then re-equilibrated to the initial condition. The detection was monitored at 210 nm; the injection volume was 2 μL.

Cryodesiccated samples were analyzed in duplicates, as follows: for each group ≈ 5 mg of cryodesiccated sample was weighed and 1 mL of acetonitrile was added prior to a 20 min ultrasonification at 25 °C. The resulting extract was filtered through a 0.2 μm PTFE filter (Agilent, Captiva EconoFilter, Santa Clara, CA, USA) and then HPLC analyzed.

For sample quantification, a standard solution of ≥99% Bisphenol A (Sigma-Aldrich, city, Germany) 1 mg/mL was prepared in LC-MS grade acetonitrile (Carlo Erba, Milano, Italy), dilutions were made at 10, 100 and 500 μg/mL concentrations in order to obtain a calibration curve (Figure 1).

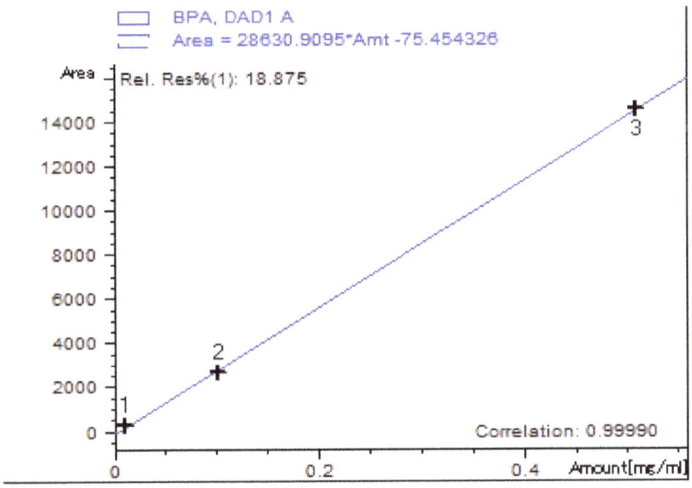

Figure 1. Bisphenol A calibration curve.

3. Results

3.1. Macroscopic Observations

Visual investigation revealed chlorosis (yellowing of leaf tissue) for the groups exposed to 200 and 100 ppm. No budding formation was observed. The 50 ppm group presented normal leaf color, but with no budding formations, while the control group showed no alteration and presented with budding formations.

The *Lemna minor* control group as well as the culture water control group were clear with no significant OD_{600} (Optical Density) increase against blank, followed by an increase at 50 ppm that peaked at 100 ppm and decreased for the 200 ppm group (Figure 2). Biofilms were observed on the roots of plants in the 100 ppm group but not in the other groups. The EMB culture plates were read after a 24 h incubation at 35 °C and revealed luxuriant colonies of non-fermenting Gram-negative bacteria, as well as punctiform colonies of *Escherichia coli*. The colonies were identified on the chromogenic media by a colorless aspect for non-fermenting Gram-negative bacteria and a pink color for *Klebsiella aerogenes* colonies, respectively blue-black with a green metallic sheen for *Escherichia coli* colonies. No colonies were present on the plates inoculated from the *Lemna minor* control group and neither from the culture media control group.

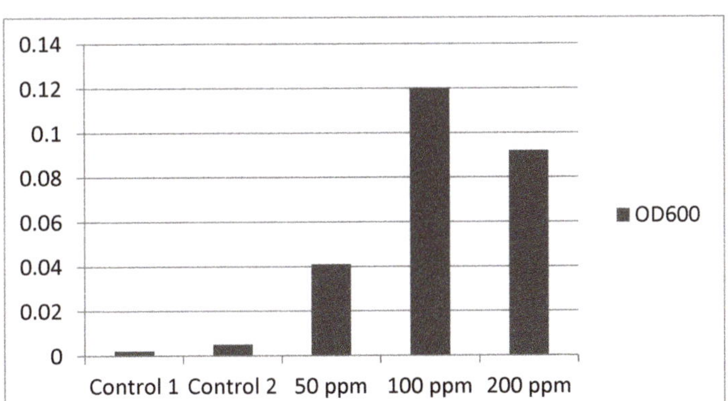

Figure 2. Spectrophotometric measurements of the culturing media turbidity at 600 nm (OD_{600}).

3.2. Total Chlorophyll Content

Chlorophyll levels were found at their lowest in the 100 ppm group, followed by the 200 and 50 ppm groups (Figures 3 and 4), the results are multiplied by a factor of 10 due to the 1:10 dilution made to perform the readings.

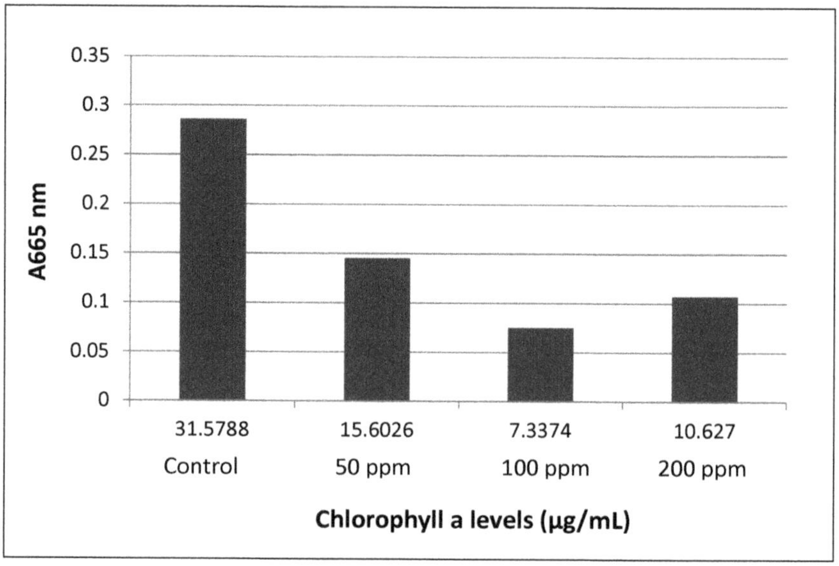

Figure 3. Chlorophyll *a* content at 665 nm absorption.

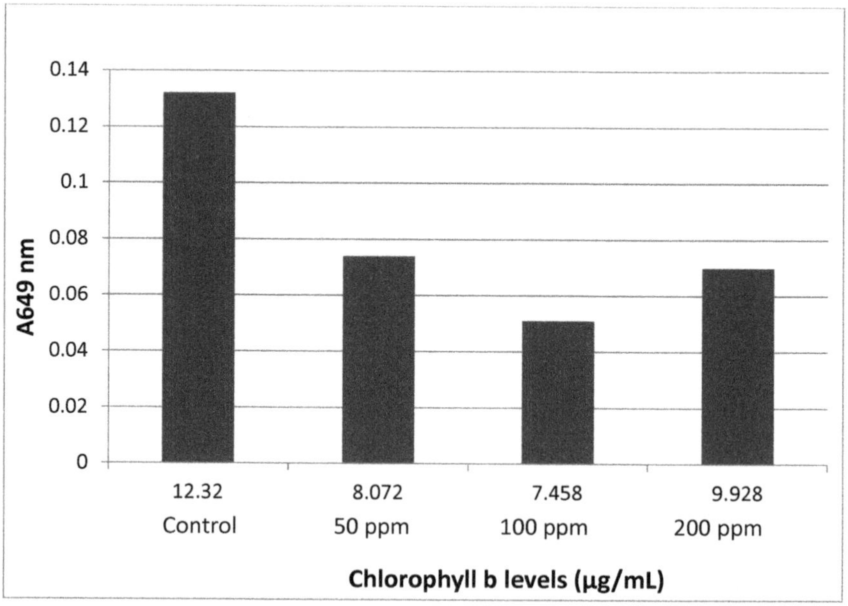

Figure 4. Chlorophyll *b* content at 649 nm absorption.

3.3. Oxidative Stress Levels

TBARS Assay results showed high absorbance at 531 nm for the control and 50 ppm groups but low absorbance with small variations for the 100 and 200 ppm groups, the measurements were made against blank, which represented all the reagents in the same ratio that underwent the same thermal treatment but without the plant tissue. MDA concentration was measured using the formula:

$$\text{MDAconcentrationinnM} = \frac{\text{Absorbanceat531nm}}{\text{molarextinctioncoeficient}(\varepsilon)}, \text{where } \varepsilon = 0.641 \times 10^{-5}.$$

The results are shown in Figure 5, converted to μM.

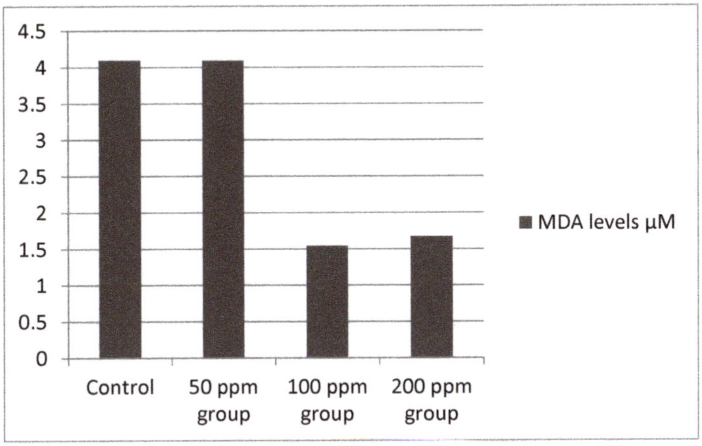

Figure 5. MDA content at 531 nm absorption.

3.4. HPLC Results

The HPLC chromatograms (Figure 6) revealed BPA absorption in all three exposed groups for P1, P2, P3 and none detected in the control group P4. The analysis was performed in duplicate where P1-1, P1-2 represent plants exposed to 50 ppm; P2-1, P2-2 represent plants exposed to 100 ppm; P3-1, P3-2 represent plants exposed to 200 ppm and P4-1, P4-2 represent the plant control group. The numerals I, II and III represent the culture water with BPA at 50, 100 and 200 ppm. (Table 1).

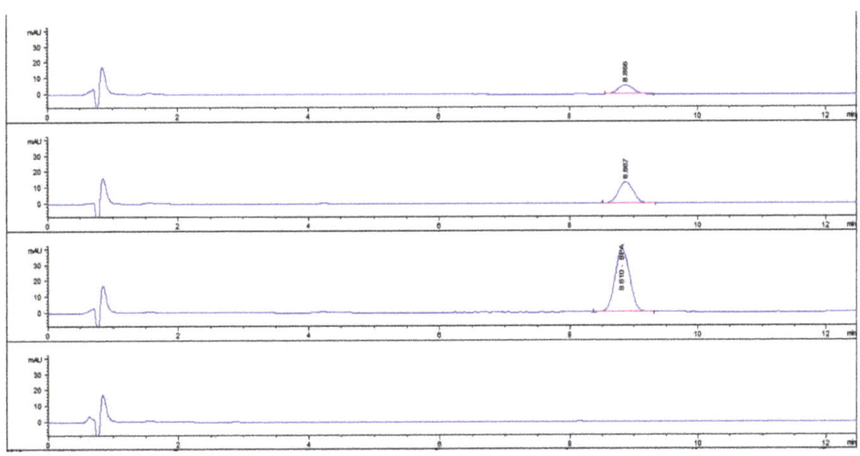

Figure 6. HPLC Chromatogram representing, from the bottom to the top, the samples P4-1, P3-1, P2-1 and P1-1.

Table 1. BPA content in *Lemna minor* samples.

Sample	Sample Mass (mg)	BPA in Sample Extract (mg/mL)	BPA in Sample Extract Expressed in Percentage (%)
P1-1	5.1	0.0057	0.11
P1-2	5.3	0.0063	0.12
P2-1	5	0.009955	0.20
P2-2	5	0.0128	0.26
P3-1	5.2	0.0243	0.47
P3-2	5.4	0.0257	0.48
P4-1	5.2	0	0
P4-2	5.3	0	0
I		0.045	
II		0.102	
III		0.201	

4. Discussion

Effects such as chlorosis and cessation of bud formation were observed macroscopically, as well as high densities of non-lactose fermenting bacteria that were confirmed by inoculation on plates with the EMB selective media. The estrogenic effect of BPA as well as its capacity to stimulate abnormal growth was first accidentally discovered after autoclaving polycarbonate culture flasks in a study to determine whether *Saccharomyces cerevisiae* produced estrogens [21,22], the culture media was prepared using distilled water autoclaved in the polycarbonate flasks from which BPA leaked. Although the microbial growth was facilitated in our study by BPA levels, chlorosis and cessation of budding could be linked to the multiple effects of BPA.

According to some authors [21–23] alterations of chlorophyll content can serve as an indicator of the physiological status of a photosynthetic organism. Therefore, it has been frequently used in ecotoxicological studies for the assessment of toxicity of inorganic and organic chemicals. It is thus possible to conclude that BPA challenged the production of photosynthetic pigments impairing photosynthesis in this study. One study found that higher microbial densities with added bacteria appeared to be cause of the degradation of chlorophyll a in two marine algae [24]. It is possible that the lower chlorophyll levels found in this study corresponding to higher optical density readings may be at least partially responsible for the results obtained. This may account for the reduced chlorophyll levels seen in the 100 ppm BPA cultures as well as the somewhat reduced levels in the 200 ppm BPA cultures. While the apparent levels of bacteria seem to be associated with a reduction in chlorophyll, the causative factors are not clear. While the increase in optical density is expected in the experimental cultures, and the reduction in chlorophyll levels probably related in part to the BPA concentrations, the effect of the BPA on bacterial growth and chlorophyll degradation needs to be fully explored. Lipid peroxidation was a distinctive and sensitive parameter showing the degree to which reactive oxygen species induced damage, as the biphenolic compound had a surprising effect on *L. minor* which was ineffective in coping with the cellular level insults of BPA for 100 and 200 ppm concentrations. This resulted in an interestingly low lipid peroxidation concentration and therefore low absorbance values at A_{531} compared to the control and 50 ppm groups.

A molecular dynamics simulation study [9] indicated that BPA can easily enter the membrane from the aqueous phase in an isolated phospholipid bilayer. With the increasing concentrations of BPA in the membrane, BPA tends to aggregate and form into cluster while dipalmitoyl phosphatidylcholine (DPPC) lipids are pulled out and adsorbed on the cluster surface, leading to membrane pore formation. Observations indicated that the lipid extraction results mainly from the dispersion interactions between BPA cluster and lipid tails, as well as weak electrostatic attractions between lipid headgroups and the two hydroxyl groups on BPA. Such lipid extraction and pore formation cause cell membrane damage that leads to chlorosis and apoptosis, revealing another cytotoxicity aspect of BPA.

5. Conclusions

Lemna minor, a macrophyte known for it's abilities in water purification processes, demonstrated a BPA absorption of ≈0.1%, 0.2% and 0.4% during the time in culture in this study. Consequences will include detrimental effects regarding its mitosis, cellular membrane integrity and viability. All of these could occur with significant environmental damage to all aquatic organisms.

The findings determined that the lack of budding formations at all experimental concentrations is an important concern from BPA pollution of aquatic systems. The low concentrations of Chlorophyl a in the group exposed to 100 ppm, and a part of the results in the 200 ppm cultures, could be explained by both the toxic effects directly, the high density of the microbial community, or both of these in concert.

Although toxic to eukaryotic organisms, the hormone-like properties of BPA appears to strongly enhance growth in some bacteria that are naturally found in freshwater environments. The detection of the increased bacterial growth of the coliform bacteria and production of a biofilm on the plants in the 100 ppm culture is a new and potentially important finding. This may occur through the epigenetic actions of BPA [25]

The implications of BPA pollution causing an increase in the growth of some bacterial species raises many important ecological questions. The changes in bacterial composition of the culture microbiome and growth not shown in other culture conditions raises other critical questions regarding ecosystem functioning and the complex effects of this endocrine disruptor.

A review [26] found that surface waters exceeded the proposed predicted no-effect concentrations in Europe 53.1% of the time, and 34.6% in North America. These data along with the new findings in this report indicate a clear and urgent need to understand the effect of BPA on microorganisms and the direct or indirect effect on chlorophyll in surface waters.

Author Contributions: Conceptualization, C.-E.P., R.W. and N.C.; methodology, C.-E.P., S.D. and R.N.; software, S.D.; investigation, N.C.; resources, C.-E.P. and N.C.; data curation, R.W.; writing—original draft preparation, R.M., C.-E.P.; writing—review and editing, R.M., C.-E.P., R.W.; funding acquisition, C.-E.P. and N.C. All authors have read and agreed to the published version of the manuscript.

Funding: This research was made possible with the support of the Non-Governmental Research Organization Biologic.

Institutional Review Board Statement: Not applicable.

Informed Consent Statement: Not applicable.

Conflicts of Interest: The authors declare no conflict of interest.

References

1. Laws, A.E. *Aquatic Pollution: An Introductory Text*, 4th ed.; Wiley: Hoboken, NJ, USA, 2017; pp. 12–13. ISBN 978-1-119-30450-0.
2. Wright, S.L.; Kelly, F.J. Plastic and Human Health: A Micro Issue? *Environ. Sci. Technol.* **2017**, *51*, 6634–6647. [CrossRef] [PubMed]
3. Markey, C.M.; Michaelson, C.L.; Sonnenschein, C.; Soto, A.M. Alkylphenols and Bisphenol A as environmental estrogens. In *The Handbook of Environmental Chemistry. Part L, Endocrine Disruptors—Part I*; Metzler, M., Ed.; Springer: Berlin/Heidelberg, Germany, 2001; Volume 3, pp. 129–153.
4. Flint, S.; Markle, T.; Thompson, S.; Wallace, E. Bisphenol A exposure, effects, and policy: A wildlife perspective. *J. Environ. Manag.* **2012**, *104*, 19–34. [CrossRef]
5. Koh, C.H.; Khim, J.S.; Villeneuve, D.L.; Kannan, K.; Giesy, J.P. Characterization of trace organic contaminants in marine sediment from Yeongil Bay, Korea: 1. Instrumental analyses. *Environ. Pollut.* **2006**, *142*, 39–47. [CrossRef] [PubMed]
6. Oehlmann, J.; Schulte-Oehlmann, U.; Kloas, W.; Jagnytsch, O.; Lutz, I.; Kusk, K.O.; Wollenberger, L.; Santos, E.M.; Paull, G.C.; Van Look, K.J.; et al. A critical analysis of the biological impacts of plasticizers on wildlife. *Philos. Trans. R. Soc. B Biol. Sci.* **2009**, *364*, 2047–2062. [CrossRef] [PubMed]

7. Canesi, L.; Fabbri, E. Environmental Effects of BPA: Focus on Aquatic Species. *Dose Response* **2015**, *13*, 1559325815598304. [CrossRef]
8. Li, L.; Wang, Q.; Zhang, Y.; Niu, Y.; Yao, X.; Liu, H. The Molecular Mechanism of Bisphenol A (BPA) as an Endocrine Disruptor by Interacting with Nuclear Receptors: Insights from Molecular Dynamics (MD) Simulations. *PLoS ONE* **2015**, *10*, e0120330. [CrossRef]
9. Chen, L.; Chen, J.; Zhou, G.; Wang, Y.; Xu, C.; Wang, X. Molecular Dynamics Simulations of the Permeation of Bisphenol A and Pore Formation in a Lipid Membrane. *Sci. Rep.* **2016**, *6*, 33399. [CrossRef] [PubMed]
10. Chen, G.; Fang, Y.; Huang, J.; Zhao, Y.; Li, Q.; Lai, F.; Xu, Y.; Xueping, T.; He, K.Z.; Jin, Y.; et al. Duckweed systems for eutrophic water purification through converting wastewater nutrients to high-starch biomass: Comparative evaluation of three different genera (*Spirodela polyrhiza, Lemna minor* and *Landoltia punctata*) in monoculture or polyculture. *RSC Adv.* **2018**, *8*, 17927–17937. [CrossRef]
11. Fekete-Kertész, I.; Kunglné-Nagy, Z.; Gruiz, K.; Magyar, Á.; Farkas, É.; Molnár, M. Assessing Toxicity of Organic Aquatic Micropollutants Based on the Total Chlorophyll Content of Lemna minor as a Sensitive Endpoint. *Period. Polytech. Chem. Eng.* **2015**, *59*, 262–271. [CrossRef]
12. Adamakis, I.S.; Malea, P.; Panteris, E. The effects of Bisphenol A on the seagrass *Cymodocea nodosa*: Leaf elongation impairment and cytoskeleton disturbance. *Ecotoxicol. Environ. Saf.* **2018**, *157*, 431–440. [CrossRef]
13. Mihaich, E.M.; Friederich, U.; Caspers, N.; Hall, A.T.; Klecka, G.M.; Dimond, S.S.; Staples, C.A.; Ortego, L.S.; Hentges, S.G. Acute and chronic toxicity testing of Bisphenol A with aquatic invertebrates and plants. *Ecotoxicol. Environ. Saf.* **2009**, *72*, 1392–1399. [CrossRef] [PubMed]
14. Garcia-Rodríguez, A.; Matamoros, V.; Fontàs, C.; Salvadó, V. The influence of Lemna sp. and Spirogyra sp. on the removal of pharmaceuticals and endocrine disruptors in treated wastewaters. *Int. J. Environ. Sci. Technol.* **2014**, *12*, 2327–2338. [CrossRef]
15. Motoyama, A.; Suzuki, A.; Shirota, O.; Namba, R. Direct determination of Bisphenol A and nonylphenol in river water by column-switching semi-microcolumn liquid chromatography/electrospray mass spectrometry. *Rapid Commun. Mass Spectrom.* **1999**, *13*, 2204–2208. [CrossRef]
16. EU; European Union Risk Assessment Report, CAS: 80-05-7 EINECS No: 201-245-8, Environment Addendum of April 2008, 4,4′-ISOPROPYLIDENEDIPHENOL (Bisphenol-A), Part1 Environment; Available from, as of Nov 13, 2012. Available online: https://publications.jrc.ec.europa.eu/repository/bitstream/111111111/15063/1/lbna24588enn.pdf (accessed on 20 June 2021).
17. Greenberg, A.E.; Trussell, R.R.; Clesceri, L.S. (Eds.) *Standard Met for the Examination of Water and Wastewater*, 20th ed.; APHA: Washington, DC, USA, 1998.
18. Brain, R.A.; Solomon, K.R. A protocol for conducting 7-day daily renewal tests with Lemna gibba. *Nat. Protoc.* **2007**, *2*, 979–987. [CrossRef]
19. Porra, R.J.; Thompson, W.A.; Kriedemann, P.E. Determination of accurate extinction coefficients and simultaneous equations for assaying chlorophylls a and b extracted with four different solvents: Verification of the concentration of chlorophyll standards by atomic absorption spectroscopy. *Biochim. Biophys. Acta* **1989**, *975*, 384–394. [CrossRef]
20. Buege, J.A.; Aust, S.D. Microsomal lipid peroxidation. *Methods Enzymol.* **1978**, *52*, 302–310. [PubMed]
21. Krishnan, A.V.; Stathis, P.; Permuth, S.F.; Tokes, L.; Feldman, D. Bisphenol-A: An estrogenic substance is released from polycarbonate flasks during autoclaving. *Endocrinology* **1993**, *132*, 2279–2286. [CrossRef]
22. Eaton, A.D.; Clesceri, L.S.; Greenberg, A.E. *Standard Methods for the Examination of Water and Wastewater*, 19th ed.; American Public Health Association, American Water Works Association, and Water Environment Federation: Washington, DC, USA, 1995.
23. Dogan, M.; Saygideger, S.D.; Ugur Colak, U. Effect of lead toxicity on aquatic macrophyte *Elodea canadensis* Michx. *Bull. Environ. Contam. Toxicol.* **2009**, *83*, 249–254. [CrossRef]
24. Szymczak-Żyla, M.; Kowalewska, G.; Louda, J.W. The influence of microorganisms on chlorophyll a degradation in the marine environment. *Limnol. Oceanogr.* **2008**, *53*, 851–862. [CrossRef]
25. Cimmino, I.; Fiory, F.; Perruolo, G.; Miele, C.; Beguinot, F.; Formisano, P.; Oriente, F. Potential Mechanisms of Bisphenol A (BPA) Contributing to Human Disease. *Int. J. Mol. Sci.* **2020**, *21*, 5761. [CrossRef]
26. Corrales, J.; Kristofco, L.A.; Steele, W.B.; Yates, B.S.; Breed, C.S.; Williams, E.S.; Brooks, B.W. Global Assessment of Bisphenol A in the Environment: Review and Analysis of Its Occurrence and Bioaccumulation. *Dose-Response* **2015**, *13*, 1559325815598308. [CrossRef] [PubMed]

materials

MDPI

Article

Enrichment Characteristics of Cr in Chromium Slag after Pre-Reduction and Melting/Magnetic Separation Treatment

Shaoyan Hu [1], Deyong Wang [1], Xianglong Li [1,*], Wei Zhao [1,*], Tianpeng Qu [1] and Yun Wang [2]

1 School of Iron and Steel, Soochow University, Suzhou 215137, China; syhu616@suda.edu.cn (S.H.); deyongwang1222@163.com (D.W.); tianpengqu8119@163.com (T.Q.)
2 China ENFI Engineering Co., Ltd., Beijing 100038, China; yunwang8901@126.com
* Correspondence: xlli@suda.edu.cn (X.L.); zhaowei0312@suda.edu.cn (W.Z.)

Abstract: Concentrating the chromium in chromium slag and improving the chromium–iron ratio is beneficial for the further utilization of chromium slag. In this paper, chromium slag obtained from a chromite lime-free roasting plant was used as the raw material. Pellets made of the chromium slag and pulverized coal were reduced at different pre-reduction temperatures and then separated by a melting separation process or magnetic separation process, respectively. The mass and composition of the metallized pellets before separation, along with the alloy and tail slag after separation, were comprehensively analyzed. The experimental results showed that the output yield of alloy, iron recovery rate, and chromium content in the alloy were all higher when using melting separation than when using magnetic separation, because of the further reduction during the melting stage. More importantly, a relatively low pre-reduction temperature and selection of magnetic separation process were found to be more beneficial for chromium enrichment in slag; the highest chromium–iron ratio in tail slag can reach 2.88.

Keywords: chromium slag; chromium–iron ratio; pre-reduction; melting separation; magnetic separation

Citation: Hu, S.; Wang, D.; Li, X.; Zhao, W.; Qu, T.; Wang, Y. Enrichment Characteristics of Cr in Chromium Slag after Pre-Reduction and Melting/Magnetic Separation Treatment. *Materials* **2021**, *14*, 4937. https://doi.org/10.3390/ma14174937

Academic Editors: Avelino Núñez-Delgado and Victor M. Prida

Received: 19 July 2021
Accepted: 25 August 2021
Published: 30 August 2021

1. Introduction

Chromium has been widely applied in many industrial processes, such as leather tanning, electroplating, and mineral extraction. It is also an important ingredient in protective coatings, especially for stainless steel.

Chromium slag is the waste residue generated by the industrial process of extracting chromium (Cr) from chromite ore. Due to different needs for chromite quality in the target product, the extraction processes for chromium (Cr) are also different, leading to diverse compositions and properties for chromium slag [1–3]. Chromium salt manufacturing is a typical extraction process, which takes chromite, sodium carbonate, and dolomite as raw materials. After high-temperature oxidation roasting, chromium oxides are transformed into water-soluble sodium chromate, and the remaining tailings become chromium slag after water leaching [4,5]. However, about 10% of the chromium remains in the chromium slag and contains water-soluble, migratory, and carcinogenic Cr^{6+}, which is harmful to the environment [6–8]. If it is not recycled or reused, not only are resources wasted, but serious pollution of the environment results, which will have a serious impact on the health of the surrounding residents because of the toxic Cr^{6+} content of the slag [9–11]. At present, the treatment method for chromium slag is mostly landfill after reduction or solidification, but there are still potential environmental hazards and resource waste to consider [12]. Thus, it is valuable to explore further treatment methods for chromium slag.

According to the flux used in the roasting process, the chromium salt manufacturing process can be divided into the lime-based roasting process and the lime-free roasting process [13]. The chromium slag used in this research was obtained from a chromite lime-free roasting plant. The main components of the lime-free roasting slag are iron oxide and chromite oxide. Based on the chemical composition characteristics of the lime-free roasting slag, the recovery and utilization methods for metallic elements mainly include the hydrometallurgical method and pyrometallurgical method. The hydrometallurgical extraction method generally uses highly corrosive acids as extractants, such as sulfuric acid and hydrochloric acid, to achieve the transfer of metallic elements from the chromium slag to the liquid phase [14]. However, the selectivity of the acid-leaching process is poor, because there is no difference in the recovery of various metallic elements from the slag. A lot of subsequent separation steps need to be added to achieve real recycling [15]. In addition, efficient and low-cost treatment of the waste acids is also a problem.

The advantage of pyrometallurgical extraction lies in its excellent selectivity. By adding the reducing agent and slagging agent and adjusting the processing temperature and operating pressure, it can achieve the selective recovery of different metallic elements. The principal elements in chromium slag are iron and chromium. Reduction by the proper amount of carbon can convert toxic Cr^{6+} into nontoxic chromium oxides, such as CrO and Cr_2O_3; meanwhile, the iron oxide can be reduced partly to metallic iron [16–18]. After the reduction treatment, the metallic iron can be separated from the chromium slag by the melting separation method or magnetic separation method because of the magnetic difference between the metal and slag [19–22]. According to extensive experimental research and productive practices, the higher the ratio of chromium to iron (Cr/Fe) in slag, the higher the economic value of the slag. In reality, only when the mass ratio of chromium to iron (Cr/Fe) in the slag is greater than 2.0 can the slag meet the requirements for further use, which is mainly the production of ferrochromium. [23]

In order to improve the mass ratio of chromium to iron (Cr/Fe) in slag, it is necessary to extract iron from chromium slag and retain chromium in chromium slag. Compared with other synthesis techniques, carbothermal reduction is the most widely used method because of its low cost and simple process. From thermodynamic data on the reduction reactions of iron oxides and chromium oxides, as shown in Table 1 [24], it is clear that iron oxides can be more easily reduced by carbon than chromium oxide. The starting temperature of chromium oxide (Cr_2O_3) reduction is 1521.0 K, which is much higher than the reduction temperature of iron oxide, which is only 994.9 K. Thus, it is practicable to control the extraction of iron and the retention of chromium by adjusting the reduction temperature.

Table 1. Thermodynamic data for reduction reactions.

No.	Equations	ΔG^{θ} [$J \cdot mol^{-1}$]	Start Temperature of Reaction [K]
1	$3Fe_2O_3 + C = 2Fe_3O_4 + CO$	124,429 − 224.2 T	555
2	$Fe_3O_4 + C = 3FeO + CO$	207,510 − 217.62 T	953.4
3	$FeO + C = Fe + CO$	149,600 − 150.36 T	994.9
4	$Cr_2O_3 + 3C = 2Cr + 3CO$	771,315 − 507.11 T	1521

However, after the controllable reduction, the reaction products need to be separated into the metallic alloy and non-metallic slag. Melting separation and magnetic separation are the most widely applied and mature separation processes. Selection of the separation process also has a significant impact on the yield and composition of the final product.

In this paper, chromium slag obtained by chromite lime-free roasting was used as the raw material, and the experimental processes of pre-reduction, followed by melting/magnetic separation, were adopted to deal with the chromium slag. The main purpose of this paper was to investigate the chromium enrichment characteristics of chromium slag under different pre-reduction temperatures and different separation processes, with the aim of achieving a higher mass ratio of chromium to iron (Cr/Fe) in the final tail slag.

2. Experiment

2.1. Raw Materials

As mentioned above, the chromium slag was obtained from a chromite lime-free roasting plant. As the key raw material of the experiment, it was necessary to obtain the specific composition of the chromium slag. The slag was ground to a particle size of less than 0.074 mm first; then, the powder slag was sent to National Analysis Center for Iron and Steel China for accurate composition analysis. T.Fe, T.Cr, Al_2O_3, MgO, CaO, and SiO_2 were clearly identified for analysis. The T.Fe and T.Cr contents were detected by oxidation–reduction titration method, and the Al_2O_3, MgO, CaO, and SiO_2 content were detected by wavelength dispersive X-ray fluorescence spectrometry method. The composition of the dry chromium slag used in the experiment is shown in Table 2. The T.Fe content and T.Cr content were relatively high, indicating resource utilization potential. Considering that the chromite was roasted in an oxidizing atmosphere, it was assumed that the iron in the slag existed in the form of Fe_2O_3, and the chromium in the slag existed in the form of Cr_2O_3. The Al_2O_3 content, MgO content, and SiO_2 content in the chromium slag were similar, while the CaO content was quite low, which is a typical feature of the chromite lime-free roasting process. In addition, it should be noted that the total content of the above components did not reach 100%. Apart from the above components, chromium slag also contains a small amount of natrium salt, titanium oxide, phosphorus oxide, and other gangue minerals. It may even include crystal water that has not been completely removed during drying. Due to their small effect on the chromium enrichment characteristics, quantitative analysis of these components was not performed.

Table 2. Composition of the chromium slag used in the experiment.

T.Fe (wt.%)	T.Cr (wt.%)	Al_2O_3 (wt.%)	MgO (wt.%)	CaO (wt.%)	SiO_2 (wt.%)	Cr/Fe
19.2	23.2	9.47	9.69	0.57	10.36	1.21

In addition to the chromium slag, another main raw material of the experiment was pulverized coal, which played the role of reducing agent. Before analyzing the composition, the pulverized coal was ground to a particle size of less than 0.074 mm, the same as the chromium slag. Proximate analysis of the pulverized coal showed that the fixed carbon content was 85.66 wt.%, the volatile content was 1.59 wt.%, and the ash content was 12.13 wt.%.

In this research, an analytically pure SiO_2 reagent was used to adjust the basicity of the pellets. The SiO_2 content in the analytical reagent was more than 99.5 wt.%.

2.2. Experimental Schemes

In order to reveal the reduction and separation behaviors of chromium and iron in slag, four experiments with different pre-reduction temperatures and different separation processes were designed and carried out in this research, as shown in Table 3. Choosing 1373 K and 1523 K as the pre-reduction temperatures were mainly based on the experiment results of Cheng G et al. [25], who found that 1373 K was the optimal reduction temperature for reducing iron from slag under their experimental conditions.

Experiment NO. 1 was pre-reduced at 1373 K, followed by a melting separation step, whose separation temperature was 1853 K, with a separation processing time of 60 min. Experiment NO. 2 was pre-reduced at 1523 K, followed by a melting separation step, whose separation temperature was also 1853K, and the separation processing time was 60 min. Experiment NO. 3 was pre-reduced at 1373 K, followed by a magnetic separation step. Experiment NO. 4 was pre-reduced at 1523 K, followed by a magnetic separation step. Uniformly, the pre-reduction time was 45 min for all experiments. In order to reduce the experimental error, each group experiment was repeated three times.

Table 3. Experimental scheme and operating parameters.

NO.	Basicity of Pellets	Pre-Reduction Temperature [K]	Pre-Reduction Time [min]	Separation Method	Separation Temperature [K]	Separation Time [min]
1	0.36	1373	45	Melting separation	1853	60
2	0.36	1523	45	Melting separation	1853	60
3	0.99	1373	45	Magnetic separation	-	-
4	0.99	1523	45	Magnetic separation	-	-

Generally, the melting point of the slag has a major influence on the separation of the alloy and slag during the melting separation process. Low-melting-point slag is beneficial for the separation of the metallic alloy from the tail slag. In order to figure out the appropriate slag composition, the $MgO–SiO_2–Al_2O_3$ phase diagram was calculated and is shown in Figure 1. The phase diagram was calculated and drawn by the thermodynamic software FactSage 7.2, which is developed by Thermfact/CRCT (Montreal, Canada) and GTT-Technologies (Aachen, Gemany). During the phase diagram calculation, the selected database was FToxide, and the operating pressure was 101,325 Pa. The isotherm lines from 1650 K to 2000 K were plotted in the phase diagram with an interval of 50 K. The calculated phase diagram was compared with the slag atlas [26] and was validated by it. According to the phase diagram of $MgO–SiO_2–Al_2O_3$ in Figure 1, the slag composition in the low-melting-point area was around m_{SiO2}:m_{Al2O3}:$m_{MgO} \approx$ 6:2:2, while the initial slag composition shown in Table 2 was m_{SiO2}:m_{Al2O3}:$m_{MgO} \approx$ 1:1:1. In order to obtain the target slag composition (m_{SiO2}:(m_{Al2O3} + m_{MgO}) = 6:4) with a low melting point, 18.38 g SiO_2 should be added to every 100 g of chromium slag. As a result, when the melting separation method was adopted, the basicity of the slag ((m_{CaO} + m_{MgO})/m_{SiO2}) was reduced to 0.36 from the initial value of 0.99 due to the addition of SiO_2. Because the melting properties of the slag have little effect on the magnetic separation process, there is no need to add SiO_2 into the pellets when using the magnetic separation method, whose basicity remains at the initial value of 0.99.

As pulverized coal is the reducing agent of chromium slag, its addition amount in the pellets is very important for reduction reactions. Since the chromium slag came from a chromate roasting process, which was in an oxidizing atmosphere, it was assumed that the iron and chromium in the chromium slag would exist in the form of Fe_2O_3 and Cr_2O_3 when they took part in the carbothermal reduction reaction and that the gaseous product of the reduction reaction would be CO. The concentration of the added carbon powder was expressed as the mole ratio of carbon to reducible oxygen (n_C/n_O), instead of mass %, in order to meaningfully represent the significance of the carbon addition. In the (n_C/n_O) value, reducible oxygen (O) is the total amount of oxygen present in the form of Fe_2O_3, while excluding the oxygen in Cr_2O_3. The reason for excluding the oxygen in Cr_2O_3 was that the purpose of this research was to reduce the iron and retain the chromium as much as possible. In order to ensure that the iron oxide was fully reduced, the addition of coal needed to be in excess, and the carbon–oxygen ratio of the experiment was n_C/n_O = 1.1.

2.3. *Experimental Procedures and Devices*

The experimental procedures are shown in Figure 2, including grinding, batching, briquetting, pre-reduction, and separation. Details of every procedure are described in combination with the relevant devices. The main devices used in the experiment were a grinder, mixer, presser, muffle furnace, and magnetic separator, as shown in Figure 3.

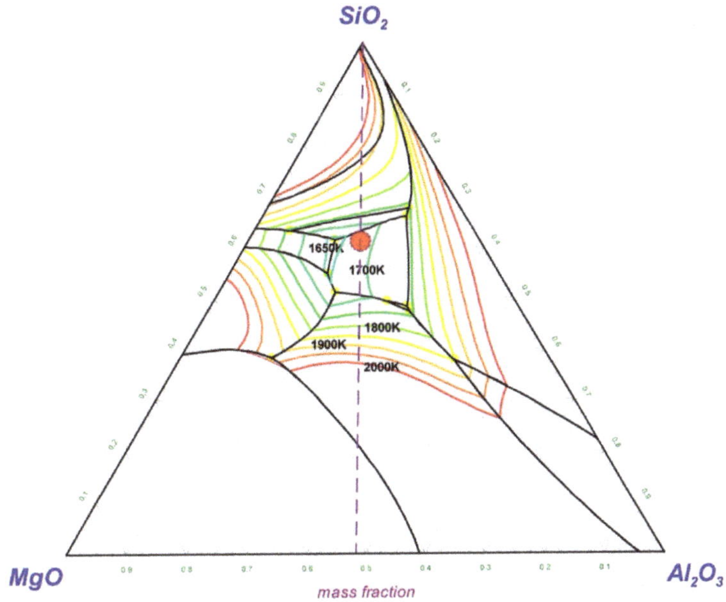

Figure 1. $MgO–SiO_2–Al_2O_3$ phase diagram with isotherm lines for p = 1 atm, drawn with FactSage 7.2.

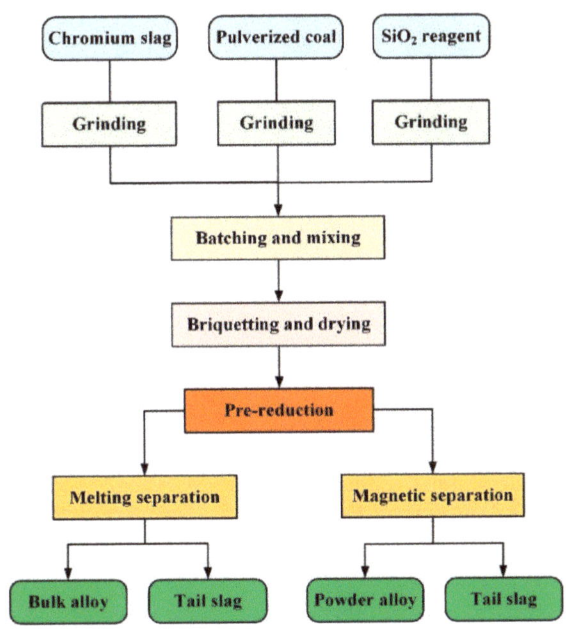

Figure 2. Experimental procedures for the pre-reduction and separation processes.

2.3.1. Grinding

As mentioned above, both the chromium slag and the pulverized coal were ground to a particle size of less than 0.074 mm first, not only for the composition analysis, but also for the subsequent batching. Crushing and grinding of the raw materials were performed in a grinder, as shown in Figure 3A. The grinder was suitable for the preparation of powdered samples of coal, ore, slag, and other raw materials. After the motor started, the eccentric block was driven to rotate at high speed to drive the exciting platform to generate the exciting vibration. After loading the material into the material bowl, the material collided with the crushing rod and the crushing ring in the material bowl strongly, then was crushed into very fine grains. After grinding, the ground powder was sieved with a 200-mesh sieve. If the particle size did not meet the requirements, the grinding time was prolonged.

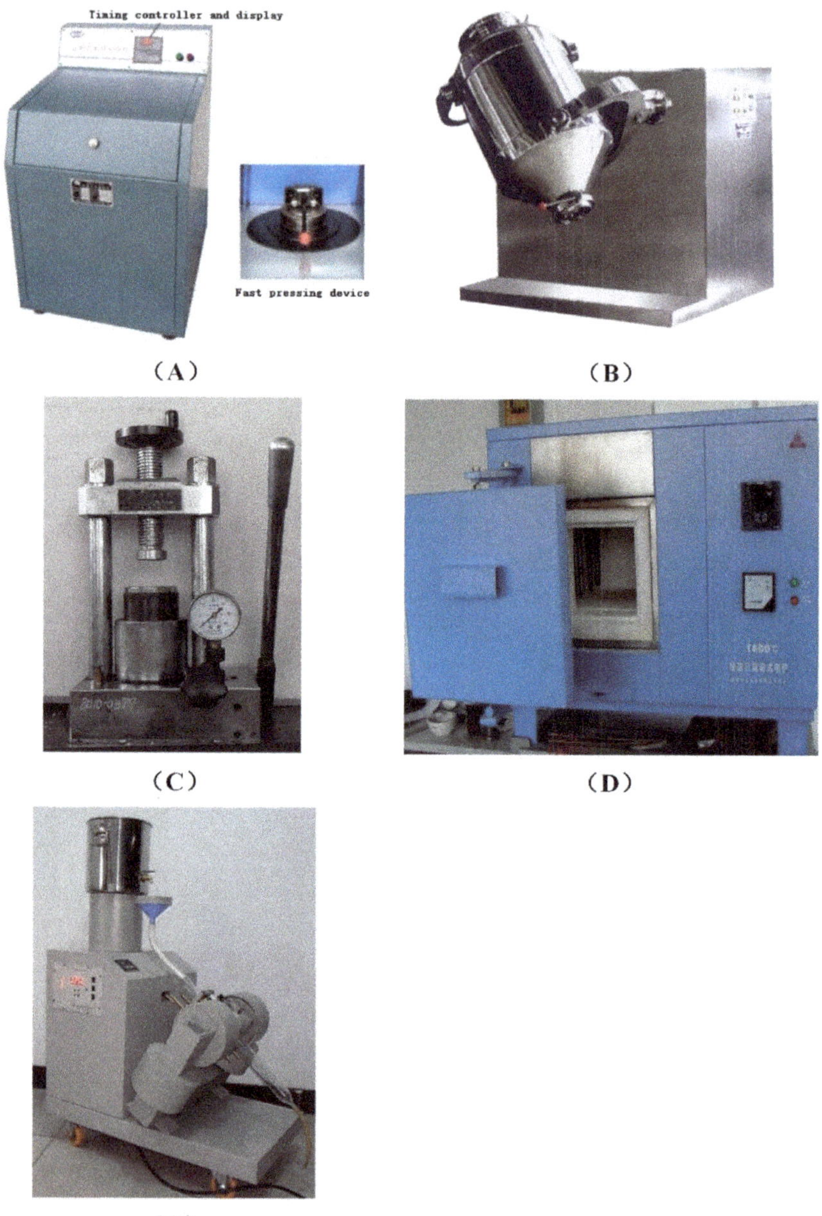

Figure 3. Experimental devices. (**A**) Grinder; (**B**) mixer; (**C**) presser; (**D**) muffle furnace; (**E**) magnetic separator.

2.3.2. Batching and Mixing

The ground chromium slag, pulverized coal, and silica were weighed and mixed in specific proportions corresponding to the respective experimental schemes. In order to achieve uniform mixing, each material was put into a 3-dimensional motion mixer, as shown in Figure 3B. During the mixing operation, various materials were well mixed due to the multi-directional rotation of the mixing tank, while avoiding the stratification phenomenon caused by gravity in general mixers.

2.3.3. Briquetting and Drying

Each group's well-mixed raw material was put into the presser for briquetting. The presser (Figure 3C) compacts powder samples. It can press granular material into tablets or granules. It is suitable for pressing tablets, catalysts, and metal powders in the laboratory. By manually pressing the lever, the oil can be pressed from the oil pool into the cylinder below the mold. The one-way valve prevents the oil from returning to the oil pool, so the high pressure of the oil cylinder can be maintained, and the maximum pressure is 60 MPa. The powder materials in the mold were molded by high pressure, and the molded pellets were obtained after demolding. The obtained pellets had an approximate diameter of 20 mm and mass of 10 g.

Before the pre-reduction step, the pellets were dried at 378 K for 4 h in an oven.

2.3.4. Pre-Reduction

Pre-reduction of the pellets was carried out in a muffle furnace, as shown in Figure 3D. This is a high-temperature resistance furnace with a maximum temperature of 1873 K. The furnace temperature was measured by a double platinum-rhodium thermocouple and controlled by the program. According to the experimental scheme, pre-reduction temperatures of 1373 K (Experiment NO. 1 and NO. 3) and 1523 K (Experiment NO. 2 and NO. 4) were used. When the muffle furnace reached the required temperature (1373 K or 1523 K), the graphite crucible containing the pellets was put into the furnace. Then, 45 min after the reduction, the graphite crucible was taken out and the pellets were covered by another graphite crucible to prevent oxidation. The cooled metallized pellets were used in subsequent experiments for melting separation or magnetic separation.

2.3.5. Melting Separation

The temperature of the muffle furnace was raised to 1853 K, and then the crucible containing the metallized pellets was put into the furnace. The temperature was maintained at 1853 K for 60 min, and then the crucible containing the sample was removed from the furnace and covered by a graphite plate to prevent oxidation during cooling. After the sample was cooled, the bulk alloy was separated from the slag, and then the slag and alloy were sent for chemical analysis.

2.3.6. Magnetic Separation

Magnetic separation was carried out with a magnetic separator, as shown in Figure 3E. The magnetic separator can generate a strong magnetic field at the middle position of the magnetic separator tube, and the maximum magnetic field intensity is 500 mT. Metallized pellets obtained by pre-reduction were ground to a particle size of less than 0.148 mm. Then, the powder was mixed into water and stirred, before being passed through the magnetic separation tube with a magnetic field intensity of 200 mT, so that the magnetic powder remained in the magnetic field in the magnetic separator tube, and the non-magnetic slag passed through the tube with the water. After all of the mixture of water and powder had been poured into the magnetic separation tube, water continued to be poured into the tube until the flowing water became clear. Then, the magnetism of the magnetic separator was switched off, and the alloy in the magnetic separation tube was flushed with water into another container until the magnetic separation tube was clean. The sample of tail slag was filtered, and the residue on the filter paper was collected and dried for 4 h to obtain dry tail slag; dry alloy powder was obtained in the same way. Finally, the dry tail slag and dry alloy powder were sent for chemical analysis.

3. Results and Discussion

After finishing the experiments, the metal distribution between the slag and the alloy, the metal recovery rate, and the mass ratio of chromium to iron (Cr/Fe) in final tail slag were compared in detail. Based on the experimental results, the proper process route and technological parameters for the recovery of chromium slag were explored.

Table 4 shows the weight and composition of the metallized pellets before separation, together with the alloy and tail slag after separation. Through pre-reduction and melting/magnetic separation, iron and chromium are distributed both in the alloy and in the tail slag. The multiple indicators, such as the output yield of alloy, the distribution ratio of metals (iron and chromium) in the slag and alloy, the recovery rate of metals (iron and chromium), and the mass ratio of chromium to iron (Cr/Fe) in the slag and alloy, can quantitatively express the effect of the process route and technological parameters on the chromium enrichment characteristics.

Table 4. Weight and composition of the metallized pellets, alloy, and tail slag.

NO.	Metallized Pellets			Alloy			Tail Slag		
	m_0 (g)	$(\%\ Fe)_0$ (wt.%)	$(\%\ Cr)_0$ (wt.%)	m_a (g)	$(\%\ Fe)_a$ (wt.%)	$(\%\ Cr)_a$ (wt.%)	m_s (g)	$(\%\ Fe)_s$ (wt.%)	$(\%\ Cr)_s$ (wt.%)
1	91.2	18.71	22.59	13.76	55.78	22.8	70.04	11.9	22.5
	87.6	18.83	22.61	12.81	56.81	23.12	71.54	11.03	21.51
	92.1	18.67	22.8	13.65	54.33	21.89	69.1	12.04	23.51
2	91.4	18.92	22.86	8.02	52.14	24.6	64.8	17.1	25.5
	92.1	19.13	22.9	8.51	53.21	24.71	64.13	17.32	25.68
	94.5	19.79	23.58	8.5	52.13	25.32	65.31	18.01	26.61
3	60	20.11	24.21	3.78	67.42	1.27	55.4	9.1	26.2
	60	20.01	24.37	3.61	66.58	1.3	54.99	9.32	26.41
	60	20.2	24.41	3.71	68.1	1.23	55.81	8.88	26.09
4	60	20.53	25.01	3.61	55.13	2.53	56.3	13.9	26.6
	60	21.51	25.89	3.69	52.78	2.68	56.59	14.01	27.19
	60	21.72	26.11	3.59	57.61	3.01	56.9	12.99	27.36

3.1. Output Yield of Alloy

The output yield of alloy is defined as φ.

$$\varphi = m_a/m_0 \times 100\% \quad (1)$$

where m_a is the mass of the obtained alloy, in g, and m_0 is the mass of the metallized pellets before separation, in g.

Previous research showed that the addition of a reducing agent had a dominant effect on the output of alloy with sufficient time and at an appropriate temperature [27,28]. As shown in Figure 4, the magnetic separation processes at different reduction temperatures have almost the same alloy output yield. However, the output of alloy from the melting separation process is significantly higher than that of the magnetic separation process. Because the melting process was carried out in the graphite crucible, the residual iron oxides and chromium oxides in the metallized pellets continued to be reduced by graphite during melting separation, and more alloy could be obtained [29,30]. Other oxide crucibles, such as alumina, magnesium oxide, zirconia, etc., would be rapidly eroded, and the slag composition would change at a melting temperature of 1853 K.

Comparing the experimental results with different pre-reduction temperatures, it should be noted that the alloy output yield at higher temperatures (1523 K) was relatively lower, especially for the melting separation process, although the thermodynamic theoretical analysis showed that the higher the temperature, the more iron oxide would be reduced [31]. However, many previous studies have found that the low-melting-point phases in the pellet will melt during pre-reduction, which decreases the permeability of the pellets and hinders the reduction reaction. As a result, the metallization rate of the pellets is relatively lower at higher temperatures [32]. According to the FeO–SiO_2 phase diagram in Figure 5, FeO and SiO_2 in slag easily form low-melting-point composite oxides, with a melting temperature of only 1451 K. As mentioned above, in Experiment NO. 2, SiO_2 was added to reduce the slag melting point to achieve a better melting separation, so the inhibition effect of a low-melting-point material on the reduction was more significant than that in Experiment NO. 4. Noting this, the FeO–SiO_2 phase diagram was calculated and drawn by the thermodynamic software FactSage 7.2. During the phase diagram calculation, the selected database was FToxide, and the operating pressure was 101,325 Pa.

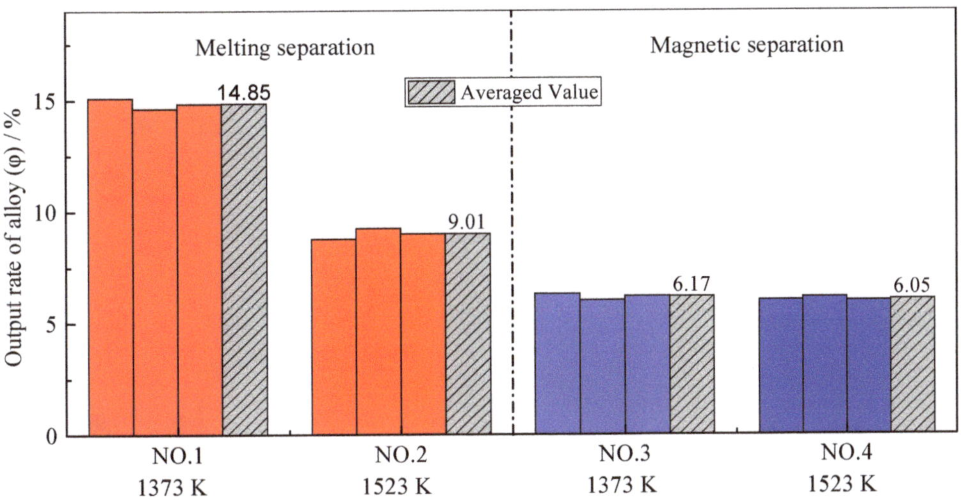

Figure 4. Output yield of alloy with different pre-reduction temperatures and different separation processes.

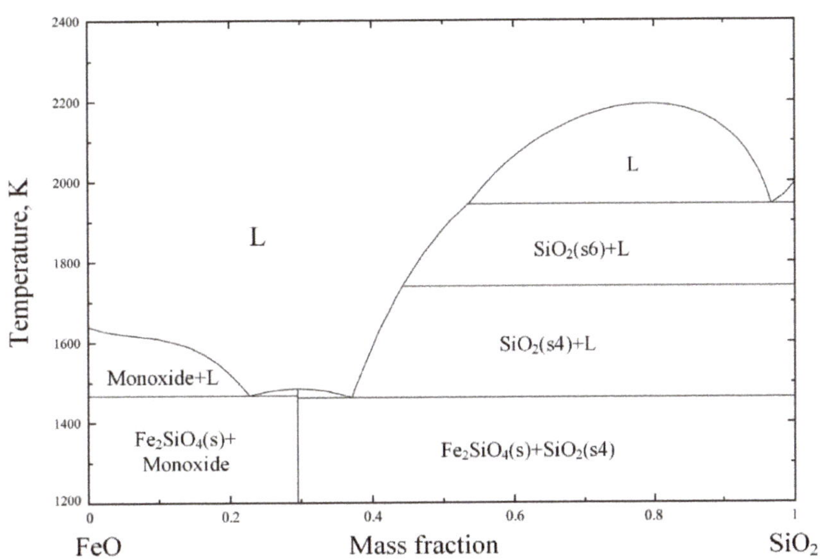

Figure 5. FeO–SiO_2 phase diagram for $p = 1$ atm, drawn with FactSage 7.2.

3.2. Iron Recovery Rate

In order to further study the transfer behaviors of iron during the processes of pre-reduction and melting/magnetic separation, the distribution of iron in the slag and alloy was assessed. The distribution ratio of iron between the slag and alloy is defined as L_{Fe}.

$$L_{Fe} = (\%\ Fe)_s/(\%\ Fe)_a \quad (2)$$

where $(\%\ Fe)_s$ is the iron content in the slag after separation and $(\%\ Fe)_a$ is the iron content in the alloy after separation, both measured as wt.%.

Figure 6 shows the distribution ratios of iron (L_{Fe}) obtained by all four group experiments. Firstly, no matter whether melting separation or magnetic separation was used, the distribution ratio of iron at higher pre-reduction temperatures was larger. Specifically, for the melting separation process, when the pre-reduction temperature increased from 1373 K to 1523 K, the average distribution ratio of iron between the slag and alloy increased from 0.21 to 0.33. For the magnetic separation process, when the pre-reduction temperature increased from 1373 K to 1523 K, the average distribution ratio of iron increased from 0.14 to 0.25. The main reason for the above phenomenon is the formation of a low-melting-point phase. Generally, the iron oxide is mainly reduced by CO in the pellets [33,34] and the melting of the low-melting-point phase will deteriorate the permeability of the pellets, thereby hindering the reduction of iron oxides, and thus more iron remains in the slag.

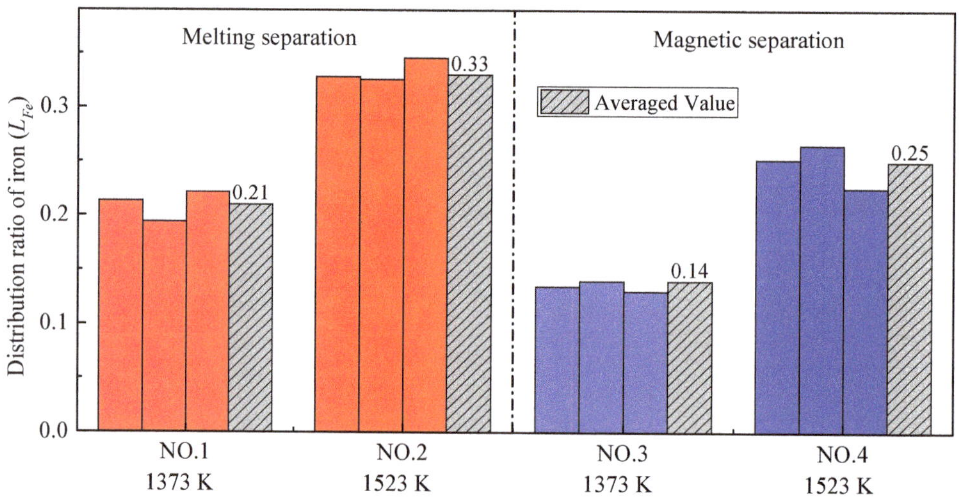

Figure 6. The distribution ratio of iron with different pre-reduction temperatures and different separation processes.

Another major concern is that, when the pre-reduction temperature is the same, the distribution ratio of iron after melting separation is significantly higher than that after magnetic separation. The reduction of chromium at melting temperature can explain these results. The reduction of chromium mainly occurs in the melting separation process, while the reduction of iron mainly occurs in the pre-reduction process. Compared with the magnetic separation process, much more metallic chromium is reduced into the alloy during melting separation, leading to the decrease in iron content in the alloy, and the increase in the iron distribution ratio between the slag and alloy.

Besides the distribution ratio, the recovery rate of iron is also an important indicator. The recovery rate of iron is defined as R_{Fe} in this paper.

$$R_{Fe} = \frac{m_a \cdot (\%\ Fe)_a}{m_0 \cdot (\%\ Fe)_0} \times 100\% \quad (3)$$

where m_a is the mass of obtained alloy, in g; $(\% Fe)_a$ is the iron content of obtained alloy, in wt.%; m_0 is the mass of metallized pellets before separation, in g; and $(\% Fe)_0$ is the iron content of the metallized pellets before separation, in wt.%.

Figure 7 shows the recovery rates of iron (R_{Fe}) obtained by all four group experiments. The recovery rate of iron after melting separation is higher than that after magnetic separation. The results validate that the reduction of iron oxide is continuously occurring in the melting stage, leading to more iron being recovered from the slag compared to the alloy and higher recovery rate of iron. In addition, it can be seen from Figure 7 that the recovery rate of iron is relatively low at higher temperatures; the reason for this is also the melting of the low-melting-point phase, as mentioned above.

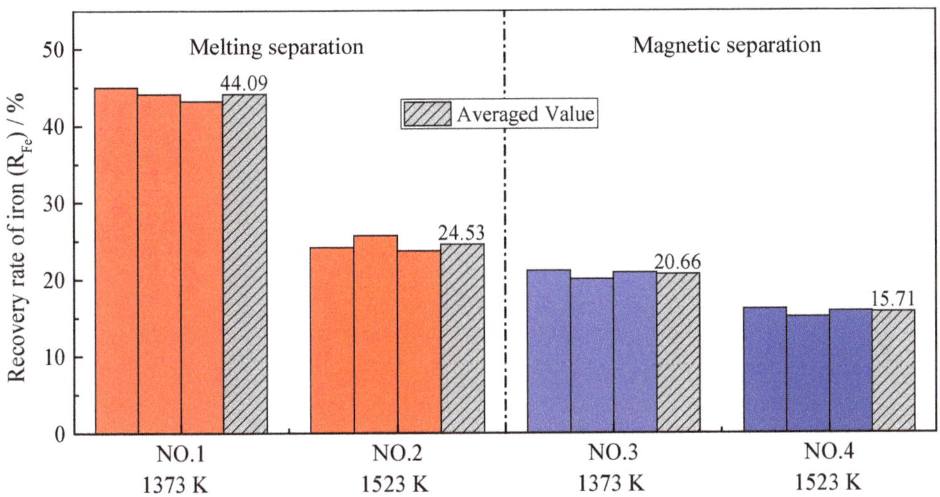

Figure 7. Recovery rate for iron with different pre-reduction temperatures and different separation processes.

3.3. Chromium Recovery Rate

The distribution ratio of chromium between the slag and alloy is defined as L_{Cr}, which is similar to the definition of the iron distribution ratio.

$$L_{Cr} = (\% Cr)_s / (\% Cr)_a \tag{4}$$

where $(\% Cr)_s$ is the chromium content in the tail slag after separation and $(\% Cr)_a$ is the chromium content in the alloy after separation, both measured as wt.%.

Because the purpose of this research was to enrich chromium in slag, the recovery rate of chromium was defined as the ratio of the mass of chromium in the tail slag to the mass of chromium in the metallized pellets, which is different from the definition of the iron recovery rate.

$$R_{Cr} = \frac{m_s \cdot (\% Cr)_s}{m_0 \cdot (\% Cr)_0} \times 100\% \tag{5}$$

where R_{Cr} is the recovery rate of chromium, in %; m_s is the mass of the tail slag after separation, in g; $(\% Cr)_s$ is the chromium content of the tail slag after separation, in wt.%; m_0 is the mass of the metallized pellets before separation, in g; and $(\% Cr)_0$ is the chromium content of the metallized pellets before separation, in wt.%.

As shown in Figures 8 and 9, the distribution ratio of chromium between the slag and alloy (L_{Cr}) after the melting separation process was far lower than that after the magnetic separation process, while the recovery rate of chromium was also significantly lower than that after the magnetic separation process, which is different from iron. In the process of magnetic separation, only strong magnetic materials, such as metallic iron and magnetite, can be retained in the magnetic field in the magnetic separation tube, while other diamagnetic or paramagnetic materials, such as slag and residual carbon, will remain with the tail slag in the flowing water [35]. According to previously published research [36–38], chromium cannot be significantly reduced into metal at a temperature of 1373–1573 K, and the products obtained from magnetic separation contain almost no chromium; that is, $(\% \, Cr)_a$ is very low and $(\% \, Cr)_s$ is high, making L_{Cr} in the magnetic separation process higher. As shown in Figure 9, the chromium recovery rates in Experiment NO. 3 and Experiment NO. 4 were close to 100%, indicating that chromium rarely enters the alloy, and almost all remains in the tail slag.

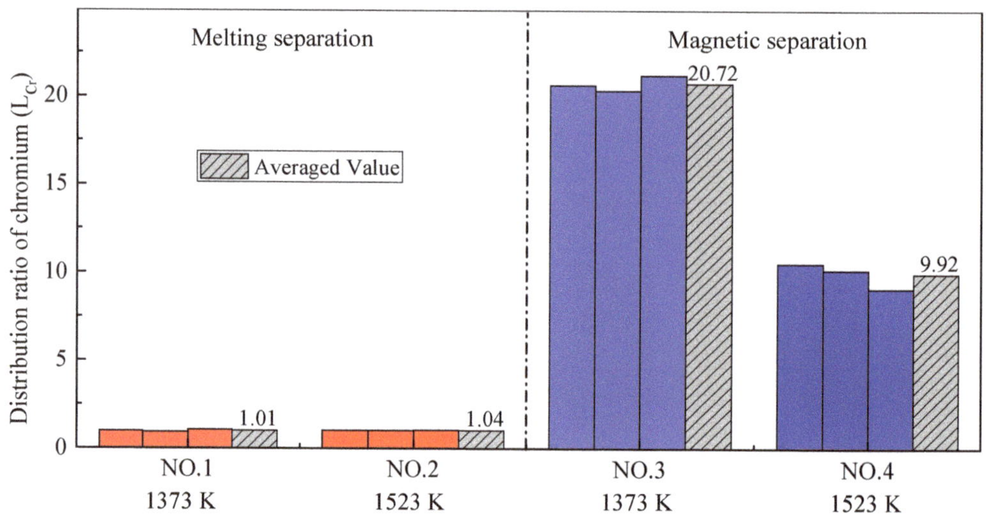

Figure 8. Distribution ratio of chromium with different pre-reduction temperatures and different separation processes.

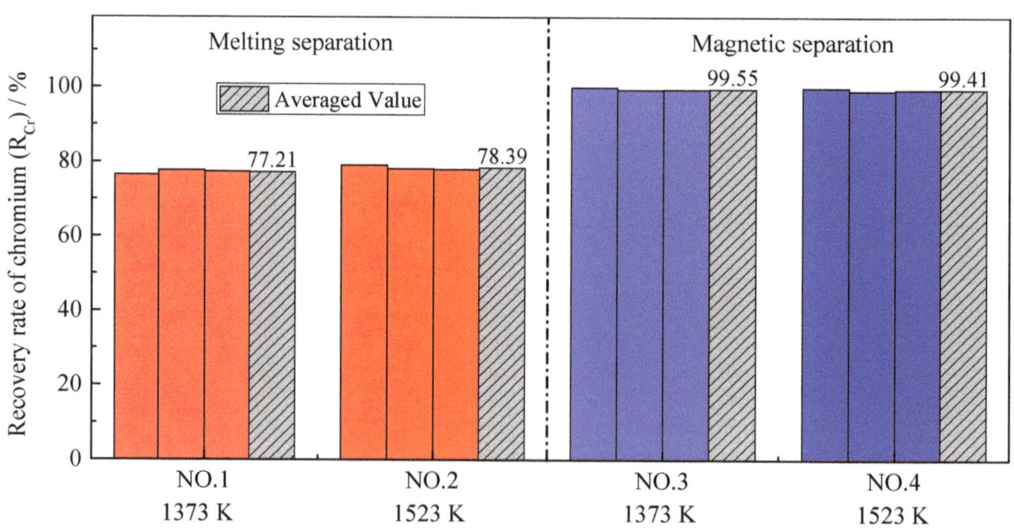

Figure 9. Recovery rate of chromium with different pre-reduction temperatures and different separation processes.

According to Figures 8 and 9, the pre-reduction temperature has little effect on the chromium distribution ratio and the chromium recovery rate in the case of melting separation. The reason for this is that reduction of chromium occurs mainly during the melting separation stage, rather than during pre-reduction. However, in the case of magnetic separation, the effect of the pre-reduction temperature is different. Although the pre-reduction temperature has little effect on the chromium recovery rate, it has an obvious effect on the chromium distribution ratio. The phenomenon can be explained by the experimental data from Table 4. For Experiment NO. 3, the average mass of alloy m_a obtained by magnetic separation was 3.70 g, and the average chromium content in the alloy $(\%\ Cr)_a$ was 1.27%. For Experiment NO. 4, the average mass of alloy m_a was 3.63 g, and the average chromium content in the alloy $(\%\ Cr)_a$ was 2.74%. Both the mass of the alloy and the chromium content in the alloy were very small, indicating that little chromium was reduced from the slag to the alloy; most of the chromium remained in the tail slag, so the recovery rate of chromium was very high and the difference was not obvious. The distribution ratio of chromium L_{Cr} is the ratio between the chromium content in the slag $(\%\ Cr)_s$ and chromium content in the alloy $(\%\ Cr)_a$. Since $(\%\ Cr)_s$ was basically unchanged, the $(\%\ Cr)_a$ increased from 1.27% to 2.74%, leading the chromium distribution ratio L_{Cr} to decrease obviously, as shown in Figure 8. This also illustrates that, as the pre-reduction temperature increases, the reduction of chromium increases slightly in the case of magnetic separation.

3.4. *Chromium–Iron Ratio in the Tail Slag (Cr/Fe)*

As mentioned above, utilization of chromium slag requires that the chromium–iron ratio (Cr/Fe) in the tail slag is higher than 2.0. The main purpose of this research was to understand the process to achieve that ratio. The Cr/Fe ratio of the metallized pellets after pre-reduction was almost 1.2, as shown in Figure 10, which is the same as the initial raw material, according to Table 2. The Cr/Fe ratio in the alloy obtained by melting separation was obviously higher than that from magnetic separation, which is mainly due to the large amount of chromium reduced by carbon during the melting stage at 1853 K. However, the Cr/Fe ratio in the tail slag obtained by melting separation was lower than that obtained by magnetic separation. According to Table 4, Figures 8 and 9, chromium mainly exists in the tail slag, and the chromium content in the tail slag changes little, although the iron content in the tail slag changes significantly.

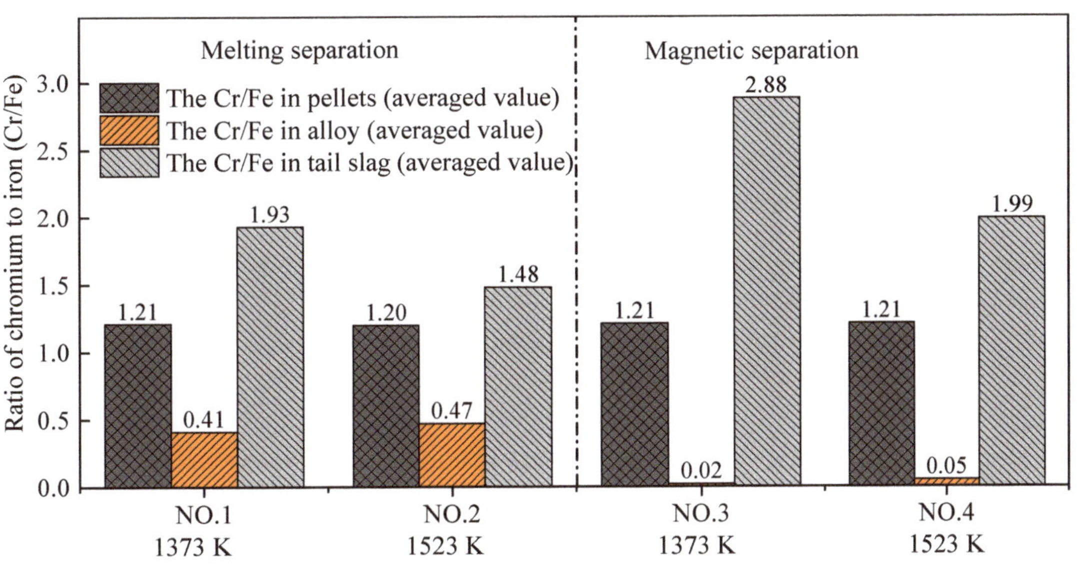

Figure 10. Chromium–iron ratio (Cr/Fe) in pellets, alloy, and tail slag.

The experimental results showed that both the melting separation and magnetic separation processes can improve the Cr/Fe ratio in tail slag. The Cr/Fe ratio in the final tail slag after the pre-reduction and separation processes decreased in the order of Experiment NO. 3, NO. 4, NO. 1, and NO. 2. Importantly, the chromium–iron ratio in the tail slag obtained by Experiment NO. 3 achieved the goal of chromium enrichment. The average Cr/Fe ratio of 2.88 was significantly higher than the critical value of 2.0.

4. Conclusions

In order to increase the chromium–iron ratio in the chromium slag generated by the chromite lime-free roasting process, carbothermal reduction, followed by the melting/magnetic separation process, was adopted and studied in this research. Experiments with different pre-reduction temperatures and different separation methods were carried out to investigate the specific distribution behaviors of iron and chromium between the metallic alloy and tail slag. The main conclusions are as follows:

(1) No matter whether melting separation or magnetic separation was selected, when the pre-reduction temperature increased from 1373 K to 1523 K, both the output yield of the alloy and the recovery rate of iron decreased. The analysis indicated that the melting of the low-melting-point phase decreased the permeability of the pellets and slowed down the reduction reaction. With the increase in the pre-reduction temperature, the amount of the molten phase increased and the reduction rate decreased.

(2) During the melting separation stage, because the processing temperature was higher than the reduction temperature of the iron oxides and chromium oxide, the reduction of iron further occurred and the chromium started to gradually reduce in the alloy. As a result, when the melting separation method was adopted, the output yield of the alloy, the chromium content in the alloy, and the iron recovery rate were all higher than when using magnetic separation. The melting separation process aggravated the transfer of chromium from the slag to alloy.

(3) The main purpose of this research was to increase the chromium–iron ratio in slag to more than 2.0 through a proper pre-reduction and separation process. Among the four experiments carried out, pre-reduction at 1373 K followed by magnetic separation obtained the highest chromium–iron ratio of 2.88 in the final tail slag. Simultaneously, the average recovery rate of chromium was as high as 99.55%, which met our requirements. The results indicated that a relatively low pre-reduction temperature and the magnetic separation method are beneficial for chromium enrichment in final tail slag.

Author Contributions: Conceptualization, S.H. and D.W.; methodology, S.H. and X.L.; software, W.Z.; formal analysis, T.Q.; investigation, T.Q.; writing—original draft preparation, S.H.; writing—review and editing, S.H. and X.L.; visualization, W.Z.; project administration, Y.W. All authors have read and agreed to the published version of the manuscript.

Funding: This research was funded by the Natural Science Foundation of Jiangsu Province, grant number BK20200869, and the China Postdoctoral Science Foundation, grant number 7114484320.

Institutional Review Board Statement: Not applicable.

Informed Consent Statement: Not applicable.

Conflicts of Interest: The authors declare no conflict of interest.

References

1. Nakasuga, T.; Nakashima, K.; Mori, K. Recovery Rate of Chromium from Stainless Slag by Iron Melts. *ISIJ Int.* **2004**, *44*, 665–672. [CrossRef]
2. Yuan, F.; Zhang, H.-N.; Li, H.; Dong, J.-H.; Xiong, H.-H.; Xu, A.-J. Recovery rates of iron, nickel, and chromium via iron-bath reduction of stainless steel dust briquettes based on corundum crucible erosion balance analysis. *J. Iron Steel Res. Int.* **2018**, *25*, 320–329. [CrossRef]
3. Mao, L.; Cui, H.; Miao, C.; An, H.; Zhai, J.; Li, Q. Preparation of $MgCr_2O_4$ from waste tannery solution and effect of sulfate, chloride, and calcium on leachability of chromium. *J. Mater. Cycles Waste Manag.* **2016**, *18*, 573–581. [CrossRef]
4. Wang, T.; He, M.; Pan, Q. A new method for the treatment of chromite ore processing residues. *J. Hazard. Mater.* **2007**, *149*, 440–444. [CrossRef]

5. Li, X.B.; Qi, T.G.; Peng, Z.H.; Liu, G.H.; Zhou, Q.S. Kinetics of chromite ore in oxidation roasting process. *Chin. J. Nonferrous Met.* **2010**, *20*, 1822–1828.
6. Matern, K.; Kletti, H.; Mansfeldt, T. Chemical and mineralogical characterization of chromite ore processing residue from two recent Indian disposal sites. *Chemosphere* **2016**, *155*, 188–195. [CrossRef] [PubMed]
7. Deakin, D.; West, L.; Stewart, D.; Yardley, B. The Leaching Characteristics of Chromite ore Processing Residue. *Environ. Geochem. Health* **2001**, *23*, 201–206. [CrossRef]
8. Thompson, C.M.; Kirman, C.R.; Proctor, D.M.; Haws, L.C.; Suh, M.; Hays, S.M.; Hixon, J.G.; Harris, M.A. A chronic oral reference dose for hexavalent chromium-induced intestinal cancer. *J. Appl. Toxicol.* **2014**, *34*, 525–536. [CrossRef] [PubMed]
9. Hori, M.; Shozugawa, K.; Matsuo, M. Hexavalent chromium pollution caused by dumped chromium slag at the urban park in Tokyo. *J. Mater. Cycles Waste Manag.* **2014**, *17*, 201–205. [CrossRef]
10. El Nemr, A.; Khaled, A.; Abdelwahab, O.; El-Sikaily, A. Treatment of wastewater containing toxic chromium using new activated carbon developed from date palm seed. *J. Hazard. Mater.* **2008**, *152*, 263–275. [CrossRef] [PubMed]
11. Liu, B.; Li, J.; Zeng, Y.; Wang, Z. Toxicity assessment and geochemical model of chromium leaching from AOD slag. *Chemosphere* **2016**, *144*, 2052–2057. [CrossRef] [PubMed]
12. Li, J.; Chen, Z.; Shen, J.; Wang, B.; Fan, L. The enhancement effect of pre reduction using zero-valent iron on the solidification of chromite ore processing residue by blast furnace slag and calcium hydroxide. *Chemosphere* **2015**, *134*, 159–165. [CrossRef] [PubMed]
13. Burke, T.; Fagliano, J.; Goldoft, M.; Hazen, R.E.; Iglewicz, R.; McKee, T. Chromite ore processing residue in Hudson County, New Jersey. *Environ. Health Perspect.* **1991**, *92*, 131–137. [CrossRef] [PubMed]
14. Tinjum, J.M.; Benson, C.H.; Edil, T.B. Mobilization of Cr(VI) from chromite ore processing residue through acid treatment. *Sci. Total Environ.* **2008**, *391*, 13–25. [CrossRef] [PubMed]
15. Zhang, Y.; Zheng, S.L.; Hao, D.U.; Xu, H.B.; Zhang, Y. Effect of mechanical activation on alkali leaching of chromite ore. *Trans. Nonferrous Met. Soc. China* **2010**, *20*, 888–891. [CrossRef]
16. Gutiérrez Paredes, J.; Romero-Serrano, A.; Plascencia-Barrera, G.; Vargas-Ramírez, M.; Zeifert, B.; Arredondo-Torres, V. Chromium oxide reduction from slag by silicon and magnesium. *Steel Res. Int.* **2005**, *76*, 764–768. [CrossRef]
17. Thacker, U.; Madamwar, D. Reduction of Toxic Chromium and Partial Localization of Chromium Reductase Activity in Bacterial Isolate DM1. *World J. Microbiol. Biotechnol.* **2005**, *21*, 891–899. [CrossRef]
18. Murugaiyan, V.; Sehar, T.; Selvaraj, S.; Selvaraj, P.K. Complete Reduction of Hazardous Cr(VI) in Chromium Ore Processing Residue Dump Site. *Asian J. Chem.* **2018**, *30*, 620–624. [CrossRef]
19. Tang, J.; Chu, M.; Feng, C.; Li, F.; Tang, Y.; Liu, Z. Melting Separation Process of High Chromium Vanadium-bearing Titanomagnetite Metallized Pellet and its Optimization by Multi-Index Synthetic Weighted Scoring Method. *ISIJ Int.* **2017**, *57*, 1156–1165. [CrossRef]
20. Tang, J.; Chu, M.; Feng, C.; Tang, Y.; Liu, Z. Melting separation behavior and mechanism of high-chromium vanadium–bearing titanomagnetite metallized pellet got from gas-based direct reduction. *ISIJ Int.* **2016**, *56*, 210–219. [CrossRef]
21. Nasr, M.; Youssef, M. Optimization of magnetizing reduction and magnetic separation of iron ores by experimental design. *ISIJ Int.* **2007**, *36*, 631–639. [CrossRef]
22. Long, H.M.; Meng, Q.M.; Wang, P.; Chun, T.J.; Yao, Y.L. Preparation of Chromium-iron metal powder from chromium slag by reduction roasting and magnetic separation. *J. Iron Steel Res. Int.* **2015**, *22*, 771–776. [CrossRef]
23. Bergeron, M.; Richer-Laflèche, M. Method for Increasing the Chrome to Iron Ratio of Chromites Products. U.S. Patent 7658894 B2, 9 February 2010.
24. Zhang, J.; Xing, X.; Song, B. *Metall Phys. Chem*; Metallurgical Industry Press: Beijing, China, 2004.
25. Cheng, G.; Quan, X.; Luo, H.; Bai, W.; Cunfang, L. Influence of Carbothermal Reduction Process on the Microscopic Properties of Lime-Free Roasting Chromite Ore Processing Residue (COPR). *IOP Conf. Ser. Mater. Sci. Eng.* **2020**, *729*, 012064. [CrossRef]
26. Verein Deutscher Eisenhüttenleute. *Slag Atlas*; Verlag Stahleisen: Düsseldorf, Germany, 1995.
27. Matsumura, T.; Takenaka, Y.; Shimizu, M. Effect of the carbon content on reduction and melting behavior of carbon composite iron orepellet. *Tetsu-Hagane* **2010**, *85*, 652–657. [CrossRef]
28. Bizhanov, A.; Kurunov, I.; Pavlov, A.; Chadaeva, O.; Chizhov, P. Study of the high-temperature reduction of ore-coal extrusion briquettes (Brex). *Metallurgist* **2014**, *57*, 871–877. [CrossRef]
29. Hino, M.; Nagasaka, T.; Higuchi, K.; Ban-Ya, S. Thermodynamic estimation on the reduction behavior of iron-chromium ore with carbon. *Met. Mater. Trans. A* **1998**, *29*, 351–360. [CrossRef]
30. Tsomondo, M.B.C.; Simbi, D.J. Kinetics of chromite ore reduction from MgO-CaO-SiO_2-FeO-Cr_2O_3-Al_2O_3 slag system by carbon dissolved in high carbon ferrochromium alloy bath. *Ironmak. Steelmak.* **2002**, *29*, 22–28. [CrossRef]
31. Sun, Y.; Gao, P.; Han, Y.; Ren, D. Reaction behavior of iron minerals and metallic iron particles growth in coal-based reduction of an oolitic iron ore. *Ind. Eng. Chem. Res.* **2013**, *52*, 2323–2329. [CrossRef]
32. Hara, Y.; Tsuchiya, M.; Kondo, S.-I. Intraparticle Temperature of Iron-Oxide Pellet during the Reduction. *Tetsu-Hagane* **1974**, *60*, 1261–1270. [CrossRef]
33. Yang, J.; Mori, T.; Kuwabara, M. Mechanism of Carbothermic Reduction of Hematite in Hematite–Carbon Composite Pellets. *ISIJ Int.* **2007**, *47*, 1394–1400. [CrossRef]

34. Gudenau, H.W.; Senk, D.; Wang, S.; Martins, K.D.M.; Stephany, C. Research in the Reduction of Iron Ore Agglomerates Including Coal and C-containing Dust. *ISIJ Int.* **2005**, *45*, 603–608. [CrossRef]
35. Long, H.; Meng, Q.; Chun, T.; Wang, P.; Li, J. Preparation of metallic iron powder from copper slag by carbothermic reduction and magnetic separation. *Can. Met. Quart* **2016**, *55*, 1–7. [CrossRef]
36. Roshchin, V.; Roshchin, A.; Akhmetov, K. Mechanism and sequence of the metal reduction in the lattice of chromospinelides. *Russ. Metall.* **2014**, *2014*, 173–178. [CrossRef]
37. Ahmed, H.M.; Viswanathan, N.N.; Björkman, B. Isothermal reduction kinetics of self-reducing mixtures. *Ironmak. Steelmak.* **2016**, *44*, 1–10. [CrossRef]
38. Zhao, Q.; Liu, C.; Zhang, B.; Jiang, M.; Qi, J.; Saxén, H.; Zevenhoven, R. Study on Extraction of Iron From Chromite. *Steel Res. Int.* **2015**, *86*, 1541–1547. [CrossRef]

materials

MDPI

Article

Use of Ashes from Lignite Combustion as Fillers in Rubber Mixtures to Reduce VOC Emissions

Miroslawa Prochon *, Dariusz Bieliński, Paulina Stepaniak, Magdalena Makowicz, Dominik Pietrzak and Oleksandra Dzeikala

Institute of Polymer and Dye Technology, Faculty of Chemistry, Lodz University of Technology, Stefanowskiego 12/16, 90-924 Lodz, Poland; dariusz.bielinski@p.lodz.pl (D.B.); 200249@edu.p.lodz.pl (P.S.); 198881@edu.p.lodz.pl (M.M.); dominik.pietrzak@p.lodz.pl (D.P.); oleksandra.dzeikala@p.lodz.pl (O.D.)

* Correspondence: miroslawa.prochon@p.lodz.pl

Abstract: This paper presents the use of ashes from brown coal combustion (BCA) as fillers in rubber mixtures, to reduce the emission of volatile organic compounds. Two types of ash, BCA1 and BCA2, were selected as fillers for styrene–butadiene rubber (SBR). The ashes were produced during the treatment of brown coal at the Bełchatów Power Plant in the years 2017 and 2018. The morphology and chemical composition of the ash were tested. Morphology studies using scanning microscopy showed differences in the grain sizes of the ashes, and EDS analysis showed a difference in their chemical compositions. Vulcanizates with different weight proportions of the individual ashes were produced. Mixtures were made with the addition of 10–30 pts. wt. ashes per 100 g of SBR. The addition of BCA1 ash at 10 and 30 pts. wt. reduced the emission of volatile organic compounds (VOC) while maintaining the good strength properties of the mixtures.

Keywords: ash after lignite combustion; filler; rubber mixtures; volatile organic compounds (VOC)

Citation: Prochon, M.; Bieliński, D.; Stepaniak, P.; Makowicz, M.; Pietrzak, D.; Dzeikala, O. Use of Ashes from Lignite Combustion as Fillers in Rubber Mixtures to Reduce VOC Emissions. *Materials* **2021**, *14*, 4986. https://doi.org/10.3390/ma14174986

Academic Editor: Avelino Núñez-Delgado

Received: 17 July 2021
Accepted: 24 August 2021
Published: 31 August 2021

1. Introduction

During the process of coal combustion, significant amounts of by-products are formed, including fly ash, furnace slag, and harmful chemical compounds in the form of gases (including CO_2, NO_x, and sulfur compounds) [1–7]. A large proportion of this furnace waste (ash and slag) is stored. The storage of ashes is not ideal due to the significant risk of atmospheric factors leading to the release of dust particles into the air. Some uses for ashes have been found. An interesting application of ashes is in the production of electrodes in lithium-ion batteries [8]. Energy waste is also used in mining (as flooring components, to strengthen the rock mass) [9], agriculture (for the production of fertilizers, soil deacidification), civil engineering (for soil stabilization, construction of embankments, flood sealing), and in environmental protection, where energy waste is used to neutralize sewage and purify exhaust gases [10].

Ashes can also be used in construction, as additives in cement and concrete [11–14]. However, not all ashes can be exploited in this way, because their chemical composition is not in line with the relevant standards. For example, A. Michalik et al. [15] developed a method of improving the properties of fly ash as an additive for cements. However, some of the ash remains and is treated as waste. Limestone ash, after the separation of the carbon fractions and silica ash, is also used in the construction industry as a concrete mass.

The possibilities of using all of an ash are quite limited, due to the problematic phase and chemical compositions of ashes, as well as their physicochemical properties. Finding a way to recycle ash could significantly reduce the amount of landfilled waste. The scale of the problem can be seen in the case of the Bełchatów Power Plant (Poland), which is the largest thermal power plant in Europe. This power plant alone produces and stores approximately three million tons of ash and slag annually [14]. The ashes produced as a result of various methods of coal combustion show significant differences in terms of their phase and chemical compositions, degrees of dispersion, and particle morphologies. The use of ashes in this form is difficult, requiring research to determine their specific properties and identify suitable applications [6].

An innovative idea may be to use ashes from brown coal combustion as fillers in elastomeric blends. Plastics are widely used in many industries, including in the automotive industry as car cabin accessories, everyday objects, interior fittings, etc., [16]. However, plastics are a potential source of volatile organic compounds (VOCs), which are released in the processes of degassing materials [17]. Volatile organic compounds are known to adversely affect human health; they constitute about 73% of carcinogenic compounds on the list of toxic compounds [18]. The highest concentrations of VOCs are recorded indoors. It is estimated that adults spend about 80% of their time indoors, which puts them in prolonged contact with these toxic compounds. Therefore, manufacturers should strive to minimize VOC emissions both during the production process and from plastic products.

Styrene–butadiene rubber (SBR) is a type of polymeric compound based on rubber that is commonly used in the tire industry [19]. SBR rubber is also used as a waste material for the growth of carbon nanomaterials, such as nanofibers or nanotubes. Small granules of rubber are subjected to pyrolysis at a temperature of 1000 °C. The pyrolysates are mixed with gases containing burned SBR and oxygen. Combustion products are used to synthesize carbon nanomaterials (CNMs) in the presence of catalysts. CNMs have a characteristic structure of approximately 30–100 microns in diameter and 10 microns in length. Due to their unique mechanical, electrical, and thermal properties, CNMs have a number of potential added value applications [20].

In a study by X. Ren and E. Sancaktar [21], by-products from the energy industry in the form of ashes were used as a fly reinforcement, replacing typical fillers such as carbon black and silica in rubbers. The partial addition of up to 10 pts. wt. of fly ash in combination of 54 and 4 pts. wt. of carbon black and silica fillers, respectively, resulted in increased elongation at break, better adhesion to the steel reinforcement cord, and improved wet grip. Lower rolling resistance was also observed, which was attributed to the more effective reinforcing effect of the silica contained in the fly ash.

Here, we present a new method of using ashes as fillers for rubber mixtures. The ashes were produced during combustion of lignite at the Bełchatów Power Plant in 2017 (BCA1] and 2018 (BCA2]. An additional aim was to reduce the emission of VOCs from the mixtures produced during the vulcanization process.

2. Materials and Methods

2.1. Materials

2.1.1. Rubber and Other Ingredients

The rubber mixtures were made using SBR butadiene–styrene rubber (KER 1500, Synthos S.A. Oświęcim, Poland) as the matrix for each mixture. The cross-linking unit included sulfur, as the cross-linking substance (S_8 orthorhombic; density 2.07 g/cm^3, Siarkopol Tarnobrzeg Sp. z o.o., Tarnobrzeg, Poland), zinc oxide as the activator (zinc white, ZnO, Huta Będzin, Poland), and CBS (Bestgum Polska Sp.z o.o., Rogowiec, Poland) as the accelerator.

2.1.2. Ash after Burning Brown Coal (BCA)

Technical carbon black N330 (75 × 10^3 m^2/kg, Fermintrade, Konin, Poland) and ash from Elektrownia Bełchatów (PGE Górnictwo i Energetyka Konwencjonalna SA, Elektrownia Bełchatów, Poland) produced in 2017 (BCA1) and 2018 (BCA2) were used as the fillers. The mixtures also included technical stearin (Torimex-Chemicals Ltd. Sp.z o.o., Konstantynów Łódzki, Poland).

2.2. Preparation of Composites

After the preparation step (see Section 2.1.2), the BCA filler was introduced at different pts. wt. into the SBR. The rubber mixtures were prepared using a mixing mill (Bridge mill, London, UK) with a roller temperature of 27–37 °C and a friction/friction of 1.1. The rolling mill parameters were as follows: roll length L = 450 mm; roll diameter D = 200 mm; rotational speed of the front roller Vp = 20 (rpm); width of the gap between rollers 1.5–3 mm. The mixtures were prepared for 6 min and then wrapped in foil and stored at 2–6 °C. Table 1 compares the tested rubber mixtures. The mixtures were made with carbon black N330 or a mixture of carbon black with 10, 20, and 30 pts. wt. ashes. Tests were carried out at room temperature under normal pressure conditions.

Table 1. Compositions of the elastomer blends.

Symbol	SBR	SBR1 (10)	SBR1 (20)	SBR1 (30)	SBR2 (10)	SBR2 (20)	SBR2 (30)
SBR [phr]				100			
BCA1 [phr]	0	10	20	30	0	0	0
BCA2 [phr]	0	0	0	0	10	20	30
N330 [phr]	50	40	30	20	40	30	20
Stearin [phr]				1			
ZnO [phr]				3			
CBS [phr]				1			
Sulphur [phr]				2			

3. Research Techniques

3.1. Researching Ashes

To determine the structure of the ashes, microscopic pictures were taken using a Keyence VHX 1000 optical microscope (Keyence International, Mechelen, Belgium). To determine the share of particles of a given size in the tested ashes, a sieve analysis was performed. For this purpose, an AS200Control device (Retsch GMBH, Haan, Germany) with EasySieve software (Retsch GMBH, Haan, Germany) was used, consisting of eight sieves with decreasing mesh sizes: 4.00 mm; 2.00 mm; 1.00 mm; 0.50 mm; 0.25 mm; 0.125 mm; 0.063 mm; 0.045 mm. The measurements lasted 3 min, during which the computer program processed the results from the testing device.

3.2. Making Mixtures

The ingredients were mixed together in the following order: rubber → ZnO → stearin → soot → CBS → sulfur. A rolling mill (Bridge type milling machine, London, UK) was used with a roll length of 150–300 mm, temperature 30 °C, rotational speed 15–13 rpm. The cross-linking kinetics of the SBR compounds were tested using an ALPHA MDR 200 rheometer with a moving nozzle (MonTech MDR 300, Buchen, Germany) at a temperature of 160 °C for 30 min, in accordance with PN-ISO 3417:1994. Hardening ($\tau 90$) was performed in a standard electrically heated hydraulic press (PH-2PW90, ZUP Nysa Sp. z o.o. Nysa, Poska) at a temperature of 160 °C, pressure 32 MPa, dimensions of heating plates 400 mm × 400 mm. After vulcanization, the samples were prepared for strength tests in accordance with the PN-EN ISO 3167:2014-09 standard. Paddle-shaped samples were cut out using a ZCP020 press (Zwick/Roell GmbH & Co. KG., Ulm, Germany).

The mechanical properties of the prepared composites were tested on a universal testing machine Zwick model 1435 (Zwick/Roell, Radeberg, Germany), in accordance with the PN-ISO 37:2007 standard. Paddle-shaped samples were subjected to stretching at room temperature at a constant speed of 500 mm/min and with a preforce of 0.3 N. The tensile strength (TSb) and the percentage elongation at break (E_b) were determined. A hardness test (H, °Sh) was performed using an electronic Shore A hardness tester (Zwick/Roell, Herefordshire, UK) with a pressing force of 12.5 Nz, in accordance with PN-80C-04238 [22].

The polymer–solvent interaction parameter was determined by the equilibrium swelling method according to PN ISO 1817:2001/ap1:2002 [23] (0.378 for SBR rubber in a toluene solvent). Samples were then allowed to evaporate in air for 72 h and reweighed.

Cross-link density was calculated as the volume fraction of rubber in the swollen material, and vs. = 106.3 mol/cm^3 for the molar volume of the solvent (toluene) solvent [24,25].

Cross-link density (v), was calculated on the basis of the Flory–Rehner equation:

$$v = \frac{\ln(1 - V_r) + V_r + \mu V_r^3}{V_0\left(V_r^{\frac{1}{3}} - \frac{V_r}{2}\right)} \quad (1)$$

where μ is the Huggins parameter for the uncrosslinked polymer–solvent system and *Vr* is the molar volume of the swelling solvent.

Rheological tests were performed using a rotational rheometer (ARES-G2 TA Instruments, New Castle, DE, USA). Stress measurements were performed regardless of the applied shear deformation, with a torque range from 0.05 μNm to 200 mNm, force range of 0.001–20 N, temperature range from −150 °C to +600 °C, cooling with liquid nitrogen, DETA module for dielectric measurements.

To test the effect of elevated temperature, the samples were placed in a thermal chamber (Binder GmbH, Tuttlingen, Germany) at 70 °C for 7 days. Three mixtures were selected and subjected to 30-day biological aging in soil. The soil samples were placed in a MEMMERT climate chamber (HPP 108 Memmert GmbH, Schwabach, Germany) for 14 days at 30 °C and 80% air humidity. The aged samples were then subjected to strength tests. The results were compared with those for samples that had not been aged. Examination of the surface morphology of the rubber mixtures was carried out using a Zeiss Ultra Plus scanning electron microscope (ZEISS, Carl Zeiss AG, Oberkochen Germany). A Nicolet 6700 FTIR spectroscope (Thermo Scientific, Waltham, MA, USA) with Fourier transform was used to analyze the chemical bonds present in the ashes and the obtained vulcanizates, using the total internal reflection method with the attenuated total reflection (ATR) attachment. Measurements were carried out in the range of 400–4000 cm^{-1} for the ashes and three selected vulcanizates.

The VOCs were studied using ion mobility spectroscopy with gas chromatography, during vulcanization in a tightly closed reactor. The heating time of the mixture at 160 °C was 2 × t90 (two times optimum cure time) in order to be sure that the vulcanization occured. The collected gases were diluted and sent to an MCC-IMS instrument (G.A.S. Gesellschaft für Analytische Sensorsysteme GmbH, Dortmund, Germany). Tests were carried out for a mixture without the addition of ash and for two mixtures with CBS1 ash (addition of 10 or 30 pts. wt. ash). Thermal analysis by differential scanning calorimetry (DSC) was carried out in order to determine the glass transition temperature (Tg) of the selected vulcanizates. The tests were carried out using a DSC_1 analyzer (Mettler Toledo, Netzsch, Switzerland) calibrated with standards (indium, zinc) at a heating rate of 10 °C/min. The SBR samples were heated from −150 °C to 350 °C under nitrogen atmosphere.

4. Results and Discussion

4.1. Characterization of BCA

It is known that the shape, particle size, and specific surface of a filler have a decisive influence on the strength of rubber–filler connections. Ash morphology was assessed on the basis of photographs taken with an optical microscope, shown in Figure 1. The ash agglomerates are visible as irregularly structured spherical particles, with a wide size distribution from several hundred nanometers to several micrometers. The grains in the BCA2 ash are much finer. The older ash contained larger particles with greater size variation.

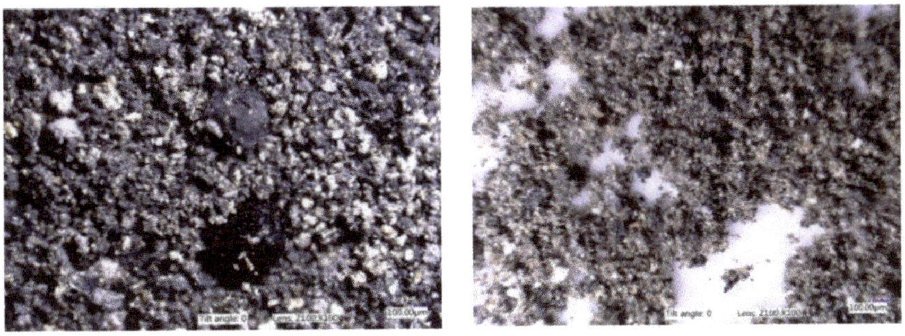

(a) (b)

Figure 1. Images of BCA magnification ×1.00: (**a**) ash BCA1 and (**b**) ash BCA2.

Table 2 shows that the largest share of BCA1 ash (74.7%) was between 63–250 μm, whereas BCA2 ash contained mostly fractions below 45–250 μm (76.9%).

Table 2. Ash sieve analysis BCA1 and BCA2.

	BCA1			BCA2		
Size Class [μm]	Δp [%]	q_3 %[μm]	Δm [g]	Δp [%]	q_3 %[μm]	Δm [g]
<45	8.1	0–18	4.30	23.6	0.52	11.79
45–63	9.2	0.51	4.59	26.1	1.45	13–05
63–125	38.2	0.62	119.00	27.2	0.44	13.63
125–250	36.5	0.29	18.18	17.3	0.14	8.68
250–500	5.8	0.02	2.91	4.3	0.02	2.15
500–1000	1.4	0.00	0.72	0.8	0.00	0.40
1000–2000	0.5	0.00	0.26	0.4	0.00	0.18
2000–4000	0.0	0.00	0.01	0.3	0.00	0.16
>4000	0.1	0.00	0.05	0.0	0.00	0.00

Legend: Δp—percentage distribution of the fraction share (%); q^3%—density of the grain class distribution (μm); Δm—particle mass distribution (g).

4.2. Characterization of SBR/BCA Composites

4.2.1. Vulcanization Kinetics

Rheometric measurements were used to determine the vulcanization time and the increase in the torsional moment ΔM. The results are presented in Table 3. The BCA fillers clearly influenced the cross-linking properties of the SBR blends. The kinetic parameters of the BCA-containing compounds were different from those of the blank compound. The addition of ash to the SBR compound resulted in lower viscosity and stiffness, which is reflected in the torque value (ML). The increase in torque decreases as the weight fraction of ash increases, leading to a visible deterioration of the mechanical parameters of vulcanizates. The unfilled sample showed a longer cure time than the filled composites.

Table 3. Influence of BCA on the rheometric properties of SBR compounds.

Symbol	M_L [dNm]	ΔM [dNm]	τ_{90} [min]
SBR	2.22	18.12	11.35
SBR1 (10)	1.90	17.11	9.67
SBR1 (20)	1.25	12.82	10.84
SBR1 (30)	1.37	12.68	12.99
SBR2 (10)	1.53	13.40	11.08
SBR2 (20)	1.52	13.03	11.36
SBR2 (30)	1.48	12.72	12.06

LL—minimum torque moment (dNm); ΔL—the decrease of torque moment (dNm) (ΔL = LHR—LL); τ_{02}—scorch time (min); τ_{90}—time of vulcanization (min).

The vulcanization times increased with increasing BCA1 concentrations, reaching the highest value at 30 phr (τ_{90} = 12.99). Subsequent studies clearly showed that the ash belongs to the group of inactive fillers, with lower increases of turbulent moments for the composition with the same content compared to the noninvasive composite. As a consequence, it leads to a visible deterioration of the mechanical parameters of vulcanizates.

4.2.2. Rheological Analysis

Rheological analysis is applicable to all materials with visco-elastic properties and describes the relationship between forces and deformation as a function of time. Figure 2 shows the dependence of changes in stress viscosity on the shear rate for individual SBR elastomer composites filled with ash. The dynamic viscosity values changed as a result of introducing ash fillers into the SBR rubber matrix. Irrespective of the type of ash used, the shear modulus decreased as the proportion of the introduced filler was increased from 10 to 20 pts. wt. This correlates with the rheometric properties (Section 4.2.1).

The use of the ash as a filler changed the viscosity of the composites. The dynamic viscosity of the mixtures was almost half that of the pure SBR sample, which may suggest the formation of connections between the mineral filler and the SBR matrix, creating a secondary structure containing mineral particles associated with the remaining components of the mixtures. As reported in [26], this may be caused by changes in deformation caused by the material microstructure, i.e., increased susceptibility to cracking and thus reduced hardness, as well as the formation of weak physical interactions connecting adjacent filler clusters.

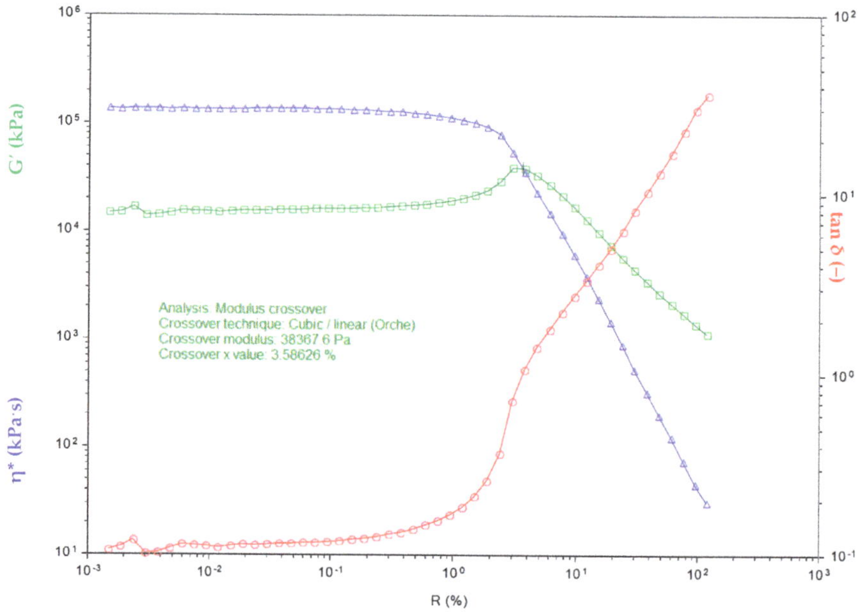

(a)

Figure 2. *Cont.*

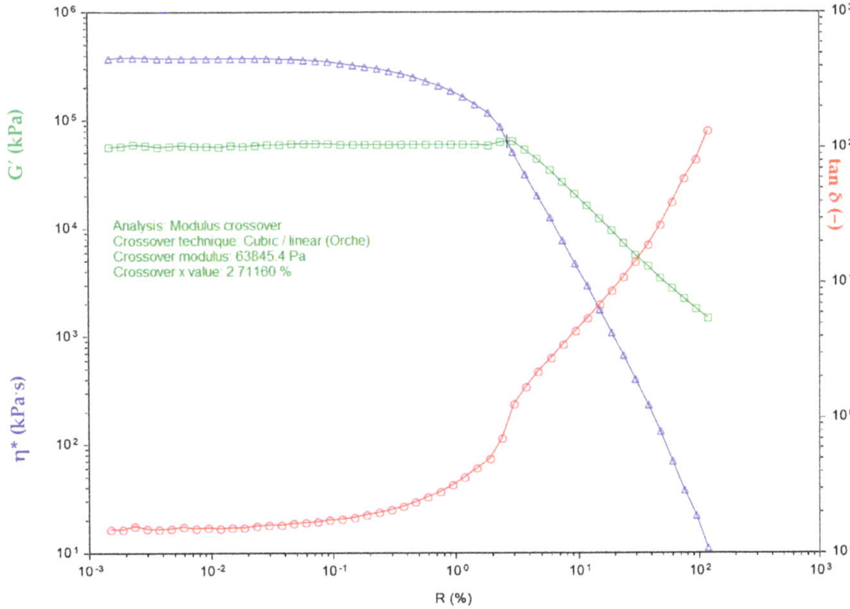
Analysis: Modulus crossover
Crossover technique: Cubic / linear (Orche)
Crossover modulus: 63845.4 Pa
Crossover x value: 2.71160 %
G′ (kPa)
η* (kPa·s)
tan δ (–)
R (%)

(b)

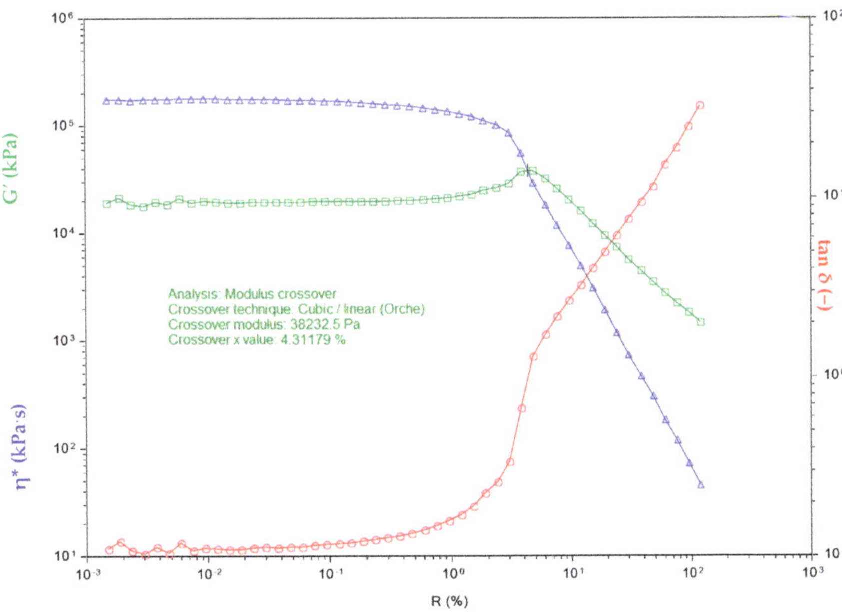
Analysis: Modulus crossover
Crossover technique: Cubic / linear (Orche)
Crossover modulus: 38232.5 Pa
Crossover x value: 4.31179 %
G′ (kPa)
η* (kPa·s)
tan δ (–)
R (%)

(c)

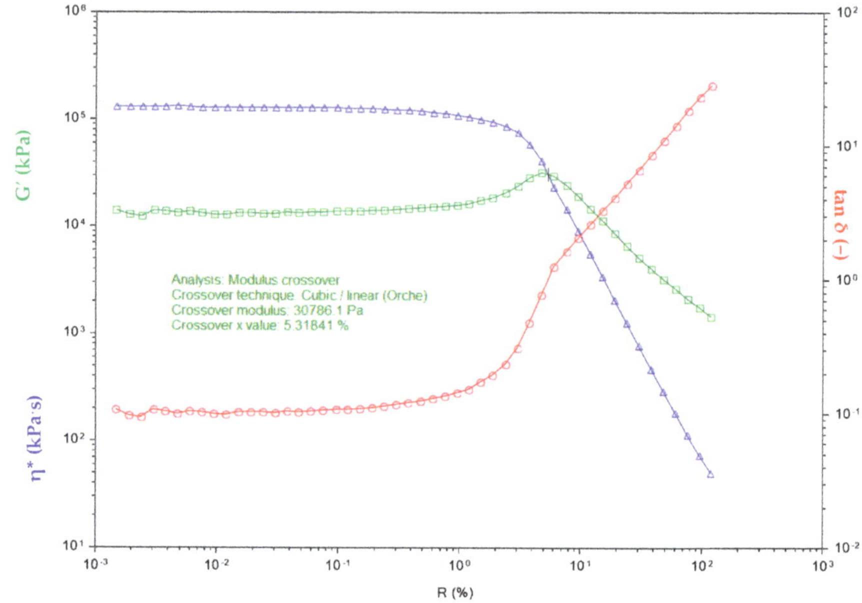

(d)

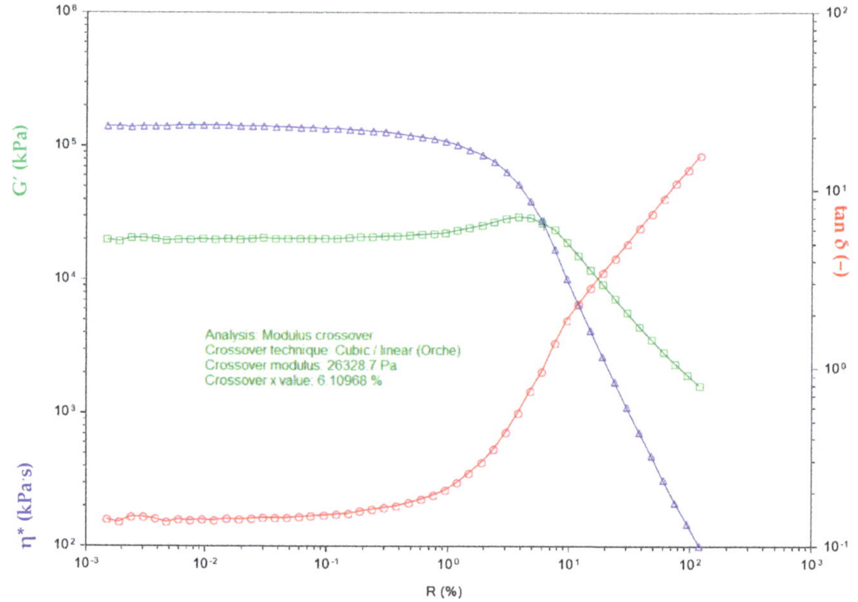

(e)

Figure 2. Dependence of viscosity and stress on the shear rate of the composites: (**a**) SBR; (**b**) SBR1 (10); (**c**) SBR2 (10); (**d**) SBR1 (20); (**e**) SBR2 (20). Sapphire curve —complex viscosity η* (kPa's) at 80 °C (processing temperature) for SBR mixtures; green curve—storage shear modulus G′ (kPa) of uncured rubber mixtures SBR at 80 °C as a function of angular frequency ω (rad's − 1) (linear viscoelastic region); red curve—mechanical loss tan δ (−) as a function of angular frequency ω (rad's − 1) (oscillation strain 100%).

4.2.3. Cross-Linking Density

The cross-linking densities of the prepared mixtures are presented in Table 4. The reference sample had the highest cross-linking density. The samples doped with BCA1 ash did not show any particular tendency, while the cross-link densities of the polymers filled with BCA2 ash increased with higher ash contents. The cross-linked density of the SBR1 (20) and SBR2 (10) samples was 53% lower than the cross-link density of the reference sample.

Table 4. The cross-linking densities of the prepared mixtures.

Symbol	SBR	SBR1 (10)	SBR1 (20)	SBR1 (30)	SBR2 (10)	SBR2 (20)	SBR2 (30)
Total network density α [mol/cm^3]	2.00×10^{-4}	1.94×10^{-4}	9.35×10^{-5}	1.39×10^{-4}	9.35×10^{-5}	1.51×10^{-5}	1.64×10^{-4}

4.2.4. FTIR

The composites were subjected to FTIR analysis to characterize the bonds present in the ash samples and in the vulcanizates. Figure 3 shows a comparison of BCA1 and BCA2 ashes, while Figures 4 and 5 show examples of spectra for SBR, SBR1 (10), and SBR1 (30) containing different amounts of ash.

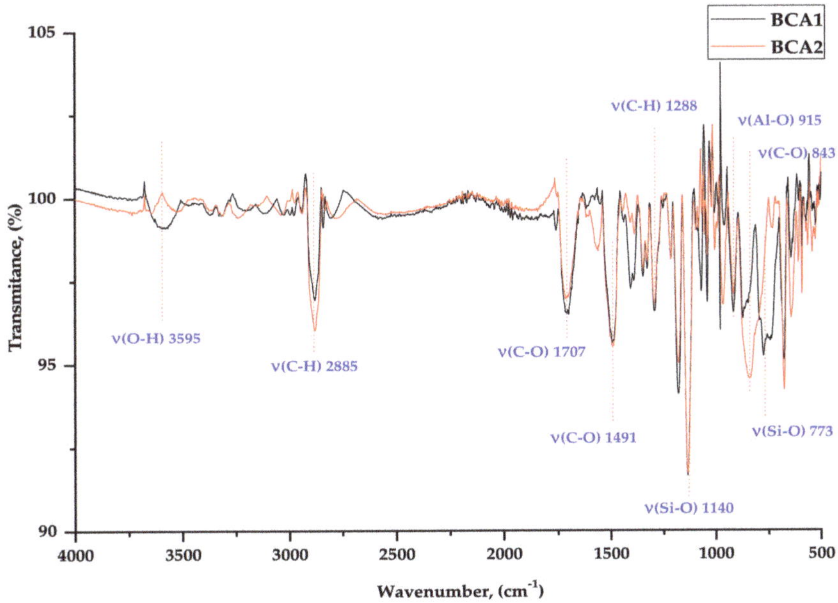

Figure 3. FTIR spectra for ashes BCA1 and BCA2.

The spectra presented in Figure 3 show valence vibrations of the bands coming from the side chains and hydroxyl (–OH) groups in the region from 3200 to 3500 cm^{-1}. At 1706 cm^{-1}, a vibration band appears (C=O). The effect of silica on interactions with other components of the BCA powder, and thus on other interactions, was manifested by shifting of the absorption bands, reducing their intensity, etc. The characteristic absorption band at 1140 cm^{-1} can be attributed to the possible interaction of Si with the protein-like system –Si–O. There is also a characteristic broad absorption band for BCA2 C-O powder at about 843 cm^{-1}. The bands between 780 and 650 cm^{-1} indicate the presence of Si–O–Si bonds. A small band was observed in the range from 3600 to 3550 cm^{-1}, which is related to the vibration of the –OH group in the BCA1 sample. The Al-O compound appears in the range of 920–910 cm^{-1} for both types of ashes.

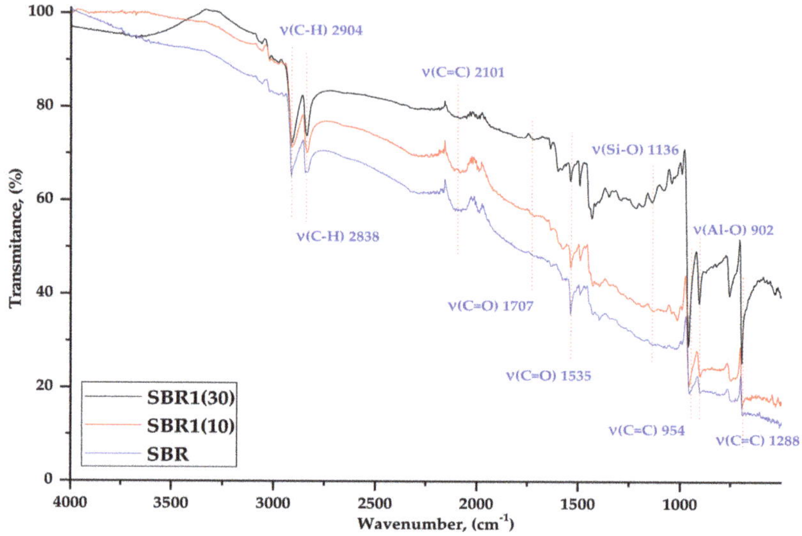

Figure 4. FTIR spectra of SBR, SBR1 (10), and SBR1 (30) composites.

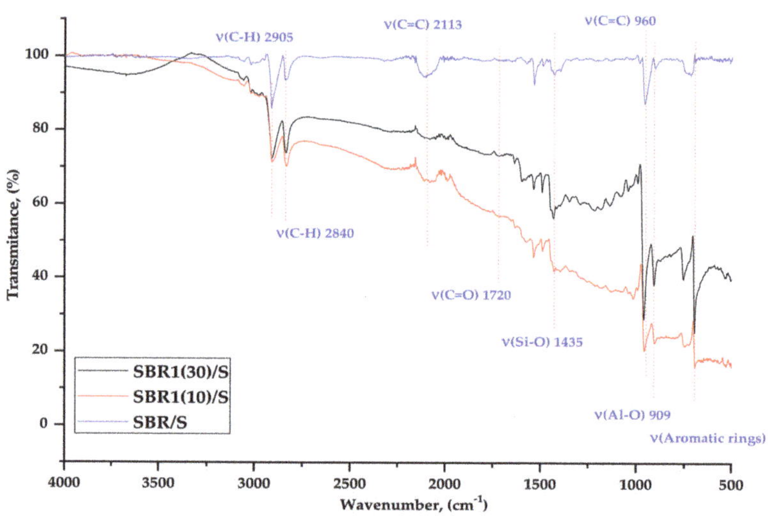

Figure 5. FTIR spectra for SBR, SBR1 (10), and SBR1 (30) after thermo-oxidative aging.

As shown in Figure 4, at 2885.07 cm^{-1} visible vibrations of the C–H groups are present in the structure of the SBR aromatic styrene ring [27]. At 2100 cm^{-1} there is a C=C bond derived from rubber butadiene. The intensity of the BCA bands changes with the amount of ash, although these changes are not as significant. Small changes are visible in the dactyloscopic area related to a slight increase in the intensity of the bands in the range of 1500–1400 cm^{-1} and 500 cm^{-1}. The first are characterized by stretching vibrations of the –C=C– aliphatic grouping and bending vibrations = C–H. However, the second is also the confirmation of the occurrence of stretching vibrations of the aromatic ring v (C=C), which come from styrene. Compared to the spectrum of BCA alone (Figures 4 and 5), the intense band at 773 cm^{-1} derived from silicates significantly reduces in intensity, which may indicate possible interaction with Si–O–C groups in the SBR aromatic rubber ring.

The band at 953 cm^{-1} is related to the absorption of stretching vibrations from the C=C groups in the aromatic ring, which become much more intense as the amount of BCA increases. This may indicate some interaction between the filler and the elastomer matrix. The prominent peak at about 901 cm^{-1} indicates the presence of Al–O groups.

In the spectra of the mixtures after temperature aging (Figure 5), a change in the intensity of certain peaks can be observed. In the case of the signal from the C=C groups, as the share of ash in the mixture increased a decrease in the band intensity was observed at the wavenumber of 2113 cm^{-1} and around 974 cm^{-1}. This was probably related to the breaking of double bonds during aging.

4.2.5. Mechanical and Hardness Tests

The influence of BCA on the mechanical properties of the SBR vulcanizates is presented in Table 5. A hardness test was performed to determine the effect of the amount of added filler on the hardness of the rubber mixtures. The samples with the addition of ash showed a decrease in hardness in the range of 8.7–14.8%. The smallest decrease was observed for the SBR2 sample (20), and the largest decrease for the SBR2 sample (30). There was no downward trend in hardness as a function of sample composition (addition of ash). This parameter is related to the cross-linking density of composites and confirms the results of the swelling measurements and rheometric tests.

Table 5. Influence of BCA filler on the mechanical properties of the SBR composites.

Symbol	**SBR**	**SBR1 (10)**	**SBR1 (20)**	**SBR1 (30)**	**SBR2 (10)**	**SBR2 (20)**	**SBR2 (30)**
H [°Sh]	60.85 ± 1.69	55.29 ± 2.63	55.39 ± 1.55	52.71 ± 1.08	54.43 ± 0.62	55.54 ± 1.12	51.83 ± 1.35
SE_{100} [MPa]	3.37 ± 0.20	2.63 ± 0.23	1.53 ± 0.06	1.49 ± 0.06	1.79 ± 0.03	1.86 ± 0.05	1.51 ± 0.06
SE_{200} [MPa]	8.47 ± 0.44	5.78 ± 0.52	2.44 ± 0.13	2.22 ± 0.13	3.48 ± 0.09	3.31 ± 0.17	2.21 ± 0.10
SE_{300} [MPa]	15.57 ± 0.75	10.66 ± 0.96	3.98 ± 0.30	3.39 ± 0.26	6.51 ± 0.18	5.98 ± 0.37	3.36 ± 0.18
T_S [MPa]	27.37 ± 0.32	18.5 ± 0.20	13.38 ± 0.66	11.86 ± 0.40	19.00 ± 0.35	15.95 ± 0.98	11.94 ± 0.54
E_b [%]	476 ± 24	441 ± 24	591 ± 25	619 ± 28	576 ± 15	535 ± 33	614 ± 14
Influence of theromoxidative and biological aging on the properties of SBR composites.							
Symbol	SBR	SBR1 (10)	SBR1 (20)	SBR1 (30)	SBR2 (10)	SBR2 (20)	SBR2 (30)
T_S [MPa]	23.16 ± 1.65	12.87 ± 1.46	10.28 ± 0.98	9.34 ± 0.54	16.98 ± 1.65	14.24 ± 0.40	6.89 ± 1.75
E_b [%]	282 ± 23	287 ± 18	370 ± 25	400 ± 16	423 ± 23	368 ± 17	422 ± 4

Legend: SE100 (MPa)—stress modulus with elongation at 100%; SE200 (MPa)—stress modulus with elongation at 200%; SE300 (MPa)—stress modulus with elongation at 300%; T_S (MPa)—tensile strength; E_b (%)—elongation at break.

The results of the tensile strength tests show a decrease in breaking strength as a result of doping the rubber mixtures with ash. The SBR control sample, without the addition of ash fillers, had the highest tensile strength. The addition of ashes caused a decrease in the tensile strength of the mixtures. The smallest decrease was observed for the SBR1 sample (10). Elongation at break increased by between 12.6% and 30% following the addition of ash fillers. The highest elongation was achieved by the SBR1 sample (30), with an increase of 30% compared to the control (without filler). The only decrease in elongation in comparison to the SBR control sample occurred in the case of SBR1 (10). The values for this sample decreased with further additions of filler. This may be related to the lack of the formation of permanent molecular interactions in the form of permanent bonds, for example covalent or ionic bonds, although FTIR analysis indicated there may be weak range metallic intermolecular interactions.

A strength test was carried out on the samples after thermo-oxidative aging. In the case of thermo-oxidative treatment, the general trend remained unchanged. The highest stresses were recorded for the SBR samples, and the lowest stresses for SBR1 (30). The decreases in the elongation at break values were comparable for each of the samples. The stress at break value for the reference sample decreased by 15% compared to the sample not subjected to thermo-oxidative aging. The average reduction in tensile strength for samples filled with ash was 23%. The elongation at break value for the reference sample decreased by 41% after temperature aging. The mean decrease in elongation at break values for the samples filled with ash was 33%. The lifetime of elastomers, including styrene–butadiene rubber, is largely dependent on the aging processes. Physical and chemical properties change, depending on the rubber content, the type of components used for the elastomer mixture and the resulting internal network structure. In the presented cases (SBR2 (10), SBR2 (20), SBR2 (30)) the decrease in tensile strength (TS) in relation to the samples not subjected to aging, and thus the observed increase in the relative elongation E_b, may be caused by changes in the viscoelastic properties caused by the introduction of the filler in the form of ash BAC. The ash is characterized by a diverse mix of micro and macro elements that influence the oxidation processes of the composite structure. In addition to elevated temperature and humidity, oxygen, ozone, and light contribute to the aging processes. During the aging processes, the cross-links are degraded, causing the relaxation of the tight and more packed elastomeric structure, which, by changing the viscoelastic interactions, leads to an increase in the elongation parameters at the moment of breaking E_b.

4.2.6. Carbonyl Index

The Carbonyl index (CI), based on changes in the carbonyl (C=O) band, is used for to predict the lifespan and development of stabilizing additives [28–31]. The CI is used specifically to monitor the absorption band of carbonyl species formed during photooxidation or thermo-oxidation processes in the range from 1850 to 1650 cm^{-1} by measuring the ratio of the carbonyl peak to the reference peak. The carbonyl index is the ratio of the band height of the carbonyl groups to the band height of the C–H groups. An increase in CI indicates more advanced degradation. One of the most common analytical techniques for monitoring oxidation reactions, including CI, is Fourier transform infrared (FTIR) spectroscopy. FTIR analysis is also able to monitor other chemical changes that occur over the lifespan of the material, by detecting functional groups present in different bands.

The CI for the tested composites was calculated after temperature aging (Table 6). The addition of ash lowered the CI, so it can be concluded that chemical reactions took place in the vulcanizate structure, leading to the remodeling of individual bonds under the influence of temperature. Increasing the amount of filler in the matrix thus caused the matrix to degrade faster during processing. The lowest value (0.05) was recorded for the SBR1 sample (10).

Table 6. Carbonyl index for samples after thermo-oxidative aging.

Symbol	Wavenumber [cm^{-1}]		Carbonyl Index (CI)
	~1700	~2800	
SBR	0.92	6.30	0.15
SBR1 (10)	0.38	7.64	0.05
SBR1 (30)	0.91	8.80	0.10

4.2.7. SEM Analysis

The morphology of the selected samples was assessed based on SEM photos taken at the breakthroughs of individual materials (Figure 6). Photographs at 250 times magnification revealed the diverse morphologies of the materials. In the case of the SBR reference sample, a small number of fine agglomerates can be observed, which may represent, e.g., the grains of the crosslinker. On the other hand, in the case of samples with the addition of ash, we can see an analogous increase in light agglomerates: the higher the addition of ash, the greater the proportion of particles with a regular structure, isolated from each other, with a wide size distribution.

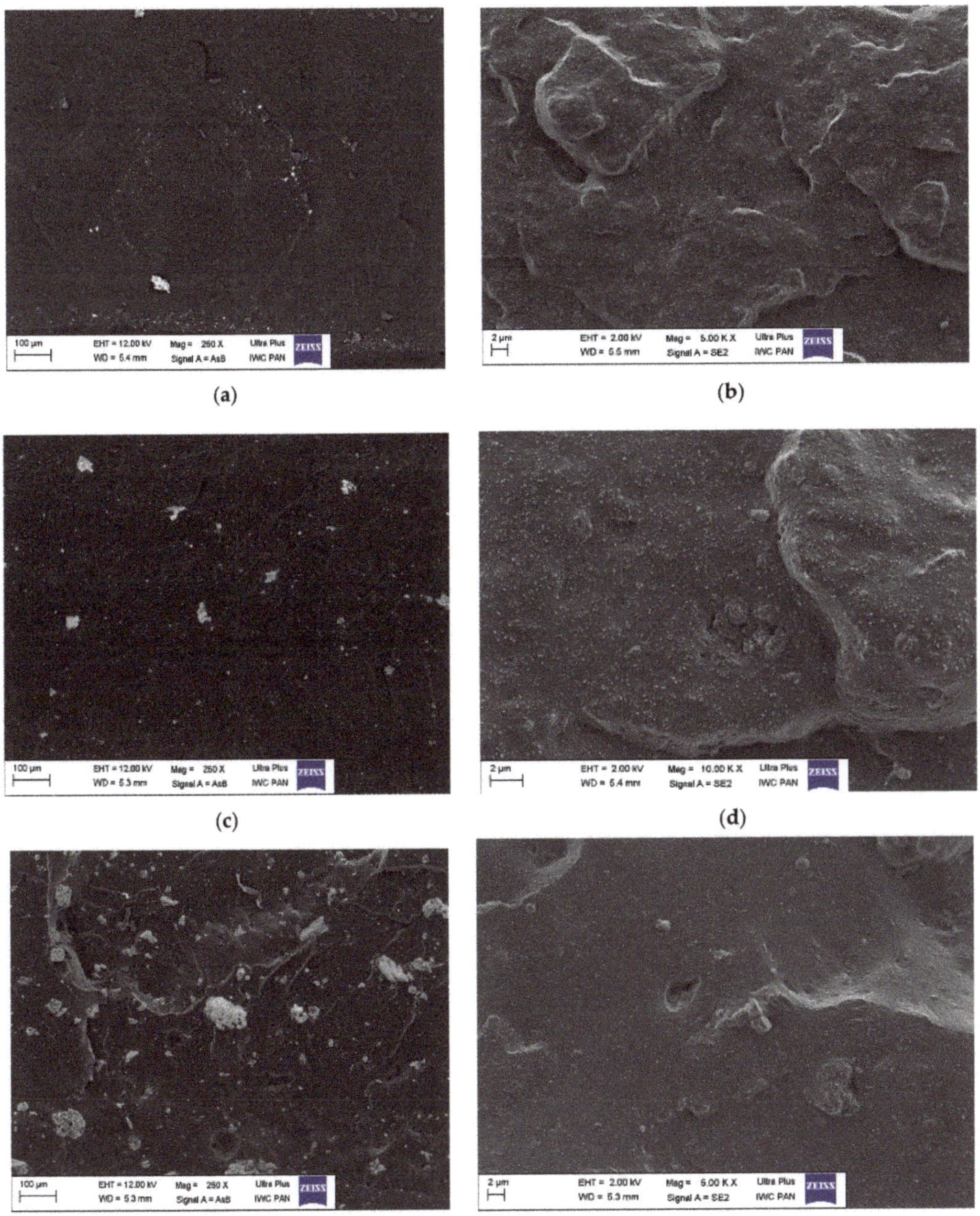

(a) (b)

(c) (d)

(e) (f)

Figure 6. SEM photographs of composites (**a**,**b**)—SBR; (**c**,**d**)—SBR1 (10); (**e**,**f**)—SBR1 (20); left: photographs with ×250 magnification, right: photographs with ×5000 magnification.

4.2.8. EDS Analysis

EDS analysis enabled qualitative and quantitative determination of the chemical compositions of the studied composites. However, it should be remembered that EDS provides only an approximate measure of individual minerals. The results are presented in Table 7. Aluminum and calcium were not observed in the SBR mix but appeared in the mixes with ash. This may indicate that these components were contained in the BCA1. The reference mixture also had a very low silicon content, whereas in SBR1 (10) and SBR1 (20) there was a significant increase in silicon content, which again leads us to the conclusion that silicon was a component of the ash.

Table 7. Compositions of analyzed elastomer blends obtained by EDS analysis.

Chemical Elements, %	Designations of Mixtures		
	SBR	**SBR1 (10)**	**SBR1 (20)**
Oxygen	5.82	9.40	10.41
Sulfur	1.46	1.65	2.03
Zinc	1.27	0.58	1.74
Silicon	0.008	0.96	1.51
Aluminum	-	0.78	0.96
Calcium	-	1.54	1.59
Iron	-	-	1.28

4.2.9. DSC & TGA Analysis

The results of thermogravimetric analysis (TGA) and differential scanning calorimetry (DSC) of selected SBR and SBR1 (10) composites are shown in Table 8, as well as in Figures 7 and 8.

The addition of BCA filler did not significantly affect the temperature of thermal decomposition of SBR. As seen from the standard deviation, the maximum decomposition temperatures of all composites were the same. The total weight loss of the elastomers tested during degradation was 97% for the reference sample and 91% for the BCA-filled composites.

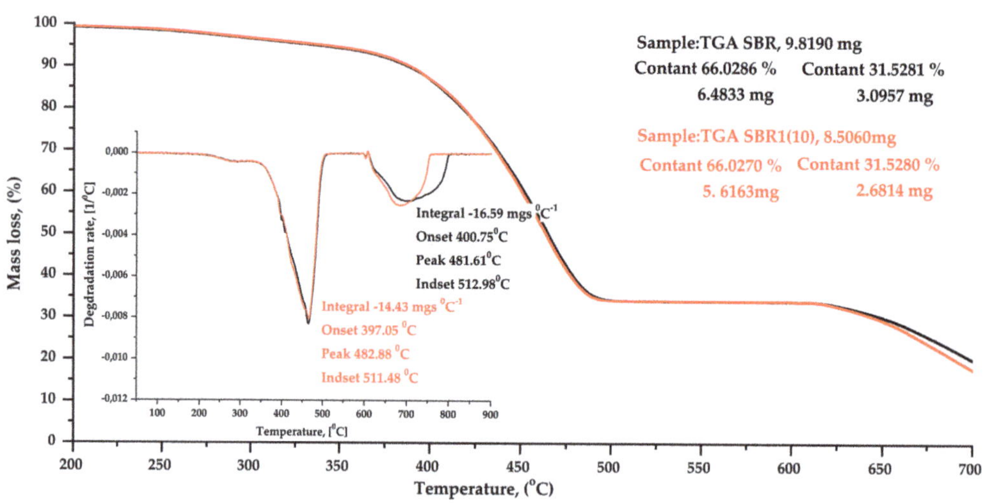

Figure 7. Thermogravimetric analysis (TGA) curves of SBR composites: reference sample SBR (black curve), SBR1 (10) sample (red curve).

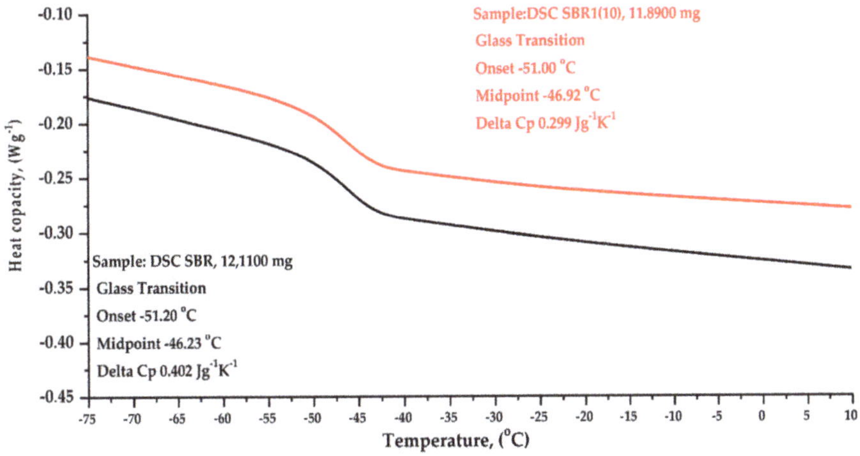

Figure 8. Comparative analysis of DSC results: unfilled SBR (black curve) and SBR10 (10 phr of BCA, red curve).

Table 8. Thermal characteristics of SBR and SBR1 (10).

Symbol	$T_{5\%}$ (°C)	$T_{peak\ 1\ (DTG)}$ (°C)	$T_{peak\ 2\ (DTG)}$ (°C)	Δm_{total} (%)	T_g (°C)
SBR	337	465	697	97.51	−46.23
SBR1 (10)	337	466.33	681	91.55	−46.92

$T_{5\%}$—decomposition temperature at 5% mass loss; $T_{p(DTG)}$—temperature of maximum conversion rate on the DTG curve; Δm_{total}—total mass loss during thermal decomposition; T_g—glass transition temperature (standard deviations: T_5, $T_{p\ (DTG)}$ ± 2 °C; Δm_{total} ± 0.6%; T_g ± 2 °C).

The glass transition temperature Tg of pure SBR is reported in the literature as 48/65 °C [32]. DSC showed that the glass transition temperature Tg starts at the beginning for the SBR composite, that is, at −46.23 °C, while for SBR1 (10) Tg starts at −46.92 °C (Figure 8). This suggests that the BCA filler acts as a plasticizer in SBR compounds. This effect is caused by certain amounts of macro- and microelements in the ashes obtained from brown coal combustion. These undoubtedly include calcium and magnesium, as well as silicon compounds. These elements, in combination with the components of the elastomer mixture, such as stearin or oxide activators, vulcanization accelerators, can impart plasticizing properties to rubber composites during cross-linking processes. The heat capacity of the SBR composite was 0.402 $Jg^{-1}K^{-1}$, whereas for SBR1 (10) the heat capacity decreased to 0.299 $Jg^{-1}K^{-1}$

4.2.10. Testing Emission of VOCs from Vulcanizates

Ashes can adsorb different types of VOC's, as well as gases or liquids in general. Ashes, depending on their physicochemical properties, can effectively adsorb various substances due to their high porosity, large surface area, suitable pore size and, for example, unburned carbon [33]. The larger surface area and carbon content with a large micropore volume results in higher adsorption capacity and longer penetration times. Additionally, the adsorption effect is related not only to the physicochemical properties of the adsorbent itself, but also to the physicochemical properties of the adsorbate, such as pore structure and surface functional groups [34].

The use of MCC-IMS allowed us to observe the effect of the ash on the release of chemical compounds from the mixture during the vulcanization process. The addition of a small amount of ash was found to reduce the amount of VOCs released during vulcanization, by approximately 48% on average (Table 9 and Figure 9). Ash as a highly porous and high-surface-area filler, can successfully trap released VOCs from rubber. The observed effect can be reduced by the polymer matrix which surround the filler particles "taking" their active groups on the surface, but also it can penetrate into the micro pores of the filler, reducing its number of free voids. With higher proportions of ash, decreases in emissions were also observed, but they were less significant. The reason for the higher VOCs emissions of the SBR1 sample (30) compared to the SBR1 sample (10) may be the amount of ash they contained.

Table 9. Comparison of the intensity of VOC signals released during the vulcanization process from the tested rubber mixtures.

	Benzothiazoles	Benzothiazole Dimer	Aniline	X1	X2
SBR	1129	399	532	973	466
SBR1(10)	961	171	178	608	264
SBR1(30)	1102	192	423	888	371

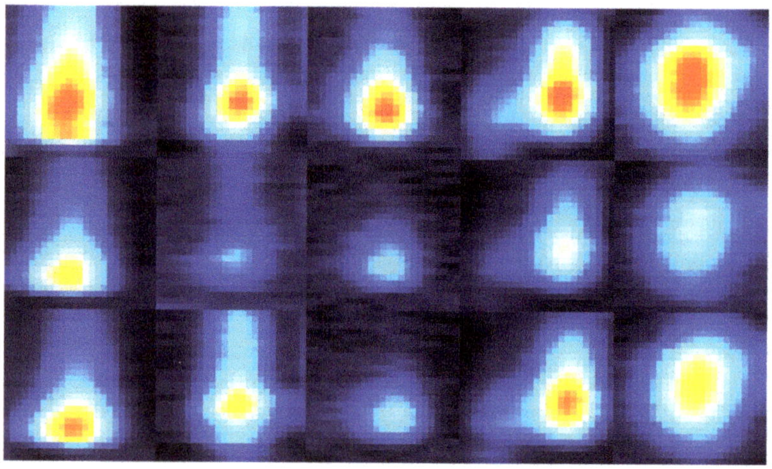

Figure 9. Graphical representation of VOCs released during the vulcanization process.

Increasing the proportion of filler in the mixture causes agglomeration, which reduces the active surface area and, as a consequence, the adsorbing properties of the filler. In addition, agglomeration of the filler reduces its degree of dispersion, which may have a direct negative impact on the amount of VOCs released.

Two VOCs that were identified in this analysis are known as decomposition by-products of accelerators: benzothiazole and aniline. There was a reduction in the emission of two other unknown chemicals, based on comparison with the reference blend.

5. Conclusions

In this study, we used ash produced during the combustion of lignite at the Bełchatów Power Plant (Poland) in 2017 and 2018 as additives in SBR mixtures. The additives reduced the emission of volatile organic compounds (VOCs) from the rubber mixtures during vulcanization. Their use to fill elastomeric mixtures could reduce production costs, while also contributing to waste management of the ash.

The incorporation of ash into the structures of the prepared mixtures was confirmed by SEM, EDS, and FTIR spectroscopy. Differences in the morphology and chemical composition of the two analyzed ashes were also noted. These differences may result from the origin of the ash, the conditions of its formation, the combustion processes used and the place of sampling.

A decrease in the mechanical properties of the vulcanizates was observed following the addition of the ashes, which proves that they are inactive fillers. The addition of a small amount of BCA1 ash (10 pts. wt.) maintained the good mechanical properties of the SBR, while also reducing VOC emissions during vulcanization. The presented solution opens the way for doping with waste ash waste from the energy and mining industries, among others. The produced vucanizates can be used in the automotive industry, among others, for example in the production of rubber products used in closed spaces, e.g., car floor mats or other elements of car cabin equipment.

Author Contributions: Conceptualization, D.B. and M.P.; methodology, M.P. and D.B.; software, D.P., D.B. and M.P.; validation, P.S., M.M. and O.D.; formal analysis, M.P. and D.B.; investigation, M.M., P.S. and D.P.; resources, M.M., P.S. and M.P.; data curation, M.P., D.B. and D.P.; writing—original draft, M.P., P.S., M.M. and O.D. All authors have read and agreed to the published version of the manuscript.

Funding: This research received no external funding.

Institutional Review Board Statement: Not applicable.

Informed Consent Statement: Not applicable.

Data Availability Statement: Not applicable.

Conflicts of Interest: The authors declare no conflict of interest.

References

1. Statistics and Balances, Wayback Machine, 2013.
2. Council of Europe. *Parliamentary Assembly, Documents de Séance*; Council of Europe, Parliamentary Assembly: Strasbourg, Germany, 2004; pp. 5–6.
3. Council of Europe. *Parliamentary, Documents: Working Papers, Ordinary Session (Fourth Part)*; Council of Europe: Strasbourg, France, 2004; Volume VI, pp. 253–259.
4. Borenstein, S. *Carbon-Emissions Culprit? Coal*; The Seattle Times Company: 2007. Available online: http://seattletimes.nwsource.com/html/nationworld/2003732690_carbon03.html (accessed on 16 July 2021).
5. US Environmental Protection Agency. Sulfur Dioxide. Available online: https://www.epa.gov/ (accessed on 16 July 2021).
6. Szponder, D.; Trybalski, K. Określanie właściwości popiołów lotnych przy użyciu różnych metod i urządzeń badawczych. *Górnictwo i Geoinżynieria* **2009**, *33*, 287–298.
7. Rajczyk, K.; Gąsior, D. Innowacyjny sposób wykorzystania niespalonych cząstek węgla zawartych w popiołach lotnych z węgla brunatnego. *Pr. Inst. Ceram. i Mater. Bud.* **2017**, *10*, 7–21.
8. Esomba, S. The Book: Twenty-First Century's Fuel Sufficiency Roadmap, Lulu. pp. 600–606. Available online: https://www.lulu.com/shop/steve-esomba-dr/shop/steve-esomba-dr/twenty-first-centurys-fuel-sufficiency-roadmap/paperback/product-19zzd294.html?page=1&pageSize=4 (accessed on 16 July 2021).
9. Tkaczewska, E.; Mróz, R.; Łój, G. Coal-biomass fly ashes for cement production of CEM II/A-V 42.5R. *Constr. Build. Mater* **2012**, *1*, 633–639. [CrossRef]
10. Formela, M.; Stryczek, S. Popioły fluidalne ze spalania węgla brunatnego jako dodatek do zaczynów uszczelniających wykorzystywanych podczas prac wypełniania pustek w górotworze. *Zesz. Nauk. Inst. Gospod. Surowcami Miner. i Energią PAN* **2017**, *97*, 117–134.
11. Stępień, M.; Białecka, B. Inwentaryzacja innowacyjnych technologii odzysku odpadów energetycznych. *Syst. Wspomagania w Inżynierii Produkcji.* **2017**, *6*, 108–123.
12. Faber, J.; Brodzik, K. Emisja LZO z tworzyw syntetycznych w komorze 1m3 - istotne źródła niepewności. *Autobusy Tech. Eksploat. Syst. Transp.* **2017**, *18*, 857–864.
13. Zabiegała, B. Jakość powietrza wewnętrznego-Lotne związki organiczne jako wskaźnik jakości powietrza wewnętrznego. *Monogr. Kom. Inżynierii Sr. PAN* **2009**, *I*, 303–315.
14. Blackley, D.C. *Polymer Latices: Science and Technology Volume 3: Applications of Latices*; Springer-Science + Business Mesia: Berlin/Heidelberg, Germany, 2012; p. 242.
15. Alves, J.O.; Zhuo, C.; Levendis, Y.A.; Tenório, J.A.S. Microstructural analysis of carbon nanomaterials produced from pyrolysis/combustion of Styrene-Butadiene-Rubber (SBR). *Mater. Res.* **2011**, *14*, 499–504. [CrossRef]
16. Ren, X.; Sancaktar, E. Use of fly ash as eco-friendly filler in synthetic rubber for tire applications. *J. Clean. Prod.* **2018**, *206*, 374–382. [CrossRef]

17. Wysokiński, A.; Skład, S.; Kalembasa. Chemiczny popiołów z węgla brunatnego i kamiennego w aspekcie ich rolniczego zagospodarowania. *Rocz. Glebozn.* **2008**, *2*, 93–97.
18. Gawlicki, M. Ostrowski, M. Aktywność wapiennych popiołów lotnych z Elektrowni 'Bełchatów' jako składnika cementów powszechnego użytku. *Pr. Inst. Ceram. i Mater. Bud.* **2012**, *5*, 66–75.
19. Hycnar, E.; Jończyk, M.W.; Ratajczak, T. Popioły lotne i iły beidellitowe z Bełchatowa jako składniki mieszanin samozestalających się. *Zesz. Nauk. Inst. Gospod. Surowcami Miner. i Energią Pol. Akad. Nauk.* **2017**, *100*, 37–48.
20. Garcia, A.B.; Cameán, I. Graphite materials prepared by HTT of unburned carbon from coal combustion fly ashes: Performance as anodes in lithium-ion batteries. *J. Power Sources* **2011**, *196*, 4816–4820.
21. Zapotoczna-Sytek, G.; Michalik, A. Badania nad zastosowaniem popiołów lotnych ze spalania węgli w kotłach fluidalnych do wytwarzania autoklawizowanego betonu komórkowego. *Pr. Inst. Ceram. i Mater. Bud.* **2010**, *3*, 123–140.
22. Gum. *Determination of Hardness according to the Shore Method;* Polish Accreditation Center: Warsaw, Poland, 2014.
23. *Liquid Resistance, Weight Method;* Polish Accreditation Center: Warsaw, Poland, 2014.
24. Lew, A.; Kardelky, C.; Haupt, P. Theory and eksperymentach Rubber Chemistry and Technology. **2003**, 533–547.
25. Przepiórkowska, A.; Prochon, M.; Zaborski, M.; Żakowska, Z.; Piotrowska, M. Biodegradable protein-containing elastomeric vulcanizates. *Rubber Chem. Technol.* **2005**, *78*, 868–878. [CrossRef]
26. Polgar, L.; Kingma, A.; Roelfs, M.; van Essen, M.; van Duin, M.; Picchioni, F. Kinetics of cross-linking and de-cross-linking of EPM rubber with thermoreversible Diels-Alder chemistry. *Eur. Polym. J.* **2017**, *90*, 150–161. [CrossRef]
27. Baeta, D.A.; Zattera, J.A.; Oliveira, M.; Oliveira, P.J. The use of styrene-butadiene rubber waste as a potential filler in nitrile rubber: Order of addition and size of waste particles. *Braz. J. Chem. Eng.* **2009**, *26*, 23–31. [CrossRef]
28. Brandt, H.; Nentwig, W.; Rooney, N.; LaFlair, R.; Wolf, U.; Duffy, J.; Puskas, J.; Kaszas, G.; Drewitt, M. *Rubber, Solution Rubbers, Encyclopedia of Industrial Chemistry;* Wiley-VCH: Hoboken, NJ, USA, 2012.
29. Mellor, D.; Moir, A.; Scott, G. The effect of processing conditions on the u.v. stability of polyolefins. *Eur. Polym. J.* **1973**, *9*, 219–225. [CrossRef]
30. Andrady, A. *Plastics and Environmental Sustainability;* John Wiley and Sons: Hoboken, NJ, USA, 2015.
31. Santos, A.S.; Agnelli, J.A.; Trevisan, D.W.; Manrich, S. Degradation and stabilization of polyolefins from municipal plastic waste during multiple extrusions under different reprocessing conditions. *Polym. Degrad. Stab.* **2002**, *77*, 441. [CrossRef]
32. Focke, W.W.; Mashele, R.P.; Nhlapo, N.S. Stabilization of low-density polyethylene films containing metal stearates as photodegradants. *J. Vinyl. Addit. Technol.* **2011**, *1*, 21–27. [CrossRef]
33. Belviso, C. State-of-the-art applications of fly ash from coal and biomass: A focus on zeolite synthesis processes and issues. *Prog. Energy Combust. Sci.* **2018**, *65*, 109–135. [CrossRef]
34. Wang, S.; Ma, Q.; Zhu, Z. Characteristics of coal fly ash and adsorption application. *Fuel* **2008**, *87*, 3469–3473. [CrossRef]

applied sciences

MDPI

Article

Using Statistical Modeling for Assessing Lettuce Crops Contaminated with Zn, Correlating Plants Growth Characteristics with the Soil Contamination Levels

Petru Cardei [1], Florin Nenciu [1,*], Nicoleta Ungureanu [2,*], Mirabela Augustina Pruteanu [1], Valentin Vlăduț [1], Dan Cujbescu [1], Iuliana Găgeanu [1] and Oana Diana Cristea [1]

1 National Institute of Research—Development for Machines and Installations Designed for Agriculture and Food Industry—INMA Bucharest, 013811 Bucharest, Romania; cardei@inma.ro (P.C.); pruteanu@inma.ro (M.A.P.); vladut@inma.ro (V.V.); cujbescu@inma.ro (D.C.); iulia.gageanu@inma.ro (I.G.); ocristea@inma.ro (O.D.C.)

2 Department of Biotechnical Systems, University Politehnica of Bucharest, 006042 Bucharest, Romania

* Correspondence: florin.nenciu@inma.ro (F.N.); nicoleta.ungureanu@upb.ro (N.U.)

Abstract: The aim of the study was to identify new mathematical models and strategies that can characterize the behavior of pollutants accumulating in the soil over time, considering the special characteristics of these chemicals that cannot be degraded or destroyed easily. The paper proposes a statistical model for assessing the accumulation of Zn in the lettuce (*Lactuca sativa* L.), based on three indicators that characterize the development of lettuce plants over time. The experimental data can be used to obtain interpolated variations of the mass increase functions and to determine several functions that express the time dependence of heavy metal accumulation in the plant. The resulting interpolation functions have multiple applications, being useful in generating predictions for plant growth parameters when they are grown in contaminated environments, determining whether pollutant concentrations may be hazardous for human health, and may be used to verify and validate dynamic mathematical contamination models.

Keywords: heavy metals; soil contamination; zinc accumulation

Citation: Cardei, P.; Nenciu, F.; Ungureanu, N.; Pruteanu, M.A.; Vlăduț, V.; Cujbescu, D.; Găgeanu, I.; Cristea, O.D. Using Statistical Modeling for Assessing Lettuce Crops Contaminated with Zn, Correlating Plants Growth Characteristics with the Soil Contamination Levels. *Appl. Sci.* **2021**, *11*, 8261. https://doi.org/10.3390/app11178261

Academic Editor: Avelino Núñez-Delgado

Received: 3 August 2021
Accepted: 3 September 2021
Published: 6 September 2021

1. Introduction

Zinc (Zn) reaches the agricultural soil through anthropogenic activities such as the application of fertilizers and pesticides, manures, sewage sludges, or various industrial activities, in many cases causing phytotoxic concentrations or contaminating agricultural products [1]. High concentrations of heavy metals in agricultural soil could generate toxic effects on human health, and their negative effect is aggravated by the property of metals to accumulate in time, producing chronic intoxication [1]. Although zinc is an important nutrient, its excessive ingestion reduces the immune functions and the level of high-density lipoproteins. Food consumption is the main pathway for human exposure to heavy metals [2], and zinc has one of the greatest accumulation potentials in plants [3]. A study [4] showed that the potential of heavy metals to be absorbed in plant parts is higher in roots, followed by stems and leaves.

Given the high-risk potential caused by the presence of heavy metals in the soil, both from the perspective of ensuring food security and human health, the accumulation of various hazardous substances in the mass of plants has been an intensely studied topic. Recent progress in the research of toxic metals and their interactions with essential elements has increased our understanding of the mechanism of toxicity at the biochemical level. Boskovic-Rakocevic studied the total Cd levels and the accumulation process in lettuce in order to evaluate the human health impact [5], determining that the uptake and accumulation of cadmium is mostly affected by soil pH, followed by Cd availability. A study [6] evaluated the biochemical effect and the physiological response of several contaminating agents in peppers raised in laboratory conditions, while other research [7] evaluated Zn phytotoxicity considering the complex biochemical reactions, and plant tolerance to Zn with regard to phytoremediation processes.

In a paper [8], the natural Cu, Cr, Zn, and Mn phytoaccumulation potential was screened for three widespread species, in the Constanta coastal area, Romania: Brassica rapa, Crambe maritima, and Lepidium draba. Samples of aboveground organs and soil were taken for each species and analyzed for heavy metal concentration values. Regarding the statistical analysis of the content of heavy metals in the soil, studies [9] evaluated the accumulation of cadmium and lead in vegetables that are grown in contaminated soils. The comparative analysis concluded that cadmium accumulation by plants grown on sewage sludge amended soils is lower than that for the inorganically contaminated soils. An extensive study on a total of 118 agricultural soil types and 43 vegetable samples have been developed in [10], assessing the spatial distribution, sources, accumulation characteristics, and potential risk of heavy metals in the agricultural soils and vegetables, by three different approaches. Contamination with heavy metals can also come from the water used to irrigate crops, as shown in a study [11] that considered the effect of wastewater irrigation on vegetables, in relation to the bioaccumulation of heavy metals and biochemical changes. The results showed a decrease in total chlorophyll and total amino acid levels in plants and an increase in the amounts of soluble sugars, total protein, and ascorbic acid in plants grown on soils irrigated with contaminated wastewater.

The statistical analysis is widely used in modern research, being the best approach for primary processing of data resulting from the investigation of phenomena. Statistical modelling is not only a way to obtain conclusive results, but also an intermediate step towards analytical, deterministic, or probabilistic models, with a greater degree of generality. In terms of plant evolution, and the bioaccumulation of contaminants in plants, a complex multivariate analysis on the presence of heavy metal contamination in agricultural soils is analyzed in a study [12]. The experimental results can be presented in statistical terms, even in terms of simple statistical models, such as the linear regressions [13]. Linear and quadratic regressions are often used in investigations envisaging the interaction of heavy metals with plants [14].

A study [15] showed a detailed analysis of the mathematical models generally used in ecology, classifying them into deterministic and statistical models. The bioaccumulation assessment uses both types of models for prediction or optimization, or for deepening various aspects. Despite the development direction that emphasizes the need to know the causality of the processes underlying the bioaccumulation processes, the use of statistical models for prediction is an extremely valuable asset [15].

To compare different types of components or the effects produced by different techniques or technologies, statistical research frequently uses the analysis of variance, or ANOVA [16–20].

Superior statistical techniques are currently employed to predict the heavy metal concentration in soil, plants, or other environments, in the frame of complex computational tools such as the neural networks [21–23].

This paper presents one-dimensional interpolations, regarding both the variation over time of the mass of lettuce plants grown on soil contaminated with zinc, and the variation of heavy metal concentration in plants. It also includes two-dimensional interpolations regarding the variation of the mass of plants (lettuce) that are grown on soil contaminated with zinc, depending on time and the amounts of zinc in the soil.

The experimental studies aimed to create a framework for a better understanding of the mechanisms of Zn toxicity in plants and to identify novel assessment perspectives based on a statistical analysis. The physiology of Zn phytotoxicity is complicated, considering Zn interference in chlorophyll biosynthesis, and other complex biochemical reactions [7]. There are different tolerances to zinc toxicity, depending on plant genotypes, with lettuce being one of the plants with higher tolerances.

The aim of this paper was to identify new tools to describe how zinc pollution affects lettuce crops, analyzing three growth characteristics: plant mass, plant height, and plant diameter. Using the least squares method on the experimental data has determined the 1 to 4 grades mathematical regression functions that describe the three growth characteristics in time. The three plant growth characteristics (mass, height, and diameter) have been correlated with the three zinc concentrations in the soil, in order to draw conclusions about the effect associated to each concentration on the lettuce experimental culture.

2. Materials and Methods

This research study has three components: the physical component, the experimental component, and the numerical component, consisting of the results of the descriptive statistical analysis of the experimental data. The research has been performed in two phases: the experimental phase and the processing of statistical data, followed by the interpretation of the results. Statistical processing led us to the elementary statistical models of the process of zinc bioaccumulation in the bio-plant bio system.

The experimental design started from the scientific problem of determining the effect that different zinc concentrations may have on lettuce plants and implicitly on human health, using statistical techniques. The evaluation method was to determine the interpolation functions, that can be used for making forecasts on the evolution over time of plant development, in soils contaminated with zinc.

The experimental unit was the plant isolated in individual pots. A single treatment was applied to the soil, followed by the monitoring of the evolution of the plant development in time. The factor has been considered that the zinc solution has three different levels (three different concentrations of zinc applied to the soil).

The response (the outcome being measured) was the plant development in the 10 weeks of monitoring. The period of 10 weeks was chosen, because it was considered the maximum until harvest should be done for human consumption. Three observational units were analyzed, namely the height of the plant, its mass, and plant thickness. The treatment stage did not change after applying the three concentrations of zinc on the soil, and the growing conditions remained the same. The resulting design considered comparisons between samples.

A completely randomized design was used, randomly assigning treatments to the experimental units in a pre-specified number, with the same number of units receiving each treatment.

The experimental method involved the use of 30 lettuce plants in individual pots, forming three groups of 10 pots. Each group was initially infested in a controlled manner with a certain level of contamination (initial concentration of zinc). Lettuce seedlings were planted on clean soil, and then contaminated with three solutions containing the following zinc concentrations: 1.5%, 3.0%, and 4.5%. Contamination was carried out by adding Zn sulphate solutions (250 mL solution, for each zinc concentration) on clean soil (1 kg/pot of fertile soil). The physicochemical properties of the control sample (i.e., of the uncontaminated soil used for the growth of all plants, which was then infested with Zn solution in each of the three groups of 10 pots) were as follows: the pH varied between 6.0–7.0; total nitrogen 1.9%; total phosphorus 0.5%; potassium 0.9%; electrical conductivity 1.2; maximum 5% soil particles over 20 mm; humidity 14.7%.

At each harvest, three plants were sacrificed, one from each group of 10 pots with different initial concentrations of Zn. For each plant harvested, the mass, height, diameter, and zinc concentration of the plants were measured for each of the three samples. At each harvest, the parameters were measured: zinc concentration in the soil, soil moisture, water content of the plant, and pH. Ten harvests were made from October to December at one-week intervals. The culture was located in a greenhouse with controlled air temperature and humidity (temperature 20 °C and humidity 65%). Four experimental parameters were measured at each harvest and introduced in the statistical analysis (mass, height, crown diameter, and zinc concentration).

The soil–plant biological system is simplified in this empirical-statistical model by the elementary dependence of the output parameters, on a single input parameter (zinc concentration in the soil in which lettuce grows), and time:

$$x = f(t, c) \tag{1}$$

where: x is one of the output parameters (plant mass, crown diameter, plant height, zinc concentration in the plant, the water content of the plant, or soil moisture, soil pH, and the amount of zinc in the soil), c is the zinc concentration from the soil in which the plant grows, and t is the time.

The physicochemical characteristics (pH, soil moisture, and plant water content) of the soil varied in relatively narrow ranges, as it can be seen in Figure 1.

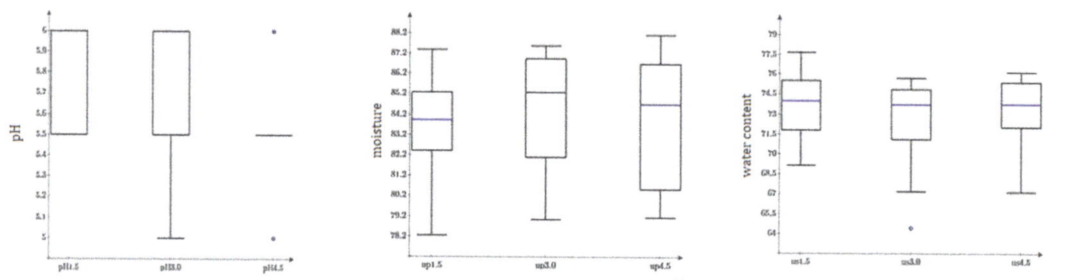

(a) (b) (c)

Figure 1. Box plot representations of the distribution of pH (**a**), soil moisture (**b**), and plant water content (**c**), during the experiment, for the three cases of soil contaminated with a solution of 1.5%, 3.0%, and 4.5% zinc concentration.

Plant and soil sampling and analysis were performed daily, up to 68 days after planting (Figure 2). The mass of the samples was determined by weighing the probes with the KERN electronic balance (0.001 g precision), and plant height from the root levels to the top of the plant was measured with a ruler.

Figure 2. Plants with different contamination concentrations, harvested at different time periods after cultivation.

Soil moisture and water content, both in the plants and soil, were determined by drying in the oven at 105 °C, in order to evaporate water both from the soil and from the plant.

Soil pH was determined using a pH determination kit. A soil sample of 20 g was dried in the oven and then placed in a bowl with 100 mL of distilled water, stirred for 30 min and then filtered and controlled with pH paper.

An overview of the evolution of the mass, height, and diameter for the last 10 harvests of lettuce grown in the three soil types (each being contaminated by adding, in the soil, three solutions having 1.5%, 3.0%, and 4.5% zinc concentrations), is given by a box plot representation in Figure 3.

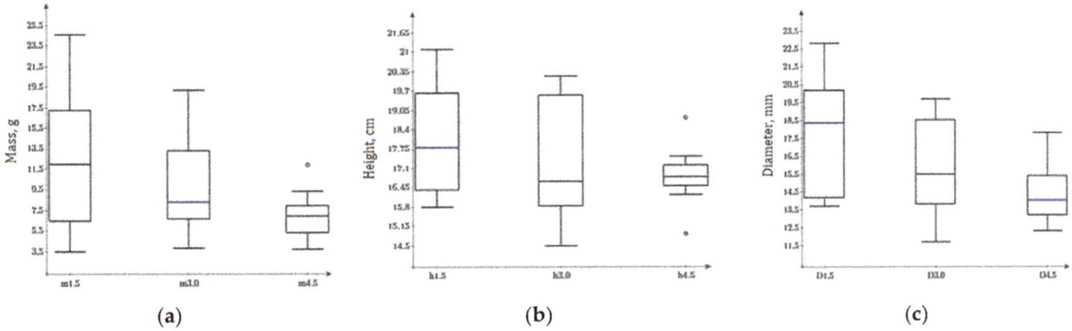

Figure 3. Box plot diagrams for mass (**a**), height (**b**), and diameter (**c**) distributions for the three cases of soil contamination with a solution of 1.5%, 3.0%, and 4.5% zinc concentration.

The determination of zinc in the contaminated soil and the lettuce plant body (root and leaves) was performed by the spectrophotometric method (flame atomic absorption), according to the procedure developed in a previous study [24].

Figure 4 shows a representation of experimental data obtained by normalized standard deviation, depending on the Zn concentration in the soil. It can be observed that the "tightest" results around the average, in terms of mass, height, and diameter of the plant, are obtained at the highest concentration of zinc in the soil.

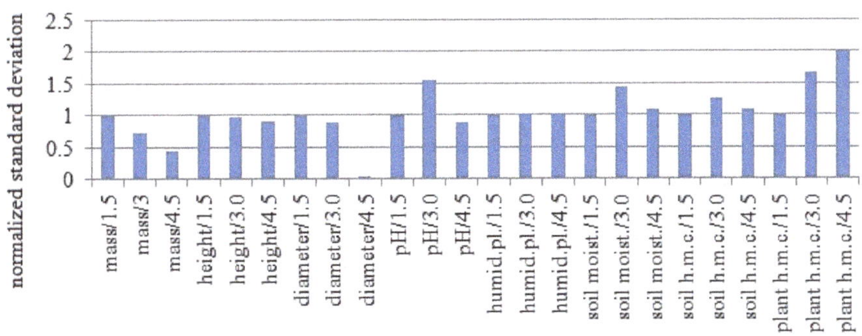

Figure 4. Variation of eight parameters of the plant after contaminating the soil, using the normalized standard deviation, in consideration of Zn accumulation (where humid.pl.: plant water content; soil moist.: soil moisture; soil h.m.c.: soil heavy metal concentration; plant h.m.c.: plant heavy metal concentration).

3. Results

3.1. Statistical Models (Polynomials Time Regression)

The statistical methods used to obtain the models are based on linear regressions (first-degree polynomials), quadratic regressions (second-degree polynomials), cubic regressions (third-degree polynomials), and the polynomials regression by fourth-degree poly.

Polynomial regressions, and non-polynomial regressions, are obtained by applying the least squares method. In the cases of the polynomial regressions, the sought coefficients are obtained by solving linear equations systems, coming from the calculation of the critical points. The calculations of the coefficients from the interpolation polynomials were made using the interpolation function of the MathCAD 15 program, whose algorithm is described in study [25]. In addition to the regression polynomial coefficients, the polyfitc function of the MathCAD 15 program also provides other important statistical features:

the standard error for the regression coefficient, the lower and upper boundary for the confidence interval of the regression coefficient, the variance inflation factor—the measure of the inflation of the regression coefficient due to multicollinearity, Student's t-test statistic if the term is significant, and *p*-value—probability of rejecting the term based on its t-statistic, when in fact it is significant. The terms used in statistical analysis are presented schematically in Table 1.

Table 1. Presentation of the main terms used in statistical analysis.

Term	Labels for Each Term Reported Upon
Coefficient	Regression coefficient for each term
Std Error	Standard error for the regression coefficient
95% CI low	Lower boundary for the confidence interval of the regression coefficient
95% CI High	Upper boundary for the confidence interval of the regression coefficient
VIF	Variance inflation factor: measure of the inflation of the regression coefficient due to multicollinearity
T	Student's t test statistic to test if the term is significant
P	*p*-value: probability of rejecting the term based on its t-statistic when in fact it is significant

As an example, Table 2 shows the final and intermediate results of the interpolation for the evolution of the mass of plants grown in soil with a zinc concentration of 4.5%. The significance and calculation algorithms of these characteristics were calculated according to [25] and the PTC MathCAD prime linear regression, (2018).

Table 2. Results of statistical analysis, for the interpolation (using first-degree polynomial) of lettuce mass variation over time, grown in soil with 4.5% zinc concentration.

Term	Coefficient	Std Error	95% CI Low	95% CI High	VIF	T	P
t^0	2.25	0.874	0.272	4.228	NaN	2.573	0.025
T	0.117	0.021	0.069	0.166	1	5.517	0.0002401

According to Tables 2–5, the equation that gives the time dependence of the plant mass of lettuces grown in soil with a concentration of 4.5%, through the polynomial function, are the following:

$$\begin{aligned} m1_{4.5}(t) &= 2.25 + 0.117t \\ m2_{4.5}(t) &= 1.656 + 0.175t - 0.0008428t^2 \\ m3_{4.5}(t) &= 0.59 + 0.424t - 0.01t^2 + 0.00009331t^3 \\ m4_{4.5}(t) &= 0.142 + 0.657t - 0.028t^2 + 0.0005041t^3 - 0.000003033t^4 \end{aligned} \quad (2)$$

where: t is the time (in days).

Table 3. Results of statistical analysis, for the interpolation (using second-degree polynomial) of lettuce diameter crowns variation over time, grown in soil with 4.5% zinc concentration.

Term	Coefficient	Std Error	95% CI Low	95% CI High	VIF	T	P
t^0	1.656	1.218	−1.153	4.466	NaN	1.359	0.152
t	0.175	0.083	−0.017	0.368	14.5	2.106	0.053
t^2	−0.0008428	0.001168	−0.003536	0.00185	14.5	−0.722	0.291

Table 4. Results of statistical analysis, for the interpolation (using third-degree polynomial) of lettuce diameter crowns variation over time, grown in soil with 4.5% zinc concentration.

Term	Coefficient	Std Error	95% CI Low	95% CI High	VIF	T	P
t^0	0.59	1.309	−2.506	3.685	NaN	0.45	0.343
t	0.424	0.176	0.008801	0.839	76.185	2.415	0.034
t^2	−0.01	0.006169	−0.025	0.004191	478.927	−1.685	0.099
t^3	0.00009331	0.00005934	−0.000047	0.0002336	200.478	1.573	0.115

Table 5. Results of statistical analysis, for the interpolation (using fourth-degree polynomial) of lettuce diameter crowns variation over time, grown in soil with 4.5% zinc concentration.

Term	Coefficient	Std Error	95% CI Low	95% CI High	VIF	T	P
t^0	0.142	1.427	−3.35	3.634	NaN	0.099	0.381
t	0.657	0.321	−0.129	1.444	246.875	2.044	0.06
t^2	−0.028	0.021	−0.079	0.023	5281	−1.331	0.155
t^3	0.0005041	0.0004744	−0.0006568	0.001665	12380	1.063	0.209
t^4	−0.000003033	0.000003474	−0.00001153	0.000005468	2948	−0.873	0.252

The interpolation curves corresponding to the polynomial Equation (2) are shown in Figure 5.

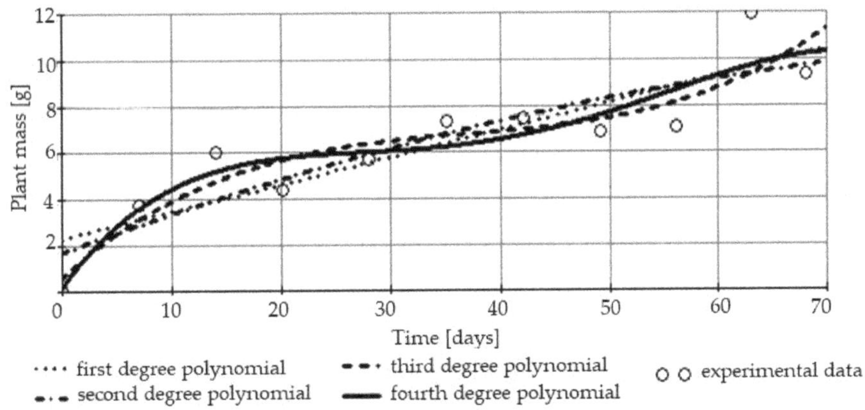

Figure 5. Time dependence of the mass of lettuces grown in soil contaminated with 4.5% zinc concentration, for 1 to 4degree polynomial interpolations.

Similarly, the polynomial regression expressions of time dependence were obtained for the other two soil contaminations, for the plant mass and their height, in accordance with the way lettuces have grown in the three types of soil. The graphical representations from Figure 6 demonstrate the regressions obtained.

The three characteristics represented graphically have a generally monotonous tendency (within the entire 68-day monitoring period). Polynomial curves of a higher degree show that there are some points of local extreme, which could deviate from reality, which should be further researched from a biological or physical perspective. It can be observed that the linear interpolation confirms the inverse proportionality between the intensity of soil contamination and the increase of the plant's mass. The second-degree interpolation curves show that towards the end of the monitoring period, the three characteristics (mass, Zn concentration, and height) cease increasing or even have decreasing tendencies. This behavior is most likely generated by the polynomial function of the second degree. However, in the case of plant mass, both the third- and fourth-degree interpolation functions indicate decreasing tendencies at the end of the monitoring period, but only for the lowest soil contamination concentration (1.5%). In the other cases, the tendency is increasing, more or less convincingly, until the end of the monitoring period.

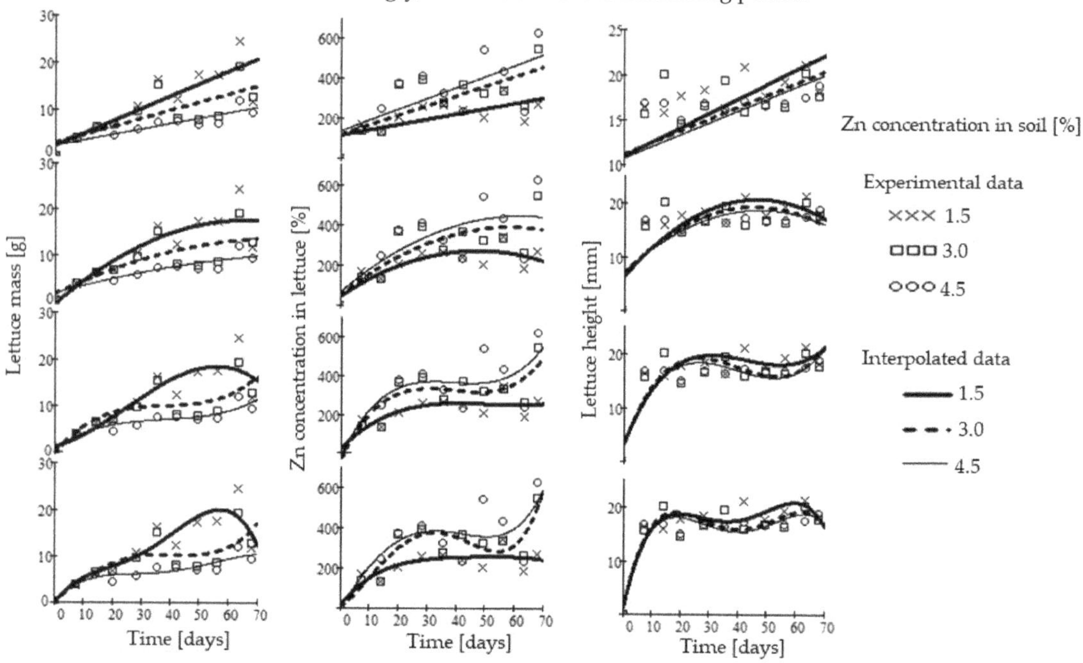

Figure 6. Graphical representation of Zn concentration, mass, and plant height over time, obtained by polynomial interpolation (1–4 degree ordered top to bottom of graph) and experimental data.

It is observed that the premises drawn for the linear interpolation case are also valid in this case (the monotony of the mass increasing and the inverse order reported to the level of Zn contamination of the soil). The parabolas depicted in Figure 6 present a maximum point (they all have the coefficient of the second-degree term negative). Theoretically, this means that there would be a time when the mass begins to fall. Since this maximum, for each of the parabolas, is outside the interpolation interval, this last statement remains subject to the validation of the extrapolation.

Starting with the interpolation by third-degree polynomials (Figure 6), the accuracy increases, the interpolation curves model increasingly better to the experimental data oscillations. In this context, this increases the challenge of explaining the origin or causes of the oscillations reported to a monotonic average curve or, relatively, to the linear regression curve (the straight line).

For example, in the case of the third- and fourth-degree interpolations (Figure 6), the maximum from the penultimate harvest, provided by the experimental data, increases in clarity.

3.2. *The Variation Obtained for the Concentration of Heavy Metals in Plants*

With the same interpolation tools, similar results can be obtained regarding the increase of Zn concentration in plants. In Figure 6, the graphs of linear interpolation of the increase of heavy metal concentration in plants are also given, at three degrees of soil contamination. The experimental data are also represented in the graph. It is observed that in the more intensely contaminated soils, the accumulation of heavy metal in plants is more intense and occurs faster (the inclination of the corresponding straight lines is more pronounced).

Figure 6 also shows the variations of Zn concentrations in lettuce grown in the soils' different Zn concentrations, in the interpolated form of grade II polynomials together with the experimental data. It can be easily observed that the same order of heavy metal concentrations is in the plants of the three categories of soil contamination, as in the case of the linear interpolation. Zinc accumulation in the plant is more intense, as the initial concentration of zinc in the soil was higher.

The order of the variation curves of heavy metal concentrations and their monotony increase in the plants is constant for the third-degree polynomial, as well as the separation of the curves (there are no intersections between the three curves). The third-degree polynomial interpolation curves, for the concentration increase in plants, are represented in Figure 6.

In general, the order and monotony of the interpolation curves of the variations of heavy metal concentration in plants are respected by the fourth-degree polynomial curves (Figure 6). With very few exceptions (the curve corresponding to soil contamination with 3.0% solution easily reaches the curve corresponding to 4.5% concentration and is very close, towards the end of the observation period, to the curve corresponding to plants grown on soil contaminated with 1.5% concentration solution), the separation of the curves is well respected.

3.3. Measurement of Plant Height at Harvest

Plant height, measured from the soil, at the time of harvest, is a significant growth parameter, and has been recorded for each plant at each stage of harvesting.

The linear interpolation leads to the regression curves in Figure 6. Plant height is a positive characteristic of evolution and, as expected, the behavior is similar to that of plant mass: the more intensely contaminated the soil is, the smaller the plant height at harvest.

3.4. Statistical Models (Polynomials Time and Zinc Concentration in Plant Regression)

Polynomial interpolations over time can be extended to the polynomial interpolations of two variables, where the second variable is the initial zinc concentration in the soil. The calculation method is similar to that used for polynomials that interpolate the temporal evolution of lettuce culture characteristics. Given the multitude of experimental data, interpolation by polynomials of two variables could be performed up to the third degree. This paper gives an example of the calculation of the interpolation polynomials of two variables of degree 1–3, for the plant mass. Similar interpolations are made for the evolution of plant height and diameter. They can also be made for any of the other characteristics measured during the experiments.

The result of the statistical interpolation calculation for the polynomials of two variables is given similarly to that for the polynomials of one variable, in Tables 6–8. The discussions made for this statistical calculation method are the same as those made for the polynomial interpolation with a single variable.

Table 6. Statistical characteristics of the linear regression by two variables for the mass of lettuces.

Term	Coefficient	Std Error	95% CI Low	95% CI High	VIF	T	P
$t^0, {c_{Zn}}^0$	4.402	2.889	−1.498	10.301	NaN	1.524	0.125
t	0.197	0.035	0.126	0.268	1.227	5.648	0.000005.271
c_{Zn}	−0.007352	0.009.692	−0.027	0.012	1.227	−0.759	0.295

The variation of plant biomass (lettuce) grown in zinc-contaminated soil is given in analytical form by Equation (3), for 1- 3-degree polynomials. Using Equation (3) and the experimental data, graphical representations of the time dependence of plant mass and the initial concentration in the soil were plotted. It is represented as portions of surfaces in three-dimensional space, the function of plant mass variation with time and the concentration of zinc, linear (Figure 7), square (Figure 8), and cubic (Figure 9), respectively. In each graph, the experimental data are represented by isolated symbols.

Table 7. Statistical characteristics of the cubic regression by two variables for the mass of the lettuces.

Term	Coefficient	Std Error	95% CI Low	95% CI High	VIF	T	P
$t^0, c_{Zn}{}^0$	3.221	11.917	−21.231	27.673	NaN	0.27	0.381
t	0.442	0.164	0.105	0.779	26.769	2.694	0.014
c_{Zn}	−0.021	0.073	−0.171	0.128	68.351	−0.293	0.378
$t \cdot c_{Zn}$	−0.0004303	0.000523	−0.001503	0.0006.428	44.753	−0.823	0.279
t^2	−0.00138	0.002.089	−0.005666	0.002905	22.054	−0.661	0.316
$c_{Zn}{}^2$	0.00004.2	0.0001039	−0.0001713	0.0002553	59.502	0.404	0.363

Table 8. Statistical characteristics of the quadratic regression by two variables for the mass of the lettuces.

Term	Coefficient	Std Error	95% CI Low	95% CI High	VIF	T	P
$t^0, c_{Zn}{}^0$	−35.198	71.732	−183.586	113.191	NaN	−0.491	0.348
t	1.07	1.294	−1.606	3.746	1470	0.827	0.278
c_{Zn}	0.317	0.772	−1.281	1.914	6780	0.41	0.362
$t \cdot c_{Zn}{}^n$	−0.00343	0.00589	−0.016	0.00876	5010	−0.582	0.331
t^2	−0.0748	0.014	−0.037	0.022	896.299	−0.528	0.342
$c_{Zn}{}^2$	−0.000914	0.00261	-6.31×10^{-3}	0.00448	33100	−0.351	0.37
$t_2 \cdot c_{Zn}$	0.000000617	0.0000654	-1.35×10^{-4}	0.000136	3020	0.009.44	0.395
$t \cdot c_{Zn}{}^2$	0.00000469	0.00000645	-8.66×10^{-6}	0.000018	1040	0.727	0.3
t^3	0.0000542	0.000170	-2.97×10^{-4}	0.0004.06	582.308	0.319	0.374
$c_{Zn}{}^3$	0.000000853	0.00000268	−0.00000468	0.00000639	9500	0.319	0.374

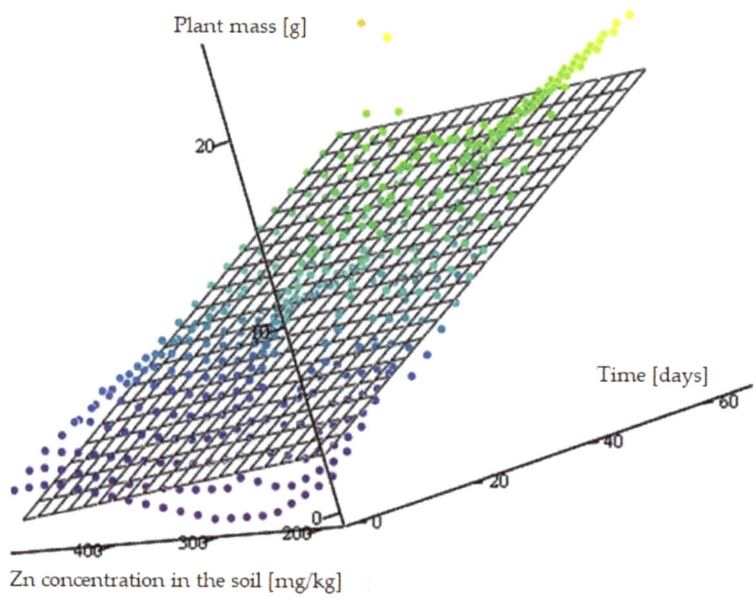

Figure 7. Variation of lettuce mass with time and the concentration of Zn in the soil, as a three-dimensional surface, corresponding to linear interpolation.

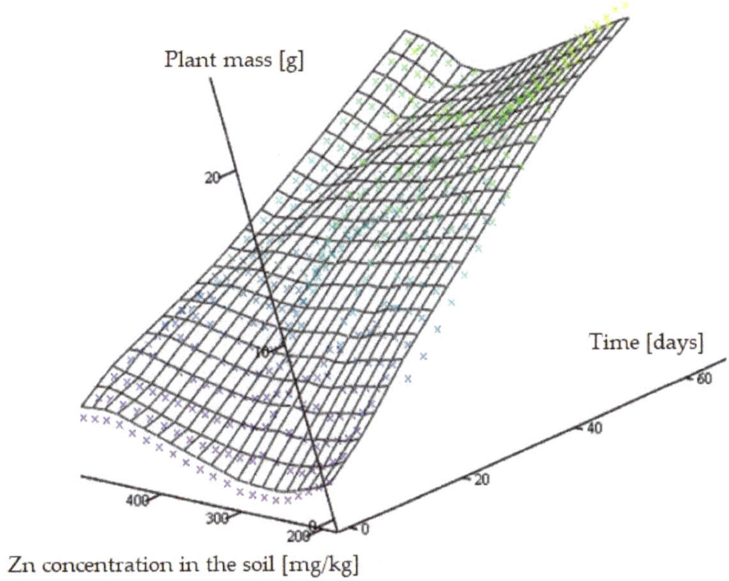

Figure 8. Variation of lettuce mass with time and the concentration of Zn in the soil, as a three-dimensional surface, corresponding to a second-degree polynomial interpolation.

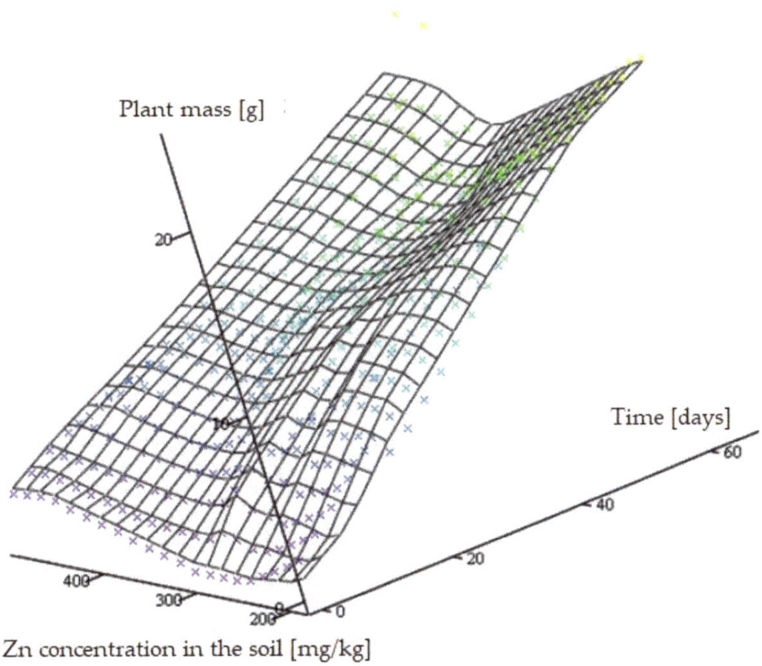

Figure 9. Variation of lettuce mass with time and the concentration of Zn in the soil, as a three-dimensional surface, corresponding to a third-degree polynomial interpolation.

The analytical expression of the three interpolation polynomials is obtained using Tables 6–8:

$$m(t, c_{Zn}) = 4.402 + 0.197 \cdot t - 0.007352 \cdot c_{Zn},$$
$$m(t, c_{Zn}) = 3.221 + 0.442 \cdot t - 0.021 \cdot c_{Zn} - 0.0004303 \cdot t \cdot c_{Zn} - 0.00138 \cdot t^2 + 0.000042 \cdot c_{Zn}^2,$$
$$m(t, c_{Zn}) = -35.198 + 1.07 \cdot t + 0.317 \cdot c_{Zn} - 0.003427t \cdot c_{Zn} - 0.007482 \cdot t^2 - 0.0009143 \cdot c_{Zn}^2 + 0.000000617 \cdot t^2 \cdot c_{Zn} + 0.000004691 \cdot t \cdot c_{Zn}^2 + 0.00005423 \cdot t^3 + 0.0000008533 \cdot c_{Zn}^3 \quad (3)$$

4. Discussion

The three areas of the surfaces drawn in Figures 7–9 show that the plant's mass increases over time; however, it tends to decrease when the initial concentration of zinc in the soil increases. This property is difficult to highlight, because:

- Zinc is also an important nutrient in plant development so at an early stage of development its rapid accumulation could promote biological growth;
- Some oscillations, which appeared in the evolution of the mass, height, or diameter of the plants, may appear due to some parameters that were not considered in the model (even if the crop was hosted in the greenhouse, lighting and temperature were greatly influenced by atmospheric conditions);
- There is a high possibility that at certain values of the accumulation of heavy metals, the plants at some point may adopt a mechanism of rejection of those elements, similar to a defense mechanism (this is a hypothesis resulting from the study and applicable and testable in deterministic models that can be built).

The variation over time of the mass increase of lettuces grown on soil contaminated by zinc (Figure 6) confirms the results of the theoretical prediction made in another study [26]. All the interpolations perform a mediation of the experimental data using the method of least squares. It can be noticed that the increase of the interpolation polynomial degree leads to a better approximation of the experimental data; however, we have no reasons of a phenomenological nature to increase the degree of polynomial interpolation over the third degree. Another reason why we are not tempted to increase the interpolation degree is that the functions found in this way are very particular and useful only in the experimental range reached for the working parameters, the extrapolation being contraindicated. For these reasons, the statistical model can be used to validate the theoretical (dynamic) model, which is generalizable not only in terms of climatic conditions, but also depends on the types of plants used.

A number of recent studies [14,27–29] are in agreement with the results we obtained and presented in this paper.

The results of the experimental data processing show that lettuces grown in soils with a higher Zn concentration have lower heights. This phenomenon is observed in many plants, for example, the authors in study [14] evaluated the presence of cadmium and zinc in sunflower and in the soil. This result is synthetically visible in the results shown in Figure 6, on the third column of the graph. It is observed that the regression curves corresponding to the plants grown in the soil with the highest concentration of Zn, are located below the regression curves corresponding to the plants grown in the soil with average zinc concentration. The regression curves corresponding to plants grown in medium contaminated soil are below the curves that represent the increase in height of the plants grown in low soil contamination.

Similar remarks on the development of the *Brassica junceea* are presented in study [27]. The influence of zinc, however, is manifested starting from a higher concentration threshold in the soil (500 ppm), that can be classified as very high. Table 4 shows the influence of increasing the concentration of zinc in the soil, that leads to a decrease in the height of the plants and a decrease in the mass. The decrease in lettuce mass caused by the increase in zinc concentration in the soil is synthetically represented in Figure 6, in the first column of the graph. The growth curve of the lettuce mass cultivated in the soil with the highest zinc concentration is located below the curve corresponding to the plants cultivated in the soil with medium zinc concentration, and is placed under the growth curve of the plant mass grown in the soil with the lowest concentration of zinc. Similar behavior to the height and mass assessment could be seen on the diameter of the crown of the plant.

The effects we noticed following the experiments performed on the lettuce are also confirmed by study [29] which mentions the effects of zinc concentration in the soil: reduction in height and biomass for cluster bean (*Cyamopsis tetragonoloba*), and development slowdown for pea (*Pisum sativum*) and rye grass (*Lolium perenne*). Another study [30,31] also concluded that the toxicity of zinc in plants is visible through the effect of the inhibition of growth and diminishes in biomass generation, while the high toxicity may be fatal. Table 9 validates the result obtained by statistical analysis study with other papers identified in the literature.

Table 9. Table summarizing key findings for the development of *Brassica junceea* in similar studies.

The Main Conclusion of the Analyzed Papers	Papers That Confirm the Conclusions of the Present Research
The mass of all plants increases over time, (with several small variations). It has not been determined what are the effects produced by high concentrations of heavy metal, with the emphasis on the doses at which the germination of the plant or its death could emerge	[14,27,29–31]
The mass of the plants at each stage of harvesting is inversely proportional to the initial concentration of zinc in the soil	[27,29–31]
The height of the plants harvested at the same time is inversely proportional to the initial concentration of zinc in the soil	[14,27,29–31]
The diameter of the plants harvested at the same time is inversely proportional to the initial concentration of zinc in the soil	[27,29–31]

All the interpolations made within the statistical models that appear in the present paper make mediation of the experimental data. It can be noticed that the increase of the interpolation polynomial degree leads to a better approximation of the experimental data; however, we have no reasons of a phenomenological nature to increase the degree of the interpolation polynomial over the value of three. Another reason why we are not tempted to increase the interpolation degree is that the functions found in this way are very particular and useful only in the experimental range reached for the working parameters, with the extrapolation not being indicated. For these reasons, the statistical model can be used to validate the theoretical (dynamic) model, which is generalizable not only in terms of climatic conditions but also for other types of plants.

The variation over time of the increase of lettuce mass grown on soils contaminated with zinc confirms the result of the theoretical prediction. More precisely, the increase of the pollutant concentration in the soil leads to a slower increase, and to the development of plant specimens with a smaller mass (also smaller height and diameter) than the plants from the same lot, grown on soil not contaminated with zinc. The plants are being studied in interaction with the soil and the growing conditions, throughout their lifetime (if we can define the lifespan, i.e., to specify the time of death of a plant). There are many more parameters that influence the development process of plants affected by excessive heavy metal content. Atmospheric parameters that play an important role in plant development (temperature, moisture, lighting, etc.), and parameters of soil structure and composition (compactness, density, porosity, nutrient content, and chemical composition in general), must be considered in experiments. Due to a large number of influential parameters, the number of experiences will increase in consequence. For this reason, wide international collaborations are recommended to carry out such experiments on portions distributed to various laboratories, under centralized monitoring.

5. Conclusions

Correlating plants' growth characteristics with the soil contamination levels using statistical modeling can obtain interpolated variations of the mass increase functions to determine several functions that express the time dependence of the heavy metal concentration in the lettuce plants. Starting from these interpolation relations, one can calculate other characteristics of the evolution process for various plants that are growing on contaminated soils.

We identified a research tool that describes how different levels of zinc contamination affects lettuce crops, analyzing three growth characteristics: plant mass, plant height, and plant diameter. We used the least squares method on the experimental data, in order to determine the 1–4 grades mathematical regression functions that describe the three growth characteristics in time. The three plant growth characteristics (mass, height, and diameter) have been correlated with the three zinc concentrations in the soil in order to draw conclusions about the effect associated to each concentration on the lettuce experimental culture.

The main conclusion of the statistical model, that was validated by experimental determinations, is the appearance of the dwarfism effect. Therefore, as the concentration of zinc in the soil increases, lettuce plants tend to have a smaller size and mass. The functions that describe the mass, height, and diameter, and variation over time, have been extended for two variables: time and zinc concentration. In this way, a better prediction was obtained regarding the plant development, under the influence of zinc concentration in the soil. All these functions can then be used for forecasting, using extrapolation.

The techniques that have been used are not necessarily new; however, they are very useful for forecasting and for studying the biodynamics of plant development.

The results obtained by the interpolation of the experimental data and their interpretation are likely to give very important indications for the following experimental step: the prolongation of the experiments over at least two to three generations of plants, which will allow drawing some conclusions on the transmission of heavy metals between generations of plants, or, probably depending on other influential parameters of the process, the decrease of the heavy metal concentration in the plants of the following generations.

Another important observation is the dependence of the contamination level on the type of soil used. Zinc contamination tests should target several soil types, given the different way in which soil structure can affect the time behavior of pollutants.

The paper addressed a specific situation where the contamination was achieved by applying a fixed dose of zinc, while the synergistic effects with other substances were not evaluated. It should be noted that the contamination process found in nature may be different: the contamination might be either continuous for long periods of time, or intermittent with maximum and minimum values.

Our experiments will continue the study with the evaluation of plant development for several other soil types, with the analysis of the synergistic effects between pollutants, and with the variation of the concentrations used for soil contamination.

Author Contributions: Conceptualization, P.C. and M.A.P.; methodology, P.C., M.A.P., F.N., V.V. and N.U.; software, P.C. and M.A.P.; validation, P.C., F.N. and N.U.; formal analysis, I.G. and O.D.C.; investigation, V.V. and D.C.; resources, N.U. and V.V.; data curation, V.V.; writing—original draft preparation, P.C. and M.A.P.; writing—review and editing, N.U. and F.N.; visualization, V.V. and N.U.; supervision, V.V. All authors are main authors and have contributed evenly to this paper. All authors have read and agreed to the published version of the manuscript.

Funding: This paper was financed by the support of the Executive Agency for Higher Education, Research, Development and Innovation Funding, Exploratory Research Program, PN-III-P4-ID-PCE-2016-0860, contr. 174/08.08.2017, "Research on the development of some mathematical models to evaluate the impact of soil contamination on fruits and vegetables—CONTAMOD" and contr. 16PFE/2018. The APC was funded by University Politehnica of Bucharest, Romania, within the PubArt Program.

Institutional Review Board Statement: Not applicable.

Informed Consent Statement: Not applicable.

Data Availability Statement: Not applicable.

Conflicts of Interest: The authors declare no conflict of interest. The funders had no role in the design of the study; in the collection, analyses, or interpretation of data; in the writing of the manuscript, or in the decision to publish the results.

References

1. Wuana, R.A.; Okieimen, F.E. Heavy Metals in Contaminated Soils: A Review of Sources, Chemistry, Risks and Best Available Strategies for Remediation. *Int. Sch. Res. Netw. ISRN Ecol.* **2011**, 402647. [CrossRef]
2. Zhou, H.; Yang, W.-T.; Zhou, X.; Liu, L.; Gu, J.-F.; Wang, W.-L.; Zou, J.-L.; Tian, T.; Peng, P.-Q.; Liao, B.-H. Accumulation of heavy metals in vegetable species planted in contaminated soils and the health risk assessment. *Int. J. Environ. Res. Public Health* **2016**, *13*, 289. [CrossRef] [PubMed]
3. Filipovic-Trajkovic, R.; Ilic, Z.S.; Sunic, L.; Andjelkovic, S. The potential of different plant species for heavy metals accumulation and distribution. *J. Food Agric. Environ.* **2012**, *10*, 959–964.
4. Sulaiman, F.R.; Hamzah, H.A. Heavy metals accumulation in suburban roadside plants of a tropical area (Jengka, Malaysia). *Ecol. Process.* **2018**, *7*, 28. [CrossRef]
5. Boskovic-Rakocevic, L.; Pavlovic, R.; Duric, M.; Pavlovic, M. Cadmium content and accumulation in lettuce grown on different types of soil. *Rev. Chim.* **2019**, *70*, 2062–2067. [CrossRef]
6. Todirascu-Ciornea, E.; Grosu, E.; Bucur, D.E.; Lobiuc, A. Biochemical and physiological effects of some organic and inorganic chemical agents in *Capsicum* spp. *Rev. Chim.* **2018**, *69*, 2703–2707. [CrossRef]
7. Chaney, R.L. Zinc Phytotoxicity. In *Zinc in Soils and Plants. Developments in Plant and Soil Sciences*; Robson, A.D., Ed.; Springer: Dordrecht, The Netherlands, 1993; Volume 55, pp. 135–150.
8. Popoviciu, D.R.; Negreanu-Pirjol, T.; Bercu, R. Copper, chromium, zinc and manganese accumulation in three common *Brassicaceae* on the Romanian littoral. *Rev. Chim.* **2016**, *67*, 670–672.
9. Apostu, M.; Tantaru, G.; Vieriu, M.; Bibire, N.; Panainte, A.D. Study of the presence of lead in a series of foods of plant origin. *Rev. Chim.* **2018**, *69*, 1223–1225. [CrossRef]
10. Alloway, B.J.; Jackson, A.P.; Morgan, H. The accumulation of cadmium by vegetables grown on soils contaminated from a variety of sources. *Sci. Total Environ.* **1990**, *91*, 223–236. [CrossRef]
11. Cai, L.; Huang, L.; Zhou, Y.; Xu, Z.; Peng, X.; Yao, L.; Zhou, Y.; Peng, P. Heavy metal concentrations of agricultural soils and vegetables from Dongguan, Guangdong. *J. Geogr. Sci.* **2010**, *20*, 121–134. [CrossRef]
12. Gupta, S.; Satpati, S.; Nayek, S.; Garai, D. Effect of wastewater irrigation on vegetables in relation to bioaccumulation of heavy metals and biochemical changes. *Environ. Monit. Assess.* **2010**, *165*, 169–177. [CrossRef] [PubMed]
13. Yang, P.; Yang, M.; Mao, R.; Shao, H. Multivariate-statistical assessment of heavy metals for agricultural soils in northern China. *Sci. World J.* **2014**, *2014*, 517020. [CrossRef]
14. Zhang, Q.; Yu, R.; Fu, S.; Wu, Z.; Chen, H.Y.H.; Liu, H. Spatial heterogeneity of heavy metal contamination in soils and plants in Hefei, China. *Sci. Rep.* **2019**, *9*, 1049. [CrossRef] [PubMed]
15. Chaves, L.H.G.; Estrela, M.A.; Sena De Souza, R. Effect on plant growth and heavy metal accumulation by sunflower. *J. Phytol.* **2011**, *3*, 4–9.
16. Cojoc, E. Modeling the Bioaccumulation of Heavy Metals in Crop Plants in the Copsa Mica Area. Ph.D. Thesis, University of Bucharest, Faculty of Biology, Section Ecology and Environmental Protection, Bucharest, Romania, 2011. Available online: http://www.cesec.ro/pdf/CojocEmilia2011.pdf (accessed on 8 July 2021).
17. Amin, H.; Arain, B.A.; Jahangir, T.M.; Abbasi, M.S.; Amin, F. Accumulation and distribution of lead (Pb) in plant tissues of guar (*Cyamopsistetragonoloba* L.) and sesame (*Sesamumindicum* L.): Profitable phytoremediation with biofuel crops. *Geol. Ecol. Landsc.* **2018**, *2*, 51–60. [CrossRef]
18. Chandra, R.; Hoduck, K. *Phytoremediation and Physiological Effects of Mixed Heavy Metals on Poplar Hybrids*; IntechOpen: London, UK, 2018. [CrossRef]
19. Eissa, M.A.; Negim, O.E. Heavy metals uptake and translocation by lettuce and spinach grown on a metal-contamined soil. *J. Soil Sci. Plant Nutr.* **2018**, *18*, 1097–1107. [CrossRef]
20. Lu, Y.; Yao, H.; Shan, D.; Jiang, Y.; Zhang, S.; Yang, J. Heavy metal residues in soil and accumulation in maize at long-term wastewater irrigation area Tongliao, China. *J. Chem.* **2015**, *2*, 1–9. [CrossRef]
21. Rouphael, Y.; Colla, G.; Bernardo, L.; Kane, D.; Trevisan, M.; Lucini, L. Zinc excess triggered polyamines accumulation in letucc root metabolome, as compared to osmotic stress under high salinity. *Front. Plant Sci.* **2016**, *7*, 842. [CrossRef] [PubMed]
22. Gharaibeh, M.A.; Bani-Hani, K.A. Prediction of phytotoxicity of metal uptake in plants using artificial neural networks. In *Proceedings of the Seventh International Conference on the Application of Artificial Intelligence to Civil and Structural Engineering*; Civil-Comp Ltd.: Stirling, UK, 2003.
23. Złoch, M.; Kowalkowski, T.; Tyburski, J.; Hrynkiewicz, K. Modelling of phytoextraction efficiency of microbially stimulated *Salix dasyclados* L. in the soil with different speciation of heavy metals. *Int. J. Phytoremediation* **2017**, *19*, 1150–1164. [CrossRef]

24. Cervantes-Ramirez, L.T.; Ramirez-López, M.; Mussali-Galante, P.; Ortiz-Hernandez, M.L.; Sanchez-Salinas, E.; Tovar-Sanchez, E. Heavy metal biomagnification and genotoxic damage in two trophic levels exposed to mine tailings: A network theory approach. *Rev. Chil. Hist. Nat.* **2018**, *91*, 6. [CrossRef]
25. Gergen, I. *Chemical and Physicochemical Methods in the Control of Vegetable Agri-Food Products Quality;* University Horizons Publishing House: Timişoara, Romania, 2003.
26. Montgomery, D.C. *Design and Analysis of Experiments*, 5th ed.; John Wiley & Sons: New York, NY, USA, 2001.
27. Cârdei, P.; Tudora, C. Theoretical research on evolution of health of plant affected by heavy metal absorption process. *Eng. Rural. Dev. Jelgava.* **2018**, 893–897. [CrossRef]
28. Cu, N. Effect of heavy metals on plant growth and ability to use fertilizing substances to reduce heavy metal accumulation by *Brassica Juncea* L. Czern. *GJSFR* **2015**, *15*, 34–40.
29. Ghani, A. Toxic effects of heavy metals on plant growth and metal accumulation in maize (*Zea mays* L.). *Iran. J. Toxicol.* **2010**, *3*, 325–334.
30. Chibuike, G.U.; Obiora, S.C. Heavy metal polluted soils: Effect on plants and bioremediation methods. *Appl. Environ. Soil Sci.* **2014**, *2014*, 752708. [CrossRef]
31. Ranjan Rout, G.; Das, P. Effect of metal toxicity on plant growth and metabolism: I. zinc. In *Sustainable Agriculture*; Springer: Dordrecht, The Netherlands, 2009; pp. 873–884. [CrossRef]

materials

MDPI

Article

Performance of Plain Concrete and Cement Blocks with Cement Partially Replaced by Cement Kiln Dust

Yasir M. Alharthi [1], Ahmed S. Elamary [1,*] and Waleed Abo-El-Wafa [2]

1 Civil Engineering Department, College of Engineering, Taif University, P.O. Box 11099, Taif 21944, Saudi Arabia; y.harthi@tu.edu.sa
2 Civil Engineering Department, Assiut University, Assiut 71515, Egypt; wam@aun.edu.eg
* Correspondence: a.elamary@tu.edu.sa; Tel.: +966-581338825

Abstract: The growth of the construction industry has led to the greater consumption of natural resources, which has a direct or indirect negative impact on the environment. To mitigate this, recycled or waste materials are being used as a partial substitute in the manufacture of concrete. Among these waste materials is cement kiln dust (CKD), which is produced during cement production. This study investigated the potential benefits of replacing part of the cement with CKD in two construction applications, i.e., plain concrete and cement blocks. This reflects positively on cost, energy, and the environment, since putting CKD in a landfill damages agricultural soil and plant respiration. In this study, an experimental program was carried out to study how replacing various percentages of ordinary portland cement (OPC) with CKD affected the compressive strengths, the tensile strengths, and the air contents of concrete and cement blocks. Although the results showed that the compressive and tensile strengths decreased as the amount of CKD increased, the air content of the concrete increased, which showed that 5% CKD was suitable for such applications. The results were used to propose two equations that approximate the concrete and cement block compressive strengths according to the CKD replacement percentage.

Keywords: cement kiln dust; compressive strength; tensile strength; concrete air content; cement blocks

Citation: Alharthi, Y.M.; Elamary, A.S.; Abo-El-Wafa, W. Performance of Plain Concrete and Cement Blocks with Cement Partially Replaced by Cement Kiln Dust. *Materials* **2021**, *14*, 5647. https://doi.org/10.3390/ma14195647

Academic Editors: Avelino Núñez-Delgado, Zhien Zhang, Elza Bontempi, Mario Coccia, Marco Race and Yaoyu Zhou

Received: 31 August 2021
Accepted: 24 September 2021
Published: 28 September 2021

1. Introduction

Over the past few decades, the demand for construction materials has increased due to the worldwide growth of construction projects. Unfortunately, this demand is fulfilled by the overconsumption of natural resources endangering the environment [1]. Accordingly, researchers are motivated to reduce the impact of construction materials on the environment and enhance the mechanical properties of construction materials. This has led to the use of waste materials (recycled aggregate and rubber) and cement kiln dust (CKD) in the manufacture of concrete and studies on how these materials affect its behavior [2–7].

Małek et al. [2] investigated the use of ferronickel slag waste aggregate (FNSWA) in place of granite aggregate and found an increase in compressive and flexural strength of 31% and 66%, respectively. Padmini et al. [3] studied the influence of parent concrete on the properties of recycled aggregate concrete. This study showed that achieving a desired compressive strength using recycled aggregate concrete requires using a lower water–cement ratio and a higher cement content than for concrete with fresh granite aggregate.

Experimentally, Chalangaran et al. [4] improved environmental noise absorption after replacing sand aggregate with recycled rubber crumbs. Fifteen percentages of fine- and coarse-grained crumbs substitutions improved sound transmission losses by 190% and 228%, respectively. They also reported [7] that the ultimate compressive strength, the tensile strength, and the modulus of elasticity of the concrete with rubber crumbs are reduced. The negative impact of the rubber crumbs was reduced by using an optimum percentage of nanosilica and metakaolin additives.

One of waste materials that researchers have used in concrete is CKD, a fine powdery material resulting from cement manufacture. Studies showed that the greatest health danger in cement production comes from dust, and several studies have been conducted by references [8–11] to assess the influence of cement dust in the workplace and how it stimulates abiotic stress responses in plants.

The environmental concerns related to portland cement (PC) production and emission and the disposal of CKD are becoming progressively significant, because 80% of CKD ends up in landfills. The extreme contamination of the surrounding environment highlights the need to find more environmentally friendly ways of disposal.

Using CKD as a partial replacement in PC is currently fraught with difficulties, and many of its applications continue to be investigated, for example, as a component in cement, an agricultural fertilizer, a soil and wastewater stabilizer, a partial replacement of soda in glass production, an antistripping agent in asphalt, and a subgrade for highway construction as stated by Shervan [12].

Maslehuddin et al. [13,14] studied properties of blended cement concretes (CKD). The mechanical properties of CKD concrete specimens were assessed using drying shrinkage and compressive strength, while the durability was assessed using electrical resistivity and chloride permeability. The compressive strength and the drying shrinkage strain of the concrete samples decreased, as the quantity of CKD increased. The chloride permeability improved, and the electrical resistance decreased when CKD was added. Finally, it is recommended that the amount of CKD in concrete be limited to 5%.

Ravindrarajah et al. [15] studied CKD as a partial replacement for cement in concrete and found that it is a cementitious substance that delays the cement setting, increases water consumption to achieve consistency and reduces the concrete strength. El-Sayed et al. [16] and Batis et al. [17] examined the impact of CKD on the cement paste compressive strength and the embedded reinforcement corrosion behavior and found that up to 5% CKD replacement by weight has no negative impact on the cement paste strength or reinforcing passivity. In addition, Batis et al. [17] reported that adding CKD and blast furnace slag (BFS) in a correct amount increases the compressive strength and the corrosion resistance of PC.

Abo-El-Nien et al. [18] studied the partial replacement of BFS cement with CKD and the effects of kiln meal and CKD on the PC paste strength. Compared to free cement, the findings revealed a high degree of hydration and a reduced compressive strength. Shoaib et al. [19] studied the effect of replacing ordinary PC and BFS cement with CKD on the splitting tensile strength of concrete mixes at various ages (1, 3, and 6 months) and concluded that it dropped as the proportion of CKD increases and it was minimal at 10% CKD.

Al-Harthy et al. [20] investigated the effect of CKD on the toughness and flexure strength of concrete mixes 3, 7, and 28 days after casting. They discovered that the mixes containing less than 5% CKD had similar toughness and flexural strength values to the control mix, especially at the water-to-binder ratio of 0.50. Mosleh et al. [21] investigated the strong growth of PC pastes with the addition of kiln meal and kiln dust, finding a higher degree of hydration and a lower compressive strength than when free cement was added.

Heikal et al. [22] studied PC clinker, BFS, and CKD composites. Three mixtures of slag cement were prepared, each mixed with 2.5%, 5.0%, 7.50%, and 10.0% CKD. The authors found that the partial replacement of BFS with CKD improves PC clinker's setting periods, electrical conductivity, and fluidity. Udoeyo and Ridnap [23] studied the characteristics of hollow sandcrete blocks with added CKD as a substitute for ordinary portland cement (OPC) and concluded that when CKD is used to replace cement, the compressive strength and density of the blocks fall as the proportion of CKD replacement rises, but the percentage of water absorption of the blocks increases.

2. Research Goal and Methodology

This research effort was designed to open new industrial areas for recycled CKD. Two main construction applications were studied in this experimental work program. The first was the effect of CKD replacement on plain concrete strength and air content. In this application, three different groups were investigated. The first group consisted of concrete samples tested for concrete compressive strength after being prepared with 0%, 2%, 5%, 8%, 10%, 15%, and 20% CKD replacements. The second group consisted of concrete samples prepared with 0% and 5% CKD replacements and tested for tensile strength. To test the effect on air content, the third group included concrete samples prepared with 0%, 5%, 10%, 15%, and 20% CKD replacements.

The second construction application was related to hollow cement blocks. Cement block specimens were prepared with 0%, 10%, 15%, 20%, and 25% CKD replacements and tested to study their effect on compressive strength and water absorption. The values are represented in the database to form equations that reflect the impact of the CKD% on the compressive strength.

3. Experimental Work

An experimental program was conducted to study the influence of replacing a percentage of OPC with CKD in 120 specimens divided into two groups: the specimens of the first group were used to measure the most important parameters of concrete, and those of the second group were used to scale the effect on cement blocks. The amount of CKD in the mix was expressed as a percentage of the total weight of cement in the mix. The types of test and specimen numbers and details are listed in Table 1. This experimental program conducted five different tests: three for plain concrete and two for the cement blocks.

Table 1. List of tests and specimens conducted.

No.	Test	Specimen Dimension	No. of Specimens
1	Chemical analysis	Cement kiln dust (CKD)	1
2	Concrete compressive strength	Cube (150 mm × 150 mm × 150 mm)	42
3	Concrete tensile strength	Cylinder (15 mm in diameter; 30 mm in height)	12
4	Concrete air content	Gilson container	15
5	Blocks compressive strength	Hollow cement blocks (200 mm × 200 mm × 400 mm)	30
6	Block absorption	Hollow cement blocks (200 mm × 200 mm × 400 mm)	20

CKD Materials

The materials used in this research were untreated raw CKD collected from electrostatic precipitators and OPC products from a Madina cement factory (Madina, KSA). A comparison between the chemical compositions of CKD produced from the Madina cement factory with the CKD limits standard in the UK is illustrated in Table 2.

The percentage of sulfate, as well as alkalinity, chloride, and silica concentrations, was within acceptable limits, according to this analysis. On the other hand, the chemical composition of cement varied based on raw materials and the manufacturing processes used. The quantity of decarburization of the calcium carbonate in the clinker-making raw materials was reflected in the level and variability of the loss on igniting the CKD.

Table 2. Typical chemical compositions of CKD and ordinary portland cement (OPC) in the UK and KSA.

Chemical Composition (%)	CKD (UK) [14]	CKD AL Madina Cement Factory (AL Madina KSA)	OPC
SiO_2	11–16	18.20	22
Al_2O_3	3–6	4.52	5
Fe_2O_3	1–4	2.92	3
CaO	38–50	49.40	64
MgO	0–2	1.21	1
SO_3	4–18	5.66	3
K_2O	3–13	2.38	<1
Na_2O	0–2	3.84	<1
Cl	0–5	5.90	<0.10
Loss on ignition	5–25	17.10	1
Free Cao	1–10	4.24	2

4. Procedures and Result Discussion

4.1. Concrete

The concrete mixture can be designed in a number of ways, but the British Standard (BS) mix was used in this study since it is concerned with cube compressive strength. Using the coefficient of variation and the lowest strength as a percentage of the mean strength, the BS technique defined the characteristic strength. This approach also took into account the influence of the fine aggregate on the determination of the fine-to-total aggregate content in the mix by ensuring that the aggregate grading followed the grading zones.

The approach, suitable for OPC (type I), uses a hypothetical concrete mix with a moderate cement content (a water/cement (w/c) ratio of 0.5) that is thoroughly compacted, correctly cured and cast with various types of cement and coarse aggregate and tested at various ages. The specified quantities are sufficient just for plain concrete.

The fine and coarse aggregates used in the mix were graded using sieve analysis in accordance with BS 882:1983. The coarse aggregate size varied from 20 to 5 mm sieve size, while the fine aggregate size ranged from 5 to 0.61 mm. The w/c ratio (0.48), the cementitious material content (250 kg/m^3), and the fine/total aggregate ratio (0.33) were all components. The relative density and the bulk density of the aggregate were measured and found to be 2.65 and 1645 kg/m^3 for coarse and 2.70 and 1750 kg/m^3 for fine aggregate.

4.1.1. Effect of CKD on Concrete Compressive Strength

The design, treatment, and control conditions of the concrete mixes, with and without CKD, were the same according to BS 1881-108 [24]. To ensure uniformity, the ingredients were combined in a dry condition for roughly one minute. Water was progressively added to the mixture, which was then blended with workability admixtures. For an additional two minutes, the materials were mechanically combined.

A traditional slump test was used to determine the consistency of fresh concrete and the temperature. Specimens of a 150 mm cube were prepared from each concrete mix to be tested in compression after 28 days. During the concrete placement, a vibrating table was used to verify that the concrete was fully compacted. After 24 h, the cube specimens were demolded and submerged in water, until they were examined. Six cubes (150 mm $\times$ 150 mm $\times$ 150 mm) were cast from each mix and cured in clean water for 28 days. Compression tests were performed on standard cubes according to BS 1881-116 [25].

The crushing load was measured using a Universal Hydraulic Testing Machine with a capacity of 2000 kN to evaluate the compressive strength of a concrete sample as illustrated in Figure 1. In this part of the study, 42 cubes were tested to examine the replacement effect of CKD on the concrete compressive strength. The proportion of cement replacement by CKD was the most important component in this experiment. Seven samples were prepared and tested with 0% (control), 2%, 5%, 8%, 10%, 15%, and 20% CKD.

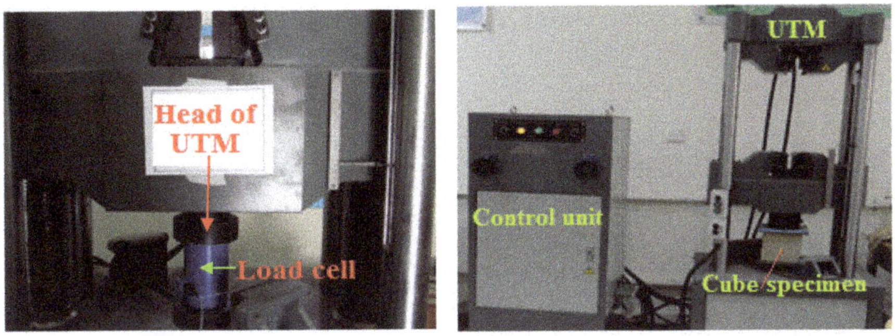

(**a**) UTM Calibration test (**b**) Cube compressive strength test

Figure 1. Concrete cubes after curing and during the test: (**a**) UTM (Universal Testing Machine) calibration test; (**b**) cube compressive strength test.

The compressive strength results found experimentally for the 42 cubes were averaged and are listed in Table 3. The relationship between the average compressive strength and the replacement percentage of CKD is shown in Figure 2. A nonlinear regression analysis was used as recommended by [26]. A fourth-degree formula was fitted and created using a numerical database, as proposed in Equation (1). The curve plotted in Figure 2 reflected the effect of CKD on the concrete compressive strength after 28 days. Equation (1) was proposed to predict the compressive strength of a concrete mix that uses CKD as a partial replacement for cement:

$$f_{cu} = -1 \times E^{-4} \times K^4 + 0.003 \times K^3 - 0.0313 \times K^2 - 0.0903 \times K + F_{cu}, \quad (1)$$

where K is the percentage of CKD; F_{cu} is the designed concrete compressive strength with 0% CKD; and f_{cu} is the anticipated concrete compressive strength after using K percentage of CKD.

Table 3. Concrete sample compressive strength results.

No.	Cube Compressive Strength after 28 Days (MPa)							
	% of CKD	**0%**	**2%**	**5%**	**8%**	**10%**	**15%**	**20%**
1	From cement portion in concrete mixes	29	28.2	28	27.6	27	24.5	19.2
2		29.3	29	27.5	27.3	26.3	25	18.5
3		28.45	28.1	27.6	26.5	26.5	23.8	18.2
4		28.41	28.3	28	27	26.8	23.7	19
5		28.66	28	27.5	26.4	26.2	24.2	18.4
6		28.5	28.1	28.2	27.2	26.2	24	18.3
Average (MPa)		28.724	28.45	27.8	27	26.5	24.2	18.6

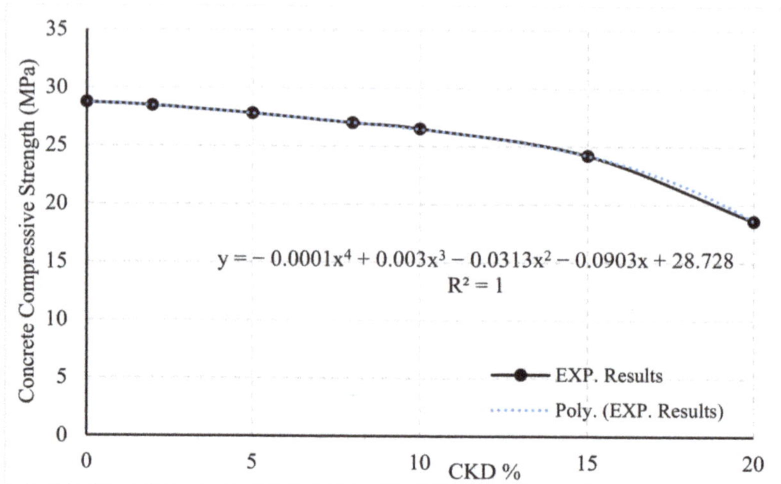

Figure 2. Compressive strength versus percentage of CKD.

The CKD describes fine-grained, extremely alkaline particulate material mostly made of oxidized, anhydrous, micron-sized particles. The bulk of prior research has found that fine CKD particles have greater sulfate and alkali concentrations and less lime content [27]. As a result, references [28] and [29] concluded that the high alkali content is responsible for the loss of compressive strength.

4.1.2. Effect of CKD on Concrete Tensile Strength

This study was performed to investigate the influence of replacing cement with CKD up to 5% on concrete tensile strength. Concrete mixes with and without CKD were mixed, treated and controlled under the same conditions. The components were combined in a dry condition for about a minute. For a further two minutes, all of the ingredients were mechanically combined to guarantee the homogeneity of the mix.

The conventional slump test measured the consistency of fresh concrete. Standard cylinders with a diameter of 150 mm and a height of 300 mm were prepared from each concrete mix according to ASTM C31 [30], as shown in Figure 3a. The ASTM C496 [31] test for splitting the tensile strength of cylindrical concrete specimens standard was used to conduct the splitting tensile test on the prepared standard cylinders.

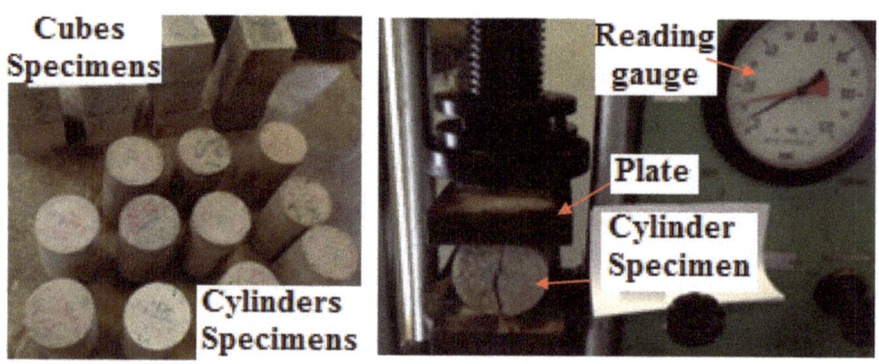

(**a**) Concrete specimens before testing (**b**) Concrete Brazilian test

Figure 3. Cylinder specimen after curing and testing: (**a**) concrete specimens before testing; (**b**) concrete Brazilian test.

All the molds were placed on a vibrating table while placing the concrete to ensure full compaction, and 6 cylinders were cast from each mix. After 48 h, all cylinder specimens were demolded and immediately submerged in potable water for 28 days. As shown in Figure 3b, all specimens were crushed in a testing machine. During the test, the load was monitored to determine the samples' tensile strengths.

In this study, a total of 12 cylinders were tested to investigate their tensile strengths after replacing cement with CKD. The main factor considered in this study was the percentage of CKD replacement (0% and 5%). The results are listed in Table 4 and showed no significant difference in strength (approximately 3%).

Table 4. Concrete tensile strength results.

No	% of CKD	Tensile Strength (MPa)						Average (MPa)
		1	2	3	4	5	6	
1	0%	2.97	3.11	2.83	2.90	3.00	2.95	2.95
2	5%	2.81	2.86	2.68	2.74	2.85	2.87	2.80

4.1.3. Effect of CKD on Air Content

The gross air contents of 15 concrete specimens were determined using the same mix that was previously used to determine compressive and tensile strength. The tests were performed using the pressurized technique according to ASTM C231 [32] for five different percentages of CKD (0% (control), 5%, 10%, 15%, and 20%). Figure 4 shows the Gilson equipment that was used to measure air content.

Figure 4. Air content testing apparatus.

The results are recorded in Table 5 and plotted in Figure 5.

Table 5 presents five groups of samples, each having three specimens as listed in column 1. Column 2 shows the percentage of CKD for each group, column 3 shows the recoded air content for each specimen, and column 4 shows the average results for each group.

These data showed that up to 5% CKD replacement can be ignored. When it was increased to 10%, the air content increased by 24%. In the last two cases, i.e., 15% and 20% CKD replacements, the air content rose by 33% and 43%, respectively, compared to the control mix.

Table 5. Concrete air content results.

No. (1)	% CKD (2)	% Air Content (3)	% Air Content Average (4)
1	0	1.05	1.05
2		1.05	
3		1.00	
1	5	1.05	1.05
2		1.05	
3		1.10	
1	10	1.30	1.30
2		1.35	
3		1.25	
1	15	1.35	1.40
2		1.45	
3		1.35	
1	20	1.45	1.50
2		1.60	
3		1.50	

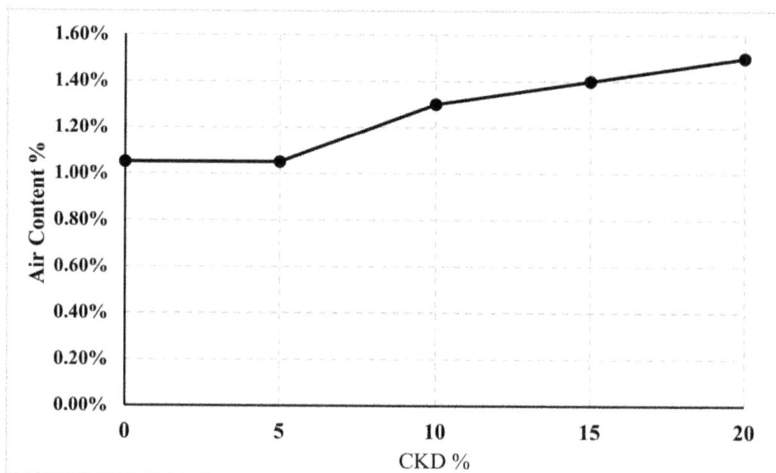

Figure 5. Relation between air content and CKD percentage.

4.1.4. Water Cement Ratio and CKD Replacement Effects

The results of this investigation for the concrete compressive strength effect were in accordance with the results of a previous study [15,33]. According to reference [15], for the control and the modified concrete, the relationship between compressive strength (f) at 28 days and the ratio of the total volume of the cementitious material to the volume of water (R) was given by F = 125.6R − 39 (R unit: N/mm^2). The expression for R is also given by $R = (V_c + kV_d)/V_w$, where V_c, V_d, and V_w are the volumes of cement, kiln dust, and water, respectively, and k is the cementing efficiency factor and written as:

$$k = \frac{C_{RQ}}{D_Q}\left[\frac{\left(\frac{w}{c}\right)_N}{\left(\frac{w}{c}\right)_{DM}} - 1\right],$$

where C_{RQ} is the weight of cement (kg/m^3); D_Q is the weight of CKD (kg/m^3); $\left(\frac{w}{c}\right)_N$ is the water/cement ratio determined from the plotted curve between strength and w/c; and $\left(\frac{w}{c}\right)_{DM}$ is the water/cement ratio in a design mix with 0% CKD.

To obtain k, $\left(\frac{w}{c}\right)_N$ was calculated for different concrete strengths of the control mixtures with different w/c ratios, and the strengths after 28 days were plotted to derive the strength and w/c relationship curve [33]. The resulting curve can be used to obtain the effective w/c ratios for mixtures with different CKD percentages.

4.2. Hollow Cement Blocks

4.2.1. Effect of CKD on Cement Block Compressive Strength

The study was conducted according to ASTM C140 [34]. Twelve hollow cement blocks (200 mm × 200 mm × 400 mm) were manufactured using mix components 1 cement: 1.66 sand and 3.5 coarse aggregate. The quantities required to produce one block were 32 *N* cement: 53.5 *N* sand: 113 *N* coarse aggregate. To reach a uniform mixture, the ingredients were mixed in a dry state for about three minutes, after which water was gradually added. The blocks were mixed and formed mechanically as shown in Figure 6a–c.

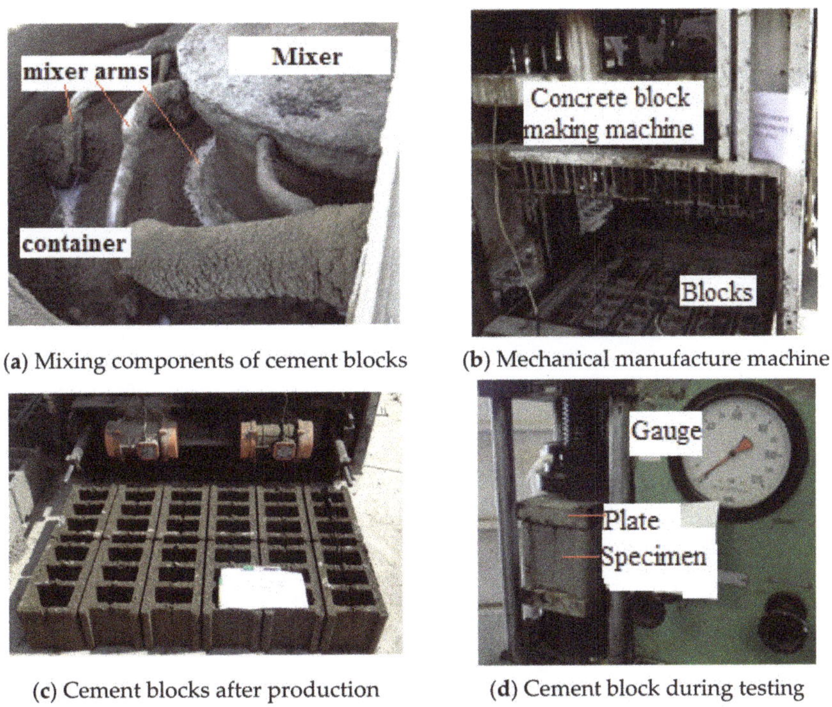

Figure 6. Hollow blocks, and stages of manufacture and under testing: (**a**) mixing components of cement blocks; (**b**) mechanical manufacture machine; (**c**) cement blocks after production; (**d**) cement block during testing.

In this part of the investigation, four different percentages of CKD were used: 10%, 15%, 20%, and 25%. The blocks with and without CKD were treated, controlled and cured under the same conditions until compression testing after 28 days, as shown in Figure 6d.

The results of the compression test are listed in Table 6. In each case, six blocks were tested, and the average compressive strength was determined. The gross and net areas were 74,600 and 44,225 mm^2, respectively.

Table 6. Hollow block compressive strength results.

% of CKD	Crushing Load (kN)	Strength (MPa) Net Area		Strength (MPa) Gross Area	
		Per Specimen	Average	Per Specimen	Average
0	55.5	12.31	11.89	7.30	7.05
	56.00	12.42		7.36	
	52.30	11.60		6.88	
	50.50	11.20		6.64	
	52.75	11.70		6.94	
	54.56	12.10		7.18	
10	48.25	10.70	10.50	6.34	6.22
	47.80	10.60		6.29	
	45.70	10.14		6.01	
	47.50	10.54		6.25	
	48.00	10.65		6.32	
	46.64	10.35		6.14	
15	37.82	8.39	9.50	4.97	5.62
	46.7	10.36		6.14	
	42.2	9.36		5.55	
	44.3	9.83		5.83	
	43.17	9.58		5.68	
	42.59	9.45		5.60	
20	35.11	7.79	7.80	4.62	4.63
	35.98	7.98		4.73	
	34.24	7.6		4.50	
	35.41	7.85		4.66	
	35.27	7.82		4.64	
	35.00	7.76		4.61	
25	27.34	6.06	6.15	3.60	3.65
	28.2	6.26		3.71	
	27.56	6.11		3.62	
	27.82	6.17		3.66	
	27.73	6.15		3.65	
	27.59	6.12		3.63	

Figure 7 shows the relation between the compressive strengths corresponding to gross area and CKD percentages.

By applying the regression analysis reported by reference [26] to the experimentally obtained numerical database, a nonlinear regression model based on a second-degree formula was fitted for this curve. Equation (2) was proposed to predict the compressive strength of blocks for the partial CKD replacement:

$$f_{cu-b} = -0.0038 \times K^2 - 0.0428 \times K + F_{cu-b}, \tag{2}$$

where K is the CKD percentage in the cement mix; F_{cu-b} is the block compressive strength with 0% CKD; and f_{cu-b} is the anticipated block compressive strength after K percentage of CKD.

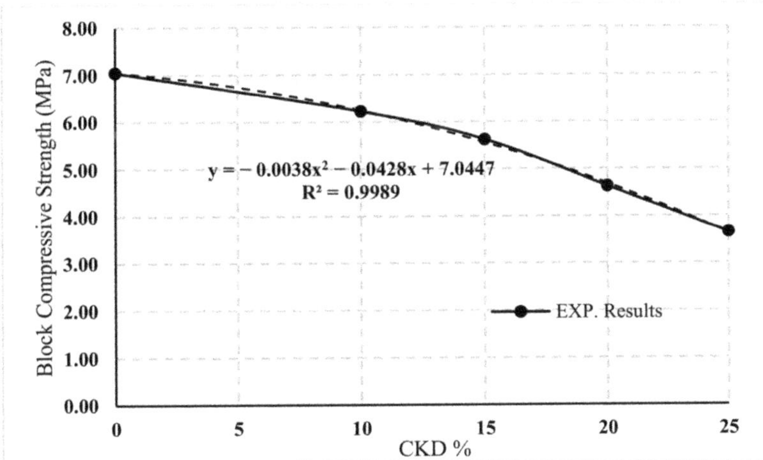

Figure 7. Relation between the block compressive strength and the CKD percentage.

4.2.2. Effect of CKD on Cement Block Absorption

The present study was concerned with an analysis of the percentage of water absorption of hollow cement blocks (200 mm × 200 mm × 400 mm) with and without CKD conducted with respect to ASTM C140 [34].

The sequence of block formation was performed using the same procedure described in the previous section. The blocks were kept in an electronic oven under 100 °C for 24 h, weighed, subjected to the same drying conditions in the oven for half an hour and then reweighed to ensure that the weight did not change. The blocks were immersed in water for 24 h and then weighed at the end of that time.

The results from both stages are shown in Table 7, which showed the average weight at the end of the dry stage and the absorbed water. The percentage of water absorption in the control specimen was 4.27%, and this percentage increased to 16%, 25%, 44%, and 47% for 10%, 15%, 20%, and 25% CKD replacement, respectively.

Table 7. Hollow blocks absorption results.

% CKD	Weight (*N*)			Average Weight (*N*)		% Absorption
	After Oven	After Submerged	Water Absorbed	Dry Weight	Water	
0	196.35	204.20	7.85	191.64	8.18	4.27
	189.70	198.15	8.45			
	191.00	198.90	7.90			
	189.50	198.00	8.50			
10	196.80	205.90	9.10	190.13	9.48	4.98
	186.20	196.10	9.90			
	186.50	196.40	9.90			
	191.00	200.00	9.00			
15	185.75	195.95	10.20	185.61	9.913	5.34
	185.40	195.15	9.75			
	185.20	195.00	9.80			
	186.10	196.00	9.90			

Table 7. *Cont.*

% CKD	Weight (*N*)			Average Weight (*N*)		% Absorption
	After Oven	After Submerged	Water Absorbed	Dry Weight	Water	
20	186.75	197.95	11.20	187.01	11.53	6.16
	187.60	199.40	11.80			
	186.60	198.20	11.60			
	187.10	198.60	11.50			
25	182.45	194.20	11.75	186.04	11.74	6.31
	188.00	199.80	11.80			
	188.30	200.00	11.70			
	185.40	197.10	11.70			

5. Assessment of the Application Tests

Table 8 shows the statistical results of the various tests: mean values for the ratio of each parameter (concrete compression and tensile strength, air content, and block compressive strength) and the CKD percentage along with the standard deviation and the minimum and maximum values.

Table 8. Assessment of various test specimens.

Test Type	% CKD	Mean	Standard Deviation	Min.	Max
Concrete compressive strength	2	0.985	0.009	0.972	0.996
	5	0.968	0.019	0.939	0.989
	8	0.940	0.014	0.921	0.954
	10	0.923	0.016	0.898	0.943
	15	0.843	0.007	0.834	0.853
	20	0.648	0.014	0.631	0.669
Concrete tensile	5	0.950	0.017	0.920	0.973
Concrete air content	5	1.033	0.058	1.000	1.100
	10	1.258	0.025	1.238	1.286
	15	1.339	0.049	1.286	1.381
	20	1.468	0.077	1.381	1.524
Block compressive strength	10	0.884	0.019	0.862	0.905
	15	0.798	0.028	0.749	0.834
	20	0.657	0.017	0.638	0.679
	25	0.517	0.016	0.501	0.541

It can be seen that having more than 5% CKD had a significant impact on the sample properties. Therefore, it is recommended that a replacement of 10% CKD can be used with an acceptable reduction in concrete and block strength.

6. Summary and Conclusions

The effects of CKD in plain concrete on compressive strength, tensile strength, and air content were studied, while the effects of CKD in cement blocks on compressive strength and water absorption were investigated. The experiments revealed the following:

- CKD has a detrimental impact on compressive strength, as evidenced by the fact that as the amount increased, the compressive strength of the concrete and cement block specimens decreased. An equation corresponding to each was presented to anticipate this reduction.

- Regarding the concrete tensile strength, there was no significant difference between 0% and 5% CKD, which was only approximately 3%.
- In the concrete mix, the percentage of air content from 0% to 5% CKD replacements made no difference; however, when CKD was increased to 10%, 15%, and 20%, it increased by 24%, 33%, and 43%, respectively.
- The percentage of water absorption due to partial replacement by CKD in cement blocks can be increased up to 25% within the allowed limits.
- This research demonstrates that CKD can be used as a primary component in concrete (e.g., plain concrete, curbs, and cement tiles) and cement block products.

Author Contributions: Conceptualization, A.S.E., W.A.-E.-W. and Y.M.A.; methodology, A.S.E., W.A.-E.-W. and Y.M.A.; software, Y.M.A.; validation, A.S.E., W.A.-E.-W. and Y.M.A.; formal analysis, A.S.E., W.A.-E.-W. and Y.M.A.; investigation, A.S.E., W.A.-E.-W. and Y.M.A.; resources, A.S.E., W.A.-E.-W. and Y.M.A.; data curation, A.S.E. and W.A.-E.-W.; writing—original draft preparation, A.S.E., W.A.-E.-W. and Y.M.A.; writing—review and editing, A.S.E., W.A.-E.-W. and Y.M.A.; visualization, A.S.E., W.A.-E.-W. and Y.M.A.; supervision, A.S.E.; project administration, A.S.E.; funding acquisition, Y.M.A. All authors have read and agreed to the published version of the manuscript.

Funding: The authors would like to acknowledge the financial support received from Taif University Researchers Supporting Project Number (TURSP-2020/276), Taif University, Taif, Saudi Arabia.

Institutional Review Board Statement: Not applicable.

Informed Consent Statement: Not applicable.

Data Availability Statement: The data presented in this study are available on request from the corresponding author.

Conflicts of Interest: The authors declare that there are no conflicts of interest.

References

1. Tam, V.W.; Soomro, M.; Evangelista, A.C. A review of recycled aggregate in concrete applications (2000–2017). *Constr. Build. Mater.* **2018**, *172*, 272–292. [CrossRef]
2. Małek, M.; Jackowski, M.; Łasica, W.; Dydek, K.; Boczkowska, A. An Experimental Study of Possible Post-War Ferronickel Slag Waste Disposal in Szklary (Lower Silesian, Poland) as Partial Aggregate Substitute in Concrete: Characterization of Physical, Mechanical, and Thermal Properties. *Materials* **2021**, *14*, 2552. [CrossRef] [PubMed]
3. Padmini, A.K.; Ramamurthy, K.; Mathews, M.S. Influence of parent concrete on the properties of recycled aggregate concrete. *Constr. Build. Mater.* **2009**, *23*, 829–836. [CrossRef]
4. Chalangaran, N.; Farzampour, A.; Paslar, N.; Fatemi, H. Experimental investigation of sound transmission loss in concrete containing recycled rubber crumbs. *Adv. Concr. Constr.* **2021**, *11*, 447–454.
5. Farzampour, A. Compressive behavior of concrete under environmental effects. In *Compressive Strength of Concrete*; Intech Open: London, UK, 2019.
6. Mansouri, I.; Shahheidari, F.S.; Hashemi, S.M.; Farzampour, A. Investigation of steel fiber effects on concrete abrasion resistance. *Adv. Concr. Constr.* **2020**, *9*, 367–374.
7. Chalangaran, N.; Farzampour, A.; Paslar, N. Nano silica and metakaolin effects on the behavior of concrete containing rubber crumbs. *CivilEng* **2020**, *1*, 264–274. [CrossRef]
8. Rahmani, A.H.; Almatroudi, A.; Babiker, A.Y.; Khan, A.A.; Alsahly, M.A. Effect of exposure to cement dust among the workers: An evaluation of health-related complications. *Open Access Maced. J. Med Sci.* **2018**, *6*, 1159. [CrossRef]
9. Omidianidost, A.; Gharavandi, S.; Azari, M.R.; Hashemian, A.H.; Ghasemkhani, M.; Rajati, F.; Jabari, M. Occupational exposure to respirable dust, crystalline silica and its pulmonary effects among workers of a cement factory in Kermanshah, Iran. *Tanaffos* **2019**, *18*, 157.
10. Shah, K.; An, N.; Ma, W.; Ara, G.; Ali, K.; Kamanova, S.; Zuo, X.; Han, M.; Ren, X.; Xing, L. Chronic cement dust load induce novel damages in foliage and buds of Malus domestica. *Sci. Rep.* **2020**, *10*, 1–2. [CrossRef]
11. Shanshal, S.A.; Al-Qazaz, H.K. Consequences of cement dust exposure on pulmonary function in cement factory workers. *Am. J. Ind. Med.* **2021**, *64*, 192–197. [CrossRef]
12. Khanna, O.S. Characterizaton and Utilization of Cement Klin Dust (CKDs) as Partial Replacements of Portland Cement. Ph.D Thesis, Department of Civil Engineering, University of Toronto, Toronto, ON, Canada, 2009.

13. Maslehuddin, O.S.B.; Al-Amoudi, M.K.; Rahman, M.R.; Ali, M.S.B. Properties of cement kiln dust concrete. *Constr. Build. Mater.* **2009**, *23*, 62357–62361. [CrossRef]
14. Maslehuddin, O.S.B.; Al-Amoudi, M.; Shameem, M.K.; Rahman, M.R.; Ali, M.I. Usage of cement kiln dust in cement products-Research review and preliminary investigations. *Constr. Build. Mater.* **2008**, *22*, 2369–2375. [CrossRef]
15. Ravindrarajah, R.S. Usage of cement kiln dust in concrete. *Int. J. Cem. Compos. Lightweight Concr.* **1982**, *4*, 95–102. [CrossRef]
16. El-Sayed, H.A.; Gabr, N.A.; Hanafi, S.; Mohran, M.A. *Reutilization of by-Pass Kiln Dust in Cement Manufacture*; International Conference on Blended Cement in Construction: Sheffield, UK, 1991.
17. Batis, G.; Rakanta, E.; Sideri, E.; Chaniotakis, E.; Papageorgiou, A. *Advantages of Simultaneous Use of Cement Kiln Dust and Blast Furnace Slag*; International Conference on Challenges of Concrete Construction; University of Dundee: Dundee, UK, 2002.
18. Abo-El-Enein, S.A. *Utilization of Cement Kiln Dust in Cement Industry and Building Products*; First international Symposium on the cement Industry: Assiut, Egypt, 1997; pp. 8–10.
19. Shoaib, M.M.; Balaha, M.M.; Abdel-Rahman, A.G. Influence of cement kiln dust substitution on the mechanical properties of concrete. *Cem. Concr.* **2000**, *30*, 337–371. [CrossRef]
20. Al-Harthy, A.S.; Al-Maamary, F.T. Effect of cement kiln dust CKD on mortar and concrete mixtures. *Constr. Build. Mater.* **2003**, *17*, 353–360. [CrossRef]
21. Mosleh, A.M. Evaluation of By-Pass Dust for the Production of Blended Cement Containing BF Slag. Master's Thesis, Institute of Environmental Studies and Research, Ain Shams University, Cairo, Egypt, 1996.
22. Heikal, M.; Aiad, I.; Helmy, I.M. Portland cement clinker, granulated slag and by bass cement dust composites. *Cem. Concr. Res.* **2002**, *32*, 1805–1812. [CrossRef]
23. Udoeyo, F.F.; Rindap, P.I. Cement kiln dust as a material for building blocks. *Glob. J. Eng. Res.* **2002**, *1*, 73–78. [CrossRef]
24. British Standards Institution (BSI). *Testing Concrete. Method for Making Test Cubes from Fresh Concrete*; BS 1881-108; British Standards Institution: London, UK, 1983.
25. British Standards Institution (BSI). *Methods for the Determination of Compressive Strength of Concrete*; BS 1881-116; British Standards Institution: London, UK, 1983.
26. Elamary, A.S.; Taha, I. Determining the Shear Capacity of Steel Beams with Corrugated Webs by Using Optimised Regression Learner Techniques. *Materials* **2021**, *14*, 2364. [CrossRef]
27. Corish, A.; Coleman, T. Cement kiln dust. *Concrete* **1995**, *29*, 40–42.
28. Bhatty, M.S.Y. Kiln dust cement blends, evaluated. *Rock Prod.* **1984**, *88*, 47–65.
29. Siddique, R.; Rajor, A. Use of cement kiln dust in cement concrete and its leachate characteristics. *Resour. Conserv. Recycl.* **2012**, *61*, 59–68.
30. American Society for Testing and Materials (ASTM). *Practice for Making and Curing Concrete Test Specimens in the Field*; ASTM C31/C31M; ASTM International: West Conshohocken, PA, USA, 2015.
31. ASTM C496-90; *Standard Test Method for Splitting Tensile Strength of Cylindrical Concrete Specimens*; Annu. Book ASTM Stand.: West Conshohocken, PA, USA, 1991; pp. 266–269.
32. ASTM C231/C231M-17a; Standard Test Method for Air Content of Freshly Mixed Concrete by the Pressure Method. In *American Society of Testing and Materials*; ASTM: West Conshohocken, PA, USA, 2017; pp. 1–10.
33. Al-Rezaiqi, J.; Alnuaimi, A.; Hago, A.W. Efficiency factors of burnt clay and cement kiln dust and their effects on properties of blended concrete. *Appl. Clay Sci.* **2018**, *157*, 51–64. [CrossRef]
34. ASTM. *Sampling and Testing Concrete Masonry Units and Related Units C140/C140M-15a*; ASTM International: West Conshohocken, PA, USA, 2016.

sustainability

MDPI

Article

Decolourisation and Biodegradation of Textile Di-azo Dye Congo Red by *Chryseobacterium geocarposphaerae* DD3

Shrabana Sarkar [1], Alex Echeverría-Vega [2,3], Aparna Banerjee [2,3] and Rajib Bandopadhyay [1,*]

1 UGC Center of Advanced Study, Department of Botany, The University of Burdwan, Purba Bardhaman 713104, India; runka.sarkar@gmail.com
2 Centro de Investigación de Estudios Avanzados del Maule, Vicerrectoría de Investigación y Posgrado, Universidad Católica del Maule, Talca 3466706, Chile; aecheverria@ucm.cl (A.E.-V.); abanerjee@ucm.cl (A.B.)
3 Centro de Biotecnología de Los Recursos Naturales (CENBio), Facultad de Ciencias Agrarias y Forestales, Universidad Católica del Maule, Talca 3466706, Chile
* Correspondence: rajibindia@gmail.com

Abstract: In the present study, *Chryseobacterium geocarposphaerae* DD3 isolated from textile industry dye effluent in West Bengal, India, displayed significant tolerance to sulfonated di-azo dye Congo red (CR), up to 500 ppm. The optimum decolourisation revealed that *C. geocarposphaerae* DD3 was capable of 96.52% decolourisation of 0.2 g L^{-1} CR within 12 h of treatment in the presence of 5 g L^{-1} glucose as supplementary carbon source. Biodegradation analysis of decolourised CR containing water was investigated by FTIR, MS and ^{1}H NMR, which confirmed the absence of azo bond as well as the toxic aromatic amines. Further, phytotoxicity analysis was performed to assess the toxicity of CR before and after bacterial treatment. Growth indexes of *Vigna radiata* L. seed confirmed that the biodegraded water was non-phytotoxic in comparison to the control CR solution. Multivariate analyses confirmed the same, showing significant differences between measured plant health indicators for CR solutions, whereas no significant differences were found between distilled and treated water. This study is novel as it is the first report of dye degradation by *C. geocarposphaerae* and may lead to a sustainable way of treating dye-contaminated water in the near future.

Keywords: wastewater; textile dye; Congo red; *Chryseobacterium*; water treatment; decolourisation; biodegradation; phytotoxicity

Citation: Sarkar, S.; Echeverría-Vega, A.; Banerjee, A.; Bandopadhyay, R. Decolourisation and Biodegradation of Textile Di-azo Dye Congo Red by *Chryseobacterium geocarposphaerae* DD3. *Sustainability* **2021**, *13*, 10850. https://doi.org/10.3390/su131910850

Academic Editor: Avelino Núñez-Delgado

Received: 19 August 2021
Accepted: 21 September 2021
Published: 29 September 2021

1. Introduction

In today's world, water pollution is an increasingly critical health issue in the context of rising industrialisation and climate change. Access to potable water is becoming rarer and more expensive worldwide. Among the plentiful causes of pollution, a major threat is the textile industry, as it often releases dye-containing effluents into natural water bodies [1]. From an economic viewpoint, although the textile industry plays an important role in any country, it is still an environmental polluter, causing more than 20% of global water pollution alone [2]. As stated by the World Health Organization (WHO), coloured effluents released after dying treatment by textile industries cause 17–20% of total water pollution worldwide. It has been found in a case study that 1.65 mg L^{-1} azo dye released into the Cristais River, São Paulo, Brazil caused a mutagenic effect in humans [3]. It has also been estimated that roughly 10–15% of the total in azo dyes used globally in textile production are lost through effluent, as they do not bind to textile fibre [4,5]. Approximately 200 billion litres of coloured wastewater are generated every year throughout the world [6]. Environmental legislation mandates that textile factories treat effluents before discharge into receiving water bodies [7]. However, synthetic textile azo dyes are unmanageable and carcinogenic [8] in nature due to the presence of −N=N− bonds [9], making them resistant to the impact of detergents, sunshine, temperatures, or other types of degradation [10]. Mainly, electron deficiency is the foremost reason why textile dyes become less susceptible to the conventional decolourisation process [11]. Nearly 2 years are required to degrade dye-derived aromatic amines in river and sea sediments [1,12]. Over 10,000 highly water soluble, commercial dye compounds are regularly used in the textile industry to dye fibres [13], causing general environmental concern.

Generally, the dye-containing effluents are intensively coloured and alkaline in pH (8.0–11.0) [14,15]. These types of effluents are also reported to be high in biochemical oxygen demand (BOD), chemical oxygen demand (COD), total suspended solids (TSS) [16], metal contamination [17] due to the presence of dye and additives (like sodium carbonate, caustic soda and other salts) [7] and temperature [18]. In this context of heavy water pollution, recycling wastewater produced by the textile industry is timely and important. Classical remediation approaches based on physical and chemical reactions are expensive and generate secondary pollutants containing amine residues, which are highly carcinogenic and show long-term effects on human as well as aquatic life. Further, chemical degradation processes are so expensive that industries often prefer not to bear that much of the cost [19]. Microbiological treatment is not that over-priced; moreover, it is an environment friendly, sustainable and efficient procedure [20]. Among different biological dye degradation processes [19], microbe-mediated azo dye degradation started back in the 1970 with the isolation of three bacterial strains, namely, *Bacillus subtilis*, *Aeromonas hydrophila* and *Bacillus cereus* [21,22]. Since then, many other bacterial strains have been reported in the field of dye bioremediation. In comparison to conventional processes, bacterial dye degradation is one of the best options for treatment of dye-containing textile effluent. While traditional processes are limited in their chemical and physical processes, environmental bacteria have the capability of rapidly adapting under adverse conditions [23], facilitating growth followed by degradation of dyes in a dye-contaminated environment. Enzymatic networks of bacteria are so well built that they are able to break down the complex chemical structures of synthetic azo dyes to use as essential nutrient sources (carbon and nitrogen sources) for their growth and development [19].

Congo red (CR) has been reported to persist in the environment for a long time—the reason why it is considered as one of the recalcitrant azo dyes. Furthermore, CR is also carcinogenic due to the presence of aromatic amine groups and is used in textile manufacturing. Hence, CR is considered as a model complex pollutant [24]. A few studies have reported on CR biodegradation using different bacteria, including *Bacillus* sp. ACT1 mediated degradation (0.05% CR) in ~30 h [25], *Pseudomonas luteola* mediated 95% degradation efficiency against 0.02% CR after 12 h [26] and *Shewanella xiamenensis* mediated 87.5% CR biodegradation in 16 h [27].

This study sought to investigate the biodegradation and detoxification of CR using a newly isolated thermotolerant bacterium *Chryseobacterium geocarposphaerae* DD3 that has not been reported earlier. The mechanism of CR biodegradation and the associated metabolites/by-products were systematically evaluated with FTIR, MS and ^{1}H NMR techniques. *C. geocarposphaerae* has been reported on earlier for its plant growth promoting activity [28], but this study elaborates a new approach to bioresource utilization in terms of azo dye bioremediation to find potential applications in textile wastewater remediation. To our knowledge, the present study is the first to report on *C. geocarposphaerae* mediated azo dye biodegradation and detoxification, and is novel in this respect. As this bacterium was isolated from contaminated textile effluent, it was hypothesized to confer elevated tolerance to synthetic azo dyes.

2. Materials and Methods

2.1. Wastewater Collection and Characteristics

Coloured effluent was collected in sterilized polythene bottle from the dye house outlet of a textile industry operation situated in West Bengal, India (latitude and longitude: 22.73718° N and 88.31861° E, respectively). QGIS 3.1 was used to prepare the map of the study area (Figure 1). Immediately after collection, some physical parameters of the wastewater, e.g., pH, temperature, colour, and smell, were recorded. Within 48 h of collection, the water sample was characterized for total suspended solids (TSS) and total dissolved solids (TDS) using a HORIBA multi-parameter meter (U-50), as well as chemical oxygen demand (COD), biological oxygen demand (BOD), and oil and grease following the standards of APHA (APHA 1999) [29].

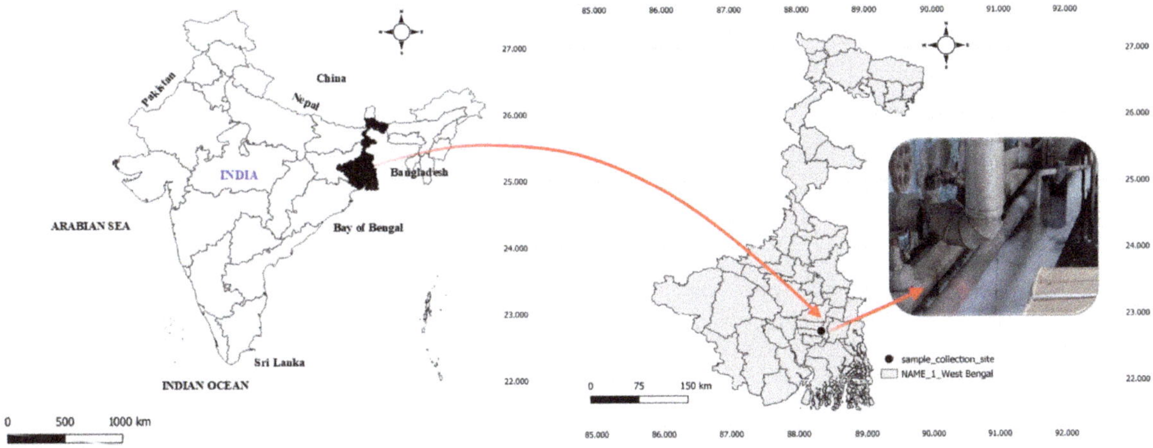

Figure 1. QGIS mapping of study site (Inset: sample collection site).

2.2. *Polyphasic Characterization of the Bacterial Isolate*

2.2.1. Bacteriological Analysis

Serially diluted collected coloured water sample was inoculated in modified nutrient agar (NA) medium (without yeast extract, pH 7.0) at 37 °C overnight. Fast growing colonies were isolated to examine their dye tolerance capacity in the modified NA medium supplemented with minimum dye concentration (0.1 g L^{-1} each of Congo red, Malachite green, Brilliant yellow, and Methyl red). Among different bacterial isolates, DD3 demonstrated growth against all the studied azo dyes (best against Congo red) other than Malachite green; thus, DD3 was chosen for further study. The isolate DD3 was primarily classified by Gram staining. Biochemical categorization of DD3 was performed on activity of extracellular enzymes (protease, amylase, catalase, and urease), substrate hydrolysis (cellulose, tributyrin, and olive oil), salt tolerance, pH, citrate and lysine utilization, motility, thermal death point, and antibiotic sensitivity tested using standard protocols [30].

For primary size and shape analysis of DD3, bacterial smear was fixed on a cover glass by 2.5% glutaraldehyde solution for 30 min and was dehydrated by passing it through 50–90% of alcohol for 5 min in each step. The bacteria on the cover glass were platinum-coated using JEOL JFC 1600 Auto fine coater (Tokyo, Japan) and observed under scanning electron microscope (JEOL JSM-6390, Tokyo, Japan) with 20 KV accelerating voltage. Further detailed morphometric analysis was performed by fixing freshly prepared overnight grown DD3 culture on carbon-coated grid followed by staining with 0.2% uranyl acetate. Then, the grid was observed under transmission electron microscope (JEM-1011 100 KV TEM, Peabody, MA, USA).

2.2.2. Molecular Characterization by 16S rRNA Amplification and Phylogenetic Analysis

Genomic DNA of DD3 was isolated using Zymo Research Fungal/Bacterial DNA MiniPrep (D6005). An approximately 1.5 kb size 16S rRNA fragment was amplified using high fidelity PCR polymerase. The PCR product was sequenced using the universal forward primer (8f) and reverse primer (1492r) with the Sanger sequencing method (Applied Biosystems 3130xl Genetic Analyzer, Foster City, CA, USA). EzTaxon platform was used for identification of the 16S rRNA sequence (http://www.ezbiocloud.net/identify, accessed on 23 May 2020). Evolutionary distances between the sequences were calculated and the phylogenetic tree was prepared by the maximum likelihood method using MEGA (version 7.0). The obtained sequence was deposited at GenBank (accession number MG905828.1).

2.3. Congo Red Decolourisation Study

Dye tolerance capacity of DD3 isolate was determined in terms of maximum tolerable concentration (MTC). In this context, mineral salts medium (MSM) was supplemented with 0.1, 0.2, 0.3, 0.4, 0.5, and 0.6 g L^{-1} of CR (BDH laboratory chemical, UAE) and incubated at 37 °C for up to 24 h. To analyse the effect of static and agitated conditions (120 rpm) on the decolourisation efficiency of DD3, the bacterial isolate was incubated up to 48 h in conical flask containing sterile mineral salt media (MSM) with maximum tolerable CR concentration (0.2 g L^{-1}). Similarly, the optimum temperature for DD3 mediated CR decolourisation was evaluated at 25, 30, 35, 37, 40, and 45 °C, respectively. Likewise, the effects of varying nutrient sources (1 g L^{-1}) (glucose and peptone as carbon source; yeast extract and beef extract as nitrogen source) were also investigated in 0.02% CR supplemented MSM. The effects of different concentrations of glucose (1–5 g L^{-1}) on DD3 mediated dye decolourisation were studied, too. For the decolourisation and biodegradation study, 0.2 mL freshly cultured inoculum (1.2×10^5 cfu/mL) was used for 100 mL media: i.e., inoculum: solution (*v/v*) = 1:500. For all the decolourisation experiments, bacteria cultured without the presence of dye and CR containing MSM incubated without bacterial inoculum were considered as biotic and abiotic controls, respectively. Cell-free supernatant was collected by centrifugation (10,000 rpm for 15 min), and absorbance of the supernatant was measured at predetermined absorbance maxima (λmax), 498 nm for CR using UV-visible spectrometer (Microprocessor Visible spectrophotometer, Model: LI-721). Percentage of decolourisation was calculated from the difference between initial (before incubation) and final (after growth) absorbance values. The best optimized condition for the decolourisation was considered for the further experiments.

To calculate decolourisation percentage the following equations was considered:

$$\text{\% of decolourisation}: \frac{Initial\ OD - Final\ OD}{Initial\ OD} \times 100 \quad (1)$$

The statistical analysis was performed using the SAS University Edition software. One-way non-parametric analysis of variance (Kruskal-Wallis test) was used to find the significance of the treatments in the final % of decolourisation of the solutions. The comparisons among treatment levels were made with Tukey's multiple comparison test. Chi-square statistic was used to test the hypothesis.

2.4. Analysis of Biodegradation Using Different Spectroscopic Methods

For Fourier transform infrared (FTIR) spectroscopic analysis, overnight-incubated DD3 culture was centrifuged (10,000 rpm for 15 min), and cell free supernatant was collected separately for the further experiments. FTIR was performed for the control (0.02% aqueous CR solution) and sample (metabolites/by-products produced after bacterial decolourisation). The scan was performed in the mid-IR region of 400–4000 cm^{-1} with two scan repeats using Perkin Elmer, UATR Two Series spectrophotometer. Peaks were analysed through OriginPro8 software. For sample extraction to perform proton nuclear magnetic resonance (^1H NMR) spectroscopy, centrifuged (15 min at 10,000 rpm at 4 °C) cell-free supernatant was vortexed thoroughly with an equal volume of ethyl acetate (*v*/*v*, 1:1). The organic layer was parted and further collected by separating funnel and air-dried at room temperature (35 °C). The collected 5–7 mg of metabolite powder was mixed with 0.5 mL deuterated solvent and analysed by ^1H NMR-300 MHZ (Bruker, Billerica, MA, USA). For mass spectrometric (MS) analysis, a millimolar concentration of the powdered by-product mixed with spectroscopic grade methanol was used.

2.5. Phytotoxicity Study

This experiment was performed by following the method described before by Rana et al. [31] with little modification. Seeds of *Vigna radiata* (L.) R. Wilkzek (mung bean) were obtained from Indian Agricultural Research Institute, PUSA, New Delhi. After washing, the healthy and uniform seeds were washed with distilled water thoroughly. Then, surface sterilisation of those seeds was performed using 0.1% $HgCl_2$ treatment, followed by rinsing 6 times with distilled water. For pre-germination treatment, seeds were soaked in 3 different conditions (3 experimental sets) for 6 h at 30 °C; set 1: Control (autoclaved, double distilled water), set 2: CR solution (0.02% aqueous CR) and set 3: DD3 treated wastewater (bacteria decolourised water). Ten seeds were placed on soaked filter paper in Petri plates. For germination purposes, the seeds were irrigated with equal volumes of autoclaved doubled distilled water. Initially, after 48 h, the number of germinated seedlings from each plate was counted and divided by the total number of seeds (10) to calculate the germination rate. At the end of the experiment (7 days), other seed growth parameters such as length of root and shoot and number, length, and width of leaves were recorded. For the assessment of phytotoxicity, the following equations were considered:

$$Relative\ germination = \frac{number\ of\ gemination\ in\ test}{number\ of\ gemination\ in\ control} \times 100 \quad (2)$$

$$\%\ of\ toxicity = \left[\frac{\{(radicle\ length\ of\ control - radicle\ length\ of\ test)\}}{radicle\ length\ of\ control}\right] \times 100 \quad (3)$$

$$Tolerance\ index = \frac{Mean\ length\ of\ longest\ root\ in\ tratment}{Mean\ length\ of\ longest\ root\ in\ control} \quad (4)$$

$$Vigour\ index = \{(mean\ root\ length + mean\ shoot\ length)\} \times \%\ of\ germination \quad (5)$$

This experiment was performed in triplicate for each experimental set. Permutational multivariate analysis of variance (PERMANOVA) test with 10,000 permutations was performed using Primer6 (PrimerE) software to determine differences between the treatments. Principal component analysis (PCA) was conducted on 3 kinds of treatments (distilled water, 0.2% CR, and bacteria mediated biodegraded CR). Varimax rotation was applied as orthogonal rotation minimizes the number of variables with high loading on each component to facilitate interpretation of the results. Statistical significance was defined as $p < 0.05$.

3. Results

3.1. Wastewater Collection and Characteristics

The waste effluent was dark greyish blue in colour with temperature of 55 °C and pH of 8.91. The high COD (4337.34 $mgO_2\ L^{-1}$), BOD (607 ppm), TSS (223 ppm), TDS (332 ppm) and oil-grease (6.32 mg L^{-1}) were recorded for the collected effluent sample (see Supplementary Materials, Table S1).

3.2. Polyphasic Characterisation and Identification of *Chryseobacterium geocarposphaerae* DD3

Based on the considerable initial tolerance of CR among the other studied dyes to DD3, this isolate was chosen for further CR decolourisation study. The DD3 isolate was Gram-negative, non-endospore forming, high-temperature tolerant (up to 95 °C), slightly halophilic (up to 3% salt/NaCl tolerant), alkaliphilic (growth until pH 12.0), aerobic, and motile in nature (see Supplementary Materials, Table S2). It was catalase-positive and was able to hydrolyse different substrates such as tributyrin and casein. Other than this, it also used lysine as an important substrate for growth. It was able to ferment citrate, D-fructose, and D-glucose, as carbon sources for growth. The bacterial isolate showed susceptibility against all the studied antibiotics (polymyxin-B, streptomycin, norfloxacin, penicillin-G, amoxicillin, chloramphenicol, ampicillin, cefotamine, ceftriaxone, mezlocillin) except the antifungal fluconazole.

16S rRNA sequencing analysis using an EZ-Taxon server revealed that isolated DD3 had 99.36% nucleotide identity with *Chryseobacterium geocarposphaerae* 91A-561. Other than this, six different *Chryseobacterium* species also exhibited more than 97% similarity with DD3 (Figure 2). Thus, based on the results of 16S rRNA analysis, DD3 was identified as *Chryseobacterium geocarposphaerae*. The electron micrographs confirmed that the bacteria *C. geocarposphaerae* DD3 are coccobacilli in shape (Figure 2). In transmission electron micrographs, the presence of outer membrane confirmed the Gram-negative nature of the bacterium.

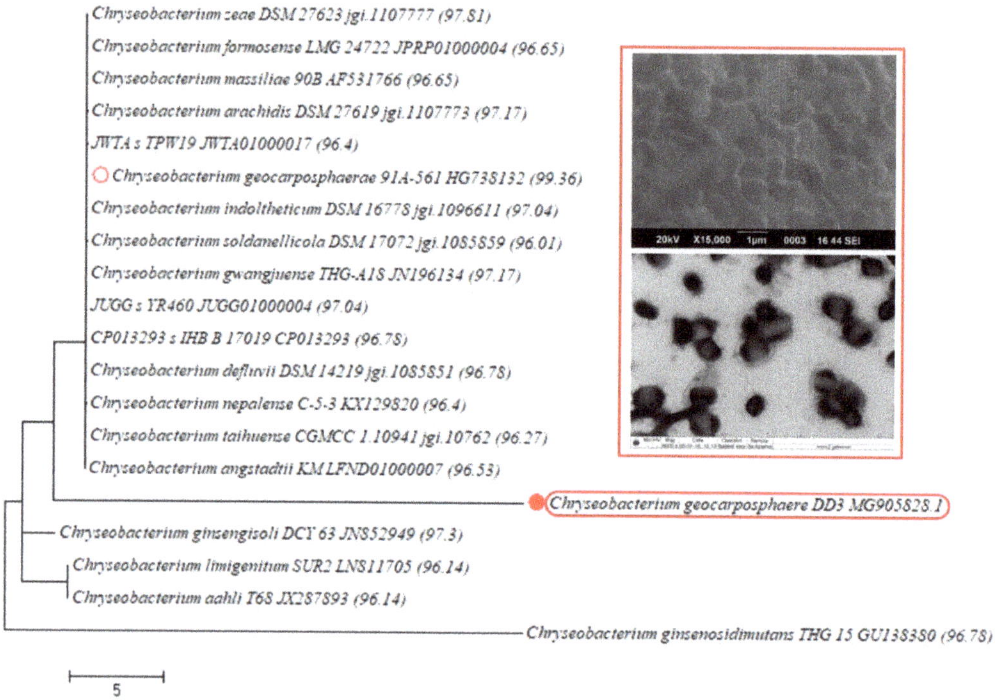

Figure 2. Phylogenetic tree of *Chryseobacterium geocarposphaerae* DD3 (Inset: scanning electron micrograph and transmission electron micrograph of *C. geocarposphaerae* DD3 representing its shape and structural characteristics).

3.3. Congo Red Decolourisation Study

Upon treating *C. geocarposphaerae* DD3 with increasing concentrations of CR, the bacteria showed MTC at the concentration of 0.5 g L^{-1}. Therefore, the bacterium showed tolerance up to 500 ppm CR concentration. The initial rate of dye decolourisation was more than 90% at 12 h of incubation, irrespective of various concentrations of CR used for the study (Figure 3A). *C. geocarposphaerae* DD3 mediated optimum decolourisation was 97.58% in the presence of 0.02% CR after 12 h. Beyond this concentration, % of decolourisation was found to be decreased with increasing initial dye concentration. Decolourisation by *C. geocarposphaerae* DD3 was studied under both static and agitated conditions (120 rpm) with 0.02% dye concentration. Maximal decolourisation was observed at 97% in the presence of agitation after 48 h of incubation. Therefore, agitation was adapted for further studies. Among a range of studied temperatures (30–45 °C), the highest percentage (96.5%) of decolourisation was found at 37 °C for *C. geocarposphaerae* DD3 mediated CR decolourisation (Figure 3B). Figure 3C shows that *C. geocarposphaerae* DD3 mediated CR decolourisation was influenced by all the studied nitrogen and carbon sources to an extent, whereas yeast extract (nitrogen source) and glucose (carbon source) elevated (70.09% for yeast extract and 96.34% for glucose) the decolourisation response. On varying the concentration of glucose as additional carbon source, the % of decolourisation increased with increasing glucose concentration, and the highest decolourisation (97.96%) was observed in the presence of 5 g L^{-1} glucose for CR supplemented (0.02%) media at 12 h. Statistical analysis showed significance for both the factors medium used for reaction and reaction time with a *p*-value 0.0041 and 0.0417, respectively (see Supplementary Materials, Figure S1). Hence, based on the paired tests between the conditions, the statistically significant experimental condition was further used for the biodegradation analysis and toxicity assessment study.

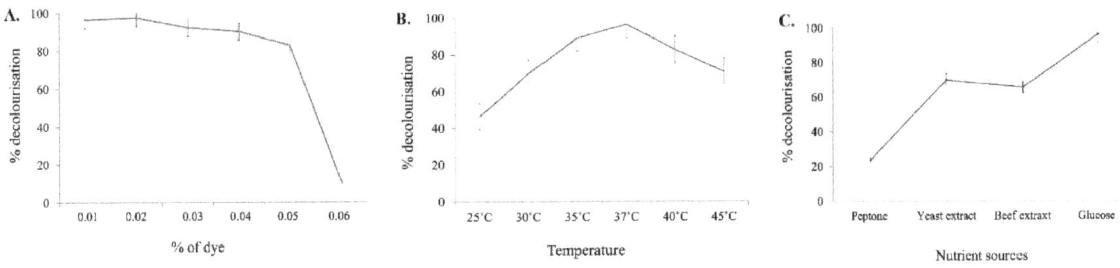

Figure 3. Study of CR decolourisation by *C. geocarposphaerae* DD3: (**A**) Effect of different percentages of CR dye; (**B**) effect of temperature (°C); (**C**) effect of different nutrient sources (nitrogen source and carbon source) on % of decolourisation.

3.4. Analysis of Biodegradation Using Different Spectroscopic Methods

Incubation with *C. geocarposphaerae* DD3 successfully resulted in decolourisation of CR. The bio-transformed metabolites related to the decolourisation were characterised by FTIR, MS, and ^{1}H NMR. The result of FTIR analysis of the CR sample and decolourised water showed differences in various peaks, indicating the change/presence of various functional groups. The result of FTIR analysis of the control CR sample showed peaks in the regions of 645 cm^{-1} (C-C bending vibration of aromatic rings), 1446 cm^{-1} (aromatic C=C stretching vibration), 1520.63 cm^{-1} (C=C stretching vibration in benzene ring), 1584 cm^{-1} (–N=N– azo bond stretching vibration), 1740 cm^{-1} (-C=O stretching vibration), and 1450–1200 cm^{-1} (phenolic C–O stretch vibration). In the FTIR spectrum of metabolites obtained after decolourisation of CR by *C. geocarposphaerae* DD3, the disappearance of the peak at 1584 cm^{-1} for –N=N- stretching vibration hinted at the reductive cleavage of azo bond of CR; the disappearance of functional groups at 1450–1200 cm^{-1} for phenol C–O stretching vibration indicated biodegradation of CR; and absence of vibrated stretch within 645–831cm^{-1} indicated absence of carcinogenic aromatic amine in the treated water sample (Figure 4A). The MS analysis of the studied sample indicated similar results (Figure 4B). By analysing the m/z spectra, a few intermediates, namely, sodium 4-amino-3-(phenyldiazenyl) naphthalene-2-sulfonate (m/z 324.2721), sodium 3,4-diaminonapthalene-2-sulfonate (m/z 260), and aniline (m/z 99.08) followed by toluene (m/z 92.00) were identified. ^{1}H NMR peaks of the treated water sample appeared at δ-1.5 ppm (-R_3CH), indicating conversion of the aromatic component to an aliphatic one [32]. The absence of ^{1}H NMR peaks at δ-7.5 (Ar-H) (Figure 5) in our study clearly indicated the absence of the carcinogenic aromatic amine in bacteria-treated aqueous CR solution [33]. Peaks in δ-6.6-8.0 are reported to be attributable to the presence of aromatic protons [32]. The absence of this stretch in this study indicated the degradation of CR without forming any carcinogenic amine as by-product (Figure 5). The peak in δ-2.3 ppm indicated the presence of the relatively non-toxic end-product toluene in the treated water.

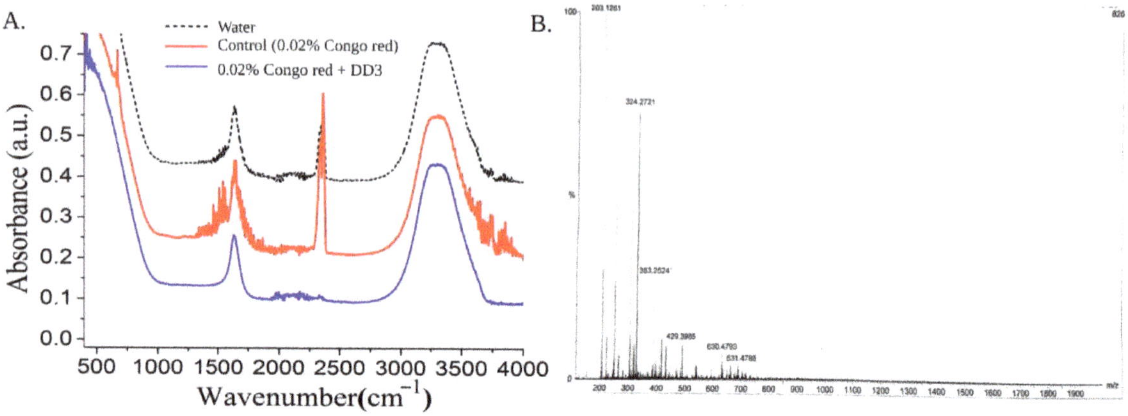

Figure 4. *C. geocarposphaerae* DD3 mediated biodegradation analysis: (**A**) FTIR of untreated and treated dye containing water sample represents the breakdown of chemical structure of CR; (**B**) mass spectrometry of bacteria treated water sample to identify the m/z ratios of intermediates produced after the treatment.

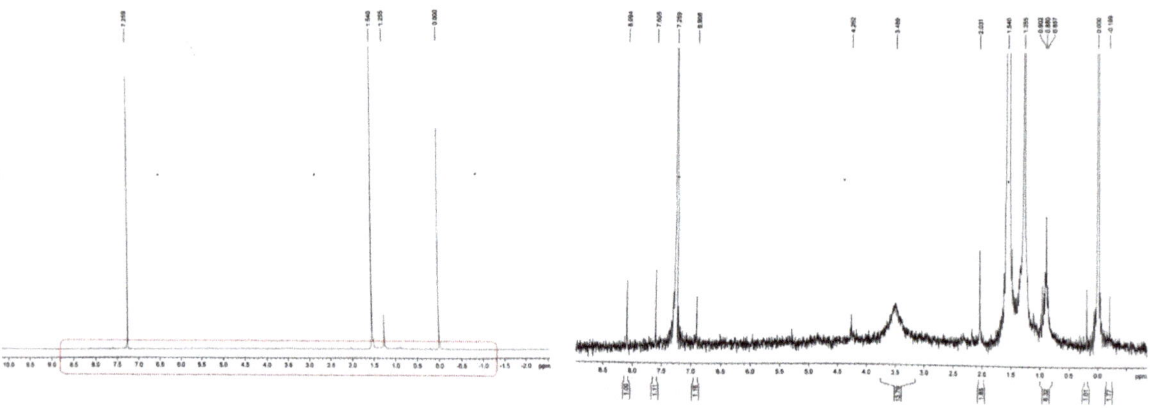

Figure 5. ^{1}H NMR spectroscopy of *C. geocarposphaerae* DD3 treated water sample.

By analysing ^{1}H NMR spectra and m/z values from MS spectrometry, probable intermediates formed after CR biodegradation were identified, and the biodegradation pathway was predicted by means of chemistry (Figure 6).

3.5. Phytotoxicity Study

The phytotoxicity of CR before and after *C. geocarposphaerae* DD3 mediated biodegradation was studied. Figure 7A,B shows germinated seedlings pre-treated with DD3-treated water and CR solution, respectively. Figure 7C shows a comparison of seedling growth (in CR solution, *C. geocarposphaerae* DD3 treated water, and control, i.e., distilled water).

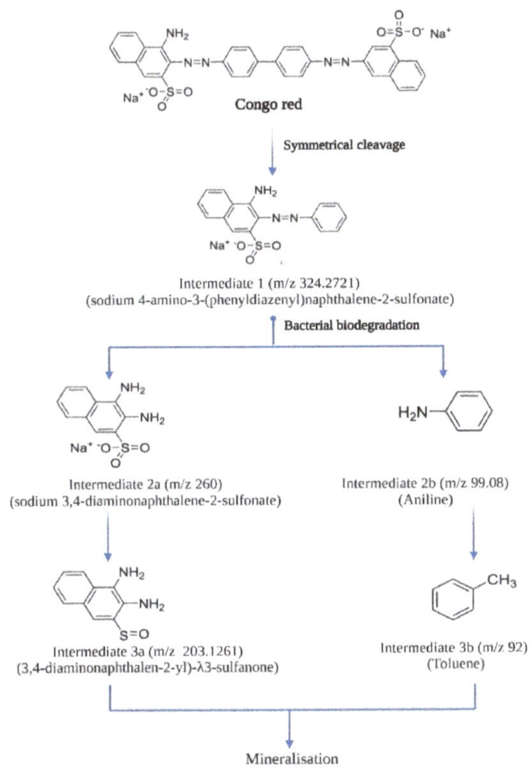

Figure 6. Prediction of CR biodegradation pathway during treatment with *C. geocarposphaerae* DD3.

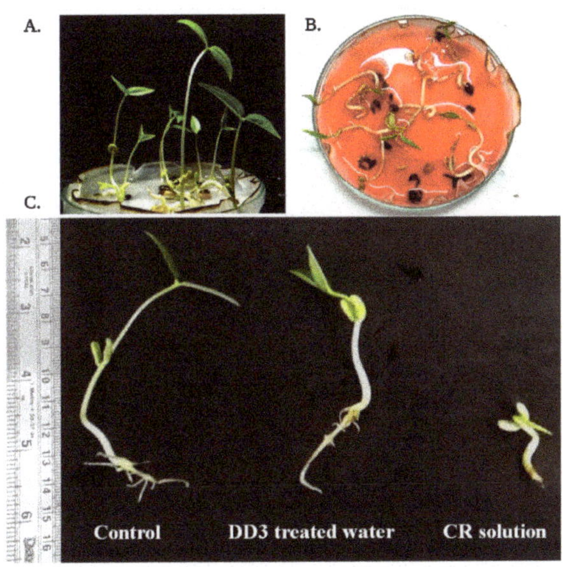

Figure 7. Phytotoxicity assessment of CR on *Vigna radiata* (L.) R. Wilkzek (mung bean) by germination of seeds pre-treated with (**A**) DD3 treated water (bacteria decolourised water) and (**B**) CR solution (0.02%CR aqueous CR); (**C**) Comparison of growth of seedlings (control, DD3 treated water, and CR solution).

Out of all the parameters measured in relation to the phytotoxicity assessment, three parameters, namely, the germination rate, % toxicity, and vigour index, showed significant differences in the treatment with CR solution compared to DD3-treated water and control (Table 1). The detailed plant growth indexes measured to assess the phytotoxicity of CR are listed in Table 1.

Table 1. Phytotoxicity assessment of CR solution and *C. geocarposphaerae* treated water sample in comparison to the control.

Characteristics	Control	DD3 Treated Water	CR Solution
Plumule length	3 cm	1.9 cm	-
Radicle length	5.6 cm	5.4 cm	-
Root length	2.4 cm	3.4 cm	1.1 cm
Number of roots	9	11	2
Shoot length	6.0 cm	4.4 cm	1.3 cm
Number of leaves	2	2	-
Length of leaf	1.9 cm	1.5 cm	-
Width of leaf	0.8 cm	0.8 cm	-
Germination rate	90%	80%	50%
Relative germination	100%	88.89%	55.56%
% toxicity	0%	3.57%	100%
Tolerance index	-	1.41	0.45
Vigour index	756	624	120

"-" signifies no growth, control denotes autoclaved double dist. water; DD3 treated water denotes bacteria decolourised water and CR solution denotes 0.02% aqueous CR solution.

From individual variable analysis, based on correlations, four groups of variables (group 1: tolerance index and root length; group 2: % toxicity, number of leaves, radicle length, and width of leaves; group 3: vigour index and shoot length; group 4: relative germination and germination rate) showed strong correlations (98%) with each other (Figure 8A). The selection was made based on correlation analysis. Multivariate analysis with PERMANOVA test was performed with the variables germination rate, plumule length, root length, number of roots, shoot length, and number and length of leaves. Significant differences were found between the different treatments ($p < 0.0033$). A principal components analysis (PCA) was performed (Figure 8B), showing 96% of total data variance explained by two principal components (PC) (PC1 was 90.9% and PC2 was 6.5%). Figure 8B shows loading of PC1 versus PC2. The relationship of CR contamination to seed germination was clearly observed as discussed before by You et al. (2019) [34]. Highest positive loading of PC1 was on 0.02% CR solution, indicating their variations may follow similar trends along with negative loading of number and length of root, which is a key index for CR contamination on seed germination. All plant germination indexes were focused on second and third quadrants, reflecting contamination with different sensitivities. The samples corresponding to the control (distilled water) and treated water showed non-significant difference, i.e., samples treated with CR showed nearly the same growth indexes as the control.

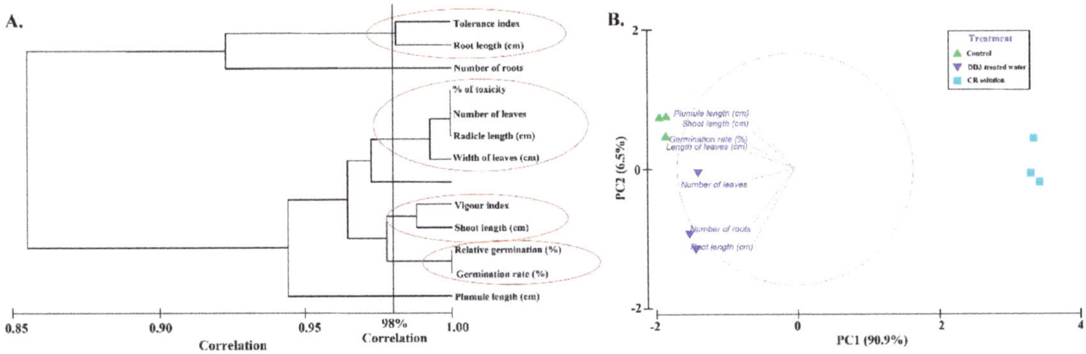

Figure 8. Multivariate analysis. (**A**) Variable Pearson correlations where red circles indicate groups of over 98% correlation, with significant differences between the different treatments ($p < 0.0033$); (**B**) Principal component analysis results, loading plot PC1 vs. PC2 showing 96% total data variance.

4. Discussion

The dye CR seemed to be toxic to the bacterium *C. geocarposphaerae* DD3 above a certain concentration, as no growth was recorded after the maximum tolerable concentration. A similar study of di-azo dye (Navitan Fast Blue) was performed previously by Nachiyar and Rajkumar [35] for *Pseudomonas aeruginosa*, which could degrade (90%) 100 mg L^{-1} Navitan Fast Blue S54, in 24 h with an MTC of 1200 mg L^{-1}; however, concentrations above this level caused acute toxicity. Better decolourisation was observed under controlled agitation, as agitation probably dominates bacterial respiration [10] and helps to utilise NADH as a cofactor to decolourise the azo dyes [36]. A similar experiment performed with *P. aeruginosa* showed better dye degradation (0.01% dye concentration) under continuous agitation (90% degradation in 48 h) than static incubation (80% in 48 h), thus supporting our findings [35]. The optimum decolourisation was found at 37 °C, but DD3 also demonstrated decolourisation capacity at 45 °C, which demonstrates the thermotolerant nature of the bacterium. A similar observation was previously reported by Kolekar et al. [36] on the effect of temperature (20–45 °C) for *Bacillus fusiformis* mediated disperse blue 79 and acid orange 10 degradation. Another study reported the effect of temperature (10–60 °C, in 10 °C increments) on CR decolourisation (94.52% of 0.01% CR at 32 ± 2 °C) [37]. According to several earlier reports [38–40], the addition of carbon and nitrogen sources increases decolourisation efficiency. As observed from this study, yeast extract (nitrogen source) and glucose (carbon source) positively influenced *C. geocarposphaerae* DD3 mediated CR decolourisation. Yeast extract as nitrogen source is probably essential for the regeneration of NADH, an electron donor that helps the microorganism to reduce the azo bond in the chemical structure of azo dye [40]. On the other hand, at the initial stage of growth, the presence of a ready carbon source (glucose) helps to produce secondary metabolites, extracellular enzymes, which in turn may influence the azo dye decolourisation [38]. According to a report by Singh et al. [41] on the effect of external nitrogen and carbon sources, *Staphylococcus hominis* RMLRT03 showed elevated decolourisation (89.81% and 93.24%, respectively) in the presence of glucose and yeast extract. On the contrary, in our study, *C. geocarposphaerae* DD3 showed better decolourisation only in the presence of glucose as a carbon source. It was clearly observed that DD3 can decolourise 0.02% CR with more proficiency (96.52% in 12 h for the media supplemented with 5 g L^{-1} glucose at 37 °C) and an adsorption efficiency of 66.63 mg/g bacterial cell. Sari et al. [42] previously described the adsorption efficiency determination. Overall, *C. geocarposphaerae* DD3 had better decolourisation efficiency than previously reported in CR decolourisation studies for *Shewanella xiamenensis* BC01 (87.5% decolourisation of 0.02% dye in 16 h) [27], *Bacillus* sp. (0.01% dye after 50 h) [25,27], or *Bacillus cohnni* RKS9 (99% degradation in 12 h for 0.01% dye) [43]. Mutant *Bacillus* sp. ACT1 (using UV and Ethidium bromide) showed 0.05%

degradation ability after a long time interval of ~30 h [25], whereas our isolate of interest took much less time, only 12 h without any strain mutation.

Bacteria isolated from dye-contaminated sites can process the bioremediation of azo dye, which is a clear indication of environmental adaptation to toxic synthetic dye [36]. In the present study, CR biodegradation was confirmed by FTIR, MS (Figure 4), and ^{1}H NMR analysis (Figure 5). Previously, the CR biodegradation pathway was studied mostly using mixed bacterial cultures [32] or fungal cultures [44], whereas comparatively little study had been pursued on monoculture-based CR degradation [45]. In this study, mainly ^{1}H NMR spectra and m/z values of intermediates (received from MS study in Figure 4B) were analysed to hypothesise a CR degradation pathway supported by FTIR. In our study, according to the FTIR data, the disappearance of functional groups at 1584 cm^{-1} for –N=N– stretching vibrations and at 1450–1200 cm^{-1} for phenolic C–O stretching vibrations in the bacteria-treated water sample indicated biodegradation of CR by reductive cleavage of azo bonds (Figure 4A), as previously studied [8,10]. Ng et al. [27] reported a similar study on CR degradation by *S. xiamenensis* BC01, but the absence of aromatic stretch was not demonstrated clearly in the treated sample. However, in this study, the absence of 645–831cm^{-1} stretching vibration for aromatic amine in the treated water sample proved the degradation of CR without forming harmful secondary pollutants. Biodegradable intermediates were analysed from MS and ^{1}HNMR. Further comparing this analytical result in a chemically integrative way allowed us to propose a CR biodegradation pathway (Figure 6). Possibly the first intermediate sodium 4-amino-3-(phenyldiazenyl) naphthalene-2-sulfonate was formed from symmetrical cleavage on the chemical structure of CR, making the molecule accessible to the active site of the enzymes for further degradation reaction. Then, sodium 3,4-diaminonapthalene-2-sulfonate was formed by reductive cleavage of azo bond. The significant disappearance of azo peaks from the FTIR spectrum of the decolourised sample also supported this reaction step. As no significant ^{1}H NMR peak of product with larger molecular weight was been found in this study, it may be hypothesised that the intermediates of the biodegraded CR are probably mineralised immediately after formation and utilised by the bacterium though the TCA cycle in their central metabolism as carbon or nitrogen sources. This could be supported by the previous work of D'Souza et al. [45] on CR biodegradation by *Alcaligenes* sp., where similar intermediates such as the intermediates 2a and 3a from our study were also formed. It was previously found by Lade et al. [10] that azo reductase is responsible for reductive cleavage of the azo bond. The intermediate is presumed to be further cleaved to form low molecular weight aniline followed by toluene (Figure 6). From a similar study previously performed by Balapure et al. [32] on reactive black 1 using a mixed bacterial consortium of *Lysinibacillus, Raoultella, Enterococcus*, and *Citrobacter* species, it can be said that the degradation of CR probably goes through a complete mineralization mechanism. Perhaps bacteria growing in wastewater containing azo dye use complex dyes as their carbon or nitrogen sources [46] by breaking them down with the help of a complex oxido-reductive enzymatic system. In this study, probably at the time of acclimatization in the CR-containing medium, that enzymatic system helped the bacterium to degrade the dye. The azo bond of CR might be reductively cleaved to form enough carbon and nitrogen compounds for the bacterium to consume as nutrient sources.

Untreated textile wastewater presents hazards when discharged into environmental sinks (water bodies) utilised for irrigation purpose [47], making it essential to discuss the risks associated with the treated water with high ecological consequences. In a previous report, Ponceau 4R showed acute toxicity to *Triticum aestivum* L. (69% germination rate on treatment with decolourised sample) and *Sorghum vulgare* L. [47]. Shoot growth was earlier reported to be inhibited in Methyl orange contamination [48]. In comparison, our results show that plants grown in CR aqueous solution probably faced a direct impact from azo dye molecules in terms of germination inhibition, seedling health, and affected shoot-root elongation, whereas toxicity of the dye was substantially reduced after bacterial biodegradation (Table 1). In a previous study by Telke et al. [49], phytotoxicity assessment revealed that textile effluent had a detrimental effect on seed germination of *Vigna radiata* (20%) compared to the treated sample (60%) and distilled water (70%). Therefore, similar observations in our study confirmed that biodegraded products of CR (bacteria-degraded water) were non-toxic and showed no negative influence on seed germination or seedling growth. In a recent study by Kishor et al. [43], similar effects were observed on seed germination and seedling growth parameters of *Phaseolus mungo* L., showing significant reduction in the toxicity of treated textile wastewater. The bacteria-degraded intermediates might be used as a source of water or plant growth nutrients for vegetation in the near future. Findings of the statistical study, i.e., principal component analysis (Figure 8) on plant germination indexes, clearly demonstrated a distant relationship between the effects of the CR solution and those of the control and bacteria-degraded water samples. This clearly indicates that *C. geocarposphaerae* DD3 is not only able to decolourise CR but also able to completely detoxify it, resulting in almost zero or insignificant phytotoxicity.

5. Conclusions

Textile dye effluent origin thermotolerant bacterium *C. geocarposphaerae* DD3 was studied for its ability to decolourise CR, followed by degradation and detoxification of CR by the bacterium. With an MTC of 0.5 g L^{-1} CR, *C. geocarposphaerae* DD3 showed the highest decolourisation efficiency of 96.52% in the presence of glucose as co-substrate within only 12 h of treatment. Based upon biodegradation analysis of CR, the process seemed to involve a complete biodegradation without forming aromatic amines as secondary pollutants. Intermediates formed after biodegradation of CR exhibited no phytotoxicity toward the germination and growth of mung bean. This biodegradation and detoxification of azo dye by *C. geocarposphaerae* DD3 is a novel approach that could lead to a sustainable bioremediation method for dye-containing wastewater treatment in the near future. In conclusion, dye effluent origin, thermotolerant *C. geocarposphaerae* DD3 degraded and detoxified recalcitrant CR in a cost-effective manner, and the recycled water could be used in the future for agricultural/irrigation purposes. This study focused on efficient bioresource utilization and future opportunities for the circular bioeconomy.

Supplementary Materials: The following are available online at https://www.mdpi.com/article/10.3390/su131910850/s1, Figure S1: Box plot representing the decolourisation percentage by statistically significant variables for DD3 mediated Congo red decolourisation. Table S1: Physico-chemical parameters of the textile effluent from collection site. Table S2: Physio-chemical characterization of bacterial isolates.

Author Contributions: Conceptualization, R.B.; formal analysis, S.S. and A.E.-V.; writing—original draft preparation, S.S.; writing—review and editing, R.B., A.B. and A.E.-V.; supervision, R.B. All authors have read and agreed to the published version of the manuscript.

Funding: This research received no external funding.

Institutional Review Board Statement: Not applicable.

Informed Consent Statement: Not applicable.

Data Availability Statement: The data presented in this study are available on request from the corresponding author.

Acknowledgments: Authors are thankful to UGC-Center of Advanced Study, Department of Botany, The University of Burdwan for pursuing research activities. S.S. and R.B. are thankful to DST-FIST (No. SR/FST/LS-1/2018/188(C) for infrastructural facilities. Authors are also thankful to Rajarshi Ghosh, Assistant Professor, Department of Chemistry, The University of Burdwan for his kind help in the analysis of data. S.S. is thankful to Swami Vivekananda Merit-Cum-Means Non Net fellowship (WBP191579671079) for pursuing PhD. All authors are also thankful to Biorender.com.

Conflicts of Interest: The authors declare no conflict of interest.

References

1. Sarkar, S.; Ponce, N.T.; Banerjee, A.; Bandopadhyay, R.; Rajendran, S.; Lichtfouse, E. Green polymeric nanomaterials for the photocatalytic degradation of dyes: A review. *Environ. Chem. Lett.* **2020**, *18*, 1569–1580. [CrossRef]
2. Agha, Y.Y.; Bahjat, S.A.; Thanoon, M.F. Assessment of Bacterial Pigments as Textile Colorants. *Indian J. Public Health Res. Dev.* **2019**, *10*, 1565–1569. [CrossRef]
3. Carneiro, P.A.; Umbuzeiro, G.A.; Oliveira, D.P.; Zanoni, M.V.B. Assessment of water contamination caused by a mutagenic textile effluent/dyehouse effluent bearing disperse dyes. *J. Hazard. Mater.* **2010**, *174*, 694–699. [CrossRef] [PubMed]
4. Baban, A.; Yediler, A.; Lienert, D.; Kemerdere, N.; Kettrup, A. Ozonation of high strength segregated effluents from a woollen textile dyeing and finishing plant. *Dye. Pigment.* **2003**, *58*, 93–98. [CrossRef]
5. Sudha, M.; Saranya, A.; Selvakumar, G.; Sivakumar, N. Microbial degradation of azo dyes: A review. *Int. J. Curr. Microbiol. Appl. Sci.* **2014**, *3*, 670–690.
6. Tkaczyk, A.; Mitrowska, K.; Posyniak, A. Synthetic organic dyes as contaminants of the aquatic environment and their implications for ecosystems: A review. *Sci. Total Environ.* **2020**, *717*, 137222. [CrossRef] [PubMed]
7. Yaseen, D.A.; Scholz, M. Textile dye wastewater characteristics and constituents of synthetic effluents: A critical review. *Int. J. Environ. Sci. Technol.* **2019**, *16*, 1193–1226. [CrossRef]
8. Mishra, S.; Nayak, J.K.; Maiti, A. Bacteria-mediated bio-degradation of reactive azo dyes coupled with bio-energy generation from model wastewater. *Clean Technol. Environ. Policy* **2020**, *22*, 651–667. [CrossRef]
9. Singh, R.L.; Singh, P.K.; Singh, R.P. Enzymatic decolorization and degradation of azo dyes—A review. *Int. Biodeterior. Biodegrad.* **2015**, *104*, 21–31. [CrossRef]
10. Lade, H.; Govindwar, S.; Paul, D. Low-cost biodegradation and detoxification of textile azo dye CI reactive blue 172 by *Providencia rettgeri* strain HSL1. *J. Chem.* **2015**, *2015*, 894109. [CrossRef]
11. Solís, M.; Solís, A.; Pérez, H.I.; Manjarrez, N.; Flores, M. Microbial decolouration of azo dyes: A review. *Process Biochem.* **2012**, *47*, 1723–1748. [CrossRef]
12. Ito, T.; Adachi, Y.; Yamanashi, Y.; Shimada, Y. Long–term natural remediation process in textile dye–polluted river sediment driven by bacterial community changes. *Water Res.* **2016**, *100*, 458–465. [CrossRef]
13. Yu, J.; Wang, X.; Yue, P.L. Optimal decolorization and kinetic modeling of synthetic dyes by *Pseudomonas* strains. *Water Res.* **2001**, *35*, 3579–3586. [CrossRef]
14. Paul, S.A.; Chavan, S.K.; Khambe, S.D. Studies on characterization of textile industrial waste water in Solapur city. *Int. J. Chem. Sci.* **2012**, *10*, 635–642.
15. Guadie, A.; Tizazu, S.; Melese, M.; Guo, W.; Ngo, H.H.; Xia, S. Biodecolorization of textile azo dye using *Bacillus* sp. strain CH12 isolated from alkaline lake. *Biotechnol. Rep.* **2017**, *15*, 92–100. [CrossRef] [PubMed]
16. Yaseen, D.A.; Scholz, M. Shallow pond systems planted with *Lemna minor* treating azo dyes. *Ecol. Eng.* **2016**, *94*, 295–305. [CrossRef]
17. Sekomo, C.B.; Rousseau, D.P.; Saleh, S.A.; Lens, P.N. Heavy metal removal in duckweed and algae ponds as a polishing step for textile wastewater treatment. *Ecol. Eng.* **2012**, *44*, 102–110. [CrossRef]
18. Shah, A.H.; Manikandan, E.; Ahmed, M.B.; Ganesan, V. Enhanced bioactivity of Ag/ZnO nanorods-A comparative antibacterial study. *J. Nanomed. Nanotechol.* **2013**, *4*, 3. [CrossRef]
19. Sarkar, S.; Banerjee, A.; Halder, U.; Biswas, R.; Bandopadhyay, R. Degradation of synthetic azo dyes of textile industry: A sustainable approach using microbial enzymes. *Water Conserv. Sci. Eng.* **2017**, *2*, 121–131. [CrossRef]
20. Zerva, I.; Remmas, N.; Kagalou, I.; Melidis, P.; Ariantsi, M.; Sylaios, G.; Ntougias, S. Effect of Chlorination on Microbiological Quality of Effluent of a Full-Scale Wastewater Treatment Plant. *Life* **2021**, *11*, 68. [CrossRef]
21. Wuhrmann, K.; Mechsner, K.L.; Kappeler, T.H. Investigation on rate—Determining factors in the microbial reduction of azo dyes. *Appl. Microbiol. Biotechnol.* **1980**, *9*, 325–338. [CrossRef]
22. Dave, S.R.; Patel, T.L.; Tipre, D.R. Bacterial Degradation of Azo Dye Containing Wastes. In *Microbial Degradation of Synthetic Dyes in Wastewaters*; Singh, S., Ed.; Springer: Cham, Switzerland, 2015; pp. 57–83. [CrossRef]
23. Manjarrez Paba, G.; Baldiris Ávila, R.; Baena Baldiris, D. Application of environmental bacteria as potential methods of azo dye degradation systems. *Glob. J. Environ. Sci. Manag.* **2021**, *7*, 131–154. [CrossRef]
24. Gharbani, P.; Tabatabaii, S.M.; Mehrizad, A. Removal of Congo red from textile wastewater by ozonation. *Int. J. Environ. Sci. Technol.* **2008**, *5*, 495–500. [CrossRef]
25. Gopinath, K.P.; Murugesan, S.; Abraham, J.; Muthukumar, K. *Bacillus* sp. mutant for improved biodegradation of Congo red: Random mutagenesis approach. *Bioresour. Technol.* **2009**, *100*, 6295–6300. [CrossRef]
26. Hsueh, C.C.; Chen, B.Y. Comparative study on reaction selectivity of azo dye decolorization by *Pseudomonas luteola*. *J. Hazard. Mater.* **2007**, *141*, 842–849. [CrossRef] [PubMed]
27. Ng, I.S.; Chen, T.; Lin, R.; Zhang, X.; Ni, C.; Sun, D. Decolorization of textile azo dye and Congo red by an isolated strain of the dissimilatory manganese-reducing bacterium *Shewanella xiamenensis* BC01. *Appl. Microbiol. Biotechnol.* **2014**, *98*, 2297–2308. [CrossRef] [PubMed]
28. Nishioka, T.; Elsharkawy, M.M.; Suga, H.; Kageyama, K.; Hyakumachi, M.; Shimizu, M. Development of culture medium for the isolation of *Flavobacterium* and *Chryseobacterium* from rhizosphere soil. *Microbes Environ.* **2016**, *31*, ME15144. [CrossRef] [PubMed]

29. American Public Health Association. *Standard Methods Barrfor the Examination of Water and Waste Water*, 20th ed.; American Public Health Association: Washington, DC, USA, 1998.
30. Pelczar, M.J.; Bard, R.C.; Burnett, G.W.; Conn, H.J.; Demoss, R.D.; Euans, E.E.; Weiss, F.A.; Jennison, M.W.; Meckee, A.P.; Riker, A.J.; et al. Manual of Microbiological Methods. In *Society of American Bacteriology*; McGraw Hill: New York, NY, USA, 1957.
31. Rana, S.; Kumar, K. Study of Phytotoxic effect of textile wastewater on seed germination and seedling growth of *Triticum aestivum*. *Int. J. Biosci. Technol.* **2017**, *10*, 58–66.
32. Balapure, K.; Aghera, P.; Bhatt, N.; Madamwar, D. Community synergism: Degradation of triazine dye reactive black 1 by mixed bacterial cultures KND_PR under microaerophilic and aerobic conditions. *Environ. Process.* **2019**, *6*, 713–739. [CrossRef]
33. Mishra, S.; Mohanty, P.; Maiti, A. Bacterial mediated bio-decolourization of wastewater containing mixed reactive dyes using jack-fruit seed as co-substrate: Process optimization. *J. Clean. Prod.* **2019**, *235*, 21–33. [CrossRef]
34. You, R.; Domínguez, C.; Matamoros, V.; Bayona, J.M.; Díez, S. Chemical characterization and phytotoxicity assessment of peri-urban soils using seed germination and root elongation tests. *Environ. Sci. Pollut. R.* **2019**, *26*, 34401–34411. [CrossRef]
35. Nachiyar, C.V.; Rajkumar, G.S. Degradation of a tannery and textile dye, Navitan Fast Blue S5R by *Pseudomonas aeruginosa*. *World J. Microbiol. Biotechnol.* **2003**, *19*, 609–614. [CrossRef]
36. Kolekar, Y.M.; Pawar, S.P.; Gawai, K.R.; Lokhande, P.D.; Shouche, Y.S.; Kodam, K.M. Decolorization and degradation of Disperse Blue 79 and Acid Orange 10, by *Bacillus fusiformis* KMK5 isolated from the textile dye contaminated soil. *Bioresour. Technol.* **2008**, *99*, 8999–9003. [CrossRef] [PubMed]
37. Hamad, M.T.; Saied, M.S. Kinetic studies of Congo red dye adsorption by immobilized *Aspergillus niger* on alginate. *Appl. Water Sci.* **2021**, *11*, 35. [CrossRef]
38. Kapdan, I.K.; Kargia, F.; McMullan, G.; Marchant, R. Effect of environmental conditions on biological decolorization of textile dyestuff by *C. versicolor*. *Enzyme Microb. Technol.* **2000**, *26*, 381–387. [CrossRef]
39. Chen, K.C.; Wu, J.Y.; Liou, D.J.; Hwang, S.C.J. Decolorization of the textile dyes by newly isolated bacterial strains. *J. Biotechnol.* **2003**, *101*, 57–68. [CrossRef]
40. Asad, S.; Amoozegar, M.A.; Pourbabaee, A.; Sarbolouki, M.N.; Dastgheib, S.M.M. Decolorization of textile azo dyes by newly isolated halophilic and halotolerant bacteria. *Bioresour. Technol.* **2007**, *98*, 2082–2088. [CrossRef] [PubMed]
41. Singh, R.P.; Singh, P.K.; Singh, R.L. Bacterial Decolorization of Textile Azo Dye Acid Orange by *Staphylococcus hominis* RMLRT03. *Toxicol. Int.* **2014**, *21*, 160–166. [CrossRef]
42. Sari, I.P.; Simarani, K. Decolorization of selected azo dye by *Lysinibacillus fusiformis* W1B6: Biodegradation optimization, isotherm, and kinetic study biosorption mechanism. *Adsorpt. Sci. Technol.* **2019**, *37*, 492–508. [CrossRef]
43. Kishor, R.; Purchase, D.; Saratale, G.D.; Ferreira, L.F.R.; Bilal, M.; Iqbal, H.M.; Bharagava, R.N. Environment friendly degradation and detoxification of Congo red dye and textile industry wastewater by a newly isolated *Bacillus cohnni* (RKS9). *Environ. Technol. Innov.* **2021**, *22*, 101425. [CrossRef]
44. Asses, N.; Ayed, L.; Hkiri, N.; Hamdi, M. Congo red decolorization and detoxification by *Aspergillus niger*: Removal mechanisms and dye degradation pathway. *BioMed Res. Int.* **2018**, *2018*, 3049686. [CrossRef]
45. D'Souza, E.; Fulke, A.B.; Mulani, N.; Ram, A.; Asodekar, M.; Narkhede, N.; Gajbhiye, S.N. Decolorization of Congo red mediated by marine *Alcaligenes* species isolated from Indian West coast sediments. *Environ. Earth Sci.* **2017**, *76*, 721. [CrossRef]
46. Barragán, B.E.; Costa, C.; Marquez, M.C. Biodegradation of azo dyes by bacteria inoculated on solid media. *Dye. Pigment.* **2007**, *75*, 73–81. [CrossRef]
47. Masarbo, R.S.; Niranjana, S.R.; Monisha, T.R.; Nayak, A.S.; Karegoudar, T.B. Efficient decolorization and detoxification of sulphonated azo dye Ponceau 4R by using single and mixed bacterial consortia. *Biocatal. Biotransform.* **2019**, *37*, 367–376. [CrossRef]
48. Haque, M.M.; Haque, M.A.; Mosharaf, M.K.; Marcus, P.K. Decolorization, degradation and detoxification of carcinogenic sulfonated azo dye methyl orange by newly developed biofilm consortia. *Saudi J. Biol. Sci.* **2021**, *28*, 793–804. [CrossRef] [PubMed]
49. Telke, A.A.; Joshi, S.M.; Jadhav, S.U.; Tamboli, D.P.; Govindwar, S.P. Decolorization and detoxification of Congo red and textile industry effluent by an isolated bacterium *Pseudomonas* sp. SU-EBT. *Biodegradation* **2010**, *21*, 283–296. [CrossRef]

sustainability

MDPI

Review

Sewage Sludge as Inhibitor of the Formation of Persistent Organic Pollutants during Incineration

Juan A. Conesa [1,2]

[1] Department of Chemical Engineering, University of Alicante, P.O. Box 99, E-03080 Alicante, Spain; ja.conesa@ua.es; Tel.: +34-965-903-400

[2] Institute of Chemical Process Engineering, University of Alicante, P.O. Box 99, E-03080 Alicante, Spain

Abstract: With the objective of suppressing dioxins and furans (PCDD/Fs) emission in municipal solid waste incineration plants (MSWI), different chemical inhibitors have been tested. Among these inhibitors, nitrogen and sulphur compounds can significantly suppress PCDD/Fs formation via de novo synthesis, which gives very interesting results with very little capital investment. In recent years, the possibility of using waste rich in nitrogen and/or sulphur as a source of inhibitor compounds has been considered, and thus has reduced the emissions of pollutants while the waste is treated. The effect of adding sludge from urban sewage treatment plants in three variants has been specially studied: directly mixing the waste, using the decomposition gas of the previously dried sludge, and using the decomposition gas of the sludge together with other inhibitors such as thiourea. Reduction of emissions in laboratory tests using model samples indicated the efficiency to be higher than 99%, using sewage sludge (SS) as an inhibitor whereas, in actual MSWI plants, the efficiency can be as high as 90%.

Keywords: PCDD/Fs; inhibition; POPs; combustion; wastes; co-combustion; co-incineration; dioxins

Citation: Conesa, J.A. Sewage Sludge as Inhibitor of the Formation of Persistent Organic Pollutants during Incineration. *Sustainability* **2021**, *13*, 10935. https://doi.org/10.3390/su131910935

Academic Editors: Avelino Núñez-Delgado, Zhien Zhang, Elza Bontempi, Mario Coccia, Marco Race and Yaoyu Zhou

Received: 10 September 2021
Accepted: 28 September 2021
Published: 1 October 2021

1. Introduction

It is a reality that the production of solid urban waste, although it has been decreasing in recent years, continues to be a problem, with a production in Europe in 2019 of 502 kg/person/year [1]. The culture of recycling of many kinds of wastes is growing more and more in society; however, landfills are still receiving greater loads of municipal wastes, even though community policy advocates a higher recycling rate and energy use of the rest fraction (not recyclable) in incineration furnaces.

In many industrialized countries, resources contained in waste are recovered by energy recovery as the main option. In this sense, it is necessary to make an effort to reuse or compost municipal solid waste, implementing a final step of energy recovery from the non-reusable and non-compostable matter.

The processes of thermal treatment of waste present a series of advantages over other methods, such as the reduction of waste (70% by mass and 90% by volume, on average), the inerting of waste (destruction of biological contamination and toxic organic compounds), the recovery of the calorific value of the waste, and the substitution of fossil fuels for the generation of Energy. However, the main problem of waste incineration is the possible production of pollutants that can cause damage to the environment.

Among the possible emitted pollutants, polycyclic aromatic hydrocarbons (PAHs) are a group of organic compounds, lipophilic in nature, solid at room temperature, and formed because of incomplete combustion. PAHs can be found in fuel-derived products such as oil, coal, and tar deposits and also as products of the use of fossil or biomass fuels. PAHs are of concern since many of them cause a variety of conditions, including cancer. PAHs have been classified into different groups, relating their toxic properties [2]. Martinez et al. [3] and Richter et al. [4], among others, conducted a review on the emission and generation of PAHs in energy-generating combustion processes and in the formation of soot. Sánchez et al. [5] presented an analytical method to determine combustion PAHs both in the gas phase and in the particles. These pollutants are present also in recycled plastic items, and have been identified as causing possible negative effects on this type of materials [6].

Other dangerous compounds are emitted during incineration, as those known as persistent organic pollutants (POPs). POPs are different groups of compounds that remain intact in nature for years once they have been released into the environment, due to a combination of physical and chemical properties; also, by their nature, POPs accumulate in the fatty tissue of living organisms.

Usually, POPs are divided into two groups (see Figure 1): "legacy" POPs, i.e., substances with the particular properties of POPs that are long-recognized as harmful and controlled under international regulation since at least 2003, and "emerging" or "new" POPs that are being considered for banning purposes nowadays [7].

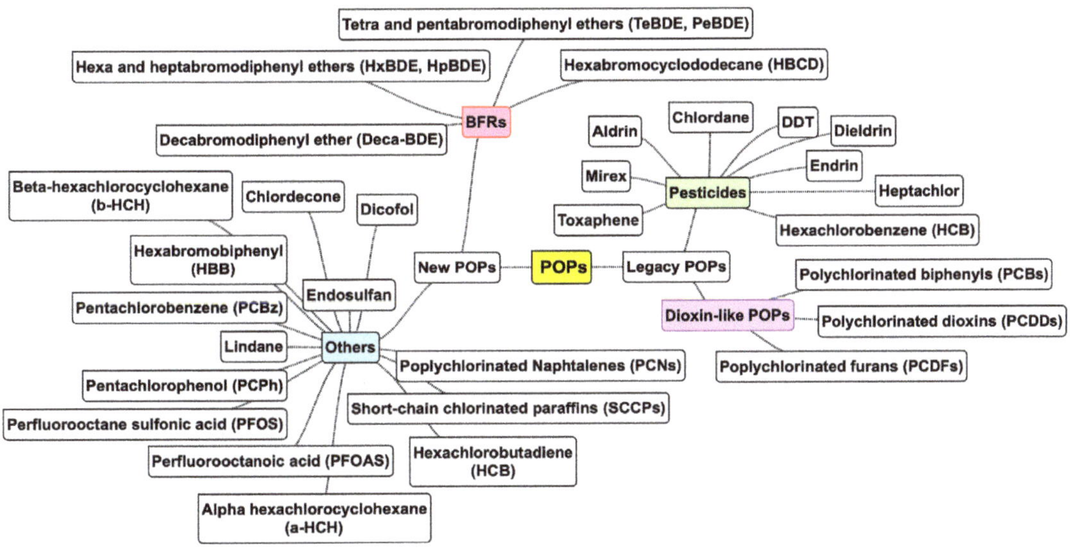

Figure 1. Persistent Organic Pollutants (POPs) considered in the Stockholm Convention [7].

In "legacy" POPs, pesticides and dioxin-like compounds are distinguished. The latter group, of particular importance given their toxicity, includes dioxins (PCDD), furans (PCDF), and biphenyls (PCB). On the other hand, the compounds classified as "new" POPs are also divided into two groups, consisting of a group of brominated compounds used as flame retardants (BFR), and another more heterogeneous group with substances that are pollutants by themselves or that can lead to very dangerous pollutants.

Among the POPs, the term 'dioxins' is applied to a set of aromatic substances, and usually refers both to PCDDs and PCDFs [8]. There are 75 chemical congeners for PCDDs and 135 congeners for PCDFs. Of all these compounds, there are 17 that develop toxic effects, and they are those that have chlorine atoms in positions 2, 3, 7, and 8. The toxicity of the 17 compounds is measured with an equivalent toxicity factor with respect to the most toxic (2,3,7,8-TetraCloroDibenzoDioxin).

PCBs (Poly Chlorinated Biphenyls) are a family of 209 compounds, of which about 130 have been identified in commercial mixtures that were industrially manufactured under different names (Aroclor, Clophen, Pyralene, etc.). Its commercialization and use have been restricted in Europe since the 85/467/EC regulation came into force. Certain PCBs, the so-called coplanar or mono-ortho congeners, are considered compounds closely related to dioxins and furans as they present similar toxicities.

There are several theories about the molecular mechanisms by which PCDDs and PCDFs are formed and subsequently emitted from combustion sources [8]. These theories are based on observations made in municipal solid waste incinerators and laboratory studies. From the studies and reviews carried out, various pathways could be considered, although the relative importance may be very different. Schematically (see Figure 2), it can be said that it can occur that the dioxins that come out through the flue gas, the fly ash, and the bottom ash come from the initial raw material, or are formed, either in the catalytic homogeneous or heterogeneous gas phase. In the homogeneous phase, they can be formed from organic compounds smaller than dioxins/furans, called precursors, that will be chlorinated derivatives, or they can be chlorinated by the radical Cl·; in the heterogeneous catalytic phase (chlorides of Cu, Zn, Fe, and others) chlorination is carried out mainly by means of Cl_2, from precursors or from carbonaceous structures (de novo synthesis). Oxygen can be incorporated into the precursors, while, in de novo synthesis, the carbonaceous structure must be oxidized [9–11].

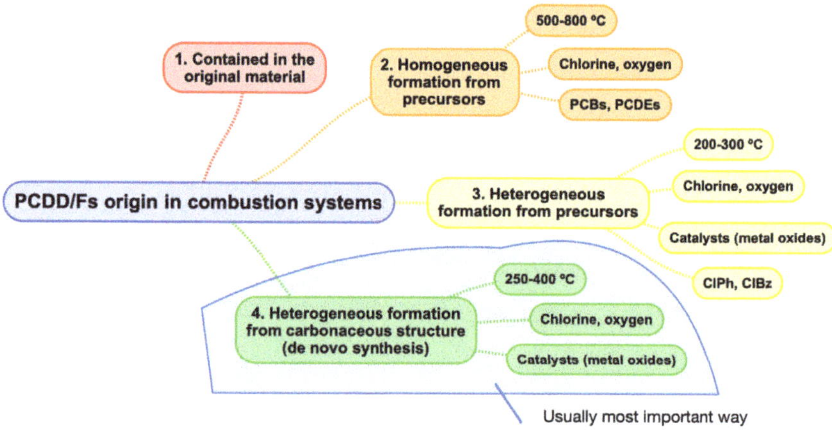

Figure 2. Possible pathways for PCDD/Fs formation in combustion systems, indicating the necessary conditions to be produced. De novo synthesis is usually the most important method, as it is the one presenting faster kinetics in the usual conditions of incineration systems [12].

Different studies [12,13] have highlighted the importance of the nature of waste (particularly, its content of Cu and Fe metals and chlorides) in the formation of dangerous compounds in the treatment processes. Likewise, the key role of the presence of oxygen in the production of PAHs, dioxins, and related compounds is highlighted, observing maximum amounts of these pollutants at intermediate oxygen levels.

Another aspect that may be interesting to analyze about the formation of dioxins in incinerators is the distribution of congeners, or what is also known as the fingerprint. By the de novo synthesis process, a distribution is obtained where the ratio between furans and dioxins is greater than unity. Fiedler et al. [10] observed that the distribution of congeners emitted in municipal waste incinerators and hazardous waste incinerators are indistinguishable. The most representative toxic congener in both cases is 23478-PeCDF, which contributes more than 30% to the total equivalent toxicity (I-TEQ).

In the present work, a review of data found in the literature about inhibition with chemicals in MSWI plants is done, with particular interest in the use of sewage sludge as a material that provides sulfur and nitrogen, and investigating the mechanism of action of these compounds.

2. Inhibition in the Dioxins and Furans Formation by S- and N-Containing Compounds

In various research, tests have been carried out showing the ability of various nitrogen and sulfur compounds to prevent the formation of dioxins and furans (PCDD/Fs) in the thermal destruction of waste. Already in 1998, Ruokojarvi et al. [14] studied the effect of four gaseous inhibitors (sulfur dioxide, ammonia, dimethylamine, and methyl mercaptan) on PCDD/PCDF formation in the combustion of liquid fuel. In a pilot plant scale, the authors observed a high decrease (98%) in the concentrations of PCDD/Fs when inhibitors were added, especially in the particle-phase PCDD/Fs concentrations.

In 2005, Pandelova et al. [15] studied the effect of the addition of more than 20 compounds to the co-combustion of coal and waste, finding that the most effective inhibitors for PCDD/F formation in flue gases were $(NH_4)_2SO_4$ and $(NH_4)_2S_2O_3$, both containing N and S. Table 1 shows a summary of the most important data noted in this Review.

Samaras et al. [16,17] mixed various sulfur and nitrogen compounds with RDF (refuse-derived-fuel) in proportions of 10% by weight. They found that the elimination of PCDD/Fs is higher than 98% using sulfur compounds, which combine with metals to form complexes and avoid the formation of pollutants. On the other hand, removal was somewhat lower using urea [16]; however, the way urea is added to the reaction medium does not affect its ability to prevent PCDD/Fs formation [17].

Thiourea [$(NH_2)_2CS$], on the other hand, with a molar ratio of (S + N)/Cl only 0.47, inhibits up to 97.3% of the PCDD/Fs content (99.8% in equivalent toxicity), although the study has not been carried out on an industrial scale as in the previous case [18]. As can be seen when comparing the results of ammonium sulfate with thiourea, the amino functional group NH_2^- tends to be more effective than NH_4^+. This fact has been confirmed by several authors, and it is attributed to the fact that the N≡C group appears in the decomposition of thiourea at 250–500 °C (in addition to ammonia, carbodiimide and H_2S), which can poison the metallic catalyst by forming some thermally stable compounds (metal complexes) [19]. The same is true of some sulfur compounds, which form sulfides or sulfates with the catalyst.

Skodras et al. [20] studied the combustion and co-combustion of natural wood, medium density fiberboard (MDF with nitrogenous additives), light poles (with a high metal content), and lignite, observing that the lower PCDD/Fs emissions are achieved using mixtures with MDF with nitrogenous additives.

Wu et al. [21] used $(NH_4)_2SO_4$ and pyrite (FeS_2) to suppress dioxin formation in a commercial incinerator, showing that removal is only partial on a practical level. However, Amand et al. [22] showed that the addition of ammonium sulfate effectively reduces the emission in a fluidized bed incinerator. For their part, Fu et al. [18] used ammonium thiosulfate, sulfamic acid (H_3NSO_3), and thiourea (CH_4N_2S) to reduce the emission of PCDD/Fs, highlighting that the combination of sulfur and nitrogen compounds provide better inhibition than separately. Thiourea has been considered by various authors, by combining both atoms. Lin et al. [23] used this in a municipal incinerator with very low doses (0.1%), showing a 91% reduction in the emission. Furthermore, these authors indicated that a reduction of NOx-type compounds takes place simultaneously. Both factors are explained by the poisoning of the catalyst metals in the presence of thiourea.

Comparing the inhibitory effect of 21 groups of substances, Pandelova et al. [15] observed that adding compounds containing only sulfur or only nitrogen achieved a relatively low inhibitory effect, while, when adding sulfur and nitrogen compounds together, the effect was much more important. For example, by adding 1% by weight of sulfamic or sulfamidic acid (H_2NSO_3H), a PCDD/Fs inhibition efficiency of around 96% was achieved [16]. Therefore, most recent studies focus on the use of compounds containing nitrogen and sulfur, where, in addition to sulfamic acid, thiourea, sulfate, and ammonium thiosulfate, among others, have been tested directly in commercial incinerators and with good results [18,21,23].

Table 1. Summary of data found in the literature.

Reference	Main Fuel	Inhibitor Used	Main Finding (PCDD/Fs Inhibition Efficiency, %)
Amand et al. [22]	SRF and wood	SS	Important decrease in the gases emission, as well as in the filter and cyclone ashes
		Ammonium sulphate	
Chen et al. [24,25]	Model samples	SS drying gases	>99%. for some congeners, especially for dioxins versus furans
Chen et al. [25]	Model fly ash	SS	97.6% (similar to TUA)
Chen et al. [25]	Model fly ash	SS	99% of inhibition in terms of toxicity (higher for dioxins than for furans)
		SO_2 and NH_3	SO_2 was more effective than NH_3 (61.9% and 38.6%, respectively)
Chen et al. [24]	Model fly ash	SS drying gases	97.6%, NH_3 and SO_2 are the most important components of the SS drying gases
Chen et al. [26]	Fly ash from MSW	SS	>96%
Fu et al. [18]	Model fly ash	sulfur–amine/ ammonium compounds	The combination of sulfur and nitrogen compounds provide better inhibition than separately
			The amino functional group NH_2^- tends to be more effective than NH_4^+.
			TUA (99:8%) > ASA (92:4%) > ATS (85:4%)
Gandon-Ros et al. [27]	PVC	SS, low O_2 presence	Increased PCDD/Fs emissions, non-inhibition observed due to the high percentage of metals in the SS
		SS, higher O_2 presence	89.2%, 71.4% and 98.8% for the inhibition ratios 0.25, 0.50 and 0.75, respectively
Hajizadeh et al. [28]	Model fly ash	SO_2 and NH_3	NH_3 was more effective at the lower temperature, while the behavior of SO_2 was the opposite (reducing a higher proportion of PCDDs than NH_3 at the higher temperature)
Lin et al. [23]	MSW	0.1% thiourea	91%
Lin et al. [29]	SS	inorganic flocculants (poly-ferric chloride, poly-aluminum chloride)	Increased PCDD/Fs emission, 30% increase
	SS	organic flocculants (polyacrylamide)	30–40% decrease
Lundin et al. [30]	Biomass and MSW	ammonium sulfate	80% (PCDDs) 50% (PCDFs) 45% (PCBs)
Ma et al. [31]	MSW	CaO and S (sulphur)	S (sulphur): HpCDD/Fs inhibition 88.1% CaO: HxCDD/Fs inhibition 85.1%
Mi Yan et al. [32]	Fly ash	ammonium sulfate and urea	both inhibitors had a higher inhibition efficiency at high temperature (650 °C) than at low temperature (350 °C)
			greater reduction in furan emissions (PDCFs)
Moreno et al. [33]	PVC containing waste	PUF	PCDD/Fs 85.7% dl-PCBs 81.2%
Ogawa et al. [34]	Wood with PVC	Coal with S (sulphur)	significant reduction in PCDD/Fs, being somewhat more important for PCDFs
Pandelova et al. [15]		1% by weight of sulfamic or sulfamidic acid	96%

Table 1. *Cont.*

Reference	Main Fuel	Inhibitor Used	Main Finding (PCDD/Fs Inhibition Efficiency, %)
Ruokojarvi et al. [14]	Liquid fuel	sulfur dioxide, ammonia, dimethylamine, and methyl mercaptan	98%
Samaras et al. [16,17]	RDF	Sulphur compounds	>98%
		Urea	Slightly lower than 98%
Soler et al. [35]	Model fly ash	TUA, TSA, ASA	the presence of the inhibitors accelerated the decomposition of the model fly ash Reactivity: TSA ≈ TUA > ASA
Wang et al. [36]	De novo runs	Different nitrogen containing compounds	$NH_4H_2PO_4 > NH_4HF_2 > (NH_4)_2SO_4 > NH_4Br$
Xiao et al. [37]	Sawdust	SS	62.9% for pellets with 10% SS 35.4% for pellets with 30% SS
Xiao et al. [37]	Wood sawdust	SS	there is a predominance of furans (PCDFs) over dioxins (PCDDs)
Zhang et al. [38]	Coal	SS	PCDD/F emissions increased from 7.00 to 32.72 pg I-TEQ/Nm^3 as the amount of SS increased from 5 to 20%
Zhan et al. [39]	Model fly ash	SS, high temperature (not de-novo synthesis)	>90%
Zhan et al. [40]	MSW	SS and TUA	>90% both
Zhong et al. [41]	MSW	5% SS	32%

Among these inhibition studies using nitrogen and sulfur compounds, research done at the University of Umeå (Sweden) used ammonium sulfate [$(NH_4)_2SO_4$] in the co-combustion of biomass and waste from the paper industry, which had a high chlorine content [30]. This compound was injected into the waste incineration plant after combustion and just before the cyclones, when the temperature was 800 °C, then the outlet gas was sampled at 130 °C. As a result, a reduction of over 80% was obtained in PCDDs, over 50% in PCDFs, and 45% in PCBs (although the latter were in a much lower proportion than the former). The inhibition was similar for all congeners.

Lin et al. [29] detected an increase in the PCDD/Fs production when adding inorganic flocculants to the incineration of SS, but an important decrease in the case of adding organic flocculants with amino substituents, ascribing the effect to the presence of such organic groups.

However, ammonia acts in a different way, since it reduces the concentration of hydrogen chloride, necessary for the formation of PCDD/Fs [42]; that is, it blocks chlorination through the formation of ammonium chloride. Larger nitrogen and sulfur molecules appear to condense and adsorb more easily in the reaction system to deactivate the catalyst than NH_3 and SO_2 [40]. In any case, as Fujimori et al. [43] found, the three inhibition pathways for dioxin formation are: (i) catalyst poisoning by sulphuration of the catalyst to form compounds of the type CuS, CuS_2, and $CuSO_4$; (ii) chlorination blocking, i.e., reaction of chlorine with nitrogen compounds that would prevent the chlorination of carbon; (iii) changing the surface of the carbonaceous matrix by sulfur and nitrogen attack to inhibit the formation of chlorinated aromatic compounds during de novo synthesis. Figure 3 shows the schematic of these three ways. Temperature plays a fundamental role in inhibition since, when using a higher temperature, thiourea tends to decompose into nitrogen and sulfur compounds instead of reacting with the metal catalyst [19,43].

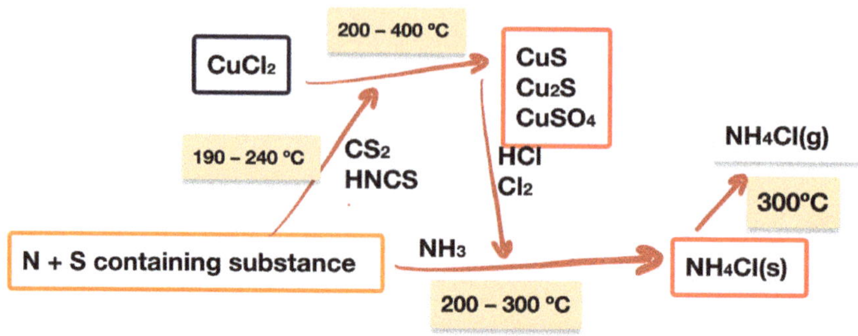

Figure 3. Pathways of inhibition of PCDD/Fs formation from N- and S- (adapted from [43]).

Regarding the difference between nitrogen and sulfur in the results of inhibition of the formation of PCDD/Fs, some authors have observed that nitrogen compounds are somewhat more effective in inhibiting the formation of PCDDs, while sulfur compounds act to a greater extent on the inhibition of PCDFs [15,28,34]. Hence, compounds containing nitrogen and sulfur at the same time are preferred. Ogawa et al. [34] studied the effect of adding SO_2 and mixing coal with an important sulfur content in the combustion of wood sawdust with PVC in a small fluidized bed incinerator, and observed a significant reduction in PCDD/Fs, being somewhat more important for PCDFs. Hajizadeh et al. [28] studied the effect of adding SO_2 and NH_3 on a laboratory scale model ash, at 225 and 375 °C (simulating the post-combustion zone), and observed, as did Zhan et al. [40], that temperature plays a very important role in the behavior of inhibitors (in addition to the concentration used and the point where they are added). NH_3 was more effective at the lower temperature, while the behavior of SO_2 was the opposite (reducing a higher proportion of PCDDs than NH_3 at the higher temperature). In the resulting solid phase (resulting fly ash), both had a greater effect than in the gas phase.

Mi Yan et al. [32] investigated the formation of PCDD/Fs produced from fly ash and its inhibition by ammonium sulfate and urea at different temperatures. The authors show that both inhibitors had a higher inhibition efficiency at high temperature (650 °C) than at low temperature (350 °C). Thus, the production of PCDD/Fs at 650 °C was quite small; a greater reduction in furan emissions (PDCFs) was observed, the most significant reductions being that of Tetra-CDF at low temperature (350 °C) and Octa-CDF at 650 °C. In a large-scale incinerator, chemical additives injected into the high-temperature section could enter the low-temperature area and continue to inhibit dioxin formation; therefore, more work is needed to evaluate different inhibitors for PCDD/Fs formation and determine the optimum injection temperature.

Marie-Rose et al. [44] suggested that the presence of NH_3 favors the formation of coke while decreasing the formation of CO_2. At a relatively low temperature (approx. 300 °C), NH_3 has been found to adsorb at strong acid sites, thereby reducing the oxidation of organic molecules to carbon dioxide. Furthermore, the formation of PCDD/Fs could also be reduced due to the blocking of these strong acid sites.

In their work, Fu et al. [18] studied three distinct –S- and $–NH_2$- or $NH^+{}_4$-containing compounds, including ammonium thiosulfate (ATS), amidosulfonic acid (ASA), and thiourea (TUA). The results revealed that the inhibition competency of the combined mixtures of S and N containing substances was strongly influenced by both the nature of the functional group of nitrogen and the value of the molar ratio (S + N)/Cl. In this way, the inhibition efficiency classifies as follows: TUA (99:8%) > ASA (92:4%) > ATS (85:4%), with the same sequence as PCDD/Fs.

On the other hand, Ma et al. [31] studied the effect of the presence of calcium oxide (CaO) and elemental sulfur (S) on the formation of PCDD/Fs during the municipal solid waste combustion process in laboratory runs performed at 800 °C. They concluded that both substances could primarily consume chlorine sources or weaken chlorination in the PCDD/F formation process to slow down PCDF formation, but the inhibition mechanisms of sulphur and calcium oxide were different. Comparing the inhibitory effect of S with that of CaO, the inhibitory effect of S on the formation of hepta-congeners (HpCDD/Fs) was more remarkable, around 88.1%; on the other hand, CaO is especially active in the inhibition of the formation of hexa-congeners of dioxins and furans more clearly than sulfur, and the inhibition was 85.1%.

Wang et al. [36] studied the inhibitory effects of four amino compounds on the formation of chlorobenzenes (CBzs) in de-novo experiments, finding that the inhibitory effects follow the order $NH_4H_2PO_4$ > NH_4HF_2 > $(NH_4)_2SO_4$ > NH_4Br under air flow, and are slightly different under nitrogen flow.

Soler et al. [35] also studied the mechanism implied in the thermal decomposition of model fly ash modified by the presence of N- and S-containing compounds. Specifically, the authors test the presence of thiourea (TUA), ammonium thiosulfate (TSA), and amidosulfonic acid (ASA) on the reactivity of fly ash air; however, the investigation was done using a thermobalance at different heating rates (5, 10, and 20 K min^{-1}) and analyzing the evolved gases. Their first conclusion is that the presence of the inhibitors, surprisingly, accelerated the decomposition of the model fly ash, i.e., the thermal decomposition took place at lower temperature (approx. 80 to 100 °C less) in the presence of TUA, TSA, or ASA. These authors assume that the increase in the decomposition rate is due to the existence of a reaction, not well defined, between the compounds containing N and S and the carbonaceous material.

3. Use of Other Sulphur and Nitrogen Containing Wastes to Reduce Emissions

In recent years, the possibility of using waste rich in nitrogen and/or sulfur as a source of inhibitor compounds has been considered, and thus the reduction of the emissions of pollutants while the waste is treated. Various authors, such as Amand et al. [22], showed the possibility of using waste mixtures to reduce pollutant emissions. The effects of the addition of urban water treatment sludge have been especially studied, since they generally have a high content of nitrogen compounds [45]. Chen et al. [24,25] used the drying gas of these sludges as a reaction atmosphere in the thermal decomposition of model samples, demonstrating that these gases are capable of suppressing the de novo synthesis of PCDD/Fs with an efficiency greater than 99% for some congeners, especially for dioxins versus furans.

Other nitrogenated wastes that have been used for inhibition are polyurethanes (mainly from mattresses). Polyurethanes are polymers made from long chain polyethers or polyesters covalently linked by a urethane bond (-NH-COO-). The manufacture of these plastics involves the reaction between polyfunctional isocyanates and alcohols [46]. Flexible polyurethane foams (FPUF) are mainly used in the manufacturing of upholstered furniture and mattresses.

Thermal degradation of different types of polyurethanes has been studied in the past. Boettner et al. [47] analyzed the volatiles produced during the combustion of four types of polyurethanes from car seats. They verified that the major products are CO and CO_2, hydrogen cyanide, acetaldehyde, and methanol, and they showed that, as the atmosphere becomes more oxidative, the amount of methanol and HCN produced decreases. In previous research, the pyrolysis and combustion of FPUF residues from mattresses was carried out, showing that the presence of oxygen effectively reduces the emission of PAHs and chlorobenzenes, as well as the NH_3 emission [48]. This decrease in ammonia has been attributed to the fact that the emission of PCDD/Fs simultaneously increases, due to the inhibitory effect shown by ammonia in the formation of these compounds [14,48].

Moreno et al. [33] tested the influence of the presence of another N-containing waste, as is polyurethane foam (PUF), in the formation of PCDD/Fs during the combustion of a PVC-containing waste. The authors also compared the presence of PUF in the waste itself with the effect of the presence of PUF pyrolysis gases in the decomposition of the PVC-containing waste. They found that the first alternative reduces the formation of

PCDD/Fs by 85.7% and dl-PCBs by 81.2% in WHO-TEQ toxicity, with the direct addition of PUF waste being the simplest one.

Inhibition by Using Thermal Decomposition Products from Sewage Sludge

The effect of adding sludge from urban sewage treatment plants in three variants has been especially studied: directly mixing the waste [22,37], using decomposition gas from previously dried sludge [25,49], and using the decomposition gas of the sludge together with other inhibitors such as thiourea [39].

Amand et al. [22] observed an important decrease in the emission of PCDD/Fs during co-combustion of solid recovered fuel (SRF) with municipal sewage sludge compared to when SRF was fired without SS as additional fuel. Also, the injection of ammonium sulphate was effective in this reduction, and the authors corroborate the presence of an 'alkali track' relating the level of alkali chlorides in the gas phase, the chlorine content in the deposits in the convection pass, and the PCDD/F formation.

Amand et al. [22] detected that direct mixing of waste can considerably reduce the levels of PCDD/Fs that are emitted. For this, they used an SS with a low dioxin content in a proportion of 8% together with 73% of wood and 19% of a waste of wood, textiles, paper, and plastic (which had a dioxin content higher than that of sludge). Experiments were carried out in a reactor on a scale close to the industrial one and under unique conditions, around 870 °C and with a slight excess of air with respect to the stoichiometric. Although the sludge contained transition metals that could favor the formation of dioxins, the dioxin emission was reduced by 86.6%. Xiao et al. [37] also observed a decrease in dioxin levels when mixing sewage sludge with sawdust during combustion; however, Zhang et al. [38] observed the opposite effect when mixing sludge with coal in a thermal power plant, which is why a more in-depth study is necessary. On the other hand, Chen et al. [49] verified that the sludge drying gas (actually slight thermal decomposition at about 300 °C of sludge previously dried at 105 °C in a different device) could inhibit the formation of dioxins in a fly ash model equivalent to those obtained in incineration plants of municipal waste, with an efficiency of up to 97.6% by weight, a result similar to that obtained with thiourea. The main compounds of the sludge drying gas were NH_3 and SO_2, although there may be other interesting nitrogen and sulfur compounds that also act as inhibitors and have not been determined. The whole experiment was carried out at 300 °C, using three types of sludge with nitrogen content around 4% and sulfur around 1%, except for one of the sludges that presented a value of 4%, although no important differences were observed with different contents in S. The doses used were 0.5–4 g of dry sludge for every 2 g of ash, and a flow of 300 mL/min of a gas was passed through them that contained 12% of O_2 in N_2. In a later work, Chen et al. [25] studied the effect of the dose of sludge, the temperature of the previous treatment of this sludge (250–350 °C), and the oxygen content (0–12%). They observed an efficiency greater than 99% of inhibition in terms of toxicity (higher for dioxins than for furans), with 300 °C being the most suitable temperature, and with a very low influence of the oxygen content. In different experiments, directly injecting SO_2 and NH_3, they established that SO_2 was more effective than NH_3 (61.9% and 38.6%, respectively). Zhan et al. [39], on the other hand, used the sludge decomposition gas obtained in a similar way to Chen et al. [24] in model fly ash, but in this case at a high temperature (650–850 °C), where the formation of PCDD/Fs could take place in the homogeneous gas phase by means of phenoxy radicals with chlorine atoms, instead of by synthesis de novo. In this case, the inhibition of dioxin formation was also observed with an efficiency of 90%. Zhan et al. [40] obtained similar inhibition results in urban solid waste incineration experiments in a pilot plant, where they added the decomposition gas of sewage sludge after combustion together with a certain amount of thiourea.

It should be noted that the experiments of Chen et al. [24] were made starting from sludge previously dried at 105 °C, which is how this residue is normally available in incineration plants, since the high moisture content of the sludge from the WWTP does not allow direct combustion.

Gomez–Rico et al. [50] studied the drying of SS at temperatures between 80 and 120 °C. In all cases, thermal drying resulted in the formation of volatile organic compounds (VOCs), including many PAHs, and prevent the need to control the emissions from the drying plants. This also indicates that the gases from SS drying can already contain a great amount of S and N gaseous compounds. In this context, Chen et al. [24] show the efficiency of SS drying gases for the inhibition of PCDD/Fs formation, in runs performed over model fly ashes, where they observed suppression of the 2,3,7,8-substitued PCDD/Fs formation. The authors calculate an efficiency up to 97.6% in wt. units and 96% in I-TEQ. Statistical treatment of the data showed that NH_3 and SO_2 are the most important components of the SS drying gases.

The sulfur compounds detected by Liu et al. [51] in pyrolysis are H_2S, SO_2, COS, CS_2, and CH_3SH (from aliphatic and aromatic organic sulfur compounds present in the sludge). Other organic compounds in the sludge such as sulfonic acid and thiophene do not influence too much, and inorganic compounds such as sulfates and sulfites also remain stable at the mentioned temperatures. Regarding the nitrogen compounds also detected during sludge pyrolysis by Tian et al. [52], NH_3, HNCO, NO, and HCN were found, as well as compounds with amino and nitrile groups, and other heterocyclic nitrogen compounds (although these at somewhat higher temperatures). These compounds come mainly from the decomposition of proteins, amines, pyridines, and pyrroles.

Zhong et al. [41] studied the PCDD/F levels and phase distributions in a full-scale municipal solid waste incinerator with co-incinerating sewage sludge. The presence of a selective reduction catalyst (SCR) was also investigated. Their primary findings are that the presence of up to 5% of SS in the fuel did not increase the PCDD/Fs emission, and that the emission in all phases decreased significantly, including the condensed water collected during the sampling.

Chen et al. [25] also studied the inhibition of PCDD/Fs by SS decomposition gases. These authors observed suppression on the formation of toxic TCDD/Fs when the decomposition takes place in the presence of gases evolved from dried SS. The efficiency was close to 100%, being slightly higher for PCDDs than for PCDFs. Chen et al. also found that SO_2 was more effective (61.9% suppression efficiency) than NH_3 (38.6%) in suppressing PCDD/Fs formation.

On the other hand, Liu et al. [51] investigated the emission of organic sulfur in the gases during SS pyrolysis, showing that both aliphatic and aromatic sulfur compounds are present and that the transformation of aliphatic sulfurs into H_2S begins at a relatively low temperature (250 °C). Also, aromatic sulfur transforms into H_2S, but the phenomenon was observed at higher temperature (350–450 °C).

Tian et al. [52] observed that the pyrolysis of SS is produced by the thermochemical decomposition of pyridine-N and pyrrole-N, and that fewer amine-N compounds are produced during pyrolysis. These authors also show that the heating rate of pyrolysis does not change the composition of the gases produced, but does produce significant changes in the species of liquid organic compounds produced.

Chen et al. [26] investigated the role of SS during the decomposition of fly ashes coming from municipal solid wastes. They found that the presence of SS inhibited the formation of PCDD/Fs by more than 96% and looked for the mechanisms of this inhibition. Treating the PCDD/Fs congener profiles of the gas emissions, they found that the SS suppressed formation through de-novo pathway, the suppression of the chlorophenol route being less important. Also, these authors determined that the main mechanism for the inhibition is the cleavage of the oxygen bridge in the molecules.

On the other hand, Zhang et al. [38] studied the emission of dioxins and furans in the combustion of coal in power plants, comparing the effect of different proportions of SS (5 to 20 wt.%). Surprisingly, the authors indicate that the amount of PCDD/Fs increased as the amount of SS increased in the fuel, when considering the whole mass balance. Nevertheless, authors demonstrate that co-combustion of SS is a sink for PCDD/Fs, as mentioned also by

Xiao et al. [37]. The main output current of these pollutants is fly ash, with a comparable amount to the output by emitted gas.

Xiao et al. [37] show that there is a predominance of furans (PCDFs) over dioxins (PCDDs) in all the output currents (stack gas, fly ash, slag) during the co-combustion of SS and wood sawdust.

Recently, a study was performed on the emissions of pollutants from the thermal decomposition of PVC and sewage sludges in different experimental conditions [27]. In this work, the materials were mixed in different proportions (ionic ratios "Ri" of (N + S) to chlorine equal to 0.25, 0.50 and 0.75) in addition to the combustion of the materials separately, and were reacted with oxygen at 850 °C. Oxygen was also present in different proportions, to study the effect of the presence of this element. Partial combustion reactions were performed with values of λ (excess oxygen) equal to 0.15 and 0.50. PAHs and PCDD/Fs were evaluated. Figure 4 shows the inhibitory effect of the presence of SS in the emission of PAHs, in all runs performed. This inhibitory effect is calculated comparing the emissions to that of the materials decomposed separately.

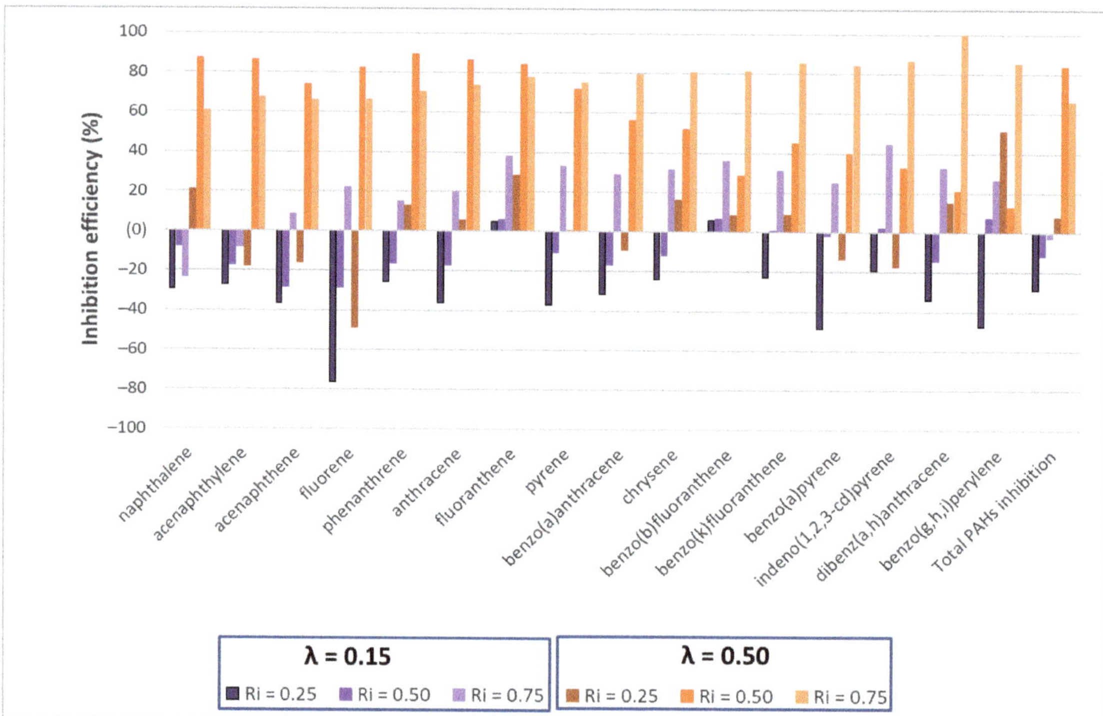

Figure 4. Inhibition efficiency (%) in PAH formation for every experiment run at 850 °C and both oxygen ratios: λ = 0.15 and 0.50.

Figure 4 shows that the production of PAHs is much higher at a low oxygen ratio (λ = 0.15), as expected, due to a less efficient combustion [53]. Concerning the presence of SS, the concentration of PAHs present in the sample decreased notably when the inhibition ratio increased, with some of the compounds being reduced by more than 80% (Figure 4). In that behavior, we found that λ = 0.50 was always better to reduce the emission of PAHs, therefore, in the presence of more oxygen, the formation of PAHs is lower, with Ri= 0.50 being the mixture that showed the best results.

For the PCDD/Fs emissions, the results obtained at different experimental conditions are outlined in Figure 5. The results obtained for PCDD/Fs at λ = 0.15 were different from those expected. Intermediate values were expected between emission of SS and PVC e-waste. For the hexa-, hepta-, and octachlorinated congeners, the mixture Ri = 0.25 presented higher dioxin emissions than expected due to the composition of the sludge, composed by around 6% of iron, among other metals in much less proportion, such as copper (0.09%). Previous studies showed that the presence of these metals catalyzes the formation reaction of PCDDs and PCDFs [12,13]. According to Fujimori et al. [54], some of the active catalysts could already be present in the sample.

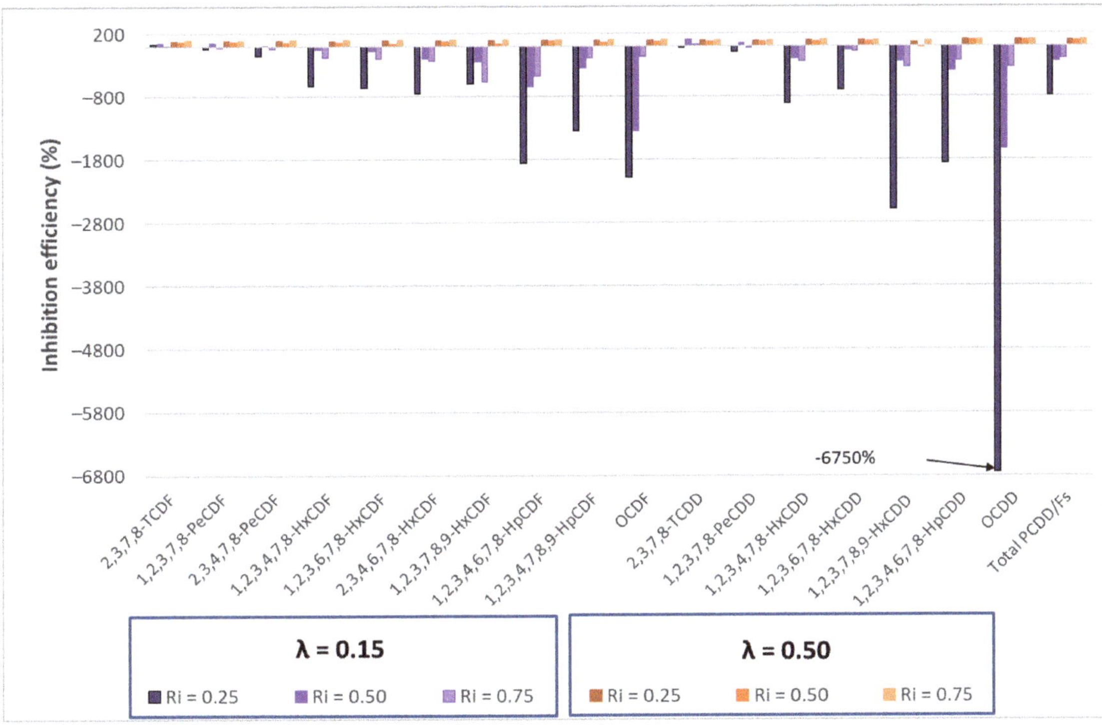

Figure 5. Inhibition efficiency (%) in PCDD/Fs formation for every experiment run at 850 °C and both oxygen ratios λ = 0.15 and 0.50.

For the low oxygen ratio, λ = 0.15, the lowest emissions were obtained for an Ri of 0.75 and the highest emissions for Ri = 0.25, i.e., a higher amount of SS produced higher inhibition. When oxygen is fed at λ = 0.50 (greater presence), the lowest emissions were obtained also for an inhibition ratio of 0.75.

It can be observed that, when working with small inhibition ratios (Ri = 0.25) or with high inhibition ratios (Ri = 0.75), λ = 0.50 was always better to reduce the concentration of PCDD/Fs. However, for intermediate inhibition ratios around 0.50, the emissions obtained were similar for both oxygen ratios.

As a conclusion, Ri = 0.75 was always better to minimize the emission of PCDD/Fs, and, therefore, in the presence of more SS, the formation of PCDD/Fs is much lower for both oxygen atmospheres, with λ = 0.50 being the oxygen ratio that clearly showed the best results in terms of the inhibition efficiency.

4. Conclusions

This review presents the main research that is being carried out regarding the use of nitrogenous and sulfurized substances in the reduction of pollutant emissions in incineration furnaces.

Different chemicals have been tested in the literature to reduce the emission of different POPs, especially those with a halogenated structure, due to their dangerousness. In this sense, substances such as thiourea, ammonium sulfate, or sulfamidic acid prevent the formation of PCDD/Fs with an efficiency higher than 90%.

Also, other substances containing nitrogen and sulfur, including wastes, appear to be a promising way to reduce persistent organic compound emissions by very high percentages. In this sense, sewage sludges seem to be a very interesting waste, as they are very abundant in urban centers and present both nitrogen and sulphur in their chemistry. In addition, the combustion of municipal waste together with sewage sludge avoids an important problem, the disposal of these sludges, which in many cases has high metal content, making them unsuitable for agricultural amendment.

The data presented in this review indicate that the joint combustion of sewage sludge and municipal waste reduces the emissions of dioxins, furans, and related species by more than 90%. This should translate into legislation requiring municipalities to add SS at MSWI plants.

Regarding the direction of future research, it is necessary to verify the efficiency of the elimination of other polluting species, particularly those that present bromine in their structure and that are generally related to the combustion of electrical and electronic materials (since bromine is used as a flame retardant in this equipment).

Funding: This research was funded by MINISTRY OF SCIENCE AND INNOVATION (SPAIN), grant number PID2019-105359RB-I00.

Institutional Review Board Statement: Not applicable.

Informed Consent Statement: Not applicable.

Conflicts of Interest: The author declares no conflict of interest.

References

1. EUROSTAT-Statistics Explained Municipal Waste Statistics—Statistics Explained. 2020; ISSN 2443-8219. Available online: https://ec.europa.eu/eurostat/statistics-explained/index.php/Municipal_waste_statistics#Municipal_waste_generation (accessed on 29 September 2021).
2. Lerda, D. *Polycyclic Aromatic Hydrocarbons (PAHs) Factsheet*, 4th ed.; European Commision, Institute for Reference of Materials and Measurements: Geel, Belgium, 2011; Available online: https://ec.europa.eu/jrc/sites/default/files/Factsheet%20PAH_0.pdf (accessed on 29 September 2021).
3. Martínez, J.D.; Puy, N.; Murillo, R.; García, T.; Navarro, M.V.; Mastral, A.M. Waste tyre pyrolysis—A review. *Renew. Sustain. Energy Rev.* **2013**, *23*, 179–213. [CrossRef]
4. Richter, H.; Benish, T.G.; Mazyar, O.A.; Green, W.H.; Howard, J.B. Formation of polycyclic aromatic hydrocarbons and their radicals in a nearly sooting premixed benzene flame. *Symp. Combust.* **2000**, *28*, 2609–2618. [CrossRef]
5. Sánchez, N.E.; Salafranca, J.; Callejas, A.; Millera, Á.; Bilbao, R.; Alzueta, M.U. Quantification of polycyclic aromatic hydrocarbons (PAHs) found in gas and particle phases from pyrolytic processes using gas chromatography-mass spectrometry (GC-MS). *Fuel* **2013**, *107*, 246–253. [CrossRef]
6. Conesa, J.A.; Nuñez, S.S.; Ortuño, N.; Moltó, J. PAH and POP Presence in Plastic Waste and Recyclates: State of the Art. *Energies* **2021**, *14*, 3451. [CrossRef]
7. Stockholm Convention—Home Page. Available online: http://chm.pops.int/Home/tabid/2121/Default.aspx (accessed on 10 May 2021).
8. Conesa, J.A. Dioxins from Agro Waste Combustion. In *Byproducts from Agriculture and Fisheries*; Wiley: Hoboken, NJ, USA, 2019; pp. 629–640.
9. McKay, G. Dioxin characterisation, formation and minimisation during municipal solid waste (MSW) incineration: Review. *Chem. Eng. J.* **2002**, *86*, 343–368. [CrossRef]
10. Fiedler, H.; Lau, C.; Eduljee, G. Statistical analysis of patterns of PCDDs and PCDFs in stack emission samples and identification of a marker congener. *Waste Manag. Res.* **2000**, *18*, 283–292. [CrossRef]

11. Conesa, J.A.; Fullana, A.; Font, R. De novo synthesis of PCDD/F by thermogravimetry. *Environ. Sci. Technol.* **2002**, *36*, 263–269. [CrossRef]
12. Palmer, D.; Pou, J.O.; Díaz-Ferrero, J.; Conesa, J.A.; Ortuño, N. Kinetics of the formation and destruction of PCDD/Fs in a laboratory tubular furnace. *Chemosphere* **2021**, *276*, 130175. [CrossRef]
13. Conesa, J.A.; Font, R.; Fullana, A.; Martín-Gullón, I.; Aracil, I.; Gálvez, A.; Moltó, J.; Gómez-Rico, M.F. Comparison between emissions from the pyrolysis and combustion of different wastes. *J. Anal. Appl. Pyrolysis* **2009**, *84*, 95–102. [CrossRef]
14. Ruokojärvi, P.H.; Halonen, I.A.; Tuppurainen, K.A.; Tarhanen, J.; Ruuskanen, J. Effect of Gaseous Inhibitors on PCDD/F Formation. *Environ. Sci. Technol.* **1998**, *32*, 3099–3103. [CrossRef]
15. Pandelova, M.E.; Lenoir, D.; Kettrup, A.; Schramm, K.-W. Primary Measures for Reduction of PCDD/F in Co-Combustion of Lignite Coal and Waste: Effect of Various Inhibitors. *Environ. Sci. Technol.* **2005**, *39*, 3345–3350. [CrossRef]
16. Samaras, P.; Blumenstock, M.; Lenoir, D.; Schramm, K.W.; Kettrup, A. PCDD/F prevention by novel inhibitors: Addition of inorganic S- and N-compounds in the fuel before combustion. *Environ. Sci. Technol.* **2000**, *34*, 5092–5096. [CrossRef]
17. Samaras, P.; Blumenstock, M.; Lenoir, D.; Schramm, K.W.; Kettrup, A. PCDD/F inhibition by prior addition of urea to the solid fuel in laboratory experiments and results statistical evaluation. *Chemosphere* **2001**, *42*, 737–743. [CrossRef]
18. Fu, J.-Y.; Li, X.-D.; Chen, T.; Lin, X.-Q.; Buekens, A.; Lu, S.-Y.; Yan, J.-H.; Cen, K.-F. PCDD/Fs' suppression by sulfur–amine/ammonium compounds. *Chemosphere* **2015**, *123*, 9–16. [CrossRef] [PubMed]
19. Zhan, M.-X.X.; Chen, T.; Fu, J.-Y.Y.; Lin, X.-Q.Q.; Lu, S.-Y.Y.; Li, X.-D.D.; Yan, J.-H.H.; Buekens, A. High temperature suppression of dioxins. *Chemosphere* **2016**, *146*, 182–188. [CrossRef] [PubMed]
20. Skodras, G.; Grammelis, P.; Samaras, P.; Vourliotis, P.; Kakaras, E.; Sakellaropoulos, G.P. Emissions monitoring during coal waste wood co-combustion in an industrial steam boiler. *Fuel* **2002**, *81*, 547–554. [CrossRef]
21. Wu, H.-L.; Lu, S.-Y.; Li, X.-D.; Jiang, X.-G.; Yan, J.-H.; Zhou, M.-S.; Wang, H. Inhibition of PCDD/F by adding sulphur compounds to the feed of a hazardous waste incinerator. *Chemosphere* **2012**, *86*, 361–367. [CrossRef]
22. Åmand, L.-E.; Kassman, H. Decreased PCDD/F formation when co-firing a waste fuel and biomass in a CFB boiler by addition of sulphates or municipal sewage sludge. *Waste Manag.* **2013**, *33*, 1729–1739. [CrossRef]
23. Lin, X.; Yan, M.; Dai, A.; Zhan, M.; Fu, J.; Li, X.; Chen, T.; Lu, S.; Buekens, A.; Yan, J. Simultaneous suppression of PCDD/F and NOx during municipal solid waste incineration. *Chemosphere* **2015**, *126*, 60–66. [CrossRef]
24. Chen, T.; Zhan, M.-X.X.; Lin, X.-Q.Q.; Li, X.-D.D.; Lu, S.-Y.Y.; Yan, J.-H.H.; Buekens, A.; Cen, K.-F.F. Inhibition of the de novo synthesis of PCDD/Fs on model fly ash by sludge drying gases. *Chemosphere* **2014**, *114*, 226–232. [CrossRef]
25. Chen, T.; Zhan, M.-X.X.; Lin, X.-Q.Q.; Fu, J.-Y.Y.; Lu, S.-Y.Y.; Li, X.-D.D.; Buekens, A.; Yan, J.-H.H. PCDD/Fs inhibition by sludge decomposition gases: Effects of sludge dosage, treatment temperature and oxygen content. *Aerosol Air Qual. Res.* **2015**, *15*, 702–711. [CrossRef]
26. Chen, Z.; Lin, X.; Zhang, S.; Xiangbo, Z.; Li, X.; Lu, S.; Yan, J. Thermal cotreatment of municipal solid waste incineration fly ash with sewage sludge for PCDD/Fs decomposition and reformation suppression. *J. Hazard. Mater.* **2021**, *416*, 126216. [CrossRef]
27. Gandon-Ros, G.; Nuñez, S.S.; Ortuño, N.; Aracil, I.; Gómez-Rico, M.F.; Conesa, J.A.; Francisca Gómez-Rico, M.; Conesa, J.A. A Win-Win Combination to Inhibit Persistent Organic Pollutant Formation via the Co-Incineration of Polyvinyl Chloride E-Waste and Sewage Sludge. *Polymers* **2021**, *13*, 835. [CrossRef] [PubMed]
28. Hajizadeh, Y.; Onwudili, J.A.; Williams, P.T. PCDD/F formation from oxy-PAH precursors in waste incinerator flyash. *Chemosphere* **2011**, *85*, 1672–1681. [CrossRef] [PubMed]
29. Lin, X.; Li, X.; Lu, S.; Wang, F.; Chen, T.; Yan, J. Influence of organic and inorganic flocculants on the formation of PCDD/Fs during sewage sludge incineration. *Environ. Sci. Pollut. Res.* **2015**, *22*, 14629–14636. [CrossRef] [PubMed]
30. Lundin, L.; Gomez-Rico, M.F.; Forsberg, C.; Nordenskjöld, C.; Jansson, S. Reduction of PCDD, PCDF and PCB during co-combustion of biomass with waste products from pulp and paper industry. *Chemosphere* **2013**, *91*, 797–801. [CrossRef] [PubMed]
31. Ma, H.; Du, N.; Lin, X.; Liu, C.; Zhang, J.; Miao, Z. Inhibition of element sulfur and calcium oxide on the formation of PCDD/Fs during co-combustion experiment of municipal solid waste. *Sci. Total Environ.* **2018**, *633*, 1263–1271. [CrossRef]
32. Yan, M.; Yang, J.; Li, X.; Hu, Y.; Yan, J.; Mi, Y.; Xiaodong, L.; Yanjun, H.; Jianhua, Y.; Jie, Y. Inhibition of PCDD/Fs formation from fly ash by ammonium sulfate and urea. *CIESC J.* **2013**, *64*, 4196–4202. [CrossRef]
33. Moreno, A.I.; Font, R.; Gomez-Rico, M.F. Inhibition effect of polyurethane foam waste in dioxin formation. *Waste Manag.* **2019**, *97*, 19–26. [CrossRef]
34. Ogawa, H.; Orita, N.; Horaguchi, M.; Suzuki, T.; Okada, M.; Yasuda, S. Dioxin reduction by sulfur component addition. *Chemosphere* **1996**, *32*, 151–157. [CrossRef]
35. Soler, A.; Conesa, J.A.; Ortuño, N. Inhibiting fly ash reactivity by adding N- and S- containing compounds. *Chemosphere* **2018**, *211*, 294–301. [CrossRef]
36. Wang, S.-J.; He, P.-J.; Lu, W.-T.; Shao, L.-M.; Zhang, H. Amino Compounds as Inhibitors of De Novo Synthesis of Chlorobenzenes. *Sci. Rep.* **2016**, *6*, 23197. [CrossRef]
37. Xiao, Z.; Yuan, X.; Jiang, L.; Chen, X.; Li, H.; Zeng, G.; Leng, L.; Wang, H.; Huang, H. Energy recovery and secondary pollutant emission from the combustion of co-pelletized fuel from municipal sewage sludge and wood sawdust. *Energy* **2015**, *91*, 441–450. [CrossRef]

38. Zhang, G.; Hai, J.; Ren, M.; Zhang, S.; Cheng, J.; Yang, Z. Emission, mass balance, and distribution characteristics of PCDD/Fs and heavy metals during cocombustion of sewage sludge and coal in power plants. *Environ. Sci. Technol.* **2013**, *47*, 2123–2130. [CrossRef] [PubMed]
39. Zhan, M.X.; Fu, J.Y.; Chen, T.; Lin, X.Q.; Li, X.D.; Yan, J.H.; Buekens, A. Suppression of dioxins by S-N inhibitors in pilot-scale experiments. *Environ. Sci. Pollut. Res.* **2016**, *23*, 16463–16477. [CrossRef]
40. Zhan, M.; Chen, T.; Lin, X.; Fu, J.; Li, X.; Yan, J.; Buekens, A. Suppression of dioxins after the post-combustion zone of MSWIs. *Waste Manag.* **2016**, *54*, 153–161. [CrossRef] [PubMed]
41. Zhong, R.; Wang, C.; Zhang, Z.; Liu, Q.; Cai, Z. PCDD/F levels and phase distributions in a full-scale municipal solid waste incinerator with co-incinerating sewage sludge. *Waste Manag.* **2020**, *106*, 110–119. [CrossRef]
42. Ooi, T.C.; Thompson, D.; Anderson, D.R.; Fisher, R.; Fray, T.; Zandi, M. The effect of charcoal combustion on iron-ore sintering performance and emission of persistent organic pollutants. *Combust. Flame* **2011**, *158*, 979–987. [CrossRef]
43. Fujimori, T.; Nakamura, M.; Takaoka, M.; Shiota, K.; Kitajima, Y. Synergetic inhibition of thermochemical formation of chlorinated aromatics by sulfur and nitrogen derived from thiourea: Multielement characterizations. *J. Hazard. Mater.* **2016**, *311*, 43–50. [CrossRef]
44. Marie-Rose, S.C.; Belin, T.; Mijoin, J.; Fiani, E.; Taralunga, M.; Chaucherie, X.; Nicol, F.; Magnoux, P. Destruction of PAH and dioxin precursors using selective oxidation over zeolite catalysts. Influence of the presence of ammonia in the flue gas. *Appl. Catal. B Environ.* **2009**, *93*, 106–111. [CrossRef]
45. Fullana, A.; Conesa, J.A.; Font, R.; Martin-Gullon, I. Pyrolysis of sewage sludge: Nitrogenated compounds and pretreatment effects. *J. Anal. Appl. Pyrolysis* **2003**, *68–69*, 561–575. [CrossRef]
46. Zhang, Y.; Xia, Z.; Huang, H.; Chen, H. Thermal degradation of polyurethane based on IPDI. *J. Anal. Appl. Pyrolysis* **2009**, *84*, 89–94. [CrossRef]
47. Agency, E.P. *Combustion Products from the Incineration of Plastics*; Environmental Protection Agency: Cincinnati, OH, USA, 1973; p. 156.
48. Moreno, A.I.; Font, R. Pyrolysis of furniture wood waste: Decomposition and gases evolved. *J. Anal. Appl. Pyrolysis* **2015**, *113*, 464–473. [CrossRef]
49. Chen, T.; Guo, Y.; Li, X.; Lu, S.; Yan, J. Emissions behavior and distribution of polychlorinated dibenzo-p-dioxins and furans (PCDD/Fs) from cement kilns in China. *Environ. Sci. Pollut. Res.* **2014**, *21*, 4245–4253. [CrossRef]
50. Gomez-Rico, M.F.; Fullana, A.; Font, R. Volatile organic compounds released from thermal drying of sewage sludge. *WIT Trans. Ecol. Environ.* **2008**, *111*, 425–433. [CrossRef]
51. Liu, S.; Wei, M.; Qiao, Y.; Yang, Z.; Gui, B.; Yu, Y.; Xu, M. Release of organic sulfur as sulfur-containing gases during low temperature pyrolysis of sewage sludge. *Proc. Combust. Inst.* **2015**, *35*, 2767–2775. [CrossRef]
52. Tian, K.; Liu, W.J.; Qian, T.T.; Jiang, H.; Yu, H.Q. Investigation on the evolution of N-containing organic compounds during pyrolysis of sewage sludge. *Environ. Sci. Technol.* **2014**, *48*, 10888–10896. [CrossRef]
53. Soler, A.; Conesa, J.A.; Iñiguez, M.E.; Ortuño, N. Pollutant formation in the pyrolysis and combustion of materials combining biomass and e-waste. *Sci. Total Environ.* **2018**, *622–623*, 1258–1264. [CrossRef] [PubMed]
54. Fujimori, T.; Takaoka, M.; Takeda, N. Influence of Cu, Fe, Pb, and Zn Chlorides and Oxides on Formation of Chlorinated Aromatic Compounds in MSWI Fly Ash. *Environ. Sci. Technol.* **2009**, *43*, 8053–8059. [CrossRef] [PubMed]

sustainability

MDPI

Article

Factors Affecting Carbonation Depth in Foamed Concrete Bricks for Accelerate CO_2 Sequestration

Abdullah Faisal Alshalif [1,*], J. M. Irwan [1,*], Husnul Azan Tajarudin [2,*], N. Othman [3], A. A. Al-Gheethi [3], S. Shamsudin [4], Wahid Ali Hamood Altowayti [3] and Saddam Abo Sabah [1]

1 Jamilus Research Centre for Sustainable Construction (JRC-SC), Faculty of Civil Engineering and Built Environment, Universiti Tun Hussein Onn Malaysia, Parit Raja 86400, Johor, Malaysia; saddam@uthm.edu.my
2 Division of Bioprocess, School of Industrial Technology, Universiti Sains Malaysia, Gelugor 11800, Pulau Pinang, Malaysia
3 Micro-Pollutant Research Centre (MPRC), Faculty of Civil Engineering and Built Environment, Universiti Tun Hussein Onn Malaysia, Parit Raja 86400, Johor, Malaysia; norzila@uthm.edu.my (N.O.); adel@uthm.edu.my (A.A.A.-G.); wahidali@uthm.edu.my (W.A.H.A.)
4 Sustainable Manufacturing and Recycling Technology, Advanced Manufacturing and Materials Center (SMART-AMMC), Universiti Tun Hussein Onn Malaysia, Parit Raja 86400, Johor, Malaysia; shazarel@uthm.edu.my
* Correspondence: faisalalshalif@gmail.com (A.F.A.); irwan@uthm.edu.my (J.M.I.); azan@usm.my (H.A.T.)

Abstract: Foamed concrete bricks (FCB) have high levels of porosity to sequestrate atmospheric CO_2 in the form of calcium carbonate $CaCO_3$ via acceleration of carbonation depth. The effect of density and curing conditions on CO_2 sequestration in FCB was investigated in this research to optimize carbonation depth. Statistical analysis using 2^k factorial and response surface methodology (RSM) comprising 11 runs and eight additional runs was used to optimize the carbonation depth of FCB for 28 days (d). The main factors selected for the carbonation studies include density, temperature and CO_2 concentration. The curing of the FCB was performed in the chamber. The results indicated that all factors significantly affected the carbonation depth of FCB. The optimum carbonation depth was 9.7 mm, which was determined at conditions; 1300 kg/m^3, 40 °C, and 20% of CO_2 concentration after 28 d. Analysis of variance (ANOVA) and residual plots demonstrated the accuracy of the regression equation with a predicted R^2 of 89.43%, which confirms the reliability of the predicted model.

Keywords: carbonation depth; chamber; temperature; statistical analysis; 2^k factorial; RSM

Citation: Alshalif, A.F.; Irwan, J.M.; Tajarudin, H.A.; Othman, N.; Al-Gheethi, A.A.; Shamsudin, S.; Altowayti, W.A.H.; Sabah, S.A. Factors Affecting Carbonation Depth in Foamed Concrete Bricks for Accelerate CO_2 Sequestration. *Sustainability* **2021**, *13*, 10999. https://doi.org/10.3390/su131910999

Academic Editor: Salman Masoudi Soltani

Received: 26 July 2021
Accepted: 26 August 2021
Published: 4 October 2021

1. Introduction

Foamed concrete is lightweight concrete made without coarse aggregate. It can either be cement or lime mortar that generates air voids in the mortar via a suitable aerated agent [1]. Foamed concrete has numerous advantages including low density, which results in a reduction of the load on the structure, especially foundations. It is also environmentally friendly and economical when compared to other types of concrete. It also provides a high degree of thermal insulation and sound-proofing [2,3]. Therefore, the applications of foamed concrete have become more popular worldwide, especially on housing constructions and insulations, road sub-based and other applications such as; old sewers, soil stabilization, trench fills earthquake purpose and storage tanks [1,2,4].

Foamed concrete has a wide range of density starting from 300 kg/m^3 to 1800 kg/m^3, which depends on the level of porosity (voids) that are introduced by the foaming agent or aluminium powder [1,5]. The reaction is initiated with water when the aluminium powder is added to the mixture. The heat of reaction under alkaline conditions generates hydrogen gas bubbles, which create air voids in the concrete to accelerate carbonation in the foamed concrete [5,6]. Several factors affect carbonation in concrete such as material chemical properties, solid physical characteristics, and curing conditions [7]. However, the concrete density, CO_2 concentration, and temperature are considered the most critical factors examined in previous research in the literature [6]. Studying the effect of each factor separately while ignoring the effect of other factors may give defective results, which in turn reduces the sustainability of the concrete. Optimization considers an appropriate solution to accelerate carbonation in foamed concrete bricks (FCB). However, the literature on the main factors affecting carbonation of foamed concrete should be studied carefully before using any optimization methods to identify the range of each factor that may cause enhancement of FCB.

Density can cause drastic changes on foamed concrete properties, particularly strength and penetration such as carbonation depth. According to [1], there is an underlining relationship between carbonation depth and density, since the increase in density increases the carbonation depth. Similarly, Namsone [8] reported that the most critical factor affecting the carbonation of foamed concrete when compared to normal concrete is the density because the change in density resulted in a change in level of porosity. It is possible to produce high pores foamed concrete with density 300 kg/m^3, which has the capability to accelerate carbonation; however, the strength of the specimens will be lower than 1 MPa [9]. On the other hand, concrete with density more than 1850 kg/m^3 will be out of the range of foamed concrete density 300 kg/m^3–1800 kg/m^3. Therefore, 1300 kg/m^3 and 1800 kg/m^3 was selected as low and high levels of densities in this study to produce FCB. Density is not the only factor affecting the carbonation of foamed concrete, particularly the curing conditions such as CO_2 concentration, as temperature and humidity also play a vital role too [10,11]. Many researchers have stated that the best range of relative humidity to accelerate carbonation in concrete is 50–70% [12,13]. However, the humidity is influenced by the changes in temperature degree and CO_2 concentration [12].

The atmospheric CO_2 concentration is the main source of sequestrated CO_2 in concrete via the carbonation process. The standard atmospheric concentration of CO_2 is between (0.03% and 0.06%), which indicates that carbonation at standard concentration is very slow [14,15]. In this case, a curing long time is required. Therefore, the chamber with different CO_2 concentration levels is used to accelerate the carbonation process in concrete [15]. The concentration of CO_2 and curing time has a strong relationship, the increase of CO_2 concentration and curing time resulted in increment of the carbonation depth and vice versa [13,16]. The researchers prefer to use a high concentration of CO_2 to accelerate carbonation in concrete with a short period of curing in the chamber [13]. Additionally, some chambers use a heater to control the temperature during curing. The increase in temperature up to 60 °C promotes the ability of the concrete to absorb CO_2 and the formation of calcium carbonate ($CaCO_3$) [17]. In contrast, higher temperatures above 60 °C may decrease the CO_2 content since solubility in water decreases at elevated temperatures, which could also decrease the carbonation rate [18]. Therefore, most researchers adopt temperatures below 60 °C to accelerate the process of concrete carbonation [19–21].

This research aims to optimize the carbonation depth in foamed concrete bricks. Hence, the effects of density, temperature and CO_2 concentration that influence CO_2 sequestration in foamed concrete via carbonation process was examined as the main factors.

2. Materials and Methods

2.1. Materials and Mix Proportion

The materials used in this study are cement, sand, water and the foaming agent. Ordinary Portland Cement (OPC) with the specific surface area 2250 cm^2/g and chemical composition details are shown in Table 1 [22].

The sand was adapted to pass through a sieve with a size of 1 mm according to BS 882-1992 [20]. According to the British Cement Association-1994, the maximum size of fine aggregate (sand) in foamed concrete is 1.18 mm. Additionally, the percentage of sand passed through 600 microns should be between 60–90% to produce foamed concrete as shown in the particle size distribution of sand in Figure 1 [21,22].

Table 1. Chemical composition of cement.

Chemical Compound	SiO_2	Al_2O_3	Fe_2O_3	SO_3	K_2O	CaO	MgO
Concentration (%)	20.6	5.4	4.2	2.2	0.6	64.8	2.2

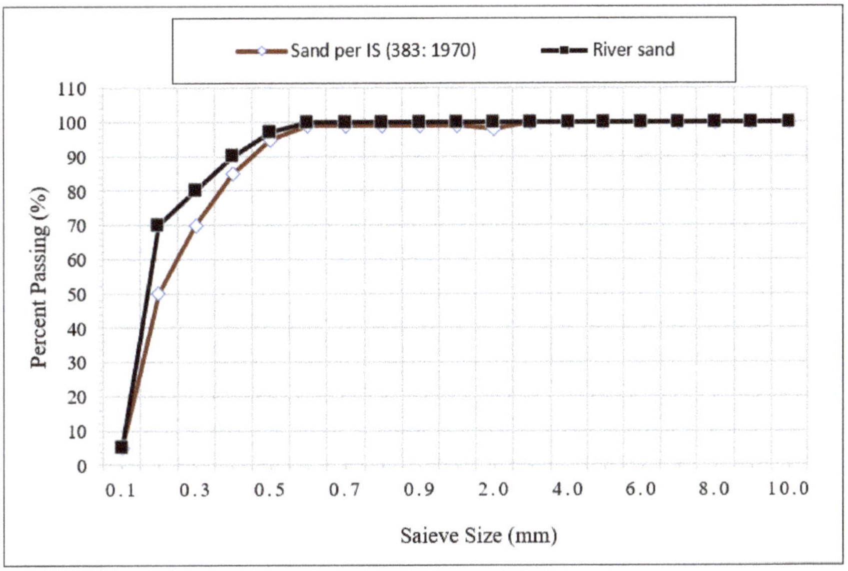

Figure 1. Grading curve of river sand.

Tap water was used for the foamed concrete mix and diluting the foaming agent. A synthetic type CF 500 foaming agent was mixed with water to produce air bubbles in the foamed concrete mixtures. The ratio of foamed agent to water was 1:20, which aerated to 65 kg/m^3 density according to the ASTM C796 Standard for foaming agents used in cellular concrete and preformed foam production [21,23]. The design of foamed concrete depends on the adjusted density. The weight of solid materials (cement/sand) was distributed in the ratio of 1:1.35 according to ACI 523.3R with the trial method of mix design [21].

For this study, 3 factors were used to optimize the carbonation depth, namely; density, temperature and CO_2 concentration using the 2^k Full Factorial and Response Surface Methodology (RSM) designs that analysed through Minitab 18 software. The software was developed at the Pennsylvania State University, USA.

The first 8 experiments were factorial runs followed by 3 centre runs for curvature analysis. The design was completed by RSM by adding 6 axials runs and 2 more runs at the centre, which resulted in a total of 19 runs. The runs were comprised of 8 factorial runs and 6 axial runs (all without repetition), while 5 runs were located at the centre. Lastly, the density of the foamed concrete was the main factor affecting the mix proportion in this study because the change of density resulted in change on materials used in the mixture proportion as shown in Table 2. Furthermore, the materials used were cement, sand and water mass which subjected to changes from run to run and in line with the changes in density.

Table 2. Mix proportions of foamed concrete by 2^k factorial and RSM designs.

Run No.	Density (kg/m^3)	Cement (kg/m^3)	Fine Sand (kg/m^3)	Water (L/m^3)	T (°C)	CO_2 (%)
1	1300	553.2	746.8	276.6	27.0	10
2	1300	553.2	746.8	276.6	27.0	20
3	1300	553.2	746.8	276.6	40.0	10
4	1300	553.2	746.8	276.6	40.0	20
5	1800	766	1034	383	27.0	10
6	1800	766	1034	383	27.0	20
7	1800	766	1034	383	40.0	10
8	1800	766	1034	383	40.0	20
9	1550	659.5	890.4	329.7	33.5	15
10	1550	659.5	890.4	329.7	33.5	15
11	1550	659.5	890.4	329.7	33.5	15
12	1550	659.5	890.4	329.7	33.5	10
13	1550	659.5	890.4	329.7	33.5	20
14	1550	659.5	890.4	329.7	27.0	15
15	1550	659.5	890.4	329.7	40.0	15
16	1300	553.2	746.8	276.6	33.5	15
17	1800	766	1034	383	33.5	15
18	1550	659.5	890.4	329.7	33.5	15
19	1550	659.5	890.4	329.7	33.5	15

2.2. *Fresh Stage Tests (Fresh Density Test/Inverted Slump Test)*

The foamed concrete was tested using the fresh density test and slump test methods. A container with 1-L capacity was used to perform the fresh density test, which was tared to zero at the balance machine before being overfilled with fresh foamed concrete. The compaction of the foamed concrete was performed by lightly tapping the sides of the container to allow consolidation of the fresh foamed concrete. The 1 litter container was weighed to obtain the fresh density of foamed concrete [24]. The inverted slump test was conducted according to the ASTM C995 (2001) standard using a slump cone and flat base plate. The slump cone was inverted and placed at the centre of the base plate and filled with fresh foamed concrete until it was filled. The inverted slump cone was lifted to 1 ft height within 3–5 s (s). The dimension of the spread was measured from four angles and recorded as shown in Figure 2 [24]. The slump flow was calculated using Equation (1).

$$\text{Slump flow} = \frac{d1 + d2}{2} \tag{1}$$

where;

d_1 = Maximum diameter of slump flow;
d_2 = Perpendicular diameter of d_1.

Figure 2. Inverted slump test.

2.3. Sample Preparation and Chamber Curing

The moulds with the size of (215 × 100 × 65 mm) were prepared to fill up by fresh foamed concrete according to the BS6073-2:2008 standard. The concrete specimens were demoulded after 24 h in moulds shown in Figure 3. The specimens were dried in the chamber at 50 °C for 72 h without supplying CO_2 in the chamber at this stage. After that, the specimens were cured in the chamber according to the conditions suggested by 2^k factorial and RSM as listed in Table 2.

Figure 3. Specimens preparation.

The curing chamber has the ability to control CO_2 concentration, temperature and sensor to monitor humidity as shown in Figure 4. The process of carbonation curing commenced after drying the specimens in the same chamber. The carbonation curing was applied for 28 d, whereas the concentration of CO_2 for each experimental run was suggested by the 2^k factorial and RSM design methods as presented in Table 2. In addition, the relative humidity was monitored along curing period for each run using a humidity sensor inside chamber. The humidity was in the range of 55–75% in all runs, which was increased and decreased within this range according to changes in temperature degree and CO_2 concentration in each run.

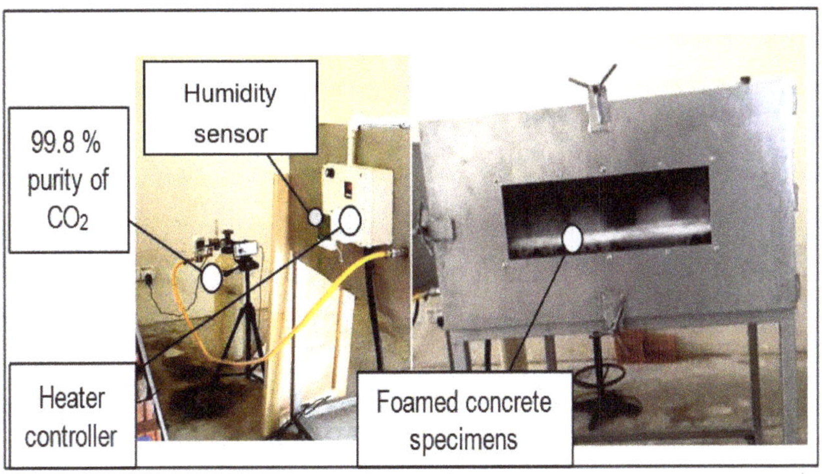

Figure 4. Setup of the carbonation curing in the chamber.

2.4. Hardened Stage Test (Carbonation Depth Test)

The depth of carbonation through the surface of FCB was measured using the simple collared dye field test for detecting carbonation. The specimens of FCB were placed in the chamber to control the CO_2 concentration and temperature according to the statistical analysis of 2^k factorial and RSM design for 28 d. The phenolphthalein solution was then diluted to indicate carbonation depth as follows; 1 g phenolphthalein dissolved in 100 mL high purity ethanol. The carbonation depth test commenced by splitting the specimen into two halves followed by spraying the freshly broken specimens with phenolphthalein indicator solution. If the colour is reddish-purple, it means the specimens are still in high alkaline condition, while a colourless edge indicates that the specimen is already carbonated and the average corresponding depth is measured. The carbonation depth was measured from the 3 sides exposed to atmospheric CO_2, whereas the average of the three sides used as the carbonation depth of the specimen was computed using Equation (2). The average of three specimens of each run of FCB was considered as carbonation depth on each run.

$$\text{Carbonation depth (mm)} = \frac{d1 + d2 + d3}{3} \quad (2)$$

whereas;

$d1$ = is the carbonation depth from the first side specimens;
$d2$ = is the carbonation depth from the second side specimens;
$d3$ = is the carbonation depth from the third side specimens.

3. Results

3.1. Fresh and Inverted Slump Tests

The fresh foamed concrete density was adjusted for each mixture via the fresh density test. The main factor for controlling the foamed concrete density is the foaming agent [25]. The three different densities used in this study as follows; 1800 kg/m^3, 1550 kg/m^3 and 1300 kg/m^3. The fresh density was measured successfully for the selected densities. Thereafter, the inverted slump test was performed to determine the workability of the foamed concrete. The results of the inverted slump test demonstrated that the spread diameter of the mixture of 1300 kg/m^3 is higher than the mixture with 1550 kg/m^3 and 1800 kg/m^3. Figure 5 depicts the increase in the spread diameter of the foamed concrete with low density compared to the foamed concrete with higher density. The foaming agent was used to produce foamed concrete with low density, therefore the spread diameter was higher.

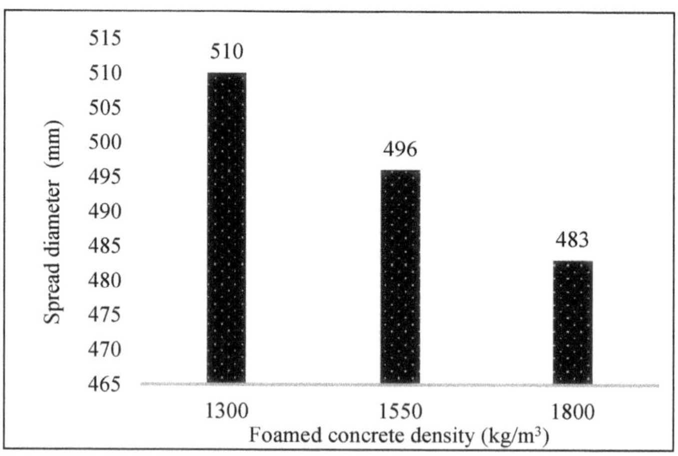

Figure 5. Spread diameter of inverted slump of foamed concrete.

3.2. Carbonation Depth of FCB

The CO_2 can be sequestrated into concrete by carbonation depth [26]. However, several factors play important roles in accelerating the sequestration of CO_2 or carbonation in concrete especially density and curing conditions such as temperature and CO_2 concentration [27]. The results of the carbonation depth of 19 runs as a response of the 2^k factorial and RSM designs were analysed. The effects of density, temperature and CO_2 concentration on the carbonation depth of FCB is presented in Figure 6.

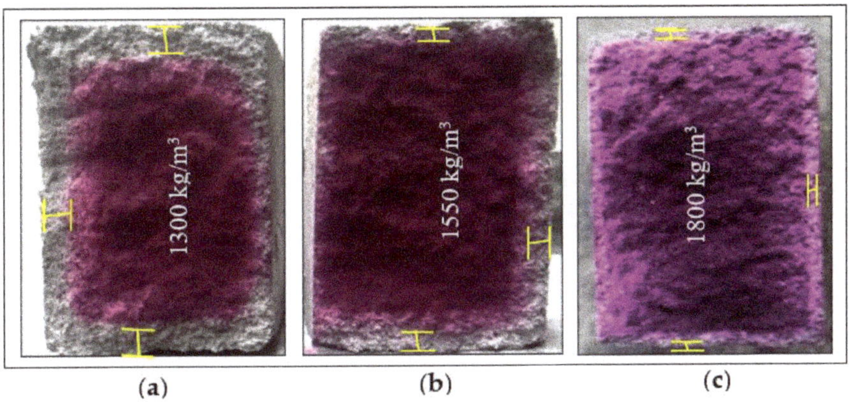

Figure 6. Carbonation depth of FCB (**a**) with 1300 kg/m^3 (**b**) with 1550 kg/m^3 (**c**) with 1800 kg/m^3 after 28 days.

The increment of carbonation depth in concrete with low density compared to concrete with a higher density is a normal effect [6]. However, the purple-red colour in the specimens with 1800 kg/m^3 was obtained due to the extreme pH value [28]. Thus, the portlandite ($Ca(OH)_2$) has the ability to control Ca and caused an expansion of the solid volume inside the concrete at pH > 12 [29,30]. Furthermore, the used of temperature between 27 °C and 40 °C help to keep H_2O in portlandite ($Ca(OH)_2$), which in turn increased CO_2 ensuing from the carbonation. In contrast, note the higher temperature corresponding to loss of H_2O as well as the solubility of CO_2 in concrete [10,31].

Consequently, the carbonation depth performance on run numbers: 4, 10 and 17 with the densities 1300 kg/m^3, 1550 kg/m^3 and 1800 kg/m^3 were 9.2 mm, 3.8 mm and 2.1 mm at 28 d, respectively as shown in Figure 7. However, the density was not the only factor that caused a significant effect on the increase or decrease of carbonation depth in FCB. Nevertheless, the change of carbonation depth on FCB that has the same density is unusual except due to some reasons. Temperature and CO_2 concentration along with curing conditions also altered the carbonation depth of FCB when the density held on some runs. For example, the highest carbonation depth was 9.2 mm at run 4 with 1300 kg/m^3, 40 °C and 20% of CO_2 concentration, while for the density at run 1, the carbonation depth was 5.6 mm when the temperature and CO_2 concentration were at 27 °C and 20%, respectively. Similarly, the carbonation depth of runs 7 and 8 are 2.1 mm and 3.2 mm at the density and temperature 1800 kg/m^3 and 40 °C, respectively. However, the concentration of CO_2 changed from 10% and 20%, respectively. This finding has demonstrated the effect of CO_2 concentration on the increase carbonation depth of FCB.

3.3. *Factorial and RSM Analyses*

3.3.1. Residual Plots of Carbonation Depth

In factorial design, the ANOVA conclusions can only be accepted when the adequacy of the underlying model has been evaluated. The primary diagnostic tool to gauge the model adequacy is residual analysis. The residual data or the measured errors should demonstrate normal distribution, independent distribution, zero mean value and constant variance σ^2 at all runs. If all residuals satisfy the aforementioned requirements, so that the F_0 ratio will follow an ***F*** distribution that will lead to accurate ANOVA results. Furthermore, the effects of nuisance factors will be excluded from the analysis [32]. In this study, the residual plots of normal probability were used to indicate whether the model meets the assumptions of the analysis or not [33]. As can be seen in Figure 8, the normal probability plot (NPP) shows the majority points cluster to a straight line and this indicates the residual distributions are likely to be a normal and hence the model meets the assumption. On top of that, the fine segregation of the points around the normal probability line demonstrates a precise prediction of the carbonation depth of FCB. Meanwhile, the versus fits in residual plots present the scattered values about zero and no obvious pattern can be observed. In addition, only two points are slightly departed from the red line in the NPP, in which the errors can be assumed as normal [32], whereas the allowable error of the findings is <5% to reflect a high level of accuracy in the data analysis [33].

3.3.2. Significance of the Factors to Carbonation Depth of FCB

The statistical significance of the factors to carbonation depth of FCB was evaluated from the results of the 19 runs of the 2^k factorial and RSM analysis. The *p*-value of each factor was below 0.05, as illustrated by ANOVA analysis in Table 3. The *p*-value of CO_2, temperature and density were; 0.003, 0.010 and 0.000, respectively. The ANOVA results reflect the highly significant effect of the factors on the response (carbonation depth). Consequently, the effect of CO_2, temperature and density were 3.67, 3.01, and −8.57, respectively. The results show that the highest effect on carbonation depth was by the density of FCB. This finding, in line with previous studies, shows that the increase or decrease of concrete density mainly affects the performance of carbonation depth [1,8]. Likewise, the CO_2 concentration and the temperature also influenced the carbonation depth of FCB. However, the increase of temperature higher than 60 °C may reduce CO_2 sequestration because the solubility of CO_2 decreases in the waste at high degree of temperature, which in turn reduces the carbonation depth in concrete [18]. Due to that, most of the researchers preferred to use temperatures lower that 60 °C to increase carbonation in concrete as practiced in this research [13].

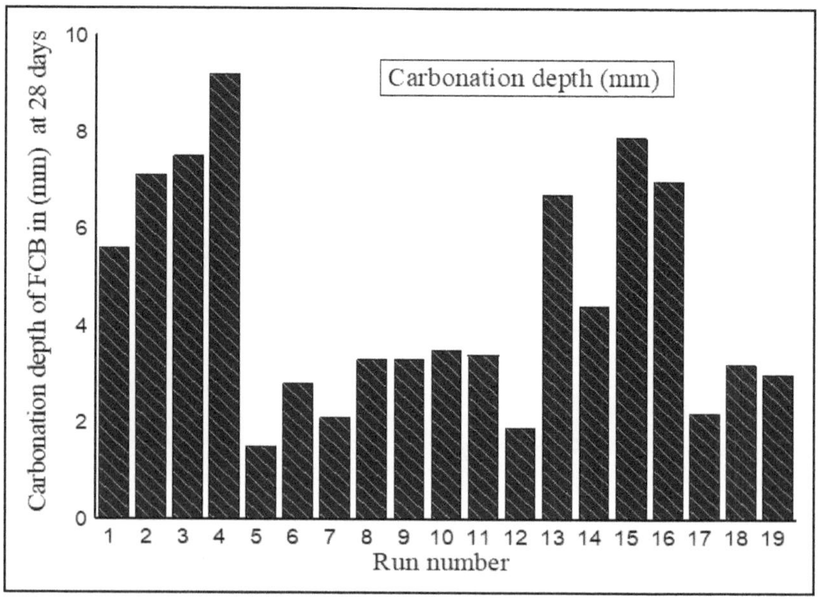

Figure 7. Carbonation depth of FCB in (mm) after 28 days.

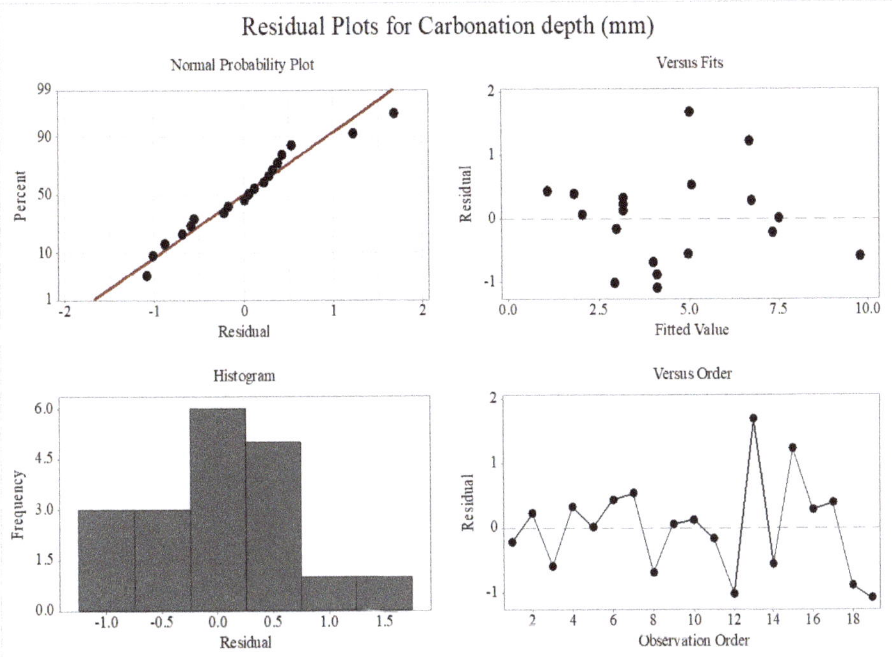

Figure 8. Residual plots of carbonation depth of FCB after 28 days.

Table 3. ANOVA analysis of the complete RSM design.

Source	DF	Adj SS	Adj MS	F-Value	*p*-Value	Effect
Model	5	89.832	17.966	22.000	0.000	-
Blocks	1	2.880	2.880	3.530	0.083	-
Linear	3	78.446	26.148	32.020	0.000	-
CO_2	1	11.025	11.025	13.500	0.003	3.670
Temperature	1	7.396	7.396	9.060	0.010	3.010
Density	1	60.025	60.025	73.490	0.000	−8.570
Square	1	11.371	11.371	13.920	0.003	3.730
Temperature * Temperature	1	11.371	11.371	13.920	0.003	3.670
Error	13	10.618	0.816	-	-	-
Lack-of-Fit	10	10.578	1.057	79.330	0.002	-
Pure Error	3	0.040	0.013	-	-	-
Total	18	100.449	-	-	-	-

The Pareto charts in Figure 9a demonstrate the significance of each input CO_2, temperature, and density. Therefore, the magnitude and the importance of the standardized effect of each factor and interactions were obtained in the statistical analysis. The horizontal bars of the factor and interaction that crosses the segmented vertical reference line is considered as statistically significant. The results show that the total number of single and double interaction terms was 9, although five of the terms were non-significant, as demonstrated in Figure 9a. Consequently, the significant terms A, B, C and BB were maintained, but the non-significant terms BC, CC, AC, AA and AB were removed from the analysis to improve the accuracy of the model as shown in Figure 9a,b. As observed, the main factors A, B and C significantly affect the carbonation of FCB. The observation from the results of C had the highest effect on the carbonation depth of FCB, followed by A and B accordingly. The curing conditions, such as temperature and CO_2 concentration, play an important role in the carbonation of concrete, as also observed by previous researchers [17,34].

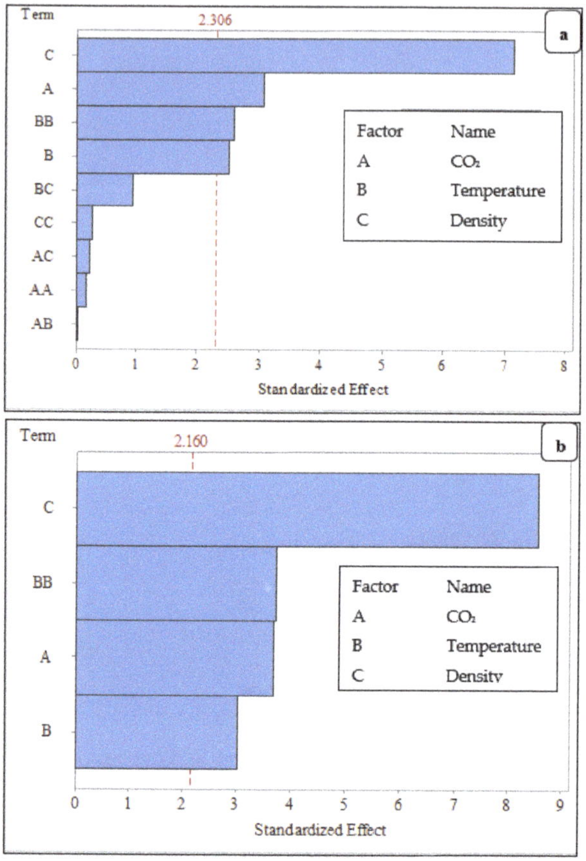

Figure 9. Pareto chart of the standardized effects at a 95% confidence interval on carbonation of FCB (**a**) before removing non-significant terms, and (**b**) after removing non-significant terms.

3.3.3. Contour Plots of Carbonation Depth of FCB

The contour plots shown in Figure 10a,b depict the effect of the parameters on carbonation depth of FCB. The contour plot is one of the most useful plots in RSM used to demonstrate the effect of two factors and holding the other factors. The plots exhibit layers with different gradually changing colours indicative of the possible independence of factors with a response. The contour plots depict the graphical relationship of two factors, i.e., density and temperature over the carbonation depth of FCB, while the CO_2 concentration is held at the centre value.

Figure 10a depicts the effect of density and CO_2 concentration on the carbonation of FCB. In general, the carbonation depth at a low level of CO_2 and temperature was very low, while it was higher at higher settings of temperature and CO_2. The lowest carbonation depth occurred when the temperature was between 28.2 °C and 35.5 °C and the CO_2 concentration was between 10% and 12%, respectively. In contrast, the highest carbonation depth occurred at 40 °C and 20% CO_2. Based on the findings, the increase in temperature and CO_2 concentration along with the curing of FCB accelerates the process of carbonation.

Figure 10b demonstrates the effect of density and CO_2 concentration on the carbonation depth of FCB. The increase in density reduced of the carbonation depth, while the increase in CO_2 concentration increased the carbonation depth. Thus, the highest carbonation depth of FCB was at 20% CO_2 for specimens with a density of 1300 kg/m^3. However, the lowest carbonation depth occurred at 10% CO_2 for the specimen 1800 kg/m^3 density.

From the above discussions, it can be surmised that carbonation depth could be enhanced at higher CO_2 concentrations and temperatures. Besides, the low density of FCB played an important role in accelerating CO_2 sequestration due to the high level of porosity.

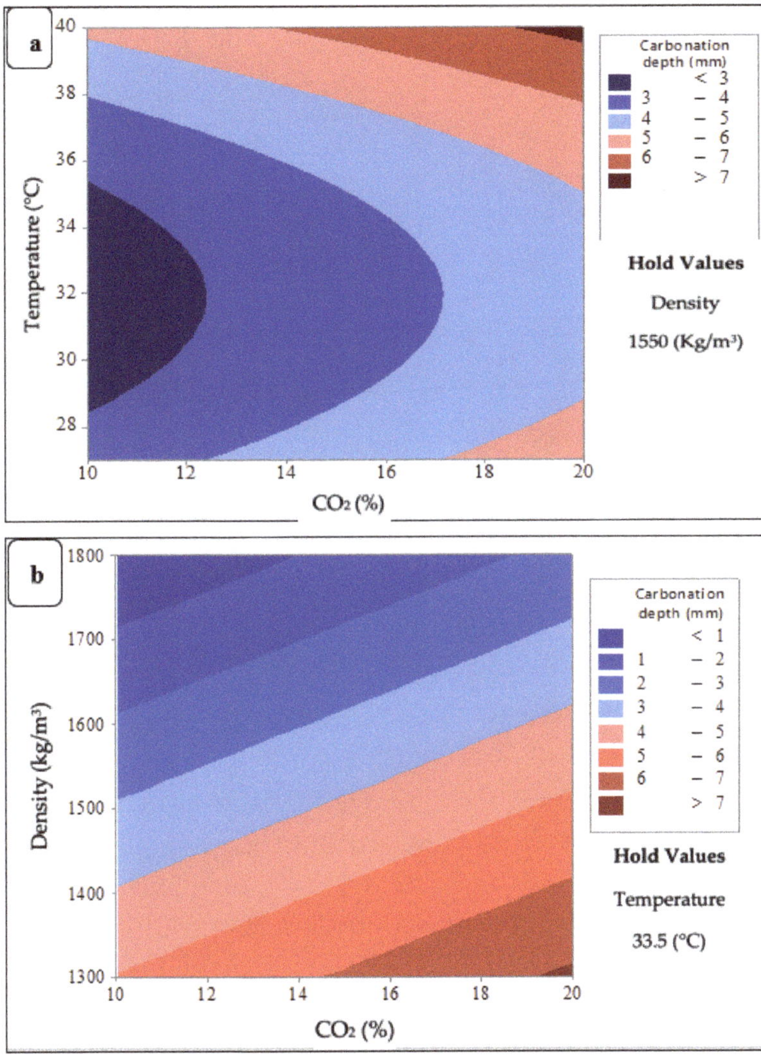

Figure 10. Contour plots for carbonation depth of FCB; (**a**) between temperature and CO_2 concentration, and (**b**) between density and CO_2 concentration.

3.3.4. Optimum Conditions of Carbonation Depth of FCB

The optimisation plot shows how different experimental settings affect the predicted carbonation depth of FCB at two targets minimum and maximum carbonation depths as shown in Figure 11a,b. The best setting of each factor is represented by the red lines, while the dotted blue line represents the highest attainment of carbonation depth of FCB. Figure 11a,b show that the single desirability (d) for the maximum and minimum carbonation depth are 1.000 and the response (y) are 9.7683 mm and 0.0458 mm, respectively.

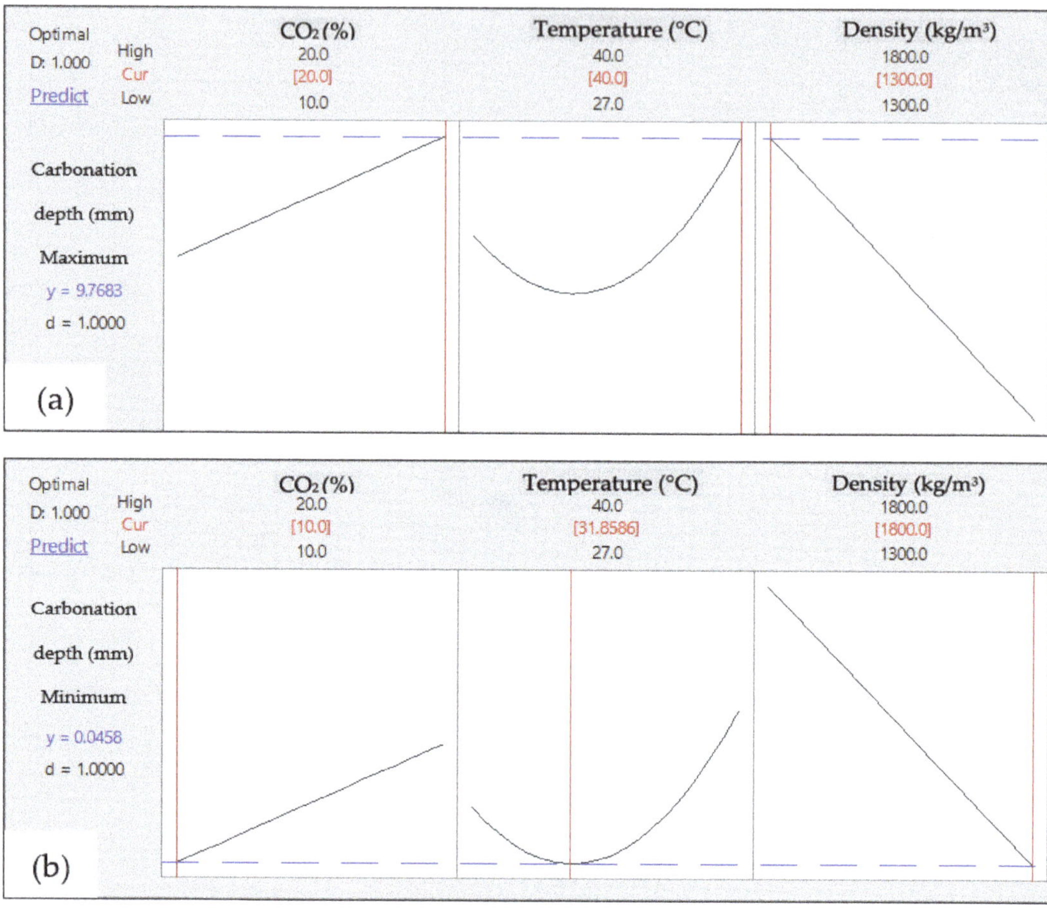

Figure 11. Optimization plot of FCB at 28 days (**a**) at the maximum target of carbonation depth, and (**b**) at the minimum target of carbonation depth.

The increase in the CO_2 concentration and temperature during the curing process increases the carbonation depth of FCB as percent in Figure 11a. Thus, the highest predicted carbonation depth of FCB was 9.7 mm, which occurred at 1300 kg/m^3, 40 °C and 20% of CO_2 concentration. The change on the factors values can make drastically change on the response value as presents in Figure 11b. The opposite trend was observed on the carbonation depth, whereby it decreased with decreasing of CO_2 concentration and temperature along curing conditions and increasing density of FCB. Therefore, the lower predicted carbonation depth was 0.0458 mm at the following conditions 10% of CO_2 concentration, 1800 kg/m^3 of FCB density and 31.8 °C of temperature.

3.3.5. Development of Initial and Final Regression Equation

The initial regression equation was developed by 2^k factorial method after the screening stage of the factors affecting carbonation depth in FCB, as shown in Equation (3). Thereafter, final regression equation in uncoded units was developed via RSM analysis after optimizing the carbonation depth of FCB as shown in Equation (4) [35].

$$\text{Carbonation depth (mm)} = 5.31 - 0.002\ CO_2 + 0.331\ \text{Temperature} - 0.00365\ \text{Density} + 0.00754\ CO_2 * \text{Temperature} + 0.000085\ CO_2 * \text{Density} - 0.000154\ \text{Temperature} * \text{Density} - 0.000005\ CO_2 * \text{Temperature} * \text{Density} - 1.4875\ \text{Ct Pt} \quad (3)$$

$$\text{Carbonation depth (mm)} = 57.9 + 0.2100\ CO_2 - 2.655\ \text{Temperature} - 0.00980\ \text{Density} + 0.0416\ \text{Temperature} * \text{Temperature} \quad (4)$$

Both equations derived from the ANOVA results illustrates the relationship between significant variables and the response of carbonation depth. The accuracy of the regression equation was further justified through the ANOVA analysis and normal probability plot. The initial equation reflects the strong effect of the factors on carbonation depth of FCB through the significant effect of the interactions between the factors. This finding confirmed by the percentage of predicted R^2 of carbonation depth, which was 99.84%. On the other hand, the predicted percentage R^2 of the carbonation depth for final regression equation was 89.43%, which is considered significant. The predicted R^2 for both equations indicates the prediction ability of the model is acceptable. Furthermore, the equations were indicated that all factors have a significant effect on the carbonation depth, which confirms the role of density and curing conditions on accelerating the sequestration of CO_2 into FCB.

3.3.6. Microstructure Analysis (SEM)

SEM images were used to identify the morphology characteristic of FCB samples that are related to the density aspects and curing conditions. Images show, after 28 days of carbonation, the formation of calcite ($CaCO_3$) in FCB, Figure 12a,b. The results revealed that a low level of calcite formation was represented in the specific surface area of carbonated FCB that cured at low temperature and CO_2 concentration 27 °C and 10%, respectively, as shown in Figure 12a. In contrast, the increment of temperature and CO_2 concentration to 40 °C and 20% were playing a vital role in the formation of calcite in FCB, as presented in Figure 12b. As expected, a great deal of hydration products mainly consisting of C-S-H formed via carbonation resulting healing of FCB pores [36]. However, the pores cannot be totally healed in 28 days due to the high level of porosity in the FCB, which has a low level of density compared to normal concrete bricks as demonstrate in SEM images. This finding confirmed the finding of the previous studies, the carbonation process is slow therefore, its takes time to heal the pores via precipitated $CaCO_3$ [37,38]. Overall, the microstructural analysis of FCB confirms that the carbonation reaction has the ability to decrease the porosity by formation of $CaCO_3$, which in turn increase with the increasing of temperature and CO_2 concentration.

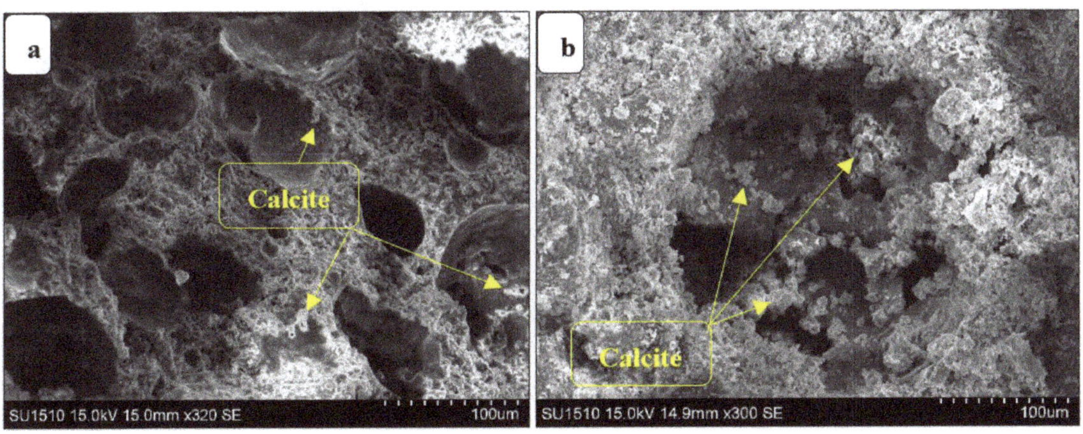

Figure 12. SEM images. (**a**) FCB specimens at 27 °C of temperature and 10% of CO_2 concentration. (**b**) FCB specimens at 40 °C of temperature and 20% of CO_2 concentration.

4. Conclusions

This study showed the use of 2^k factorial and RSM as statistical analysis tools to optimize the carbonation depth of FCB. The optimization was carried out to investigate the effect of the parameters (density, temperature and CO_2 concentration) on the carbonation depth of FCB. Based on the desirability optimization approach, the optimal carbonation depth was 9.7 mm, which was achieved with 1300 kg/m^3, 40 °C and 20% CO_2 concentration. The density of FCB is considered the most significant factor on the carbonation depth followed by CO_2 concentration and temperature with the effective values −8.57, 3.67, and 3.01, respectively. In contrast, the minimum carbonation depth could be achieved when the density, temperature and CO_2 concentration are at the following levels of 1800 kg/m^3, 31.8 °C and 10% CO_2 concentration, respectively. The significance of the factors used to accelerate the carbonation depth of FCB presents novel feedback. Notably, a single parameter may accelerate the carbonation depth, but to reach the optimum point, the other factors cannot be neglected. Therefore, the statistical analysis and optimization of the carbonation depth are required to sequester large quantities of CO_2 into FCB.

Author Contributions: A.F.A.; conceptualization, data curation, formal analysis, resources, writing—original draft, funding acquisition, J.M.I.; conceptualization, supervision, validation, funding acquisition, project administration, writing—review and editing, H.A.T.; methodology, microstructure analysis, visualization, N.O.; data curation and review, A.A.A.-G.; investigation and software, S.S.; chamber fabrication, curing conditions, funding, W.A.H.A.; writing—review and editing and S.A.S.; data curation, investigation and analysis. All authors have read and agreed to the published version of the manuscript.

Funding: The research was funded by MDR (VOT No. H487), Fundamental Research Grant Scheme (FRGS/1/2019/WAB05/UTHM/02/1) and Universiti Sains Malaysia.

Institutional Review Board Statement: Not applicable.

Informed Consent Statement: Not applicable.

Data Availability Statement: The results of the study are not placed in any publicly archived datasets.

Acknowledgments: This research was supported by Universiti Tun Hussein Onn Malaysia through MDR (Vot H487) and Ministry of Higher Education (MOHE) Malaysia through Fundamental Research Grant Scheme (FRGS)(FRGS/1/2019/WAB05/UTHM/02/1). The research was also supported by Universiti Sains Malaysia.

Conflicts of Interest: The authors declare no conflict of interest.

References

1. Narayanan, N.; Ramamurthy, K. Structure and properties of aerated concrete: A review. *Cem. Concr. Compos.* **2000**, *22*, 321–329. [CrossRef]
2. Amran, Y.H.M.; Farzadnia, N.; Ali, A.A.A. Properties and applications of foamed concrete; A review. *Constr. Build. Mater.* **2015**, *101*, 990–1005. [CrossRef]
3. Lim, S.K.; Tan, C.S.; Zhao, X.; Ling, T.C. Strength and toughness of lightweight foamed concrete with different sand grading. *KSCE J. Civ. Eng.* **2015**, *19*, 2191–2197. [CrossRef]
4. Mindess, S. *Developements in the Formulation and Reinforcement of Concrete*; Woodhead Publishing: Sawston, UK, 2019; ISBN 9781845692636.
5. Song, Y.; Li, B.; Yang, E.; Liu, Y.; Ding, T. Cement & Concrete Composites Feasibility study on utilization of municipal solid waste incineration bottom ash as aerating agent for the production of autoclaved aerated concrete. *Cem. Concr. Compos.* **2015**, *56*, 51–58.
6. Costa, B.L.D.S.; Freitas, J.C.D.O.; Melo, D.M.D.A.; Araujo, R.G.D.S.; de Oliveira, Y.H.; Simão, C.A. Evaluation of density influence on resistance to carbonation process in oil well cement slurries. *Constr. Build. Mater.* **2019**, *197*, 331–338. [CrossRef]
7. Kellouche, Y.; Boukhatem, B.; Ghrici, M.; Tagnit-Hamou, A. Exploring the major factors affecting fly-ash concrete carbonation using artificial neural network. *Neural Comput. Appl.* **2019**, *31*, 969–988. [CrossRef]
8. Namsone, E.; Šahmenko, G.; Korjakins, A. Durability Properties of High Performance Foamed Concrete. *Procedia Eng.* **2017**, *172*, 760–767. [CrossRef]
9. Jiang, J.; Lu, Z.; Niu, Y.; Li, J.; Zhang, Y. Study on the preparation and properties of high-porosity foamed concretes based on ordinary Portland cement. *Mater. Des.* **2016**, *92*, 949–959. [CrossRef]

10. Wang, T.; Huang, H.; Hu, X.; Fang, M.; Luo, Z.; Guo, R. Accelerated mineral carbonation curing of cement paste for CO_2 sequestration and enhanced properties of blended calcium silicate. *Chem. Eng. J.* **2017**, *323*, 320–329. [CrossRef]
11. Zhang, D.; Shao, Y. Effect of early carbonation curing on chloride penetration and weathering carbonation in concrete. *Constr. Build. Mater.* **2016**, *123*, 516–526. [CrossRef]
12. Ekolu, S.O. A review on effects of curing, sheltering, and CO_2 concentration upon natural carbonation of concrete. *Constr. Build. Mater.* **2016**, *127*, 306–320. [CrossRef]
13. Alshalif, A.F.; Irwan, J.M.; Othman, N.; Al-Gheethi, A.A.; Shamsudin, S. A systematic review on bio-sequestration of carbon dioxide in bio-concrete systems: A future direction. *Eur. J. Environ. Civ. Eng.* **2020**, 1–20. [CrossRef]
14. Yoon, I.S.; Çopuroğlu, O.; Park, K.B. Effect of global climatic change on carbonation progress of concrete. *Atmos. Environ.* **2007**, *41*, 7274–7285. [CrossRef]
15. Pacheco Torgal, F.; Miraldo, S.; Labrincha, J.A.; De Brito, J. An overview on concrete carbonation in the context of eco-efficient construction: Evaluation, use of SCMs and/or RAC. *Constr. Build. Mater.* **2012**, *36*, 141–150. [CrossRef]
16. Lovato, P.S.; Possan, E.; Molin, D.C.C.D.; Masuero, Â.B.; Ribeiro, J.L.D. Modeling of mechanical properties and durability of recycled aggregate concretes. *Constr. Build. Mater.* **2012**, *26*, 437–447. [CrossRef]
17. Drouet, E.; Poyet, S.; Le Bescop, P.; Torrenti, J.M.; Bourbon, X. Carbonation of hardened cement pastes: Influence of temperature. *Cem. Concr. Res.* **2019**, *115*, 445–459. [CrossRef]
18. Liu, L.; Ha, J.; Hashida, T.; Letters, S.T. Development of a CO_2 solidification method for recycling autoclaved lightweight concrete waste. *J. Mater. Sci. Lett.* **2001**, 1791–1794. [CrossRef]
19. Ji, L.; Yu, H.; Yu, B.; Zhang, R.; French, D.; Grigore, M.; Wang, X.; Chen, Z.; Zhao, S. Insights into Carbonation Kinetics of Fly Ash from Victorian Lignite for CO_2 Sequestration. *Energy Fuels* **2018**, *32*, 4569–4578. [CrossRef]
20. Muthu Kumar, E.; Ramamurthy, K. Effect of fineness and dosage of aluminium powder on the properties of moist-cured aerated concrete. *Constr. Build. Mater.* **2015**, *95*, 486–496. [CrossRef]
21. Kumar, R.; Lakhani, R.; Tomar, P. *A Simple Novel Mix Design Method and Properties Assessment of Foamed Concretes with Limestone Slurry Waste*; Elsevier B.V.: Amsterdam, The Netherlands, 2018; Volume 171, ISBN 9113322834. [CrossRef]
22. Alshalif, A.F.; Irwan, J.M.; Tajarudin, H.A.; Othman, N.; Al-Gheethi, A.A.; Shamsudin, S.; Altowayti, W.A.H.; Abo Sabah, S. Optimization of Bio-Foamed Concrete Brick Strength via Bacteria Based Self-Healing and Bio-Sequestration of CO_2. *Materials* **2021**, *14*, 4575. [CrossRef]
23. Chen, B.; Wu, Z.; Liu, N. Experimental Research on Properties of High-Strength Foamed Concrete. *J. Mater. Civ. Eng.* **2012**, *24*, 113–118. [CrossRef]
24. Zambrano Leal, A. Sociedad de control y profesión docente. Las imposturas de un discurso y la exigencia de una nueva realidad. *Antimicrob. Agents Chemother.* **2012**, *95*, 45–52. [CrossRef]
25. Panesar, D.K. Cellular concrete properties and the effect of synthetic and protein foaming agents. *Constr. Build. Mater.* **2013**, *44*, 575–584. [CrossRef]
26. Possan, E.; Thomaz, W.A.; Aleandri, G.A.; Felix, E.F.; dos Santos, A.C.P. CO_2 uptake potential due to concrete carbonation: A case study. *Case Stud. Constr. Mater.* **2017**, *6*, 147–161. [CrossRef]
27. Lo, T.Y.; Nadeem, A.; Tang, W.C.P.; Yu, P.C. The effect of high temperature curing on the strength and carbonation of pozzolanic structural lightweight concretes. *Constr. Build. Mater.* **2009**, *23*, 1306–1310. [CrossRef]
28. Chang, C.F.; Chen, J.W. The experimental investigation of concrete carbonation depth. *Cem. Concr. Res.* **2006**, *36*, 1760–1767. [CrossRef]
29. De Ceukelaire, L.; Van Nieuwenburg, D. Accelerated carbonation of a blast-furnace cement concrete. *Cem. Concr. Res.* **1993**, *23*, 442–452. [CrossRef]
30. Meima, J.A.; Comans, R.N.J. Geochemical modelling of weathering reactions in MSWI bottom ash. *Environ. Sci. Technol.* **1997**, *31*, 1269–1276. [CrossRef]
31. Villain, G.; Thiery, M.; Platret, G. Measurement methods of carbonation profiles in concrete: Thermogravimetry, chemical analysis and gammadensimetry. *Cem. Concr. Res.* **2007**, *37*, 1182–1192. [CrossRef]
32. Shamsudin, S.; Lajis, M.A.; Zhong, Z.W.; Ahmad, A.; Wagiman, A. Weld strength in solid–state recycling of aluminum chips: Schweißnahtfestigkeit im Festkörper-Recycling von Aluminium-Spänen. *Mater. Werkst.* **2017**, *48*, 290–298. [CrossRef]
33. Biglarijoo, N.; Nili, M.; Hosseinian, S.M.; Razmara, M.; Ahmadi, S.; Razmara, P. Modelling and optimisation of concrete containing recycled concrete aggregate and waste glass. *Mag. Concr. Res.* **2017**, *69*, 306–316. [CrossRef]
34. Yang, K.-H.; Seo, E.-A.; Tae, S.-H. Carbonation and CO_2 uptake of concrete. *Environ. Impact Assess. Rev.* **2014**, *46*, 43–52. [CrossRef]
35. Talebi, A.; Razali, Y.S.; Ismail, N.; Rafatullah, M.; Azan Tajarudin, H. Selective adsorption and recovery of volatile fatty acids from fermented landfill leachate by activated carbon process. *Sci. Total Environ.* **2020**, *707*, 134533. [CrossRef]
36. Qin, L.; Gao, X. Recycling of waste autoclaved aerated concrete powder in Portland cement by accelerated carbonation. *Waste Manag.* **2019**, *89*, 254–264. [CrossRef]
37. Sharma, D.; Goyal, S. Accelerated carbonation curing of cement mortars containing cement kiln dust: An effective way of CO_2 sequestration and carbon footprint reduction. *J. Clean. Prod.* **2018**, *192*, 844–854. [CrossRef]
38. Branch, J.L.; Kosson, D.S.; Garrabrants, A.C.; He, P.J. The impact of carbonation on the microstructure and solubility of major constituents in microconcrete materials with varying alkalinities due to fly ash replacement of ordinary Portland cement. *Cem. Concr. Res.* **2016**, *89*, 297–309. [CrossRef]

materials

MDPI

Article

Research on the Preparation Parameters and Basic Properties of Premelted Calcium Aluminate Slag Prepared from Secondary Aluminum Dross

Shaoyan Hu [1], Deyong Wang [1], Dong Hou [1,*], Wei Zhao [1,*], Xianglong Li [1], Tianpeng Qu [1] and Qingde Zhu [2]

[1] School of Iron and Steel, Soochow University, Suzhou 215137, China; syhu616@suda.edu.cn (S.H.); deyongwang1222@163.com (D.W.); xlli202005@163.com (X.L.); tianpengqu8119@163.com (T.Q.)

[2] Hongxing Iron and Steel Co., Ltd., Jiuquan Iron and Steel Group, Jiayuguan 735100, China; zhuqingde@jiugang.com

* Correspondence: houdong0702@suda.edu.cn (D.H.); zhaowei0312@suda.edu.cn (W.Z.)

Abstract: Secondary aluminum dross is a byproduct of the electrolytic aluminum industry, whose main components are Al_2O_3, AlN and Na_3AlF_6. Secondary aluminum dross is a type of hazardous waste, with a tremendous yield every year. Realizing the harmless treatment or resource utilization of secondary aluminum dross has important economic and social benefits. In the present research, the process of preparing premelted calcium aluminate slag used for molten steel refining from secondary aluminum dross was studied in detail. Firstly, the chemical composition and phase component of secondary aluminum dross were analyzed systematically. Then, according to phase diagram analysis and melting point measurement, the appropriate mixing ratio of CaO and secondary aluminum dross and the appropriate calcination temperature were determined. On this basis, an experiment of premelted calcium aluminate slag preparation was carried out in a tubular resistance furnace. The phase component and micromorphology of the premelted slag were analyzed by XRD and SEM. The results show that the main component of the premelted calcium aluminate slag is $11CaO{\cdot}7Al_2O_3{\cdot}CaF_2$ phase with a low melting point. The original Na_3AlF_6 phase, which is the cause of leachable fluoride in secondary aluminum dross, disappears totally, and there is no water-soluble fluoride detected in the leaching toxicity detection. The research indicates that the process of preparing premelted calcium slag from secondary aluminum dross is feasible, which provides a helpful reference for the resource utilization of secondary aluminum dross.

Keywords: secondary aluminum dross; premelted calcium aluminate slag; thermodynamic calculation; lime-based calcination; leaching toxicity

Citation: Hu, S.; Wang, D.; Hou, D.; Zhao, W.; Li, X.; Qu, T.; Zhu, Q. Research on the Preparation Parameters and Basic Properties of Premelted Calcium Aluminate Slag Prepared from Secondary Aluminum Dross. *Materials* **2021**, *14*, 5855. https://doi.org/10.3390/ma14195855

Academic Editors: Avelino Núñez-Delgado, Zhien Zhang, Elza Bontempi, Mario Coccia, Marco Race and Yaoyu Zhou

Received: 17 September 2021
Accepted: 3 October 2021
Published: 6 October 2021

1. Introduction

Aluminum is a nonferrous metal, with the largest production and consumption in the world [1]. It is widely used in construction, transportation, electric power, aerospace and other fields, playing an important role in the national economic construction and national defense industries as a basic raw material [2]. At present, metallic aluminum is generally produced by the process of cryolite-alumina molten salt electrolysis, which is a multiphase electrolyte system composed of cryolite-based fluoride as solvent and alumina as solute [3]. Hence, the Na_3AlF_6-Al_2O_3 binary system and Na_3AlF_6-AlF_3-Al_2O_3 ternary system are the basis of the electrolytic aluminum industry [4,5].

Aluminum dross is slag floating on the surface of molten aluminum during the electrolytic process, which is composed of unreacted alumina, cryolite and other raw materials, as well as a small amount of other impurities generated from chemical reactions and the falling-off of anode and cathode materials [6]. It is estimated that 30–50 kg of aluminum dross is generated with the production of 1 ton of metallic aluminum. Considering the tremendous yield of metallic aluminum, millions of tons of aluminum dross are newly generated every year in the world [7]. The aluminum dross generated from the electrolytic process is generally called primary aluminum dross, which usually contains more than 50 wt.% metallic aluminum. Most of the metallic aluminum in primary aluminum dross can be easily separated and recovered by various methods, such as ash frying, ball milling, the rotary kiln process, etc. [8–12]. The residue after metallic aluminum extraction is called secondary aluminum dross. The main components of secondary aluminum dross are Al_2O_3, AlN and a small amount of metallic aluminum, in which the AlN can react with water and generate ammonia [13,14]. Moreover, secondary aluminum dross contains a small amount of fluoride and cyanide that are soluble in water [15,16]. If secondary aluminum dross is stacked or landfilled without appropriate treatment, it will cause pollution of water resources, atmosphere and soil, which cause serious harm to the environment [17]. At present, secondary aluminum dross has been included in the list of hazardous wastes [12]. Therefore, it is of great significance to achieve harmless treatment for resource utilization of secondary aluminum dross.

Due to the low content of metallic aluminum, it is not economical to extract aluminum from secondary aluminum dross. Much research on the utilization of secondary aluminum dross has been carried out, as described below. As the content of Al_2O_3 in secondary aluminum dross is high, mixing secondary aluminum dross with other raw materials to prepare refractories is an important utilization approach. Adeosun et al. [18] prepared refractory bricks by mixing kaolin and secondary aluminum dross and studied the specific effects of mixing ratio and sintering temperature on refractory properties. Ewais et al. [19] utilized aluminum sludge and aluminum dross to manufacture calcium aluminate cement. Aluminum sludge is a byproduct of aluminum profile processing, and it is composed of calcium aluminate hydrate, calcium carbonate and hydroxide. Mailar et al. [20] tried to replace some sintering raw materials with secondary aluminum dross to make refractory bricks. Although the performance of the final product met the relevant basic standards, its oxidation resistance property was far inferior to refractory sintered bricks without aluminum dross addition. In a subsequent study, Li et al. [21] found that reducing the content of salt impurities in secondary aluminum dross can effectively improve the oxidation resistance level of the refractories. In recent years, with the development and wide application of flocculants, using secondary aluminum dross to produce flocculants has also been studied by many researchers [22–24]. Shi et al. [25] prepared a polyaluminum chloride flocculant by pickling secondary aluminum dross with hydrochloric acid. The new flocculant showed good performance in removing impurities from wastewater and other waste liquid. Du et al. [26] optimized the operation method of preparing a flocculant from secondary aluminum dross, which helped to reduce the dosage of hydrochloric acid. Chao et al. [27] studied the influence of hydrochloric acid concentration, leaching temperature, leaching time and additives and obtained optimum conditions for the preparation of polyaluminum chloride by secondary aluminum dross acid leaching. Additionally, Kang et al. [28] used secondary aluminum dross as a raw material to produce low-iron aluminum sulfate by the method of co-deposition, whose product is expected to be widely used in papermaking, textile, water purification and other fields. David et al. [29] proposed a simple method with high efficiency for generating high pure hydrogen by hydrolysis in tap water of highly activated aluminum dross.

As mentioned above, many approaches to produce industrial products from secondary aluminum dross have been studied. However, few approaches have been applied to large-scale production. On the one hand, the existence of AlN and fluorine salt in secondary aluminum dross has an unstable, negative impact on the quality of industrial products. On the other hand, the consumption amount of those industrial products is too small compared with the generation amount of secondary aluminum dross, and it is difficult to solve the problem of aluminum dross accumulation.

As mentioned above, the main components of secondary aluminum dross are Al_2O_3 and AlN. Coincidentally, Al_2O_3 is also the main component of molten steel refining slag, and AlN has reducibility and deoxidation abilities, which are beneficial to molten steel desulfurization. Therefore, the application of secondary aluminum dross in molten steel refining has attracted attention and research [30]. Li et al. [31] used secondary aluminum dross to modify the composition of molten steel refining slag. Lime, fluorite and secondary aluminum dross were mixed and added into molten steel as refining slag, and as a result, fine deoxidation and desulfurization performance were achieved. Wang et al. [32] carried out an experiment to study the effect of aluminum dross on the desulphurization of pipe line steel, using secondary aluminum dross to replace aluminum oxide in the refining slag. Similarly, better desulfurization results than traditional desulfurization agents were achieved. Both the above studies verified the excellent performance of secondary aluminum dross in molten steel desulfurization. The theoretical reason for that advantage happens to be the AlN and fluoride contained in the secondary aluminum dross. AlN has reducibility and deoxidation abilities, which can reduce the oxygen content both in refining slag and molten steel, thereby improving the thermodynamic conditions of desulfurization [33]. Fluoride helps to reduce the viscosity of refining slag and improve the kinetic conditions of desulfurization [34].

In previous research on molten steel refining slag, aluminum dross was mixed with another flux and then directly added into molten steel without premelting treatment. On the one hand, it is not conducive to the rapid melting of slag. On the other hand, cryolite in aluminum dross will volatilize while adding it into molten steel and pollutes the steelmaking environment seriously, which is one of the key factors limiting its industrial application [35,36]. In addition, because secondary aluminum dross is hazardous waste, if it is not pretreated properly, its transportation and storage are subject to limitations. Therefore, the process of preparing calcium aluminate premelted slag from secondary aluminum slag is proposed in this study. Secondary aluminum dross needs to be mixed with lime firstly, then the mixture is calcinated at high temperature. During the calcination stage, Al_2O_3 in secondary aluminum dross reacts with CaO in lime, forming calcium aluminate with a low melting point, and those leachable, water-soluble fluorides are transformed into harmless calcium fluoride [37–39].

In order to check the feasibility of the process and find out the appropriate operating parameters, this research took industrial secondary aluminum dross generated by an aluminum electrolysis plant as raw material. Composition analysis and phase characterization of the secondary aluminum dross were carried out firstly. On this basis, thermodynamic calculation and melting point measurement were performed to obtain the appropriate material mixing ratio and calcination temperature. Based on that, lab-scale experiments preparing premelted calcium aluminate slag were carried out. Mineralogical characteristics and leaching toxicity of the premelted slag were detected.

2. Analysis of Secondary Aluminum Dross

As mentioned above, the secondary aluminum dross was obtained from an aluminum electrolysis plant, which was the residue left after extracting metallic aluminum from primary aluminum dross by the ash-frying method. As the key raw material of the present research, it was essential to obtain the basic properties of the secondary aluminum dross, including chemical composition and phase component. The secondary aluminum dross was ground to a particle size smaller than 0.074 mm with a ball mill pulverizing machine; then, the pulverized dross was sent for detailed inspection and analysis. Firstly, qualitative chemical composition analysis of the secondary aluminum dross was carried out by the X-ray fluorescence spectrometry method, and the results are shown in Table 1. Major elements in the secondary aluminum dross are Al and O, indicating that the main components are Al_2O_3 and other aluminum-containing compounds, which conform to the typical properties of secondary aluminum dross [6]. Apart from the Al and O, contents of the F, Na, Cl and Mg are relatively high, more than 1%.

Table 1. Qualitative analysis of the secondary aluminum dross composition by X-ray fluorescence spectrometry.

Element	Al	O	F	Na	Mg	Cl	Si	V	Ca	S	Ti	K	Fe
Content, wt.%	50.89	23.4	3.6	2.73	1.34	1.19	0.725	0.529	0.491	0.363	0.322	0.302	0.229

Based on the results of the qualitative analysis, a subsequent quantitative analysis was carried out to obtain an accurate chemical composition. Referring to a previous report [6] and the present qualitative analysis results, contents of the total aluminum (T.Al), metallic aluminum (M.Al), total fluorine (F), total nitrogen (N) and total sodium (Na) were detected purposefully. The results of the quantitative analysis are shown in Table 2. Comparing the detection results shown in Tables 1 and 2, it can be seen that the deviation of the main elements is small, indicating that the detection results are accurate and acceptable. As can be seen from Table 2, the content of M.Al is only 4.58%, indicating that most of the metallic aluminum in the primary aluminum dross was extracted and recovered.

Table 2. Quantitative analysis of the secondary aluminum dross composition.

Element	T.Al	M.Al	F	N	Na
Content, wt.%	55.49	4.58	2.61	9.92	2.25

Both Tables 1 and 2 show only the element contents. In order to know the specific components of the secondary aluminum dross well, it was necessary to analyze its phase component by the X-ray diffraction method (XRD). The XRD analysis result of the secondary aluminum dross is shown in Figure 1. The main phases in the secondary aluminum dross are Al_2O_3, AlN, $MgAl_2O_4$, Na_3AlF_6 and $Al_2O_3{\cdot}H_2O$.

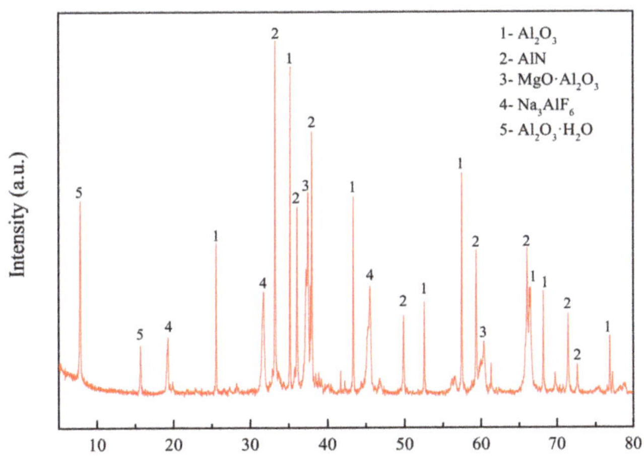

Figure 1. X-ray diffraction phase analysis of the secondary aluminum dross.

According to the chemical composition and phase component, it can be concluded that the element of N mainly exists in the form of AlN, the element of Mg mainly exists in the form of $MgO{\cdot}Al_2O_3$ and the element of F mainly exists in the form of Na_3AlF_6. Based on the element contents shown in Table 2 and the phase components, the content of different phases can be estimated. It is assumed that all the N exists in the form of AlN. Since the content of N is 9.92%, the content of AlN should be 29.05%. It is assumed that all the F exists in the form of Na_3AlF_6. Since the content of F is 2.61%, the content of Na_3AlF_6 should be 4.81%. Similarly, it is assumed that all the Mg exists in the form of $MgO{\cdot}Al_2O_3$. Because the content of Mg was not quantitatively detected, the content of $MgO{\cdot}Al_2O_3$ was estimated based on the qualitative analysis result. Since the content of Mg is 1.34%, the content of $MgO{\cdot}Al_2O_3$ should be 6.15%. Finally, it is assumed that all the rest of the aluminum elements, apart from those contained in M.Al, AlN, Na_3AlF_6 and $MgO{\cdot}Al_2O_3$, exist in the form of Al_2O_3. Based on the above assumption, the content of Al_2O_3 could be estimated, and the calculated content of Al_2O_3 was found to be 54.44%.

3. Thermodynamic Analysis and Melting Point Measurement

The purpose of this research was to prepare premelted calcium aluminate slag from secondary aluminum dross. It was necessary to determine the appropriate proportion of lime and secondary aluminum dross, as well as the calcination temperature. At high temperature, CaO and Al_2O_3 can form various calcium aluminate compounds with different melting points. Finding out the appropriate proportion with a low melting point is not only beneficial to reducing the calcination temperature but also to the rapid melting of the calcium aluminate slag during the molten steel refining process.

3.1. Phase Diagram Analysis

Considering that the main components of secondary aluminum dross are Al_2O_3 and AlN, a CaO-Al_2O_3-AlN ternary phase diagram was calculated and drawn and is shown in Figure 2. The ternary phase diagram was calculated and drawn by thermodynamic software FactSage 7.2, which was developed by Thermfact/CRCT (Montreal Canada) and GTT-Technology (Aachen, Germany). During the ternary phase diagram calculation, the selected database was FToxide, and the operating pressure was 101,325 Pa. Isotherm lines from 1273 K to 2273 K were plotted in the diagram with an interval of 50 K. According to the CaO-Al_2O_3-AlN ternary phase diagram, it is obvious that the addition of AlN will significantly increase the melting point temperature of the mixture. CaO and Al_2O_3 can form a low-melting point phase of calcium aluminate, while AlN can hardly form a new phase with CaO and Al_2O_3. The XRD analysis results shown in Figure 1 also illustrate that the AlN exists as an independent phase. Thus, in order to analyze the formation thermodynamics of calcium aluminate more clearly, a CaO-Al_2O_3 binary phase diagram was obtained from the slag atlas [40] and is shown in Figure 3, ignoring the influence of the AlN phase temporarily.

As shown in Figure 3, there exist five binary phases: $3CaO{\cdot}Al_2O_3{\cdot}$(C3A), $12CaO{\cdot}7Al_2O_3$ (C12A7), $CaO{\cdot}Al_2O_3$ (CA), $CaO{\cdot}2Al_2O_3$ (CA2) and $CaO{\cdot}6Al_2O_3$ (CA6). The liquidus temperature drops rapidly while adding Al_2O_3 to CaO. The peritectic temperature between CaO and $3CaO{\cdot}Al_2O_3$ is 1812 K. The minimum melting compositions of the CaO-Al_2O_3 binary system are the eutectics between $12CaO{\cdot}7Al_2O_3$ and either $3CaO{\cdot}Al_2O_3{\cdot}$or $CaO{\cdot}Al_2O_3$, which are located at 1673 K and 1668 K, respectively. Once beyond the $12CaO{\cdot}7Al_2O_3$ eutectic with $CaO{\cdot}Al_2O_3$, the liquidus temperature begins to rise rapidly with the increase in aluminum content. The eutectic temperature between $CaO{\cdot}Al_2O_3$ and $CaO{\cdot}2Al_2O_3$ is 1875 K. The eutectic temperature between $CaO{\cdot}2Al_2O_3$ and $CaO{\cdot}6Al_2O_3$ is 2035 K. The peritectic temperature between $CaO{\cdot}6Al_2O_3$ and Al_2O_3 is 2103 K.

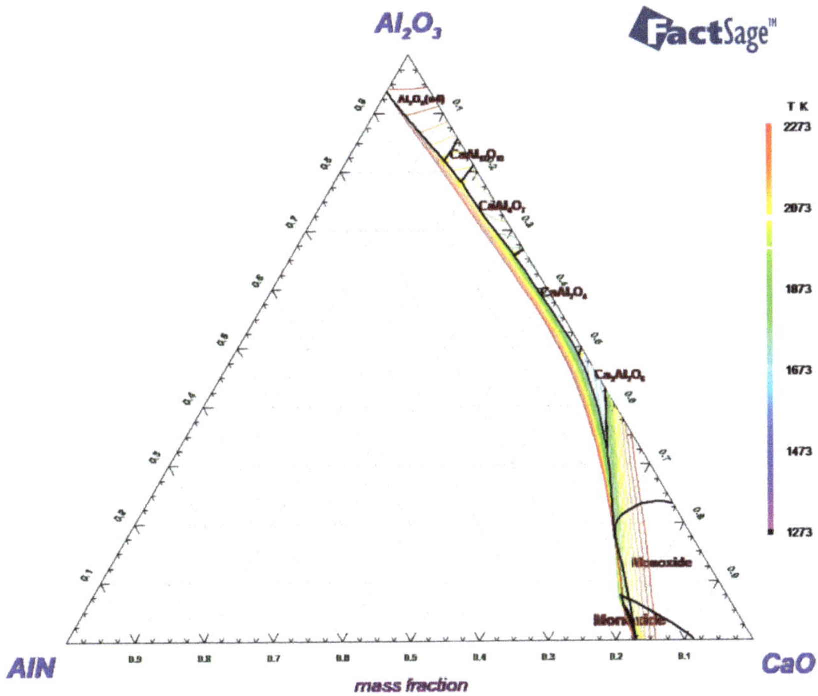

Figure 2. CaO-Al_2O_3-AlN ternary phase diagram for p = 1 atm, drawn with FactSage 7.2.

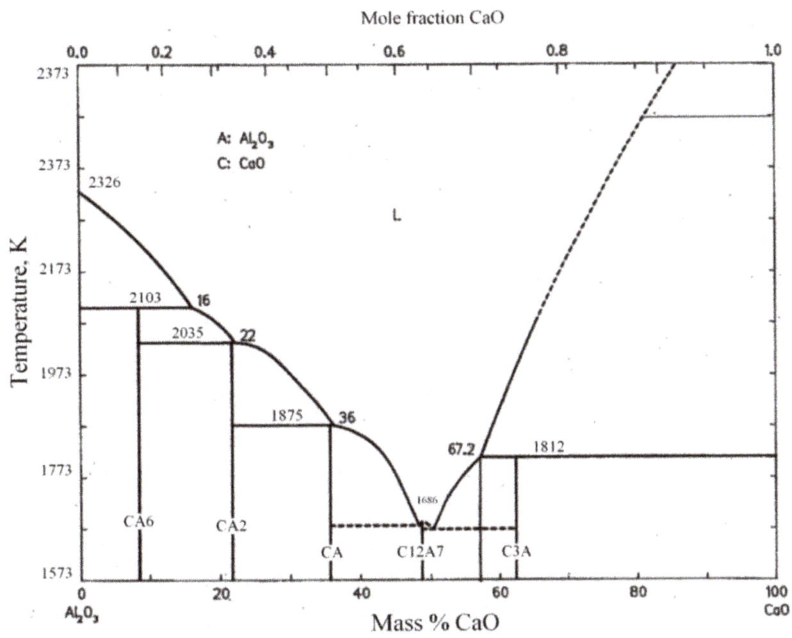

Figure 3. CaO-Al_2O_3 binary phase diagram, obtained from the slag atlas [40].

Considering the existence of Na_3AlF_6 in secondary aluminum dross, it is necessary to analyze the effect of Na_3AlF_6 on the CaO-Al_2O_3 system. However, there is no CaO-Al_2O_3-

Na_3AlF_6 ternary phase diagram currently. As reported by Huang et al. [39], Na_3AlF_6 can react with CaO and Al_2O_3 according to the formula shown in Equation (1). Na_3AlF_6 can be transformed into CaF_2 at high temperature. Due to the lack of a CaO-Al_2O_3-CaF_2-Na_2O phase diagram, the CaO-Al_2O_3-CaF_2 ternary phase diagram was adopted for analyzing the influence of fluoride, as shown in Figure 4. Considering the low content of Na in secondary aluminum dross, the analysis is considered acceptable. As shown in Figure 4, adding CaF_2 into the CaO-Al_2O_3 system is helpful to reduce the melting temperature. If the mixing ratio of CaO and Al_2O_3 is near the $12CaO{\cdot}7Al_2O_3$ phase, adding a small amount of CaF_2 can form a new phase of $11CaO{\cdot}7Al_2O3{\cdot}CaF_2$, whose melting point is around 1773 K.

$$17CaO + 6Al_2O_3 + 2Na_3AlF_6 = 5CaF_2 + 3Na_2O + 11CaO{\cdot}7Al_2O_3{\cdot}CaF_2 \quad (1)$$

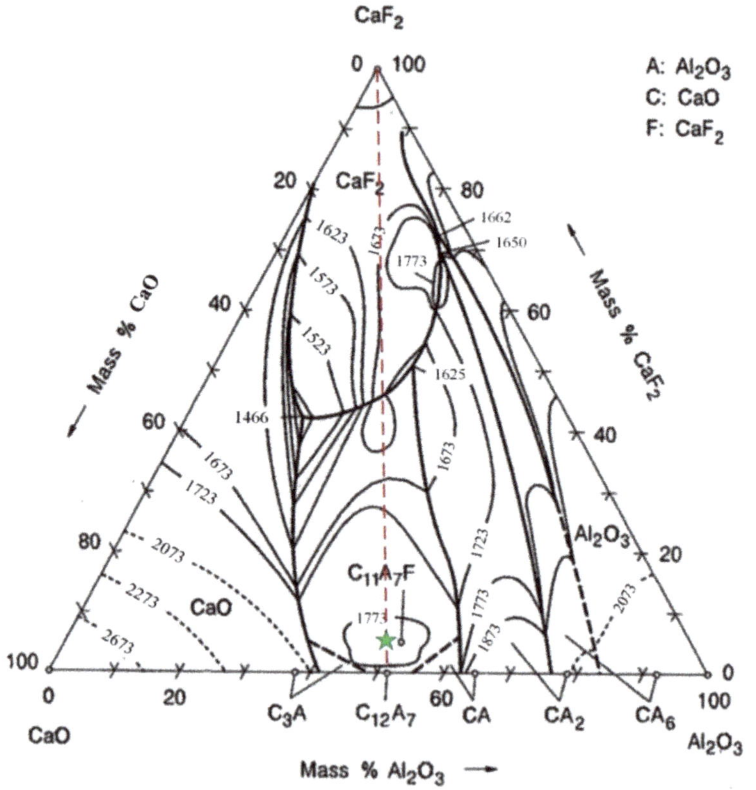

Figure 4. CaO-Al_2O_3-CaF_2 ternary phase diagram, obtained from the slag atlas [40].

As a summary, in order to obtain a low-melting point mixture of lime and secondary aluminum dross, the ratio of CaO to Al_2O_3 should be adjusted near to $12CaO{\cdot}7Al_2O_3$, whose mass ratio is 0.94.

3.2. Melting Point Measurement Experiment

3.2.1. Experimental Scheme

Based on the above phase diagrams' analysis, melting point measurement experiments were carried out to study the effect of the raw material mixing ratio. An analytical reagent of CaO was used for raw material mixing, whose CaO content was more than 98.0%. According to the analyzed composition of secondary aluminum dross, the studied mass ratio of CaO to secondary aluminum dross was changed from 0.4:1 to 1.2:1, with an interval of 0.2.

3.2.2. Experimental Procedures and Devices

The analytical reagent of CaO and the secondary aluminum dross were mixed firstly according to the experimental scheme. Then, the mixed raw material was placed into a ball milling machine for further crushing and mixing. All of the raw materials were ground to a particle size smaller than 0.074 mm. Then, the mixed powdery raw material was pressed into a tablet and pushed into the melting point measuring instrument. The heating rate of the instrument was about 10 K/min, and the maximum temperature was about 1723 K. An image of the sample can be observed by a camera so as to judge the softening temperature, hemisphere temperature and flowing temperature of the sample. Pictures of the ball milling machine and the melting point measuring instrument are shown in Figure 5.

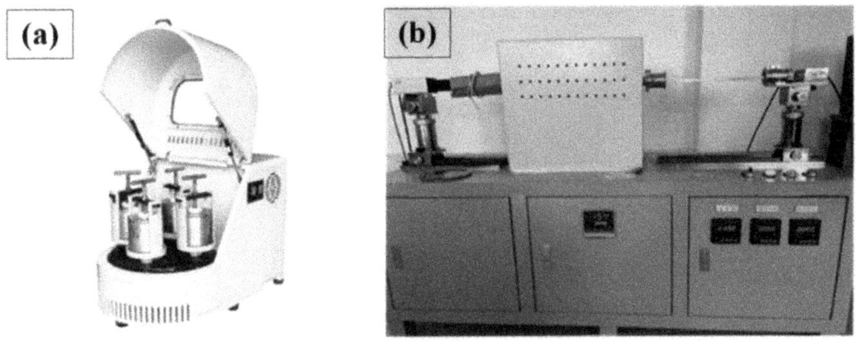

Figure 5. Devices used in melting point measurement experiment. (**a**) Ball milling machine; (**b**) melting point measuring instrument.

3.2.3. Results and Discussion

Results of the melting point measurement are shown in Table 3. Within the temperature limit of the instrument, only the samples with mass ratios of 0.6:1, 0.8:1 and 1.0:1 melt. Pictures of the three samples in hemispherical shape and after being taken out from the instrument are shown in Figure 6. Other samples could not even soften. When the mass ratio of CaO to secondary aluminum dross changes from 0.6:1 to 1.0:1, the melting temperature decreases slightly. The flowing temperatures of all three samples are lower than 1723 K. The measured results indicate that as long as the mass ratio varies from 0.6:1 to 1.0:1, the flowable liquid phase can be obtained at 1723 K. Considering that the difference in melting temperature between the above three samples is small, and the mass ratio of 0.6:1 is closer to the composition of $12CaO{\cdot}7Al_2O_3$, the mass ratio of 0.6:1 and the calcination temperature of 1723 K were selected for subsequent research.

Table 3. Melting point of raw material mixture measured by the experiment.

Mass Ratio of CaO to Secondary Aluminum Dross	0:1	0.4:1	0.6:1	0.8:1	1.0:1	1.2:1
Softening temperature, K	—	—	1700	1672	1653	—
Hemispherical temperature, K	—	—	1709	1686	1667	—
Flowing temperature, K	—	—	1721	1705	1699	—

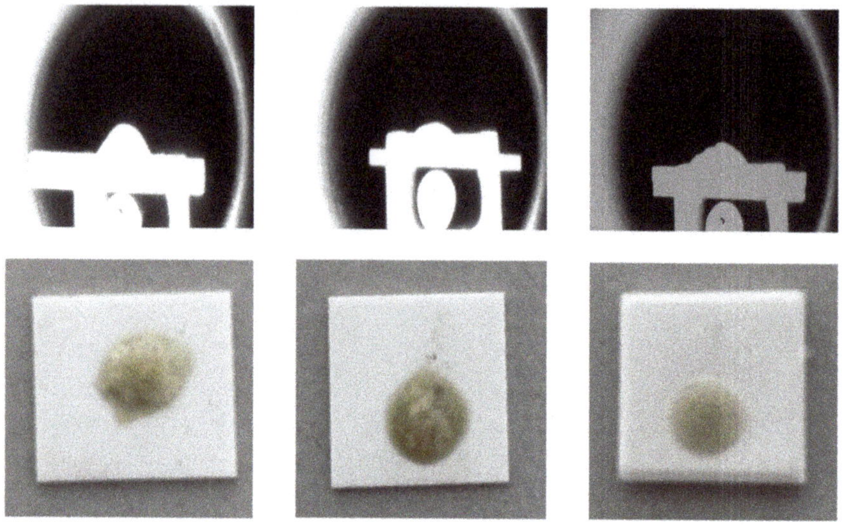

Figure 6. Pictures of the melted samples during melting point measurement.

3.3. *Thermodynamic Calculation of the Calcination Product*

In order to understand the reaction product of CaO and secondary aluminum dross at high temperature in advance, thermodynamic calculation was performed before the premelted slag preparation experiment. The reaction product was calculated by the Equilib module of FactSage software. The composition of the secondary aluminum dross was simplified as 55 wt.% Al_2O_3, 5 wt.% $MgO{\cdot}Al_2O_3$, 30 wt.% AlN, 5 wt.% Na_3AlF_6, and 5 wt.% Al. As the input parameter of the Equilib module, the total weight of the secondary aluminum dross was 100 g, and the weight of CaO was 60 g. In order to compare the influence of ambient atmosphere on the reaction products, 10 g of N_2 or 10 g of Ar was added as a reactant, respectively. Because there was fluoride in the reactants, the database of FToxid-OXFL was selected, in which CaF_2, AlF_3, Na_3AlF_6, MgF_2 were all included. The operating pressure of the atmosphere was set as 101,325 Pa.

Results of the thermodynamic calculation are shown in Table 4. As shown by the calculation results, the reaction products contain a gas phase, liquid phase, $11CaO{\cdot}7Al_2O_3{\cdot}CaF_2$ phase, AlN phase and metallic Al. The mass proportion of $11CaO{\cdot}7Al_2O_3{\cdot}CaF_2$ in the reaction products is the largest. Phase transformations between the raw material and reaction products are discussed in detail. The Na_3AlF_6 phase in the original secondary aluminum dross disappears totally in the reaction products. Most of the Na_3AlF_6 is transformed into the $11CaO{\cdot}7Al_2O_3{\cdot}CaF_2$ phase, part of the fluoride melts into the liquid phase and little volatile fluoride is in the gas phase. The $MgO{\cdot}Al_2O_3$ phase also disappears totally in the reaction products. Part of the Mg element melts into the liquid phase in the form of MgO, and the rest volatilizes into the gas phase in the form of Mg vapor. Ambient atmosphere has a big influence on the distribution of the Mg element. In N_2 atmosphere, most of the Mg element is in the liquid phase. However, in Ar atmosphere, most of the Mg element is in the gas phase. Ambient atmosphere also has a big influence on the transformation of metallic Al. In N_2 atmosphere, metallic Al reacts with N_2 and forms AlN, leading to the total disappearance of metallic Al and the increase in AlN. In Ar atmosphere, metallic Al cannot be transformed into AlN, thus the weight of AlN is unchanged. However, the weight of metallic Al decreases from the initial 5 g to 3.6615 g in Ar atmosphere. This is because part of the metallic Al has a displacement reaction with MgO or CaO, which also leads to the increase in Mg and Ca vapor in the gas phase. As the main components in reactants, CaO reacts with Al_2O_3 to form a liquid phase and an $11CaO{\cdot}7Al_2O_3{\cdot}CaF_2$ phase, which is consistent with the analysis of the CaO-Al_2O_3-CaF_2 ternary phase diagram.

Table 4. Thermodynamic calculation results of the CaO and secondary aluminum dross calcination product.

Phase Components	Calculation Results in N_2 Atmosphere		Calculation Results in Ar Atmosphere	
	Weight	Specific Composition	Weight	Composition
Gas	9.3122 g	7.7302 g N_2; 1.5469 g Na; 0.031709 g Mg; et al.	12.657 g	10 g Ar; 1.6393 g Na; 0.82349 g Mg; 0.13539 g Ca; et al.
Liq-Oxyfluoride	41.473 g	20.282 g CaO; 18.7433 g Al_2O_3; 1.3459 g MgO; 0.41321 g AlF_3; 0.37792 g CaF_2; 0.24021 g $NaAlO_2$; et al.	26.643 g	13.409 g CaO; 13.039 g Al_2O_3; 0.071320 g AlF_3; 0.061993 g CaF_2; 0.050781 g MgO; et al.
$11CaO{\cdot}7Al_2O_3{\cdot}CaF_2$	82.573 g	—	97.0039 g	—
AlN	36.642 g	—	30 g	—
Metallic Al	0	—	3.6615 g	—

4. Experiment of Premelted Slag Preparation

4.1. Experimental Scheme

Based on the selected raw material mixing ratio and calcination temperature, the experiment of premelted calcium aluminate slag preparation was carried out. The experiment was finished in a modified tubular resistance furnace, as shown in Figure 7. A gas supply device and an exhaust device were installed on the furnace. The gas supply device was used to control the ambient atmosphere in the furnace, including top blowing and bottom blowing. In the present research, N_2 or Ar was used as a gas source to study the influence of ambient atmosphere on the calcination products, just like the condition of the thermodynamic calculation. In addition, the mass ratio of CaO to secondary aluminum dross was set to 0.6:1, the calcination temperature was set to 1723 K and the calcination time was set to 2 h.

4.2. Experimental Procedures

The experimental procedure is as follows: The analytical reagent of CaO and the secondary aluminum dross were mixed according to the mass ratio of 0.6:1. The mixed raw material was ground to a particle size smaller than 0.074 mm by a ball milling machine. Then, 200 g of the raw material mixture was placed into an alumina crucible, which was placed into the tubular resistance furnace subsequently. The furnace cover was closed, the gas supply device was turned on and N_2 or Ar were blown according to the experimental scheme. The top blowing gas was blown through a quartz tube, the distance between whose exit and the raw material surface was set to 200 mm to avoid blowing out the powdery raw material. The flow rate of the gas was set to 10 mL/min. Then, the heating system of the furnace was turned on. The furnace started to heat up at a rate of about 10 K/min, and the temperature was maintained at 1723 K for 2 h. After that, the heating system was turned off, and the crucible cooled naturally with the furnace. Due to the thermal insulation effect of the furnace, the cooling process lasted about 12 h. After cooling to room temperature, the crucible was taken out from the furnace. The premelted slags were analyzed by XRD, SEM and leaching toxicity.

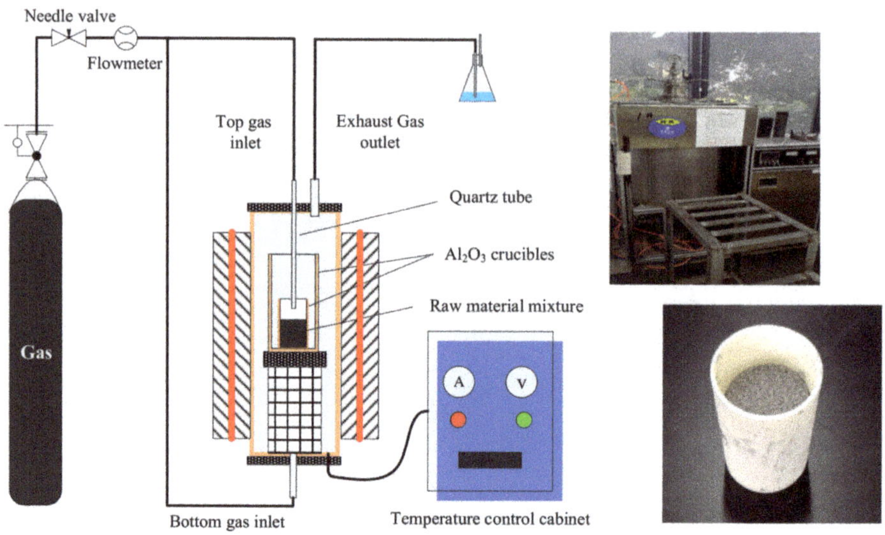

Figure 7. Schematic diagram and physical diagram of the experimental furnace.

4.3. *Results and Discussion*

4.3.1. XRD Analysis

Figure 8 shows the X-ray diffraction patterns of the two premelted slags. The XRD pattern of premelted slag calcinated in N_2 atmosphere is similar with that calcinated in Ar atmosphere. The phase components in both premelted slags are $11CaO·7Al_2O3·CaF_2$ and AlN and $MgO·Al_2O_3$, among which $11CaO·7Al_2O_3·CaF_2$ is the major phase. The detected phase components are consistent with the results of the thermodynamic calculation. There is a little amount of $MgO·Al_2O_3$ phase in the premelted slag, which should be formed by liquid-phase solidification during the cooling process. No metallic Al is found in the premelted slag calcinated in Ar atmosphere, which is different from the thermodynamic calculation. It is estimated that the metallic Al content may be too low to be detected, or the metallic Al may be oxidized by the residual air in the furnace. Comparing the XRD pattern of the original secondary aluminum dross shown in Figure 1 and that of premelted slags shown in Figure 8, it can be found that the Na3AlF6 phase disappears totally after calcination. All of the residual F exists in the form of $11CaO·7Al_2O_3·CaF_2$. $11CaO·7Al_2O_3·CaF_2$ has good thermal and chemical stability and is insoluble in water. The transformation of F is expected to solve the problem of leaching toxicity induced by water-soluble fluoride.

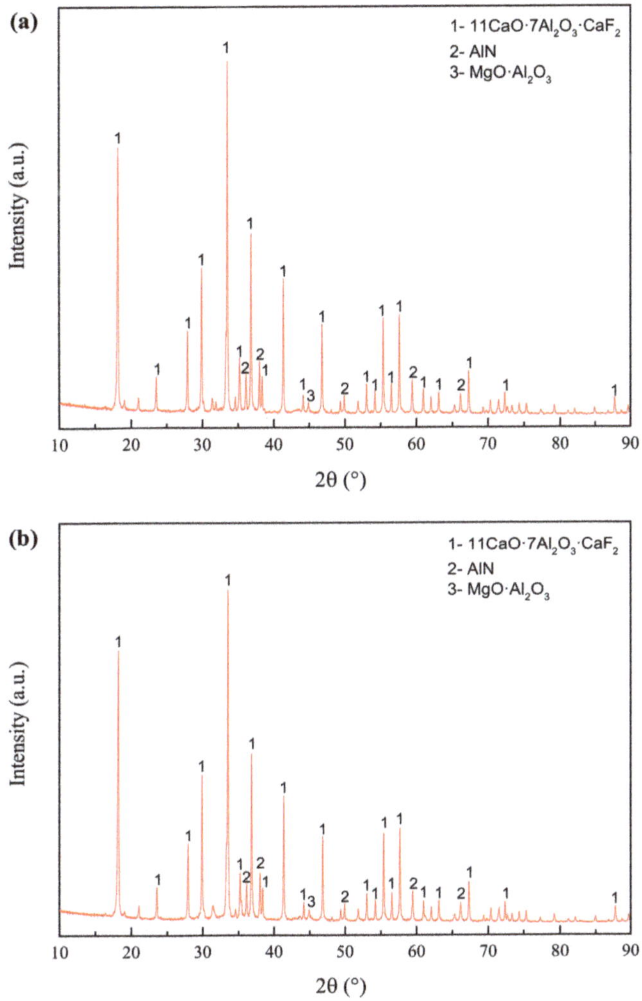

Figure 8. XRD patterns of the premelted calcium aluminate slags. (**a**) Calcinated in N_2 atmosphere; (**b**) calcinated in Ar atmosphere.

4.3.2. Micromorphology Analysis

Scanning electron microscope (SEM) and energy-dispersive X-ray spectrometry (EDS) were used to observe the difference of micromorphology between the original secondary aluminum dross and the premelted calcium aluminate slag. Figure 9 shows the BSE micrographs of typical phases in the original secondary aluminum dross. The morphologies of the different phases are quite different, and the distribution is uneven. As shown in Figure 9a, EDS was used to analyze the element distribution of the bright phase. According to the elemental mapping distribution, the bright phase contains Al_2O_3 and AlN, which coexisted and aggregated in flakes with a relatively compact structure. At the same time, a flocculent phase and bulk phase were observed in local areas, just as shown in Figure 9b. According to the elemental mapping distribution, the flocculent phase should be the Na_3AlF_6 phase. Two spots on the white bulk phase were selected for element analysis, the main element compositions of which are Cl and Na, indicating that the white bulk phase should be NaCl. However, there was no NaCl phase detected in the XRD pattern. It is supposed that the content of the NaCl phase is too low to be detected. From the observation of SEM, the white bulk phase is only scattered in some local areas, and the amount is very small.

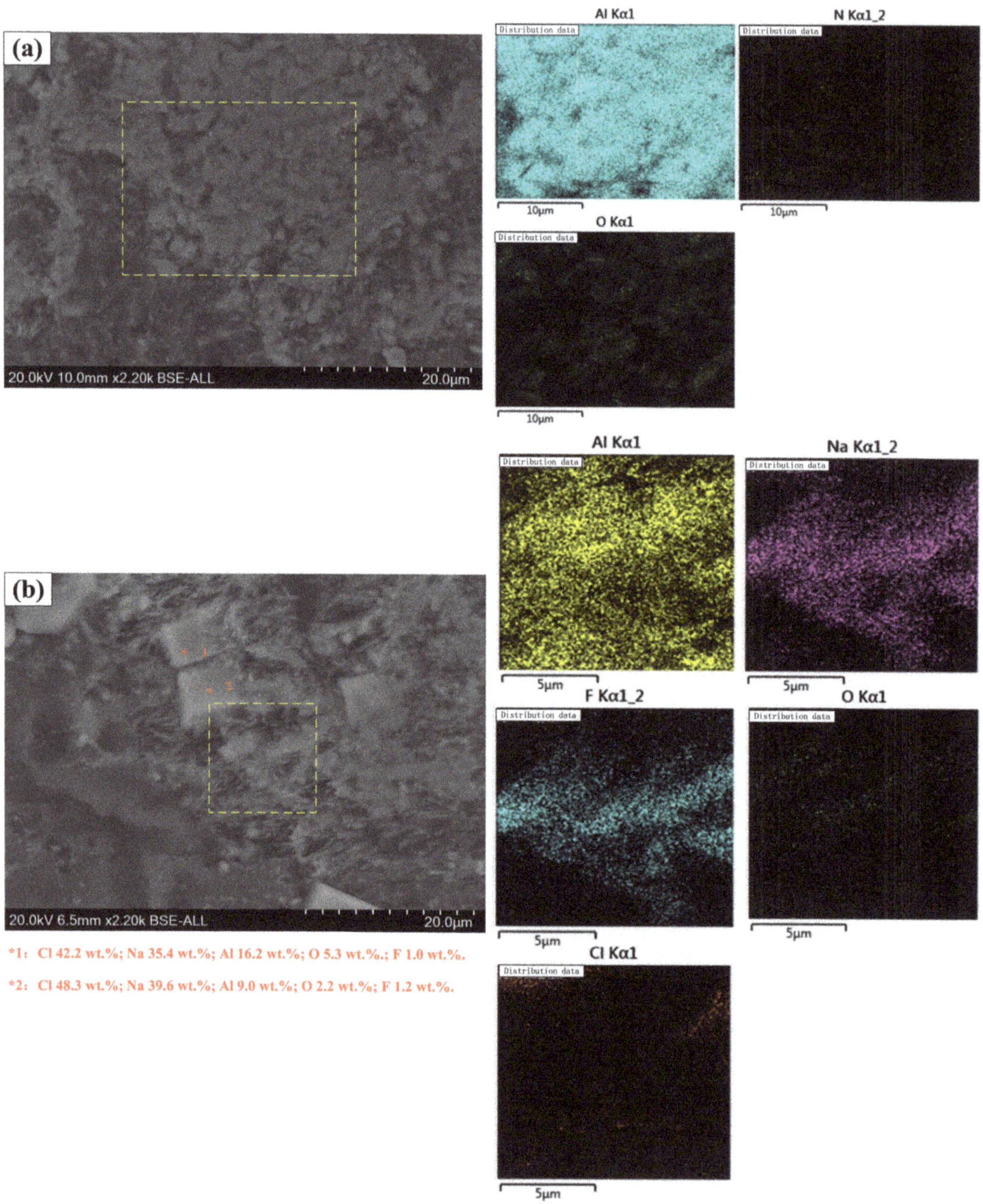

Figure 9. BSE micrographs of typical phases in original secondary aluminum dross, (**a**) phases of Al_2O_3 and AlN coexistence, (**b**) Na_3AlF_6 phase and NaCl phase.

As a contrast, the micrographs of typical phases in premelted calcium aluminate slag are shown in Figure 10. The micromorphology of the premelted slag is more compact and uniform, mainly due to the melting process. According to the elemental mapping distribution shown in Figure 10a, the main components of the bright phase are Al-Ca-O, and the main components of the gray phase are Al-Mg-O. Thus, it is inferred that the gray phase is $MgO \cdot Al_2O_3$. In order to reveal the accurate component of the bright phase, a magnified image of the bright phase is shown in Figure 10b. Area scan and spot scan were performed on the bright phase. The spot scan result shows that the bright phase contains 38.9 wt.% Ca, 26.5 wt.% Al and 1.8 wt.% F. According to the theoretical calculation, the $11CaO \cdot 7Al_2O_3 \cdot CaF_2$ phase theoretically contains 34.1 wt.% Ca, 26.8 wt.% Al and 2.7 wt.% F. Therefore, the bright phase should be $11CaO \cdot 7Al_2O_3 \cdot CaF_2$. Overall, the phases observed in the micrographs match well with the XRD analysis results.

4.3.3. Leaching Toxicity Assessment

According to the hazardous waste identification standard in China (GB5085.3-2007), contents of the water-soluble fluoride in the original secondary aluminum dross and premelted slag were detected to evaluate their leaching toxicity. The used detection method was ion chromatography, whose detection lower limit was 0.74 mg/L. Results of the detection are shown in Table 5. "ND" in the table indicates "not detected" because the content is lower than 0.74 mg/L. It can be seen that the leachable fluoride content in original secondary aluminum dross reaches 866 mg/L, which is far higher than the national standard limit. This is the important reason why secondary aluminum dross is included in the list of hazardous wastes. Fortunately, after the calcination with CaO, the leachable fluoride content in the premelted slag is not detected, regardless of calcination in N_2 atmosphere or in Ar atmosphere. There is almost no leachable fluoride in the premelted calcium aluminate slag prepared from secondary aluminum dross, which meets the national environmental safety standard.

Table 5. Contents of water-soluble fluoride in original secondary aluminum dross and premelted slag.

Sample	Standard Limit, mg/L	Detection Result, mg/L
Original secondary aluminum dross	100	866
Premelted slag calcinated in N_2 atmosphere		ND
Premelted slag calcinated in Ar atmosphere		ND

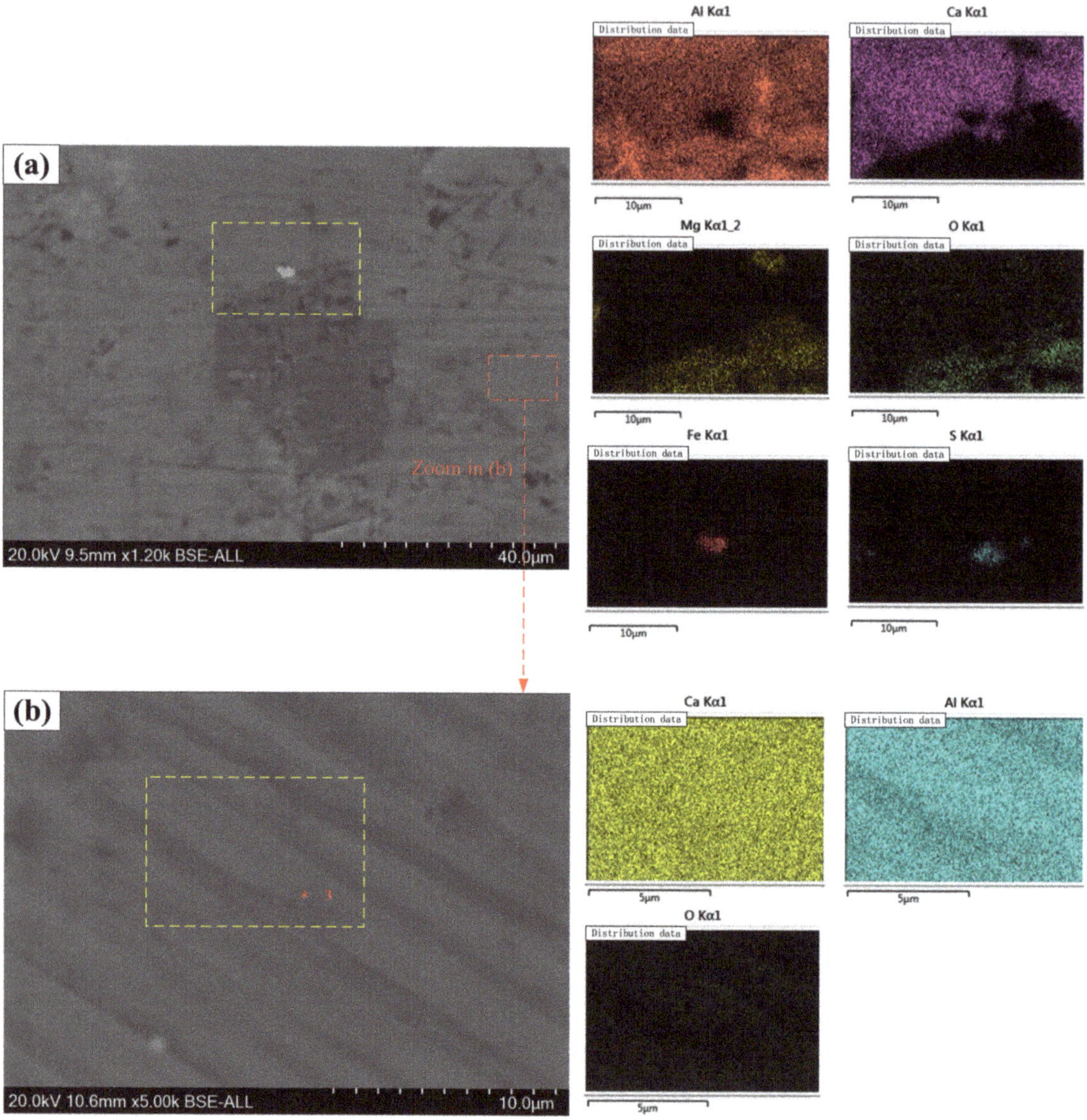

Figure 10. BSE micrographs of premelted calcium aluminate slag, (**a**) typical phases in slag, (**b**) enlarged view of the $11CaO{\cdot}7Al_2O_3{\cdot}CaF_2$ phase.

5. Conclusions

In the present research, the process of preparing premelted calcium aluminate slag from secondary aluminum dross was studied in detail. The appropriate mixing ratio of CaO and secondary aluminum dross and appropriate calcination temperature were determined by thermodynamic analysis and calcination experiments. The main conclusions are as follows:

1. According to quantitative chemical analysis and XRD pattern analysis, specific phase components of the used secondary aluminum dross should be: 54.44 wt.% Al_2O_3, 29.05 wt.% AlN, 6.15 wt.% $MgO{\cdot}Al_2O_3$, 4.81 wt.% Na_3AlF_6, 4.58 wt.% Al.

2. Phase diagram analysis indicates that the AlN phase can hardly form a new phase with CaO and Al_2O_3, existing as an independent phase. The low-melting point composition of the CaO-Al_2O_3 binary system is near to the $12CaO{\cdot}7Al_2O_3$ phase. Adding CaF_2 to the $12CaO{\cdot}7Al_2O_3$ phase can form a new phase of $11CaO{\cdot}7Al_2O_3{\cdot}CaF_2$.
3. When the mass ratio of CaO to secondary aluminum dross varies in the range of 0.6:1 to 1.0:1, the mixture of CaO and secondary aluminum dross can melt within 1723 K. Moreover, the melting point decreases slightly with the increase in the mass ratio in the above range.
4. Premelted calcium aluminate slag was obtained by calcinating the mixture of CaO and secondary aluminum dross with a mass ratio of 0.6:1 at 1723 K for 2 h. The premelted slag contains phases of $11CaO{\cdot}7Al_2O_3{\cdot}CaF_2$, AlN, and $MgO{\cdot}Al_2O_3$, in which $11CaO{\cdot}7Al_2O_3{\cdot}CaF_2$ is the major phase. The original Na_3AlF_6 phase disappears completely, leading to undetectable water-soluble fluoride during the leaching toxicity detection. The experimental results agree well with the thermodynamic calculation results.

Although the composition and leaching toxicity of the premelted calcium aluminate slag meets the requirement of the molten steel refining slag, the specific metallurgical performance of the premelted slag needs further research in a follow-up study.

Author Contributions: Conceptualization, S.H. and D.W.; methodology, S.H. and D.H.; software, W.Z.; formal analysis, S.H. and D.H.; investigation, X.L. and T.Q.; writing—original draft preparation, S.H.; writing—review and editing, S.H., D.H. and W.Z.; visualization, W.Z.; project administration, Q.Z. All authors have read and agreed to the published version of the manuscript.

Funding: This research was funded by National Natural Science Foundation of China (No. 52104337), the Natural Science Foundation of Jiangsu Province (BK20200869) and the China Postdoctoral Science Foundation (2020M681709).

Institutional Review Board Statement: Not applicable.

Informed Consent Statement: Not applicable.

Conflicts of Interest: The authors declare no conflict of interest.

References

1. Ma, Q.; Zhang, J.X. Analysis and development trend of the global primary aluminum market. *Light Metals* **2016**, *11*, 1–7.
2. Redkin, A.; Apisarov, A.; Dedyukhin, A.; Kovrov, V. Recent Developments in Low-Temperature Electrolysis of Aluminum. *ECS Trans.* **2013**, *50*, 205–213. [CrossRef]
3. Yu, B.Y.; Zhao, Z.H.; Zhang, S.; An, R.Y.; Chen, J.M.; Li, R.; Zhao, G.P. Technological development pathway for a low-carbon primary aluminum industry in China. *Technol. Forecast. Soc. Chang.* **2021**, *173*, 121052. [CrossRef]
4. Yang, Y.C. Effect factors of sodium-cryolite electrolysis system in electrolytic aluminum process. *Hydrometall. China* **2012**, *31*, 89–91.
5. Thonstad, J. Alternative electrolyte compositions for aluminum electrolysis. *Miner. Process. Extr. Metall.* **2005**, *114*, 188–191. [CrossRef]
6. David, E.; Kopac, J. Aluminum recovery as a product with high added value using aluminum hazardous waste. *J. Hazard. Mater.* **2013**, *261*, 316–324. [CrossRef] [PubMed]
7. Xu, S.M. The situation of generation, treatment and supervision of common industrial solid wastes in China. *IOP Conf. Ser. Earth Environ. Sci.* **2018**, *113*, 1–4. [CrossRef]
8. Guo, R.; Liu, X.Z.; Li, Q.D.; Yi, X.M. Present situation of high value recycling technology of aluminum ash. *Inorg. Chem. Ind.* **2017**, *49*, 12–15, 25.
9. Tsakiridis, P.E. Aluminum salt slag characterization and utilization: A review. *J. Hazard. Mater.* **2012**, *217–218*, 1–10. [CrossRef]
10. Hwang, J.Y.; Huang, X.; Xu, Z. Recovery of metals from aluminum dross and salt cake. *J. Miner. Mater. Charact. Eng.* **2006**, *5*, 47–62.
11. Du, Z.C.; Li, H.Q.; Bao, W.J.; Li, S.P.; Cai, W.Q. Reaction mechanism of desilification process of high aluminum fly ash by alkali solution. *Chin. J. Process Eng.* **2011**, *11*, 442–447.
12. Xing, X.J.; Wu, Y.D. Review on development on the utilization of aluminum dross. *Environ. Eng.* **2021**, *39*, 148–152.
13. Yang, H.; Shen, S.F.; Liu, H.Y.; Zheng, X.J.; Li, W.G.; Zhao, Q.C.; Wang, J.L.; Luo, Y.F. Study on mineralogical characteristics of secondary aluminum ash. *Nonferrous Metals Eng.* **2019**, *9*, 117–124.
14. Jiang, L.; Qiu, M.F.; Ding, Y.D. Hydrolysis behavior of AlN in aluminum dross. *Chin. J. Nonferrous Metals* **2012**, *22*, 3555–3561.
15. Shen, H.; Forssberg, E. An overview of recovery of metals form slags. *Waste Manag.* **2003**, *23*, 933–949. [CrossRef]
16. Hong, J.P.; Wang, J. Process of aluminum dross recycling and life assessment for Al-Si alloys and brown fused alumi. *Trans. Nonferrous Metal Soc. China* **2010**, *11*, 2155–2161. [CrossRef]

17. Navid, H.J.; Stark, T.D.; Roper, R. Classification and reactivity of secondary aluminum production waste. *J. Hazard. Toxic Radioact. Waste* **2014**, *18*, 1–11.
18. Adeosun, S.O.; Akpan, E.I.; Dada, M.O. Refractory characteristics of aluminum dross-kaolin composite. *JOM* **2014**, *66*, 2253–2261. [CrossRef]
19. Ewais, E.M.M.; Khalil, N.M.; Amin, M.S.; Ahmed, Y.M.Z.; Barakat, M.A. Utilization of aluminum sludge and aluminum slag (dross) for the manufacture of calcium aluminate cement. *Ceram. Int.* **2009**, *35*, 3381–3388. [CrossRef]
20. Mailar, G.; Sujay, R.N.; Sreedhara, B.M.; Manu, D.S.; Parameshwar, H.; Jayakesh, K. Investigation of concrete produced using recycled aluminium dross for hot weather concreting conditions. *Resour.-Effic. Technol.* **2016**, *2*, 68–80. [CrossRef]
21. Li, A.P.; Zhang, H.J.; Yang, H.M. Evaluation of aluminum dross as raw material for high-alumina refractory. *Ceram. Int.* **2014**, *40*, 12585–12590. [CrossRef]
22. Li, F.T.; Jiang, J.Q.; Wu, S.J. Preparation and performance of a high purity poly-aluminum chloride. *Chem. Eng. J.* **2015**, *156*, 64–69. [CrossRef]
23. Yang, Z.; Gao, B.; Yue, Q. Coagulation performance and residual aluminum speciation of $Al_2(SO_4)_3$ and poly-aluminum chloride (PAC) in Yellow River water treatment. *Chem. Eng. J.* **2010**, *165*, 122–132. [CrossRef]
24. Zarchi, I.; Friedler, E.; Rebhun, M. Poly-aluminum chloride as an alternative to alum for the direct filtration of drinking water. *Environ. Technol.* **2013**, *34*, 1199–1209. [CrossRef] [PubMed]
25. Shi, J.L.; Huang, Z.L.; Qin, Q.W.; Yuan, C.G.; Chen, J.Z.; Zhou, Y. Experiment study on the preparation of polyaluminum chloride with secondary aluminum dross. *Metal Mine* **2021**, *50*, 206–210.
26. Du, K.F.; Lv, S.S.; Ni, H.J.; Wang, X.X.; Chen, L.F.; Li, Z.Y. Research on preparation of poly-aluminum chloride by aluminum ash and waste hydrochloric acid. *J. Funct. Mater.* **2020**, *51*, 1207–1213.
27. Chao, X.; Zhang, T.A.; Zhang, Y.B.; Lv, G.Z.; Chen, Y. Study on the preparation of polyaluminum chloride by acid leaching of secondary aluminum dross. *Nonferrous Metals Sci. Eng.* **2021**, *12*, 1–6.
28. Kang, W.T.; Li, J.J.; Li, X.Y.; Liu, Y.M. Study on the new process of manufacturing low-iron aluminum sulfate. *J. Hebei Univ. Sci. Technol.* **2001**, *20*, 65–67, 79.
29. David, E.; Kopac, J. Hydrolysis of aluminum dross material to achieve zero hazardous waste. *J. Hazard. Mater.* **2012**, *209*, 501–509. [CrossRef]
30. Yuasa, G.; Yajima, T.; Ukal, A. Refining practice and application of the Ladle Furnace (LF) Process in Japan. *ISIJ Int.* **1984**, *24*, 412–418. [CrossRef]
31. Li, Y.L.; Zhang, L.F.; Yang, W.; Wang, L.S. Development of ladle slag reducer using aluminum dross. *Iron Steel* **2014**, *49*, 17–23.
32. Wang, D.Y.; Liu, C.J.; Min, Y.; Yu, F.Z.; Jiang, M.F. Effect of Al ash on desulfurization of pipe line steel. *China Metall.* **2007**, *17*, 14–16.
33. Hassall, G.J.; Jackaman, D.P.; Hawkins, R.J. Phosphorus and sulfur removal from liquid steel in ladle steelmaking processes. *Ironmak. Steelmak.* **1991**, *18*, 359–369.
34. Pezzin, R.O.; Berger, A.P.; Grillo, F.F.; Junca, E.; Furtado, H.S.; Oliveira, J.R. Analysis of the influence of the solid and liquid phases on steel desulfurization with slags from the CaO-Al_2O_3 systems using computational thermodynamics. *J. Mater. Res. Technol.* **2020**, *9*, 838–846. [CrossRef]
35. Zhang, Y.Z.; Zhu, X.F.; Yang, J.W.; Zhao, Y.J.; Wang, S.N. Study on harmless application of industrial aluminum dross in steelmaking. *Gansu Metall.* **2020**, *42*, 115–117, 120.
36. Li, S.; Liu, W.C.; Liu, Z.K.; Yan, K. Technical state and prospect on processing of aluminum dross. *Nonferrous Metals (Extr. Metall.)* **2018**, *10*, 25–30.
37. Park, J.H.; Min, D.J. Thermodynamics of fluoride vaporisation from slags containing CaF_2 at 1773 K. *Steel Res. Int.* **2004**, *75*, 807–811. [CrossRef]
38. Li, X.Q.; Shen, S.F.; Wang, L.; Zheng, X.J. Study on roasting characteristics and removal of harmful elements of secondary aluminum dross. *Nonferrous Metals (Extr. Metall.)* **2020**, *58*, 69–74.
39. Huan, S.X.; Wang, Y.W.; Di, Y.Z.; You, J.; Peng, J.P.; Hong, Y.M. Experimental study on alumina extraction by calcination of secondary aluminum ash. *Conserv. Util. Miner. Resour.* **2020**, *40*, 34–39.
40. Verein Deutscher Eisenhüttenleute. *Slag Atlas;* Verlag Stahleisen: Dusseldorf, Germany, 1995.

processes

Article

A Small-Scale Study on Removal of Heavy Metals from Contaminated Water Using Water Hyacinth

An The Huynh [1,2], Yi-Ching Chen [1,*] and Bich Ngoc Thi Tran [3]

1 Department of Environmental Engineering, Da-Yeh University, Changhua 51591, Taiwan; anht@tdmu.edu.vn
2 Department of Management Sciences, Thu Dau Mot University, Binh Duong 590000, Vietnam
3 Institute of Environmental Sciences and Technology, Tra Vinh University, Tra Vinh 940000, Vietnam; ngocbich@tvu.edu.vn
* Correspondence: yiching@mail.dyu.edu.tw; Tel.: +88-69-3910-7419

Abstract: The aim of this research was to determine whether water hyacinth can be used to remove heavy metals, such as cadmium, arsenic, lead, zinc, and copper, from industrial wastewater. Investigations of the pollution removal or prevention potential of aquatic macrophytes, such as heavy metal bio-indicators in aquatic habitats, can prove to be advanced field studies. Water hyacinth is one of the aquatic plant species that has been effectively utilized for the treatment of wastewater. It is extremely effective in removing stains, suspended solids, BOD, organic matter, and heavy metals. This research focused on the use of water hyacinth to treat wastewater from heavy metals. Water hyacinths can grow in sewage, absorbing and digesting contaminants and transforming sewage effluents into comparatively clean water in the process. As a result, the plants have the potential to be used as natural water purification systems at a fraction of the cost of a standard sewage treatment facility. The experiment was performed using healthy, young, and acclimatized water hyacinths. Containment water with a cadmium concentration of 0.5 mg/L, arsenic concentration of 0.5 mg/L, lead concentration of 2 mg/L, zinc concentration of 5 mg/L, and copper concentration of 5 mg/L was added to five different polyethylene pots with 100 g of water hyacinth in each pot. After 30 days, the removal efficiency for heavy metals (Cd, As, Pb, Zn, and Cu) reached 59–92%, and the results were within the permitted limits according to the National Technical Regulation on Industrial Wastewater in Vietnam. Based on this information, it is possible to deduce that water hyacinth can be utilized to remove cadmium, arsenic, lead, zinc, and copper from industrial wastewater effluents efficiently.

Keywords: heavy metals; water hyacinth; cadmium; arsenic; lead; zinc; copper

Citation: Huynh, A.T.; Chen, Y.-C.; Tran, B.N.T. A Small-Scale Study on Removal of Heavy Metals from Contaminated Water Using Water Hyacinth. *Processes* **2021**, *9*, 1802. https://doi.org/10.3390/pr9101802

Academic Editors: Avelino Núñez-Delgado, Zhien Zhang, Elza Bontempi, Mario Coccia, Marco Racea and Yaoyu Zhou

Received: 7 September 2021
Accepted: 8 October 2021
Published: 11 October 2021

Publisher's Note: MDPI stays neutral with regard to jurisdictional claims in published maps and institutional affiliations.

1. Introduction

Water pollution is a burning problem for developing countries. As societies grow, the amount of domestic and industrial waste also increases exponentially. Amounts of waste are increasing but the centralized treatment systems are not enough to deal with them along with unconcentrated discharge habits, which have caused extremely serious water pollution. In particular, heavy metal pollution involving cadmium (Cd), arsenic (As), lead (Pb), zinc (Zn), and copper (Cu) is a serious problem due to the particularly dangerous toxicity of these elements affecting human health, organisms, and the environment [1].

Traditional methods, including physical and chemical processes used to treat heavy metals, are being applied, most of which have complex processes, are quite expensive in terms of economics, and have high technical requirements [2]. A study on removing arsenic contamination in soil by phytoremediation has been published [3]. Additionally, wastewater treatment using aquatic plants has been applied in many parts of the world, which has the advantages of low costs, easy operation, and high pollution treatment level [4–6]. This is a technology for the treatment of wastewater in natural and environmentally friendly conditions while increasing biodiversity and improving the landscape, environment, and local ecosystems [7].

The popular application of water hyacinth (*Eichhornia crassipes*) for wastewater treatment has gained worldwide acclaim [8]. Several metals have extraordinarily high attraction and accretion capacities for water hyacinths [9,10]. A typical marshland colonized by water hyacinth can act, in a way, as "nature's kidney" for proper wastewater treatment, preserving the earth's valuable water resources [8]. Water hyacinth has gained much attention for its capacity to be cultivated in extremely polluted water, in addition to its tendency to accumulate metal ions [11]. Water scarcity is becoming more prevalent around the biosphere and in numerous countries it might be unavoidable by the end of the year. With the realization that surface water contamination is a worldwide problem, the situation becomes much more alarming. Several methods for sustainable water resource utilization have been developed to solve this problem, with wastewater recovery and reuse currently being one of the most important targets. The two most common anthropogenic sources of metal in aquatic ecosystems, according to research, are domestic and industrial waste [12]. Heavy metals in water can be dangerous even in small quantities. Since the beginning of the industrial revolution, hazardous metal contamination in the biosphere has increased drastically. Water hyacinth is an aquatic plant that can effectively remove a variety of pollutants from water, making it important in wastewater treatment [4,5].

Water hyacinth is a major material in the handicrafts industry in Vietnam. It is cheap and commonly planted. Furthermore, water hyacinth can be a good tool for the removal of heavy metal contamination through phytoremediation technology. For economic reasons, the use of water hyacinth in wastewater treatment should be promoted in Vietnam. In addition, water hyacinth is also used in Vietnam as fodder for livestock, compost for mushrooms, and manure. Dried water hyacinth can be processed for use in braided ropes, weaving mats, crafts, or furniture. Therefore, this study was conducted to evaluate the ability of water hyacinth to absorb heavy metals (Pb, Cd, Zn, Cu, As) from wastewater. The results indicated good performance in the reduction of concentrations of heavy metals. The removal efficiency for heavy metals (Cd, As, Pb, Zn, and Cu) reached 59–92% after 30 days, notably for Pb, which was reduced by 92.4%.

2. Characteristics of Water Hyacinth

2.1. *Water Hyacinth*

Water hyacinth is a rapidly growing, floating aquatic plant from the South American Amazon Bay [13]. It is well-known for its propensity to multiply, with the plant's population being able to double in just 12 days. Its propensity to flourish in severely contaminated waters is also well-known. As an aquatic plant capable of improving oxidation pond effluent quality and a key factor for single, integrated, and advanced treatment systems for urban, agricultural, and industrial waste streams, water hyacinth has been intensively explored. Beside wastewater treatment, water hyacinth can also be used in valuable products like biogas, bioethanol, biohydrogen, biofertilizers, and fish feed [14].

2.2. *Ecological Factors*

The ability of ecological technology to recover and reuse resources is an environmentally friendly feature. In aquatic ecosystems, for example, nutrients from phosphorous and nitrogenous wastewater components are recycled into usable biomass via ecological food chains. Warm, nutrient-rich water is ideal for water hyacinth growth. The ideal H^+ ion potential for this aquatic plant's growth is neutral; however, it can potentially withstand hydrogen values ranging from 4 to 10. Water hyacinth can be used to treat a variety of types of wastewater thanks to this key feature. The ideal water temperature for growth is between 28 °C and 30 °C. Temperatures exceeding 33 °C stifle further development. The ideal air temperature is between 21 °C and 30 °C. Water hyacinth can also thrive in both nutrient-rich and nutrient-poor water, as well as in water that has been heavily contaminated with diverse biological and inorganic industrialized effluents containing metal ions [15–17]. Water hyacinth is a common and prolific plant that is free-floating, highly tolerant of harsh settings, and capable of rapid vegetative reproduction and survival in a variety of environments. The maximum reproductive rate is 54.4 g dry weight/m^2/day [15,16]. Plants grow more quickly in the summer, covering 15% more surface area every day.

2.3. Socio-Economic Potential of Water Hyacinth in Vietnam

Water hyacinths were introduced into Vietnam around 1905 and developed throughout the Southern Rivers on inland canals, affecting agricultural production, waterway traffic, and drainage [18]. Water hyacinths also hinder fishing and seriously affect water supply. They grow rapidly in the rainy season, and they are the residences of mosquitoes and disease-causing insects. Water hyacinth seems useless but it is considered a valuable and precious raw material for handicrafts and a "new discovery of the 21st century" because it has been exploited and used since the flood season in 2000. Recently, people in the area of the Mekong Delta have been interested in handicraft products made from this plant. Furthermore, according to the statistics of the General Department of Vietnam Customs (GDVNC), these products are also exported to foreign countries to meet the needs of fastidious markets such as those of the US, China, Japan, Europe, and South Korea, [19].

Thanks to this handicraft industry, many localities have solved unemployment problems and helped farmers increase their incomes, contributing to the eradication of hunger and poverty reduction. Since 2000, water hyacinth knitting has developed strongly in the Mekong Delta, especially in the provinces of Dong Thap, Long An, An Giang, and Vinh Long, and the development of this profession has created more jobs and increased income for farming households. Therefore, water hyacinth is also heavily exploited to provide raw materials for production.

2.4. Mechanism of Wastewater Treatment Using Water Hyacinth

Contaminants and stockpiles can be found in water hyacinth biomass. As they store pollutants in their tissues, these plants are known as bioaccumulators. They have a high tolerance for pollutants such as heavy metals and may absorb significant quantities of them. Phytoextraction is a technique for the removal of heavy metals from contaminated water sources [20]. The routes for pollutant uptake are described below.

Root absorption: The roots absorb contaminants in the aquatic environment. A large cation change throughout the cell membrane is caused by the presence of carboxyl groups in the root system. This acts as a mechanism for heavy metal transport within the root system, where active absorption takes place. Aerobic bacteria develop well in the water systems due to the root structures of water hyacinths (as well as other aquatic floras). Aerobic bacteria acquire nutrients and harvest inorganic compounds, which plants consume as nourishment. The plants improve rapidly and can be collected as rich and profitable fertilizer.

Foliar absorption: In addition to root absorption, foliar absorption can provide plants with small amounts of some pollutants. They are absorbed passively by stoma cells and cuticle fissures.

In this type of absorption, fibrous or feathery roots are trapped in floating particles and germs as bacterial and fungal growth attachment points. Pollutants on the root surface are absorbed by the bacteria that live on it. An ionic imbalance throughout the cell membrane also occurs.

3. Methods

3.1. Instruments

The instruments used in the study included a micropipette (Eppendorf, Hamburg, Germany), flask, test tube, electric stove, and an analytical balance from Adam (UK) with an accuracy of 0.0001 mg.

ICP-MS equipment (Perkin Elmer, Waltham, MA, USA, ELAN 9000) was used to analyze samples with the following parameters: RF power: 1000 (W); sample injector (Perkin Elmer, Waltham, MA, USA): 26 (rpm); auxiliary gas (Perkin Elmer, Waltham, MA, USA): 2 (L/min); plasma gas (Perkin Elmer, Waltham, MA, USA): 0.85 (L/min); nebulizer gas (Perkin Elmer, Waltham, MA, USA): 2.4 (L/min).

3.2. Chemicals

Standard solutions of cadmium, arsenic, lead, copper, and zinc were prepared in a standard solution of 100 μg/mL.

The chemicals used were all pure chemicals manufactured by Merck Chemical (Darmstadt, Germany). The solutions were prepared with double-distilled water.

3.3. Analysis Sample

The sample collection and processing period started on 20 May 2021. After being planted, the sample was monitored and the heavy metal content in the water was analyzed three times, after 10, 20, and 30 days, to observe the metal content in water treated with water hyacinth over time. The heavy metals (Cd, As, Pb, Zn, and Cu) in the plants were determined using inductively coupled plasma-mass spectrometry.

3.4. Experimental Setup

The experiments lasted 30 days and 100 g of water hyacinth was used in the pot. For planting, 30-L foam containers were used, which were washed of dust and soil with distilled water beforehand. The plants were fixed with stones (inexpensive, poorly absorbing substrates). Concentration selection experiments were undertaken based on the permitted threshold levels for Cd, As, Pb, Zn, and Cu in water environments according to the QCVN 40:2011/BTNMT National Technical Regulation on Industrial Wastewater in Vietnam. The experimental concentration was three to five times higher than the allowable threshold. Water hyacinth was grown in irrigation water containing the heavy metals Cd, As, Pb, Zn, and Cu according to selected concentrations:

1. Plant water containing 0.5 mg/L of Cd (II);
2. Plant water containing 0.5 mg/L of As (III);
3. Plant water containing 2 mg/L of Pb (II);
4. Plant water containing 5 mg/L of Zn (II);
5. Plant water containing 5 mg/L of Cu (II).

As a matching plant control sample, plants were planted in distilled water with stone media.

The analytical parameters were Cd, As, Pb, Zn, and Cu in water.

3.5. Water Properties

When simplifying the natural environment, the possibility of the accumulation of heavy metals in the plants had to be taken into consideration whilst at the same time ensuring the accuracy of the experiment. The water composition helped us to identify the nutrient content and determine whether additional nutrients had to be added to the plants [21]. Further, the metal content in the water was also an aspect that we were interested in. Identifying the presence of metal ions and the amount of accumulation helped us to assess whether there was competition for these metal ions in the accumulation in the plants and to identify the research plant that could handle Cd, As, Pb, Zn, and Cu metal ions [2]. The water samples were collected in Phu Giao district, Binh Duong, Vietnam, and the levels of water quality are listed in Table 1.

Based on the results of the water quality analysis shown in Table 1 and the Vietnam National Technical Regulation QCVN 08-MT:2015/BTNMT (Column B1), the parameters for water quality and other heavy metals content were within the allowable thresholds of the national standards. The appropriate quantities of nutrients and mineral ions required for proper plant development were used and there were no metal elements that could obstruct the accumulating process. As a result, the uptake of the tested elements should have been faster.

Table 1. Results of water quality analysis.

No.	Parameters	Unit	Content	QCVN 08-MT:2015/BTNMT (Column B1)
1	Temperature	°C	25.1	-
2	pH	-	6.7	5.5–9
3	EC	μS/cm	58	-
4	DO	mg/L	8.77	≥4
5	TDS	mg/L	27	-
6	BOD_5	mg/L	12.2	15
7	COD	mg/L	23.3	30
8	TSS	mg/L	7.3	50
9	As	mg/L	<0.005	0.05
10	Cd	mg/L	<0.0005	0.01
11	Pb	mg/L	<0.005	0.05
12	Cr	mg/L	<0.005	0.04
13	Hg	mg/L	<0.0005	0.001
14	Zn	mg/L	0.051	1.5
15	Mn	mg/L	0.158	0.5
16	Fe	mg/L	0.772	1.5
17	Total nitrogen	mg/L	2.92	-
18	$N–NO_3^-$	mg/L	0.97	10
19	$N–NO_2^-$	mg/L	<0.005	0.05
20	$N–NH_4^+/NH_3$	mg/L	<0.006	0.9
21	Total phosphorus	mg/L	<0.05	-
22	$P–PO_4^{3-}$	mg/L	<0.05	0.3
23	S^{2-}	mg/L	<0.04	-
24	CN^-	mg/L	-	0.05
25	Coliform	MPN/100mL	1600	7500

3.6. Data Analysis

The analyzed data were assessed and compared with the current QCVN 40:2011/BTNMT Vietnamese standards. The data were processed using Excel and Statgraphics software.

4. Results and Discussions

4.1. The Height Growth of Water Hyacinth after 30 Days

By definition, plant growth is associated with the process of increasing the mass, height, and size of cells or cell organelles. Growth can also represent the development and reproduction of a plant [22]. Height is one of the important indicators for assessing the growth of plants containing different environmental heavy metals (Cd, As, Pb, Zn, and Cu). In addition to the dependence of height on the genetic characteristics of a variety, it also depends on external conditions such as climate, temperature, oxygen, mineral nutrition, fertilizer, water, etc. [23]. The growth of plants in polluted water is important for the absorption and accumulation of heavy metals in plants. The height results for the water hyacinths are presented in Table 2. There was a significant difference in height growth for these samples with $p < 0.05$, which could validly explain their differences in plant growth.

Table 2. Effects of heavy metal concentrations on the height growth of water hyacinths after 30 days.

Element	Original Height(cm)	Height after 30 Days (cm)
Cd	43	$44.0^{a} \pm 0.5$
As	43	$44.3^{ab} \pm 0.5$
Pb	43	$45.7^{bc} \pm 0.8$
Zn	43	$46.3^{c} \pm 1.5$
Cu	43	$46.7^{c} \pm 0.6$
	$p = 0.0105$ *	

(1) a, b, and c represent statistically different values. (2) * means significant with $p < 0.05$.

The results in Table 2 show that the height growth rate of the water hyacinths depended on the absorption capacity of the plants. The decreasing height corresponded to the heavy metals contained in the water in the order Cu > Zn > Pb > As > Cd. The growth condition can be described as follows:

- Plant growth in water environment with 0.5 mg/L Cd and As content. Water hyacinth showed good growth signs, strong and green stems and leaves. The height did not change much;
- Plant growth in a water environment contaminated with Pb continued after 30 days; however, the stems and leaves had yellowed, and some stems and leaves were wilted;
- The plants in water with a Zn concentration of 5 mg/L grew well, increasing in height and standing strong, but in the last days of the cycle, some leaves were wilted and yellowed;
- The plants in water with 5 mg/L of Cu showed good growth: there was an increase in height in some of the canopies and the appearance of extra branches in young trees, lush green stems, and leaves.

The height of the water hyacinth in each pot was considered as height growth. As water hyacinth is a wild plant, it grows very well under natural conditions [24]. However, the height growth of the water hyacinths differed depending on the different metals; some plants even withered, died, and did not develop young plants. The biomass of the pot was then significantly reduced.

4.2. *The Ability of Water Hyacinth to Accumulate Heavy Metals*

Phytoremediation refers to the use of plants to partially or substantially remediate selected contaminants in contaminated soil, sludge, sediment, groundwater, surface water, and wastewater [25]. Phytoremediation is popular because of its cost-effectiveness, aesthetic advantages, and long-term applicability [26].

The objective of utilizing water hyacinth in phytoremediation technique testing was to assess its efficiency for heavy metal treatment. Therefore, after being planted, the sample was monitored and the heavy metal content in the water was analyzed three times, after 10, 20, and 30 days, to observe the metal content in water treated with water hyacinth over time. The results are presented in Table 3.

Table 3. The concentrations of heavy metals in water treated with water hyacinth over time.

Days	Cadmium (Cd) Concentration			Arsenic (As) Concentration			Lead (Pb) Concentration			Zinc (Zn) Concentration			Copper (Cu) Concentration		
	Cd (mg/L)	CV (%)	H (%)	As (mg/L)	CV (%)	H (%)	Pb (mg/L)	CV (%)	H (%)	Zn (mg/L)	CV (%)	H (%)	Cu (mg/L)	CV (%)	H (%)
0	0.5			0.5			2			5			5		
10	0.406 ± 0.0006	0.14	18.8	0.394 ± 0.0006	0.15	21.2	1.498 ± 0.001	0.07	25.1	3.891 ± 0.002	0.05	22.2	3.859 ± 0.002	0.05	22.8
20	0.301 ± 0.002	0.51	39.8	0.285 ± 0.001	0.35	43.0	0.618 ± 0.002	0.32	69.1	2.686 ± 0.001	0.04	46.3	2.527 ± 0.003	0.12	49.5
30	0.203 ± 0.002	0.75	59.4	0.196 ± 0.002	1.02	60.8	0.153 ± 0.005	3.27	92.4	1.989 ± 0.004	0.18	60.2	1.963 ± 0.0006	0.03	60.7
QCVN 40:2011/BTNMT (Column B)	0.1			0.1			0.5			3			2		

H (%): removal efficiency percentage, CV (%): coefficient of variation.

The results show that the ability of water hyacinth to accumulate heavy metals gradually decreased with the remaining heavy metal content in the water in the order Cd < Zn < Cu < As < Pb. In addition, Table 3 also shows that, with regard to the experimental conditions of the sample pots, the concentrations of Cd, As, Pb, Zn, and Cu in the water decreased gradually with increasing time of treatment with water hyacinth. Specifically, the Cd, As, Pb, Zn, and Cu in the water had initial concentrations of 0.5 mg/L, 0.5 mg/L, 2 mg/L, 5 mg/L, and 5 mg/L, respectively. Ten days after planting the water hyacinths, the concentrations of Cd, As, Pb, Zn, and Cu in the water were 0.406 mg/L, 0.394 mg/L, 1.498 mg/L, 3.891 mg/L, and 3.859 mg/L, respectively. By day 30 of the experiment, the remining Pb content in the solution reached the lowest level (0.153 mg/L) compared to the other elements, and it was within the permitted limits according to Column B of the QCVN 40:2011/BTNMT National Technical Regulation on Industrial Wastewater in Vietnam.

The experiments on growing water hyacinths in water containing heavy metals (Cd, As, Pb, Zn, and Cu) showed that this plant can still grow and develop at a certain level of pollution. The analysis results for the heavy metal content (Cd, As, Pb, Zn, and Cu) in the water showed a trend of decreasing concentration over time. Thus, water hyacinth has the ability to clean water contaminated with heavy metals (Cd, As, Pb, Zn, and Cu) very well.

4.3. Removal Efficiency Percentage and Remaining Percentages of Heavy Metals in Water

The removal efficiency percentage for the plants' ability to absorb heavy metals (Cd, As, Pb, Zn, and Cu) in the water over time is shown in Table 3.

The treatment efficiencies of water hyacinth for Cd, As, Pb, Zn, and Cu over the 30 days of the experiment were 59.4%, 60.8%, 92.4%, 60.2%, and 60.7% respectively, corresponding to the initial concentrations of 0.5 mg/L, 0.5 mg/L, 2 mg/L, 5 mg/L, and 5 mg/L. By the end of the 30-day survey period, the cleaning rates of the water hyacinths for all heavy metals (Cd, As, Pb, Zn, and Cu) were mostly high (59–92%).

The results show that the remaining heavy metal content in the water treated with water hyacinths decreased significantly after 30 days compared to the original concentrations. The remaining percentage of Pb especially was only 7.65%, corresponding to an original concentration of 2 mg/L.

The findings of the pot trials reveal that this plant has the potential to absorb heavy elements such as cadmium, arsenic, lead, zinc, and copper from wastewater. When the plants were added to the pots, the cadmium, arsenic, lead, zinc, and copper concentrations in the pots with the plants were significantly reduced. As a result, we can conclude that evaporation and settlement caused very little loss. The results show that water hyacinth is an effective plant capable of removing heavy metals from wastewater.

Previous bench-scale tests [27] using water hyacinth treatment for polluted river water and synthetic solutions demonstrated up to 63% removal of Al, 62% Zn, 47% Cd, 22% Mn, and 23% As in just seven hours of exposure to the plant. The results demonstrated very good removal efficiency in a very short time compared with this study, which only achieved removals of 18.8% Cd and 21.2% As in 10 days. Nevertheless, the initial metal concentrations were very low (0.00623 mg/L Cd and 0.00211 mg/L As) in the bench-scale tests compared to the much higher initial metal concentrations (0.5 mg/L Cd and 0.5 mg/L As) used in this study. Additionally, a study of water hyacinth as a biosorbent has been undertaken by using dry water hyacinth biomass [28]. This study involved a test that achieved 93% As(III) removal efficiency with 120 min shaking time and initial metal concentrations of 0.2 mg/L in solution. However, it is hard to make an objective comparison since the treatment mechanism was different to this study.

5. Conclusions

The efficiency of wastewater treatment for the removal of Cd, As, Pb, Zn, and Cu was assessed according to the concentrations of heavy metals in water. The elimination of contaminants from the water was quite effective when the plants were grown together. According to the results of the experiments, water hyacinth was able to reduce Cd concentration by 59.4%, As concentration by 60.8%, Pb concentration by 92.4%, Zn concentration by 60.22%, and Cu concentration by 60.74%. The total heavy metal concentration was reduced by 66.7%. This treatment system proved cost-effective because of the low installation and maintenance costs. The system could be used alone or in conjunction with other wastewater treatment systems. In conclusion, the current study revealed the viability of employing the aquatic plant *Eichhornia* to treat wastewater in a "sustainable" and environmentally benign manner. As this was merely a laboratory-size baseline study, more research on a larger scale is needed in the future, with a focus on phytoremediation and resource utilization.

Author Contributions: Conceptualization, A.T.H.; methodology, A.T.H.; software, A.T.H.; Formal analysis, A.T.H. and Y.-C.C.; resources, data curation, A.T.H.; writing—original draft preparation, B.N.T.T. and A.T.H.; writing—review and editing, Y.-C.C. All authors have read and agreed to the published version of the manuscript.

Funding: This research received no external funding.

Institutional Review Board Statement: Not applicable.

Informed Consent Statement: Not applicable.

Data Availability Statement: The data presented in this study are available on request from the corresponding author. The data are not publicly available.

Conflicts of Interest: The authors declare no conflict of interest.

References

1. Zheng, N.; Wang, Q.; Zheng, D. Health risk of Hg, Pb, Cd, Zn, and Cu to the inhabitants around Huludao Zinc Plant in China via consumption of vegetables. *Sci. Total Environ.* **2007**, *383*, 81–89. [CrossRef]
2. Dong, T.M.H.; Hoang, T.T.T.; Dao, Q.P. Research and selection of some plants capable of absorbing heavy metals (Cr, Cu, Zn) in mud dredging of Tan Hoa-Lo Gom canal (In Vietnamese: Nghiên cứu và lựa chọn một số thực vật có khả năng hấp thu các kim loại nặng (Cr, Cu, Zn) trong bùn nạo vét kênh Tân Hóa-Lò Gốm). *Sci. Technol. Dev. J.* **2008**, *11*, 59–67.
3. Le, H.Q.; Chen, Y.C.; Huynh, T.A.; Thai, V.L. A Study on Removing Arsenic Contamination in Soil by Phytoremediation. *Key Eng. Mater.* **2019**, *818*, 113–117. [CrossRef]
4. Zimmels, Y.; Kirzhner, F.; Roitman, S. Use of naturally growing aquatic plants for wastewater purification. *Water Environ. Res.* **2004**, *76*, 220–230. [CrossRef]
5. Dar, S.H.; Kumawat, D.M.; Singh, N.; Wani, K.A. Sewage Treatment Potential of Water Hyacinth (Eichhornia crassipes). *Res. J. Environ. Sci.* **2011**, *5*, 377–385. [CrossRef]
6. Vo, T.H.; Truong, P.K.D.; Tran, P.K.M.; Le, H.T.; Nguyen, M.T.; Pham, T.M.T. Survey the efficiency of domestic wastewater handling of water hyacinth and buffalo spinach (In Vietnamese: Khảo sát hiệu quả xử lý nước thải sinh hoạt của lục bình và ngổ trâu). *J. Thu Dau Mot Univ.* **2014**, *14*, 25–30.
7. Le, T.T.; Nguyen, T.M. Research on handling heavy metal pollution in water-by-Water Hyacinth (Eichhornia Crassipes) and Reed (Phragmites Australis) (In Vietnamese: Nghiên cứu sử dụng bèo tây (Eichhornia Classical) và cây sậy (Phragmites australis) xử lý nước bị ô nhiễm các kim loại nặng, cadimi (Cd), chì (Pb), kẽm (Zn) và đồng (Cu)). *Sci. J. Hong Duc Univ.* **2020**, *5*, 133–142.
8. Tchobanoglous, G.; Maitski, F.; Thoson, K.; Chadwick, T.H. Evaluation and performance of city of San Diego plot scale aquatic wastewater management system using water hyacinth. *Res. J. Water Pollut. Control Fed.* **1989**, *61*, 1625–1635.
9. Zaranyika, M.F.; Mutoko, F.; Murahwa, H. Uptake of Zn, Co, Fe and Cr by water hyacinth (Eichhorniacrassipes) in Lake Chivero, Zimbabwe. *Sci. Total Environ.* **1994**, *153*, 117–121. [CrossRef]
10. Zhu, Y.L.; Zayed, A.M.; Qian, J.H.; Souza, M.D.; Terry, N. Phytoaccumulation of trace elements by wetland plants: 11. Water hyacinth. *J. Environ. Qual.* **1999**, *28*, 339–344. [CrossRef]
11. Anushee, M. Environmental challenge vis a vis opportunity: The case of water hyacinth. *J. Environ. Qual* **2005**, *32*, 122–138.
12. Häder, D.-P.; Banaszak, A.T.; Villafañe, V.E.; Narvarte, M.A.; González, R.A.; Helbling, E.W. Anthropogenic pollution of aquatic ecosystems: Emerging problems with global implications. *Sci. Total Environ.* **2020**, *713*. [CrossRef] [PubMed]
13. Bolenz, S.; Omran, H.; Gierschner, K. Treatments of water hyacinth tissue to obtain useful products. *J. Biol. Wastes* **1990**, *33*, 63–74. [CrossRef]
14. Dwivedi, M.; Dwivedi, A.K. Valuable Product from Water Hyacinth—Review Paper. *Int. Res. J. Eng. Technol.* **2018**, *5*, 838–843.
15. Gopal, B. Water Hyacinth. *Elsevier Aquat. Plant Stud.* **1987**, *471*, 11–19.
16. Trivedy, R.K. Water hyacinth-based systems for waste treatment. *Adv. Wastewater Treat. Technologies. Glob. Sci.* **1998**, *1*, 463–486.
17. Granato, M. Cyanide degradation by water hyacinths, Eichhornia crassipes (Mart) Solms. *Biotechnol. Lett.* **1993**, *15*, 1085–1090. [CrossRef]
18. Flower. Learn about Water Hyacinth (In Vietnamese: Tìm hiểu về hoa lục bình—Water Hyacinth). Available online: https://vietflower.info/2018/tim-hieu-ve-hoa-luc-binh-water-hyacinth (accessed on 1 August 2021).
19. General Department of Vietnam Customs (GDVNC). Statistics of import and export goods (In Vietnamese: Báo cáo thống kê của Tổng cục Hải quan về các thị trường xuất khẩu sản phẩm thủ công mỹ nghệ của việt nam 2017). *Vietnam* **2017**, *2*, 1–22.
20. Garbisu, C.; Alkorta, I. Phytoextraction: A cost-effective plant-based technology for the removal of metals from the environment. *Bioresour. Technol.* **2001**, *77*, 229–236. [CrossRef]
21. Dang, D.K. Research on the use of plants to improve heavy metal contaminated soil in areas of mineral exploitation (In Vietnamese: Nghiên cứu sử dụng thực vật để cải tạo đất bị ô nhiễm kim loại nặng tại các vùng khai thác khoáng sản). *J. Environ.* **2011**, *5*, 201–207.
22. Nguyen, M.C. *Plant Growth Regulators Applied in Agriculture (In Vietnamese: Các chất điều hòa sinh Trưởng Thực vật Ứng Dụng Trong nông Nghiệp)*; Agriculture Publishing House: Hanoi, Vietnam, 2012.

23. Bui, T.K.A. Study on the Use of Plants (Ferns) to Treat Arsenic Contamination in Soil in Mineral Mining Areas (In Vietnamese: Nghiên cứu sử dụng thực vật (dương xỉ) để xử lý ô nhiễm Asen Trong đất Vùng Khai Thác Khoáng Sản). Ph.D. Thesis, University of Natural Sciences-Vietnam National University, Hanoi, Vietnam, 2011.
24. Le, T.P.; Le, T.D. Studying on absorbing possibility of lead Pb^{+2} in aquatic environments of Hyacinth (In Vietnamese: Nghiên cứu hấp thụ Pb^{2+} trong nước của cây lục bình). *Sci. J. TDMU* **2016**, 3, 42–49.
25. Sukha, R.V.; Srivastava, P.N. Phytoremediation—Green for Environmental Clean. Proceedings of Taal2007: The 12th World Lake Conference, Rajasthan, India, 28 October–2 November 2007; pp. 1016–1021.
26. Annette, C.D.; Jerald, L.S. Advances in Phytoremediation. *Environ. Health Perspect.* **2001**, *109*, 163–168.
27. Jones, J.L.; Jenkins, R.O.; Haris, P.I. Extending the geographic reach of the water hyacinth plant in removal of heavy metals from a temperate Northern Hemisphere River. *Sci. Rep.* **2018**, *8*, 1–15. [CrossRef] [PubMed]
28. Al Rmalli, S.W.; Harrington, C.F.; Ayub, M.; Haris, P.I. A biomaterial based approach for arsenic removal from water. *J. Environ. Monit.* **2005**, *7*, 279–282. [CrossRef] [PubMed]

materials

MDPI

Article

Preparation of TiO_2/Black Talc Composite Photocatalyst and the Research on Its Adsorption-Degradation Coupling Effects

Huan Shuai [1,2], Yuxin Wang [1], Jiao Wang [3,*], Gaoxiang Du [1,2,*], Daimei Chen [1,*] and Yu Liang [4]

1 School of Materials Science and Technology, China University of Geosciences, Beijing 100083, China; shuaihuan@email.cugb.edu.cn (H.S.); 2103190011@email.cugb.edu.cn (Y.W.)
2 Beijing Yiyi Star Technology Co., Ltd., Beijing 100089, China
3 School of Basic Education, Beijing Polytechnic College, Beijing 100042, China
4 School of Materials Science and Technology, Shenyang University of Chemical Technology, Shenyang 110142, China; liangyu1234@syuct.edu.cn
* Correspondence: wj@bgy.edu.cn (J.W.); dgx@cugb.edu.cn (G.D.); chendaimei@cugb.edu.cn (D.C.)

Abstract: In this paper, a TiO_2/black talc composite photocatalyst was prepared by the sol-gel method using TBOT as titanium source and black talc as carrier. Rhodamine B was used as the targeted pollutant to study the adsorption role of carbon in black talc. The results showed that with the adsorption-degradation cycles, the illumination time can be reduced by 40%. The adsorption rate and degradation rate of the composite photocatalyst was also increased. The degradation rate of Rhodamine B reached more than 95%, which fully shows the synergistic effect between TiO_2 nanoparticles and black talc. In this way, the adsorption-degradation coupling of the photocatalyst could be realized.

Keywords: nano-TiO_2 particles; black talc; characteristic adsorption; photodegradation; synergistic effect

Citation: Shuai, H.; Wang, Y.; Wang, J.; Du, G.; Chen, D.; Liang, Y. Preparation of TiO_2/Black Talc Composite Photocatalyst and the Research on Its Adsorption-Degradation Coupling Effects. *Materials* **2021**, *14*, 6038. https://doi.org/10.3390/ma14206038

Academic Editors: Marco Race, Elza Bontempi, Avelino Núñez-Delgado, Zhien Zhang, Mario Coccia and Yaoyu Zhou

Received: 1 September 2021
Accepted: 24 September 2021
Published: 13 October 2021

1. Introduction

With the development of industry in recent years, organic pollutants in air and water are emerging and increasing with astonishing speed, causing a series of environmental issues [1,2] and health problems [3] that affect our lives seriously. Therefore, it is very important to investigate an environmentally friendly, low-cost, simple, facile, and efficient method [4] to deal with the above-mentioned problem. Among the various solutions, physical adsorption [5,6] and photocatalytic degradation [7–11] are regarded as two common, effective solutions for pollution treatment. Titanium dioxide (TiO_2) has been well known as a star photocatalyst for organic pollution in air and water for its superiority of low price, stable chemical properties [12–16], suitable conduction band and valence band potential, no secondary pollution, and low photocorrosion [17–19]. However, TiO_2 has some disadvantages, such as low specific surface area, high photogenerated electron-hole recombination rate, narrow light utilization wavelength range, high agglomeration rate, and difficulty to be recycled [20], which limit its further applications. Therefore, it is urgent to find a carrier with porous structure and stable properties to load TiO_2 to overcome the above-mentioned problems [21].

Non-metallic minerals, such as diatomite, zeolite, and sepiolite, have been widely used as carriers [22,23] due to their low cost, excellent stability, and large specific area. Biochar, such as biomass bamboo fiber [24], macroalgae [25], coconut shell biochar [26], microalgae, and nut shells [27], has also been shown to promote the catalytic effect of photocatalysts and enhance the adsorption of organic molecules [28]. Pinna et al. [27] produced a biochar-decorated TiO_2 photocatalyst through a simple drop casting method. The composite Biochar -TiO_2 material has a better catalytic effect than pure titanium dioxide, which proves that BC NPs have the ability to act as a promoter. Moreover, the enhanced adsorption of organic molecules, coupled with the improved charge carrier separation provided by BC NPs, resulted in a nearly two-fold increase in photocatalytic performance. Zhang et al. [29] synthesized TiO_2 supported acid activated sepiolite (TiO_2/AAS) fibers under low temperature conditions and it was found that the specific surface area of the composite material was much larger than that of the original sepiolite. The photocatalytic activity of the composite was also excellent and superior to that of the TiO_2/sepiolite (raw sepiolite) and pure TiO_2. The addition of sepiolite significantly improved the photocatalytic activity of TiO_2. Suá et al. added TiO_2 to zeolite and found that the photocatalytic activity of the composite was ten times better than that of the pure TiO_2 particles. The more uniform the morphology of TiO_2 particles, the better the oxidation ability [30]. Liu et al. [31] prepared TiO_2/zeolite composite material by the sol-gel method and studied its adsorption and photocatalytic degradation performance for sulfadiazine (SDZ) under ultraviolet light irradiation. The results showed that, under neutral pH value, 90% of SDZ can be removed by TiO_2/zeolite within 120 min. Black talc is a kind of non-metallic mineral with excellent surface affinity, chemical stability, and thermal stability [32]. Compared with other minerals, it has a unique carbon layer between the black talc layers and the crystal grains of talc, which is beneficial to gather the surrounding organic pollutants, thus improving the catalytic efficiency of the photocatalyst.

In this study, a TiO_2/black talc composite photocatalyst was prepared by the sol-gel method. Using black talc as carrier can not only increase the contact area, but also facilitate the dispersion of TiO_2, resulting in the improvement of the photodegradation efficiency of the photocatalyst. The innovation of this research lies in the use of the adsorption degradation cycle test method, which can make full use of the adsorption properties of black talc and the degradation performance of TiO_2 in the composite photocatalyst, thus improving the overall rate of adsorption degradation. Compared with traditional degradation method, it can reduce the light time by 60% and reduce the energy consumption. This study is of great significance for the comprehensive utilization of black talc.

2. Materials and Methods

2.1. Materials

Black talc was purchased from Guangfeng District, Jiangxi, China. Tetrabutyl titanate (TBOT) was purchased from Aladdin Reagent Co., Ltd. (Shanghai, China). The sulfuric acid, sodium hydroxide, and analytical pure ethanol were purchased from Beijing Chemical Plant (Beijing, China)). Deionized water was used throughout the experiment.

2.2. Synthesis of the Composite Material

In a typical synthesis, excess 10% sulfuric acid solution was added to 300 mesh black talc powder and the suspension was stirred for 2 h. Then, the mixture was filtered and washed with DI H_2O for three times. Finally, it was dried at 105 °C for 24 h prior to use and was labelled as BT (black talc). BT-OC (Oxygen-calcined black talc) was prepared by the calcination of BT sample at 550 °C for 2 h in air. The composite photocatalyst was prepared as follows. A certain amount of TBOT was dissolved in ethanol. Then, BT was added, and the mixture was stirred for one hour and dried at 105 °C. Then, it was calcined at 550 °C for 2 h under N_2 atmosphere. The obtained composite photocatalyst was labelled as BT-xT, where x is the mass ratio of TiO_2 to black talc.

2.3. Photodegradation Experiment

Photocatalytic activities of samples were evaluated by their performance as catalysts in the photocatalytic oxidation of RhB in water. In this experiment, 1.5 mg photocatalyst was dispersed in 50 mL of RhB solution (30 mg/L). As a control group, 50 mL of RhB solution (30 mg/L) was taken without adding anything. Prior to degradation, the suspension was magnetically stirred in the dark for a period of time to establish adsorption–desorption equilibrium between the pre-irradiation photocatalyst and RhB. High-pressure mercury lamp with a power of 300 W was chosen as the light source of photodegradation and its dominant wavelength was 365 nm. At given intervals of illumination, a specimen (3 mL of the suspension was collected and centrifuged. The filtrates were analyzed by UV-vis spectroscopy at 554 nm.

2.4. Characterization Methods

X-ray diffraction (XRD) patterns of the samples were recorded using a Bruker D8 Advance diffractometer (Bruker, Germany), with CuKαradiation (λ = 0.15418) at 40 kV and 30 mA. Scanning electron microscopy (SEM) (JSM7500F, JEOL, Tokyo, Japan) was used to observe the microstructure of the samples. X-ray photoelectron spectra (XPS) were obtained using the radiation of Al Kα line (1486.6 eV, 300 W) as the excitation source. Binding energies were referenced to the C1s peak at 284.8 eV. The BET surface area of the samples was determined by N_2 adsorption by using NOVA4000 equipment (Quantachrome, Boynton Beach, FL, USA). Prior to N2 adsorption, the samples were evacuated at 473 K under vacuum for 4 h.

3. Results

3.1. The Structure, Morphologies and Composition of the Prepared Catalysts

X-ray diffraction analysis was used to analyze the phase structure and composition of the prepared samples. Figure 1 shows the XRD patterns for BT, BT-10%T, and BT-20%T samples. The characteristic diffraction peaks for the anatase phase are observed in both BT-10%T and BT-20%T samples, suggesting that the anatase phase of the TiO_2 has been successfully synthesized on black talc [33]. The intensities of the characteristic peaks of talc decreased with the TBOT amount and all the characteristic peaks of talc are observed in the composite photocatalyst, which indicates that calcination at 550 °C cannot damage the black talc structure.

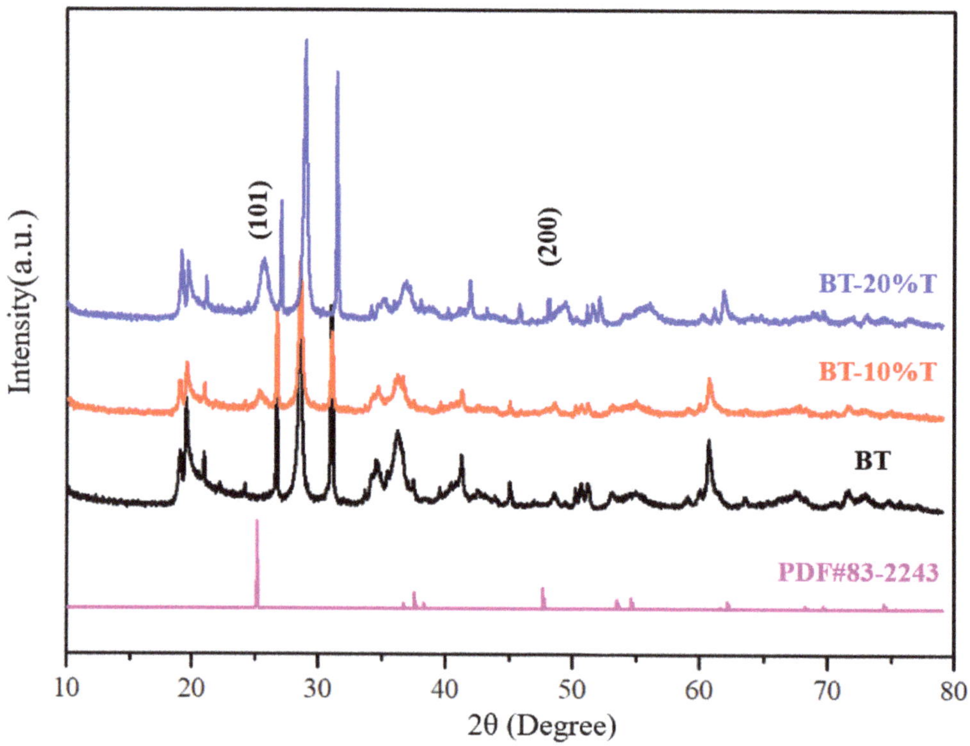

Figure 1. XRD diffraction patterns for photocatalytic material.

Figure 2 shows SEM images of black talc and BT samples. It can be seen that black talc exhibits a layered structure with different sizes (Figure 2a,b). Most of the particle sizes are less than 2 microns, but a minority can reach hundreds of microns. In the prepared photocatalyst composite, a large amount of TiO_2 particles with uniform size of 50 nm are agglomerated on the surface of the black talc (Figure 2c,d).

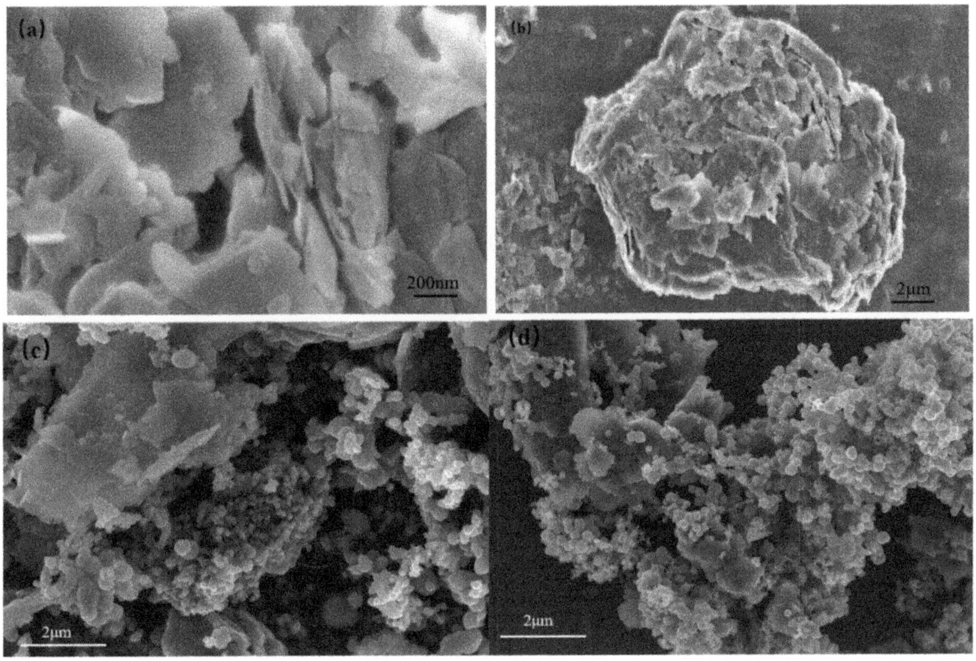

Figure 2. (**a**,**b**) SEM images of black talc; (**c**,**d**) SEM images of the photocatalyst.

In order to further analyze the interaction between the TiO_2 and black talc, X-ray photoelectron spectroscopy (XPS) was used to test and analyze BT and BT-T (Figure 3). Figure 3a shows that BT-T has a characteristic peak of Ti 2p, which is not existed in BT. This indicated that TiO_2 was successfully loaded on black talc. The analysis of the chemical state of O 1s is shown in Figure 3b. There are two characteristic peaks emerging at 532.6 eV and 529.8 eV for BT-T, which can be attributed to the existence of Si-O-Si and Ti-O-Si, respectively [34,35], while only one characteristic peak can be observed in BT sample. From this comparison, it can be seen that there is a chemical bonding between TiO_2 and black talc, which indicates the good stability of the composite.

3.2. Nitrogen Adsorption-Desorption Isotherms of the Prepared Catalysts

N_2 adsorption-desorption isotherms of black talc and calcined talc are presented in Figure 4. Both the black talc and calcined talc have almost the same value of specific surface area, i.e., 11.96 m^2/g and 10.17 m^2/g, respectively. This means that the pore structure of talc is not affected by the calcination at 550 °C, which is consistent with the analysis of XRD results. Since both the nitrogen adsorption and desorption curves of the two samples have hysteresis loops, which fit in typical IV according to the classification of the International Union of Pure Theory and Applied Chemistry (IUPAC), it can be seen that both BT and BT-OC samples have mesoporous structures.

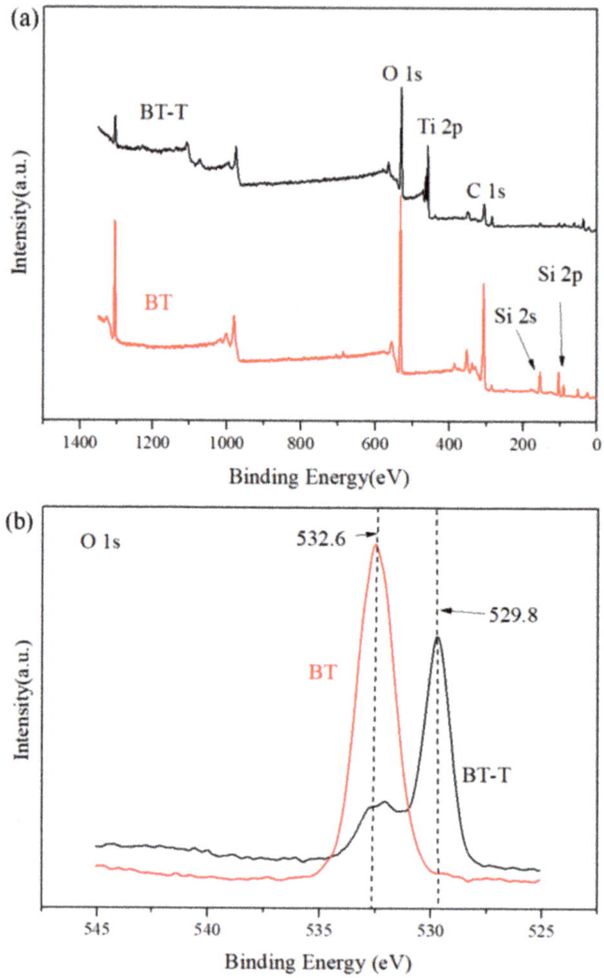

Figure 3. (**a**) XPS survey spectra of BT and BT-T, (**b**) O 1s high-resolution energy spectrum.

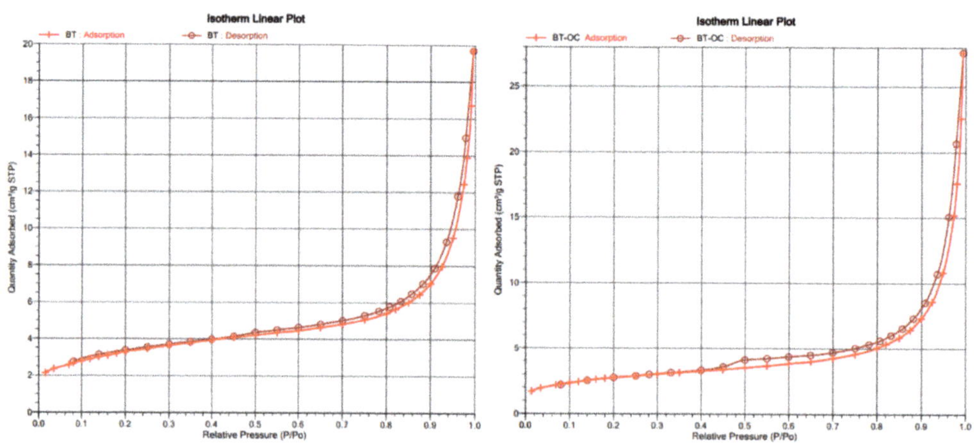

Figure 4. N_2 adsorption-desorption isotherms of BT and BT-OC samples.

3.3. The Adsorption Behavior and Photodegradation Behavior of the Prepared Photocatalysts

Figure 5 shows the adsorption effect curves of different samples for Rhodamine B. Under dark conditions, the compound Rhodamine B itself in control group is relatively stable with only 2.4% decay in the concentration. The adsorption rate of Rhodamine B by uncalcined black talc is gradually improved from 5.3% to 21.0% with the increase of dose amount from 10 mg to 90 mg. However, compared with BT sample (50 mg), the adsorption rate of Rhodamine B by BT-OC calcined by oxygen decreased sharply to 4.3%, which was only a little higher than the control group. This indicated that oxygen calcination could remove organic carbon that can be used to adsorb Rhodamine B. It also shows that the adsorption rate of BT-T (50 mg) was 17.8%, which was slightly lower than the adsorption rate of 50 mg BT (19.7%). This is caused by the adhesion of TiO_2 on the surfaces of black talc, which can affect part of the role of carbon, resulting in a decrease in adsorption rate. After 120 min, the adsorption rate remained basically unchanged, meaning that the absorption equilibrium was achieved.

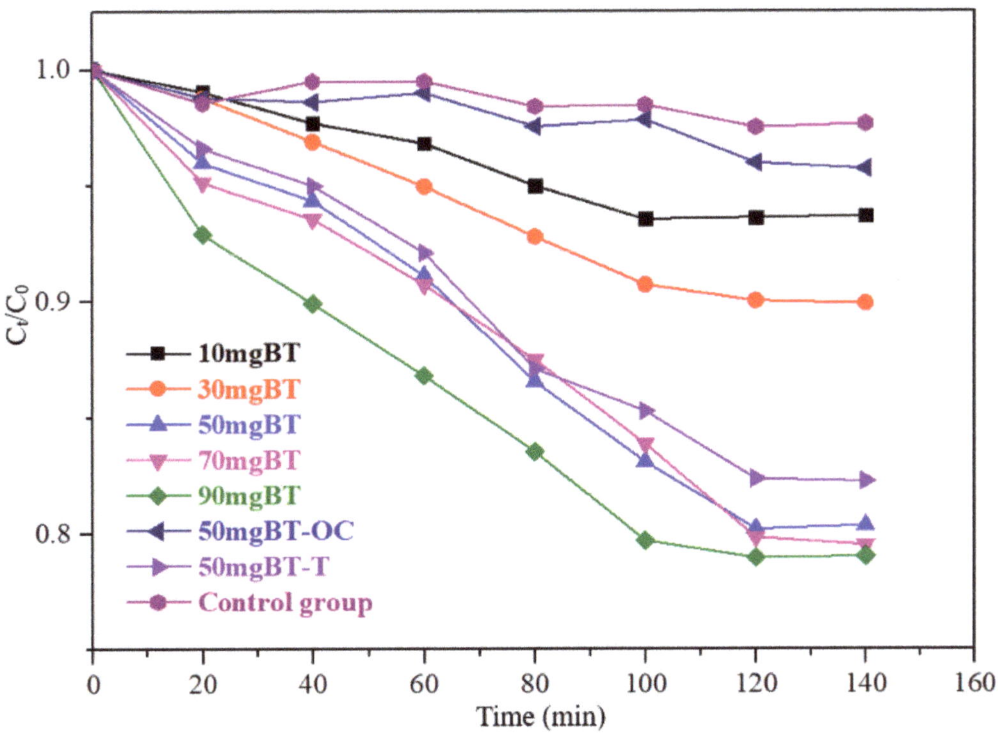

Figure 5. The adsorption effect curves of RhB on different samples.

The physical adsorption of Rhodamine B solution to black talc follows the quasi-first-order kinetic equation, as shown in Figure 6, where the adsorption performance for each sample can be quantitatively evaluated through the apparent rate constant k.

$$-\ln(C/C_0) = kt \quad (1)$$

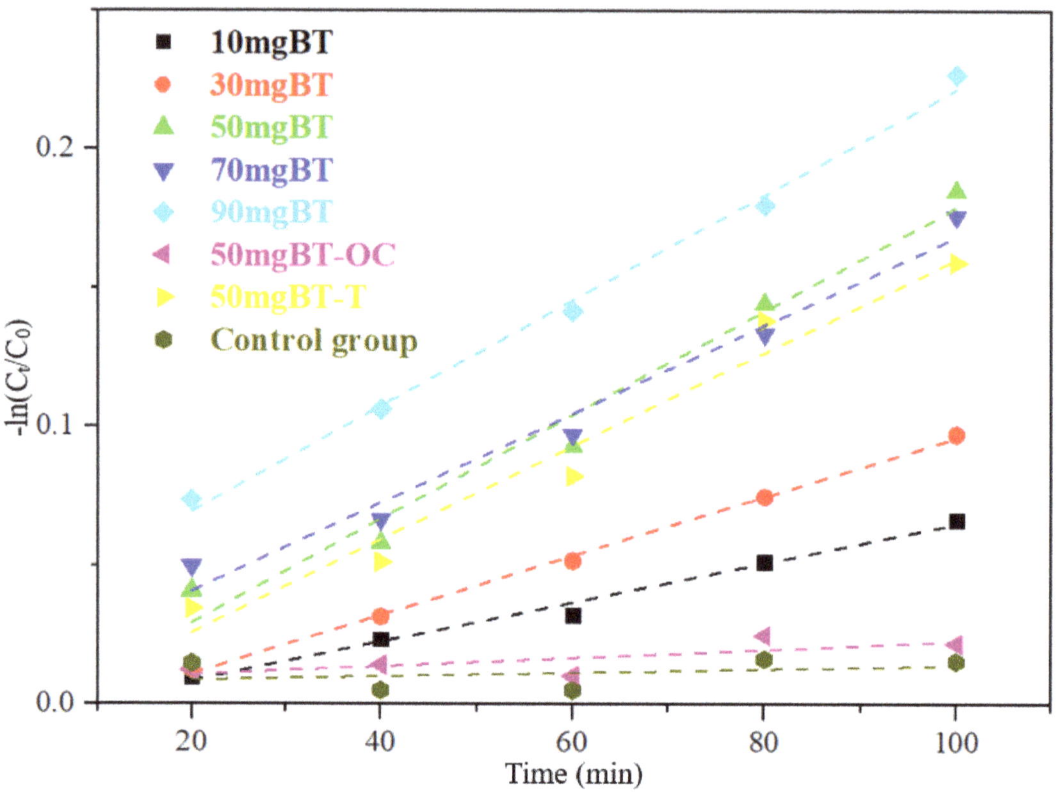

Figure 6. Kinetics fitting curves for the adsorption of RhB.

Table 1 below shows the value of the correlation coefficient R^2 of the kinetic equation and the apparent rate constant k. It can be seen that the correlation coefficient of the kinetic equation of BT-OC and the control group is not high. This is due to the small adsorption capacity of these two groups. The test data fluctuate greatly, causing the R^2 value to be small, and the correlation coefficients of the other groups of kinetic equations are higher, which can be fitted by the kinetic equation. The maximum value of the apparent rate constants 50 mg BT and 90 mg BT is 0.0019, which shows that the adsorption rate of 50 mg BT and 90 mg BT is the fastest, followed by 50 mg BT-T. In order to maximize the reaction efficiency, 50 mg sample was used in the degradation stage of this experiment, which had the highest adsorption efficiency.

Table 1. The value of the correlation coefficient R^2 and constant k of the kinetic equation.

Time	10 mgBT	30 mgBT	50 mgBT	70 mgBT	90 mgBT	50 mgBT-OC	50 mgBT-T	Control Group
R^2	0.98873	0.99967	0.98733	0.99865	0.99944	0.17109	0.94654	0.1593
K	0.0007	0.0011	0.0019	0.0016	0.0019	0.0001	0.0017	0.00006

After the adsorption experiment, the photocatalytic degradation experiment was conducted on the BT and BT-T samples. At this stage, a set of adsorption degradation cycle groups was added to compare with constant light degradation experiments. The cycle process contained 20 min degradation followed by 20 min adsorption and it would be continued until rhodamine B was completely degraded.

Figure 7 shows the degradation effect curves of different samples on RhB. Under UV irradiation, the concentration of RhB in BT and the control group changed slightly (both less than 4%), indicating that Rhodamine B was quite stable under UV irradiation. The degradation rate of BT-T after 200 min irradiation reached 95.1% and the degradation rate of BT-T in the adsorption-degradation cycle group was 96.2%. Although the results were very similar, the total irradiation time of BT-T in the adsorption-degradation cycle group was 120 min, which was only 60% of the total irradiation time in the continuous light group, thus proving the excellent effect of the adsorption degradation cycle groups. The degradation cycle method can greatly reduce the use of light and achieve a degradation efficiency slightly higher than that of constant lightening.

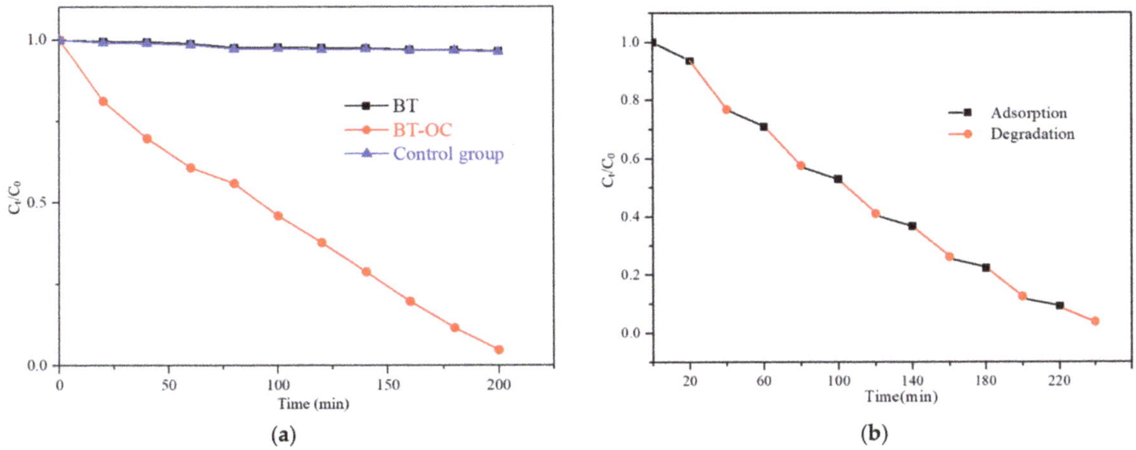

Figure 7. (**a**) Degradation effect curves of RhB with the different samplesunder UV irradiation. (**b**) The adsorption-degradation effect curve of RhB with BT-T.

In order to compare the coupling performance of BT-T and BT-T cycle on the adsorption and degradation of Rhodamine B, we performed the kinetic fitting of the adsorption and degradation of Rhodamine B for 120 min.

As shown in Figure 8, the apparent rate constants of adsorption and degradation of BT-T cycle are 0.00274 and 0.00967, respectively. The apparent rate constants of adsorption and degradation of BT-T are 0.00146 and 0.00764, respectively. The adsorption and degradation rates of the former are respectively 1.88 times and 1.27 times of the latter, showing that the adsorption and degradation recycling can increase the rate of adsorption and degradation and reduce the time required for the reaction. This is because in the stage of pure adsorption of BT-T, the adsorption capacity will reach saturation. After the pores are full, the adsorption capacity will be significantly weakened, but the cyclic method can degrade the adsorbed RhB in time, making the adsorption capacity decrease to a small extent.

In order to further test the stability of the prepared photocatalyst, cyclic degradation tests were conducted, and the results are shown in Figure 9. The experimental results show that the adsorption-degradation rate for RhB being catalyzed by BT-T was still around 90% after five cycles, indicating good stability and reusability.

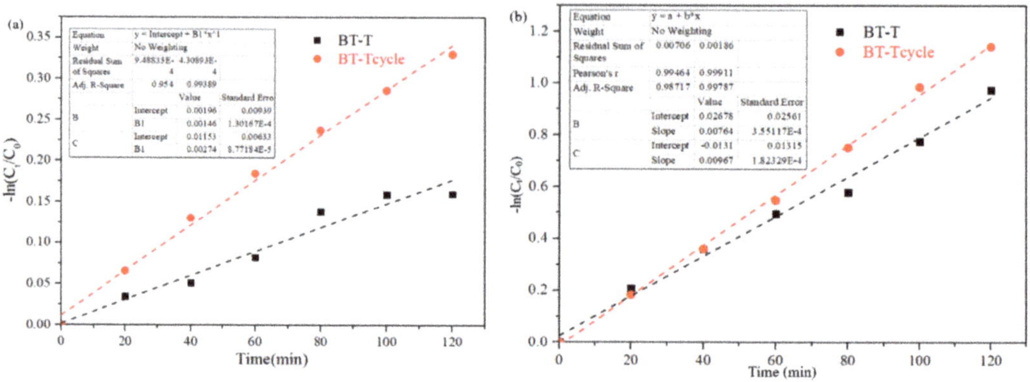

Figure 8. (**a**) Kinetics fitting plots for the adsorption of RhB. (**b**) Kinetics fitting plots for the degradation of RhB.

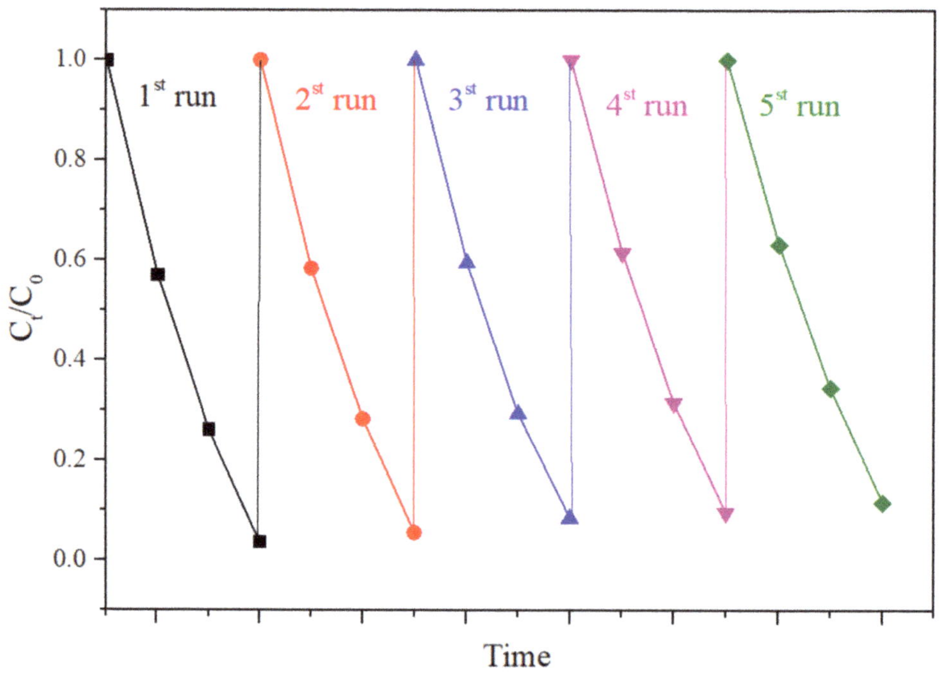

Figure 9. Cycling degradation test of RhB on BT-T sample.

4. Discussion

Based on the above analysis, the photocatalytic mechanism of BT-T composite was proposed. As shown in Scheme 1, TiO_2 nanoparticles adhered to the surfaces of black talc and formed stable Si-O-Ti bonds. This can avoid agglomeration and ensure that most of TiO_2 can be exposed to contact with pollutants. Moreover, there is a graphene-like carbon layer in black talc, which has a characteristic adsorption effect on organic pollutants. Thus, organic pollutants can be gathered around black talc. TiO_2 nanoparticles that were adhered on the surfaces of black talc can degrade pollutants and reduce the concentration of the pollutants, which in turn can promote the adsorption of the pollutants. Therefore, both TiO_2 and black talc have synergic effects on each other, thus improving the photocatalytic activity of the prepared composite.

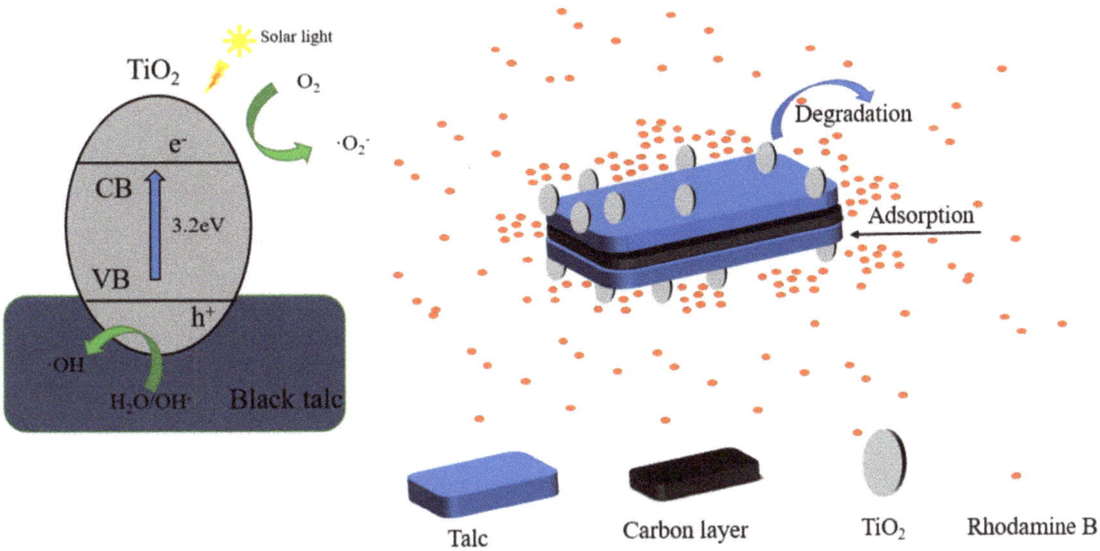

Scheme 1. Schematic diagram of photocatalytic performance enhancement mechanism of BT-T composite material.

5. Conclusions

In this study, a TiO_2/black talc photocatalyst was prepared through the sol-gel method. Black talc combined with TiO_2 via chemical bonds and the composite demonstrated good stability. The adsorption-degradation cycle can significantly improve the coupling performance of the photocatalyst for rhodamine B. Compared with the constant light degradation group, it can reduce irradiation time by 60% and reduce energy consumption. The degradation rate of Rhodamine B by the composite photocatalyst could reach more than 95% after 120 min and the degradation rate of the composite photocatalyst was more than 90% after five cycles. Black talc can adsorb and enrich the pollutants around the photocatalyst, which helps to improve its catalytic efficiency. Therefore, black talc can be used as an effective carrier for the improvement of semiconductor photocatalyst performance and cost reduction.

Author Contributions: Conceptualization, J.W. and G.D.; methodology, H.S., J.W. and G.D.; validation, H.S., J.W. and D.C.; formal analysis, H.S. and D.C.; investigation, H.S., Y.W. and D.C.; data curation, H.S. and Y.W.; writing—original draft preparation, H.S., Y.W. and D.C.; writing—review and editing, J.W., G.D. and Y.L; visualization, H.S., Y.W. and Y.L.; supervision, J.W.; project administration, J.W.; funding acquisition, J.W. All authors have read and agreed to the published version of the manuscript.

Funding: This work was partly supported by the Beijing Municipal Education Commission Science and Technology Plan Key Project (No. KZ201910853043) and Liaoning Provincial Natural Science Foundation (No. 2021-MS-260).

Institutional Review Board Statement: Not applicable.

Informed Consent Statement: Not applicable.

Data Availability Statement: The data presented in this study are available on request from the corresponding author.

Conflicts of Interest: The authors declare no conflict of interest.

References

1. Qiao, B.; Chen, Y.; Tian, M.; Wang, H.; Yang, F.; Shi, G.; Zhang, L.; Peng, C.; Luo, Q.; Ding, S. Characterization of water soluble inorganic ions and their evolution processes during PM 2.5 pollution episodes in a small city in southwest China. *Sci. Total Environ.* **2019**, *650*, 2605–2613. [CrossRef] [PubMed]
2. Xue, J.; Kannan, K. Mass flows and removal of eight bisphenol analogs, bisphenol A diglycidyl ether and its derivatives in two wastewater treatment plants in New York State, USA. *Sci. Total Environ.* **2019**, *648*, 442–449. [CrossRef] [PubMed]
3. Li, X.; Yu, J.; Jaroniec, M.; Chen, X. Cocatalysts for selective photoreduction of CO_2 into solar fuels. *Chem. Rev.* **2019**, *119*, 3962–4179. [CrossRef]
4. Gao, W.; Tian, J.; Fang, Y.; Liu, T.; Zhang, X.; Xu, X.; Zhang, X. Visible-light-driven photo-Fenton degradation of organic pollutants by a novel porphyrin-based porous organic polymer at neutral pH. *Chemosphere* **2020**, *243*, 125334. [CrossRef] [PubMed]
5. Kameda, T.; Ito, S.; Yoshioka, T. Kinetic and equilibrium studies of urea adsorption onto activated carbon: Adsorption mechanism. *J. Dispers. Sci. Technol.* **2016**, *38*, 1063–1066. [CrossRef]
6. GilPavas, E.; Dobrosz-Gómez, I.; Gómez-García, M.-A. Optimization and toxicity assessment of a combined electrocoagulation, $H_2O_2/Fe^{2+}/UV$ and activated carbon adsorption for textile wastewater treatment. *Sci. Total Environ.* **2018**, *651*, 551–560. [CrossRef] [PubMed]
7. Yan, X.; Yuan, X.; Wang, J.; Wang, Q.; Zhou, C.; Wang, D.; Tang, H.; Pan, J.; Cheng, X. Construction of novel ternary dual Z-scheme Ag_3VO_4/C_3N_4/reduced TiO_2 composite with excellent visible-light photodegradation activity. *J. Mater. Res.* **2019**, *34*, 2024–2036. [CrossRef]
8. Ding, X.; Wang, W.; Zhang, A.; Zhang, L.; Yu, D. Efficient visible light degradation of dyes in wastewater by nickel–phosphorus plating–titanium dioxide complex electroless plating fabric. *J. Mater. Res.* **2019**, *34*, 999–1010. [CrossRef]
9. Prakash, J.; Sun, S.; Swart, H.C.; Gupta, R.K. Noble metals-TiO_2 nanocomposites: From fundamental mechanisms to photocatalysis, surface enhanced Raman scattering and antibacterial applications. *Appl. Mater. Today* **2018**, *11*, 82–135. [CrossRef]
10. Selcuk, S.; Zhao, X.; Selloni, A. Structural evolution of titanium dioxide during reduction in high-pressure hydrogen. *Nat. Mater.* **2018**, *17*, 923–928. [CrossRef]
11. Ullattil, S.G.; Narendranath, S.B.; Pillai, S.C.; Periyat, P. Black TiO_2 Nanomaterials: A Review of Recent Advances. *Chem. Eng. J.* **2018**, *343*, 708–736. [CrossRef]
12. Ortega-Méndez, J.A.; Herrera-Melián, J.A.; Araña, J.; Espino-Estévez, M.R.; Doña-Rodríguez, J.M. Performance and Economic Assessment of the Treatment of Phenol with TiO_2 Photocatalysis, Photo-Fenton, Biological Aerated Filter, and Wetland Reactors. *Chem. Eng. Technol.* **2017**, *40*, 1165–1175. [CrossRef]
13. Pal, U.; Ghosh, S.; Chatterjee, D. Effect of sacrificial electron donors on hydrogen generation over visible light–irradiated nonmetal-doped TiO_2 photocatalysts. *Transit. Met. Chem.* **2012**, *37*, 93–96. [CrossRef]
14. Tsoukleris, D.S.; Maggos, T.; Vassilakos, C.; Falaras, P. Photocatalytic degradation of volatile organics on TiO_2 embedded glass spherules. *Catal. Today* **2007**, *129*, 96–101. [CrossRef]
15. Castro, R.H.R.; Wang, B. The Hidden Effect of Interface Energies in the Polymorphic Stability of Nanocrystalline Titanium Dioxide. *J. Am. Ceram. Soc.* **2011**, *94*, 918–924. [CrossRef]
16. Fujishima, A.; Rao, T.N.; Tryk, D.A. Titanium dioxide photocatalysis. *J. Photochem. Photobiol. C Photochem. Rev.* **2000**, *1*, 1–21. [CrossRef]
17. Chen, D.; Zhu, Q.; Zhou, F.; Deng, X.; Li, F. Synthesis and photocatalytic performances of the TiO_2 pillared montmorillonite. *J. Hazard. Mater.* **2012**, *235*, 235–236. [CrossRef] [PubMed]
18. Gaya, U.I.; Abdullah, A.H. Heterogeneous photocatalytic degradation of organic contaminants over titanium dioxide: A review of fundamentals, progress and problems. *J. Photochem. Photobiol. C Photochem. Rev.* **2007**, *9*, 1–12. [CrossRef]
19. Chatterjee, D.; Moulik, S.K.; Giribabu, L.; Kanaparthi, R.K. Dye sensitization of a large band gap semiconductor by an iron(III) complex. *Transit. Met. Chem.* **2014**, *39*, 641–646. [CrossRef]
20. Lavanya, T.; Satheesh, K.; Dutta, M.; Jaya, N.V.; Fukata, N. Superior photocatalytic performance of reduced graphene oxide wrapped electrospun anatase mesoporous TiO_2 nanofibers. *J. Alloy. Compd.* **2014**, *615*, 643–650. [CrossRef]
21. Srikanth, B.; Goutham, R.; Narayan, R.B.; Ramprasath, A.; Gopinath, K.P.; Sankaranarayanan, A.R. Recent advancements in supporting materials for immobilised photocatalytic applications in waste water treatment. *J. Environ. Manag.* **2017**, *200*, 60–78. [CrossRef]
22. Rasalingam, S.; Peng, R.; Koodali, R.T.; Dong, F. Removal of Hazardous Pollutants from Wastewaters: Applications of TiO_2-SiO_2 Mixed Oxide Materials. *J. Nanomater.* **2014**, *2014*, 10. [CrossRef]
23. Low, J.; Cheng, B.; Yu, J. Surface modification and enhanced photocatalytic CO_2 reduction performance of TiO_2: A review. *Appl. Surf. Sci.* **2017**, *392*, 658–686. [CrossRef]
24. Sun, J.; Lin, X.; Xie, J.; Zhang, Y.; Wang, Q.; Ying, Z. Facile synthesis of novel ternary g-CN_4/ferrite/biochar hybrid photocatalyst for efficient degradation of methylene blue under visible-light irradiation. *Colloids Surf. A Physicochem. Eng. Asp.* **2020**, *606*, 125556. [CrossRef]
25. Fazal, T.; Razzaq, A.; Javed, F.; Hafeez, A.; Rashid, N.; Amjad, U.S.; Rehman, M.S.U.; Faisal, A.; Rehman, F. Integrating adsorption and photocatalysis: A cost effective strategy for textile wastewater treatment using hybrid biochar-TiO_2 composite. *J. Hazard. Mater.* **2020**, *390*, 121623. [CrossRef]

26. Zhang, S.; Lu, X. Treatment of wastewater containing Reactive Brilliant Blue KN-R using TiO_2/BC composite as heterogeneous photocatalyst and adsorbent. *Chemosphere* **2018**, *206*, 777–783. [CrossRef]
27. Pinna, M.; Binda, G.; Altomare, M.; Marelli, M.; Dossi, C.; Monticelli, D.; Spanu, D.; Recchia, S.B. Nanoparticles over TiO_2 Nanotube Arrays: A Green Co-Catalyst to Boost the Photocatalytic Degradation of Organic Pollutants. *Catalysts* **2021**, *11*, 1048. [CrossRef]
28. Kim, J.R.; Kan, E. Heterogeneous photocatalytic degradation of sulfamethoxazole in water using a biochar-supported TiO_2 photocatalyst. *J. Environ. Manag.* **2016**, *180*, 94–101. [CrossRef] [PubMed]
29. Zhang, G.; Xiong, Q.; Xu, W.; Guo, S. Synthesis of bicrystalline TiO_2 supported sepiolite fibers and their photocatalytic activity for degradation of gaseous formaldehyde. *Appl. Clay Sci.* **2014**, *102*, 231–237. [CrossRef]
30. Suárez, S.; Jansson, I.; Ohtani, B.; Sánchez, B. From Titania nanoparticles to Decahedral Anatase Particles: Photocatalytic activity of TiO_2 /zeolite hybrids for VOC oxidation. *Catal. Today* **2018**, *326*, 2–7. [CrossRef]
31. Liu, X.; Liu, Y.; Lu, S.; Guo, W.; Xi, B. Performance and mechanism into TiO_2/Zeolite composites for sulfadiazine adsorption and photodegradation. *Chem. Eng. J.* **2018**, *350*, 131–147. [CrossRef]
32. Li, C.; Wang, R.; Lu, X.; Zhang, M. Mineralogical characteristics of unusual black talc ores in Guangfeng County, Jiangxi Province, China. *Appl. Clay Sci.* **2013**, *74*, 37–46. [CrossRef]
33. Jaiswal, R.; Bharambe, J.; Patel, N.; Dashora, A.; Kothari, D.C.; Miotello, A. Copper and Nitrogen co-doped TiO_2 photocatalyst with enhanced optical absorption and catalytic activity. *Appl. Catal. B Environ.* **2015**, *168*, 333–341. [CrossRef]
34. Wang, B.; Zhang, G.; Leng, X.; Sun, Z.; Zheng, S. Characterization and improved solar light activity of vanadium doped TiO_2/diatomite hybrid catalysts. *J. Hazard. Mater.* **2015**, *285*, 212–220. [CrossRef] [PubMed]
35. Xia, Y.; Li, F.; Jiang, Y.; Xia, M.; Xue, B.; Li, Y. Interface actions between TiO_2 and porous diatomite on the structure and photocatalytic activity of TiO_2-diatomite. *Appl. Surf. Sci.* **2014**, *303*, 290–296. [CrossRef]

materials

MDPI

Article

Activated Graphene Oxide-Calcium Alginate Beads for Adsorption of Methylene Blue and Pharmaceuticals

Burcu Gunes [1], Yannick Jaquet [2], Laura Sánchez [3], Rebecca Pumarino [3], Declan McGlade [1], Brid Quilty [1], Anne Morrissey [4], Zahra Gholamvand [1], Kieran Nolan [5] and Jenny Lawler [1,6,*]

1 DCU Water Institute, School of Biotechnology, Dublin City University, Glasnevin, D09 NA55 Dublin, Ireland; burcu.gunes@dcu.ie (B.G.); declan.mcglade2@mail.dcu.ie (D.M.); brid.quilty@dcu.ie (B.Q.); z.gholamvand@gmail.com (Z.G.)
2 Institut Technologie du vivant, University of Applied Sciences and Arts Western Switzerland, Rte de Moutier 14, 2800 Delémont, Switzerland; jaquet.yannick@gmail.com
3 Faculty of Biology, University of Oviedo Calle Catedrático Valentín Andrés Álvarez, 33006 Oviedo, Spain; lauusnchzbna@gmail.com (L.S.); rebecacgp@gmail.com (R.P.)
4 DCU Water Institute, School of Mechanical and Manufacturing Engineering, Dublin City University, Glasnevin, D09 NA55 Dublin, Ireland; anne.morrissey@dcu.ie
5 DCU Water Institute, School of Chemical Sciences, Dublin City University, Glasnevin, D09 NA55 Dublin, Ireland; kieran.nolan@dcu.ie
6 Qatar Environment and Energy Research Institute (QEERI), Hamad Bin Khalifa University, Doha 34110, Qatar
* Correspondence: jlawler@hbku.edu.qa; Tel.: +974-445-48116

Abstract: The remarkable adsorption capacity of graphene-derived materials has prompted their examination in composite materials suitable for deployment in treatment of contaminated waters. In this study, crosslinked calcium alginate–graphene oxide beads were prepared and activated by exposure to pH 4 by using 0.1M HCl. The activated beads were investigated as novel adsorbents for the removal of organic pollutants (methylene blue dye and the pharmaceuticals famotidine and diclofenac) with a range of physicochemical properties. The effects of initial pollutant concentration, temperature, pH, and adsorbent dose were investigated, and kinetic models were examined for fit to the data. The maximum adsorption capacities q_{max} obtained were 1334, 35.50 and 36.35 mg g^{-1} for the uptake of methylene blue, famotidine and diclofenac, respectively. The equilibrium adsorption had an alignment with Langmuir isotherms, while the kinetics were most accurately modelled using pseudo- first-order and second order models according to the regression analysis. Thermodynamic parameters such as $\Delta G°$, $\Delta H°$ and $\Delta S°$ were calculated and the adsorption process was determined to be exothermic and spontaneous.

Keywords: adsorption; graphene oxide; methylene blue; pharmaceuticals; kinetics; isotherms and thermodynamics

Citation: Gunes, B.; Jaquet, Y.; Sánchez, L.; Pumarino, R.; McGlade, D.; Quilty, B.; Morrissey, A.; Gholamvand, Z.; Nolan, K.; Lawler, J. Activated Graphene Oxide-Calcium Alginate Beads for Adsorption of Methylene Blue and Pharmaceuticals. *Materials* **2021**, *14*, 6343. https://doi.org/10.3390/ma14216343

Academic Editor: Avelino Núñez-Delgado

Received: 16 September 2021
Accepted: 18 October 2021
Published: 23 October 2021

Publisher's Note: MDPI stays neutral with regard to jurisdictional claims in published maps and institutional affiliations.

1. Introduction

Micropollutants such as pharmaceuticals, personal care products, surfactants and pesticides [1], as well as synthetic dyes [2], have been found virtually ubiquitously in environmental matrices over the past decade. One major source of organic micropollutants is the effluents from wastewater treatment plants (WWTPs), since most of these emerging contaminants are poorly biodegradable [3]. In addition, the hazards presented to human health and the ecosystem by thousands of trace contaminants in a "cocktail effect" are not yet well understood, although advances in effect-based biomonitoring aim to address this [4,5].

Methylene blue (MB) is a heterocyclic aromatic chemical dye used in textile, paper and cosmetic industries [6]. It is not highly toxic but has significant adverse impacts on aquatic ecosystems [2], retarding the photosynthetic activity of aquatic plants by affecting the light penetration, consuming dissolved oxygen or isolating metal ions, producing microtoxicity to organisms [7–10]. It can also be harmful to human health, causing heart rate increase, nausea and vomiting [11]. Methylene blue is widely used as an indicator pollutant to demonstrate the efficiency of novel adsorbent materials in the literature. Famotidine (FMTD) is a histamine H_2-receptor antagonist used for treating gastroesophageal reflux disease and Zollinger–Ellison syndrome [12]. Famotidine has been shown to persist in WWTP effluents [13–15]. Diclofenac (DFC) is a non-steroidal anti-inflammatory drug prescribed to reduce inflammation, pain and dysmenorrhea; consumption is associated with serious dose-dependent gastrointestinal, renal and hepatic adverse effects, and increases vascular and coronary risks by about 33% [16,17]. Diclofenac is monitored in European surface waters under the watch list mechanism for the Water Framework Directive, and has been found almost ubiquitously in wastewater influent, wastewater effluent and surface waters [18]. Diclofenac exposure in trout has been shown to induce severe glomerulonephritis, resulting in kidney failure [19], and it has been implicated in the collapse of Asian vulture populations [20].

The removal of micropollutants and synthetic dyes using membrane-based technologies, ozonation, photolysis, photocatalysis [21–23], electrolysis, Fenton [24], photo-Fenton oxidation and electrochemical oxidation [25] has been extensively investigated in the literature. In particular, adsorption technology holds a lot of advantages, such as easy operation, fast decolorization and chemical oxygen demand removal efficiency; however, the main limitation of adsorption technology is the low and non-selective adsorption capacity of traditional adsorbents. The enhancement of adsorption capacity by increasing surface area and optimizing pore size has received much research attention [26–31].

Graphene-based materials have received increasing attention as potential candidates for composite preparation due to their high specific surface area and adsorption capacity. Graphene oxide (GO) is a two-dimensional complex of carbon atoms decorated with a multitude of oxygen-containing functional groups densely packed in a honeycomb framework [32]. GO has unique properties, such as a large theoretical surface area, high thermal and chemical stability, high conductivity and good mechanical flexibility [33], showing a great potential as an adsorbent for the removal of pharmaceuticals [34], heavy metals [35,36] or dyes [37]. However, using bare GO as an adsorbent agent causes the agglomeration of GO, which requires a complex high-speed centrifuge for GO separation from the solution [38]. Therefore, in this study, to stabilize the GO [39] and maximize the ease of recovery, GO was incorporated into an alginate matrix (an anionic polysaccharide used in paints, inks or pharmaceuticals). Alginate forms a hydrogel when mixed with divalent cations, such as Ca^{2+}, giving good mechanical properties. Acid-activation of the beads provides an enhancement in the surface, area including micro- and mesopores [40,41]. In fact, the adsorption capacity of GO-montmorillonite/sodium alginate beads was recently investigated [42].

In this work, the acid-activated (0.1 M HCl pH 4) adsorption capacity of calcium alginate graphene oxide beads was evaluated as novel adsorbents for MB, FMTD and DFC removal. In addition, the influence of initial pollutant concentration, adsorbent dose, adsorption temperature and pH on adsorption capacity was investigated, along with an examination of the kinetic and thermodynamic modelling of the reactions.

2. Materials and Methods

2.1. Materials

Graphite flakes (GF) were purchased from Asbury Carbons. Diclofenac sodium (DFC, 99%), famotidine (FMTC), methylene blue (MB) and alginic acid sodium salt (Na-Alg) were purchased from Sigma Aldrich. The structure of analytes is given in Appendix A. Calcium chloride dihydrate, sodium hydroxide, potassium permanganate and absolute ethanol were purchased from Fischer Chemicals. Sulfuric acid (H_2SO_4, 95–98%) and hydrogen peroxide (H_2O_2, 30%) was purchased from Merck Millipore. Hydrochloric acid (37%) was provided by Acros Organics Dublin, Ireland.

2.2. Methods

Preparation of Graphene oxide (GO) Solution and Ca-Alg2/GO Beads

GO was prepared according to a modified version of Hummer's method [43]. In this method, expanded graphite derived from graphite flasks is treated with H_2SO_4 to be mixed with H_2O_2 in order to produce GO particles. To establish the concentration, 1 g of GO suspension was placed in a dried, weighed beaker, dried overnight at 60 °C and weighed again, and the concentration was then adjusted to 1% GO in DI water on a dry mass basis. The details of the method and Ca-Alg2/GO bead preparation as well as dry and wet images of Ca-Alg and Ca-Alg/GO beads are given in Appendix B.

2.3. Acid Activation of the Beads Activation

Beads for activation were placed into 600 mL beakers of DI water adjusted to pH 4 using 0.1M HCl, which were agitated for 3 h. Afterwards, the beads were collected, rinsed three times with 300 mL of DI water and stored in a closed bottle at RT.

2.4. Characterization

The surface morphological structure of the beads was examined using scanning electron microscopy (SEM) analysis using a Hitachi 3400 SEM, following gold coating. Functional groups of the GO sheets as well as of the Ca-Alg2 and Ca-Alg2/GO beads were identified by Fourier transform infrared spectroscopy (Appendix C). In addition, the GO used for the beads' preparation was characterized by X-ray diffraction and Raman spectroscopy (Appendix D).

2.5. Adsorption Measurements

All adsorption measurements were carried out in 250 mL flasks with 0.05 g of adsorbent (Ca-Alg2, Ca-Alg_2/GO5, Ca-Alg2/GO10 or Ca-Alg2/GO20 dried beads) over 24 h on a shaker table operating at 125 rpm at room temperature (22 °C), unless otherwise specified. Equilibrium for all pollutants was established by 24 h. In total, 75 mL of pollutant solution at a concentration of 10 mg L^{-1} was added in each case, with a pH of 7 for MB and FMTD and a pH of 2 for DFC, unless otherwise specified. The pollutant concentration was determined using a UV-VIS spectrophotometer (Varian) at a wavelength of 660, 286 and 274 nm for MB, FMTD and DFC, respectively. Experiments were carried out in triplicate and the average values reported along with the error bars represent the standard deviation.

2.5.1. Initial Pollutant Concentration

The initial pollutant concentrations tested were 10, 100, 500 and 1000 mgL^{-1} for MB, 10, 25, 100 and 250 mgL^{-1} for FMTD and 1, 5, 10 and 25 mgL^{-1} for DFC. The absorbed amount at equilibrium (q_{eq} (mg g^{-1})) was calculating using Equation (1):

$$q_{eq} = \frac{(C_0 - C_{eq}) \cdot V}{m_{ads}} \tag{1}$$

where C_0 (mg L^{-1}) is the initial pollutant concentration, C_{eq} (mg L^{-1}) the equilibrium pollutant concentration, V (L) the solution volume and m_{ads} (g) the adsorbent mass.

2.5.2. Adsorbent Dose

The effect of the adsorbent dose was studied using 0.01, 0.025, 0.05 and 0.1 g of Ca-Alg2, Ca-Alg2/GO5, Ca-Alg2/GO10 or Ca-Alg2/GO20 dried beads.

2.5.3. pH

The adsorption was performed at pH 7, 9, 10 and 11 for MB and FMTD whereas the adsorption for DFC was at pH 2, 3.5, 5 and 7.

2.5.4. Temperature

The influence of the temperature was studied by performing the adsorption process at 4, 22 and 30 °C.

2.5.5. Thermodynamics

The thermodynamic parameters of adsorption were determined at 4, 22 and 30 °C in order to evaluate the feasibility and the spontaneous nature of the adsorption. The adsorption distribution coefficient K_d is calculated using Equation (2):

$$K_d = \frac{C_0 - C_{eq}}{C_{eq}} \tag{2}$$

where C_0 (mg L^{-1}) is the initial concentration of the solution and C_{eq} (mg L^{-1}) the equilibrium concentration in solution. A plot of $\ln(K_d)$ versus $1/T$ gives a straight line where the enthalpy change ΔH° ($J{\cdot}mol^{-1}$) and the entropy change ΔS° ($J{\cdot}K^{-1}{\cdot}mol^{-1}$) can be calculated using (Equation (3)):

$$\ln(K_d) = \frac{\Delta S^\circ}{R} - \frac{\Delta H^\circ}{RT} \tag{3}$$

where R is the ideal gas constant (8.345 $J{\cdot}mol^{-1}{\cdot}K^{-1}$) and T (K) is the temperature of the solution during the adsorption process. The standard Gibbs free energy change can be obtained from Equation (4):

$$\Delta G^\circ = \Delta H^\circ - T\Delta S^\circ \tag{4}$$

2.5.6. Kinetics

Kinetic parameters were studied using 0.05 g of Ca-Alg2, Ca-Alg2/GO5, Ca-Alg2/GO10 or Ca-Alg2/GO20 dried beads. The three most common models were examined as to their fit to the experimental data [44]. The adsorbate capacity q_t (mg g^{-1}) at time t was calculated using Equation (5):

$$q_t = \frac{(C_0 - C_t){\cdot}V}{m_{ads}} \tag{5}$$

where C_0 (mgL^{-1}) is the initial concentration, C_t (mgL^{-1}) the concentration at time t, V (L) the volume of pollutant solution and m_{ads} (g) the adsorbent mass.

The linearized integral form of the pseudo-first-order Lagergren equation is given by Equation (6):

$$\ln\left(q_{eq} - q_t\right) = \ln\left(q_{eq}\right) - k_1{\cdot}t \tag{6}$$

where k_1 (min^{-1}) is the Lagergren rate constant, q_{eq} (mg g^{-1}) is the maximum adsorbed amount at equilibrium, and q_t (mg g^{-1}) is the amount of adsorption at time t (min). The values of k_1 and q_{eq} were determined from the intercept and the slope of a plot of $\ln(q_{eq} - q_t)$ versus t.

The linearized integral form of the pseudo-second-order model is shown in Equation (7):

$$\frac{t}{q_t} = \frac{1}{k_2{\cdot}q_{eq}^2} - \frac{1}{q_{eq}}{\cdot}t \tag{7}$$

where k_2 ($g{\cdot}mg^{-1}{\cdot}min^{-1}$) is the pseudo second-order rate constant of adsorption. The parameters k_2 and q_{eq} were determined from the intercept and the slope of a plot of t/q_t versus t.

The intraparticle diffusion model is represented in Equation (8):

$$q_t = k_{ip}{\cdot}t^{1/2} + C_{ip} \tag{8}$$

where k_{ip} (mg $g^{-1}\cdot min^{-0.5}$) is an intraparticle diffusion rate constant and C_{ip} (mg g^{-1}) is related to the thickness of the diffusion boundary layer. These parameters were obtained from a plot of q_t versus $t^{1/2}$.

2.5.7. Adsorption Isotherms

The Langmuir model and the Freundlich model were examined for their utility in describing the adsorption process. The Langmuir equation is detailed in Equation (9): [45]

$$q_{eq} = q_{max}\cdot\frac{C_{eq}}{k_L + C_{eq}} \tag{9}$$

where q_{max} (mg g^{-1}) is the maximum adsorption capacity corresponding to complete monolayer coverage, C_{eq} (mgL^{-1}) is the concentration at equilibrium in the solution and k_L (Lg^{-1}) is a constant related to adsorption capacity and the energy of adsorption.

The Freundlich equation is an empirical model based on the adsorption on a heterogeneous surface [46], and is given in Equation (10):

$$q_{eq} = k_F\cdot C^n \tag{10}$$

where k_F (L$\cdot g^{-1}$) and n (-) are the Freundlich constants, indicating the adsorption capacity and the adsorption intensity, respectively. In order to determine the Langmuir and Freundlich constants, Excel Solver was used to fit the adsorption isotherm models with the experimental data. The sum of squared differences between experimental q_{eq} and calculated q_{eq} was minimized by changing the constants of the models with the solver to find the best non-linear regression.

2.6. Desorption Studies

After the concentration at equilibrium was determined, the beads were removed from the solution and were washed three times with DI water. The desorption of pollutants from the beads was examined using three different desorption systems, 0.1 M HCl, 1 M NaCl and ethanol 1% *v*/*v*. The desorption process was carried out in 250 mL conical flasks with 75 mL of desorption solution at RT. The conical flasks were agitated for 24 h at 125 rpm. Then, the final concentration in solution was determined using UV-VIS and the percentage desorption was calculated using Equation (11):

$$\text{Desorption} = \frac{(q_{eq,a} - q_{eq,d})}{q_{eq,a}}\cdot 100 \tag{11}$$

where $q_{eq,d}$ (mg g^{-1}) is the adsorbed amount at equilibrium after 24 h of desorption, and $q_{eq,a}$ (mg g^{-1}) is the adsorbed amount at equilibrium after 24 h of adsorption.

3. Results and Discussion

3.1. Characterisation of Beads

SEM analysis was carried out in order to characterize the morphological structure of the beads, and the results are given Figures 1 and 2 at 500× and 5000× magnification, respectively.

The SEM images show that increased graphene oxide concentration altered the morphological structure of the beads by providing increased porosity and roughness. Due to that increase, the beads had a greater surface available for interactions between adsorbate and adsorbent. The Ca-Alg2/GO20 was typically carbonaceous with similarities to the structure of activated carbon. The FTIR spectrum of GO sheets, Ca-Alg2 and Ca-Alg2/GO beads is given in S4. No significant difference was observed between Ca-Alg2 and Ca-Alg2/GO beads, as the functional groups of the alginate overlap with GO.

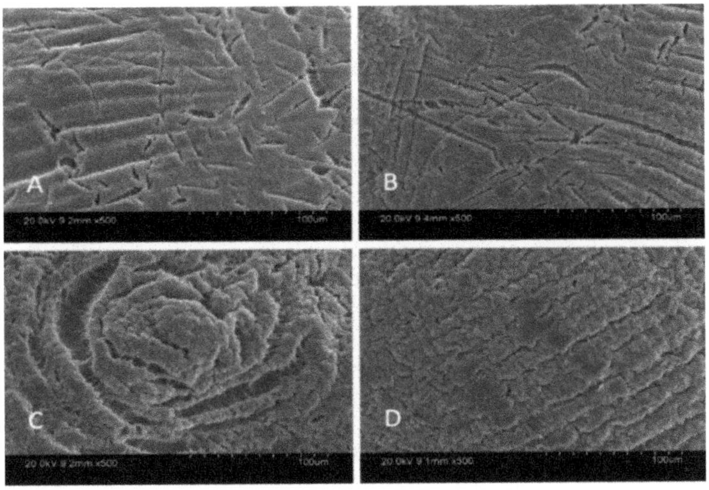

Figure 1. SEM images of beads at 500× magnification: Ca-Alg2 (**A**), Ca-Alg2/GO5 (**B**), Ca-Alg2/GO10 (**C**), Ca-Alg2/GO20 (**D**).

Figure 2. SEM images of beads at 5000×; Ca-Alg2 (**A**), Ca-Alg2/GO5 (**B**), Ca-Alg2/GO10 (**C**), Ca-Alg2/GO20 (**D**).

3.2. *Effect of Operating Parameters on the Adsorption*

3.2.1. Contact Time

The effect of the contact time on q_t was examined by taking samples over 24 h. The average of the results obtained for the adsorption of MB, FMTD and DFC is given in Figure 3A–C, respectively.

The adsorption gradually increased with the contact time and slowed down progressively to reach an equilibrium after 24 h. An increased adsorbate capacity was achieved with an increased concentration of GO in the composites. Ca-Alg2/GO20 beads were the most efficient for the adsorption of each compound, as expected. For the adsorption of MB, Ca-Alg2/GO10 and Ca-Alg2/GO20 beads caused a significant improvement in the adsorbate capacity by increasing it from 6.91 ± 0.83 mg g^{-1} to 9.18 ± 0.08 and 10.63 ± 0.17 mg g^{-1}, respectively, in comparison to Ca-Alg2 (control), with the p values of 0.034 and 0.023. For the adsorption of FMTD, regardless the level of GO, Ca-Alg2/GO composites showed a significant improvement ($p < 0.05$) in adsorbate capacity from 2.78 ± 0.34 mg g^{-1} to a maximum of 7.95 ± 0.54 mg g^{-1}. On the other hand, for the adsorption of DFC, no significant enhancements were seen (high level of error bars), which is attributed to the molecular limitations of DFC.

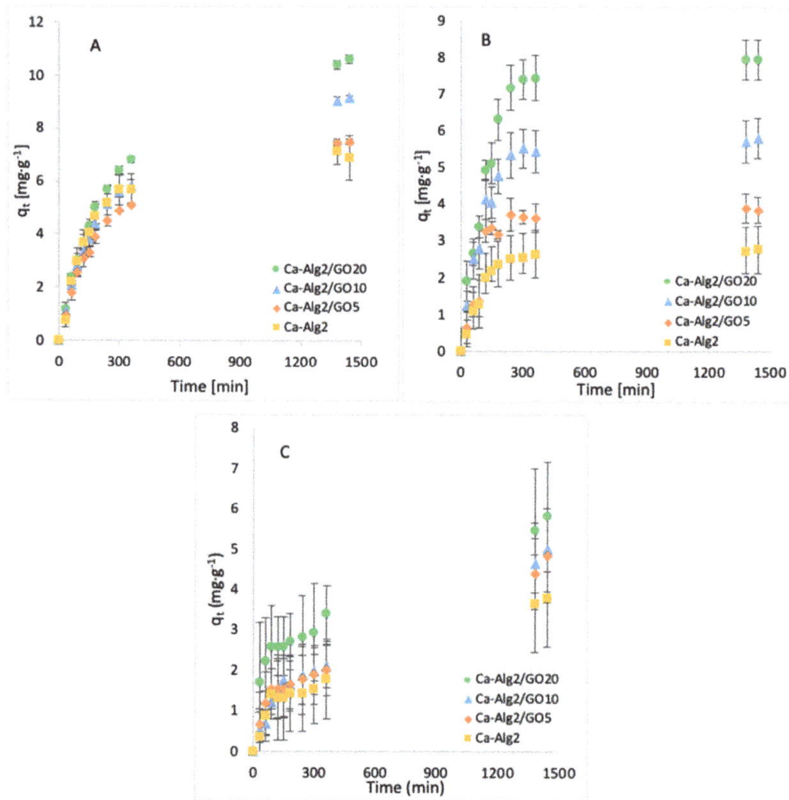

Figure 3. Effect of the contact time on q_{eq} of MB (**A**), FMTD (**B**) and DFC (**C**).

3.2.2. Pollutant Concentration

The effects of the different concentrations of methylene blue, famotidine and diclofenac on the adsorption density (q_{eq}) and the percentage removal are shown in Figures 4–6.

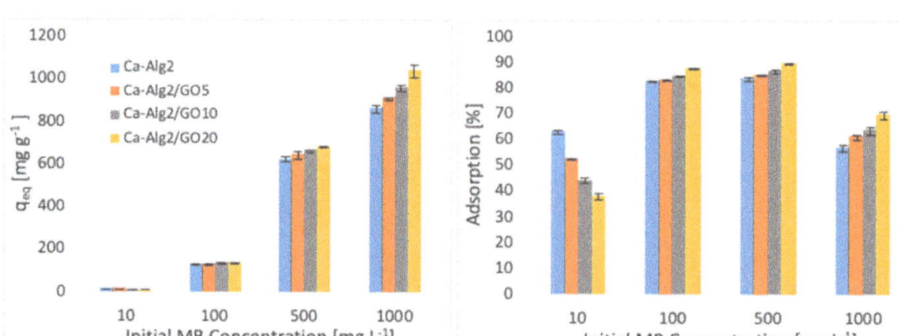

Figure 4. Effect of the initial MB concentration on q_{eq} and adsorption percentage.

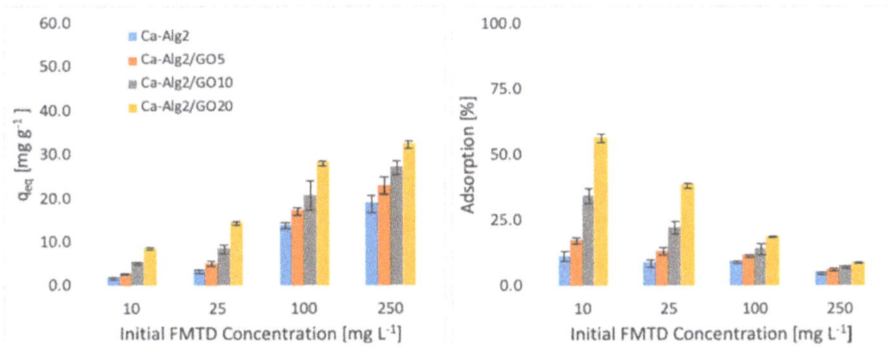

Figure 5. Effect of the initial concentration of FMTD on q_{eq} and adsorption percentage.

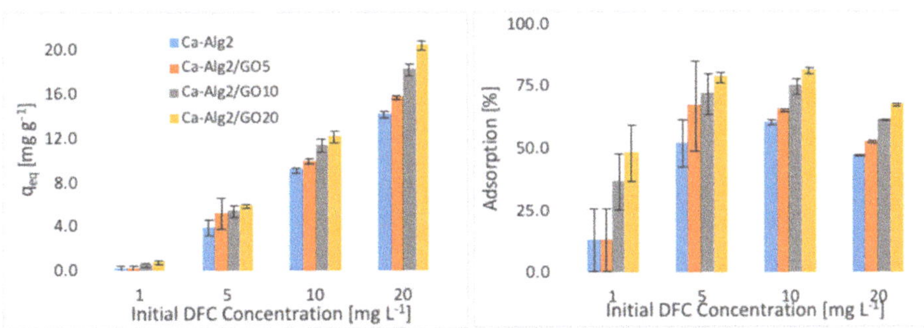

Figure 6. Effect of the initial concentration of DFC on q_{eq} and adsorption percentage.

Higher adsorbed amounts of pollutant at equilibrium (q_{eq}) were achieved with an increased GO level in the composites when the initial concentration of the pollutants (MB, FMTD and DFC) was higher. For the highest initial pollutant concentrations, which were 1000, 250 and 20 mgL^{-1} for MB, FMTD and DFC, respectively, all Ca-Alg2/GO beads (regardless of the GO level) showed a significant increase in q_{eq} value ($p < 0.05$) for DFC, whereas Ca-Alg2/GO10 and Ca-Alg2/GO20 beads caused a significant improvement in the q_{eq} value of MB and FMTD. The greatest enhancements were achieved with Ca-Alg2/GO20 beads, and were from 856.4 ± 16.8 to 1036.2 ± 30.7mg g^{-1} (p: 0.022) for MB, from 18.6 ± 1.9 to 32.2 ± 0.8 mg g^{-1} (p: 0.004) for FMTD and from 14.1 ± 0.329 to 20.4 ± 0.427 mg g^{-1} (p: 0.0001) for DFC at the highest pollutant concentrations.

Ca-Alg2/GO20 beads showed the highest adsorption percentage for each pollutant, as expected. The highest adsorption percentages were found to be 89.4 ± 0.25, 56.0 ± 1.7 and 80.9 ± 1.35 for MB, FMTD and DFC, respectively, which were significantly higher than the corresponding control.

3.2.3. Adsorbent Dose

The effects of the adsorbent dose on adsorption density and adsorption percentage were observed by using four different masses of beads varying from 0.01 to 1.0. The adsorption densities (q_{eq}) of four different types of beads and the percentages of adsorption of MB, FMTD and DFC on Ca-Alg2/GO20 beads are given in Figures 7–9, respectively, as a function of the amount of adsorbent.

A decreased adsorption density has been observed with the increased adsorbent dose regardless of the type of beads and of pollutant. On the other hand, adsorption percentage increased significantly ($p < 0.05$) when using Ca-Alg2/GO20 beads from 52.9 ± 1.7 to 76 ± 0.4% for MB, from 12.2 ± 0.9 to 58.2 ± 0.5% for FMTD and from 39.7 ± 4.1

to 96.1 ± 1.7% for DFC, due to the increase in adsorbent dose from 0.1 to 1.0. On the contrary, the adsorption density decreased significantly ($p < 0.05$) from 38.9 ± 1.6 to 5.7 ± 0.1 mg g^{-1} for MB, from 9.2 ± 1.3 to 4.3 ± 0. mg g^{-1} for FMTD and from 29.5 ± 2.5 to 7.2 ± 0.1 mg g^{-1} for DFC due to the increase in adsorbent dose from 0.1 to 1.0. The reduction in the adsorption density was attributed to a lower quantity adsorbed per unit weight of the adsorbent, causing the presence of unsaturated adsorption sites [47] when the adsorbent dose was increased [48,49]. Ca-Alg2/GO20 beads were shown to exhibit significantly better adsorption than Ca-Alg2 beads.

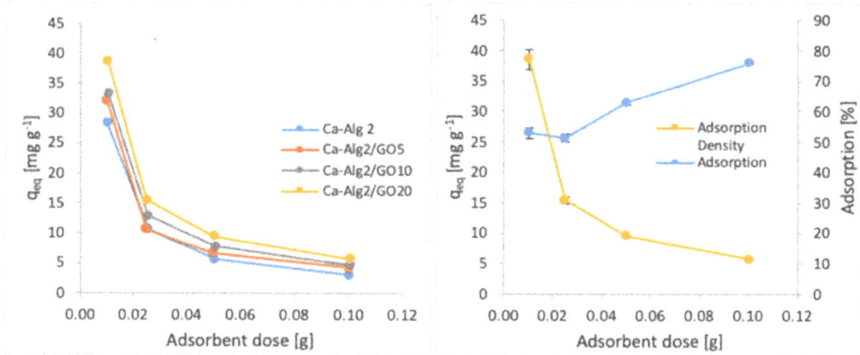

Figure 7. Effect of the adsorbent dose on qeq MB.

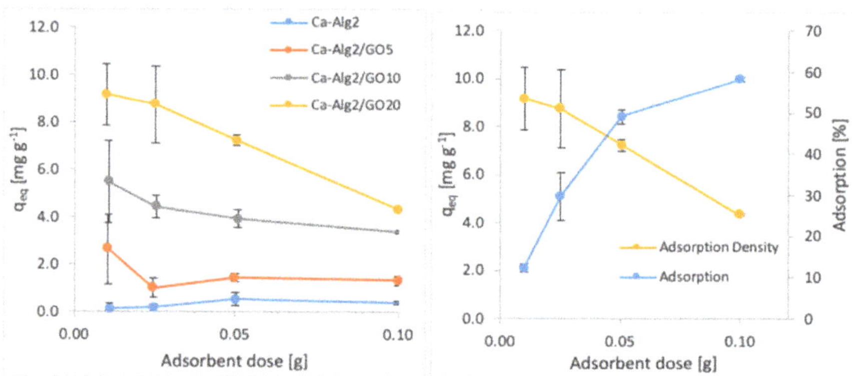

Figure 8. Effect of the adsorbent dose on qeq FMTD.

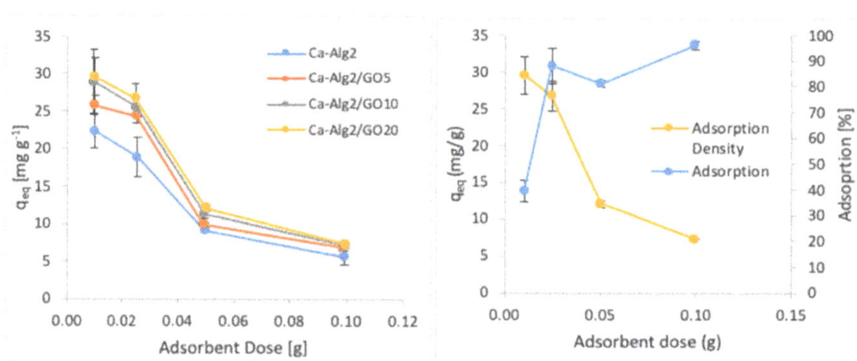

Figure 9. Effect of the adsorbent dose on qeq DFC.

3.2.4. pH

The impact of the pH on adsorption was observed by using four pH values (7, 9, 10, and 11.5 for cationic molecules and 2, 3.5, 5 and 7 for anionic molecules). The averages of the results obtained for the adsorption of MB, FMTD and DFC are shown in the figures below (Figure 10A–C).

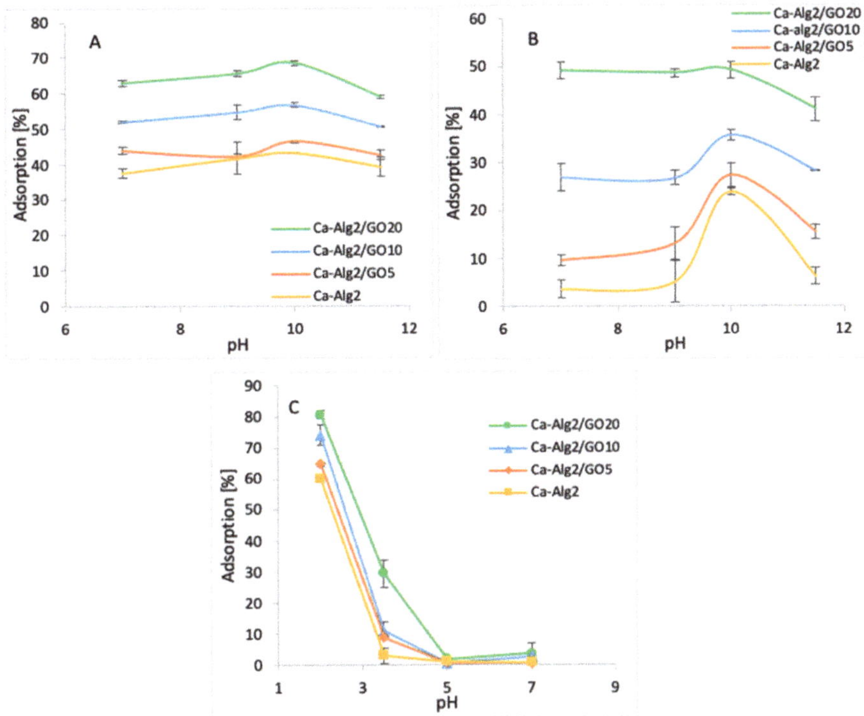

Figure 10. Effect of the pH on q_{eq} MB (**A**), FMTD (**B**), DFC (**C**).

The adsorption percentage of MB at pH 7 was 62.9 ± 0.75%, which slightly increased up to 68.6 ± 0.7% at pH 10 and attained a maximum value (Figure 10A) when Ca-Alg2/GO20 beads were used. Similarly, the highest MB adsorptions were obtained at pH 10 with other beads. The adsorption percentage of FMTD showed a peak at pH 10 with for different types of beads, with the maximum of 49.2 ± 1.6% (Figure 10B) when Ca-Alg2/GO20 was used. Therefore, the adsorption of cationic molecules, MB and FMTD, increases with higher pH solutions until they reach a pH of 10, then it starts to decrease. This observation can be explained by looking at the pK_a values of the analytes and Ca-Alg2/GO20 beads. At pH 10, FMTD is in the neutral form, since its pK_a value is 6.98 [50], and it possess lower water solubility, thereby enhancing the adsorption process at this pH. MB still possesses a positive charge at pH 10; however, GO has an increased negative charge at pH 10 since the phenolic groups of GO now becoming ionized (GO pK_a = 4.3; 6.6; 9.8 all acid groups and 50% of GO phenolic groups will be ionized) [51], enhancing the charge attraction between MB and the adsorbent, which explains the larger adsorption capacity between Ca-Alg2/GO20 and Ca-Alg2 in Figure 10A.

On the other hand, for the anionic molecules, DFC, the highest adsorption percentage was observed as 96.1 ± 1.7% (Figure 10C) when Ca-Alg2/GO20 beads were used at pH 2. The adsorption percentage showed a sudden drop when the pH level increased to 3.5, and further increases in pH level had a negative impact on adsorption for four different bead types. The observed decrease in the adsorption of diclofenac at higher pH is a consequence of the non-ionized to ionized form of diclofenac (pK_a = 4.0), alginate (pK_a mannuronic = 3.38 and guluronic acid = 3.65 [52]) and GO. At pH 2, the ratio of non-ionized to ionized diclofenac is 100:1, that to alginate is 45:1 (pK_a 3.65), and that to GO is 200:1 (see Appendix E). These ratios change at pH 3.5 to 3.16:1, 1.4:1 and 6.3:1 for diclofenac, alginate and GO, respectively. Thus, diclofenac, alginate and GO gain negative charge, and as a consequence repulsion occurs. Furthermore, as diclofenac becomes negatively charged, its water solubility is significantly enhanced, consequently reducing adsorption [53,54]. The calculation of the pKa values of the pollutants and adsorbents is given in Appendix E.

3.2.5. Temperature

Adsorption studies were performed across three temperatures ranging from 4 to 30 °C (4, 22 and 30 °C). The averages of the results obtained for the adsorption of MB, FMTD and DFC are shown in Figure 11A–C.

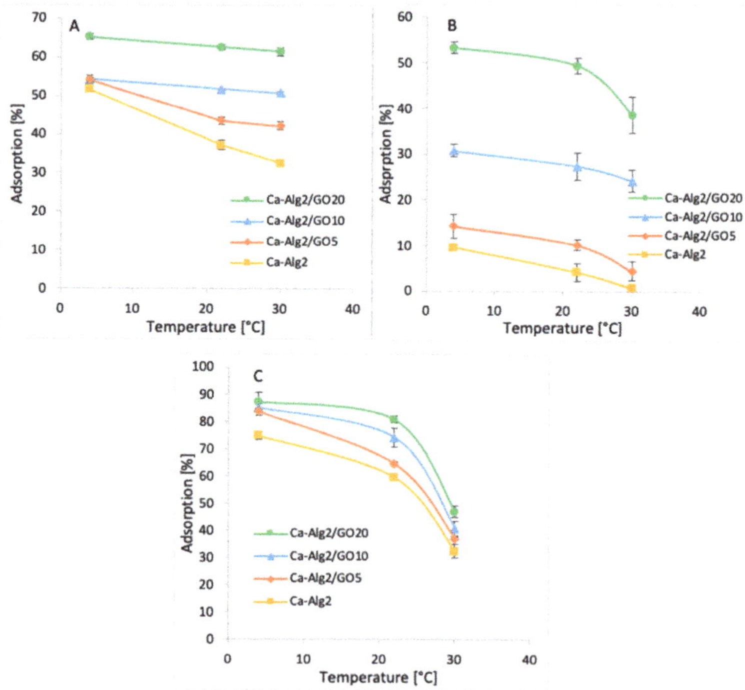

Figure 11. Effect of the temperature on q_{eq} of MB (**A**), FMTD (**B**), DFC (**C**).

The coldest temperature (4 °C) showed the highest adsorption percentages of 65.3 ± 0.9% (Figure 11A), 53.3 ± 1.2% (Figure 11B) and 87.3 ± 3.4% (Figure 11C) for MB, FMTD and DFC, respectively when Ca-Alg2/GO20 beads were used. The adsorption of MB, FMTD and DFC decreased significantly due to stepwise increases in temperature to 22 and 30 °C, and it reached 61.1 ± 1.0, 38.6 ± 4.0 and 47.1 ± 2.0%, respectively, under the same conditions. This may be explained by an exothermic adsorption process [55].

3.3. Thermodynamics

Thermodynamic studies were conducted based on the feasibility and the spontaneous nature of adsorption [44]. The distribution coefficients for the adsorption K_d, enthalpy change ΔH°, entropy change ΔS° and Gibbs free energy change ΔG° were calculated using Equations (2)–(4). The results are given in the Tables 1–3.

Table 1. Thermodynamic parameters for MB adsorption on Ca-Alg2, Ca-Alg2/GO5, Ca-Alg2/GO10 and Ca-Alg2/GO20 beads.

Adsorbent	K_d			ΔG° (kJ·mol^{-1})			ΔH° (kJ·mol^{-1})	ΔS° (J·K^{-1}·mol^{-1})
	4 °C	22 °C	30°C	4 °C	22 °C	30 °C		
Ca-Alg2	1.07	0.59	0.48	−0.144	1.240	1.855	−21.44	−76.9
Ca-Alg2/GO5	1.18	0.77	0.73	−0.341	0.504	0.879	−13.34	−46.9
Ca-Alg2/GO10	1.19	1.08	1.04	−0.401	−0.186	0.091	−3.712	−12.0
Ca-Alg2/GO20	1.88	1.68	1.60	−1.459	−1.272	−1.189	−4.338	−10.4

Table 2. Thermodynamic parameters for FMTD adsorption on Ca-Alg2, Ca-Alg2/GO5, Ca-Alg2/GO10 and Ca-Alg2/GO20 beads.

Adsorbent	K_d			ΔG° (kJ·mol^{-1})			ΔH° (kJ·mol^{-1})	ΔS°(J·K^{-1}·mol^{-1})
	4 °C	22 °C	30 °C	4 °C	22 °C	30 °C		
Ca-Alg2	0.11	0.04	0.01	4.747	9.465	11.56	−67.85	−262
Ca-Alg2/GO5	0.17	0.11	0.05	3.926	6.226	7.248	−31.46	−128
Ca-Alg2/GO10	0.44	0.38	0.32	1.847	2.505	2.797	−8.277	−36.6
Ca-Alg2/GO20	1.14	0.98	0.63	−0.418	0.473	0.868	−14.13	−49.5

Table 3. Thermodynamic parameters for DFC adsorption on Ca-Alg2, Ca-Alg2/GO5, Ca-Alg2/GO10 and Ca-Alg2/GO20 beads.

Adsorbent	K_d			ΔG° (kJ·mol^{-1})			ΔH° (kJ·mol^{-1})	ΔS° (J·K^{-1}·mol^{-1})
	4 °C	22 °C	30 °C	4 °C	22 °C	30 °C		
Ca-Alg2	2.90	1.55	0.55	−2.758	−0.042	1.165	−44.56	−151
Ca-Alg2/GO5	4.96	1.80	0.58	−3.982	−0.704	0.753	−54.43	−182
Ca-Alg2/GO10	5.48	2.61	0.74	−4.352	−1.342	−0.004	−50.67	−167
Ca-Alg2/GO20	6.88	3.81	0.93	−4.896	−2.047	−0.781	−48.74	−158

A significant decrease ($p < 0.05$) in the distribution coefficient (K_d) was observed in all cases when the adsorption temperature was increased from 4 to 30 °C, indicating better adsorption at lower temperatures. For example, the K_d value for the adsorption of MB, FMTD and DFC on Ca-Alg2/GO20 beads decreased from 1.88 to 1.60, 1.14 to 0.63 and 6.88 to 0.93, respectively, as a result of increasing the temperature from 4 to 30 °C. Furthermore, negative enthalpy (ΔH°) and entropy (ΔS°) changes were seen in the adsorption of MB, FMTD and DFC on Ca-Alg2 and Ca-Alg2/GO beads (Tables 1–3). Negative enthalpy change indicates that the adsorption process is exothermic in nature, while negative entropy change suggests a reduction in randomness at the solid–solute interface during adsorption [56]. Moreover, the adsorption of MB and DFC on all beads was found to be spontaneous at 4 °C and 22 °C, respectively, due to the negative Gibbs free energy changes (ΔG°); however, the spontaneity decreased with the increasing temperature. Spontaneity was achieved when Ca-Alg2/GO20 beads were used as the adsorbent. The adsorption of FMDT on Ca-Alg2/GO20 was only found to be spontaneous at 4 °C. Several studies indicate that the absolute magnitude of the change in Gibbs free energy for physisorption is between −20 kJ·mol^{-1} and 0 kJ·mol^{-1}, and chemisorption occurs between −80 kJ·mol^{-1} and −400 kJ·mol^{-1} [57,58]. Thus, the adsorption process observed seems to be physisorption.

3.4. Kinetics

Three models, pseudo-first-order Lagergren, the pseudo-second-order model and intraparticle diffusion, were fitted to the experimental data, and the models are given in Figure 12. All the kinetic parameters for the adsorption of MB, FMTD and DFC are given in Tables 4–6.

The adsorbed amounts of MB and FMTD predicted using the pseudo-first-order model are lower than the experimental data, and the values of R^2 are better under the pseudo-second-order model. The experimental values for MB are 6.91 mg g^{-1} and 10.63 mg g^{-1} for Ca-Alg2 and Ca-Alg2/GO20, respectively, but the results for the pseudo-first-order model are 6.27 mg g^{-1} and 9.79 mg g^{-1} for Ca-Alg2 and Ca-Alg2/GO20, respectively; for the pseudo-second-order model, the results are 7.84 mg g^{-1} and 12.66mg g^{-1} higher than in the experimental data; however, the R^2 values are 0.9907 instead of 0.9657 for the case of Ca-Alg2, and 0.9993 instead of 0.9991 for Ca-Alg2/GO20. For FMTD, the experimental results are 2.78 mg g^{-1} and 7.95 mg g^{-1}, but the predictions of the pseudo-first-order model are 2.57 mg g^{-1} and 8.50 mg g^{-1} for Ca-Alg2 and Ca-Alg2/GO20, respectively; for the pseudo-second-order model, the predictions are 2.93 mg g^{-1} and 8.50 mg g^{-1}; as we can see, the experimental data are higher than those from the pseudo-first-order model and lower than those from the pseudo-second-order model only for the Ca-Alg2 beads, while for the Ca-Alg2/GO20 beads the experimental results are lower than the predicted ones. However, the R^2 results are better for the pseudo-second-order model. This indicates that the adsorption process does not fit the pseudo-first-order model, and shows applicability to the pseudo-second-order model for describing the adsorption of methylene blue and famotidine onto Ca-Alg2 and Ca-Alg2/GO beads [59].

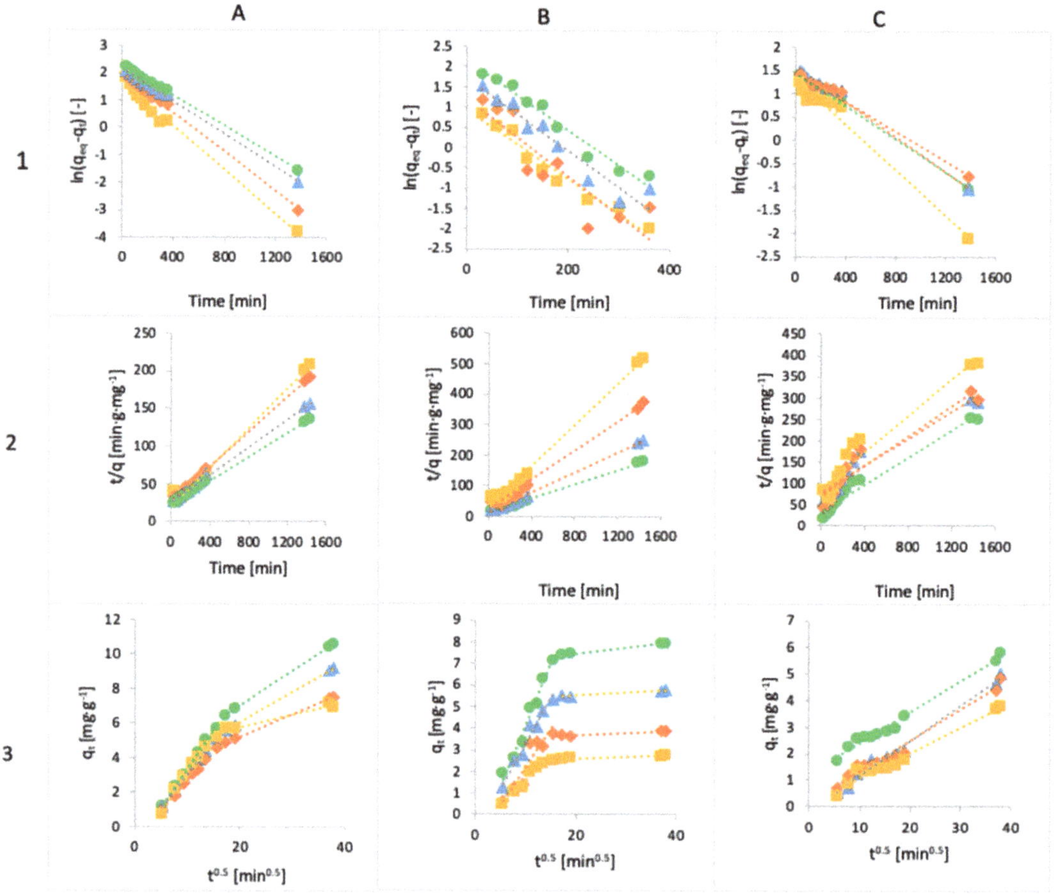

Figure 12. The linearized integral form of the pseudo-first-order Lagergren equation of methylene blue (**A1**), famotidine (**B1**) and diclofenac (**C1**), linearized integral form of the pseudo-second-order model of methylene blue (**A2**), famotidine (**B2**) and diclofenac (**C2**) and intraparticle diffusion model of methylene blue (**A3**), famotidine (**B3**) and diclofenac (**C3**) using Ca-Alg2 (■), Ca-Alg2/GO5 (♦), Ca-Alg2/GO10 (▲) and Ca-Alg2/GO20 (•) beads.

Table 4. Kinetic parameters for MB adsorption onto Ca-Alg2, Ca-Alg2/GO5, Ca-Alg2/GO10 and Ca-Alg2/GO20 dried beads.

Adsorbent	Pseudo-First-Order			Pseudo-Second-Order			Intraparticle Diffusion		
	q_{eq} (mg g^{-1})	k_1 (min^{-1})	R^2 (-)	q_{eq} (mg g^{-1})	k_2 (g·mg^{-1}·min^{-1})	R^2 (-)	k_{id} (mg g^{-1}·min$^{-0.5}$)	C (mg g^{-1})	R^2 (-)
Ca-Alg2	6.27	5.1 × 10^{-3}	0.9657	7.84	8.3 × 10^{-4}	0.9907	0.479	−1.68	0.9580
Ca-Alg2/GO5	7.22	3.6 × 10^{-3}	0.9974	8.70	5.0 × 10^{-4}	0.9995	0.358	−0.957	0.9997
Ca-Alg2/GO10	8.55	3.0 × 10^{-3}	0.9979	10.81	3.5 × 10^{-4}	0.9991	0.409	−1.10	0.9996
Ca-Alg2/GO20	9.79	2.8 × 10^{-3}	0.9991	12.66	2.7 × 10^{-4}	0.9993	0.457	−1.28	0.9998

Table 5. Kinetic parameters for FMTD adsorption onto Ca-Alg2, Ca-Alg2/GO5, Ca-Alg2/GO10 and Ca-Alg2/GO20 dried beads.

Adsorbent	Pseudo-First-Order			Pseudo-Second-Order			Intraparticle Diffusion		
	q_{eq} (mg g^{-1})	k_1 (min^{-1})	R^2 (-)	q_{eq} (mg g^{-1})	k_2 (g·mg^{-1}·min^{-1})	R^2 (-)	k_{id} (mg g^{-1}·min$^{-0.5}$)	C (mg g^{-1})	R^2 (-)
Ca-Alg2	3.70	2.4 × 10^{-3}	0.9756	4.64	5.5 × 10^{-4}	0.9405	0.248	−0.882	0.8809
Ca-Alg2/GO5	4.28	1.6 × 10^{-3}	0.9820	5.72	4.4 × 10^{-4}	0.9147	0.340	−1.23	0.9355
Ca-Alg2/GO10	4.74	1.9 × 10^{-3}	0.9865	6.31	3.3 × 10^{-4}	0.9422	0.409	−0.821	0.8817
Ca-Alg2/GO20	4.26	1.8 × 10^{-3}	0.9879	6.33	7.7 × 10^{-4}	0.9712	0.556	−1.44	0.9992

Table 6. Kinetic parameters for DFC adsorption onto Ca-Alg2, Ca-Alg2/GO5, Ca-Alg2/GO10 and Ca-Alg2/GO20 dried beads.

Adsorbent	Pseudo-First-Order			Pseudo-Second-Order			Intraparticle Diffusion		
	q_{eq} (mg g^{-1})	k_1 (min^{-1})	R^2 (-)	q_{eq} (mg g^{-1})	k_2 (g·mg^{-1}·min^{-1})	R^2 (-)	k_{id} (mg g^{-1}·min$^{-0.5}$)	C (mg g^{-1})	R^2 (-)
Ca-Alg2	3.70	2.4 × 10^{-3}	0.9756	4.64	5.5 × 10^{-4}	0.9405	0.095	0.107	0.9684
Ca-Alg2/GO5	4.28	1.6 × 10^{-3}	0.9820	5.72	4.4 × 10^{-4}	0.9147	0.117	0.110	0.9740
Ca-Alg2/GO10	4.74	1.9 × 10^{-3}	0.9865	6.31	3.3 × 10^{-4}	0.9422	0.131	−0.130	0.9858
Ca-Alg2/GO20	4.26	1.8 × 10^{-3}	0.9879	6.33	7.7 × 10^{-4}	0.9712	0.117	1.190	0.9835

The diffusion mechanism during the adsorption process was studied using the intra-particle diffusion model. The plot of q_t versus $t^{1/2}$ shows a non-linear form, indicating that the adsorption process occurs in more than one step, as there are two distinct linear regions. The first straight region is attributed to macro-pore diffusion, and the second linear region to micro-pore diffusion. The first portion characterizes the instantaneous utilization of the adsorbing sites on the adsorbent surface. On the other hand, the second region is attributed to the slow diffusion of the methylene blue from the surface film into the micro-pores [60].

The predictions for the adsorbed amount of diclofenac obtained using the pseudo-first-order model fit the experimental data better than those obtained using the pseudo-second-order model. The experimental results are 3.78 mg g^{-1} and 5.81 mg g^{-1} for Ca-Alg2 and Ca-Alg2/GO20, respectively; for the pseudo-first-order model, the predictions are 3.70 mg g^{-1} and 4.64 mg g^{-1} for Ca-Alg2 and Ca-Alg2/GO20, respectively; for the pseudo-second-order model, the predictions are 4.26 mg g^{-1} and 6.33 mg g^{-1} for Ca-Alg2 and Ca-Alg2/GO20, respectively. The predictions are higher in every case except for the pseudo-first-order model and the Ca-Alg2/GO20 beads. Moreover, the values of the correlation coefficient R^2 are higher for the pseudo-first-order model, meaning that the adsorption process of diclofenac into Ca-Alg2 and Ca-Alg2/GO beads can be described by the Lagergren model. The intraparticle diffusion model shows a straight line, indicating the adsorption process, because the intercept is close to 0 [61].

3.5. Adsorption Isotherms

The adsorption isotherms for methylene blue were built by testing nine different concentrations, namely, 1, 5, 10, 25, 50, 100, 250, 500 and 1000 mg·L^{-1}. The adsorption was carried out over 24 h at 125 rpm, room temperature, and pH 7, with 0.05 g of beads. The isotherms obtained for each kind of bead are shown in Figure 13, and the data fit results are given in Table 7. The results show that an increase in GO concentration of the beads improves the adsorbed amount of dye at equilibrium. The Langmuir model fits the experimental data better than the Freundlich model, as indicated by the goodness-of-fit tests (Table 7).

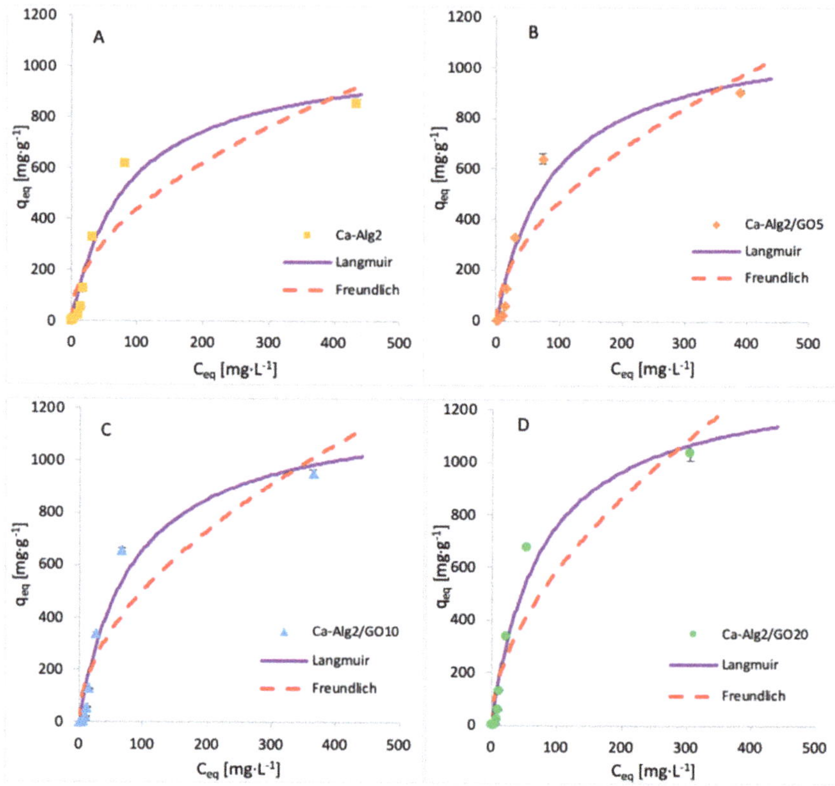

Figure 13. Adsorption isotherms for MB using (**A**) Ca-Alg2, (**B**) Ca-Alg2/GO5, (**C**) Ca-Alg2/GO10 and (**D**) Ca-Alg2/GO20 dried beads.

Table 7. Langmuir and Freundlich isotherm constants for MB adsorption onto Ca-Alg2 and Ca-Alg2/GO beads.

Adsorbent	**Langmuir**			**Freundlich**		
	q_{max} (mg·g^{-1})	K_L (L·g^{-1})	R^2 (-)	K_F (L·g^{-1})	n (-)	R^2 (-)
Ca-Alg2	1064	85.56	0.9778	43.26	0.503	0.9270
Ca-Alg2/GO5	1153	88.93	0.9716	41.02	0.530	0.9109
Ca-Alg2/GO10	1212	84.21	0.9782	42.64	0.537	0.8941
Ca-Alg2/GO20	1334	76.21	0.9894	45.10	0.558	0.8541

The adsorption isotherms for FMTD were performed by using 1, 5, 10, 25, 50, 100 and 250 mgL^{-1} of solution, whereas for diclofenac, the concentrations were 1, 5, 10, 15 and 20 mgL^{-1}. The adsorption process was carried out over 24 h at 125 rpm, room temperature, and pH 7, with 0.05 g of beads. The adsorption models of FMTD and DFC are given in Figures 14 and 15, while the data fit results are given in Tables 8 and 9, respectively.

The results for the adsorption isotherms of FMTD and DFC indicate similar behaviour to MB. Indeed, the Langmuir model better fits the experimental data between each compound and each different kind of bead. Tables 7–9 shows the Langmuir and Freundlich isotherm constants and correlation coefficients for MB, FMTD and DFC adsorption.

The values of R^2 are higher with the Langmuir model, indicating that this model better fits the experimental data than the Freundlich isotherm for each kind of bead and each pharmaceutical. The Langmuir model presumes that the adsorption process occurs on a homogenous surface via monolayer adsorption.

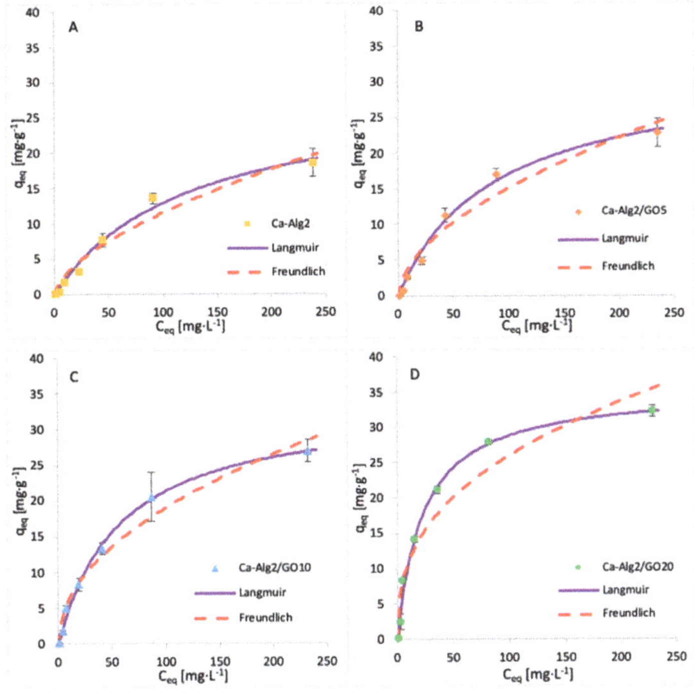

Figure 14. Adsorption isotherms of FMTD using (**A**) Ca-Alg2, (**B**) Ca-Alg2/GO5, (**C**) Ca-Alg2/GO10 and (**D**) Ca-Alg2/GO20 dried beads.

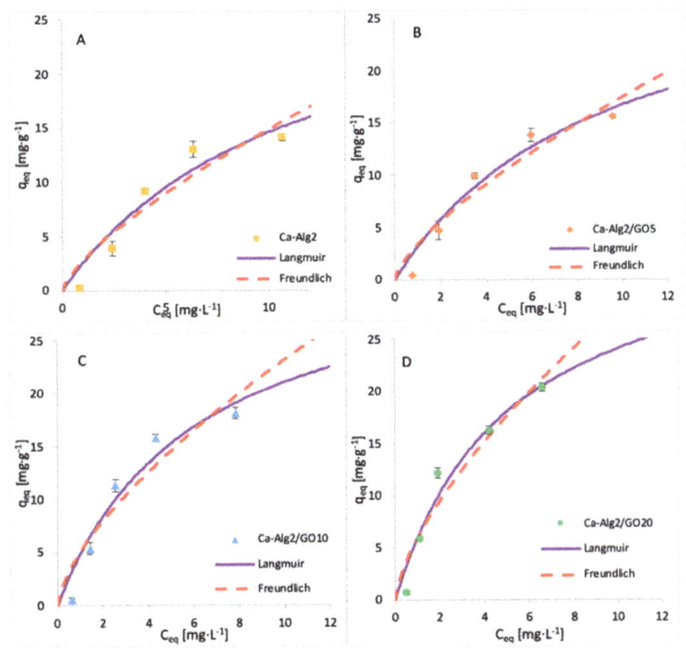

Figure 15. Adsorption isotherms of DFC using (**A**) Ca-Alg2, (**B**) Ca-Alg2/GO5, (**C**) Ca-Alg2/GO10 and (**D**) Ca-Alg2/GO20 dried beads.

Table 8. Langmuir and Freundlich isotherm constants for FMTD adsorption onto Ca-Alg2 and Ca-Alg2/GO beads.

Adsorbent	Langmuir			Freundlich		
	q_{max} (mg·g^{-1})	K_L (L·g^{-1})	R^2 (-)	K_F (L·g^{-1})	n (-)	R^2 (-)
Ca-Alg2	28.96	123.2	0.9809	0.680	0.615	0.9351
Ca-Alg2/GO5	31.69	85.39	0.9733	1.190	0.552	0.9173
Ca-Alg2/GO10	33.57	57.02	0.9611	2.099	0.479	0.8593
Ca-Alg2/GO20	35.50	23.10	0.9214	4.647	0.374	0.7491

Table 9. Langmuir and Freundlich isotherm constants for DFC adsorption onto Ca-Alg2 and Ca-Alg2/GO beads.

Adsorbent	Langmuir			Freundlich		
	q_{max} (mg·g^{-1})	K_L (L·g^{-1})	R^2 (-)	K_F (L·g^{-1})	n (-)	R^2 (-)
Ca-Alg2	30.74	11.10	0.9457	2.795	0.725	0.8937
Ca-Alg2/GO5	31.81	9.020	0.9175	3.441	0.705	0.8707
Ca-Alg2/GO10	33.72	5.988	0.8886	5.055	0.662	0.8401
Ca-Alg2/GO20	36.35	5.066	0.8872	5.992	0.672	0.8322

The constants K_F and n indicate the adsorption capacity and the adsorption intensity, respectively. As indicated by the experimental data, the adsorption capacity K_F increases gradually with graphene oxide concentration, and the constant n is lower than 1, meaning the adsorption isotherm is favourable. The maximum adsorption capacities q_{max} obtained are 1334, 35, 50 and 36.35 mg g^{-1} for the uptake of methylene blue, famotidine and diclofenac, respectively. This means that Ca-Alg2/GO beads are an efficient adsorbent for the removal of these contaminants, particularly for methylene blue as, to our knowledge, this is the highest adsorption capacity for MB that has been reported in the literature.

3.6. Desorption

The desorption of each compound adsorbed onto Ca-Alg2 and Ca-Alg2/GO beads was studied using HCl/NaOH 0.1M, NaCl 1M and ethanol 1% v:v. The results of the percentage desorbed after 24 h are shown in Figure 16.

The desorption of MB from the beads was higher when using HCl 0.1 M than with NaCl or ethanol. The results show that 89 ± 9.9% and 44 ± 0.6% of MB are desorbed for Ca-Alg2 and Ca-Alg2/GO20 beads, respectively. Indeed, excesses of H^+ protons seem to be able to force the cationic dye to be released by arising on the adsorption sites on the surfaces of the beads. It is more difficult to cause the release of MB from beads with graphene oxide due to the stronger affinity. NaCl 1 M also showed good results for the desorption of MB. However, the ionic strength of sodium chloride 1 M destabilizes the structure of calcium alginate beads, making them soft, fragile, and crumbly. As such, a high-concentration salt solution cannot be use as a desorbent due to the inability to reuse the beads. Ethanol solution showed low desorption of MB, as the main interactions between adsorbate and adsorbent are typically ionic bonds, and ethanol has a low ability to remove the dye from the beads with van der Waals' forces.

On the other hand, the desorption of famotidine using HCl 0.1 M showed less satisfactory results. This might be due to the poor solubility in a low pH solution or to the presence of hydrogen bonds between the adsorbent and the adsorbate, making hydrochloric acid unable to release the drug into the solution. Hydrogen bonds could also explain why the ionic strength of NaCl is insufficient to remove the cationic pharmaceutical from the beads, in addition to destroying the stability of the beads. As for MB, ethanol has little effect on the van der Waals' interaction for the removal of famotidine because of the stability of hydrogen bonds.

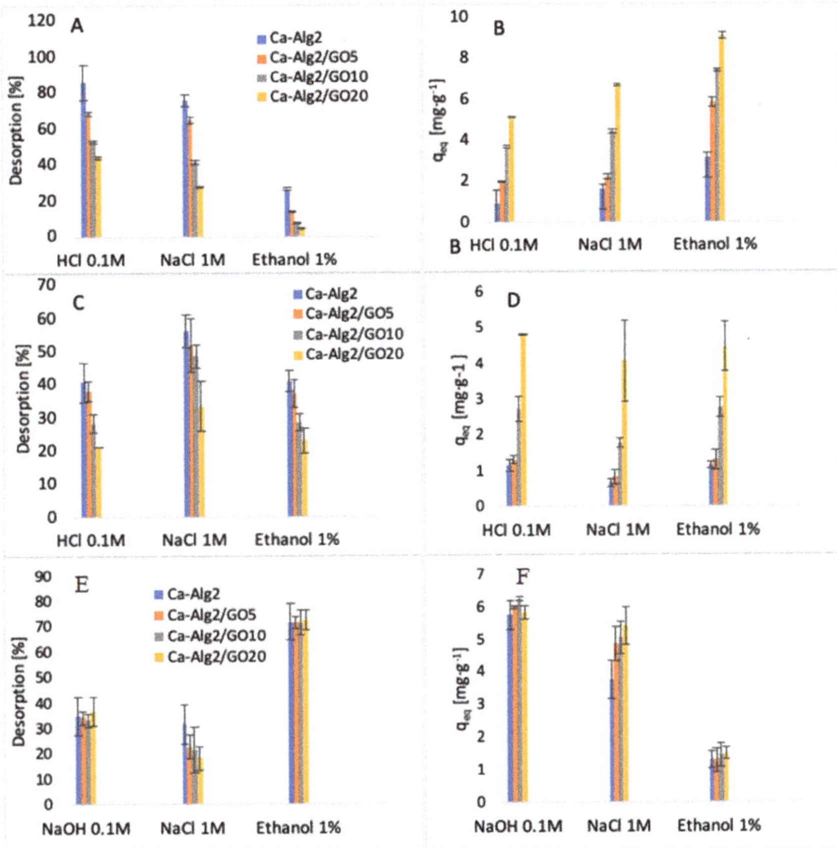

Figure 16. Desorption (%) of MB (**A**), desorption q_{eq} of MB (**B**), desorption (%) of FMTD (**C**), desorption q_{eq} of FMTD (**D**), desorption (%) of DFC (**E**) and desorption q_{eq} of DFC (**F**).

The results regarding the desorption of diclofenac from the beads showed that ethanol 1% v:v is able to release the pharmaceutical into solution. The percentages of desorption are high, around 70%, and the results present little difference between each of the kinds of beads. Diclofenac was desorbed from the beads with NaOH 0.1 M, mainly due to a change in the pH of the solution. As the results of the effect of pH showed, the adsorption of diclofenac was low in high pH solution. Similarly, to other compounds, NaCl is not an efficient desorbent, and it damages the stability of the beads by interacting with the structure of the polymer.

4. Conclusions

The SEM analysis showed that an increase in graphene oxide modified the morphological structure of the beads. Indeed, they become more porous and rougher with a higher surface available for the interactions between adsorbate and adsorbent. As expected, during the adsorption process, Ca-Alg2/GO20 beads present the best adsorption for each compound. The effect of initial concentration, adsorbent dose, pH and temperature all play an important role in the adsorption. The results show that a higher concentration of pharmaceuticals increases the force of the diffusion of the drug adsorbed by the beads. With a lower concentration of beads, the adsorbed amount at equilibrium q_{eq} increases because of a higher amount adsorbed per unit of weight of the adsorbent. On the other hand, the percentage of removal decreases due to the fewer adsorption sites available.

The pH can modify the structure of the beads along with the pharmaceuticals, causing a change in the interactions between adsorbate and adsorbent. The adsorption process is better at low temperature than high temperature, meaning the adsorption mechanism is exothermic (as confirmed by H°), and thermodynamic studies show that the physisorption is spontaneous. The pseudo-second-order model is the best fit for the experimental data concerning the kinetics of adsorption for methylene blue and famotidine; however, the Lagergren pseudo-first order model is better suited to diclofenac. Furthermore, each compound follows the Langmuir model for isotherm adsorption, with a maximum adsorption capacity of 1334, 35.50 and 36.35 $mg{\cdot}g^{-1}$ for methylene blue, famotidine and diclofenac, respectively. It should be noted that the adsorption capacity of MB onto calcium alginate graphene oxide beads, particularly Ca-Alg2/GO20 composites, was found to be superior in comparison to other adsorbents, ranging from commercial activated carbon adsorption capacity (980.3 $mg{\cdot}g^{-1}$) to bioadsorbents adsorption capacity, such as modified biomass of baker's yeast (869.6 $mg{\cdot}g^{-1}$), and many other natural adsorbents, such as biomass and coal-derived activate carbon, as reviewed in detail by [62]. In addition, GO and alginate incorporation outperformed a graphene-derived nanocomposite ($Fe_3O_4^-$ graphene at mesoporous SiO_2), wherein the adsorption capacity of this graphene-derived nanocomposite was reported to be 178.49 $mg{\cdot}g^{-1}$ in terms of MB removal [32]. Furthermore, the adsorption capacity of MB was recently published as 150.66 $mg{\cdot}g^{-1}$ when GO was incorporated with sodium alginate to produce aero gel beads. Therefore, it can be concluded that graphene oxide calcium alginate composite is a superabsorbent useful for MB removal from water, in addition to being superior to the previously researched compounds [42,63]. Furthermore, by treating the Alg2/GO20 composites that had come in contact with the absorbates MB or diclofenac with 0.1 M HCl and ethanol 1% v:v, the absorbates could be efficiently removed/desorbed, and the Alg2/GO20 composite beads were regenerated without damage to bead integrity. Further investigations need to be performed for famotidine, as little desorption was observed with the desorption candidates examined.

These beads appear to be an efficient adsorbent for dyes and pharmaceuticals, particularly for methylene blue. This novel technology could be applied as a polishing step in water treatment in order to reduce the concentration of these micropollutants, as well as of the synthetic dyes that negatively impact the environment, human health and aquatic life. Methylene blue is widely used a model pollutant in adsorption studies. It is interesting to note that while the performance of the beads assessed in this study is excellent for MB, it is less than ideal for FMTD and DFC. This calls into question the validity of the common approach of using a single-component pollutant for novel adsorbent testing. However, the treatment of an MB, FMTD and DFC mixture via Alg2/GO beads might have an influence on the adsorption/desorption capacities. Therefore, conducting an experimental study targeting the removal of mixed pollutants via Alg2/GO beads is suggested as a future direction.

Author Contributions: Conceptualization, Z.G., B.Q., A.M., K.N. and J.L.; methodology, B.G., D.M., Z.G., B.Q., A.M., K.N. and J.L.; validation, D.M., K.N. and J.L.; formal analysis, Z.G. and K.N; investigation, B.G., Y.J., L.S., R.P., D.M., Z.G., B.Q., A.M., K.N. and J.L.; resources, K.N. and J.L.; data curation, Y.J., B.G., K.N. and J.L.; writing—original draft preparation, B.G., Y.J., K.N. and J.L.; writing—review and editing, B.G., K.N. and J.L.; visualization, Y.J., L.S., R.P. and B.G.; supervision, B.G., B.Q., A.M., K.N. and J.L.; project administration, K.N., B.Q., A.M. and J.L.; funding acquisition, B.Q., A.M., K.N. and J.L. All authors have read and agreed to the published version of the manuscript.

Funding: This research was funded by Environmental Protection Agency, grant number 2011-W-MS-8, and the APC was funded by Qatar Environment and Energy Research Institute.

Acknowledgments: The authors acknowledge the immense support of Environmental Protection Agency Ireland for funding the project and for unwavering support. The authors are most grateful to Qatar Environment and Energy Research Institute for covering the publication charge. Yannick Jaquet, Laura Sánchez, Rebecca Pumarino was funded with the support of the Erasmus+ program of the European Union.

Conflicts of Interest: The authors declare no conflict of interest.

Disclaimer: The European Commission's support for the production of this publication does not constitute an endorsement of the contents, which reflect the views only of the authors, and the Commission cannot be held responsible for any use that may be made of the information contained therein.

Appendix A

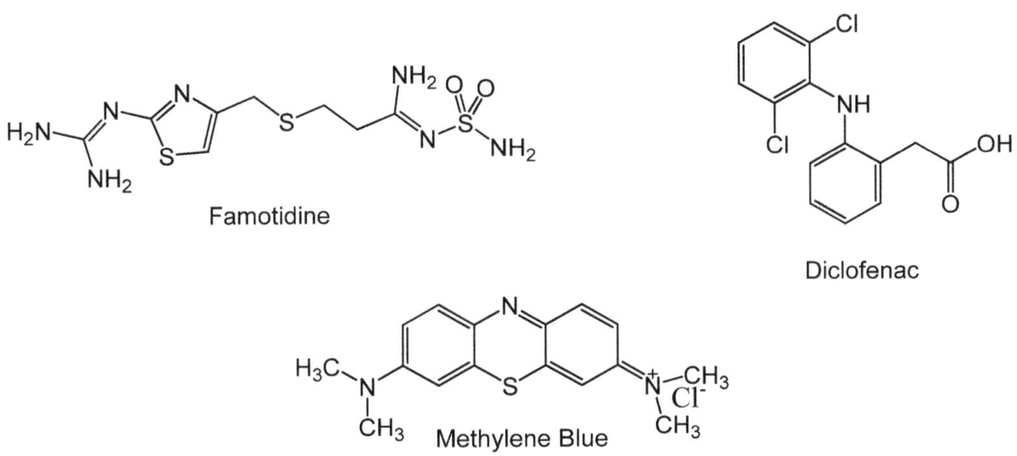

Figure A1. Structure of Analytes.

Appendix B

Appendix B.1. Preparation of GO

Graphite flakes were microwaved in small portions in a glass beaker in a 700 W microwave (modified, vented) on maximum power three times for 10 s each time to produce expanded graphite; 2 g of expanded graphite, 10 g of dried potassium permanganate and 250 mL of H_2SO_4 were added to a 2 L round bottomed flask, and stirred overnight at room temperature (RT). After that, 500 mL of deionized (DI) water was added into the flask, maintaining the temperature below 80 °C using an ice bath.

In total, 100 mL of H_2O_2 was added slowly to the mixture (color change to golden brown), after that the mixture was stirred for 30 min and the resulting particles were washed with 750 mL of 10% HCl solution and centrifuged three times. To remove any residual traces of preparatory chemicals, the obtained GO was resuspended in 750 mL of DI water and centrifuged three times, and the washed GO was resuspended in 50 mL of DI water to achieve a thick paste. To establish the concentration, 1 g of GO suspension was spread in a dried, weighed beaker, dried overnight at 60 °C and weighed again, and the concentration was then adjusted to 1% GO in DI water on a dry mass basis.

Appendix B.2. Preparation of Ca-Alg2/GO Beads

Na-Alg/GO solutions were prepared by weighing 5, 10 or 20 g of 1% *w/w* GO concentrate into a conical flask, then 50 mL of DI water was added to the flask. The suspension was ultrasonicated for 20 min to ensure good dispersion of GO, and was then added to 250 mL of a 2% *w/w* Na-Alg solution and stirred until a homogenous mixture was obtained.

Ca-Alg2 beads were prepared using the Na-Alg solution without GO by using a syringe pump with a flow rate of 20 mL min^{-1} and two 10 mL syringes with a 1 mm needle. The solution was dropped into an aqueous coagulation bath of $CaCl_2$ 6% *w/v*, and the bath was continuously agitated with a magnetic stirrer in order to prevent beads agglomeration. The beads were left for 24 h without agitation to achieve the complete formation of Ca-

Alg2 gel beads, and the beads were collected and washed three times with 500 mL of DI water. The same procedure was used to prepare Ca-Alg2/GO5, Ca-Alg2/GO10 and Ca-Alg2/GO20 beads. Activated beads were then dried at 60 °C over 2 days in order to reach a constant weight. The beads were stored in plastic vials at room temperature.

Figure A2. Images of Ca-Alg/GO and Ca-Alg dried beads (average diameter 0.46 ± 0.02 cm).

Figure A3. Images of (**a**) Ca-Alg and (**b**) Ca-Alg/GO wet gel beads. The scale bar = 0.5 cm.

Appendix C.

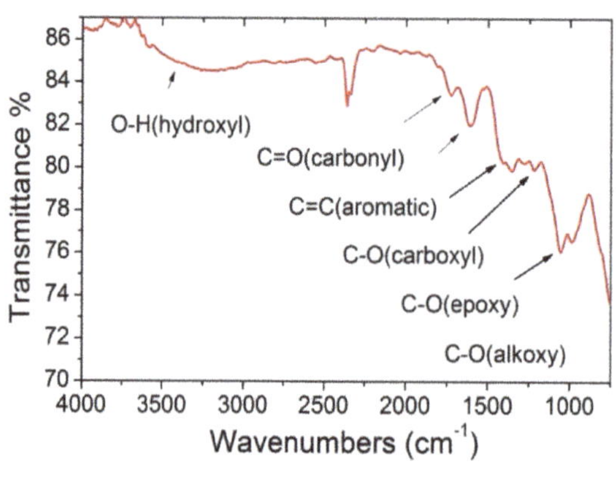

Figure A4. GO FTIR spectrum of GO sheets used in the preparation of alginate beads.

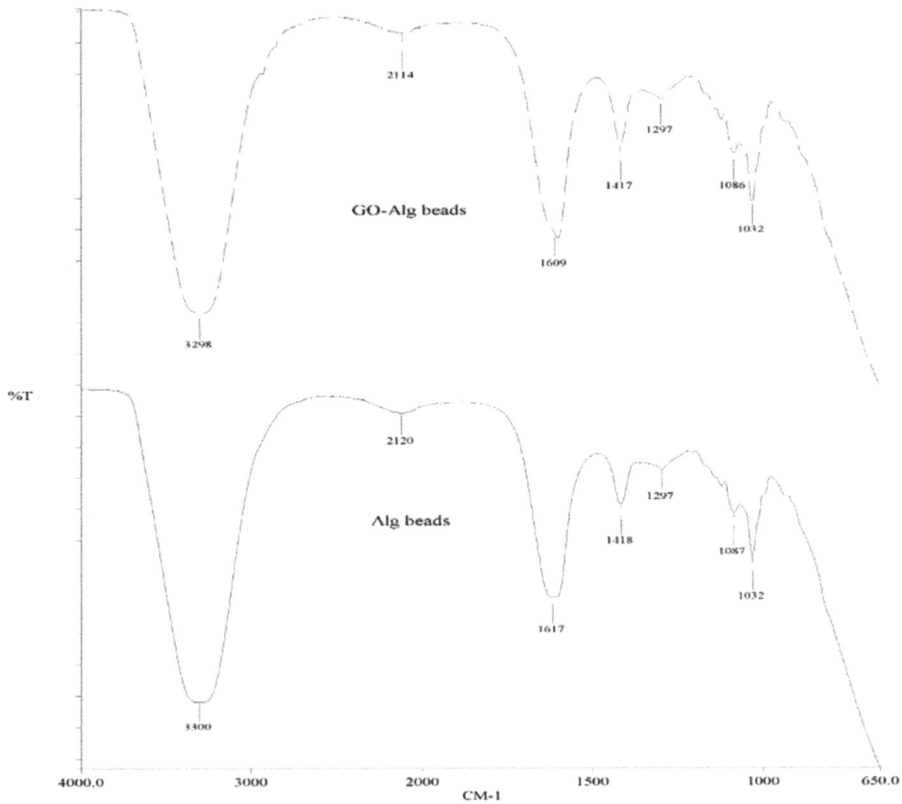

Figure A5. FTIR spectrum of Ca-Alg2 and Ca-Alg2/GO beads.

Appendix D.

XRD pattern of GO used in the preparation of the alginate beads

It can be seen that the as-prepared GO shows a distinct peak at 10° attributed to the (002) plane and corresponding to a layer-to-layer distance (d-spacing) of about 0.82 nm. The crystallite size of graphene oxide sheets is about 7.5nm based on the calculations from the half width at maximum (HWFM) of the X-ray diffraction peak using Scherrer's equation.

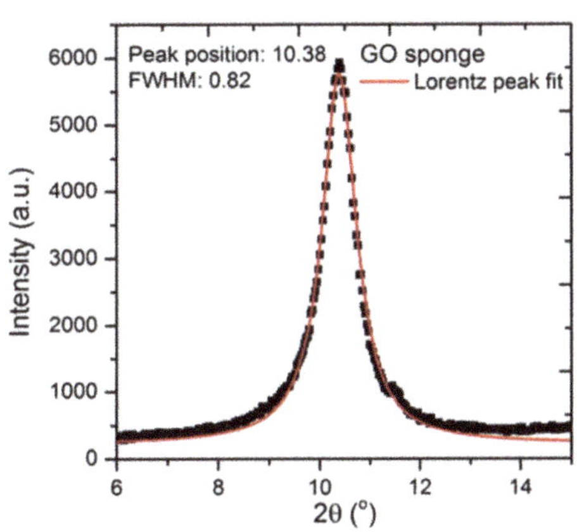

Figure A6. Raman spectrum of GO used in the preparation of the alginate beads.

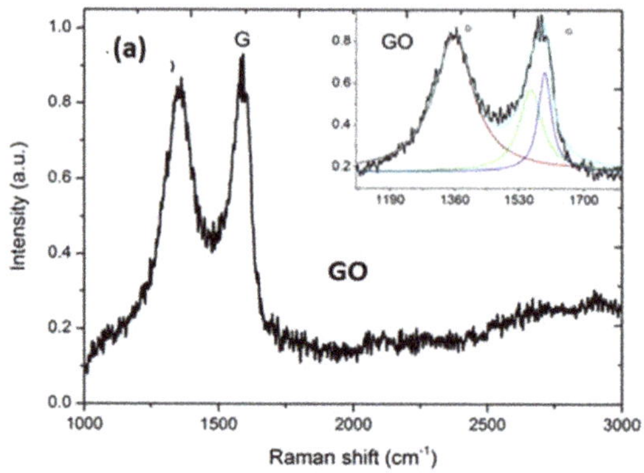

Figure A7. GO displayed a strong D band at 1357 cm^{-1}. Raman spectra can be deconvoluted (inset spectrum) to three Lorentzian peaks (with adj. R-Squared $\geq$ 92%). D/IG is 1.27 for GO.

Appendix E.

Detailed calculation of pKa values for pollutants and adsorbents, Henderson–Hasselbach equation.

Table A1. Reported pKa values from the literature.

Alginate		Graphene Oxide			Famotidine	Diclofenac
pK_a = 3.38	pK_a = 3.65	pK_a = 4.3	pK_a = 6.6	pK_a = 9.8	pK_a=6.98	pK_a = 4.15

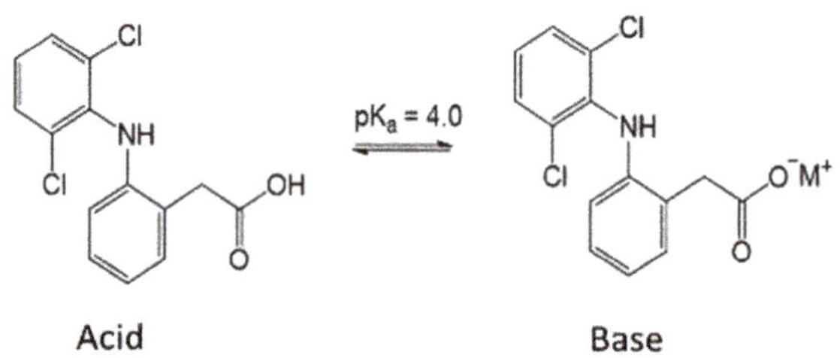

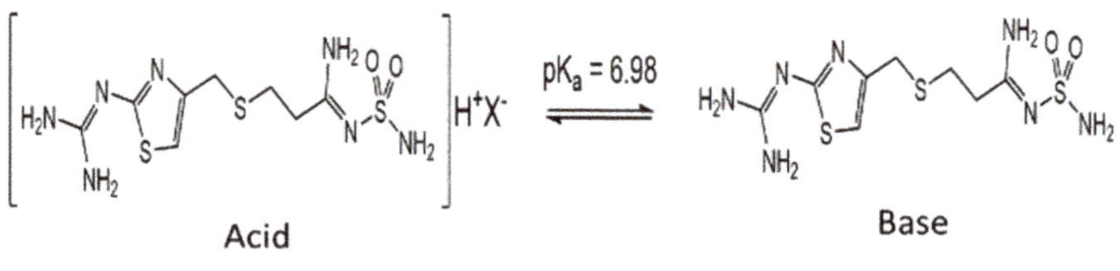

Henderson–Hasselbach Equation:

$$pH = pKa + \log [acid]/[base]$$

Table A2. Example of [acid]/[base] ratio for Diclofenac at various pH values.

pH of Solution	pH = 2	pH = 3.5	pH = 5	pH = 7
Acid/base	100/1	3.16:1	0.1:1	0.001:1

Appendix F.

SEM, TEM and AFM images of bare GO in various forms

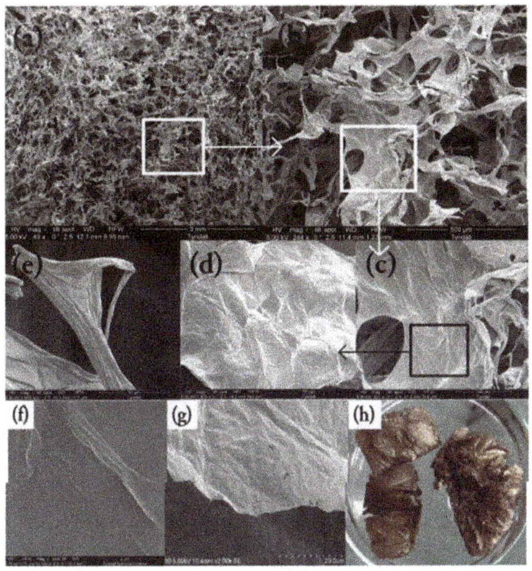

Figure A8. SEM images of (**a**–**e**) GO sponge at different magnifications, (**f**) GO sheet surface, (**g**) GO single sheet edge, (**h**) GO sponge photo.

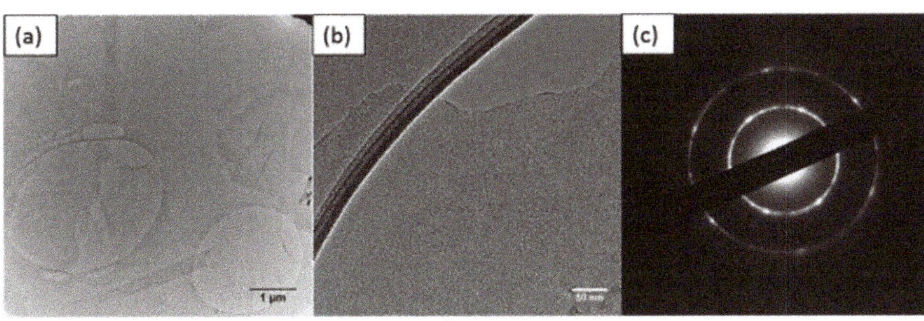

Figure A9. TEM images of (**a**) thin GO sheet, (**b**) HETEM of single-layer GO sheet and (**c**) SAED pattern of GO.

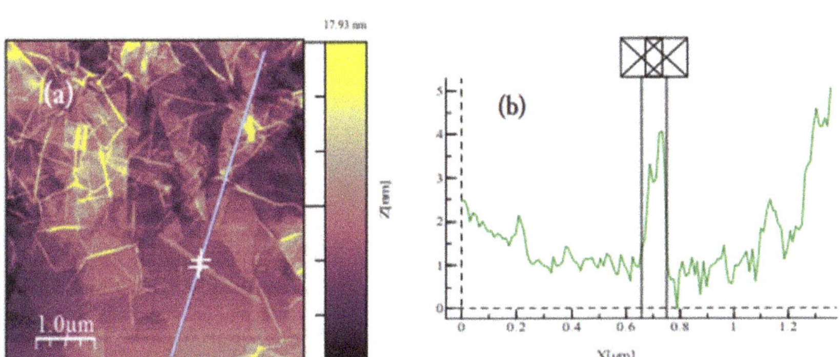

Figure A10. (**a**) AFM image of GO dispersion deposited on SiO_2/Si substrate. GO sheets are expected to be around 1 nm thick due to the presence of covalently bonded oxygen and the displacement of the sp3 hybridized carbon atoms slightly above and below the original graphene basal plane 304. (**b**) Average thickness of ~0.8 nm according to the height profile.

References

1. Luo, Y.; Guo, W.; Ngo, H.H.; Nghiem, L.D.; Hai, F.I.; Zhang, J.; Liang, S.; Wang, X.C. A review on the occurrence of micropollutants in the aquatic environment and their fate and removal during wastewater treatment. *Sci. Total Environ.* **2014**, *473–474*, 619–641. [CrossRef]
2. Guo, J.-Z.; Li, B.; Liu, L.; Lv, K. Removal of methylene blue from aqueous solutions by chemically modified bamboo. *Chemosphere* **2014**, *111*, 225–231. [CrossRef]
3. Benstoem, F.; Becker, G.; Firk, J.; Kaless, M.; Wuest, D.; Pinnekamp, J.; Kruse, A. Elimination of micropollutants by activated carbon produced from fibers taken from wastewater screenings using hydrothermal carbonization. *J. Environ. Manag.* **2018**, *211*, 278–286. [CrossRef]
4. Schwarzenbach, R.P.; Escher, B.I.; Fenner, K.; Hofstetter, T.B.; Johnson, C.A.; von Gunten, U.; Wehrli, B. The Challenge of Micropollutants in Aquatic Systems. *Science* **2006**, *313*, 1072–1077. [CrossRef] [PubMed]
5. Miller, T.H.; Bury, N.R.; Owen, S.; MacRae, J.I.; Barron, L.P. A review of the pharmaceutical exposome in aquatic fauna. *Environ. Pollut.* **2018**, *239*, 129–146. [CrossRef]
6. Jeon, Y.S.; Lei, J.; Kim, J.-H. Dye adsorption characteristics of alginate/polyaspartate hydrogels. *J. Ind. Eng. Chem.* **2008**, *14*, 726–731. [CrossRef]
7. Keskinkan, O.; Göksu, M.Z.L. Assessment of the dye removal capability of submersed aquatic plants in a laboratory-scale wetland system using anova. *Braz. J. Chem. Eng.* **2007**, *24*, 193–202. [CrossRef]
8. Mittal, A.K. Stdie or Soretio of Dycr by Sulfomlcd Col dA Gamdema fucidutn. *Indian J. Environ. Health* **1989**, *31*, 105–111.
9. El Qada, E.N.; Allen, S.J.; Walker, G. Adsorption of Methylene Blue onto activated carbon produced from steam activated bituminous coal: A study of equilibrium adsorption isotherm. *Chem. Eng. J.* **2006**, *124*, 103–110. [CrossRef]
10. Bulut, Y.; Aydın, H. A kinetics and thermodynamics study of methylene blue adsorption on wheat shells. *Desalination* **2006**, *194*, 259–267. [CrossRef]
11. Hassan, A.; Abdel-Mohsen, A.-M.; Fouda, M.M. Comparative study of calcium alginate, activated carbon, and their composite beads on methylene blue adsorption. *Carbohydr. Polym.* **2014**, *102*, 192–198. [CrossRef] [PubMed]
12. Lin, S.-Y. An Overview of Famotidine Polymorphs: Solid-State Characteristics, Thermodynamics, Polymorphic Transformation and Quality Control. *Pharm. Res.* **2014**, *31*, 1619–1631. [CrossRef] [PubMed]
13. Kristofco, L.A.; Brooks, B.W. Global scanning of antihistamines in the environment: Analysis of occurrence and hazards in aquatic systems. *Sci. Total. Environ.* **2017**, *592*, 477–487. [CrossRef]
14. Murphy, S.; Saurel, C.; Morrissey, A.; Tobin, J.; Oelgemoeller, M.; Nolan, K. Photocatalytic activity of a porphyrin/TiO2 composite in the degradation of pharmaceuticals. *Appl. Catal. B Environ.* **2012**, *119–120*, 156–165. [CrossRef]
15. Rad, T.S.; Khataee, A.; Kayan, B.; Kalderis, D.; Akay, S. Synthesis of pumice-TiO2 nanoflakes for sonocatalytic degradation of famotidine. *J. Clean. Prod.* **2018**, *202*, 853–862. [CrossRef]
16. Bort, R.; Ponsoda, X.; Jover, R.; Gómez-Lechón, M.J.; Castell, J.V. Diclofenac toxicity to hepatocytes: A role for drug metabolism in cell toxicity. *J. Pharmacol. Exp. Ther.* **1999**, *288*, 65–72.
17. McGettigan, P.; Henry, D. Cardiovascular Risk and Inhibition of Cyclooxygenase. *JAMA* **2006**, *296*, 1633–1644. [CrossRef]
18. Gros, M.; Petrović, M.; Ginebreda, A.; Barceló, D. Removal of pharmaceuticals during wastewater treatment and environmental risk assessment using hazard indexes. *Environ. Int.* **2010**, *36*, 15–26. [CrossRef]
19. Triebskorn, R.; Casper, H.; Scheil, V.; Schwaiger, J. Ultrastructural effects of pharmaceuticals (carbamazepine, clofibric acid, metoprolol, diclofenac) in rainbow trout (*Oncorhynchus mykiss*) and common carp (*Cyprinus carpio*). *Anal. Bioanal. Chem.* **2007**, *387*, 1405–1416. [CrossRef]
20. Nambirajan, K.; Muralidharan, S.; Roy, A.A.; Manonmani, S. Residues of Diclofenac in Tissues of Vultures in India: A Post-ban Scenario. *Arch. Environ. Contam. Toxicol.* **2017**, *74*, 292–297. [CrossRef]
21. Starukh, G. Photocatalytically Enhanced Cationic Dye Removal with Zn-Al Layered Double Hydroxides. *Nanoscale Res. Lett.* **2017**, *12*, 1–8. [CrossRef]
22. Huang, R.; He, L.; Zhang, T.; Li, D.; Tang, P.; Zhao, Y.; Feng, Y. Fabrication and Adsorption Behavior of Magnesium Silicate Hydrate Nanoparticles towards Methylene Blue. *Nanomaterials* **2018**, *8*, 271. [CrossRef] [PubMed]
23. Basha, S.; Keane, D.; Nolan, K.; Oelgemoeller, M.; Lawler, J.; Tobin, J.M.; Morrissey, A. UV-induced photocatalytic degradation of aqueous acetaminophen: The role of adsorption and reaction kinetics. *Environ. Sci. Pollut. Res.* **2014**, *22*, 2219–2230. [CrossRef]
24. Zhao, L.; Yang, S.-T.; Feng, S.; Ma, Q.; Peng, X.; Wu, D. Preparation and Application of Carboxylated Graphene Oxide Sponge in Dye Removal. *Int. J. Environ. Res. Public Health* **2017**, *14*, 1301. [CrossRef]
25. Cunha, D.L.; Kuznetsov, A.; Achete, C.A.; Machado, A.; Marques, M. Immobilized TiO2 on glass spheres applied to heterogeneous photocatalysis: Photoactivity, leaching and regeneration process. *PeerJ* **2018**, *6*, e4464. [CrossRef]
26. Feng, Z.; Odelius, K.; Rajarao, G.K.; Hakkarainen, M. Microwave carbonized cellulose for trace pharmaceutical adsorption. *Chem. Eng. J.* **2018**, *346*, 557–566. [CrossRef]
27. Lin, L.; Jiang, W.; Xu, P. Comparative study on pharmaceuticals adsorption in reclaimed water desalination concentrate using biochar: Impact of salts and organic matter. *Sci. Total Environ.* **2017**, *601–602*, 857–864. [CrossRef]
28. Paunovic, O.; Pap, S.; Maletic, S.; Taggart, M.; Boskovic, N.; Sekulic, M.T. Ionisable emerging pharmaceutical adsorption onto microwave functionalised biochar derived from novel lignocellulosic waste biomass. *J. Colloid Interface Sci.* **2019**, *547*, 350–360. [CrossRef]

29. Shan, D.; Deng, S.; Zhao, T.; Wang, B.; Wang, Y.; Huang, J.; Yu, G.; Winglee, J.; Wiesner, M.R. Preparation of ultrafine magnetic biochar and activated carbon for pharmaceutical adsorption and subsequent degradation by ball milling. *J. Hazard. Mater.* **2016**, *305*, 156–163. [CrossRef]
30. Yan, Y.; Li, J.; Kong, F.; Jia, K.; He, S.; Wang, B. L-Lysine-grafted graphene oxide as an effective adsorbent for the removal of methylene blue and metal ions. *Beilstein J. Nanotechnol.* **2017**, *8*, 2680–2688. [CrossRef]
31. Khan, M.A.; Hameed, B.H.; Lawler, J.; Kumar, M.; Jeon, B.-H. Developments in activated functionalized carbons and their applications in water decontamination: A review. *DESALINATION Water Treat.* **2014**, *54*, 422–449. [CrossRef]
32. Wu, X.-L.; Shi, Y.; Zhong, S.; Lin, H.; Chen, J.-R. Facile synthesis of Fe3O4-graphene@mesoporous SiO_2 nanocomposites for efficient removal of Methylene Blue. *Appl. Surf. Sci.* **2016**, *378*, 80–86. [CrossRef]
33. Wu, S.; Zhao, X.; Li, Y.; Du, Q.; Sun, J.; Wang, Y.; Wang, X.; Xia, Y.; Wang, Z.; Xia, L. Adsorption Properties of Doxorubicin Hydrochloride onto Graphene Oxide: Equilibrium, Kinetic and Thermodynamic Studies. *Materials* **2013**, *6*, 2026–2042. [CrossRef] [PubMed]
34. Deng, X.; Lü, L.; Li, H.; Luo, F. The adsorption properties of Pb(II) and Cd(II) on functionalized graphene prepared by electrolysis method. *J. Hazard. Mater.* **2010**, *183*, 923–930. [CrossRef]
35. Repo, E.; Warchoł, J.K.; Bhatnagar, A.; Sillanpää, M. Heavy metals adsorption by novel EDTA-modified chitosan–silica hybrid materials. *J. Colloid Interface Sci.* **2011**, *358*, 261–267. [CrossRef]
36. Kadirvelu, K. Removal of heavy metals from industrial wastewaters by adsorption onto activated carbon prepared from an agricultural solid waste. *Bioresour. Technol.* **2001**, *76*, 63–65. [CrossRef]
37. Algothmi, W.M.; Bandaru, N.M.; Yu, Y.; Shapter, J.; Ellis, A. Alginate–graphene oxide hybrid gel beads: An efficient copper adsorbent material. *J. Colloid Interface Sci.* **2013**, *397*, 32–38. [CrossRef] [PubMed]
38. Li, K.; Yu, J.; Wang, Q.; Li, L.; Zhang, W.; Ma, J.; Zhang, J.; Liu, P.; Li, D. Improved sodium storage properties of nickel sulfide nanoparticles decorated on reduced graphene oxide nanosheets as an advanced anode material. *Nanotechnology* **2021**, *32*, 195406. [CrossRef]
39. Chandy, T.; Mooradian, D.L.; Rao, G.H. Evaluation of modified alginate-chitosan-polyethylene glycol microcapsules for cell encapsulation. *Artif. Organs* **1999**, *23*, 894–903. [CrossRef]
40. Dutta, D.; Borah, B.J.; Saikia, L.; Pathak, M.G.; Sengupta, P.; Dutta, D.K. Synthesis and catalytic activity of Ni°-acid activated montmorillonite nanoparticles. *Appl. Clay Sci.* **2011**, *53*, 650–656. [CrossRef]
41. Oladipo, A.A.; Gazi, M. Enhanced removal of crystal violet by low cost alginate/acid activated bentonite composite beads: Optimization and modelling using non-linear regression technique. *J. Water Process. Eng.* **2014**, *2*, 43–52. [CrossRef]
42. Tao, E.; Ma, D.; Yang, S.; Hao, X. Graphene oxide-montmorillonite/sodium alginate aerogel beads for selective adsorption of methylene blue in wastewater. *J. Alloys Compd.* **2020**, *832*, 154833. [CrossRef]
43. Kumar, M.; McGlade, D.; Ulbricht, M.; Lawler, J. Quaternized polysulfone and graphene oxide nanosheet derived low fouling novel positively charged hybrid ultrafiltration membranes for protein separation. *RSC Adv.* **2015**, *5*, 51208–51219. [CrossRef]
44. Zhou, M.; Li, Q.; Zhong, S.; Chen, J.; Lin, H.; Wu, X.-L. Facile large scale fabrication of magnetic carbon nano-onions for efficient removal of bisphenol A. *Mater. Chem. Phys.* **2017**, *198*, 186–192. [CrossRef]
45. Allen, S.; Mckay, G.; Porter, J. Adsorption isotherm models for basic dye adsorption by peat in single and binary component systems. *J. Colloid Interface Sci.* **2004**, *280*, 322–333. [CrossRef] [PubMed]
46. Liu, T.; Li, Y.; Du, Q.; Sun, J.; Jiao, Y.; Yang, G.; Wang, Z.; Xia, Y.; Zhang, W.; Wang, K.; et al. Adsorption of methylene blue from aqueous solution by graphene. *Colloids Surf. B Biointerfaces* **2012**, *90*, 197–203. [CrossRef]
47. Li, Y.; Liu, F.; Xia, B.; Du, Q.; Zhang, P.; Wang, D.; Wang, Z.; Xia, Y. Removal of copper from aqueous solution by carbon nanotube/calcium alginate composites. *J. Hazard. Mater.* **2010**, *177*, 876–880. [CrossRef] [PubMed]
48. Popuri, S.R.; Vijaya, Y.; Boddu, V.M.; Abburi, K. Adsorptive removal of copper and nickel ions from water using chitosan coated PVC beads. *Bioresour. Technol.* **2009**, *100*, 194–199. [CrossRef]
49. Hameed, B. Evaluation of papaya seeds as a novel non-conventional low-cost adsorbent for removal of methylene blue. *J. Hazard. Mater.* **2009**, *162*, 939–944. [CrossRef] [PubMed]
50. Islam, N.M.; Narurkar, M.M. Solubility, Stability and Ionization Behaviour of Famotidine. *J. Pharm. Sci.* **1993**, *45*, 682–686. [CrossRef]
51. Konkena, B.; Vasudevan, S. Understanding Aqueous Dispersibility of Graphene Oxide and Reduced Graphene Oxide through pKa Measurements. *J. Phys. Chem. Lett.* **2012**, *3*, 867–872. [CrossRef]
52. Francis, N.L.; Hunger, P.M.; Donius, A.E.; Riblett, B.W.; Zavaliangos, A.; Wegst, U.G.K.; Wheatley, M.A. An ice-templated, linearly aligned chitosan-alginate scaffold for neural tissue engineering. *J. Biomed. Mater. Res. Part A* **2013**, *101*, 3493–3503. [CrossRef]
53. Rocher, V.; Siaugue, J.-M.; Cabuil, V.; Bee, A. Removal of organic dyes by magnetic alginate beads. *Water Res.* **2008**, *42*, 1290–1298. [CrossRef]
54. Beltrán, F.J.; Pocostales, P.; Alvarez, P.; Oropesa, A.L. Diclofenac removal from water with ozone and activated carbon. *J. Hazard. Mater.* **2009**, *163*, 768–776. [CrossRef]
55. Al-Degs, Y.; El-Barghouthi, M.; El-Sheikh, A.; Walker, G. Effect of solution pH, ionic strength, and temperature on adsorption behavior of reactive dyes on activated carbon. *Dye. Pigment.* **2008**, *77*, 16–23. [CrossRef]
56. Kumar, M.; Tripathi, B.P.; Shahi, V.K. Crosslinked chitosan/polyvinyl alcohol blend beads for removal and recovery of Cd(II) from wastewater. *J. Hazard. Mater.* **2009**, *172*, 1041–1048. [CrossRef]
57. Weng, C.-H.; Lin, Y.-T.; Tzeng, T.-W. Removal of methylene blue from aqueous solution by adsorption onto pineapple leaf powder. *J. Hazard. Mater.* **2009**, *170*, 417–424. [CrossRef]

58. Han, S.S.; Lee, H.M. Adsorption properties of hydrogen on (10,0) single-walled carbon nanotube through density functional theory. *Carbon* **2004**, *42*, 2169–2177. [CrossRef]
59. Wang, P.; Cao, M.; Wang, C.; Ao, Y.; Hou, J.; Qian, J. Kinetics and thermodynamics of adsorption of methylene blue by a magnetic graphene-carbon nanotube composite. *Appl. Surf. Sci.* **2014**, *290*, 116–124. [CrossRef]
60. Doğan, M.; Özdemir, Y.; Alkan, M. Adsorption kinetics and mechanism of cationic methyl violet and methylene blue dyes onto sepiolite. *Dye. Pigment.* **2007**, *75*, 701–713. [CrossRef]
61. Cheung, A.; Szeto, Y.S.; McKay, G. Intraparticle diffusion processes during acid dye adsorption onto chitosan. *Bioresour. Technol.* **2007**, *98*, 2897–2904. [CrossRef]
62. Rafatullah, M.; Sulaiman, O.; Hashim, R.; Ahmad, A. Adsorption of methylene blue on low-cost adsorbents: A review. *J. Hazard. Mater.* **2010**, *177*, 70–80. [CrossRef] [PubMed]
63. Ali, S.A.; Yaagoob, I.Y.; Mazumder, M.A.; Al-Muallem, H.A. Fast removal of methylene blue and Hg(II) from aqueous solution using a novel super-adsorbent containing residues of glycine and maleic acid. *J. Hazard. Mater.* **2019**, *369*, 642–654. [CrossRef]

materials MDPI

Article

An Acylhydrazone-Based Fluorescent Sensor for Sequential Recognition of Al^{3+} and $H_2PO_4^-$

Donghwan Choe and Cheal Kim *

Department of Fine Chemistry, Seoul National University of Science and Technology (SNUT), Seoul 136-742, Korea; ehdghksdl_@naver.com
* Correspondence: chealkim@snut.ac.kr; Tel.: +82-2-972-6673; Fax: +82-2-981-9147

Abstract: A novel acylhydrazone-based fluorescent sensor **NATB** was designed and synthesized for consecutive sensing of Al^{3+} and $H_2PO_4^-$. **NATB** displayed fluorometric sensing to Al^{3+} and could sequentially detect $H_2PO_4^-$ by fluorescence quenching. The limits of detection for Al^{3+} and $H_2PO_4^-$ were determined to be 0.83 and 1.7 μM, respectively. The binding ratios of **NATB** to Al^{3+} and **NATB**-Al^{3+} to $H_2PO_4^-$ were found to be 1:1. The sequential recognition of Al^{3+} and $H_2PO_4^-$ by **NATB** could be repeated consecutively. In addition, the practicality of **NATB** was confirmed with the application of test strips. The sensing mechanisms of Al^{3+} and $H_2PO_4^-$ by **NATB** were investigated through fluorescence and UV–Visible spectroscopy, Job plot, ESI-MS, ^1H NMR titration, and DFT calculations.

Keywords: aluminum ion; dihydrogen phosphate; acylhydrazone; fluorescent chemosensor; sequential detection; calculations

Citation: Choe, D.; Kim, C. An Acylhydrazone-Based Fluorescent Sensor for Sequential Recognition of Al^{3+} and $H_2PO_4^-$. *Materials* **2021**, *14*, 6392. https://doi.org/10.3390/ma14216392

Academic Editors: Marco Race, Elza Bontempi, Avelino Núñez-Delgado, Zhien Zhang, Mario Coccia and Yaoyu Zhou

Received: 23 September 2021
Accepted: 20 October 2021
Published: 25 October 2021

1. Introduction

Al^{3+}, the third most abundant metallic element in nature [1,2], is broadly employed in daily life in packaging materials, pharmaceuticals, food additives, machinery, clinical medicines, and water purification [3,4]. Owing to its widespread usage, Al^{3+} could be readily accumulated in the body, which leads to the development of diverse diseases such as Parkinson's and Alzheimer's disease [5,6]. Dihydrogen phosphate ($H_2PO_4^-$) is an important component related to many intercellular activities, such as signaling mediation, protein phosphorylation, enzymatic reactions, ion-channel regulation, and so on [7–9]. However, excessive agricultural use of phosphate causes eutrophication or massive algal growth, leading to a deficiency in oxygen levels [10–12]. For these reasons, there has been a strong demand for the development of sensing and detection methods for Al^{3+} and $H_2PO_4^-$.

The traditional analytical methods reported for the analysis of cations and anions, such as ICP-AES, AAS, and electrochemical methods, have been largely restricted due to their expensive instruments, complicated procedures, and the need for highly trained operators [13–15]. In contrast, fluorescence methods have shown the advantages of cost-effectiveness, simplicity, easy operation, and high sensitivity [16–18]. While numerous fluorescent chemosensors for a single analyte have been reported, fluorescent chemosensors that allow the sequential sensing of multiple analytes with great selectivity and sensitivity are still needed [19–21] because they are more cost-effective, recyclable and practical [22–24]. Several fluorescent sensors have been addressed for consecutive sensing of Al^{3+} and several anions [25–28] or several cations and $H_2PO_4^-$ [29–31]. In addition, Kumar et al. reported a fluorescent sensor for sequential sensing of Al^{3+} and $H_2PO_4^-/HSO_4^-$ [32]. The practical importance of sequential sensing may have potential applications such as logic gates and molecular switches. Nevertheless, a sequential fluorescent sensor that can exclusively detect Al^{3+} and $H_2PO_4^-$ has not been reported to date.

As Al^{3+} is a hard cation, chemosensors containing hard base units, such as nitrogen or oxygen atoms, prefer to coordinate with Al^{3+} [33–35]. In this regard, acylhydrazone derivatives, having oxygen and nitrogen atoms, are expected to be a suitable functional group to design an Al^{3+} chemosensor [36–38]. Naphthalene moieties have been widely applied for the design of fluorescent sensors because of their excellent photophysical properties as a fluorophore [39–41]. Hence, we expected that a compound including both acylhydrazone and naphthalene may operate as a fluorescence chemosensor for Al^{3+}.

In the current study, we designed an acylhydrazone-based fluorescent sensor, **NATB**, which showed green fluorescence emissions with Al^{3+} and could sequentially detect $H_2PO_4^-$ through fluorescence quenching with high sensitivity and selectivity. A sensing mechanism of **NATB** to Al^{3+} and $H_2PO_4^-$ was illustrated by fluorescence and UV–Vis spectroscopy, Job plot, ESI-MS, ^{1}H NMR titration, and calculations.

2. Experimental Section

2.1. Materials and Equipment

All solvents and reagents were commercially obtained from TCI (TCI, Nihonbashi-Honcho, Tokyo, Japan) and Sigma-Aldrich (MilliporeSigma, Burlington, MA, USA). NMR experiments were conducted using DMSO-d_6 as the solvent, and the data were recorded on a Varian spectrometer (Varian, Palo Alto, CA, USA). Fluorescence and UV–Visible spectra were measured with Perkin Elmer machines (Perkin Elmer, Waltham, MA, USA). The quantum yields of **NATB** and **NATB**-Al^{3+} were relatively determined with quinine ($\Phi = 0.54$ in 1×10^{-1} M H_2SO_4) as a reference. ESI-MS data were recorded on a Thermo Finnigan machine (Thermo Finnigan LLC, San Jose, CA, USA).

2.2. Synthesis of N′-[(E)-(3-tert-butyl-2-hydroxyphenyl)methylidene]-3-hydroxynaphthalene-2-carbohydrazide (NATB)

The intermediate compound, 3-hydroxy-2-naphthohydrazide (**2**), was synthesized following a previously reported method [42]. The excess amounts of 3-(*tert*-butyl)-2-hydroxybenzaldehyde (**1**, 1.8 mmol) and 3-hydroxy-2-naphthohydrazide (**2**, 0.3 mmol) were mixed in absolute EtOH (10 mL) with a catalytic amount of HCl and stirred at room temperature for 1 day. A yellow precipitate was filtered, rinsed with cold absolute EtOH, and dried (77.2 mg, 70.1%); ^{1}H NMR in DMSO-d_6: δ 12.42 (s, 1H), 12.24 (s, 1H), 11.19 (s, 1H), 8.63 (s, 1H), 8.45 (s, 1H), 7.93 (d, 1H), 7.77 (d, 1H), 7.53 (t, 1H), 7.38 (t, 1H), 7.35 (s, 1H), 7.32 (d, 1H), 7.30 (d, 1H), 6.91 (t, 1H), 1.43 (s, 9H). ^{13}C NMR in DMSO-d6: δ 163.23 (1C), 156.90 (1C), 153.75 (1C), 151.45 (1C), 136.36 (1C), 135.84 (1C), 130.31 (1C), 129.53 (1C), 128.59 (1C), 128.54 (1C), 128.20 (1C), 126.69 (1C), 125.75 (1C), 123.75 (1C), 119.86 (1C), 118.70 (1C), 117.54 (1C), 110.54 (1C), 34.39 (1C), 29.16 (3C). ESI-MS (*m/z*): [**NATB** + H^+]$^+$ calcd 363.17, found 363.04.

2.3. Preparation of Spectroscopic Experiments

An **NATB** stock (10 mM) was prepared in DMSO. The stock solutions (20 mM) of varied cations were prepared using their nitrate salts (Al^{3+}, Na^+, Cr^{3+}, Fe^{2+}, Ca^{2+}, Cd^{2+}, Zn^{2+}, Pb^{2+}, Co^{2+}, Cu^{2+}, In^{3+}, Mn^{2+}, Ga^{3+}, Ni^{2+}, Mg^{2+}, Ag^+, Hg^{2+} and K^+) or perchlorate salt (Fe^{3+}). The concentrated solutions (20 mM) of varied anions were prepared using their tetrabutylammonium salts ($H_2PO_4^-$, SCN^-, BzO^-, N_3^-, OAc^- and NO_2^-), tetraethylammonium salts (F^-, Cl^-, Br^-, I^- and CN^-), sodium salts (S^{2-} and ClO^-), or potassium salts (HPO_4^{2-}, PO_4^{3-}, HSO_4^- and $P_2O_7^{4-}$(PPi)). All spectroscopic experiments were conducted in MeOH.

2.4. Competitive Experiments

For Al^{3+}, 6 μL (10 mM) of an **NATB** stock in DMSO was mixed into MeOH (2 mL) to make 30 μM. A total of 4.5 μL of various cations (20 mM) in DMF was diluted in **NATB** to make 45 μM. Finally, 4.5 μL (20 mM) of an Al^{3+} stock in DMF was mixed into each solution to produce 45 μM, and their fluorescent spectra were measured.

For $H_2PO_4^-$, 6 μL (10 mM) of an **NATB** stock in DMSO and 4.5 μL (20 mM) of an Al^{3+} stock in DMF were diluted into MeOH (2 mL) to produce 30 μM of **NATB**-Al^{3+}. We added 4.5 μL of various anions (20 mM) in H_2O to **NATB**-Al^{3+} to produce 45 μM. A total of 4.5 μL (20 mM) of an $H_2PO_4^-$ stock was diluted into each solution to produce 45 μM. Their fluorescent spectra were measured.

2.5. Determination of Association Constant (K)

The association constant (*K*) was calculated using Li's method [43]. If the ligand (*L*) and the analyte (*M*) form an m-n complex, M_mL_n, the equilibrium constant of the corresponding complex, *K*, can be expressed by the following equation:

$$[M]^m = \frac{1}{nK}\frac{1}{[L]_T^{n-1}}\frac{1-\alpha}{\alpha^n}$$

where,

$$[M] = the\ concentration\ of\ analyte$$

$$[L\}_T = the\ total\ concentration\ of\ ligand$$

and α could be described as:

$$\alpha = \frac{I - I_{max}}{I_{min} - I_{max}}$$

where,

$$I = the\ fluorescence\ intensity\ of\ complex$$

2.6. Calculations

Calculations were achieved with the Gaussian 16 program [44]. Optimal geometries of **NATB** and **NATB**-Al^{3+} were provided with the DFT method [45,46]. B3LYP was selected as the hybrid functional basis set. The 6–31G(d,p) basis set was implemented to all atoms except Al^{3+} [47,48], and the LANL2DZ basis set was employed for applying ECP to Al^{3+} [49–51]. No imaginary frequency was found in the optimized states of **NATB** or **NATB**-Al^{3+}, indicating their local minima. The solvent effect of MeOH was considered with IEFPCM [52]. Based on the energy-optimized structures of **NATB** and **NATB**-Al^{3+}, the plausible UV–Vis transition states were calculated by the TD-DFT method with 20 lowest singlet states.

3. Results and Discussion

The synthesis of **NATB** was conducted as depicted in Scheme 1. The condensation reaction of 3-(*tert*-butyl)-2-hydroxybenzaldehyde (**1**) and 3-hydroxy-2-naphthohydrazide (**2**) produced the desired product, *N*′-[(E)-(3-tert-butyl-2-hydroxyphenyl)methylidene]-3-hydroxynaphthalene-2-carbohydrazide (**NATB**), which was verified with ^1H NMR, ^{13}C NMR (Figures S1 and S2), and ESI-MS.

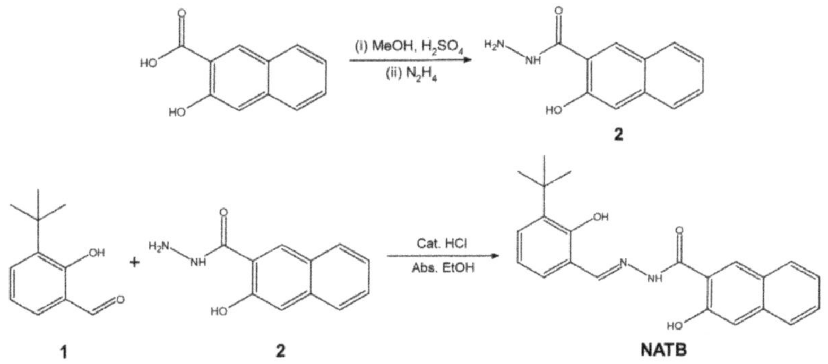

Scheme 1. Synthesis of NATB.

3.1. Spectroscopic Examination of NATB to Al^{3+}

To confirm the fluorescence selectivity of **NATB**, the fluorescence emission was studied with a variety of cations in MeOH (Figure 1). As a result, **NATB** exhibited notable fluorescence emission at 526 nm with Al^{3+}, while **NATB** and **NATB** with other cations showed negligible or no fluorescence emission (λ_{ex} = 358 nm). These outcomes demonstrated that **NATB** could be utilized as a fluorescent probe for the selective sensing of Al^{3+}. On the other hand, **NATB** was soluble in aqueous media, but it did not show any selectivity to Al^{3+}. In addition, the fluorescence emission of **NATB** was examined with various anions including dihydrogen phosphate. **NATB** had no selectivity for the anions.

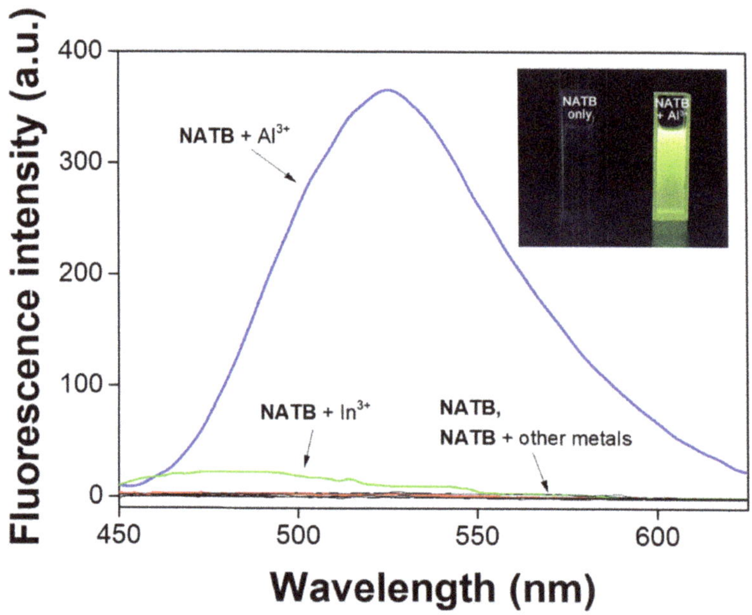

Figure 1. Fluorescence changes of **NATB** (30 μM) with a variety of cations (1.5 equiv) in MeOH. Photograph: the fluorescent images of **NATB** and **NATB**-Al^{3+} (λ_{ex}: 358 nm).

To check the concentration-dependent properties of **NATB** to Al^{3+}, fluorescence titration was carried out (Figure 2). **NATB** exhibited little fluorescence with a tiny quantum yield (Φ = 0.008). However, the continuous increase in Al^{3+} up to 1.5 equiv significantly enhanced the green fluorescence emission at 526 nm (Φ = 0.162). UV–Vis spectrometry was also conducted with Al^{3+} to examine its photophysical characteristics (Figure 3). Upon the addition of Al^{3+}, the absorption of 310 nm clearly decreased, while a new absorption of 325 nm constantly increased up to 1.5 equiv. An explicit isosbestic point was observed at 315 nm, verifying that the coordination of **NATB** with Al^{3+} produced a stable complex.

A 1:1 stoichiometric coordination between **NATB** and Al^{3+} was suggested by the Job plot experiment (Figure S3), which was explicitly supported by ESI-MS analysis (Figure S4). The positive ion mass displayed a large peak of 596.16 (*m/z*), which was correlated to [**NATB**($-H^+$) + Al^{3+} + 2 DMF + NO_3^-]$^+$ (calcd. 596.23). The association constant (K) of **NATB**-Al^{3+} was confirmed to be 3.6×10^4 M^{-1} (Figure S5) based on Li's method [43]. The detection limit of **NATB** toward Al^{3+} was 0.83 μM, based on 3σ/slope (Figure S6).

The ^1H NMR titrations were achieved to investigate the binding mechanism of **NATB** toward Al^{3+} (Figure 4). Upon the addition of Al^{3+} to **NATB**, the proton H_{14} continually disappeared and the protons H_5 and H_6 were deshielded. These results indicate that the deprotonated oxygen on the *tert*-butylphenol group and the oxygen and nitrogen on the acylhydrazone group may be coordinated to Al^{3+} (Scheme 2).

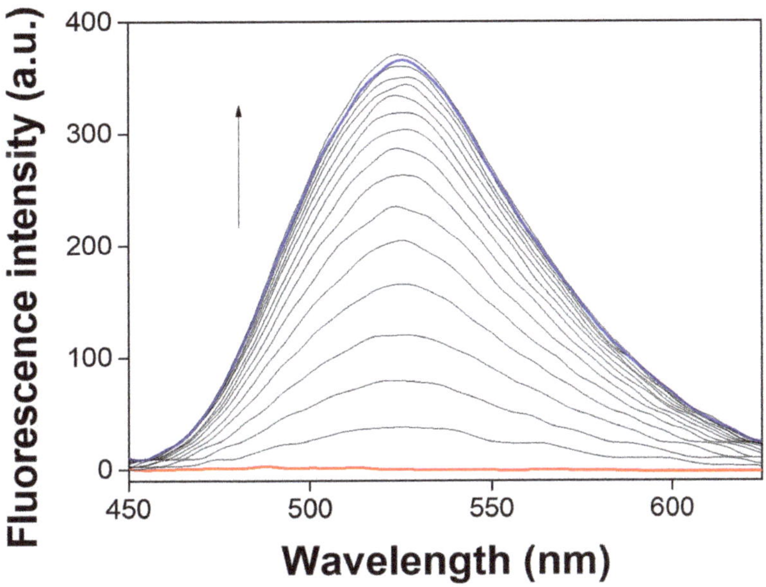

Figure 2. Fluorescence titration of **NATB** (30 μM) with varied amounts of Al^{3+} (0–1.5 equiv) in MeOH.

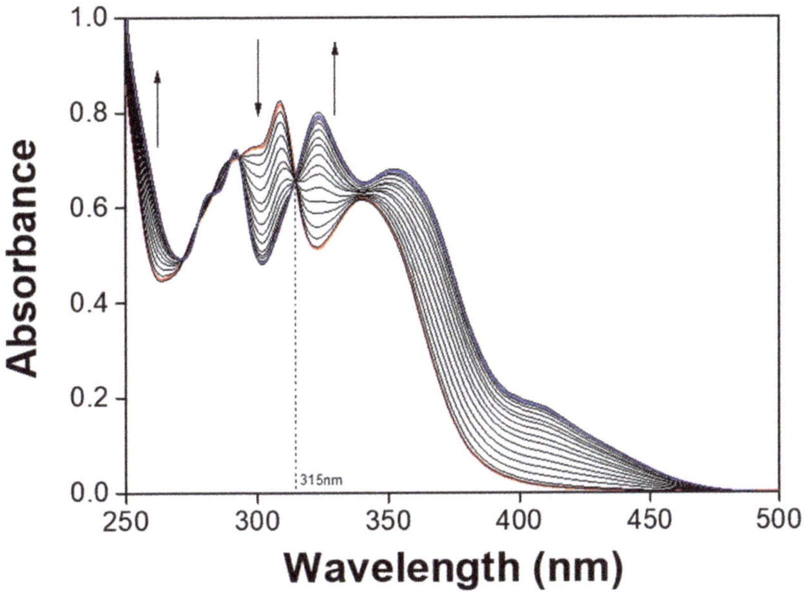

Figure 3. UV–Vis changes of **NATB** (30 μM) with varied amounts of Al^{3+} (0–1.5 equiv) in MeOH.

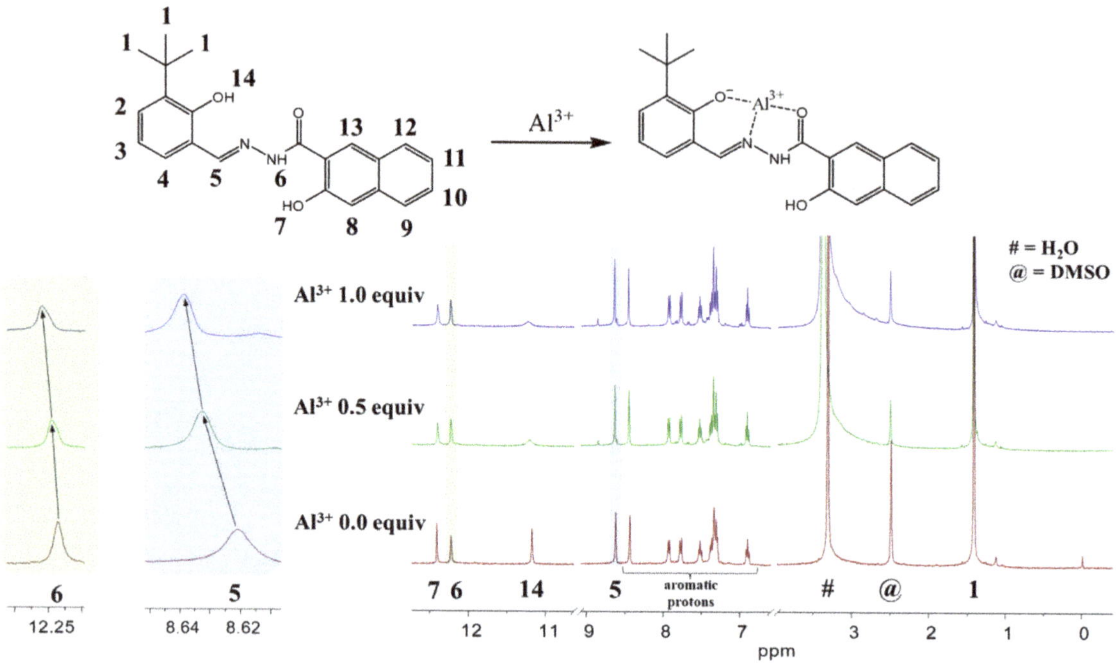

Figure 4. ^{1}H NMR titration of **NATB** with Al^{3+} in DMSO-d_6.

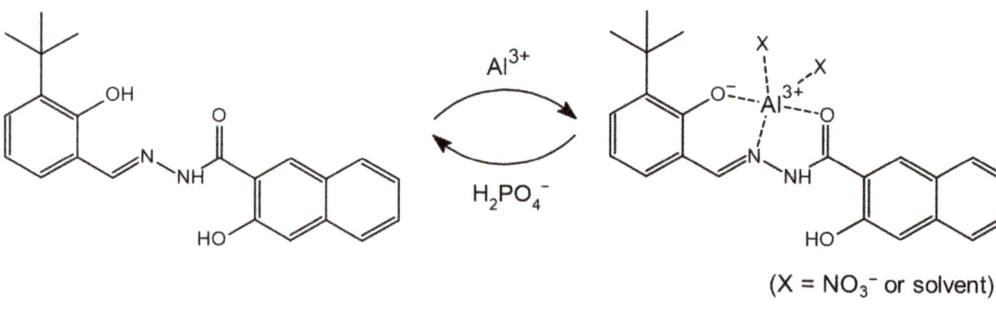

Scheme 2. Sequential recognition mechanism of Al^{3+} and $H_2PO_4^-$ by **NATB**.

To verify the practicability of **NATB** as a probe for Al^{3+}, an interference experiment was conducted (Figure S7). **NATB** could detect Al^{3+} with other cations without significant interferences, except for In^{3+}, Fe^{3+} and Cu^{2+}. These three cations bound more tightly to **NATB** than Al^{3+}. For the practical application of **NATB**, test kits were prepared by immersing filter paper strips in the **NATB** solution. When **NATB**-coated test kits were immersed in a range of concentrations of Al^{3+} solutions, the obvious green fluorescence emission showed up above 2 mM of Al^{3+} under UV light (Figure 5a). However, the fluorescence was not displayed when those strips were applied to the same concentration of other cations (Figure 5b). These results indicate the potential applications of **NATB** in easily recognizing Al^{3+} without any complicated tools.

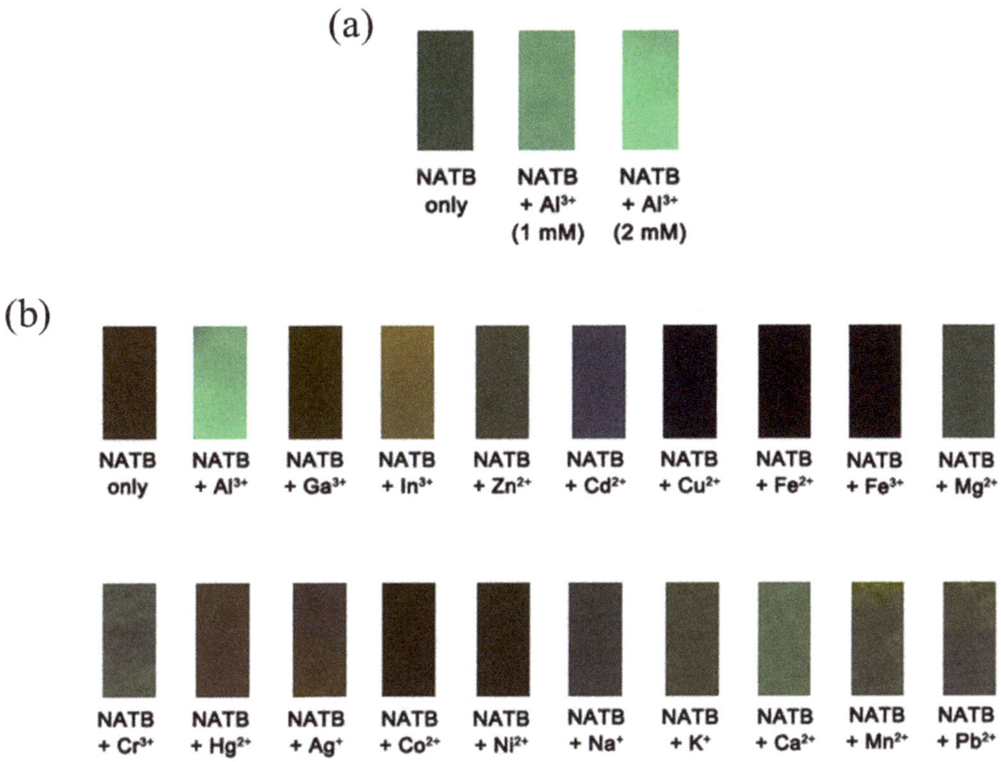

Figure 5. Detection of Al^{3+} by **NATB**-coated test kits (10 mM). (**a**) **NATB**-coated test kits immersed in the solution of different Al^{3+} concentrations; (**b**) **NATB**-coated test kits immersed in 2 mM of various cation solutions.

3.2. Calculations

To comprehend the Al^{3+}-sensing property of **NATB**, DFT calculations were performed with the Gaussian 16 program (Figure 6). As the Job plot, ESI-MS, and ^{1}H NMR titration implied a 1:1 stoichiometric coordination of **NATB** with Al^{3+}, all calculations were conducted with 1:1 stoichiometry. **NATB** showed a dihedral angle of 0.013° (1O, 2C, 3N, and 4C) with a planar structure (Figure 6a). The coordination of **NATB** with Al^{3+} distorted its structure, showing a dihedral angle of 98.875° (Figure 6b).

Based on the energy-minimized structures of **NATB** and **NATB**-Al^{3+}, TD-DFT calculations were conducted to inspect the transition energies and molecular orbitals. **NATB** featured the main absorption induced from the HOMO → LUMO (347.28 nm), showing intra-charge transfer (ICT) transition from the *tert*-butylphenol to the naphthol (Figure S8). The major absorption of **NATB**-Al^{3+} derived from the HOMO-1 → LUMO transition (412.27 nm) also showed a similar ICT transition (Figures S9 and S10). The reduction in the energy gap was consistent with the red-shift of the experimental absorption. These outcomes led us to conclude that the fluorescence turn-on mechanism of **NATB** to Al^{3+} may be a chelation-enhanced fluorescence (CHEF) effect [53]. Based on experimental data and theoretical calculations, an appropriate binding structure of **NATB**-Al^{3+} is proposed in Scheme 2.

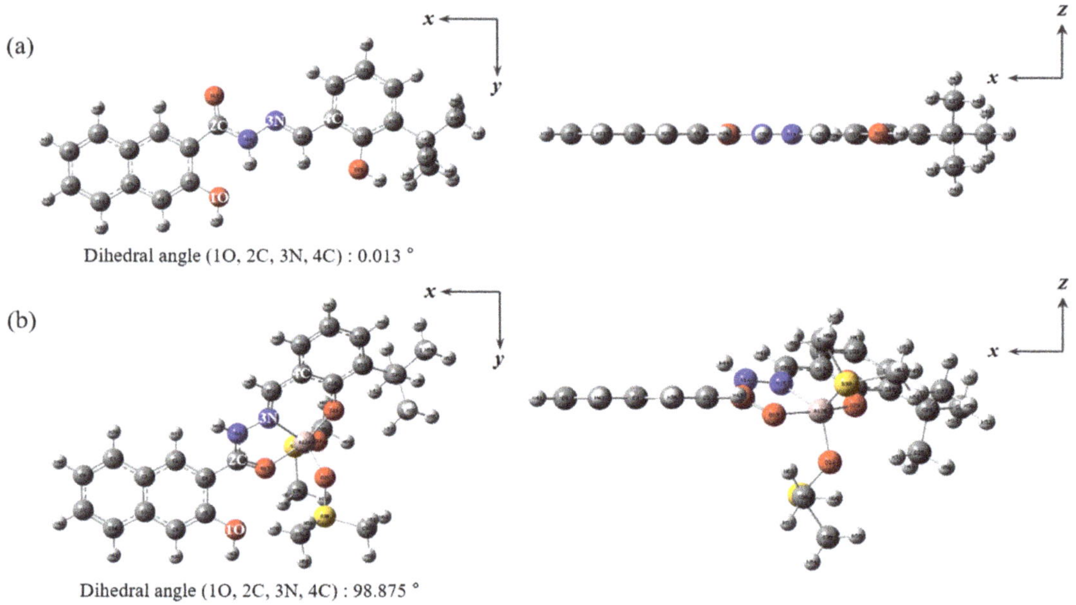

Figure 6. Energy-optimized forms of (**a**) **NATB** and (**b**) **NATB**-Al^{3+}.

3.3. Spectroscopic Examination of NATB-Al^{3+} to $H_2PO_4^-$

We studied the fluorescence selectivity of **NATB**-Al^{3+} to a range of anions such as $H_2PO_4^-$, Cl^-, CN^-, OAc^-, F^-, ClO^-, I^-, N_3^-, BzO^-, SCN^-, Br^-, NO_2^-, S^{2-}, HPO_4^{2-}, PO_4^{3-}, HSO_4^-, and PPi in MeOH (Figure 7). Most of the anions did not affect the fluorescence emission of **NATB**-Al^{3+}, while the addition of $H_2PO_4^-$ toward **NATB**-Al^{3+} resulted in significant fluorescence quenching (λ_{ex} = 358 nm). The result demonstrated that **NATB**-Al^{3+} could be used as a chemosensor for $H_2PO_4^-$ with fluorescence quenching.

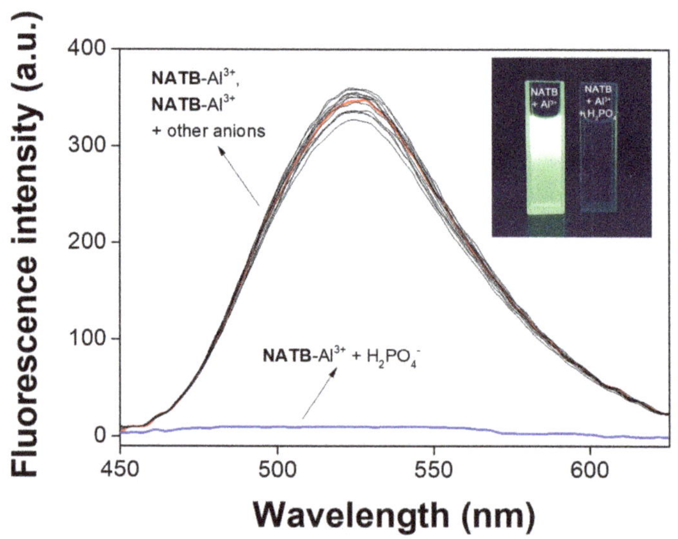

Figure 7. Fluorescence changes of **NATB**-Al^{3+} (30 μM) with various anions (45 μM) in MeOH (λ_{ex}: 358 nm). Photograph: the fluorescent images of **NATB**-Al^{3+} and **NATB**-Al^{3+}-$H_2PO_4^-$ (λ_{ex}: 358 nm).

The fluorescence titration experiments were conducted to verify the fluorescence quenching ability of $H_2PO_4^-$ toward **NATB**-Al^{3+} (Figure 8). The fluorescence of **NATB**-Al^{3+} consistently diminished with the addition of $H_2PO_4^-$ up to 1.5 equiv (Φ = 0.005). UV–Vis spectroscopy showed that the continuous addition of $H_2PO_4^-$ increased the absorbance at 310 nm, while those at 270 and 325 nm decreased with the explicit isosbestic points at 253 and 315 nm (Figure 9). The UV–Vis spectrum of $H_2PO_4^-$ with **NATB**-Al^{3+} is analogous to that of free **NATB**, implying that the addition of $H_2PO_4^-$ released Al^{3+} from the **NATB**-Al^{3+} complex (Figure S11).

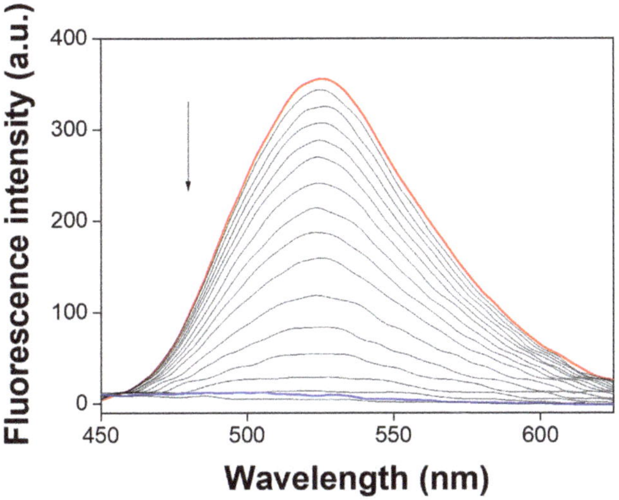

Figure 8. Fluorescence titration of **NATB**-Al^{3+} (30 μM) with various amounts of $H_2PO_4^-$ (0–1.5 equiv) in MeOH.

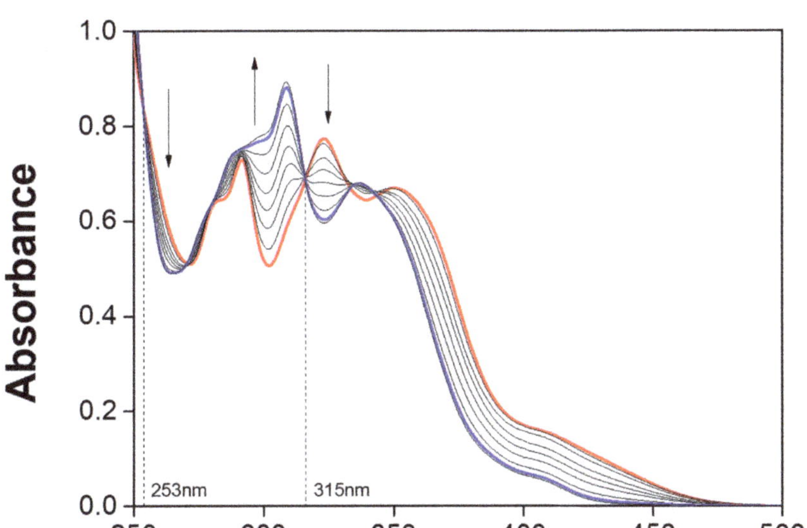

Figure 9. UV–Vis changes of **NATB**-Al^{3+} (30 μM) with various amounts of $H_2PO_4^-$ (0–1.5 equiv) in MeOH.

The stoichiometry of $H_2PO_4^-$ toward **NATB**-Al^{3+} was determined by the Job plot experiment (Figure S12), which exhibited a 1:1 stoichiometry. The mass spectral analysis displayed a peak of 395.06 (m/z), which demonstrated the regeneration of **NATB** ([**NATB** + H^+ + MeOH]$^+$; calcd. 395.20) (Figure S13). These outcomes supported the mechanism that the addition of $H_2PO_4^-$ released Al^{3+} from **NATB**-Al^{3+}, which resulted in the loss of fluorescence. Based on Li's method [43], the association constant (K) for $H_2PO_4^-$ with **NATB**-Al^{3+} was calculated as 1.2×10^4 M^{-1} (Figure S14). The detection limit of **NATB**-Al^{3+} toward $H_2PO_4^-$ was determined as 1.7 μM, based on 3σ/slope (Figure S15). Importantly, **NATB** is the first fluorescent sensor for the consecutive sensing of Al^{3+} and $H_2PO_4^-$ (Table S1). On the other hand, **NATB** showed higher detection limits for Al^{3+} and $H_2PO_4^-$ compared to Kumar's work [32], but it could solely detect $H_2PO_4^-$ without the interference of HSO_4^-.

The reversibility in the response of **NATB** was verified through the alternative additions of Al^{3+} and $H_2PO_4^-$ (Figure 10). The fluorescence emission of **NATB** repeated its enhancing and quenching processes several times without fluorescence efficiency loss. To verify that **NATB**-Al^{3+} is an effective fluorescence probe for $H_2PO_4^-$, the interference of other anions was tested (Figure S16). The results indicated that the presence of other anions (1.5 equiv) did not interfere with the fluorescence quenching of **NATB**-Al^{3+} toward $H_2PO_4^-$.

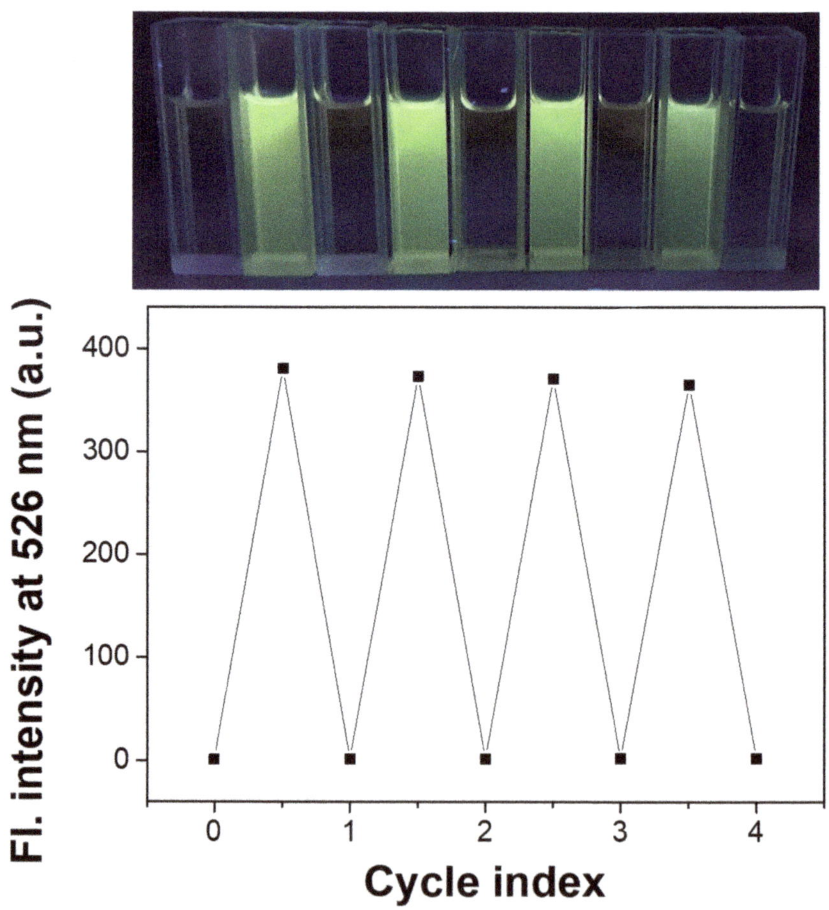

Figure 10. Change in fluorescence of **NATB** (30 μM) upon alternate addition of Al^{3+} and $H_2PO_4^-$ in MeOH (λ_{ex}: 358 nm).

4. Conclusions

An acylhydrazone-based chemosensor **NATB** was developed and its sequential recognition of Al^{3+} and $H_2PO_4^-$ was studied. **NATB** showed a strong fluorescence increase with Al^{3+}, and its complex **NATB**-Al^{3+} sequentially detected $H_2PO_4^-$ by releasing Al^{3+} with turn-off fluorescence. Importantly, **NATB** is the first sequential fluorescent probe for selective sensing of Al^{3+} and $H_2PO_4^-$. Detection limits of **NATB** for Al^{3+} and $H_2PO_4^-$ were calculated as 0.83 and 1.7 μM, respectively, based on 3σ/slope. **NATB** could repeat sequential recognition of Al^{3+} and $H_2PO_4^-$ several times and could be applied to detect Al^{3+} in test strips. The sensing mechanism of **NATB** toward Al^{3+} and $H_2PO_4^-$ was demonstrated with a Job plot, ESI-MS, ^{1}H NMR spectroscopy, and theoretical calculations. The detection mechanism of **NATB** toward Al^{3+} is suggested to be a CHEF effect through DFT calculations.

Supplementary Materials: The following are available online at https://www.mdpi.com/article/10.3390/ma14216392/s1, Table S1: Examples of chemosensors for successive detection related to Al^{3+} or $H_2PO_4^-$ or both; Figure S1: ^{1}H NMR spectrum of **NATB** in DMSO-d_6; Figure S2: ^{13}C NMR spectrum of **NATB** in DMSO-d_6; Figure S3: Job plot for the binding of **NATB** with Al^{3+} (50 μM) in MeOH. Fluorescence intensity at 526 nm is plotted as a function of the molar ratio of $[Al^{3+}]/([Al^{3+}] + [$**NATB**$])$; Figure S4: Positive-ion ESI mass spectrum of **NATB** (100 μM) in MeOH upon the addition of 1 equiv of Al^{3+} in DMF; Figure S5: Li's equation plot (at 526 nm) of **NATB** (30 μM) in MeOH, based on fluorescence titration, assuming 1:1 stoichiometry for association between **NATB** and Al^{3+}; Figure S6: Calibration curve of **NATB** as a function of Al^{3+} concentration in MeOH. [**NATB**] = 30 μM and $[Al^{3+}]$ = 0–18 μM (λ_{ex} = 358 nm); Figure S7: Competitive experiments of **NATB** (30 μM) toward Al^{3+} (45 μM) in the presence of other metal ions (45 μM, λ_{ex} = 358 nm) in MeOH; Figure S8: (a) The theoretical excitation energies and the experimental UV–Vis spectrum of **NATB.** (b) The major electronic transition energies and molecular orbital contributions of **NATB**; Figure S9: (a) The theoretical excitation energies and the experimental UV–Vis spectrum of **NATB**-Al^{3+}. (b) The major electronic transition energies and molecular orbital contributions of **NATB**-Al^{3+}; Figure S10: The major molecular orbital transitions and excitation energies of **NATB** and **NATB**-Al^{3+}; Figure S11: UV–Vis spectra of **NATB** and **NATB**-Al^{3+} with $H_2PO_4^-$ in MeOH, respectively; Figure S12: Job plot for the stoichiometry of **NATB**-Al^{3+} with $H_2PO_4^-$ (30 μM) in MeOH. Fluorescence intensity at 526 nm is plotted as a function of the molar ratio of [**NATB**-Al^{3+}]/([**NATB**-Al^{3+}] + $[H_2PO_4^-]$); Figure S13: Positive-ion ESI mass spectrum of **NATB**-Al^{3+} (100 μM) in MeOH upon the addition of 1 equiv of $H_2PO_4^-$ in H_2O; Figure S14: Li's equation plot (at 526 nm) of **NATB**-Al^{3+} (30 μM) based on fluorescence titration in MeOH, assuming 1:1 stoichiometry for association between **NATB**-Al^{3+} and $H_2PO_4^-$; Figure S15: Calibration curve of **NATB**-Al^{3+} in MeOH as a function of $H_2PO_4^-$ concentration. [**NATB**-Al^{3+}] = 30 μM and $[H_2PO_4^-]$ = 0.0–18.0 μM (λ_{ex} = 358 nm); Figure S16: Interference studies of **NATB**-Al^{3+} (30 μM) toward $H_2PO_4^-$ (45 μM) in the presence of other anions (45 μM, λ_{ex} = 358 nm) in MeOH.

Author Contributions: D.C. and C.K. provided the initial idea for this work; D.C. contributed to the collection and analysis of field test data; D.C. and C.K. wrote the paper. All authors have read and agreed to the published version of the manuscript.

Funding: National Research Foundation of Korea (2018R1A2B6001686) is acknowledged.

Conflicts of Interest: The authors declare no conflict of interest.

References

1. Pungut, N.A.S.; Saad, H.M.; Sim, K.S.; Tan, K.W. A turn on fluorescent sensor for detecting Al^{3+} and colorimetric detection for Cu^{2+}: Synthesis, cytotoxicity and on-site assay kit. *J. Photochem. Photobiol. A Chem.* **2021**, *414*, 113290. [CrossRef]
2. Maity, D.; Govindaraju, T. Pyrrolidine constrained bipyridyl-dansyl click fluoroionophore as selective Al^{3+}sensor. *Chem. Commun.* **2010**, *46*, 4499–4501. [CrossRef]
3. Kim, S.Y.; Lee, S.Y.; Kang, J.H.; Kim, M.S.; Kim, A.; Kim, C. Colorimetric detection of $Fe^{3+/2+}$ and fluorescent detection of Al^{3+} in aqueous media: Applications and DFT calculations. *J. Coord. Chem.* **2018**, *71*, 2401–2414. [CrossRef]
4. Kaur, R.; Saini, S.; Kaur, N.; Singh, N.; Jang, D.O. Rhodamine-based fluorescent probe for sequential detection of Al^{3+} ions and adenosine monophosphate in water. *Spectrochim. Acta Part A Mol. Biomol. Spectrosc.* **2020**, *225*, 117523. [CrossRef]

5. Cheah, P.W.; Heng, M.P.; Izati, A.; Ng, C.H.; Tan, K.W. Rhodamine B conjugate for rapid colorimetric and fluorimetric detection of aluminium and tin ions and its application in aqueous media. *Inorg. Chim. Acta* **2020**, *512*, 119901. [CrossRef]
6. Maity, D.; Govindaraju, T. A differentially selective sensor with fluorescence turn-on response to Zn^{2+} and dual-mode ratiometric response to Al^{3+} in aqueous media. *Chem. Commun.* **2012**, *48*, 1039–1041. [CrossRef]
7. Zhou, J.; Yuan, Y.-F.; Zhuo, J.-B.; Lin, C.-X. Synthesis and characterization of cyclophane: The highly selective recognition of Fe^{3+} in aqueous solution and $H_2PO_4^-$ in acetonitrile solution. *Tetrahedron Lett.* **2018**, *59*, 1059–1064. [CrossRef]
8. Singh, A.; Nishith, U.; Trivedi, D.R. Spectroscopic studies of colorimetric receptors for detection of biologically important inorganic F^-, AcO^- and $H_2PO_4^-$ anions in organo-aqueous medium: Real-life application. *Inorg. Chem. Commun.* **2020**, *115*, 107874. [CrossRef]
9. Goswami, S.; Maity, S.; Maity, A.C.; Das, A.K.; Khanra, K.; Mandal, T.K.; Bhattacharyya, N. A macrocyclic piperazine linked extremely Zn^{2+} selective fluorescent chemosensor with bio-imaging and for $H_2PO_4^-$ sensing. *Tetrahedron Lett.* **2014**, *55*, 5993–5997. [CrossRef]
10. Cao, C.; You, X.; Feng, L.; Luo, G.; Yue, G.; Ji, X. Synthesis of new chromogenic sensors containing thiourea and selective detection for F^-, $H_2PO_4^-$, and Ac^- anions. *Can. J. Chem.* **2020**, *98*, 659–666. [CrossRef]
11. Choi, J.H.; Pandith, A.; Chakradhar, D.; Kim, H.R.; Kim, H.S. Al^{3+}-Morpholine-appended Anthracene Ensemble as a Dual Photonic Switch for $H_2PO_4^-$ and CN^- Ions and its Biological Applications. *Bull. Korean Chem. Soc.* **2019**, *40*, 138–145. [CrossRef]
12. Pandith, A.; Uddin, N.; Choi, C.H.; Kim, H.S. Highly selective imidazole-appended 9,10-*N,N'*-diaminomethylanthracene fluorescent probe for switch-on Zn^{2+} detection and switch-off $H_2PO_4^-$ and CN^- detection in 80% aqueous DMSO, and applications to sequential logic gate operations. *Sens. Actuators B Chem.* **2017**, *247*, 840–849. [CrossRef]
13. Theetharappan, M.; Neelakantan, M.A. A Water-Soluble Schiff Base Turn-on Fluorescent Chemosensor for the Detection of Al^{3+} and Zn^{2+} Ions at the Nanomolar Level: Application in Live-Cell Imaging. *J. Fluoresc.* **2021**, *31*, 1277–1290. [CrossRef]
14. Chandra, R.; Manna, A.K.; Sahu, M.; Rout, K.; Patra, G.K. Simple salicylaldimine-functionalized dipodal *bis* Schiff base chromogenic and fluorogenic chemosensors for selective and sensitive detection of Al^{3+} and Cr^{3+}. *Inorg. Chim. Acta* **2020**, *499*, 119192. [CrossRef]
15. Jeong, H.Y.; Lee, S.Y.; Han, J.; Lim, M.H.; Kim, C. Thiophene and diethylaminophenol-based "turn-on" fluorescence chemosensor for detection of Al^{3+} and F^- in a near-perfect aqueous solution. *Tetrahedron* **2017**, *73*, 2690–2697. [CrossRef]
16. Wang, M.; Lu, L.; Song, W.; Wang, X.; Sun, T.; Zhu, J.; Wang, J. AIE-active Schiff base compounds as fluorescent probe for the highly sensitive and selective detection of Al^{3+} ions. *J. Lumin.* **2021**, *233*, 117911. [CrossRef]
17. Park, S.M.; Saini, S.; Park, J.E.; Singh, N.; Jang, D.O. A benzothiazole-based receptor for colorimetric detection of Cu^{2+} and S^{2-} ions in aqueous media. *Tetrahedron Lett.* **2021**, *73*, 153115. [CrossRef]
18. So, H.; Cho, H.; Lee, H.; Tran, M.C.; Kim, K.T.; Kim, C. Detection of zinc (II) and hypochlorite by a thiourea-based chemosensor via two emission channels and its application in vivo. *Microchem. J.* **2020**, *155*, 104788. [CrossRef]
19. Feng, X.; Fu, Y.; Jin, J.; Wu, J. A highly selective and sensitive fluorescent sensor for relay recognition of Zn^{2+} and $HSO_4^-/H_2PO_4^-$ with "on-off" fluorescent responses. *Anal. Biochem.* **2018**, *563*, 20–24. [CrossRef]
20. Du, K.; Niu, S.; Qiao, L.; Dou, Y.; Zhu, Q.; Chen, X.; Zhang, P. A highly selective ratiometric fluorescent probe for the cascade detection of Zn^{2+} and $H_2PO_4^-$ and its application in living cell imaging. *RSC Adv.* **2017**, *7*, 40615–40620. [CrossRef]
21. Hwang, S.M.; Kim, M.S.; Lee, M.; Lim, M.H.; Kim, C. Single fluorescent chemosensor for multiple targets: Sequential detection of Al^{3+} and pyrophosphate and selective detection of F^- in near-perfect aqueous solution. *N. J. Chem.* **2017**, *41*, 15590–15600. [CrossRef]
22. Liu, Y.; Wang, X.; Feng, E.; Fan, C.; Pu, S. A highly selective sequential recognition probe for Zn^{2+} and $HSO_4^-/H_2PO_4^-$ based on a diarylethene chemosensor. *Spectrochim. Acta Part A Mol. Biomol. Spectrosc.* **2021**, *246*, 119052. [CrossRef] [PubMed]
23. Jain, H.; Deswal, N.; Joshi, A.; Ramachandran, C.N.; Kumar, R. Triazole-appended pyrano[2,3-*c*]pyrazolone based colorimetric chemosensors for recognition of Fe^{3+} ions and their molecular logic gate behavior. *Anal. Methods* **2019**, *11*, 3230–3243. [CrossRef]
24. Qu, W.J.; Yan, G.T.; Ma, X.L.; Wei, T.B.; Lin, Q.; Yao, H.; Zhang, Y.M. "Cascade recognition" of Cu^{2+} and $H_2PO_4^-$ with high sensitivity and selectivity in aqueous media based on the effect of ESIPT. *Sens. Actuators B Chem.* **2017**, *242*, 849–856. [CrossRef]
25. Li, S.; Cao, D.; Meng, X.; Hu, Z.; Li, Z.; Yuan, C.; Zhou, T.; Han, X.; Ma, W. A novel schiff base fluorescent probe based on coumarin and benzothiazole for sequential detection of Al^{3+} and PPi and its applicability in live cell imaging. *J. Photochem. Photobiol. A Chem.* **2020**, *392*, 112427. [CrossRef]
26. Fu, J.; Li, B.; Mei, H.; Chang, Y.; Xu, K. Fluorescent schiff base probes for sequential detection of Al^{3+} and F^- and cell imaging applications. *Spectrochim. Acta Part A Mol. Biomol. Spectrosc.* **2020**, *227*, 117678. [CrossRef] [PubMed]
27. Sun, X.J.; Liu, T.T.; Fu, H.; Li, N.N.; Xing, Z.Y.; Yang, F. A Naphthalimide-Based Fluorescence "Off-on-Off" Chemosensor for Relay Detection of Al^{3+} and ClO^-. *Front. Chem.* **2019**, *7*, 549. [CrossRef]
28. Huang, M.X.; Lai, J.P.; Sun, H.; Wu, W.Z. A simple, highly selective and ultra-sensitive "off-on-off" fluorescent chemosensor for successive detection of aluminum ion and phosphate in water samples. *Microchem. J.* **2019**, *151*, 104195. [CrossRef]
29. Zhang, Y.M.; Chen, X.P.; Liang, G.Y.; Zhong, K.P.; Yao, H.; Wei, T.B.; Lin, Q. A water-soluble fluorescent chemosensor based on Asp functionalized naphthalimide for successive detection Fe^{3+} and $H_2PO_4^-$. *Can. J. Chem.* **2018**, *96*, 363–370. [CrossRef]
30. Meng, X.; Li, S.; Ma, W.; Wang, J.; Hu, Z.; Cao, D. Highly sensitive and selective chemosensor for Cu^{2+} and $H_2PO_4^-$ based on coumarin fluorophore. *Dyes Pigments* **2018**, *154*, 194–198. [CrossRef]

31. Purkait, R.; Das Mahapatra, A.; Chattopadhyay, D.; Sinha, C. An azine-based carbothioamide chemosensor for selective and sensitive turn-on-off sequential detection of Zn(II) and $H_2PO_4^-$, live cell imaging and INHIBIT logic gate. *Spectrochim. Acta Part A Mol. Biomol. Spectrosc.* **2019**, *207*, 164–172. [CrossRef] [PubMed]
32. Kumar, A.; Kumar, V.; Upadhyay, K.K. An Al^{3+} and $H_2PO_4^-/HSO_4^-$ selective conformational arrest and bail to a pyrimidine-naphthalene anchored molecular switch. *Analyst* **2013**, *138*, 1891–1897. [CrossRef] [PubMed]
33. Anu, D.; Naveenm, P.; Rajamanikandan, R.; Kaveri, M.V. Development of hydrazide based fluorescence probe for detection of Al^{3+} ions and application in live cell image. *J. Photochem. Photobiol. A Chem.* **2021**, *405*, 112921. [CrossRef]
34. Sun, J.; Li, Y.; Shen, S.; Yan, Q.; Xia, G.; Wang, H. A squaraine-based fluorescence turn on chemosensor with ICT character for highly selective and sensitive detection of Al^{3+} in aqueous media and its application in living cell imaging. *Spectrochim. Acta Part A Mol. Biomol. Spectrosc.* **2020**, *228*, 117590. [CrossRef]
35. Xu, Y.; Kong, L.; Bai, L.; Chen, A.; Li, N.; Cheng, L.; Liu, W.; Sun, X.; Tao, F.; Wang, L.; et al. A new water-soluble polymer fluorescent chemosensor with thiophene Schiff base site for selectively sensing Al^{3+} ions. *Tetrahedron* **2021**, *79*, 131888. [CrossRef]
36. Li, Z.; Wang, Q.; Wang, J.; Jing, X.; Wang, S.; Xiao, L.; Li, L. A fluorescent light-up probe for selective detection of Al^{3+} and its application in living cell imaging. *Inorg. Chim. Acta* **2020**, *500*, 119231. [CrossRef]
37. Liao, Z.; Liu, Y.; Han, S.F.; Wang, D.; Zheng, J.Q.; Zheng, X.J.; Jin, L.P. A novel acylhydrazone-based derivative as dual-mode chemosensor for Al^{3+}, Zn^{2+} and Fe^{3+} and its applications in cell imaging. *Sens. Actuators B Chem.* **2017**, *244*, 914–921. [CrossRef]
38. Zhu, G.; Huang, Y.; Wang, C.; Lu, L.; Sun, T.; Wang, M.; Tang, Y.; Shan, D.; Wen, S.; Zhu, J. A novel coumarin-based fluorescence chemosensor for Al^{3+} and its application in cell imaging. *Spectrochim. Acta Part A Mol. Biomol. Spectrosc.* **2019**, *210*, 105–110. [CrossRef]
39. Das, A.K.; Goswami, S. 2-Hydroxy-1-naphthaldehyde: A versatile building block for the development of sensors in supramolecular chemistry and molecular recognition. *Sens. Actuators B Chem.* **2017**, *245*, 1062–1125. [CrossRef]
40. Immanuel David, C.; Bhuvanesh, N.; Jayaraj, H.; Thamilselvan, A.; Parimala Devi, D.; Abiram, A.; Prabhu, J.; Nandhakumar, R. Experimental and Theoretical Studies on a Simple S-S-Bridged Dimeric Schiff Base: Selective Chromo-Fluorogenic Chemosensor for Nanomolar Detection of Fe^{2+} & Al^{3+} Ions and Its Varied Applications. *ACS Omega* **2020**, *5*, 3055–3072. [CrossRef]
41. Das, B.; Dolai, M.; Dhara, A.; Ghosh, A.; Mabhai, S.; Misra, A.; Dey, S.; Jana, A. Solvent-Regulated Fluorimetric Differentiation of Al^{3+} and Zn^{2+} Using an AIE-Active Single Sensor. *J. Phys. Chem. A* **2021**, *125*, 1490–1504. [CrossRef]
42. Singh, R.; Das, G. "Turn-on" Pb^{2+} sensing and rapid detection of biothiols in aqueous medium and real samples. *Analyst* **2019**, *144*, 567–572. [CrossRef]
43. Yang, R.; Li, K.; Wang, K.; Zhao, F.; Li, N.; Liu, F. Porphyrin assembly on β-cyclodextrin for selective sensing and detection of a zinc ion based on the dual emission fluorescence ratio. *Anal. Chem.* **2003**, *75*, 612–621. [CrossRef] [PubMed]
44. Frisch, M.J.; Trucks, G.W.; Schlegel, H.B.; Scuseria, G.E.; Robb, M.A.; Cheeseman, J.R.; Scalmani, G.; Barone, V.; Petersson, G.A.; Nakatsuji, H.; et al. *Gaussian 16, Revision C.01*; Gaussian, Inc.: Wallingford, CT, USA, 2016.
45. Becke, A.D. Density-functional thermochemistry. III. The role of exact exchange. *J. Chem. Phys.* **1993**, *98*, 5648–5652. [CrossRef]
46. Lee, C.; Yang, W.; Parr, R.G. Development of the Colle-Salvetti correlation-energy formula into a functional of the electron density. *Phys. Rev. B* **1988**, *37*, 785–789. [CrossRef] [PubMed]
47. Hariharan, P.C.; Pople, J.A. The influence of polarization functions on molecular orbital hydrogenation energies. *Theor. Chim. Acta* **1973**, *28*, 213–222. [CrossRef]
48. Francl, M.M.; Pietro, W.J.; Hehre, W.J.; Binkley, J.S.; Gordon, M.S.; DeFrees, D.J.; Pople, J.A. Self-consistent molecular orbital methods. XXIII. A polarization-type basis set for second-row elements. *J. Chem. Phys.* **1982**, *77*, 3654–3665. [CrossRef]
49. Hay, P.J.; Wadt, W.R. Ab initio effective core potentials for molecular calculations. Potentials for the transition metal atoms Sc to Hg. *J. Chem. Phys.* **1985**, *82*, 270–283. [CrossRef]
50. Wadt, W.R.; Hay, P.J. Ab initio effective core potentials for molecular calculations. Potentials for main group elements Na to Bi. *J. Chem. Phys.* **1985**, *82*, 284–298. [CrossRef]
51. Hay, P.J.; Wadt, W.R. Ab initio effective core potentials for molecular calculations. Potentials for K to Au including the outermost core orbitale. *J. Chem. Phys.* **1985**, *82*, 299–310. [CrossRef]
52. Klamt, A.; Moya, C.; Palomar, J. A Comprehensive Comparison of the IEFPCM and SS(V)PE Continuum Solvation Methods with the COSMO Approach. *J. Chem. Theory Comput.* **2015**, *11*, 4220–4225. [CrossRef] [PubMed]
53. Jo, T.G.; Bok, K.H.; Han, J.; Lim, M.H.; Kim, C. Colorimetric detection of Fe^{3+} and Fe^{2+} and sequential fluorescent detection of Al^{3+} and pyrophosphate by an imidazole-based chemosensor in a near-perfect aqueous solution. *Dyes Pigments* **2017**, *139*, 136–147. [CrossRef]

processes

MDPI

Article

Synthesis of ZnO-CuO and ZnO-Co_3O_4 Materials with Three-Dimensionally Ordered Macroporous Structure and Its H_2S Removal Performance at Low-Temperature

Tao Yu, Zhuo Chen *, Yundong Wang and Jianhong Xu *

The State Key Laboratory of Chemical Engineering, Department of Chemical Engineering, Tsinghua University, Beijing 100084, China; yut19@mails.tsinghua.edu.cn (T.Y.); wangyd@tsinghua.edu.cn (Y.W.)
* Correspondence: chenz2020@mail.tsinghua.edu.cn (Z.C.); xujianhong@mail.tsinghua.edu.cn (J.X.)

Abstract: H_2S is a common but hazardous impurity in syngas, biogas, or natural gas. For some advanced power generation technologies, such as integrated gasification combined cycle (IGCC), solid oxide fuel cells, H_2S content needs to be reduced to an acceptable level. In this work, a series of highly porous Zn-Cu and Zn-Co composites with three-dimensionally ordered macropores (3DOM) structure were synthesized via the colloidal crystal template method and used to remove H_2S at 150 °C and one atm. Scanning electron microscopy (SEM), transmission electron microscopy (TEM), nitrogen adsorption studies, X-ray diffraction (XRD), and X-ray photoelectron spectroscopy (XPS) were carried out to analyze the fresh and spent adsorbents. The results show that all the adsorbents possess well-ordered macropores, a large surface area, and a highly dispersed active phase. The relative content of Zn and (Cu or Co) has a significant influence on the desulfurization performance of adsorbents. The addition of CuO significantly increases the sulfur capacity and 3DOM-$Zn_{0.5}Cu_{0.5}$ shows the largest sulfur capacity of all the adsorbents, reaching up to 102.5 mg/g. The multiple adsorption/regeneration cycles of 3DOM-$Zn_{0.5}Cu_{0.5}$ and 3DOM-$Zn_{0.5}Co_{0.5}$ indicate that the as-prepared adsorbents are stable, and the sulfur capacity can still exceed 65% of the fresh adsorbents after six cycles.

Keywords: porous materials; H_2S removal; low-temperature desulfurization; metal oxide; regeneration ability

Citation: Yu, T.; Chen, Z.; Wang, Y.; Xu, J. Synthesis of ZnO-CuO and ZnO-Co_3O_4 Materials with Three-Dimensionally Ordered Macroporous Structure and Its H_2S Removal Performance at Low-Temperature. *Processes* **2021**, *9*, 1925. https://doi.org/10.3390/pr9111925

Academic Editor: Avelino Núñez-Delgado

Received: 30 September 2021
Accepted: 24 October 2021
Published: 27 October 2021

1. Introduction

Hydrogen sulfide (H_2S) is a major impurity that originates from varieties of processes such as coke ovens, coal or natural gas manufacturing, and oil refining [1,2]. H_2S impurity must be removed since it is highly hazardous to both the environment and industrial processes. For example, H_2S can corrode industrial equipment and pipelines, and sulfur dioxide generated by H_2S combustion is a cause of acid rain. The olfactory threshold of H_2S is 0.005 ppm and the acceptable environment limit for H_2S is 0.02–0.1 ppm [3].

Many methods, including alcohol amine absorption, adsorption, ionic liquid absorption, etc., have been developed to remove H_2S from gas streams [4–6]. Among these methods, adsorption has attracted more and more attention from scholars because of its low cost and high desulfurization activity. In 1976, various metal oxides were thermodynamically screened to remove H_2S, and 11 metal oxides were found to have high-temperature desulfurization potential by Westmoreland et al. [7], including Zn, Fe, Mn, Co, V, W, Mo, Ca, Sr, Ba, and Cu. The adsorption reaction mechanism is based on the formation of metal sulfides from metal oxides and hydrogen sulfide [8,9]. Compared with high-temperature (>500 °C) desulfurization, mid-(200–500 °C) and low-temperature (<200 °C) desulfurization can simplify operations and reduce costs. Because of its thermal stability and desulfurization activity, ZnO has attracted much attention in low and mid-temperature desulfurization [10–12].

Bulk ZnO has a large particle size, insufficient porosity, and low surface area, which results in low sulfurization kinetics and grains agglomeration. To solve this problem, porous ZnO and ZnO nanoparticles supported on meso-/microporous materials have been extensively studied. Tran et al. [13] prepared a porous ZnO with high porosity using agarose gels as templates. The result showed that porous ZnO exhibited a much higher sulfur capacity (457 $mg \cdot g^{-1}$) than that of the commercial ZnO (245 $mg \cdot g^{-1}$). Yang et al. [14] synthesized a series of mesoporous ZnO/SiO_2 sorbents with the aid of activated carbon. The presence of activated carbon made ZnO highly dispersed and produced more oxygen vacancies in ZnO. The highest H_2S adsorption capacity of this material could reach up to 160.95 $mg \cdot g^{-1}$ and corresponding ZnO utilization was 69%. Moreover, with the increase of ZnO loading, the adsorption capacity decreased due to ZnO agglomeration. Furthermore, researchers investigated the desulfurization performance of ZnO/mesoporous silica prepared in various ways. Wang et al. [15,16] prepared mesoporous SBA-15 with different loading of ZnO nanoparticles via incipient wetness impregnation and investigated H_2S removal ability. The prepared materials could remove H_2S from 1000 ppm to less than 0.1 ppm at a zinc loading of 3.04 wt% and a sulfur capacity of 436.6 $mg \cdot g^{-1}$. They also found that as the reaction proceeded, the formation of zinc sulfide would condense in the pores and limit the gas diffusion, thereby reducing adsorption ability. Geng et al. [17] adopted a melt infiltration method to prepare ZnO supported on SBA-15, MCM-41, and MCM-48 and conducted desulfurization experiments at room temperature. Compared with the impregnation method, melt infiltration could make ZnO nanoparticles more uniform in the pores. The optimal ZnO loading on SBA-15 and MCM-41 was 20 wt%, and on MCM-48 was 30 wt%, and the sulfur capacities were 41.0, 54.9, 53.2 $mg \cdot g^{-1}$, respectively.

However, pure ZnO has some drawbacks, such as grain agglomeration and zinc evaporation. The addition of another metal oxide as an active additive was investigated [18–23]. For instance, a series of ZnO/TiO_2 composites with different Zn/Ti ratios were prepared from amorphous citrate precursors by Lew et al. [19]. The desulfurization result indicated that the addition of TiO_2 could stabilize ZnO adsorbents but raise operation temperature. Tran et al. [13] found that Ni-doped ZnO adsorbent showed a much higher desulfurization capacity compared with pure ZnO as Ni provided additional active sites. Moreover, The Ni-doped ZnO adsorbent was easily and completely regenerated by heat treatment in air. Furthermore, Shangguan et al. [20] combined Al_2O_3 and K_2CO_3 additives with ZnO adsorbents. The prepared materials showed an improved performance and a high resistance to generating COS. Balsamo et al. [21] loaded mixed oxides of Zn and Cu on activated carbon. The utilization ratio of the active phase increased along with CuO content and CuO could obviously promote the reactivity of ZnO. Generally, incorporating zinc oxide with another metal oxide produces a synergistic effect during desulfurization processes and the surface reactions are more complex.

Recently, 3DOM materials with interconnected pores have been extensively studied and have attracted widespread attention in the fields of photonics, batteries, and catalyst support [24–26]. 3DOM materials have an ordered skeleton structure that consists of uniform close-packed macropores, ranging from several nanometers to several hundred nanometers. The materials have some advantages including high surface area, excellent internal diffusion, large porosity, and large pore volume due to their unique structure, making them attractive as catalysts, especially for diffusion-controlling processes. In earlier studies, ZnO- [27], Fe_2O_3- [28], CuO- [29], Co_3O_4- [30] based 3DOM have been prepared and their desulfurization abilities were investigated. The results showed that the structure provided open and interconnected pores and improved metal oxides dispersion, resulting in better reaction kinetics and higher H_2S breakthrough capacity. However, few reports of the materials with mixed metal oxides used as H_2S removal adsorbents have been found.

According to the above findings, 3DOM structure can promote gas diffusion inside the adsorbent and improve adsorption-reaction kinetics. Copper oxide or cobalt oxide, as additives, have a synergic effect with ZnO-based adsorbent in the desulfurization processes. In this work, a series of 3DOM ZnO-CuO and 3DOM ZnO-Co_3O_4 adsorbents were synthesized for the first time. The total metal oxide's loading was 30 wt.% but with different Zn/Cu or Zn/Co atomic ratios. The H_2S removal performance of the adsorbents was evaluated under dry conditions. The initial and spent adsorbents were characterized by XRD, SEM, TEM, N_2 sorption, and thermogravimetry-mass spectrometry (TG-MS). The regeneration ability was also studied. This study intends to provide a new insight for a rational design of efficient ZnO-based adsorbents.

2. Materials and Methods

2.1. Chemicals

Zinc nitrate hexahydrate (AR, 99%), Copper dinitrate (AR, 99%), and Cobalt nitrate hexahydrate (AR, 99%) were obtained from Shanghai Macklin Biochemical Co. Ltd (Shanghai, China). Styrene (AR, 99.5%), tetraethyl orthosilicate (AR, 99%) and citric acid (AR, 99%) were provided by Energy Chemical (Anhui, China). Potassium persulfate (AR, 99.5%) was purchased from Modern Oriental Technology development Co. Ltd (Beijing, China). Methanol (AR, 99.5%), ethylene glycol (AR, 99.5%), ethanol (AR, 99.7%) were obtained from Shanghai Titan Scientific Co. Ltd (Shanghai, China). Hydrochloric acid (36–38 wt%) was purchased from Beijing lanyi chemical products Co. Ltd. (Beijing, China).

2.2. Preparation of Adsorbents

3DOM adsorbents were prepared by the colloidal crystal template method as described elsewhere [31,32]. Before preparation, the monodispersed polystyrene (PS) microspheres and well-arrayed assembly template are required. The typical preparation process of 3DOM adsorbents is as follows. First, tetraethyl orthosilicate (TEOS), anhydrous ethanol, hydrochloric acid, and distilled water at a molar ratio of 1:3.9:0.3:1.8 were mixed to obtain silicon sol. Then a certain amount of zinc nitrate hexahydrate and copper dinitrate (or a certain amount of zinc nitrate hexahydrate and cobalt nitrate hexahydrate) were dissolved in a mixture of methanol and ethylene glycol (2:3, V/V) to achieve a metal nitrate solution with a concentration of 1 M. Then the silicon sol and metal nitrate solution were mixed to get a final precursor solution with different metal contents. The well-arrayed PS hard templates were thoroughly immersed in the precursor solution for 6h and the excessive solution was removed by vacuum filtration. After that, the wet template was air-dried at room temperature overnight. Finally, the obtained material was calcined under flowing air in tube finance. The temperature was raised from room temperature to 300 °C at a heating rate of 1 °C/min and held for 2 h, then raised to 500 °C at 1 °C/min and held for 3 h. In this work, the nominal metal oxides content (Zn + Cu, Zn + Co) was set to 30 wt%, whereas the atomic ratio of Zn:Cu or Zn:Co was varied. The adsorbents were denoted as 3DOM-Zn_xM_{1-x}, where M stands for Cu or Co; x indicates the atom fraction of Zn with respect to Zn + M. Letters S and R, represent the spent and regenerated adsorbents, respectively.

2.3. Adsorbents Characterization

Nitrogen adsorption-desorption isotherms were determined by the auto surface area and pore size analyzer (Quadrasorb SI) at liquid nitrogen boiling temperature (77 K). The specific area of adsorbents was evaluated by the Brunauer–Emmett–Teller (BET) method and the pore characteristics were evaluated by the Barrett–Joyner–Halenda (BJH) method. The phase structure of the adsorbents was determined by X-ray diffraction (XRD, D8 Advance, Bruker, Germany). X-ray photoelectron spectroscopy (XPS) measurements were performed using ESCALAB 250Xi spectrometer equipped with a monochromated Al Kα source. The morphology of adsorbents was observed by scanning electron microscope (SEM, SU8010, Hitachi, Japan) and high-resolution transmission electron microscope (HRTEM, JEM2100F, JEOL, Akishima-shi, Japan).

2.4. H_2S Adsorption Experiments

H_2S dynamic adsorption tests were carried out at 150 °C to determine the breakthrough capacities of adsorbents under a dry atmosphere on a quartz tubular fixed bed (6 mm inner diameter × 380 mm long). The apparatus is shown in Figure 1. For desulfurization experiments, ~0.04 g as-prepared adsorbent was ground to 40–80 mesh and placed in the middle of a quartz tube, supported by quartz wool. The required gas composition was supplied by separated gas cylinders using mass flow controllers. Prior to the desulfurization test, the air inside the apparatus was removed by N_2 for 20 min at the flow rate of 50 mL/min. Then the fixed bed was heated to 150 °C. When the temperature was reached, the sulfur breakthrough capacity was measured using an H_2S/N_2 mixture with 1000 ppm H_2S as a sulfur source. The initial concentration of H_2S was 200 ppm and the breakthrough point was set when the tail gas contained 20 ppm H_2S, which was determined by a gas chromatograph (huaai, GC-9560) equipped with a flame photometric detector (FPD). The detection limitation of GC was 0.5 ppm. After desulfurization, the spent adsorbents were regenerated under flow air in a tube furnace at 650 °C for 3 h and the desulfurization performance of regenerated adsorbents was evaluated under the same conditions as fresh adsorbents.

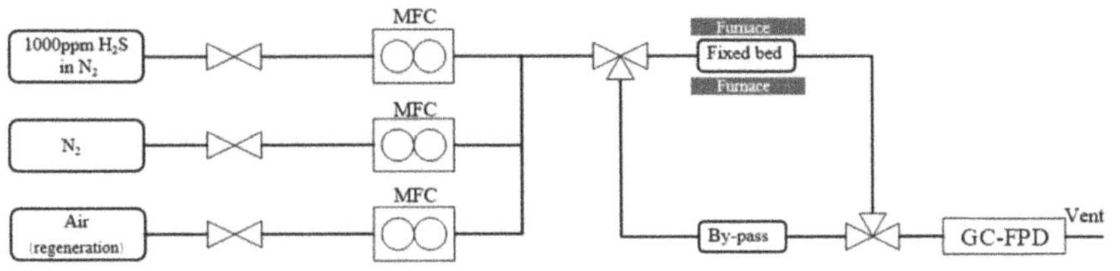

Figure 1. Apparatus for desulfurization test.

The breakthrough capacity (BC) was calculated by the following equation:

$$BC = \frac{v \times \int_0^t (C_0 - C_t)dt \times M_s \times 10^{-3}}{V_{mol} \times m_{ads}} \quad (1)$$

where BC is the breakthrough sulfur capacity (mg S/g adsorbent), v is the gas flow rate (L/min), t is the breakthrough time (min), C_0 and C_t are initial and outlet concentration of H_2S (ppm), respectively, M_s is the molar weight of S (32 g/mol), V_{mol} is the molar volume (24.45 L/mol at room temperature, 1 atm), m_{ads} is the weight of adsorbent (g).

3. Results and Discussion

3.1. Characterization of Adsorbents

3.1.1. Morphology and Porous Structure of Adsorbents

Figure 2 shows the SEM images of PS templates and as-prepared materials. From Figure 2a,b, it can be seen that the as-synthesized PS microspheres are approximately spherical with a diameter of about 320 nm, and they are monodispersed. After assembly, the template with a close-packed and well-arrayed structure was obtained. Figure 2c–i shows the prepared adsorbents with different metal oxides content. Obviously, all the materials had uniform, interconnected pore structures, resembling a honeycomb. The pore size of 3DOM materials was about 210 nm, indicating a shrinkage of about 35% relative to the initial size of PS microspheres. The shrinkage was due to the melting of the PS template and sintering of metal oxides. Similar phenomena have been observed in other literature [26,28,33]. Furthermore, the macropores were connected by "windows" (the diameter is ~40 nm), which was caused by the removal of tightly packed PS spheres during calcination. These windows were important for the diffusion of gas molecules inside the adsorbents. Among the above materials, silica acted as a solidification agent and carrier for metal oxides, which stabilized 3DOM structures. From Figure 2d–i, it can be concluded that when the total metal oxides content was fixed, 3DOM structures can basically remain unchanged when adjusting the ratio of Zn/M.

Figure 2. SEM images of PS template (**a**,**b**); 3DOM-Zn (**c**); 3DOM-$Zn_{0.5}Cu_{0.5}$ (**d**); 3DOM-$Zn_{0.33}Cu_{0.66}$ (**e**); 3DOM-$Zn_{0.66}Cu_{0.33}$ (**f**); 3DOM-$Zn_{0.5}Co_{0.5}$ (**g**); 3DOM-$Zn_{0.33}Co_{0.66}$ (**h**); 3DOM-$Zn_{0.66}Co_{0.33}$ (**i**).

Figure 3 shows the high-resolution transmission electron microscopy (HRTEM) images of typical as-prepared adsorbents and the metal elements mapping of 3DOM-$Zn_{0.5}Cu_{0.5}$ and 3DOM-$Zn_{0.5}Co_{0.5}$. As shown in Figure 3, the 3DOM materials contained highly periodically arrayed uniform macropores, which was consistent with SEM results. The walls of the adsorbents were formed by SiO_2 and nanocrystallized particles, and the shrinkage during calcination led to significant mesoporous within the wall [28]. In Figure 3b,c, EDX elemental mappings demonstrated a uniform distribution of Zn, Cu, or Zn, Co, indicating metal oxides were successfully loaded on the skeleton and the 3DOM structure could prevent the agglomeration of metal oxide nanoparticles.

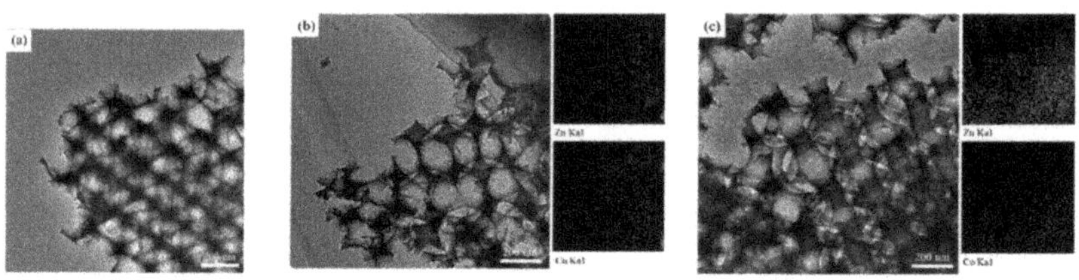

Figure 3. HRTEM images and EDX mapping of 3DOM-Zn (**a**); 3DOM-$Zn_{0.5}Cu_{0.5}$ (**b**); 3DOM-$Zn_{0.5}Co_{0.5}$ (**c**).

N_2 adsorption-desorption analysis was carried out to study the physical properties of prepared adsorbents. The nitrogen adsorption-desorption isotherms and pore size distribution are presented in Figure 4. It can be seen that all the 3DOM materials displayed similar curves, conforming to the type II isotherm according to the IUPAC classification [34,35]. Owing to the unrestricted monolayer or multilayer adsorption, the middle section of isotherms was approximately linear. The nitrogen adsorption increased significantly when P/P_0 was close to one, which indicated the materials possessed a macroporous structure [36]. Moreover, in the P/P_0 range of 0.8–1.0, H_3 hysteresis hoops could be observed in all materials, indicating the presence of mesoporous in the wall of macropores [36,37], which could be seen in the SEM and TEM images. Moreover, the inflection points of all 3DOM materials appeared at relative pressure below 0.1. This demonstrated the existence of a certain number of micropores [38].

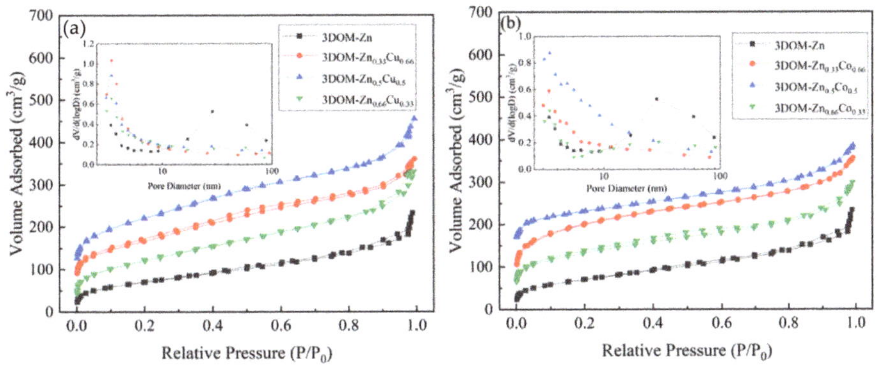

Figure 4. Nitrogen adsorption–desorption isotherms of 3DOM-Zn and 3DOM-Zn_xCu_{1-x} (**a**), 3DOM-Zn and 3DOM-Zn_xCo_{1-x} (**b**) (insert is pore size distribution calculated by BJH method).

The textural parameters are summarized in Table 1. Compared with 3DOM-Zn, the addition of copper oxide or cobalt oxide could increase the BET specific surface area. Since the formation and growth of CuO/Co_3O_4 grains are different from those of ZnO under the same condition, inconsistent shrinkage occurred during the calcination process [39,40]. The shrinkage led to the formation of micropores, which was consistent with the results of insert images in Figure 4 [27], and the increase in the number of micropores increased the specific surface area. As the Zn/Cu ratio decreases, the BET specific area increased. 3DOM-$Zn_{0.33}Co_{0.66}$ also possessed the largest specific area among the three adsorbents containing cobalt oxide, while the specific area of 3DOM-$Zn_{0.5}Co_{0.5}$ was slightly less than that of 3DOM-$Zn_{0.66}Co_{0.33}$. This may be caused by the sintering of different grains during the calcination process. Furthermore, the proportion of micropores increased with the addition of copper/cobalt oxide, while the average pore sizes calculated by the Barret–Joyner–Halenda (BJH) method almost remained unchanged.

Table 1. Textural parameters of 3DOM materials [1].

Name	S_{BET} (m^3/g)	V_t (cm^3/g)	d_{BJH} (nm)
3DOM-Zn	187.1	0.467	3.361
3DOM-$Zn_{0.33}Cu_{0.66}$	213.2	0.335	3.387
3DOM-$Zn_{0.5}Cu_{0.5}$	209.1	0.379	3.423
3DOM-$Zn_{0.66}Cu_{0.33}$	198.4	0.337	3.437
3DOM-$Zn_{0.33}Co_{0.66}$	237.1	0.288	3.405
3DOM-$Zn_{0.5}Co_{0.5}$	217.5	0.264	3.347
3DOM-$Zn_{0.66}Co_{0.33}$	221.7	0.442	3.389

[1] S_{BET}, BET specific area; V_t, total pore volume; d_{BJH}, pore size determined from BJH data.

In summary, the prepared adsorbents possess a well-arrayed and interconnected porous framework which contains hierarchically pores composed of macro-, meso-, and micropores. The adsorbents have a high surface area, which can be affected by the Zn/Cu or Zn/Co ratio.

3.1.2. XRD Analysis

Figure 5 shows the XRD patterns of as-prepared adsorbents and SiO_2 with a 3DOM structure. It can be seen that both the 3DOM-Zn and SiO_2 exhibited a typical amorphous structure. However, as shown in Figure 5a, the peaks of 3DOM-Zn at ~24° shifted slightly towards the larger angles compared to those of SiO_2 but with no obvious peaks [41]. This might be attributed to the following reasons, the ZnO nanoparticles were too small to be detected by the XRD method and the ZnO was well dispersed in the SiO_2 matrix [42,43]. Compared with the standard card of CuO (JCPDS PDF no. 48-1548), all the adsorbents containing CuO showed CuO diffraction peaks, which could be indexed to (1,1,-1), (1,1,1), (2,0,-2), (0,2,0), (1,1,-3), (3,1,-1), (2,2,0) crystal faces, and the peak intensity decreased with the increase of ZnO content. 3DOM-$Zn_{0.66}Cu_{0.33}$, which possessed the lowest content of copper, had very weak diffraction peaks. Figure 5b shows the XRD patterns of 3DOM-Zn and adsorbents with different Zn/Co ratios. The relatively weak but clear reflections at 31.3°, 36.9°, 38.5°, 59.4°, and 65.2° appeared in the XRD pattern, assigning to the cobalt oxide phase (JCPDS PDF no. 42-1467). Furthermore, adsorbents with different Zn/Cu or Zn/Co ratios can be regarded as Zn-doped, Cu-doped, or Co-doped adsorbents, and doping will affect lattice parameters to some extent [8], and this is also the reason why some samples have weak diffraction peaks. However, the very weak diffraction peaks of ZnO appeared in both the 3DOM-Zn_xCu_{1-x} and 3DOM-Zn_xCo_{1-x} XRD patterns rather than 3DOM-Zn. The results indicated that zinc oxide and copper oxide (or cobalt oxide) combined together partially through intra-grain and inter-granular coupling, and the latter made the faint diffraction peak of zinc oxide appear [44,45].

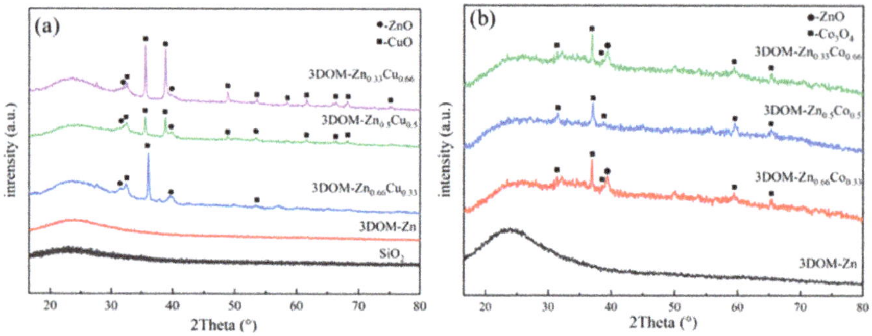

Figure 5. XRD patterns of 3DOM-Zn_xCu_{1-x} (**a**) and 3DOM-Zn_xCo_{1-x} (**b**).

3.1.3. XPS Analysis

Combining the XRD analysis and previous studies, the phase composition of as-prepared adsorbents could be roughly determined. XPS was further performed to analyze amorphous species. Figure 6a shows the Zn 2p spectra of all the prepared adsorbents, two characteristic peaks in the pattern at binding energy (BE) values of 1022.5 eV and 1045.7 eV were observed, corresponding to the Zn 2p3/2 and Zn 2p1/2, respectively [18]. The two characteristic peaks shifted slightly towards higher binding energy with the addition of copper oxide or cobalt oxide but still maintained the original shape of the peaks. A part of copper oxide or cobalt oxide would interact with zinc oxide to form CuO/ZnO or Co_3O_4/ZnO heterogeneous composite [46]. Moreover, the electronegativity of Cu or Co was higher than Zn, and ZnO acted as an electron donor. Thus, the characteristic peaks shifted slightly. The results indicated that all the adsorbents contained ZnO species. Cu 2p scan spectra are shown in Figure 6b. The appearance of two peaks (933.2 eV and 953.5 eV), as well as two satellite peaks between 940–945 eV and 960–965 eV illustrated the presence of CuO (Cu^{2+}) [47]. In the XPS spectra of Co, as shown in Figure 6c, the peaks at 781.3 eV and 797.6 eV illustrated the presence of Co^{3+} and Co^{2+}, and no metallic Co could be identified by XPS spectra. Similarly, with the increase of Cu or Co content, the peak intensity increased, which was consistent with XRD results.

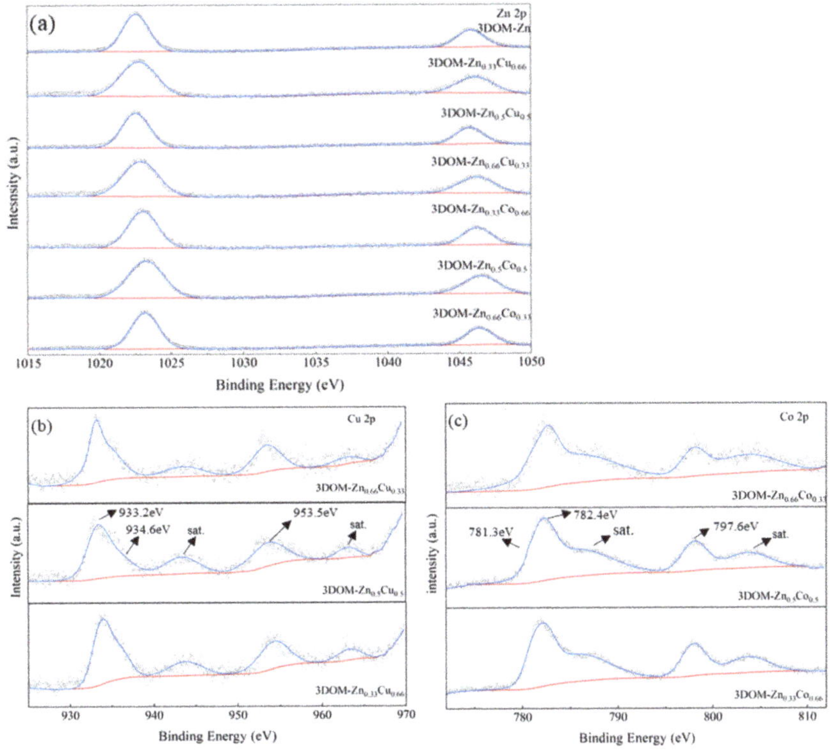

Figure 6. XPS spectra of Zn 2p (**a**), Cu 2p (**b**) and Co 2p (**c**) of the adsorbents.

3.2. H_2S Removal of Synthesized Adsorbents

The H_2S removal performance of all the prepared adsorbents was carried out via a fixed bed at 150 °C, 1atm. Figure 7a shows the breakthrough curves and the corresponding breakthrough sulfur capacities are displayed in Figure 7b. Clearly, adsorbents with different active phase contents possess remarkable differences in desulfurization ability. Under the same test condition, 3DOM-Zn had the shortest breakthrough time (about 170 min), and it also had the lowest sulfur capacity (45.9 mg/g) and active phase utilization (32.5%). The addition of Cu/Co into the adsorbents could improve desulfurization performance at different levels. Among the three adsorbents containing Co, both the sulfur capacity and active phase utilization increased with the increase of Co content, and the sulfur capacity of 3DOM-$Zn_{0.33}Co_{0.66}$ reached up to 72.7 mg/g. The addition of copper had a greater impact on the adsorption performance and the breakthrough times were much longer than others. The sulfur capacity and active phase utilization of 3DOM-$Zn_{0.5}Cu_{0.5}$ reached up to 102.5 mg/g and 72.7%, respectively. As demonstrated in the characteristic results of XRD, HRTEM and nitrogen adsorption, the inter-connected macropores, large specific area, and well-dispersed nanoparticles were beneficial to improve desulfurization performance due to excellent mass transfer inside the adsorbents and easy access to the active phase. Pore structure plays an important role in gas-solid non-catalytic reactions, especially reactions which produce large amounts of products, such as the reaction of metal oxide and H_2S [27]. However, the difference of specific area between 3DOM-Zn_xCo_{1-x} and 3DOM-Zn_xCu_{1-x} (such as 3DOM-$Zn_{0.33}Co_{0.66}$ and 3DOM-$Zn_{0.33}Cu_{0.66}$, 237.1 vs 213.2 m^2/g) might illustrate that the specific area was not the major cause of the improvement of adsorption ability (72.7 vs. 102.5 mg/g). The copper oxide/cobalt oxide activity and size of nanoparticles might also be the reason for the difference in desulfurization performance. Further studies of fresh and spent adsorbents are needed.

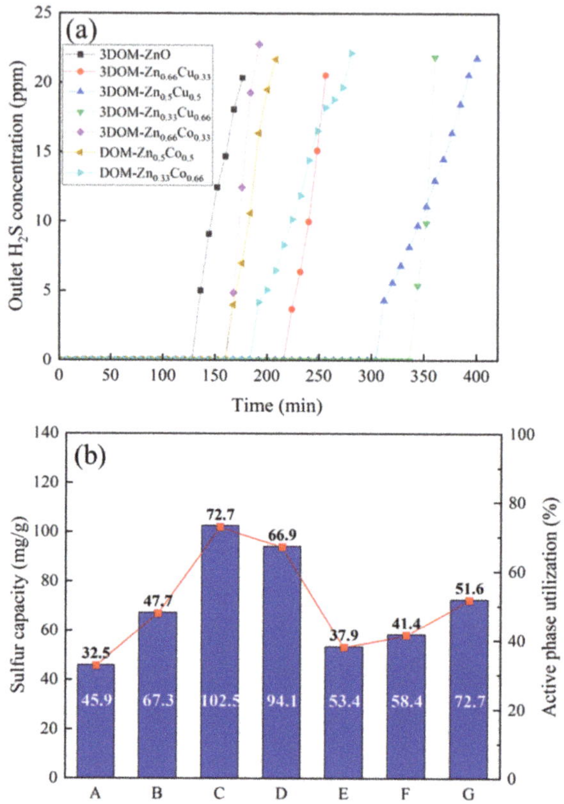

Figure 7. Breakthrough curves (**a**) and breakthrough capacities (**b**) of prepared adsorbents (A: 3DOM-Zn, B: 3DOM-$Zn_{0.66}Cu_{0.33}$, C: 3DOM-$Zn_{0.5}Cu_{0.5}$, D: 3DOM-$Zn_{0.33}Cu_{0.66}$, E: 3DOM-$Zn_{0.66}Co_{0.33}$, F: 3DOM-$Zn_{0.5}Co_{0.5}$, G: 3DOM-$Zn_{0.33}Co_{0.66}$).

The H_2S removal performance of as-prepared adsorbents was also compared to the reported results of metal oxide-based adsorbents, and the results are shown in Table 2. Clearly, the prepared adsorbents, especially 3DOM-$Zn_{0.5}Cu_{0.5}$ and 3DOM-$Zn_{0.33}Co_{066}$, showed much better sulfur capacities even at low metal oxide proportion, indicating the bi-metal oxide had a synergistic effect on desulfurization. In conclusion, the bi-metal oxide materials demonstrate promising potential in H_2S removal.

3.3. *Analysis of Spent Adsorbents*

The spent 3DOM-$Zn_{0.5}Cu_{0.5}$ and 3DOM-$Zn_{0.5}Co_{0.5}$ were selected to analyze physical properties and the desulfurization process. As mentioned above, the inter-connected macropores can enhance the diffusion kinetics and increase the specific area, thereby improving the dispersion of the active phase and desulfurization ability of adsorbents. It is necessary to figure out the structure change of adsorbents during the desulfurization process. Figure 8 shows the SEM images of spent 3DOM-$Zn_{0.5}Cu_{0.5}$ and 3DOM-$Zn_{0.5}Co_{0.5}$. Compared to Figure 2d,g, the spent adsorbents still maintained a relatively complete three-dimensional and well-ordered structure but the surface of adsorbents changed from smooth to rough, with granules appearing. Moreover, it can be seen that some mesopores between macropores were blocked by the products, even local collapse occurred. The reason for the phenomenon is that the desulfurization of hydrogen sulfide is a process of volume expansion. The adsorbents become denser and denser, and pore block even occurs in some serious cases during the process [28]. Figure 9 shows the HRTEM images of the above two adsorbents as well as EDX mapping. Similar to Figure 8, their adsorbents still maintained a well-ordered macroporous structure. EDX mapping images of Zn and Cu (or Co) depicted that metal oxide was evenly distributed in the adsorbents. The results were in good agreement with the aforementioned results. Furthermore, the mapping of S also showed the uniform distribution of sulfur species and corresponded to the metal mapping as well, indicating that sulfur was mainly adsorbed by the metal oxide active phase. Sulfur species of adsorption products need to be further determined.

Table 2. Summary of the performance of metal oxide-based adsorbents for H_2S removal.

Adsorbents	Proportion of Metal Oxide (%)	Feed Gas Composition	Synthesis Temperature (°C)	Desulfurizaion Temperature (°C)	Breakthrough Capacity (mg/g Sorbent)	Ref.
3DOM-CuO/SiO_2	50	500 mg/m^3 H_2S, N_2 balance	500	Room temperature	97	[29]
3DOM-Co_3O_4/SiO_2	57	3% H_2O, 500 mg/m^3 H_2S, N_2 balance	500	30	75	[30]
3DOM-Fe_2O_3/SiO_2	71.8–72.3	300 mg/m^3 H_2S, 5 % H_2, N_2 balance	500	350	38.92	[28]
Z30/K6	30	800 ppm H_2S, N_2 balance	300	Room temperature	37.6	[48]
ZnO/SBA-15-F	15	200 ppm H_2S, N_2 balance	550	Room temperature	21.8	[49]
20Cu/MSU-1	20	5% H_2S, 95% CH_4	450	Room temperature	18.3	[50]
SZ-30-400	30	100 ppm H_2S, N_2 balance	400	Room temperature	90.7	[43]
3DOM-Zn	30	200 ppm H_2S, N_2 balance	500	150	45.9	This work
3DOM-$Zn_{0.5}Cu_{0.5}$	30	200 ppm H_2S, N_2 balance	500	150	102.5	This work
3DOM-$Zn_{0.33}Co_{066}$	30	200 ppm, H_2S, N_2 balance	500	150	72.7	This work

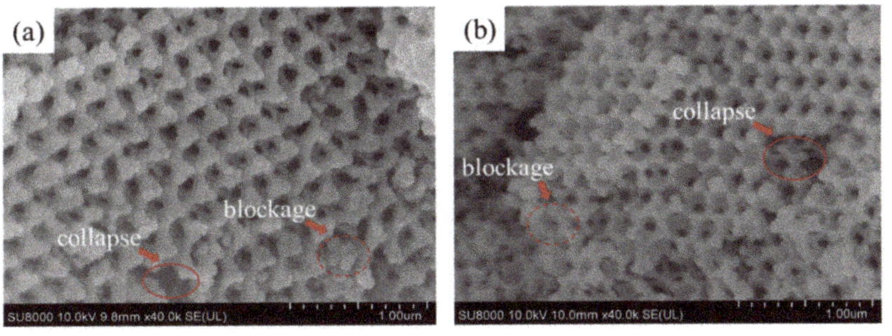

Figure 8. SEM images of spent 3DOM-$Zn_{0.5}Cu_{0.5}$ (**a**) and 3DOM-$Zn_{0.5}Co_{0.5}$ (**b**).

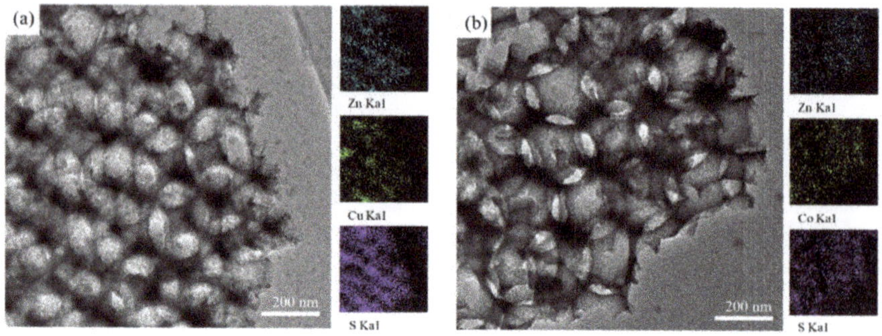

Figure 9. HRTEM images and EDX mapping of spent 3DOM-$Zn_{0.5}Cu_{0.5}$ (**a**); 3DOM-$Zn_{0.5}Co_{0.5}$ (**b**).

XRD and XPS analysis was performed to study sulfur species after the adsorption test. The XRD patterns of spent and regenerated 3DOM-$Zn_{0.5}Cu_{0.5}$ and 3DOM-$Zn_{0.5}Co_{0.5}$ are shown in Figure 10. As shown in Figure 10a, there was almost no copper oxide and zinc oxide in 3DOM-$Zn_{0.5}Cu_{0.5}$-R after adsorption and the XRD pattern was in good agreement with the characteristic of copper sulfide (JCPDS PDF no. 06-0464), indicating copper sulfide was the main product during the adsorption process. However, in the pattern of 3DOM-$Zn_{0.5}Co_{0.5}$-R, after the adsorption of H_2S, the peaks of Co_3O_4 became weak and only a few peaks corresponded to cobalt sulfide in the pattern of 3DOM-$Zn_{0.5}Co_{0.5}$-R. The relatively high kinetics and low limitation of thermodynamics of cobalt oxide and H_2S made the reaction occur almost exclusively on the solid surface, and a quantity of cobalt oxide remained unreacted [30]. The XPS spectra of spent adsorbents are shown in Figure 11. Similar to the fresh adsorbents, two distinguished peaks could be noticed in Zn 2p spectra, which was consistent with characteristic Zn 2p3/2 and Zn 2p1/2. This indicated that the oxidation state of Zn in both fresh and spent adsorbents were Zn^{2+} and no metallic Zn was produced in the adsorption process. Figure 11b shows Cu 2p spectra of 3DOM-$Zn_{0.5}Cu_{0.5}$-R. Compared to Figure 6b, the Cu 2p main peaks were shifted to lower binding energies (933.2 to 932.5 eV, 953.5 to 952.3 eV), which could be attributed to the appearance of CuS [47,51]. Besides, the weak peaks at 933.2 eV and 953.5 eV, together with a satellite peak indicated few CuO still existed. As shown in Figure 11c, S 2p spectrum curve could be fitted by three characteristic peaks, with the binding energies at 161.8 eV, 162.7 eV, and 169.5 eV, respectively, which suggested the formation of ZnS, CuS, and metal sulfate [52,53]. In the XPS spectrum of Co, multiple peaks could be observed, and the two peaks around 782.5 eV and 798.0 eV were ascribed to Co 2p1/2 and Co 2p3/2, respectively. The curve was fitted with four peaks with a binding energy of 778.8 eV, 782.5 eV, 786.1 eV, and 798.0 eV, respectively. The peaks at 778.8 eV and the satellite peak at 804.3 eV were characteristic of CoO [54]. In addition, two peaks at 782.5 eV and 798.0 eV were regarded to be CoS. Also, no metallic Co could be found from the spectrum. For sulfur species of 3DOM-$Zn_{0.5}Co_{0.5}$-R, as shown in Figure 11e, the spectrum had a great difference from that in Figure 11c. The curve could be fitted into three peaks with the binding energy of 162.7 eV, 164.8 eV, and 169.0 eV, corresponding to CoS, element S, and sulfate, respectively [55].

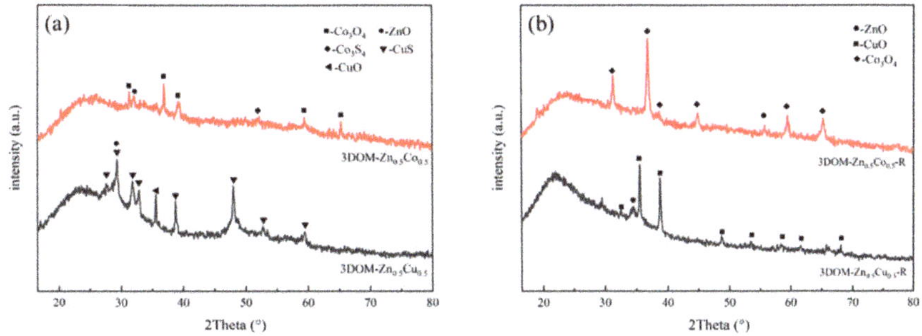

Figure 10. XRD patterns of spent (**a**) and regenerated (**b**) 3DOM-$Zn_{0.5}Cu_{0.5}$ and 3DOM-$Zn_{0.5}Co_{0.5}$.

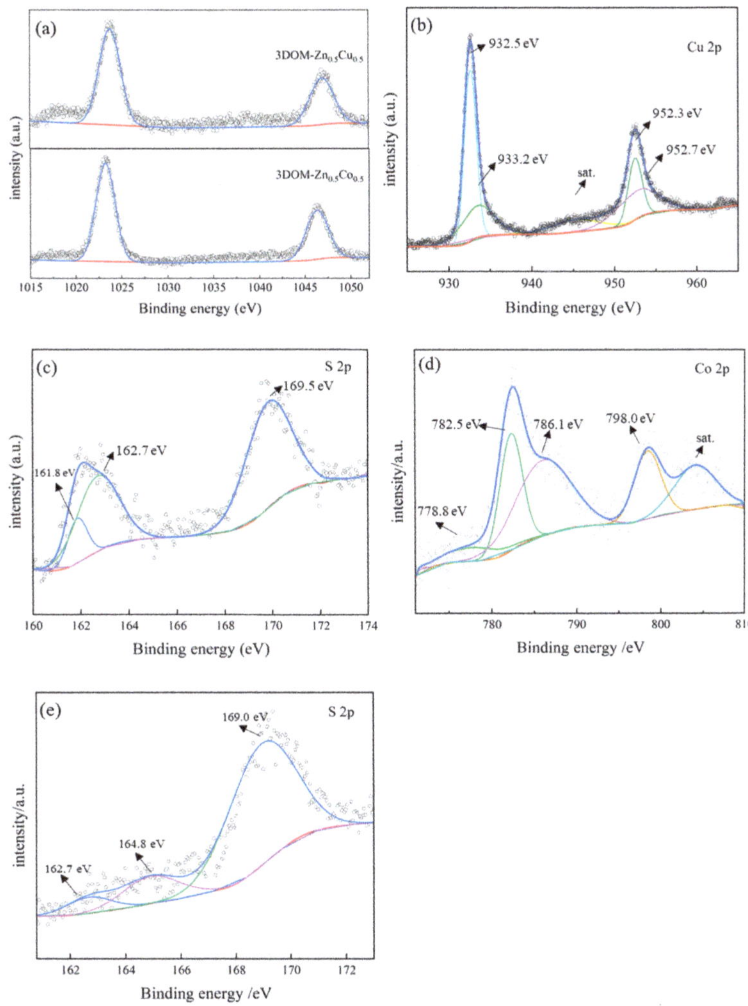

Figure 11. XPS spectra of different element: Zn 2p of 3DOM-$Zn_{0.5}Cu_{0.5}$ and 3DOM-$Zn_{0.5}Co_{0.5}$ (**a**), Cu 2p (**b**) and S 2p (**c**) of 3DOM-$Zn_{0.5}Cu_{0.5}$; Co 2p (**d**) and S 2p (**e**) of 3DOM-$Zn_{0.5}Co_{0.5}$.

Based on the XPS analysis of 3DOM-$Zn_{0.5}Cu_{0.5}$-R and 3DOM-$Zn_{0.5}Co_{0.5}$-R and previous literature, both the copper oxide and cobalt oxide have good desulfurization kinetics [7]. 3DOM structures could avoid pore blockage caused by rapid grain growth. In the desulfurization of 3DOM-Zn_xCo_{1-x}, a quantity of elemental S was produced, and S would attach to the surface of the adsorbents, which will slow down the reaction and prevent H_2S from reacting with the active phase inside. This resulted in a smaller sulfur capacity than that of 3DOM-Zn_xCu_{1-x}.

3.4. Successive Sulfidation-Regeneration Performance

Taking into account the economy and convenience of the adsorbents, the regeneration ability of 3DOM-$Zn_{0.5}Cu_{0.5}$ and 3DOM-$Zn_{0.5}Co_{0.5}$ was investigated. The spent adsorbents were regenerated in a tube furnace at 650 °C with the flow air at the rate of 100 mL/min for 3 h. Figure 12 shows the sulfur capacities of the above two adsorbents within six adsorption/regeneration cycles. It can be clearly seen that the sulfur capacities of 3DOM-$Zn_{0.5}Cu_{0.5}$ significantly decreased by more than 30% after the 1ST cycle. From the 2nd to 6th cycles, sulfur capacities almost remained unchanged. A similar phenomenon occurred on 3DOM-$Zn_{0.5}Co_{0.5}$, but the decline rate was about 25%, less than that of 3DOM-$Zn_{0.5}Cu_{0.5}$. Based on the SEM, XRD, and XPS analysis above, the reason for the sulfur capacity decline might be the formation of metal sulfate during the adsorption, which occupied the active phase site, destroyed well-ordered structure and it was hard to be reduced. However, a quantity of elemental S was produced during the adsorption process of 3DOM-$Zn_{0.5}Co_{0.5}$, which will attach to the surface of the adsorbent. In the regeneration process, the attached S was removed by reducing to sulfur oxide and the active phase was exposed again. Sulfur capacity was restored to a certain extent. However, the degree of regeneration is related to the regeneration conditions, such as temperature, moisture, and so on. To find a proper regeneration condition, further work needs to be done.

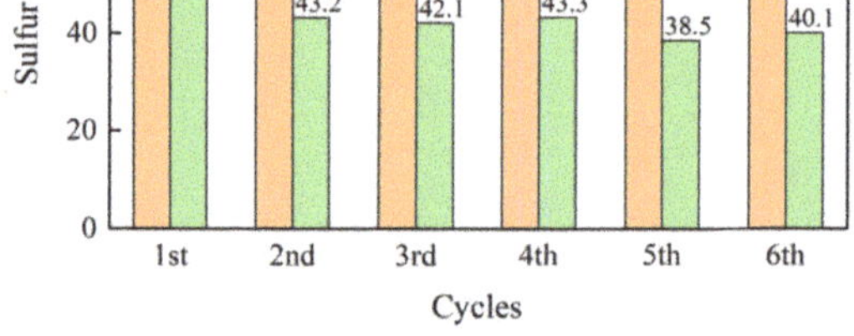

Figure 12. Sulfur capacity of 3DOM-$Zn_{0.5}Cu_{0.5}$ and 3DOM-$Zn_{0.5}Co_{0.5}$ during cycles.

4. Conclusions

In summary, a series of 3DOM adsorbents containing two active phases (ZnO, CuO or ZnO, Co_3O_4) were synthesized via a colloidal crystal template method. These adsorbents possessed uniform and inter-connected macropores, large surface area, and highly dispersed active phase. The results of the adsorption test showed that the relative content of Zn and (Cu or Co) had a significant influence on desulfurization performance. Among the three adsorbents containing Cu, the 3DOM-$Zn_{0.5}Cu_{0.5}$ showed the highest sulfur capacity (102.5 mg/g), while 3DOM-$Zn_{0.33}Co_{0.66}$ performed the largest sulfur capacity (72.7 mg/g) among adsorbents containing Co. In addition, the results of successive adsorption/regeneration indicated that the 3DOM-$Zn_{0.5}Cu_{0.5}$ and 3DOM-$Zn_{0.5}Co_{0.5}$ could retain most of the desulfurization ability after regeneration. The sulfur capacity still remained more than 65% of the original value after six times regeneration. This work will provide some reference for the preparation of novel and efficient adsorbents in the future.

Author Contributions: Conceptualization, T.Y.; methodology, T.Y.; formal analysis, T.Y.; investigation, T.Y.; resources, Y.W.; data curation, T.Y.; writing—original draft preparation, T.Y.; writing—review and editing, Z.C. and J.X.; supervision, Y.W. and J.X.; project administration, Y.W.; funding acquisition, Y.W. All authors have read and agreed to the published version of the manuscript.

Funding: This research was funded by National Natural Science Foundation of China, grant number 22178196, 21991100, 21991101 and 22108147; China Postdoctoral Science Foundation, grant number 2021M691761; Shui Mu Xue Zhe of Tsinghua University, grant number 2020SM056.

Institutional Review Board Statement: Not applicable.

Informed Consent Statement: Not applicable.

Acknowledgments: This research was funded by National Natural Science Foundation of China, grant number 22178196, 21991100, 21991101, 22108147; Shui Mu Xue Zhe of Tsinghua University, grant number 2020SM056; China Postdoctoral Science Foundation, grant number 2021M691761.

Conflicts of Interest: The authors declare no conflict of interest.

References

1. Yang, C.; Yang, S.; Fan, H.; Wang, Y.; Shangguan, J. Tuning the ZnO-Activated Carbon Interaction through Nitrogen Modification for Enhancing the H_2S Removal Capacity. *J. Colloid Interface Sci.* **2019**, *555*, 548–557. [CrossRef]
2. Okonkwo, C.N.; Lee, J.J.; De Vylder, A.; Chiang, Y.; Thybaut, J.W.; Jones, C.W. Selective Removal of Hydrogen Sulfide from Simulated Biogas Streams Using Sterically Hindered Amine Adsorbents. *Chem. Eng. J.* **2020**, *379*, 122349. [CrossRef]
3. Meng, F.N.; Di, X.P.; Dong, H.W.; Zhang, Y.; Zhu, C.L.; Li, C.; Chen, Y.J. Ppb H_2S Gas Sensing Characteristics of Cu_2O/CuO Sub-Microspheres at Low-Temperature. *Sensors Actuators B* **2013**, *182*, 197–204. [CrossRef]
4. Lu, J.-G.; Zheng, Y.-F.; He, D.-L. Selective Absorption of H_2S from Gas Mixtures into Aqueous Solutions of Blended Amines of Methyldiethanolamine and 2-Tertiarybutylamino-2-Ethoxyethanol in a Packed Column. *Sep. Purif. Technol.* **2006**, *52*, 209–217. [CrossRef]
5. Xue, M.; Chitrakar, R.; Sakane, K.; Ooi, K. Screening of Adsorbents for Removal of H_2S at Room Temperature. *Green Chem.* **2003**, *5*, 529–534. [CrossRef]
6. Liu, X.; Wang, R. H_2S Removal by Peroxo Heteropoly Compound/Ionic Liquid Solution. *Fuel Process. Technol.* **2017**, *160*, 78–85. [CrossRef]
7. Westmoreland, P.R.; Harrison, D.P. Evaluation of Candidate Solids for High-Temperature Desulfurization of Low-Btu Gases. *Environ. Sci. Technol.* **1976**, *10*, 659–661. [CrossRef]
8. Yang, C.; Yang, S.; Fan, H.-L.; Wang, J.; Wang, H.; Shangguan, J.; Huo, C. A Sustainable Design of ZnO-Based Adsorbent for Robust H_2S Uptake and Secondary Utilization as Hydrogenation Catalyst. *Chem. Eng. J.* **2020**, *382*, 122892. [CrossRef]
9. Yang, C.; Wang, Y.; Fan, H.; de Falco, G.; Yang, S.; Shangguan, J.; Bandosz, T.J. Bifunctional ZnO-MgO/Activated Carbon Adsorbents Boost H_2S Room Temperature Adsorption and Catalytic Oxidation. *Appl. Catal. B Environ.* **2020**, *266*, 118674. [CrossRef]
10. Sun, J.; Modi, S.; Liu, K.; Lesieur, R. Kinetics of Zinc Oxide Sulfidation for Packed-Bed Desulfurizer Modeling. *Energy Fuels* **2007**, *21*, 1863–1871. [CrossRef]
11. Garces, H.F.; Galindo, H.M.; Garces, L.J.; Hunt, J.; Morey, A.; Suib, S.L. Low Temperature H_2S Dry-Desulfurization with Zinc Oxide. *Microporous Mesoporous Mater.* **2010**, *127*, 190–197. [CrossRef]
12. Abdullah, A.H.; Mat, R.; Somderam, S.; Abd Aziz, A.S.; Mohamed, A. Hydrogen Sulfide Adsorption by Zinc Oxide-Impregnated Zeolite (Synthesized from Malaysian Kaolin) for Biogas Desulfurization. *J. Ind. Eng. Chem.* **2018**, *65*, 334–342. [CrossRef]
13. Tran, D.T. Synthesis of Porous ZnO Based Materials Using an Agarose Gel Template for H_2S Desulfurization. *RSC Adv.* **2016**, *6*, 1339–1345. [CrossRef]
14. Yang, C.; Wang, J.; Fan, H.; Hu, Y.; Shen, J.; Shangguan, J.; Wang, B. Activated Carbon-Assisted Fabrication of Cost-Efficient ZnO/SiO_2 Desulfurizer with Characteristic of High Loadings and High Dispersion. *Energy Fuels* **2018**, *32*, 6064–6072. [CrossRef]
15. Wang, X.; Sun, T.; Yang, J.; Zhao, L.; Jia, J. Low-Temperature H_2S Removal from Gas Streams with SBA-15 Supported ZnO Nanoparticles. *Chem. Eng. J.* **2008**, *142*, 48–55. [CrossRef]
16. Wang, X.H. Chemical Characterization of Mesoporous Material Supported ZnO Nanoparticles for Hydrogen Sulfide Capture from Gas Streams. *Adv. Mater. Res.* **2010**, *129–131*, 143–148. [CrossRef]
17. Geng, Q.; Wang, L.-J.; Yang, C.; Zhang, H.-Y.; Zhao, Y.-R.; Fan, H.-L.; Huo, C. Room-Temperature Hydrogen Sulfide Removal with Zinc Oxide Nanoparticle/Molecular Sieve Prepared by Melt Infiltration. *Fuel Process. Technol.* **2019**, *185*, 26–37. [CrossRef]

18. Dhage, P.; Samokhvalov, A.; Repala, D.; Duin, E.C.; Tatarchuk, B.J. Regenerable Fe–Mn–ZnO/SiO_2 Sorbents for Room Temperature Removal of H_2S from Fuel Reformates: Performance, Active Sites, Operando Studies. *Phys. Chem. Chem. Phys.* **2011**, *13*, 2179–2187. [CrossRef]
19. Lew, S.; Sarofim, A.F.; Flytzani-Stephanopoulos, M. Sulfidation of Zinc Titanate and Zinc Oxide Solids. *Ind. Eng. Chem. Res.* **1992**, *31*, 1890–1899. [CrossRef]
20. Shangguan, J.; Zhao, Y.; Fan, H.; Liang, L.; Shen, F.; Miao, M. Desulfurization Behavior of Zinc Oxide Based Sorbent Modified by the Combination of Al_2O_3 and K_2CO_3. *Fuel* **2013**, *108*, 80–84. [CrossRef]
21. Balsamo, M.; Cimino, S.; de Falco, G.; Erto, A.; Lisi, L. ZnO-CuO Supported on Activated Carbon for H_2S Removal at Room Temperature. *Chem. Eng. J.* **2016**, *304*, 399–407. [CrossRef]
22. De Falco, G.; Montagnaro, F.; Balsamo, M.; Erto, A.; Deorsola, F.A.; Lisi, L.; Cimino, S. Synergic Effect of Zn and Cu Oxides Dispersed on Activated Carbon during Reactive Adsorption of H_2S at Room Temperature. *Microporous Mesoporous Mater.* **2018**, *257*, 135–146. [CrossRef]
23. Huang, Z.B.; Liu, B.S.; Wang, F.; Amin, R. Performance of Zn–Fe–Mn/MCM-48 Sorbents for High Temperature H_2S Removal and Analysis of Regeneration Process. *Appl. Surf. Sci.* **2015**, *353*, 1–10. [CrossRef]
24. Kanamura, K.; Akutagawa, N.; Dokko, K. Three Dimensionally Ordered Composite Solid Materials for All Solid-State Rechargeable Lithium Batteries. *J. Power Sources* **2005**, *146*, 86–89. [CrossRef]
25. Wei, Y.; Liu, J.; Zhao, Z.; Duan, A.; Jiang, G. The Catalysts of Three-Dimensionally Ordered Macroporous $Ce_{1-x}Zr_xO_2$-Supported Gold Nanoparticles for Soot Combustion: The Metal–Support Interaction. *J. Catal.* **2012**, *287*, 13–29. [CrossRef]
26. Xu, J.; Liu, J.; Zhao, Z.; Xu, C.; Zheng, J.; Duan, A.; Jiang, G. Easy Synthesis of Three-Dimensionally Ordered Macroporous $La_{1-x}K_xCoO_3$ Catalysts and Their High Activities for the Catalytic Combustion of Soot. *J. Catal.* **2011**, *282*, 1–12. [CrossRef]
27. Wang, L.-J.; Fan, H.-L.; Shangguan, J.; Croiset, E.; Chen, Z.; Wang, H.; Mi, J. Design of a Sorbent to Enhance Reactive Adsorption of Hydrogen Sulfide. *ACS Appl. Mater. Interfaces* **2014**, *6*, 21167–21177. [CrossRef]
28. Fan, H.-L.; Sun, T.; Zhao, Y.-P.; Shangguan, J.; Lin, J.-Y. Three-Dimensionally Ordered Macroporous Iron Oxide for Removal of H_2S at Medium Temperature. *Environ. Sci. Technol.* **2013**, *47*, 4859–4865. [CrossRef]
29. Wang, J.; Wang, L.; Fan, H.; Wang, H.; Hu, Y.; Wang, Z. Highly Porous Copper Oxide Sorbent for H_2S Capture at Ambient Temperature. *Fuel* **2017**, *209*, 329–338. [CrossRef]
30. Wang, J.; Yang, C.; Zhao, Y.-R.; Fan, H.-L.; Wang, Z.-D.; Shangguan, J.; Mi, J. Synthesis of Porous Cobalt Oxide and Its Performance for H 2 S Removal at Room Temperature. *Ind. Eng. Chem. Res.* **2017**, *56*, 12621–12629. [CrossRef]
31. Yang, H.; Deng, J.; Xie, S.; Jiang, Y.; Dai, H.; Au, C.T. Au/MnO_x/3DOM SiO_2: Highly Active Catalysts for Toluene Oxidation. *Appl. Catal. A Gen.* **2015**, *507*, 139–148. [CrossRef]
32. Sawangphruk, M.; Limtrakul, J. Effects of Pore Diameters on the Pseudocapacitive Property of Three-Dimensionally Ordered Macroporous Manganese Oxide Electrodes. *Mater. Lett.* **2012**, *68*, 230–233. [CrossRef]
33. Li, L.; Zhang, H.; Zhou, P.; Meng, X.; Liu, L.; Jia, J.; Sun, T. Three Dimensional Ordered Macroporous Zinc Ferrite Composited Silica Sorbents with Promotional Desulfurization and Regeneration Activity at Mid-High Temperature. *Appl. Surf. Sci.* **2019**, *470*, 177–186. [CrossRef]
34. Wei, Y.; Zhao, Z.; Li, T.; Liu, J.; Duan, A.; Jiang, G. The Novel Catalysts of Truncated Polyhedron Pt Nanoparticles Supported on Three-Dimensionally Ordered Macroporous Oxides (Mn, Fe, Co, Ni, Cu) with Nanoporous Walls for Soot Combustion. *Appl. Catal. B Environ.* **2014**, *146*, 57–70. [CrossRef]
35. Li, L.; Huang, X.; Hu, T.; Wang, J.; Zhang, W.; Zhang, J. Synthesis of three-dimensionally ordered macroporous composite Ag/Bi_2O_3–TiO_2 by dual templates and its photocatalytic activities for degradation of organic pollutants under multiple modes. *New J. Chem.* **2014**, *38*, 5293. [CrossRef]
36. Li, H.; Zhang, L.; Dai, H.; He, H. Facile Synthesis and Unique Physicochemical Properties of Three-Dimensionally Ordered Macroporous Magnesium Oxide, Gamma-Alumina, and Ceria−Zirconia Solid Solutions with Crystalline Mesoporous Walls. *Inorg. Chem.* **2009**, *48*, 4421–4434. [CrossRef]
37. Chen, D.; Li, Z.; Wan, Y.; Tu, X.; Shi, Y.; Chen, Z.; Shen, W.; Yu, C.; Tu, B.; Zhao, D. Anionic Surfactant Induced Mesophase Transformation to Synthesize Highly Ordered Large-Pore Mesoporous Silica Structures. *J. Mater. Chem.* **2006**, *16*, 1511. [CrossRef]
38. Li, W.-C.; Lu, A.-H.; Weidenthaler, C.; Schüth, F. Hard-Templating Pathway To Create Mesoporous Magnesium Oxide. *Chem. Mater.* **2004**, *16*, 5676–5681. [CrossRef]
39. Yarbrough, R.; Davis, K.; Dawood, S.; Rathnayake, H. A Sol–Gel Synthesis to Prepare Size and Shape-Controlled Mesoporous Nanostructures of Binary (II–VI) Metal Oxides. *RSC Adv.* **2020**, *10*, 14134–14146. [CrossRef]
40. Taufik, A.; Albert, A.; Saleh, R. Sol-Gel Synthesis of Ternary CuO/TiO2/ZnO Nanocomposites for Enhanced Photocatalytic Performance under UV and Visible Light Irradiation. *J. Photochem. Photobiol. A Chem.* **2017**, *344*, 149–162. [CrossRef]
41. Cannas, C.; Casu, M.; Lai, A.; Musinu, A.; Piccaluga, G. XRD, TEM and 29Si MAS NMR Study of Sol-Gel ZnO-SiO2 Nanocomposites. *J. Mater. Chem.* **1999**, *9*, 1765–1769. [CrossRef]
42. Machala, L.; Zboril, R.; Gedanken, A. Amorphous Iron(III) Oxide—A Review. *J. Phys. Chem. B* **2007**, *111*, 4003–4018. [CrossRef]
43. Liu, G.; Huang, Z.-H.; Kang, F. Preparation of ZnO/SiO_2 Gel Composites and Their Performance of H_2S Removal at Room Temperature. *J. Hazard. Mater.* **2012**, *215–216*, 166–172. [CrossRef]
44. Habibi, M.H.; Karimi, B. Application of Impregnation Combustion Method for Fabrication of Nanostructure CuO/ZnO Composite Oxide: XRD, FESEM, DRS and FTIR Study. *J. Ind. Eng. Chem.* **2014**, *20*, 1566–1570. [CrossRef]

45. Sujinnapram, S.; Onreabroy, W.; Nantawisarakul, T.; Chia, S.-P.; Ratnavelu, K.; Muhamad, M.R. XRD, Photoluminescence and Optical Absorption Investigations of Cobalt-Doped ZnO. In Proceedings of the AIP Conference; AIP: Kuala Lumpur, Malaysia, 2009; pp. 340–343.
46. Zhu, L.; Li, H.; Liu, Z.; Xia, P.; Xie, Y.; Xiong, D. Synthesis of the 0D/3D CuO/ZnO Heterojunction with Enhanced Photocatalytic Activity. *J. Phys. Chem. C* **2018**, *122*, 9531–9539. [CrossRef]
47. Lee, Y.-H.; Ju, H.; Rha, S.-K.; Lee, S.-H.; Lee, Y.-S. Anode Effects in Electroplated Cu Film. *J. Surf. Anal.* **2011**, *17*, 282–286. [CrossRef]
48. Li, L.; Sun, T.H.; Shu, C.H.; Zhang, H.B. Low Temperature H2S Removal with 3-D Structural Mesoporous Molecular Sieves Supported ZnO from Gas Stream. *J. Hazard. Mater.* **2016**, *311*, 142–150. [CrossRef]
49. Hussain, M.; Abbas, N.; Fino, D.; Russo, N. Novel Mesoporous Silica Supported ZnO Adsorbents for the Desulphurization of Biogas at Low Temperatures. *Chem. Eng. J.* **2012**, *188*, 222–232. [CrossRef]
50. Montes, D.; Tocuyo, E.; González, E.; Rodríguez, D.; Solano, R.; Atencio, R.; Ramos, M.A.; Moronta, A. Reactive H2S Chemisorption on Mesoporous Silica Molecular Sieve-Supported CuO or ZnO. *Microporous Mesoporous Mater.* **2013**, *168*, 111–120. [CrossRef]
51. Gebhardt, J.E.; McCarron, J.J.; Richardson, P.E.; Buckley, A.N. The Effect of Cathodic Treatment on the Anodic Polarization of Copper Sulfides. *Hydrometallurgy* **1986**, *17*, 27–38. [CrossRef]
52. Nefedov, V.I. A Comparison of Results of an ESCA Study of Nonconducting Solids Using Spectrometers of Different Constructions. *J. Electron. Spectrosc. Relat. Phenom.* **1982**, *25*, 29–47. [CrossRef]
53. Strohmeier, B.R.; Levden, D.E.; Field, R.S.; Hercules, D.M. Surface Spectroscopic Characterization of CuAl2O3 Catalysts. *J. Catal.* **1985**, *94*, 514–530. [CrossRef]
54. Biesinger, M.C.; Payne, B.P.; Grosvenor, A.P.; Lau, L.W.M.; Gerson, A.R.; Smart, R.S.C. Resolving Surface Chemical States in XPS Analysis of First Row Transition Metals, Oxides and Hydroxides: Cr, Mn, Fe, Co and Ni. *Appl. Surf. Sci.* **2011**, *257*, 2717–2730. [CrossRef]
55. Madec, L.; Xia, J.; Petibon, R.; Nelson, K.J.; Sun, J.-P.; Hill, I.G.; Dahn, J.R. Effect of Sulfate Electrolyte Additives on LiNi1/3Mn1/3Co1/3O2/Graphite Pouch Cell Lifetime: Correlation between XPS Surface Studies and Electrochemical Test Results. *J. Phys. Chem. C* **2014**, *118*, 29608–29622. [CrossRef]

processes

MDPI

Article

Fast and Efficient Removal of Uranium onto a Magnetic Hydroxyapatite Composite: Mechanism and Process Evaluation

Tao Ou [1,2], Hairong Peng [1,2], Minhua Su [1,2,*], Qingpu Shi [1,2], Jinfeng Tang [2,*], Nan Chen [1,2] and Diyun Chen [1,2]

1 Guangdong Provincial Key Laboratory of Radionuclides Pollution Control and Resources, Guangzhou University, Guangzhou 510006, China; outao1997@163.com (T.O.); drogba11phr@foxmail.com (H.P.); L60882969@163.com (Q.S.); nancychen@gzhu.edu.cn (N.C.); cdy@gzhu.edu.cn (D.C.)

2 School of Environmental Science and Engineering, Guangzhou University, Guangzhou 510006, China

* Correspondence: mhsu@gzhu.edu.cn (M.S.); jinfeng@gzhu.edu.cn (J.T.)

Abstract: The exploration and rational design of easily separable and highly efficient sorbents with satisfactory capability of extracting radioactive uranium (U)-containing compound(s) are of paramount significance. In this study, a novel magnetic hydroxyapatite (HAP) composite (HAP@ $CoFe_2O_4$), which was coupled with cobalt ferrite ($CoFe_2O_4$), was rationally designed for uranium(VI) removal through a facile hydrothermal process. The U(VI) ions were rapidly removed using HAP@ $CoFe_2O_4$ within a short time (i.e., 10 min), and a maximum U(VI) removal efficiency of 93.7% was achieved. The maximum adsorption capacity (Q_{max}) of the HAP@$CoFe_2O_4$ was 338 mg/g, which demonstrated the potential of as-prepared HAP@$CoFe_2O_4$ in the purification of U(VI) ions from nuclear effluents. Autunite $[Ca(UO_2)_2(PO_4)_2(H_2O)_6]$ was the main crystalline phase to retain uranium, wherein U(VI) was effectively extracted and immobilized in terms of a relatively stable mineral. Furthermore, the reacted HAP@$CoFe_2O_4$ can be magnetically recycled. The results of this study reveal that the suggested process using HAP@$CoFe_2O_4$ is a promising approach for the removal and immobilization of U(VI) released from nuclear effluents.

Keywords: cobalt ferrite; adsorption; hydroxyapatite; mineralization; uranium(VI)

Citation: Ou, T.; Peng, H.; Su, M.; Shi, Q.; Tang, J.; Chen, N.; Chen, D. Fast and Efficient Removal of Uranium onto a Magnetic Hydroxyapatite Composite: Mechanism and Process Evaluation. *Processes* **2021**, *9*, 1927. https://doi.org/10.3390/pr9111927

Academic Editor: Andrea Petrella

Received: 19 September 2021
Accepted: 21 October 2021
Published: 28 October 2021

1. Introduction

Uranium (U), a long-persisting and highly hazardous radionuclide, is generally released from uranium mining, nuclear weapon testing, and nuclear accidents. Uranium released from untreated wastewater poses serious threats to aquatic life and human health [1–4]. Because of the high mobility of uranium(VI) ions under oxidizing conditions and the consequent environmental risks, the retardation and immobilization of UO_2^{2+}, which is the most stable uranium specie, have attracted considerable attention [5]. Adsorption is the most facile and promising approach to the treatment of metal ions because of its low cost, high efficiency, and ease of operation [6–10].

Many cost-effective and ecofriendly adsorbents with components ubiquitous in the environment have been fabricated to remove U(VI) or other heavy metal ions, including iron and manganese oxides or hydroxides, organic matter, silicates, phosphate minerals, clay minerals and their modified forms [11–17]. Hydroxyapatite $[Ca_{10}(PO_4)_6(OH)_2$, HAP] is a promising absorbent material because it can effectively immobilize considerable amounts of U(VI) due to its unique physical, chemical, mechanical, and biological properties [18,19]. Because UO_2^{2+} exhibits a strong affinity for PO_4^{3-}, and their product uranyl phosphate generally has low solubility under most conditions, the phosphate group in the HAP structure dominates the transport and transformation behavior of U(VI) in nature [5,20–22]. Autunite, meta-autunite, torbernite, meta-torbernite, and uranyl selenite are the principal uranyl minerals, and phosphorus can be used to promote the mineralization and retention of U(VI) to form stable secondary uranium minerals to mitigate U pollution in the environment [23]. Numerous studies have been performed on the morphology control, synthesis, and adsorption applications of HAP. Zhou et al. [24] reported the synthesis of strontium (Sr) doped hydroxyapatite (HAP) for the increased Cr(VI) adsorption through a hydrothermal method. Ma et al. [25] found that the addition of a certain amount of alendronate in the synthesis of hydroxyapatite helped to form a loose porous nanospheres with low crystallinity, showing good adsorption capacity for Pb^{2+}, Cu^{2+}, and Cd^{2+}. Xiong et al. [26] and Su et al. [27] revealed that HAP with a porous structure can remove U(VI) ions from an aqueous solution. Although such HAP materials can remove pollutants, a key problem is the separation of granular absorbents. Therefore, the rational design of a fast, efficient, and easily separable emergency material is critical for the purification of U(VI)-containing wastewater from the nuclear industry or in case of accidents.

$CoFe_2O_4$, a compound with excellent magnetic and unique properties, has been widely used in applications, such as permanent magnets, storage devices, magnetic recording, electronics, and pigments [28–31]. Compared with Fe_3O_4, $CoFe_2O_4$ has more advantages, such as simpler synthesis, more hydroxyl groups on the surface, and excellent chemical inertness [32–35]. Many researchers have verified the potential of $CoFe_2O_4$ and its magnetic recovery [36–38]. To effectively recover the adsorbed material and maintain the stability of the target material, nano $CoFe_2O_4$ and HAP can be coupled through a facile hydrothermal process. The as-synthesized HAP@$CoFe_2O_4$ exhibits superior composition uniformity, narrow particle size distribution, and can be therefore magnetically separated from reaction systems [39]. To the best of our knowledge, the use of a rationally designed HAP@$CoFe_2O_4$ composite as a sorbent to abate and fix U(VI) has been rarely reported.

This study aims to (I) synthesize easily separable and highly efficient HAP@$CoFe_2O_4$ through a facile hydrothermal method, (II) investigate the interaction between HAP@$CoFe_2O_4$ and U(VI) ions, (III) determine the mechanisms toward U(VI) removal using magnetic HAP@$CoFe_2O_4$ and (IV) develop a preliminary process for the U(VI) removal. The U(VI) adsorption behavior by HAP@$CoFe_2O_4$ was discussed in detail based on the outcomes from equilibrium and kinetics analysis. The findings of this study provide a promising and economic material for the treatment of uranium-contaminated sites.

2. Materials and Methods

2.1. Materials

Citric acid ($C_6H_8O_7$, 99%) was obtained from Tianjin Fuchen Chemical Reagent Factory (Tianjin, China). Salts of calcium (II) ($Ca(NO_3)_2$, 99%), iron (III) ($FeCl_3 \cdot 6H_2O$, 99%), and cobalt (II) ($CoCl_2 \cdot 6H_2O$, 98%) were purchased from Tianjin Fuchen Chemical Reagent Factory (Tianjin, China) and Shanghai Aladdin Biochemical Technology Co., Ltd. (Shanghai, China). Both diammonium hydrogen phosphate ($(NH_4)_2HPO_4$) and aqua ammonia ($NH_3 \cdot H_2O$, 25–28%) were purchased from Sinopharm Chemical Reagent Co., Ltd. (Shanghai, China). Ethylene glycol ($(CH_2OH)_2$, $\geq$99.5%) were purchased from Shanghai Aladdin Biochemical Technology Co., Ltd. A stock solution containing 1 g/L of uranium was prepared by dissolving $UO_2(NO_3)_2 \cdot 6H_2O$ in ultrapure water.

2.2. Synthesis and Characterizations of HAP@$CoFe_2O_4$

Nanosized HAP that was used as a precursor was prepared using a chemical precipitation method described in our previous study [40]. The HAP@$CoFe_2O_4$ composite was hydrothermally fabricated (Text S1 in Supplementary Materials). The initial mole ratio of HAP/$CoFe_2O_4$ is 0.5/1. First, 30 mL of ethylene glycol and 20 mL of ultrapure water were added to the beaker, and subsequently, 0.5 mmol of HAP precursors were added to the beaker and sonicated for 15 min to ensure uniform dispersion. Then, 2 mmol of ferric chloride and 1 mmol of cobalt chloride were added to the beaker and stirred until they were dissolved. Aqua ammonia was titrated to increase the pH of the solution to 10.0 with magnetic stirring at room temperature for 1 h. The mixed solution was transferred into a 100 mL Teflon-lined, stainless-steel autoclave and stored at 180 °C for 24 h to facilitate reactions in an oven. Then, the autoclave was cooled to room temperature naturally, and the precipitates were obtained. The precipitates were centrifuged and washed with alcohol and ultrapure water 3–4 times, and finally, the obtained solid was placed in a constant-temperature oven and dried at 60 °C for 4 h.

The prepared adsorbent samples' phases and crystal structures were determined through an X-ray diffraction (XRD) instrument with Cu K_α radiation (λ = 1.5418 Å) operated at a 2000 W power and 10°/min scanning rate in the 2θ range from 10° to 90°. Additionally, the specific surface area, magnetic properties, and functional groups of the as-prepared samples were analyzed using the Brunauer–Emmett–Teller specific surface analyzer (ASAP 2020, Micromeritics, Norcross, GA, USA), magnetometer (VSM, Dexing, Model: 250), and Bruker Tensor 27 FT-IR spectrometer (Bruker Tensor27, Germany), respectively. The surface morphology and elemental compositions of the HAP@$CoFe_2O_4$ samples were characterized by a transmission electron microscope (JEOLJSM-2100F, Japan) and X-ray photoelectron spectroscopy (Thermo Fisher Scientific ESCALAB250Xi spectrometer, Waltham, MA, USA). The point of zero charge (pzc) of the HAP@$CoFe_2O_4$ was measured by a Zeta potentiometer (NanoBrook Zeta PALS Potential Analyzer, Brookhaven Instruments, Holtsville, NY, USA).

2.3. *U(VI) Removal Experiments*

A series of batch sorption experiments were performed to investigate the U(VI) removal performance of HAP@$CoFe_2O_4$. The solution volume was 50 mL. The influence of pH was investigated in the range of 2.0–6.0. The dosage of the sorbent varied from 0.1 to 0.3 g/L. A dosage of 0.2 g/L of HAP@$CoFe_2O_4$ exhibited excellent U(VI) removal ability and was thus selected as the optimal dosage for the experiments, while the initial U(VI) concentration and contact time varied. The residual U in the solution after filtration was determined using a uranium microanalyzer. The amount of U(VI) ions adsorbed by the adsorbent at time t was estimated using Equation (1):

$$Q_t = \frac{(C_0 - C_t)V}{m} \quad (1)$$

In Equation (1), Q_t, C_0, C_t, V, and m denote the adsorption capacity of HAP@$CoFe_2O_4$ for U(VI) at time t, initial U(VI) concentration (mg/L), U(VI) concentration measured at time t, volume of the reaction system, and the adsorbent's mass, respectively.

3. Results and Discussion

3.1. *Characterization*

The X-ray diffraction (XRD) patterns of the HAP@$CoFe_2O_4$ are illustrated in Figure 1a. The diffraction characteristic peaks of the HAP@$CoFe_2O_4$ located at 25.338°, 31.740°, and 49.463° corresponded to the (201), (211), and (213) planes, respectively, of typical hydroxyapatite (PDF no. 09-0432). The characteristic peaks of the HAP@$CoFe_2O_4$ were located at 29.946°, 35.270°, 42.862°, 56.672°, and 65.422°, which corresponded to the planes of (220), (311), (400), (511), and (531) of $CoFe_2O_4$ (PDF no. 22-1086), respectively. The average microcrystalline size of the $CoFe_2O_4$ calculated by Scherer equation (Equation S1) is 10.60 nm. The HAP@$CoFe_2O_4$ almost preserved the crystal structure of the HAP during the preparation process. The XRD results indicated that a hybrid material composed of HAP and $CoFe_2O_4$ was obtained.

Figure 1b indicated that the HAP@$CoFe_2O_4$ exhibited a specific surface area of 12.161 m^2/g, which suggested that the as-obtained HAP@$CoFe_2O_4$ could be beneficial for interface reactions because of the sufficient active sites. The N_2 adsorption-desorption isotherm of HAP@$CoFe_2O_4$ was a type IV curve and belonged to the H3 type hysteresis loop, which indicated that HAP@$CoFe_2O_4$ is a flat slit structure. The average pore size of the product calculated according to the Barrett–Joyner–Halenda (BJH) method was 15.292 nm.

A broad peak near 3448.3 cm^{-1}, which can be attributed to the stretching vibration of the –OH groups, was observed in the Fourier-transform infrared (FT-IR) spectra of the HAP@$CoFe_2O_4$ (Figure 1c). This spectrum, which exhibited bands at 2360.99 cm^{-1} and between 1476.14 and 1383.85 cm^{-1} could be explained by the adhesion of CO_2 from the atmosphere during the synthesis of the HAP in highly alkaline conditions [41–43]. The peak at 1638.06 cm^{-1} was related to the –COO group, possibly because of the citric acid added during synthesis. The peaks at approximately 572.89 and 602.94 cm^{-1} represented asymmetric bending vibration, and the peak at 1044.82 cm^{-1} represented symmetric stretching vibration; these were attributed to the tetrahedron PO_4^{3-} groups of HAP. For FT-IR, Co/Fe–O bonds are typical in the 580–598 banding [44–46]. In this case, the peak of the Co/Fe–O bond may have overlapped with the typical vibration of the PO_4^{3-} group of HAP. The charge on the surface of the material affects its adsorption performance. As displayed in Figure 1d, the pHpzc of the HAP@$CoFe_2O_4$ was approximately 2.8, even in the solution with 15 ppm U(VI).

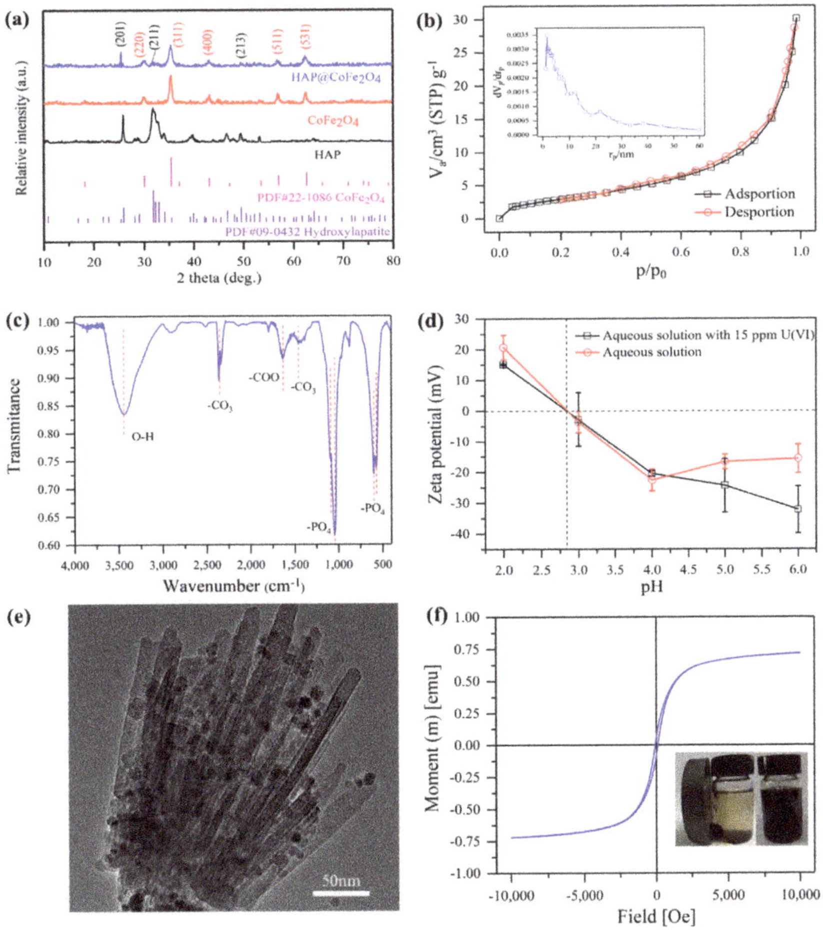

Figure 1. XRD patterns (**a**) of HAP, $CoFe_2O_4$ as well as HAP@$CoFe_2O_4$, (**b**) the nitrogen adsorption-desorption isotherm (calculated from the Brunauer–Emmett–Teller [BET method] and pore size distribution (calculated from the BJH method), (**c**) Fourier transform infrared [FT-IR] spectrum, (**d**) zeta-potential and (**e**) transmission electron microscopy (TEM) image, and (**f**) the hysteresis loop of HAP@$CoFe_2O_4$ (inset of Figure 1f displays the separation of HAP@$CoFe_2O_4$ with and without an additional magnet).

Because the morphology including size and shape of nanomaterials considerably affected the physical and chemical properties of the product, the morphology of the

$HAP@CoFe_2O_4$ was investigated through transmission electron microscopy (TEM). As presented in Figure 1e, $HAP@CoFe_2O_4$ morphology showed a typical bundle-like structure with numerous $CoFe_2O_4$ nanoparticles. $CoFe_2O_4$ was evenly dispersed and tightly adhered to the surface of the HAP. The results from hysteresis loop analysis showed that the as-prepared $HAP@CoFe_2O_4$ possessed excellent magnetic properties, suggesting that $HAP@CoFe_2O_4$ could be easily separated using an extra magnet within a short time after the reaction (Figure 1f). Compared with other separation technology (e.g., filtering and centrifugal separation) applying an external magnetic field and using magnetic material(s) allows the adsorbent to be more quickly recovered from the reaction system and thus saving time and cost. These results revealed that the $HAP@CoFe_2O_4$ composites were successfully synthesized and can be applied in the treatment process of eliminating U(VI) from the aqueous solution.

3.2. *U(VI) Removal by Magnetic $HAP@CoFe_2O_4$*

Dosage experiments were conducted for confirming the optimal dosage of adsorbents. As shown in Figure 2a, with an increase in the dosage, the adsorption sites increased in the solution, which were more conducive to the binding of UO_2^{2+}, leading to the improvement of removal efficiency. The $HAP@CoFe_2O_4$ maintained the high efficiency adsorption of U(VI). The HAP considerably dominated the adsorption process, whereas the adsorption contribution of the $CoFe_2O_4$ to U(VI) was insignificant.

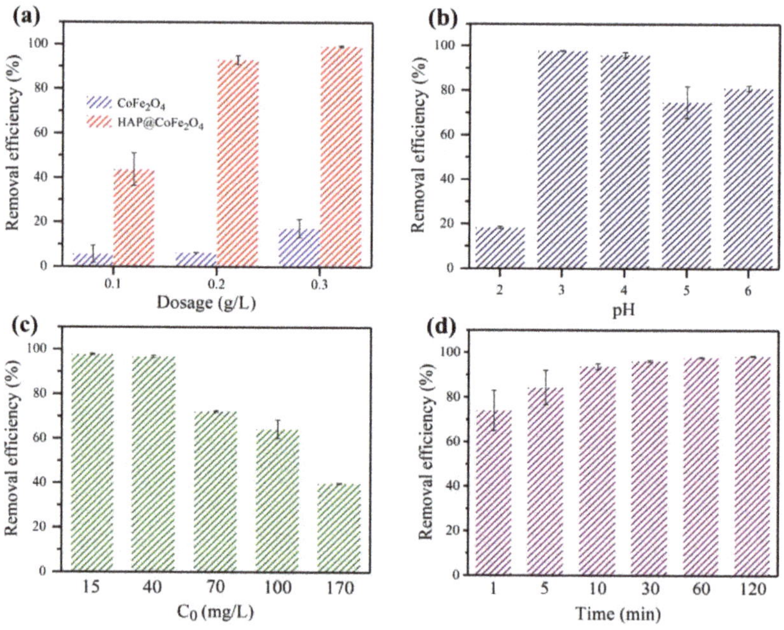

Figure 2. (**a**) Effect of dosage on the U(VI) adsorption of $CoFe_2O_4$ and $HAP@CoFe_2O_4$ at various dosages (Reaction conditions: initial U(VI) concentration = approximately 15 mg/L, pH = 3.0, adsorbent dosage = 0.1–0.3 g/L, T = 298 K); (**b**) Removal rate of U(VI) by magnetic $HAP@CoFe_2O_4$ under various pH (Reaction conditions: initial U(VI) concentration = approximately 15 mg/L, pH = 2.0–6.0, adsorbent dosage = 0.2 g/L, T = 298 K); (**c**) Effect of initial U(VI) concentration (reaction conditions: initial U(VI) concentration = 15–170 mg/L, pH = 3.0, adsorbent dosage = 0.2 g/L, T = 298 K); (**d**) Effect of contact time (reaction conditions: initial U(VI) concentration = approximately 15 mg/L, t = 0–120 min, pH = 3.0, adsorbent dosage = 0.2 g/L, T = 298 K).

Because pH is a critical factor affecting the interaction between HAP@$CoFe_2O_4$ particles and U(VI), experiments were performed to investigate the effect of the solution pH on U(VI) adsorption by changing the pH value (2.0–6.0) of the solution (Figure 2b). When the pH value was less than 3.0, U(VI) was mainly present in the form of uranyl ions (UO_2^{2+}) in the solution [47]. Moreover, HAP@$CoFe_2O_4$ is not stable under the conditions with a pH below 3.0 because the HAP might dissolve. A higher pH reduced protonation in the solution, and the U(VI) changed. At a low pH, the high H^+ concentration of the solution competed intensely with the UO_2^{2+}, which limited the UO_2^{2+} adsorption by the HAP@ $CoFe_2O_4$. At pH = 3.0, the adsorption capacity considerably increased and reached the maximum. As the pH increased, the adsorption of U(VI) by HAP@$CoFe_2O_4$ weakened marginally. When the pH was 3.0 or higher, the charge of as-fabricated HAP@$CoFe_2O_4$ composite decreased and became negative [19]. Meanwhile, under high pH conditions, the UO_2^{2+} began to hydrolyze and form new species, such as $(UO_2)_2(OH)_2^{2+}$, $(UO_2)_3(OH)^{5+}$, $(UO_2)_4(OH)^{7+}$, $UO_2(OH)^+$, and $(UO_2)_3(OH)^{7-}$. The affinity of these species is generally lower than that of the UO_2^{2+} [48]. Therefore, the HAP@$CoFe_2O_4$ composite does not easily absorb these species. Nevertheless, the HAP@$CoFe_2O_4$ particles still maintained high adsorption of U(VI), which indicated that the composite material could efficiently adsorb U(VI) in a wide pH range, making it capable of removing uranium from contaminated wastewater under various conditions.

The reaction system comprising 0.2 g/L of sorbent and approximately 15 mg/L U(VI) exhibited the highest removal efficiency (Figure 2c). Figure 2d indicated that the adsorption efficiency of the HAP@$CoFe_2O_4$ was as high as 93% in 10 min. Initially, sufficient active surface sites were present, which resulted in a high adsorption efficiency and intensive interaction between the HAP@$CoFe_2O_4$ and U(VI). On prolonging the contact time, U(VI) ions occupied most of the active sites of the HAP@$CoFe_2O_4$, and the adsorption reached equilibrium at 30 min. Even when the time was increased, stable adsorption efficiency was maintained.

3.3. Adsorption Isotherms and Kinetics

To investigate the adsorption behavior, Langmuir (Equation (S2)) and Freundlich (Equation (S3)) isotherm models were used to simulate the adsorption process of the HAP@$CoFe_2O_4$ for U(VI). The correlation coefficients (R^2) of the HAP@$CoFe_2O_4$ were 0.9390 and 0.8663 for the Langmuir and Freundlich isotherm models, respectively (Figure 3a and Table 1). This result indicated that the adsorption process of U(VI) onto HAP@$CoFe_2O_4$ could be better interpreted using the Langmuir isotherm model than the Freundlich isotherm. On the basis of the Langmuir isotherm model, the Q_{max} of the HAP@$CoFe_2O_4$ for U(VI) was 338 mg/g, which was almost consistent with the experimental results. The two lower correlation coefficients may be due to the combined effects of physical adsorption and surface precipitation. In the process of the HAP@$CoFe_2O_4$ absorbing uranium, the dominant adsorption mechanism is related to the initial uranium concentration. Surface adsorption may contribute considerably to the total uranium adsorption at low uranium concentrations [49].

Table 1. Freundlich and Langmuir isotherm constants for the U(VI) adsorption onto the as-fabricated HAP@$CoFe_2O_4$ composite.

Materials	Langmuir Isotherm			Freundlich Isotherm		
	q_m(mg/g)	K_L(L/mg)	R^2	K_F(L/mg)	1/n	R^2
HAP@$CoFe_2O_4$	338	0.91	0.9981	145.64	0.20	0.8465

The fitting constants for pseudo-first-order (Equation (S4)) and pseudo-second-order (Equation (S5)) kinetic models are displayed in Figure 3b and Table 2. The pseudo-second-order kinetic model exhibited greater R^2 (0.9922) than that of the pseudo-first-order kinetic model (0.9790), which indicated that the kinetic process of the HAP@$CoFe_2O_4$ for U(VI) was well described by the pseudo-second-order kinetic model. Furthermore, the adsorption capacity (67.61 mg/g) calculated using the pseudo-second-order kinetic model was consistent with the experimental result (68.85 mg/g), which indicated that the adsorption process of U(VI) by the HAP@$CoFe_2O_4$ was controlled by chemical adsorption.

Table 2. Kinetic parameters for the adsorption of U(VI) onto HAP@$CoFe_2O_4$.

Kinetics Model	Parameter	Values
	k_1 (L mg^{-1})	1.539
Pseudo-first-order	R^2	0.9790
	q_e	65.83
	k_2 (L mg^{-1})	0.44
Pseudo-second-order	R^2	0.9922
	q_e	67.61

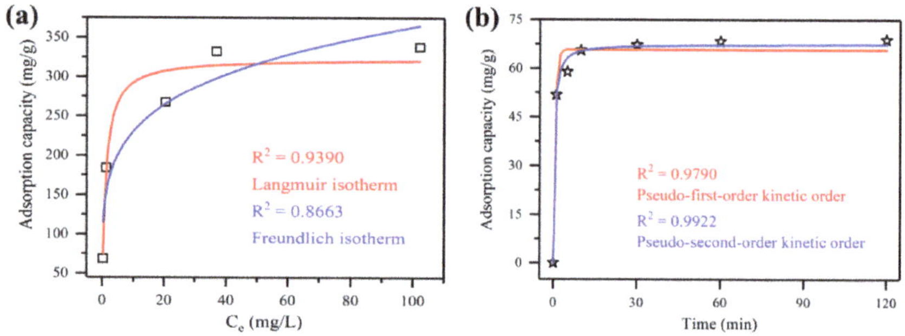

Figure 3. Isotherms and kinetics of U(VI) adsorption onto the as-fabricated HAP@$CoFe_2O_4$ composite: (**a**) Langmuir isotherm and Freundlich isotherm models, (**b**) pseudo-first-order and pseudo-second-order kinetics.

3.4. U(VI) Removal Mechanisms

XRD, FT-IR, and XPS were performed to understand the adsorption mechanism of U(VI) by the HAP@$CoFe_2O_4$ composite. The XRD patterns (Figure 4a) revealed that a uranium-containing compound was formed. After adsorption, numerous diffraction peaks were observed at 2θ = 10.523°, 18.009°, 20.903°, 24.738°, 25.576°, 27.724°, 35.880°, 40.972°, and 44.581°, which corresponded to the crystal planes of (001), (110), (111), (102), (200), (201), (212), (310), and (302) of the meta-autunite 9A [($Ca(UO_2)_2(PO_4)_2(H_2O)_6$, PDF no. 72-2117]. These results were in good agreement with the standard pattern of autunite, which could be explained by the dissolution of the HAP and then precipitation of the autunite on the surface of the HAP [50]. The presence of autunite indicated a successful fixation of uranium on the adsorbent. The XRD pattern also illustrated the presence of $CoFe_2O_4$ in the adsorbed material, which proved that the adsorbent was still magnetic after the reaction.

FT-IR analysis (Figure 4b) revealed that both the adsorbents before and after the reaction exhibited a distinct single and wide peak at approximately 3445 cm^{-1}, which was attributed to the –OH groups. This result indicated that the existence of –OH groups in the HAP@$CoFe_2O_4$ is a typical feature of HAP [51]. The peak at 1616 cm^{-1} is characteristic of –COO, which could be attributed to the absorption of CO_2 in air during the test [40,41]. The peaks at 1044.82 and 602.94 cm^{-1}, (before adsorption) and 1007.83 and 600.90 cm^{-1} (after adsorption) were attributable to the PO_4^{3-} group in the adsorbent [40,52], and the position and intensity of the characteristic peaks in the adsorbent changed considerably before and after the reaction. The results revealed that the PO_4^{3-} in the HAP@$CoFe_2O_4$ is involved in the reaction. After the adsorption reaction, a new peak appeared at 546.327 cm^{-1}, which corresponded to Co/Fe–O bonding [53,54]. A new absorption peak of the HAP@$CoFe_2O_4$ after the reaction appeared at 911 cm^{-1}, which was characteristic of UO_2^{2+}, providing strong evidence of uranium loading. The essence of removing U(VI) by using the HAP@$CoFe_2O_4$ is the chemical reaction between UO_2^{2+} and the HAP. The UO_2^{2+} was adsorbed and eventually incorporated into the Ca^{2+}–PO_4^{3-}–OH according to Equation (2):

$$Ca^{2+} + 2PO_4{}^{3-} + H_2O + 2UO_2^{2+} \rightarrow Ca(UO_2)_2(PO_4)_2(H_2O)_6 \quad (2)$$

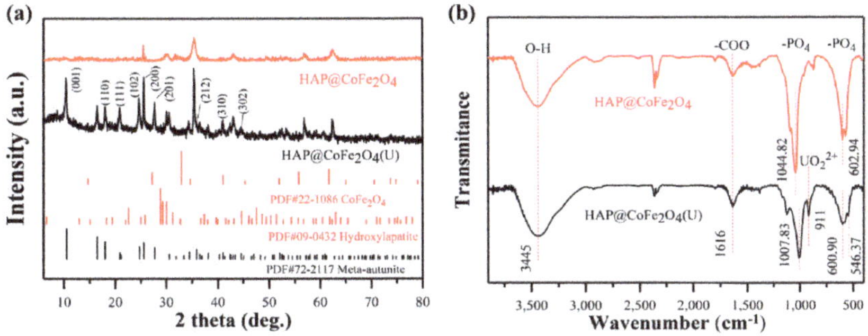

Figure 4. X-ray diffraction patterns (**a**) and FT-IR spectrum (**b**) of HAP@$CoFe_2O_4$ and U(VI)-loaded HAP@$CoFe_2O_4$.

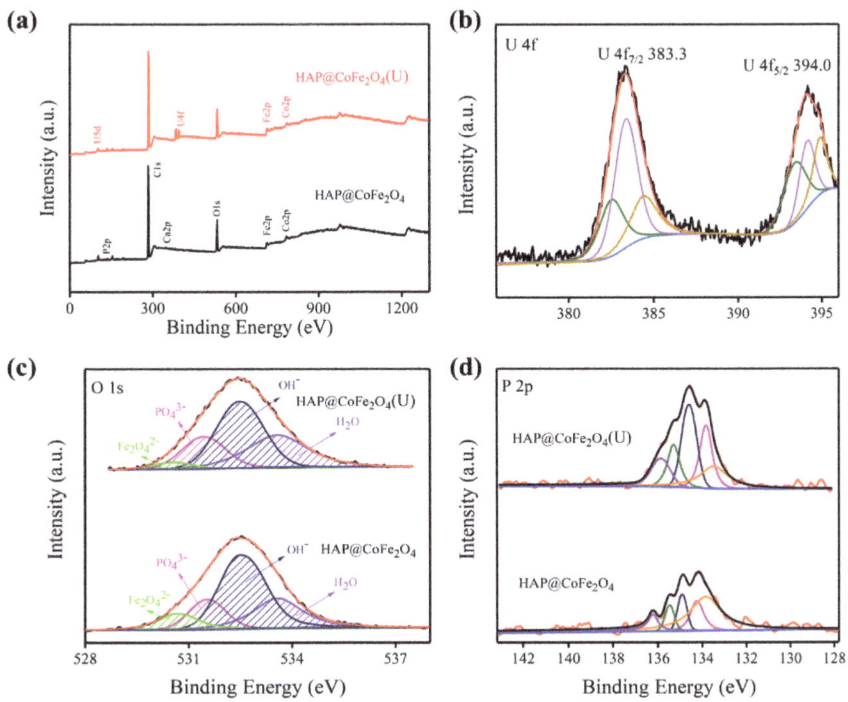

Figure 5. X-ray spectroscopy (XPS) full-survey spectra (**a**) as well as the (**b**) U 4f, (**c**) O 1s, (**d**) P 2p.

The XPS survey spectra (Figure 5a) revealed the peaks of C 1s, O 1s, Ca 2p, P 2p, Fe 2p, and Co 2p, which indicated that the $CoFe_2O_4$ was successfully loaded on the HAP. A new strong double peak for the antisymmetric vibration of $[O{=}UVI{=}O]^{2+}$ appeared in the XPS spectra of the HAP@$CoFe_2O_4$ reacted with U(VI), which was consistent with the EDS elemental mapping results (Figure S1), suggesting that uranium was successfully adsorbed on the surface of the material. The high resolution of the U 4f spectrum (Figure 5b) can be well fitted at binding energies of 383.3 (U $4f_{7/2}$) and 394.0 eV (U $4f_{5/2}$), indicating that the valence of uranium did not change during the adsorption process [55,56]. The Co spectrum (Figure S2a) revealed two peaks at binding energies 782.58 and 786.84 eV, which were attributed to Co $2p_{3/2}$ and its shake-up satellites, respectively, whereas peaks at higher binding energies (~797.81 and 804.44 eV) corresponded to Co $2p_{1/2}$ and their shake-up satellites, respectively [57]. The Fe 2p spectrum (Figure S2b) displayed into two peaks at binding energies of 712.13 and 726.02 eV, which respectively correspond to Fe $2p_{3/2}$ and Fe $2p_{1/2}$, evidencing the existence of Fe^{3+} [58–60]. As displayed in Figure 5d, after reaction with U(VI), the P 2p peaks shifted slightly towards higher binding energy, which indicated that the presence of phosphorus considerably affected the adsorption of uranium [19]. The spectra of O 1s (Figure 5c) can be divided into four peaks, namely, anion oxide (530.64 eV, $Fe_2O_4^{2-}$), phosphate group (531.53 eV, PO_4^{3-}), hydroxyl groups (532.50 eV, −OH), and adsorbed groups (H_2O, 533.57 eV), respectively. The peak area ratio of –OH varied considerably (from 50.33% to 40.26%) before and after U(VI) adsorption, which evidenced that the surface hydroxyl groups (–OH) played a key role in the adsorption of U(VI) through the sharing of electrons to form U–O bonds [61,62].

3.5. Preliminary Evaluation of the U(VI) Removal Process

The suggested schematic of the removal of the U(VI) from wastewater using HAP@$CoFe_2O_4$ is presented in Figure 6, 93.7%, where the U(VI) was removed within 10 min. As a magnetic separable material, the advantage of using HAP@$CoFe_2O_4$ is that the adsorption/separation is a simple and efficient process, U(VI) can be effectively removed from wastewater and the U(VI) loaded can be easily separated from water using magnetic field, which is beneficial for the minimization of secondary waste. The process using HAP@$CoFe_2O_4$ as an adsorbent for the removal of U(VI) has economic feasibility. The use of HAP@$CoFe_2O_4$ for the U(VI) removal/separation could result in a significant reduction of costs due to easier separation of the solid phase and energy savings. HAP@$CoFe_2O_4$ can be considered to be an effective and environmental material for the elimination of U(VI) from mining wastewater.

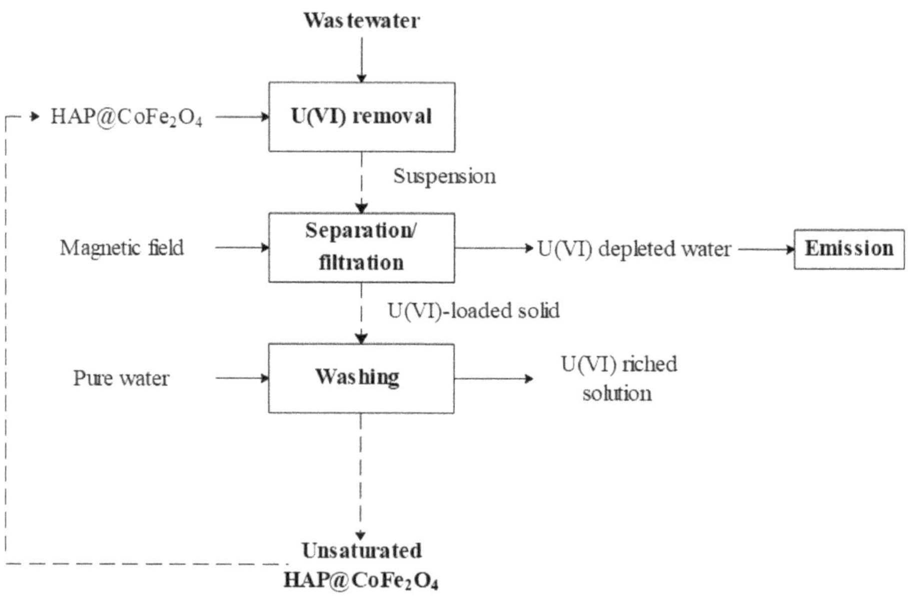

Figure 6. Schematic of the removal of U(VI) using HAP@$CoFe_2O_4$. The dashed lines present the solid phases.

4. Conclusions

This study developed a promising and easily separable composite material, HAP@$CoFe_2O_4$, for removing uranium ions from radioactive wastewater. A U(VI) removal efficiency of 93.7% can be achieved by the HAP@$CoFe_2O_4$ in 10 min, and the maximum adsorption capacity was 338 mg/g. Therefore, HAP@$CoFe_2O_4$ can be used as an emergency material to treat uranium-containing radioactive wastewater generated in nuclear accidents. The UO_2^{2+} were transferred to the HAP@$CoFe_2O_4$ by reacting with dissolved calcium and phosphate, thereby forming a more stable compound-autunite [$Ca(UO_2)_2(PO_4)_2(H_2O)_6$]. The adsorption and incorporation are the main ways for U(VI) removal. Simulation of the interaction process between the HAP@$CoFe_2O_4$ and UO_2^{2+} by using adsorption isotherm models and adsorption kinetic models revealed that the adsorption process followed the Langmuir isotherm and pseudo-second-order dynamic model, which indicated that the adsorption of U(VI) by the HAP@$CoFe_2O_4$ was mainly controlled by chemical adsorption. Moreover, the magnetic property of the as-prepared HAP@$CoFe_2O_4$ was measured and found to be used for magnetic separation. Therefore, the developed process using HAP@$CoFe_2O_4$ as an adsorbent is promising to mitigate uranium pollution generated in a nuclear accident, and a pilot scale test for process optimization is planned in the future.

Supplementary Materials: The following are available online at https://www.mdpi.com/article/10.3390/pr9111927/s1, Figure S1: EDS image of U(VI) loaded HAP@$CoFe_2O_4$. Figure S2: (a) Co 2p, and (b) Fe 2p XPS spectra of the HAP@$CoFe_2O_4$ composite before and after the removal of U(VI).

Author Contributions: Conceptualization, T.O. and Q.S.; Methodology, T.O. and Q.S.; Writing-Original Draft, T.O. and H.P.; Resources, M.S. and D.C.; Funding acquisition, M.S., N.C. and D.C.; Formal analysis, T.O. and J.T.; Writing—Review & Editing, T.O., M.S. and J.T.; Supervision, M.S. and D.C. All authors have read and agreed to the published version of the manuscript.

Funding: This research was funded by the National Natural Science Foundation of China (22076034, 41877290), the Natural Science Foundation of Guangdong Province of China (2021A1515010067) and Foundation of Department of Education of Guangdong Province of China (2018KTSCX176) and Guangzhou University Intramural Scientific Research Project (YG2020012).

Institutional Review Board Statement: Not applicable.

Informed Consent Statement: Not applicable.

Data Availability Statement: The data presented in this study are available in this article.

Acknowledgments: This work was supported by the National Natural Science Foundation of China (22076034, 41877290), the Natural Science Foundation of Guangdong Province of China (2021A1515010067), and Foundation of Department of Education of Guangdong Province of China (2018KTSCX176) and Guangzhou University Intramural Scientific Research Project (YG2020012).

Conflicts of Interest: The authors declare no conflict of interest.

References

1. Yang, Y.; Saiers, J.E.; Barnett, M.O. Impact of interactions between natural organic matter and metal oxides on the desorption kinetics of uranium from heterogeneous colloidal suspensions. *Environ. Sci. Technol.* **2013**, *47*, 2661–2669. [CrossRef]
2. Lu, S.H.; Zhu, K.R.; Hayat, T.; Alharbi, N.S.; Chen, C.L.; Song, G.; Chen, D.Y.; Sun, Y.B. Influence of carbonate on sequestration of U(VI) on perovskite. *J. Hazard. Mater.* **2019**, *364*, 100–107. [CrossRef]
3. Pan, Q.J.; Odoh, S.O.; Asaduzzaman, A.M.; Schreckenbach, G. Adsorption of uranyl species onto the rutile (110) surface: A periodic DFT study. *Chemistry* **2012**, *18*, 1458–1466. [CrossRef] [PubMed]
4. Kong, L.J.; Zhang, H.M.; Ji, W.; Shih, K.M.; Su, M.H.; Diao, Z.H.; Xu, R.M.; Hou, L.A.; Song, G.; Chen, D.Y. Recovery of phosphorus rich krill shell biowaste for uranium immobilization: A study of sorption behavior, surface reaction, and phase transformation. *Environ. Pollut.* **2018**, *243*, 630–636. [CrossRef] [PubMed]
5. Niu, Z.W.; Wei, X.Y.; Qiang, S.R.; Wu, H.Y.; Pan, D.Q.; Wu, W.S.; Fan, Q.H. Spectroscopic studies on U(VI) incorporation into $CaCO_3$: Effects of aging time and U(VI) concentration. *Chemosphere* **2019**, *220*, 1100–1107. [CrossRef] [PubMed]
6. Tang, J.F.; Su, M.H.; Wu, Q.H.; Wei, L.Z.; Wang, N.N.; Xiao, E.Z.; Zhang, H.G.; Wei, Y.J.; Liu, Y.K.; Ekberg, C.; et al. Highly efficient recovery and clean-up of four heavy metals from MSWI fly ash by integrating leaching, selective extraction and adsorption. *J. Clean. Prod.* **2019**, *234*, 139–149. [CrossRef]

7. Wang, D.; Xu, Y.B.; Xiao, D.F.; Qiao, Q.G.; Yin, P.; Yang, Z.L.; Li, J.X.; Winchester, W.; Wang, Z.; Hayat, T. Ultra-thin iron phosphate nanosheets for high efficient U(VI) adsorption. *J. Hazard. Mater.* **2019**, *371*, 83–93. [CrossRef]
8. Li, X.; Liu, Y.; Zhang, C.L.; Wen, T.; Zhuang, L.; Wang, X.X.; Song, G.; Chen, D.Y.; Ai, Y.J.; Hayat, T.; et al. Porous Fe_2O_3 microcubes derived from metal organic frameworks for efficient elimination of organic pollutants and heavy metal ions. *Chem. Eng. J.* **2018**, *336*, 241–252. [CrossRef]
9. Xia, X.; Shen, J.; Cao, F.; Wang, C.J.; Tang, M.; Zhang, Q.Y.; Wei, S.S. A facile synthesis of hydroxyapatite for effective removal strontium ion. *J. Hazard. Mater.* **2019**, *368*, 326–335. [CrossRef]
10. Liu, X.; Huang, Y.; Duan, S.; Wang, Y.; Li, J.; Chen, Y.; Hayat, T.; Wang, X. Graphene oxides with different oxidation degrees for Co(II) ion pollution management. *Chem. Eng. J.* **2016**, *302*, 763–772. [CrossRef]
11. Zhang, H.M.; Ruan, Y.; Liang, A.P.; Shih, K.M.; Diao, Z.H.; Su, M.H.; Hou, L.A.; Chen, D.Y.; Lu, H.; Kong, L.J. Carbothermal reduction for preparing nZVI/BC to extract uranium: Insight into the iron species dependent uranium adsorption behavior. *J. Clean. Prod.* **2019**, *239*, 117873. [CrossRef]
12. Wu, W.Y.; Chen, D.Y.; Li, J.W.; Su, M.H.; Chen, N. Enhanced adsorption of uranium by modified red muds: Adsorption behavior study. *Environ. Sci. Pollut. Res.* **2018**, *25*, 18096–18108. [CrossRef]
13. Su, M.H.; Liao, C.Z.; Chan, T.S.; Shih, K.M.; Xiao, T.F.; Chen, D.Y.; Kong, L.J.; Song, G. Incorporation of cadmium and nickel into ferrite spinel solid solution: X-ray diffraction and X-ray absorption fine structure analyses. *Environ. Sci. Technol.* **2018**, *52*, 775–782. [CrossRef] [PubMed]
14. Wu, H.Y.; Li, P.; Pan, D.Q.; Yin, Z.X.; Fang, Q.H.; Wu, W.S. Interactions between silicon oxide nanoparticles (SONPs) and U(VI) contaminations: Effects of pH, temperature and natural organic matters. *Plos One* **2016**, *11*, e0149632. [CrossRef] [PubMed]
15. Li, D.E.; Kaplan, D.I.; Chang, H.S.; Seaman, J.C.; Jaffe, P.R.; van Groos, P.K.; Scheckel, K.G.; Segre, C.U.; Chen, N.; Jiang, D.T.; et al. Spectroscopic evidence of uranium immobilization in acidic wetlands by natural organic matter and plant roots. *Environ. Sci. Technol.* **2015**, *49*, 2823–2832. [CrossRef] [PubMed]
16. Yuvaraja, G.; Su, M.; Chen, D.Y.; Pang, Y.; Kong, L.J.; Subbaiah, M.V.; Wen, J.C.; Reddy, G.M. Impregnation of magnetic—Momordica charantia leaf powder into chitosan for the removal of U(VI) from aqueous and polluted wastewater. *Int. J. Biol. Macromol.* **2020**, *149*, 127–139. [CrossRef]
17. Krajňák, A.; Viglašová, E.; Galamboš, M.; Krivosudský, L. Application of HDTMA-intercalated bentonites in water waste treatment for U(VI) removal. *J. Radioanal. Nucl. Chem.* **2017**, *314*, 2489–2499. [CrossRef]
18. Wen, H.; Pan, Z.Z.; Giammar, D.; Li, L. Enhanced uranium immobilization by phosphate amendment under variable geochemical and flow conditions: Insights from reactive transport modeling. *Environ. Sci. Technol.* **2018**, *52*, 5841–5850. [CrossRef] [PubMed]
19. Zheng, N.C.; Yin, L.Y.; Su, M.H.; Liu, Z.Q.; Tsang, D.C.W.; Chen, D.Y. Synthesis of shape and structure-dependent hydroxyapatite nanostructures as a superior adsorbent for removal of U(VI). *Chem. Eng. J.* **2020**, *384*, 123262. [CrossRef]
20. Kong, L.J.; Ruan, Y.; Zheng, Q.Y.; Su, M.H.; Diao, Z.H.; Chen, D.Y.; Hou, L.A.; Chang, X.Y.; Shih, K.M. Uranium extraction using hydroxyapatite recovered from phosphorus containing wastewater. *J. Hazard. Mater.* **2020**, *382*, 120784. [CrossRef]
21. Chen, B.D.; Wang, J.; Kong, L.J.; Mai, X.X.; Zheng, N.C.; Zhong, Q.H.; Liang, J.Y.; Chen, D.Y. Adsorption of uranium from uranium mine contaminated water using phosphate rock apatite (PRA): Isotherm, kinetic and characterization studies. *Colloids Surf. A* **2017**, *520*, 612–621. [CrossRef]
22. Cumberland, S.A.; Douglas, G.; Grice, K.; Moreau, J.W. Uranium mobility in organic matter-rich sediments: A review of geological and geochemical processes. *Earth-Sci. Rev.* **2016**, *159*, 160–185. [CrossRef]
23. Burns, P.C. U^{6+} minerals and inorganic compounds: Insights into an expanded structural hierarchy of crystal structures. *Can. Mineral.* **2005**, *43*, 1839–1894. [CrossRef]
24. Zhou, Y.; Li, W.; Jiang, X.; Sun, Y.; Yang, H.; Liu, Q.; Cao, Y.; Zhang, Y.; Cheng, H. Synthesis of strontium (Sr) doped hydroxyapatite (HAp) nanorods for enhanced adsorption of Cr (VI) ions from wastewater. *Ceram. Int.* **2021**, *47*, 16730–16736. [CrossRef]
25. Ma, J.; Xia, M.; Zhu, S.; Wang, F. A new alendronate doped HAP nanomaterial for Pb^{2+}, Cu^{2+} and Cd^{2+} effect absorption. *J. Hazard. Mater.* **2020**, *400*, 123143. [CrossRef]
26. Xiong, T.; Li, Q.; Liao, J.; Zhang, Y.; Zhu, W. Highly enhanced adsorption performance to uranium(VI) by facile synthesized hydroxyapatite aerogel. *J. Hazard. Mater.* **2022**, *423*, 127184. [CrossRef] [PubMed]
27. Su, M.H.; Tsang, D.C.W.; Ren, X.Y.; Shi, Q.P.; Tang, J.F.; Zhang, H.G.; Kong, L.J.; Hou, L.A.; Song, G.; Chen, D.Y. Removal of U(VI) from nuclear mining effluent by porous hydroxyapatite: Evaluation on characteristics, mechanisms and performance. *Environ. Pollut.* **2019**, *254*, 112891. [CrossRef]
28. Ghaffarian, F.; Ghasemzadeh, M.A.; Aghaei, S.S. An efficient synthesis of some new curcumin based pyrano[2,3-d] pyrimidine-2,4(3H)-dione derivatives using $CoFe_2O_4$@OCMC@Cu(BDC) as a novel and recoverable catalyst. *J. Mol. Struct.* **2019**, *1186*, 204–211. [CrossRef]
29. Dey, C.; Baishya, K.; Ghosh, A.; Goswami, M.M.; Ghosh, A.; Mandal, K. Improvement of drug delivery by hyperthermia treatment using magnetic cubic cobalt ferrite nanoparticles. *J. Magn. Magn. Mater.* **2017**, *427*, 168–174. [CrossRef]
30. Tartaj, P.; Morales, M.D.; Veintemillas-Verdaguer, S.; Gonzalez-Carreno, T.; Serna, C.J. The preparation of magnetic nanoparticles for applications in biomedicine. *J. Phys. D: Appl. Phys.* **2003**, *36*, R182–R197. [CrossRef]
31. Chandra, G.; Srivastava, R.C.; Reddy, V.R.; Agrawal, H.M. Effect of sintering temperature on magnetization and Mossbauer parameters of cobalt ferrite nanoparticles. *J. Magn. Magn. Mater.* **2017**, *427*, 225–229. [CrossRef]

32. Zhu, H.; Shen, Y.; Wang, Q.; Chen, K.; Wang, X.; Zhang, G.; Yang, J.; Guo, Y.; Bai, R. Highly promoted removal of Hg(II) with magnetic $CoFe_2O_4@SiO_2$ core–shell nanoparticles modified by thiol groups. *RSC Adv.* **2017**, *7*, 39204–39215. [CrossRef]
33. Coutinho, T.C.; Malafatti, J.O.D.; Paris, E.C.; Tardioli, P.W.; Farinas, C.S. Hydroxyapatite-$CoFe_2O_4$ magnetic nanoparticle composites for industrial enzyme immobilization, use, and recovery. *ACS Appl. Nano Mater.* **2020**, *3*, 12334–12345. [CrossRef]
34. Ansari, S.M.; Ghosh, K.C.; Devan, R.S.; Sen, D.; Sastry, P.U.; Kolekar, Y.D.; Ramana, C.V. Eco-friendly synthesis, crystal chemistry, and magnetic properties of manganese-substituted $CoFe_2O_4$ nanoparticles. *ACS Omega* **2020**, *5*, 19315–19330. [CrossRef] [PubMed]
35. Cao, Z.; Zuo, C. Direct synthesis of magnetic $CoFe_2O_4$ nanoparticles as recyclable photo-fenton catalysts for removing organic dyes. *ACS Omega* **2020**, *5*, 22614–22620. [CrossRef] [PubMed]
36. Dos Santos, J.M.N.; Pereira, C.R.; Pinto, L.A.A.; Frantz, T.; Lima, E.C.; Foletto, E.L.; Dotto, G.L. Synthesis of a novel $CoFe_2O_4$/chitosan magnetic composite for fast adsorption of indigotine blue dye. *Carbohyd. Polym.* **2019**, *217*, 6–14. [CrossRef]
37. Foroughi, F.; Hassanzadeh-Tabrizi, S.A.; Amighian, J.; Saffar-Teluri, A. A designed magnetic $CoFe_2O_4$–hydroxyapatite core–shell nanocomposite for Zn(II) removal with high efficiency. *Ceram. Int.* **2015**, *41*, 6844–6850. [CrossRef]
38. Elizalde, M.L.M.; Acha, C.; Molina, F.V.; Antonel, P.S. Composites of poly(3,4-ethylenedioxythiophene) and $CoFe_2O_4$ nanoparticles: Composition influence on structural, electrical, and magnetic properties. *J. Phys. Chem. C* **2020**, *124*, 6884–6895. [CrossRef]
39. Manikandan, V.; Mirzaei, A.; Vigneselvan, S.; Kavita, S.; Mane, R.S.; Kim, S.S.; Chandrasekaran, J. Role of ruthenium in the dielectric, magnetic properties of nickel ferrite (Ru-$NiFe_2O_4$) nanoparticles and their application in hydrogen sensors. *ACS Omega* **2019**, *4*, 12919–12926. [CrossRef]
40. Shi, Q.P.; Su, M.H.; Yuvaraja, G.; Tang, J.F.; Kong, L.J.; Chen, D.Y. Development of highly efficient bundle-like hydroxyapatite towards abatement of aqueous U(VI) ions: Mechanism and economic assessment. *J. Hazard. Mater.* **2020**, *394*, 122550. [CrossRef] [PubMed]
41. Champeau, M.; Thomassin, J.M.; Jerome, C.; Tassaing, T. In situ FT-IR micro-spectroscopy to investigate polymeric fibers under supercritical carbon dioxide: CO2 sorption and swelling measurements. *J. Supercrit. Fluids* **2014**, *90*, 44–52. [CrossRef]
42. Tsuru, K.; Yoshimoto, A.; Kanazawa, M.; Sugiura, Y.; Nakashima, Y.; Ishikawa, K. Fabrication of carbonate apatite block through a dissolution–precipitation reaction using calcium hydrogen phosphate dihydrate block as a precursor. *Materials* **2017**, *10*, 374. [CrossRef]
43. Tchoffo, R.; Ngassaa, G.B.P.; Tonlé, I.K.; Ngameni, E. Electroanalysis of diquat using a glassy carbon electrode modified with natural hydroxyapatite and β-cyclodextrin composite. *Talanta* **2021**, *222*, 121550. [CrossRef] [PubMed]
44. Crocella, V.; Cavani, F.; Cerrato, G.; Cocchi, S.; Comito, M.; Magnacca, G.; Morterra, C. On the role of morphology of $CoFeO_4$ spinel in methanol anaerobic oxidation. *J. Phys. Chem. C* **2012**, *116*, 14998–15009. [CrossRef]
45. Debnath, B.; Bansal, A.; Salunke, H.G.; Sadhu, A.; Bhattacharyya, S. Enhancement of magnetization through interface exchange interactions of confined NiO nanoparticles within the mesopores of $CoFe_2O_4$. *J. Phys. Chem. C* **2016**, *120*, 5523–5533. [CrossRef]
46. He, Q.M.; Rui, K.; Che, C.H.; Yang, J.H.; Wen, Z.Y. Interconnected $CoFe_2O_4$–polypyrrole nanotubes as anode materials for high performance sodium ion batteries. *Acs Appl. Mater. Inter.* **2017**, *9*, 36927–36935. [CrossRef] [PubMed]
47. Camacho, L.M.; Deng, S.G.; Parra, R.R. Uranium removal from groundwater by natural clinoptilolite zeolite: Effects of pH and initial feed concentration. *J. Hazard. Mater.* **2010**, *175*, 393–398. [CrossRef]
48. Li, X.L.; Wu, J.J.; Liao, J.L.; Zhang, D.; Yang, J.J.; Feng, Y.; Zeng, J.H.; Wen, W.; Yang, Y.Y.; Tang, J.; et al. Adsorption and desorption of uranium (VI) in aerated zone soil. *J. Environ. Radioact.* **2013**, *115*, 143–150. [CrossRef] [PubMed]
49. Guan, D.X.; Ren, C.; Wang, J.; Zhu, Y.; Zhu, Z.; Li, W. Characterization of lead uptake by nano-sized hydroxyapatite: A molecular scale perspective. *ACS Earth Space Chem.* **2018**, *2*, 599–607. [CrossRef]
50. Fuller, C.C.; Bargar, J.R.; Davis, J.A.; Piana, M.J. Mechanisms of uranium interactions with hydroxyapatite: Implications for groundwater remediation. *Environ. Sci. Technol.* **2002**, *36*, 158–165. [CrossRef]
51. Han, M.N.; Kong, L.J.; Hu, X.L.; Chen, D.Y.; Xiong, X.Y.; Zhang, H.M.; Su, M.H.; Diao, Z.H.; Ruan, Y. Phase migration and transformation of uranium in mineralized immobilization by wasted bio-hydroxyapatite. *J. Clean. Prod.* **2018**, *197*, 886–894. [CrossRef]
52. El-Maghrabi, H.H.; Younes, A.A.; Salem, A.R.; Rabie, K.; El-shereafy, E.-S. Magnetically modified hydroxyapatite nanoparticles for the removal of uranium (VI): Preparation, characterization and adsorption optimization. *J. Hazard. Mater.* **2019**, *378*, 120703. [CrossRef]
53. Teng, Y.; Liu, Z.Y.; Yao, K.; Song, W.B.; Sun, Y.J.; Wang, H.L.; Xu, H.H. Preparation of attapulgite/$CoFe_2O_4$ magnetic composites for efficient adsorption of tannic acid from aqueous solution. *Int. J. Environ. Res. Public Health* **2019**, *16*, 2187. [CrossRef]
54. Deng, L.; Shi, Z.; Peng, X.X. Adsorption of Cr(VI) onto a magnetic $CoFe_2O_4$/MgAl-LDH composite and mechanism study. *RSC Adv.* **2015**, *5*, 61. [CrossRef]
55. Guo, Y.; Gong, Z.; Li, C.; Gao, B.; Li, P.; Wang, X.; Zhang, B.; Li, X. Efficient removal of uranium (VI) by 3D hierarchical Mg/Fe-LDH supported nanoscale hydroxyapatite: A synthetic experimental and mechanism studies. *Chem. Eng. J.* **2020**, *392*, 123682. [CrossRef]
56. Feng, Y.; Ma, B.; Guo, X.; Sun, H.; Zhang, Y.; Gong, H. Preparation of amino-modified hydroxyapatite and its uranium adsorption properties. *J. Radioanal. Nucl. Chem.* **2018**, *319*, 437–446. [CrossRef]
57. Tan, L.C.; Liu, Q.; Jing, X.Y.; Liu, J.Y.; Song, D.L.; Hu, S.X.; Liu, L.H.; Wang, J. Removal of uranium(VI) ions from aqueous solution by magnetic cobalt ferrite/multiwalled carbon nanotubes composites. *Chem. Eng. J.* **2015**, *273*, 307–315. [CrossRef]

58. Zhou, L.C.; Ji, L.Q.; Ma, P.C.; Shao, Y.M.; Zhang, H.; Gao, W.J.; Li, Y.F. Development of carbon nanotubes/ $CoFe_2O_4$ magnetic hybrid material for removal of tetrabromobisphenol A and Pb(II). *J. Hazard. Mater.* **2014**, *265*, 104–114. [CrossRef]
59. Xiong, S.Q.; Ye, S.D.; Hu, X.H.; Xie, F.Z. Electrochemical detection of ultra-trace Cu(II) and interaction mechanism analysis between amine-groups functionalized $CoFe_2O_4$/reduced graphene oxide composites and metal ion. *Electrochim. Acta* **2016**, *217*, 24–33. [CrossRef]
60. Dong, Y.C.; Chui, Y.S.; Ma, R.G.; Cao, C.W.; Cheng, H.; Li, Y.Y.; Zapien, J.A. One-pot scalable synthesis of Cu-$CuFe_2O_4$/graphene composites as anode materials for lithium-ion batteries with enhanced lithium storage properties. *J. Mater. Chem. A* **2014**, *2*, 13892–13897. [CrossRef]
61. Chen, L.; Zhang, K.S.; He, J.Y.; Xu, W.H.; Huang, X.J.; Liu, J.H. Enhanced fluoride removal from water by sulfate-doped hydroxyapatite hierarchical hollow microspheres. *Chem.Eng. J.* **2016**, *285*, 616–624. [CrossRef]
62. Wu, F.C.; Pu, N.; Ye, G.; Sun, T.X.; Wang, Z.; Song, Y.; Wang, W.Q.; Huo, X.M.; Lu, Y.X.; Chen, J. Performance and mechanism of uranium adsorption from seawater to poly(dopamine)-inspired sorbents. *Environ. Sci. Technol.* **2017**, *51*, 4606–4614. [CrossRef] [PubMed]

materials

MDPI

Article

Construction of BiOCl/Clinoptilolite Composite Photocatalyst for Boosting Formaldehyde Removal

Yonghao Di, Xiangwei Zhang, Xinlin Wang and Shuilin Zheng *

School of Chemical and Environmental Engineering, China University of Mining and Technology (Beijing), Beijing 100083, China; 18910320633@163.com (Y.D.); zhangxiangwei0613@126.com (X.Z.); wangxinlin8426@163.com (X.W.)
* Correspondence: zhengsl@cumtb.edu.cn; Tel.: +86-10-62339920

Abstract: Binary composite was synthesized via coupling BiOCl with alkali leached natural clinoptilolite (40B0/CN), which showed retarded recombination of photo-generated carriers. The clinoptilolite was pretreated with alkali leaching, resulting in a larger pore size and high cation exchange capacity. The modified clinoptilolite was more feasible for the growth of BiOCl and to promote the adsorption ability for formaldehyde (HCHO). In addition, the cation exchange capacity was conducive to anchor Bi^{3+}, further leading to the reduction of the particle size of BiOCl. The carrier effect of alkali leached natural clinoptilolite promoted the amorphous transformation of BiOCl at low temperature, which simultaneously produced more distortions and defects in the BiOCl lattice. The 40B0/CN composite exhibited the superior light absorption ability with a narrower band gap. The photocatalytic degradation rate for HCHO of 40B0/CN under solar light reached 87.7%, and the reaction rate constant was 0.0166 min^{-1}, which was 1.6 times higher than that of BiOCl. This paper gave a deep insight into photocatalytic technology to efficiently degrade formaldehyde.

Keywords: BiOCl; clinoptilolite; formaldehyde; photocatalysis

Citation: Di, Y.; Zhang, X.; Wang, X.; Zheng, S. Construction of BiOCl/Clinoptilolite Composite Photocatalyst for Boosting Formaldehyde Removal. *Materials* **2021**, *14*, 6469. https://doi.org/10.3390/ma14216469

Academic Editors: Avelino Núñez-Delgado, Zhien Zhang, Elza Bontempi, Mario Coccia, Marco Race and Yaoyu Zhou

Received: 7 October 2021
Accepted: 21 October 2021
Published: 28 October 2021

1. Introduction

Owning to the increasing attention to the living environment, indoor air quality (IAQ) has traditionally been regarded as an important factor that affects the health of humans [1–3]. In particular, formaldehyde, as one of the most concerned volatile organic compounds (VOCs), has emerged as being one of the main reasons causing cancer due to its widespread source and high toxicity [4–6]. To date, massive research efforts have been devoted to the development of adsorption [7], biodegradation [8], thermal catalysis [9], and photocatalysis [10] technologies to remove formaldehyde. Due to uncomplicated processes and an efficient treatment effect, adsorption technology has recently been widely studied. As a kind of natural zeolite and an efficient and environmental adsorption material, natural clinoptilolite (NC) has received widespread attention owing to its abundant resources, low price, and its massive mesoporous structure, which could promote the diffusion and mass transportation of formaldehyde and active species [11,12]. However, its formaldehyde adsorption capacity is limited. On the other hand, when external environmental conditions change, formaldehyde molecules are easily desorbed from the surface of NC, resulting in secondary pollution. Therefore, to make NC more feasible for practical application, it is very necessary to endow NC adsorbents with formaldehyde degradation function to ensure the continuous purification ability of materials to formaldehyde.

Among various degradation technologies for formaldehyde purification, photocatalysis has been proven to be effective in degrading or mineralizing formaldehyde. Thereinto, BiOCl is an inexpensive and promising photocatalyst with the advantages of a suitable band gap (3.40 eV), its non-toxicity, and being cost-effective and environmentally friendly [13,14]. Moreover, BiOCl displays the typical layered structure, which is conducive to the separation and migration of electron and hole pairs, and so it shows better photocatalytic activity.

Massive reports have been proved that formaldehyde could be degraded by BiOCl under ultraviolet or solar light irradiation [15,16]. However, it is currently constrained by the shortcoming of lower degradation efficiency under solar light irradiation and tendency to form larger particle size, which inevitably decreases the number of active sites [17]. In view of the large specific surface area, NC is a promising carrier candidate for BiOCl to construct a coupling system with a higher formaldehyde adsorption capacity, adsorption selectivity, and efficient formaldehyde degradation performance. However, the smaller pore size of NC may hinder the diffusion of BiOCl into the holes. Previous studies have been proved that alkali leaching process could enlarge the pore size significantly [18,19]. Predictably, the NC after alkali leaching treatment would be more feasible for the deposition of BiOCl. Based on the above assumptions, combining alkali leached natural clinoptilolite and BiOCl is a satisfactory strategy to enhance the removal of formaldehyde, which would not only possess superior adsorption performance, but also exhibit formaldehyde degradation performance.

Overall, the BiOCl/alkali leached natural clinoptilolite was successfully prepared via facile liquid-phase hydrolysis method and well characterized by various instruments. The catalytic performance was examined by degrading formaldehyde in different systems. Meanwhile, the effect of preparation parameters such as reaction temperature and mass ratio of BiOCl were studied as well. Besides, the mechanism of the composite was fully explored and proposed.

2. Experimental

2.1. Materials

Sodium hydroxide (NaOH), bismuth nitrate ($Bi(NO_3)_3$), glycol ($C_2H_6O_2$), sodium chloride (NaCl), and formaldehyde solution (37 wt%) were purchased from Beijing Reagent Co. (Beijing, China). All the above reagents were analytical reagent grade without any further purification. The natural clinoptilolite was provided by Bayannur City, Inner Mongolia Province, China. Deionized water was used throughout the experiments.

2.2. Composites Preparation

Alkali leached natural clinoptilolite (labeled as NC-Na-3.0) was prepared via impregnation method. In a typical synthesis, the NC and NaOH (3.0 mol/L) solutions were transformed into deionized water and the solid/liquid ratio was 1:10. The above suspension was further stirred for 6 h in a water bath at 60 °C. Then, the suspension was filtrated, dried in an oven (DHG-9240A, Huiyi Sifang Technical Service Co., Beijing, China) and collected for further use.

BiOCl/alkali leached natural clinoptilolite was synthesized through liquid-phase hydrolysis. Typically, 2.0 g NC-Na-3.0 was added into 61.42 mL $Bi(NO_3)_3$ ethylene glycol solution with the molar concentration of (m = 0.0125, 0.025, 0.0375, 0.05, 0.0625 mmol/L) under magnetic stirring to form a homogeneous suspension, then the suspension was continuously stirred in a water bath under different temperatures. After that, 92.13 mL sodium chloride solution with the same molar concentration of $Bi(NO_3)_3$ was added into the above suspension through the peristaltic pump (BT100M, Chuangrui Pump Co., Baoding, China), and finally the suspension was further reacted for 1 h. Then, the powder was collected after washing and drying. The composites with a different mass ratio of BiOCl were marked as XBY/CN (X and Y present the mass ratio of BiOCl and reaction temperature, respectively. X = 10%, 20%, 30%, 40%, and 50%, Y = 0, 25, 50, 75, and 90 °C). Moreover, pure BiOCl (labeled as B25 and B0) was obtained according to the abovementioned method at 25 °C and 0 °C water bath, respectively, without using NC-Na-3.0.

2.3. Characterizations

D8 advance X-ray diffractometer (Bruker, Karlsruhe, Germany) equipped with Cu-Kα radiation (λ = 0.154056 nm) was applied to investigate the phase structure of as-prepared composite. The range of 2θ was from 5° to 80° with a scanning rate at 8°/min.

The morphology was studied on a scanning electron microscopy (SEM) (S-4800, Tokyo, Hitachi, Japan). The optical performance of composite was recorded through a UV-vis spectrophotometer (UV-9000s, Metash Instruments Co., Shanghai, China). The specific surface area and pore distributions were obtained by JW-BK (JWGB Sci. & Tech., Beijing, China) at liquid nitrogen temperature (77 K). Photoluminescence (PL) spectra was recorded through a fluorescence spectrophotometer (F-7000 PL, Tokyo, Hitachi, Japan) at 360 nm emission wavelength.

2.4. *Evaluation of Photocatalytic Activity*

One gram of as-prepared composite was evenly dispersed at the bottom of a glass-surface vessel (φ = 10 cm). Then, the glass-surface vessel was put on the support platform in the reaction chamber with a volume of 60 L. After that, the reaction chamber was closed and 20 μL formaldehyde diluent (18.5 wt%) was injected through a micro syringe. The electric heating plate and fan were employed to accelerate the volatilization of formaldehyde and finally the formaldehyde was evenly dispersed in the reaction chamber. After 2 min, the electric heating plate was turned off. The temperature was maintained at room temperature and the relative humidity maintained at around 50% (when testing under high humidity conditions, the relative humidity was maintained at about 75%). Before the simulated sunlight irradiation, 45 min of dark reaction was conducted to achieve the adsorption-desorption equilibrium on the surface of the composites. Then, the catalytic process was examined with a light intensity of 50 mW/cm^2 at the sample surface. Generally, the degradation rate of formaldehyde was equal to the formation rate of CO_2, thus the degradation rate of formaldehyde could be calculated according to the following formula:

$$D = \frac{M_1 \times \triangle CO_2}{M_2 \times C_0}. \quad (1)$$

M_1 and M_2 represent the relative molecular weights of formaldehyde and carbon dioxide, respectively. $\triangle CO_2$ is the increment of CO_2 (mg/m^3). C_0 represents the initial concentration of formaldehyde (mg/m^3), and the theoretical calculation value is 66.8mg/m^3. The concentration of CO_2 and formaldehyde was monitored through a photoacoustic spectrum.

3. Results and Discussion

3.1. *Cation Exchange Effect of Catalyst*

Since NC possessed a superior cation exchange effect, it is necessary to further investigate whether the cation exchange performance is conducive to the growth of BiOCl, which is important for understanding the mechanism of the catalyst. A certain amount of NC-Na-3.0 was transferred into $Bi(NO_3)_3$ ethylene glycol solutions and the suspension was dispersed evenly under vigorous stirring. Then, the suspension was transferred to a water bath at 0 °C and 90 °C, respectively. After stirring for 2 h, the suspension was filtrated and dried, then the as-prepared composites were marked as CN-0 and CN-90, respectively. The chemical components of the above two samples were further analyzed. As shown in Table 1, the content of Bi in CN-0 and CN-90 was higher than NC-Na-3.0, while the content of Na and Ca in CN-0 and CN-90 are significantly lower than NC-Na-3.0, indicating that ion-exchangeable reaction between the Na^+, Ca^{2+} in NC-Na-3.0 and Bi^{3+} was carried out. Moreover, there is little difference in the chemical components between CN-0 and CN-90, indicating that a low temperature could hardly inhibit the ion exchange between Na^+, Ca^{2+} in NC-Na-3.0 and Bi^{3+}.

Table 1. X-ray fluorescence spectroscopy (XRF) results of clinoptilolite supports before and after ion exchange with bismuth nitrate at different temperatures (wt%).

Sample	SiO_2	Al_2O_3	CaO	MgO	K_2O	Na_2O	Fe_2O_3	Bi_2O_3
NC-Na-3.0	61.39	21.46	5.35 ± 0.11	2.23	2.38	6.32 ± 0.13	0.83	0.01 ± 0.00
CN-0	64.31	22.37	3.39 ± 0.08	2.28	2.13	1.58 ± 0.06	0.63	2.86 ± 0.06
CN-90	64.25	21.92	3.23 ± 0.08	2.21	2.26	1.54 ± 0.06	0.85	2.76 ± 0.07

3.2. *Phase Structure of Catalysts*

The phase structure of XBY/CN composites and the contrastive samples were investigated by XRD spectra. As shown in Figure 1a, the spectrum of B25 was basically consistent with the standard spectrum of BiOCl (JCPDS: 06-0249), indicating that the purity and crystallinity of BiOCl in B25 synthesized at room temperature were high [20]. The peaks appeared at 26.0°, 32.5°, and 46.8°, and could be indexed as (101), (110), and (200) planes of BiOCl in XB25/CN, respectively, and the peak intensity gradually enhanced with the increase of the loading ratio of BiOCl. Meanwhile, the peaks located at 26.0°, 32.5°, and 46.8°, which were attributed to (101), (102), and (200) planes of clinoptilolite, were gradually decreased. This implies that more BiOCl were successfully grafted on the surface of NC-Na-3.0. Moreover, the peak relative intensity of (110) crystal face of BiOCl in XBY/CN was higher than (101) crystal face of B25 and standard spectrum of BiOCl, which revealed that BiOCl presented an evident (110) preferred orientation on the surface of NC-Na-3.0.

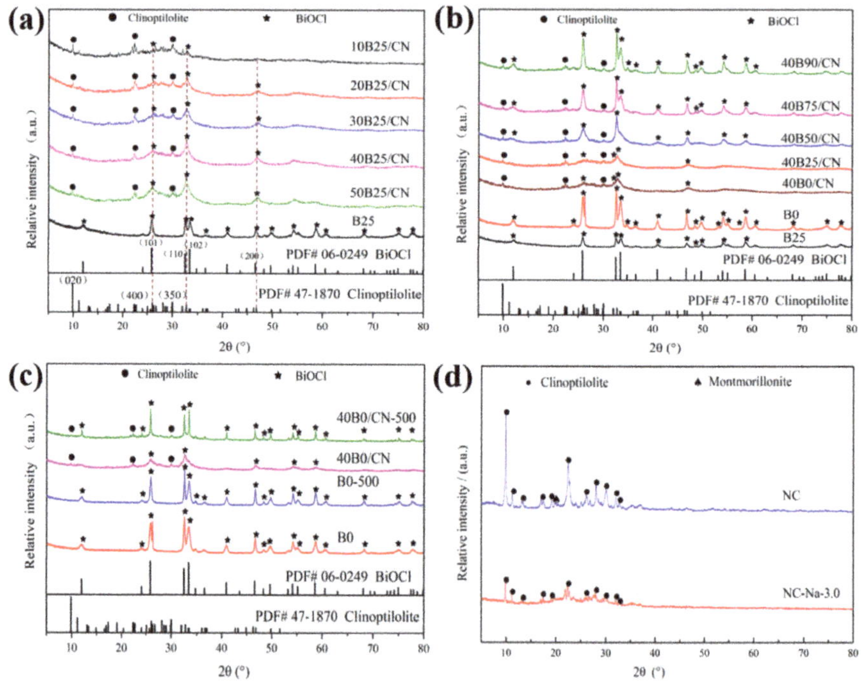

Figure 1. (**a**) XRD spectra of B25 and XBY/CN with different BiOCl loading amounts; (**b**) different reaction temperature; (**c**) XRD spectra of B0 and 40B0/CN before and after calcination; (**d**) XRD spectra of NC and NC-Na-3.0.

The XRD spectra of 40BY/CN composites under different reaction temperatures were shown in Figure 1b. As can be seen, compared with B25, the diffraction peaks of BiOCl in B0 were significantly sharpened and the intensity was improved as well, indicating that the lower temperature was beneficial to increasing the crystallinity of BiOCl. However, the peaks' intensity was enhanced gradually with the increased reaction temperature, indicating that the existence of NC-Na-3.0 exhibited a significant impact on the crystallization and actual loading amount of BiOCl.

Furthermore, the XRD spectra of B0 and 40B0/CN composites after calcination at 500 °C for 1 h (B0-500, 40B0/CN-500) was further studied to confirm the carrier effect of NC-Na-3.0. As shown in Figure 1c, the diffraction peaks of B0 did not change significantly after calcination, while the peak of BiOCl diffraction peaks in 40B0/CN were significantly sharpened and enhanced after calcination. This phenomenon could be attributed to the amorphous transformation of BiOCl at low temperature caused by the introduction of NC-Na-3.0. The induced transformation of NC-Na-3.0 might produce more distortions and defects in the BiOCl lattice, which would effectively promote the photocatalytic performance of the composites under solar light [21].

3.3. Pore Structure and Specific Surface Area of Catalysts

The specific surface area, pore volume, pore distributions of composites, and the average crystallite sizes of BiOCl in composites (calculated based on the Debye-Scherrer equation) are summarized in Table 2. NC-Na-3.0 possesses the largest specific surface area, total pore volume, and average pore diameter, which are beneficial for the adhesion and dispersion of BiOCl nanoparticles. Compared with B25 and B0, the specific surface area and pore volume of XBY/CN is improved, and the average pore size decreases apparently, suggesting that a large number of microporous pores were formed. When the mass ratio of BiOCl is small, the crystallite size of BiOCl loaded on NC-Na-3.0 carrier increases compared with B25, which may be due to the acid-base characteristics of the carrier surface. After alkali leaching treatment, more active centers that can react with acid remained on the mineral surface of natural clinoptilolite. These active centers would react with HNO_3, which was produced by the hydrolysis of $Bi(NO_3)_3$ in NaCl aqueous solution, and finally the neutralization reaction could be realized. Therefore, the pH would be maintained to neutral, and the higher pH value promoted the rapid nucleation and growth of BiOCl. With the increase of the mass ratio of BiOCl, the crystallite size of BiOCl in XB25/CN gradually decreases as well as the total pore volume and average pore size (compared with NC-Na-3.0), indicating that the loading of BiOCl particles on the carrier may be completed preferentially in the pores of the carrier.

Table 2. Specific surface area, total pore volume, average pore diameter, and crystallite size of different samples.

Sample	Crystallite Size of BiOCl (nm)	S_{BET} (m^2/g)	Total Pore Volume (cm^3/g)	Average Pore Diameter (nm)
NC-Na-3.0	——	61.43	0.310	16.67
B25	19.5	29.29	0.160	15.63
10B25/CN	19.9	56.63	0.283	15.43
20B25/CN	12.5	45.29	0.181	13.61
30B25/CN	12.3	43.68	0.176	12.97
40B25/CN	12.2	42.55	0.162	12.38
50B25/CN	12.9	36.75	0.158	11.57
B0	18.6	23.89	0.148	15.84
40B0/CN	11.3	65.54	0.254	12.2
40B25/CN	12.2	42.55	0.162	12.38
40B50/CN	18.1	45.78	0.206	12.72
40B75/CN	23.3	50.82	0.209	13.52
40B90/CN	26.8	55.47	0.254	15.76

When the mass ratio of BiOCl was 40%, the smallest crystallite size of BiOCl and smaller average pore size of 40B0/CN appeared, which not only improved the photocatalytic activity of the composites but also facilitated the diffusion and mass transportation of reactants and active species. Besides, the uniformly BiOCl significantly improved the specific surface area and pore volume of 40B0/CN composite, and the surface area of 40B0/CN was 2.24 and 2.79 times as high as that of B0 and B25, respectively. It can also be seen from Table 2 that a higher reaction temperature is unfavorable to the formation of a smaller crystallite size of BiOCl.

3.4. Morphology and Structure of Catalysts

The morphology and structure of as-prepared composites were investigated by SEM images. As shown in Figure 2, the B25 sample prepared at 25 °C exhibited a flake structure with a diameter of about 110~160 nm and a thickness of about 20~40 nm. The aggregate exhibit flake structure with a diameter of about 2~3 μm. The B0 sample prepared at 0 °C was mainly composed of flake structure with a diameter of about 60~80 nm and a thickness of about 20 nm. The BiOCl nanosheets were agglomerated into flower-like structure with a diameter of about 500 nm. As shown in Figure 2a,b, the particle size of B0 was significantly smaller than B25, and the size of B0 aggregates was small and uniform as well, which further indicated that a lower temperature could inhibit the growth of BiOCl and contribute to the dispersion on the surface of NC-Na-3.0. As shown in Figure 2c, the NC-Na-3.0 exhibited massive macropores with the pore size of 70~400 nm, while most of the original macropores of NC-Na-3.0 disappeared after loading of BiOCl (Figure 2d), which was ascribed to the growth of BiOCl in the macropores. As shown in Figure 2d, massive aggregates of BiOCl with the diameter of 100~200 nm and nanosheets of BiOCl with the diameter of 20~40 nm and a thickness of about 10 nm were dispersed on the surface of NC-Na-3.0. The smaller size of BiOCl might be caused by the carrier effect of NC-Na-3.0.

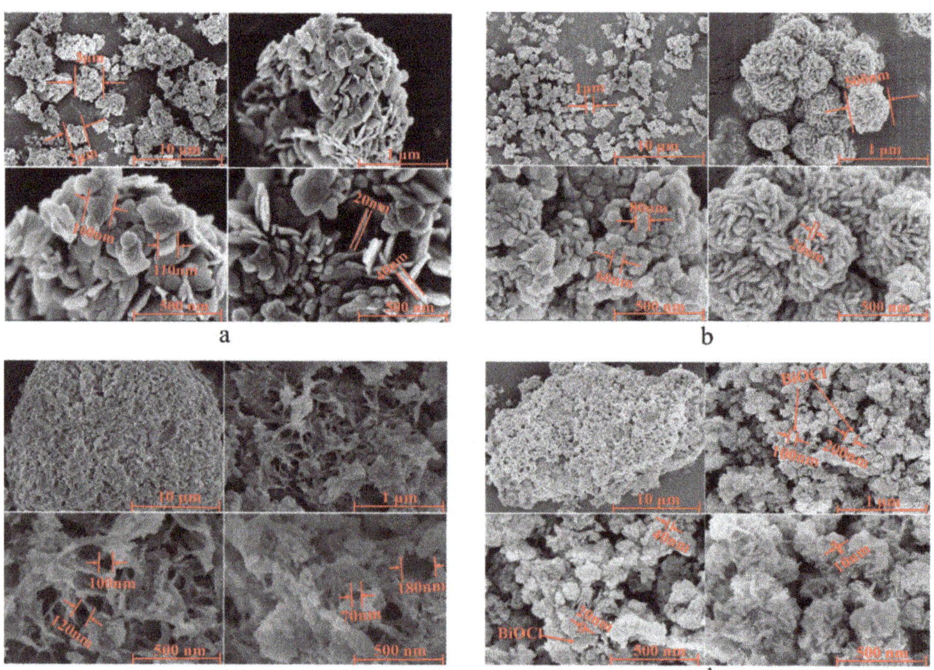

Figure 2. SEM images of pure BiOCl ((**a**) B25, (**b**) B0), clinoptilolite support ((**c**) NC-Na-3.0) and 40B0/CN (**d**).

3.5. Optical Properties and Photoelectrochemical Performance of Catalysts

The optical performance of the as-prepared composites was measured via UV-Vis DRS spectra. The light absorption intensity of B25, B0, and 40B0/CN in the UV region, especially in the UV region of 230~315 nm, increased in turn, indicating that the smaller size of BiOCl was helpful to improve the utilization of ultraviolet light. In the region of >355 nm, especially in the region of 355~565 nm, the light absorption ability of B0, B25, and 40B0/CN increased in turn, indicating that the lower crystallinity of BiOCl might facilitate improving the utilization of visible light. In addition, the band gap values (Eg) were calculated using the Kubelka–Munk method [22], and the results were displayed in Figure 3b. The band gap of B25, B0 and 40B0/CN were estimated as 3.41 eV, 3.46 eV, and 3.37 eV, respectively. The narrower band gap of 40B0/CN further revealed that more visible light could be harvested.

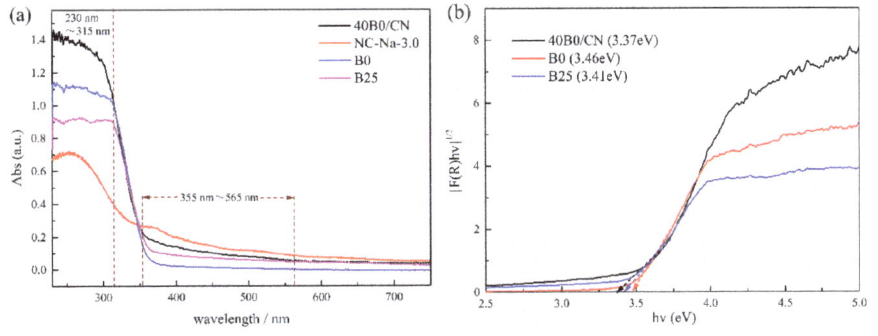

Figure 3. UV-Vis DRS (**a**) and band gaps (**b**) of different samples.

3.6. Photocatalytic Activity and Stability of Photocatalysts

Since the mass ratio of BiOCl could directly affect the catalytic performance of XB25/CN composites, the degradation rate of formaldehyde with a different loading amount of BiOCl was studied systemically. As depicted in Figure 4a, the concentration of HCHO showed a declining trend with the extension of time under dark condition, and gradually reached adsorption equilibrium within 45 min. Moreover, the concentration of CO_2 remained almost unchanged during the dark adsorption process, which demonstrated that the decrease of HCHO was mainly attributed to the materials' adsorption rather than degradation. B25 exhibited the smallest reduction of HCHO, resulting from poor adsorption ability. The XB25/CN composites possessed the higher adsorption of HCHO, which revealed that the introduction of NC-Na-3.0 significantly enhanced the adsorption performance. As shown in Figure 4b, the concentration of CO_2 under different systems increased significantly, and the formation efficiency increased with the increase of BiOCl loading from 10% to 40%, and decreased when the BiOCl loading was excessive (X > 40%), suggesting that higher BiOCl loading amount was not conducive to the degradation of formaldehyde by composites. The corresponding first-order kinetics plot shown in Figure 4c,d demonstrated that the 40B25/CN composite shows superior formaldehyde purification performance under sunlight and the degradation efficiency was higher than of pure BiOCl. The K value of 40B25/CN (0.0133 min^{-1}) was 1.32 times higher than that of B25 (0.0101 min^{-1}) [23,24], further indicating that the introduction of NC-Na-3.0 significantly improves the photocatalytic performance of BiOCl.

It can be seen from Figure 5, that the formation efficiency of CO_2 gradually increased with the decrease of reaction temperature. Moreover, the reaction rate constant gradually increased as well, which indicated that the lower reaction temperature was conducive to the improvement of the catalytic performance of B/CN. Specifically, when the reaction temperature decreased to 0 °C, the composite displayed the faster formation rate of CO_2, and the K value of 40B0/CN (0.0166 min^{-1}) was 1.61 times higher than that of B0 (0.0103 min^{-1}). Thus, 40B0/CN was selected for further use in this work.

The stability of photocatalytic performance is of great significance to the practical application of photocatalyst. The stability mentioned in this work includes the reusability of 40B0/CN and the photocatalytic performance of HCHO under high humidity conditions. It can be seen from Figure 6 that the degradation rate of HCHO and the formation rate of CO_2 have no obvious reduction after three reuse experiments of 40B0/CN, indicating that the composite possessed the superior sustainable degradation performance for HCHO under solar light. Additionally, there was a minor reduction of HCHO and CO_2 in the high humidity environment, which further indicated that 40B0/CN displayed good moisture resistance and had the application conditions in the high humidity environment.

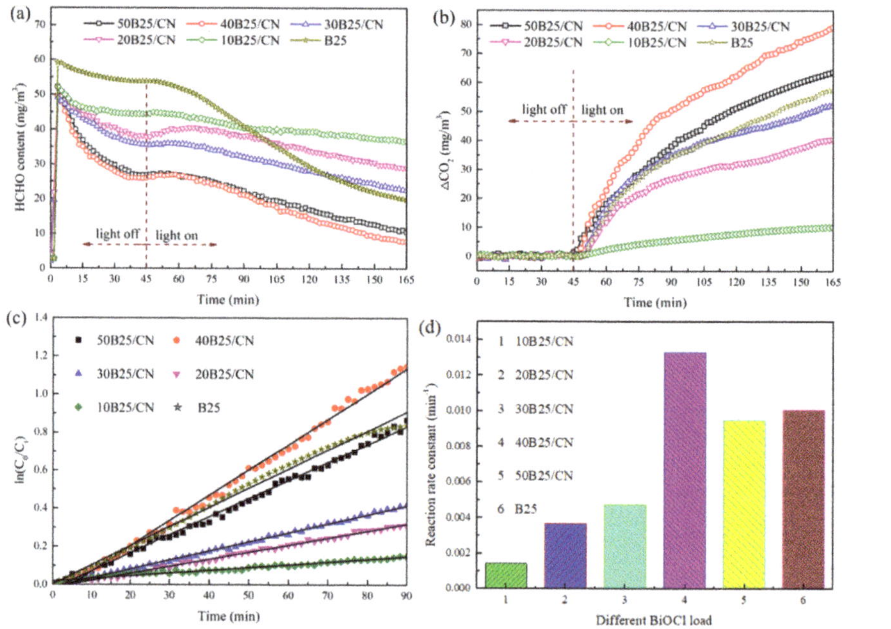

Figure 4. Variations of HCHO concentration (**a**), corresponding increased CO_2 concentration ($\triangle CO_2$) (**b**), the kinetic curves (**c**) and the reaction rate constant values (**d**) over different samples under solar light.

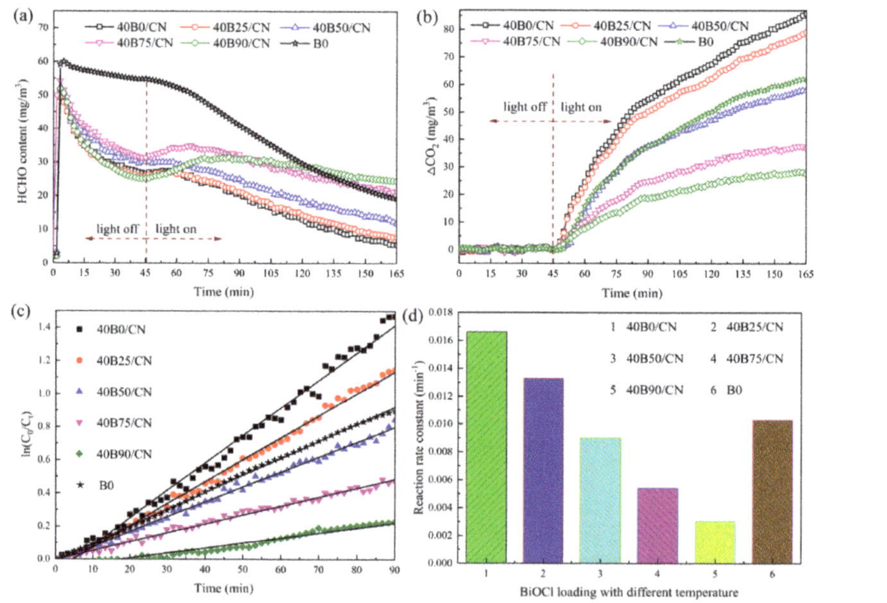

Figure 5. Variations of HCHO concentration (**a**), corresponding increased CO_2 concentration ($\triangle CO_2$) (**b**), the kinetic curves (**c**) and the reaction rate constant values (**d**) over different samples under solar light.

The XRD spectra of 40B0/CN, B0 and B25 before and after photocatalytic reaction were depicted in Figure 7. Obviously, the peak of BiOCl in B25 and B0 after the photocatalytic reaction was significantly sharpened and the intensity was enhanced, suggesting that the stability of the pure BiOCl was poor. The pure BiOCl synthesized by precipitation method had a smaller particle size. Moreover, the structural stability and crystallinity of it were generally lower than those of BiOCl after calcination. Therefore, it is also understandable that the structure and crystallinity changed after receiving light radiation energy. However, only slight changes were observed in 40B0/CN, which might be attributed to the existence of NC-Na-3.0, which could effectively maintain the stability of BiOCl.

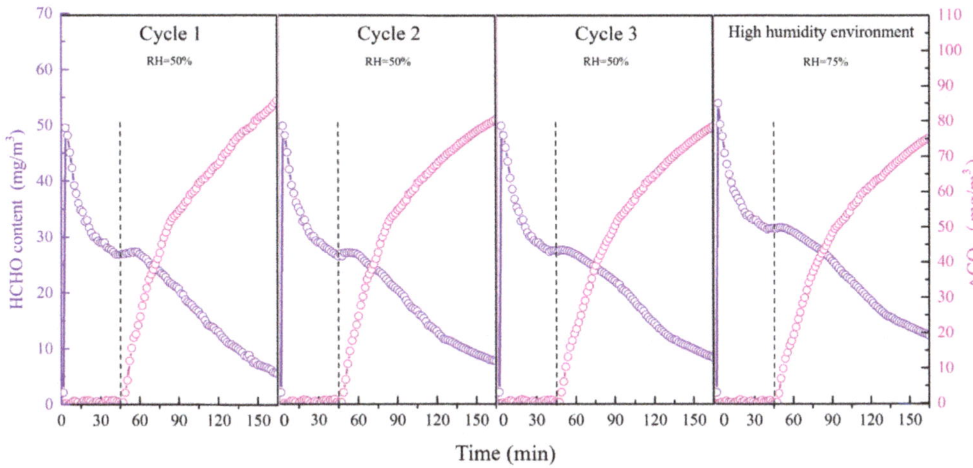

Figure 6. Repeated formaldehyde degradation performance of 40B0/CN in sunlight and its formaldehyde degradation performance in a high humidity environment.

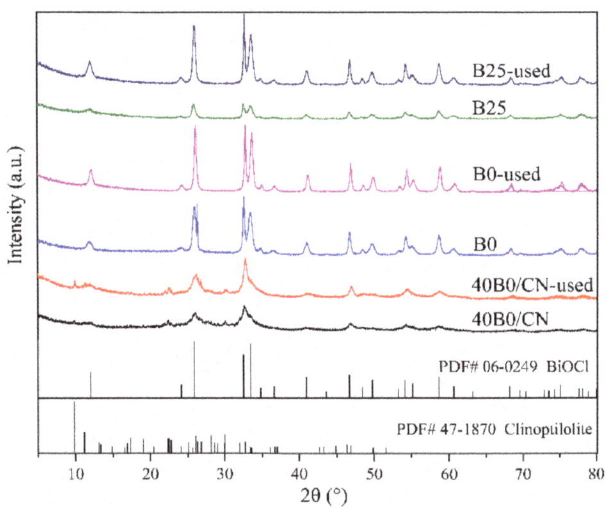

Figure 7. XRD spectra of B0, B25, and B0/CN before and after photocatalytic reaction under sunlight.

3.7. Reaction Mechanism

For the photocatalyst, the separation efficiency of photogenerated carriers was closely related to the photocatalytic activity. Based on the above analysis and discussion, it could be proved that the adsorption performance and photocatalytic performance of the composites were improved compared with BiOCl. The PL spectra was carried out to further investigate the separation efficiency of photogenerated carriers. As depicted in Figure 8a, the composites were excited under 290 nm and all of them revealed a wide emission peak from 350 to 800 nm. It is clearly observed that 40B0/CN showed weaker intensity than B25 and B0, which demonstrated that 40B0/CN possessed a lower recombination rate of electron-hole pairs [25,26]. This is consistent with the experimental results of the HCHO degradation rate in Figure 8b.

In conclusion, the preparation mechanism of 40B0/CN composite was as follows: After alkali leaching, NC-Na-3.0 obtained more large pores, and due to the existence of a strong electric field inside and around pores, more equilibrium ions Na^+ and Ca^{2+} could be gathered. When NC-Na-3.0 was added into $Bi(NO_3)_3$ ethylene glycol solution, NC-Na-3.0 would carry out the ion exchange between Bi^{3+} and Na^+, Ca^{2+} in the pores of NC-Na-3.0. Then, with the addition of NaCl aqueous solution, BiOCl preferentially nucleated and deposited around Bi^{3+} in the pores. Due to the confinement effect of the carrier, the growth of BiOCl would be inhibited, and the size of BiOCl would be reduced and the dispersion would be improved [27]. Moreover, the lower reaction temperature further decreased the deposition size of BiOCl on NC-Na-3.0 and further promoted its dispersion. The existence of NC-Na-3.0 also induced the transformation of BiOCl nanocrystals to an amorphous state.

The possible degradation mechanism was discussed, and the conclusion was depicted as three aspects: (1) Compared with pure BiOCl, the specific surface area of the 40B0/CN increased significantly and the adsorption of HCOH was significantly enhanced. The probability of collision between BiOCl particles and HCHO molecules could be greatly increased, which promoted the efficiency of photocatalytic degradation of HCHO; (2) More nano-scale BiOCl particles were evenly distributed on the surface of the carrier, and these particles have the tendency of amorphous transformation, resulting in increased distortion in the lattice. Thus, more visible light could be harvested by 40B0/CN, and the utilization of solar light was improved; (3) The combination of BiOCl and NC-Na-3.0 effectively promoted the separation and migration of photogenerated carriers, further improving photocatalytic performance.

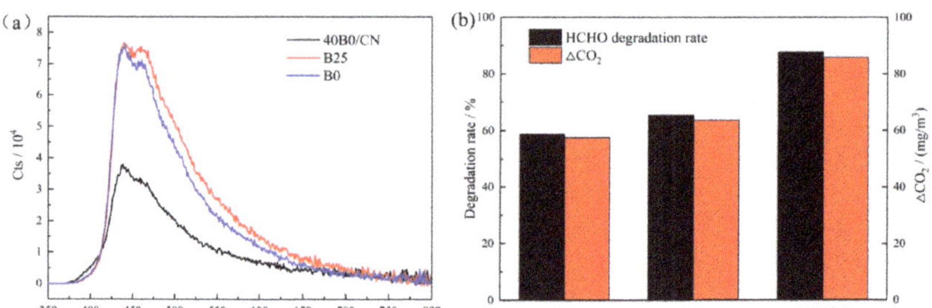

Figure 8. Photoluminescence spectra (**a**) and formaldehyde degradation rate under sunlight (**b**) of B25, B0 and 40B0/CN composites.

4. Conclusions

The BiOCl/alkali leached natural clinoptilolite with superior photodegradation performance of gaseous formaldehyde were successfully prepared via facile liquid-phase hydrolysis method. There were a large number of BiOCl nanosheets with a diameter of 20~40 nm and a thickness of about 10 nm growth in the pores of NC-Na-3.0. The photocatalytic degradation rate of HCOH under solar light reached 87.7%, and the reaction rate constant was 0.0166 min^{-1}, which was 1.6 times higher than that of BiOCl. Moreover, the composites possessed superior adsorption properties, good reusability, and moisture resistance. The introduction of NC-Na-3.0 as the carrier enhanced the adsorption performance for HCHO, promoted the directional deposition of BiOCl, and reduced the particle size of BiOCl. On the other hand, the improvement of dispersion and the induced transformation of BiOCl effectively reduced the band gap value, which enhanced the solar light absorption and boosted the separation and migration of photogenerated carriers.

Author Contributions: Conceptualization, Y.D. and S.Z; data curation, Y.D. and X.Z.; funding acquisition, Y.D.; investigation, Y.D., X.Z. and X.W.; writing—original draft, Y.D., X.Z. and X.W.; writing—review and editing, S.Z. All authors have read and agreed to the published version of the manuscript.

Funding: This research is funded by the Natural Science Foundation of China (52074307) and the Fok Ying Tung Education Foundation, China (171042).

Institutional Review Board Statement: Not applicable.

Informed Consent Statement: Not applicable.

Data Availability Statement: The data presented in this study are available on request from the corresponding author.

Acknowledgments: The authors gratefully acknowledge the financial support provided by the Natural Science Foundation of China (52074307) and the Fok Ying Tung Education Foundation, China (171042).

Conflicts of Interest: The authors declare no conflict of interest.

References

1. Baures, E.; Blanchard, O.; Mercier, F.; Surget, E.; le Cann, P.; Rivier, A.; Gangneux, J.P.; Florentin, A. Indoor air quality in two French hospitals: Measurement of chemical and microbiological contaminants. *Sci. Total Environ.* **2018**, *642*, 168–179. [CrossRef]
2. Bian, Y.; Wang, R.; Wang, S.; Yao, C.; Ren, W.; Chen, C.; Zhang, L. Metal–organic framework-based nanofiber filters for effective indoor air quality control. *J. Mater. Chem. A* **2018**, *6*, 15807–15814. [CrossRef]
3. Cheek, E.; Guercio, V.; Shrubsole, C.; Dimitroulopoulou, S. Portable air purification: Review of impacts on indoor air quality and health. *Sci. Total Environ.* **2021**, *766*, 142585. [CrossRef] [PubMed]
4. Li, J.; Cui, W.; Chen, P.; Dong, X.A.; Chu, Y.; Sheng, J.; Zhang, Y.; Wang, Z.; Dong, F. Unraveling the mechanism of binary channel reactions in photocatalytic formaldehyde decomposition for promoted mineralization. *Appl. Catal. B Environ.* **2020**, *260*, 118130. [CrossRef]
5. Na, C.J.; Yoo, M.J.; Tsang, D.C.W.; Kim, H.W.; Kim, K.H. High-performance materials for effective sorptive removal of formaldehyde in air. *J. Hazard Mater.* **2019**, *366*, 452–465. [CrossRef]
6. Suresh, S.; Bandosz, T.J. Removal of formaldehyde on carbon-based materials: A review of the recent approaches and findings. *Carbon* **2018**, *137*, 207–221. [CrossRef]
7. Liu, L.; Liu, J.; Zeng, Y.; Tan, S.J.; Do, D.D.; Nicholson, D. Formaldehyde adsorption in carbon nanopores—New insights from molecular simulation. *Chem. Eng. J.* **2019**, *370*, 866–874. [CrossRef]
8. Zvulunov, Y.; Ben-Barak-Zelas, Z.; Fishman, A.; Radian, A. A self-regenerating clay-polymer-bacteria composite for formaldehyde removal from water. *Chem. Eng. J.* **2019**, *374*, 1275–1285. [CrossRef]
9. Wang, Z.; Xiong, F.; Zhang, Z.; Sun, G.; Xu, H.; Chai, P.; Huang, W. Surface Chemistry of Formaldehyde on Rutile TiO_2 (011)-(2 × 1) Surface: Photocatalysis Versus Thermal-Catalysis. *J. Phys. Chem. C* **2017**, *121*, 25921–25929. [CrossRef]
10. Feng, Y.; Ling, L.; Nie, J.; Han, K.; Chen, X.; Bian, Z.; Li, H.; Wang, Z.L. Self-Powered Electrostatic Filter with Enhanced Photocatalytic Degradation of Formaldehyde Based on Built-in Triboelectric Nanogenerators. *ACS Nano* **2017**, *11*, 12411–12418. [CrossRef] [PubMed]
11. Rashidi, R.; Yousefinejad, S.; Mokarami, H. Catalytic ozonation process using CuO/clinoptilolite zeolite for the removal of formaldehyde from the air stream. *Int. J. Environ. Sci. Technol.* **2018**, *16*, 6629–6636. [CrossRef]
12. Rahimi, E.; Nazari, F.; Javadi, T.; Samadi, S.; da Silva, J.A.T. Potassium-enriched clinoptilolite zeolite mitigates the adverse impacts of salinity stress in perennial ryegrass (*Lolium perenne* L.) by increasing silicon absorption and improving the K/Na ratio. *J. Environ. Manag.* **2021**, *285*, 112142. [CrossRef] [PubMed]
13. Hu, X.; Li, C.; Sun, Z.; Song, J.; Zheng, S. Enhanced photocatalytic removal of indoor formaldehyde by ternary heterogeneous BiOCl/TiO_2/sepiolite composite under solar and visible light. *Build. Environ.* **2020**, *168*, 106481. [CrossRef]
14. Liu, M.Y.; Lin, G.L.; Liu, Y.M.; Lin, X.Y.; Wang, L.J.; Xu, Y.F.; Song, X.C. Ternary heterojunction Ag/AgIO3/ BiOCl(CMC) by a biomass template for photodegradation of tetracycline hydrochloride and gaseous formaldehyde. *Solid State Sci.* **2021**, *112*, 106517. [CrossRef]
15. Cui, D.; Wang, L.; Xu, K.; Ren, L.; Wang, L.; Yu, Y.; Du, Y.; Hao, W. Band-gap engineering of BiOCl with oxygen vacancies for efficient photooxidation properties under visible-light irradiation. *J. Mater. Chem. A* **2018**, *6*, 2193–2199. [CrossRef]
16. Jia, Z.; Li, T.; Zheng, Z.; Zhang, J.; Liu, J.; Li, R.; Wang, Y.; Zhang, X.; Wang, Y.; Fan, C. The BiOCl/diatomite composites for rapid photocatalytic degradation of ciprofloxacin: Efficiency, toxicity evaluation, mechanisms and pathways. *Chem. Eng. J.* **2020**, *380*, 122422. [CrossRef]

17. Hou, W.; Xu, H.; Cai, Y.; Zou, Z.; Li, D.; Xia, D. Precisely control interface OVs concentration for enhance 0D/2D $Bi_2O_2CO_3$/BiOCl photocatalytic performance. *Appl. Surf. Sci.* **2020**, *530*, 147218. [CrossRef]
18. Joorasty, M.; Hemmati, A.; Rahbar-Kelishami, A. NaOH/clinoptilolite-Fe_3O_4 as a novel magnetic catalyst for producing biodiesel from Amygdalus scoparia oil: Optimization and kinetic study. *Fuel* **2021**, *303*, 121305. [CrossRef]
19. de Souza, V.; Villarroel-Rocha, J.; de Araújo, M.; Sapag, K.; Pergher, S. Basic Treatment in Natural Clinoptilolite for Improvement of Physicochemical Properties. *Minerals* **2018**, *8*, 595. [CrossRef]
20. Cao, J.; Li, J.; Chu, W.; Cen, W. Facile synthesis of Mn-doped BiOCl for metronidazole photodegradation: Optimization, degradation pathway, and mechanism. *Chem. Eng. J.* **2020**, *400*, 125813. [CrossRef]
21. Mao, C.; Cheng, H.; Tian, H.; Li, H.; Xiao, W.-J.; Xu, H.; Zhao, J.; Zhang, L. Visible light driven selective oxidation of amines to imines with BiOCl: Does oxygen vacancy concentration matter? *Appl. Catal. B Environ.* **2018**, *228*, 87–96. [CrossRef]
22. Boukhatem, H.; Khalaf, H.; Djouadi, L.; Gonzalez, F.V.; Navarro, R.M.; Santaballa, J.A.; Canle, M. Photocatalytic activity of mont-La (6%)-$Cu_{0.6}Cd_{0.4}S$ catalyst for phenol degradation under near UV visible light irradiation. *Appl. Catal. B Environ.* **2017**, *211*, 114–125. [CrossRef]
23. Liu, M.; Zheng, L.; Lin, G.; Ni, L.; Song, X. Synthesis and photocatalytic activity of BiOCl/diatomite composite photocatalysts: Natural porous diatomite as photocatalyst support and dominant facets regulator. *Adv. Powder Technol.* **2020**, *31*, 339–350. [CrossRef]
24. Hu, X.; Li, C.; Song, J.; Zheng, S.; Sun, Z. Multidimensional assembly of oxygen vacancy-rich amorphous TiO_2-BiOBr-sepiolite composite for rapid elimination of formaldehyde and oxytetracycline under visible light. *J. Colloid Interface Sci.* **2020**, *574*, 61–73. [CrossRef] [PubMed]
25. Hao, L.; Huang, H.; Guo, Y.; Zhang, Y. Multifunctional $Bi_2O_2(OH)(NO_3)$ Nanosheets with {001} Active Exposing Facets: Efficient Photocatalysis, Dye-Sensitization, and Piezoelectric-Catalysis. *ACS Sustain. Chem. Eng.* **2018**, *6*, 1848–1862. [CrossRef]
26. Abd-Elnaiem, A.; Abdel-Rahim, M.; Abdel-Latief, A.; Mohamed, A.; Mojsilovic, K.; Stepniowski, W. Fabrication, Characterization and Photocatalytic Activity of Copper Oxide Nanowires Formed by Anodization of Copper Foams. *Materials* **2021**, *14*, 5030. [CrossRef] [PubMed]
27. Ma, Z.; Li, P.; Ye, L.; Zhou, Y.; Su, F.; Ding, C.; Xie, H.; Bai, Y.; Wong, P.K. Oxygen vacancies induced exciton dissociation of flexible BiOCl nanosheets for effective photocatalytic CO_2 conversion. *J. Mater. Chem. A* **2017**, *5*, 24995–25004. [CrossRef]

MDPI
St. Alban-Anlage 66
4052 Basel
Switzerland
www.mdpi.com

MDPI Books Editorial Office
E-mail: books@mdpi.com
www.mdpi.com/books

MDPI